World Scientific – Nobel Laureate Series: Vol 1

# HARRY MARKOWITZ

## Selected Works

**World Scientific — Nobel Laureate Series**

---

Vol. 1 Harry Markowitz: Selected Works
*edited by Harry M Markowitz*

World Scientific – Nobel Laureate Series: Vol 1

# HARRY MARKOWITZ

## Selected Works

edited by

**HARRY M MARKOWITZ**

University of California, San Diego, USA

World Scientific

NEW JERSEY • LONDON • SINGAPORE • BEIJING • SHANGHAI • HONG KONG • TAIPEI • CHENNAI

*Published by*

World Scientific Publishing Co. Pte. Ltd.
5 Toh Tuck Link, Singapore 596224
*USA office:* 27 Warren Street, Suite 401-402, Hackensack, NJ 07601
*UK office:* 57 Shelton Street, Covent Garden, London WC2H 9HE

**British Library Cataloguing-in-Publication Data**
A catalogue record for this book is available from the British Library.

**World Scientific — Nobel Laureate Series: Vol. 1**
**HARRY MARKOWITZ**
**Selected Works**



ISBN-13 978-981-283-363-1
ISBN-10 981-283-363-3
ISBN-13 978-981-283-364-8 (pbk)
ISBN-10 981-283-364-1 (pbk)

Printed in Singapore by World Scientific Printers

## *In Memory of Alan S Manne*

Alan Manne was about my age and played polo. He did not die in a polo accident but blacked out while on his horse. He woke up briefly and was told that he would not be able to use his arms or legs. He said he would rather go. He asked his family whether his horse had stepped on him. He was assured that his horse had been a lady and just stood there. He asked his children to take care of their mother. He said that he had been working on a joint paper, told in which drawer the latest draft was located, and asked one of his children to give it to his coauthor.

This book is also dedicated to George B. Dantzig who was Alan and my mentor, and to all our spouses who waited patiently while we played our favourite competitive game: research.

# Foreword

I remember very clearly the day I met Harry. I walked into his office where he sat writing a response to a letter that arrived on his desk. I knew nothing about financial economics. He was generous with his time. After a brief conversation we headed to dinner to meet his wife, Barbara. That dinner lasted for many hours as he patiently explained the finer details and mathematics of diversification, utility functions, optimization algorithms, and of course, portfolio selection. As we finished dinner, he listed a few articles he thought would help improve my understanding. I read all those references soon after that dinner and many times in the years since.

New students to the field of financial economics will find the foundation of our discipline in these works. *Portfolio Section* in Chapter 2 explains "risk", "diversification", and "portfolio selection". The companion article, *The Utility of Wealth*, speaks to the idea of measuring wealth. More advanced readers will find support for efficient market theory and behavioral finance in these classic articles. Students and practitioners in the field of Operations Research will find the "Markowitz Rule" which today often aids the speed of determination of larhge matrix inversion probldms in Chapter 3. Practitioners will find not only supporting logic for the pursuit of diversification in their client's portfolios through optimizations in this chapter but a detailed discussion of a solution algorithm. Chapter 4 again reaches into the field of Operations Research with its articles about the SIMSCRIPT language one of several computer languages that traces its origin to Harry. Readers will also find lessons on research methodology in these articles.

The first works in Chapter 5 speak to ideas either directly mentioned in or alluded to in Harry's first book. Several of these ideas remain a source of contention. Economic theorists still argue that all forms of utility functions hold value for empirical uses and practitioners spend countless hours trying to select a representative utility function in their portfolio construction problems. *Approximating Expected Utility by a Function of Mean and Variance* serves as the touch point for this argument while *Mean-Variance Versus Direct Utility Maximization* provides good empirical evidence to reject many utility maximization exercises as unnecessary. Later articles in this chapter return to the topic of computer programming languages with application to portfolio problems. This chapter concludes with entries about the topic of long run investment.

Chapters 6 and 7 are a collection of papers that comfortably are at rest in the world of theoretical financial economics or the world of the practitioner.

Several of articles in these chapters offer good guidance to practitioners as well as sound theoretical arguments.

A few weeks ago Harry, Barbara and I again sat at dinner discussing this volume. Once again he was patient and generous with his time. As he explained which articles he included in the different chapters, I found myself thinking about the concepts contained in many of these classic pieces — about how freely the ideas in these papers move between theory and practice. And just what these ideas mean to so many people in the operations research field, simulation teams, the financial services industry as well as different academy departments. The ideas in these pages have spawned industries, helped companies and governments address complex problems, and launched the careers of many professionals. What more can be said, enjoy!

Bernard V. Tew
October 2007

# Contents


| | | |
|---|---|---|
| **Foreword** | | **vii** |
| **Acknowledgements** | | **ix** |
| **Chapter 1** | **Overview** | **1** |
| | Trains of Thought | 3 |
| **Chapter 2** | **1952** | **11** |
| | Portfolio Selection | 15 |
| | The Early History of Portfolio Theory 1600–1960 | 31 |
| | The Utility of Wealth | 43 |
| **Chapter 3** | **Rand [I] and The Cowles Foundation** | **51** |
| | Industry-wide, Multi-industry and Economy-wide Process Analysis | 53 |
| | Alternate Methods of Analysis | 85 |
| | The Elimination Form of the Inverse and its Application to Linear Programming | 99 |
| | The Optimization of a Quadratic Function Subject to Linear Constraints | 115 |
| | The General Mean-variance Portfolio Selection Problem | 139 |
| **Chapter 4** | **Rand [II] and CACI** | **147** |
| | Simulating with SIMSCRIPT | 155 |
| | Programming by Questionnaire | 165 |
| | SIMSCRIPT | 211 |
| | Barriers to the Practical Use of Simulation Analysis | 271 |
| **Chapter 5** | **IBM's T. J. Watson Research Center** | **279** |
| | Approximating Expected Utility by a Function of Mean and Variance | 281 |
| | Mean-variance Versus Direct Utility Maximization | 293 |
| | The Value of a Blank Check | 309 |
| | The "Two beta" Trap | 319 |
| | Portfolio Analysis with Factors and Scenarios | 329 |
| | Sparsity and Piecewise Linearity in Large Portfolio Optimization Problems | 337 |

| | | |
|---|---|---|
| | The ER and EAS Formalisms for System Modeling and the EAS-E Language | 357 |
| | EAS-E: An Integrated Approach to Application Development | 377 |
| | The System Architecture of EAS-E: An Integrated Programming and Database Language | 405 |
| | Samuelson and Investment for the Long Run | 417 |
| | Investment for the Long Run: New Evidence for an Old Rule | 429 |
| **Chapter 6** | **Baruch College (CUNY) and Daiwa Securities** | **443** |
| | Investment Rules, Margin and Market Volatility | 445 |
| | Risk Adjustment | 453 |
| | Normative Portfolio Analysis: Past, Present and Future | 467 |
| | Individual versus Institutional Investing | 473 |
| | Foundations of Portfolio Theory | 481 |
| | Fast Computation of Mean-variance Efficient Sets Using Historical Covariances | 491 |
| | Computation of Mean-semivariance Efficient Sets by the Critical Line Algorithm | 507 |
| | Data Mining Corrections | 519 |
| **Chapter 7** | **Harry Markowitz Company** | **529** |
| | The Likelihood of Various Stock Market Return Distributions: Part 1: Principles of Inference | 531 |
| | The Likelihood of Various Stock Market Return Distributions: Part 2: Empirical Results | 545 |
| | Resampled Frontiers Versus Diffuse Bayes: | 573 |
| | An Experiment On Socks Ties and Extended Outcomes | 591 |
| | Single-Period Mean-variance Analysis in a Changing World | 601 |
| | Financial Market Simulation | 615 |
| | Portfolio Optimization with Factors, Scenarios and Realistic Short Positions | 627 |
| | Market Efficiency: A Theoretical Distinction and So What? | 641 |
| | Efficient Portfolios, Sparse Matrices, and Entities: A Retrospective | 661 |
| | DeFinetti Scoops Markowitz | 669 |
| | CAPM Investors Do Not Get Paid for Bearing Risks: *A Linear Relation Does not Imply Payment for Risk* | 697 |

# Acknowledgements

I would like to thank the publishers of the various articles that are reproduced herein for permission to do so. I would also like to express my gratitude to my many coauthors for the fun and interesting times we have had collaborating on these works. Further, I would like to express special thanks to Sandhya of World Scientific Publishing for guidance and encouragement in preparing the enclosed collection and my introductory comments on them.

Almost last but definitely not least, I wish to acknowledge the tireless and good-spirited labors of my secretary Mary Schultz a.k.a. Midge.

Finally I want to warmly acknowledge the role of Bernie Tew who was personally authorized by my wife, Barbara Markowitz, to bug me until I finished my part of all this.

**Chapter 1**

**Trains of Thought**
Reprinted with permission from Markowitz, H. M. (1993). *Trains of Thought.* The American Economist, 37(1), 3–9.

**Chapter 2**

**Portfolio Selection**
Reprinted with permission from Markowitz, H. M. (1952). *Portfolio Selection.* The Journal of Finance, 7(1), 77–91.

**The Early History of Portfolio Theory: 1600–1960**
Reproduced and republished from Markowitz, H. M. (1999). *The Early History of Portfolio Theory: 1600–1960.* Financial Analysts Journal, with permission from CFA Institute. All rights reserved.

**The Utility of Wealth**
Reprinted with permission from Markowitz, H. M. (1952). *The Utility of Wealth.* The Journal of Political Economy, 60, 151–158.

**Chapter 3**

**Industry-wide, Multi-industry and Economy-wide Process Analysis**
The publisher has attempted to identify the rightful owner of this paper. Anyone else who claims rights is requested to contact World Scientific Publishing Co. at editor@wspc.com.sg.

**The Elimination Form of the Inverse and its Application to Linear Programming**
Reprinted with permission, Markowitz, H. M. (1957). *The Elimination Form of the Inverse and its Application to Linear Programming.* Management Science, 3, 1957, 255–269. Copyright 2008, the Institute for Operations Research and the Management Sciences, 7240 Parkway Drive, Suite 310, Hanover, MD 21076 USA.

**Alternate Methods of Analysis**
Reprinted with permission from Manne, A., Markowitz, H. M., (1963). *Alternate Methods of Analysis* in *Studies in Process Analysis*: *Economy-Wide Production Capabilities*, pp. 8–20. Reprinted with permission of the Cowles Foundation, which retains all rights under the original copyright.

**The Optimization of a Quadratic Function Subject to Linear Constraints**
Reprinted with permission from Markowitz, H. M. (1956). Naval Research Logistics Quarterly, Vol. 3, pp. 111–33.

**The General Mean-Variance Portfolio Selection Problem**
Reprinted with permission, Philosophical Transactions of the Royal Society A, Markowitz, H. M. (1994). *The General Mean-Variance Portfolio Selection Problem*, 347A, 1994, 543–549.

**Chapter 4**

**Barriers to the Practical use of Simulation Analysis**
© 1981 IEEE. Reprinted, with permission, from Markowitz, H. M. *Barriers to the Practical use of Simulation Analysis.* 1981 Winter Simulation Conference Proceedings, 1, 3–9.

**Simulating with Simscript**
Reprinted with permission, Markowitz, H. M. (1966). *Simulating with Simscript.* Management Science, 12(10), 1966, B396–B405. Copyright 2008, the Institute for Operations Research and the Management Sciences, 7240 Parkway Drive, Suite 310, Hanover, MD 21076 USA.

**Simscript**
Reprinted with permission from Markowitz, H. M. (1979). *Simscript.* Encyclopedia of Computer Science and Technology, 13, 79–136.

**Programming by Questionnaire**
Reprinted with permission of the RAND Corporation, Ginsberg, A. S., Markowitz, H. M. and Oldfather, P. M. (1965). *Programming by Questionnaire.* Memo RM-4460-PR, 1–42.

**Chapter 5**

**Approximating Expected Utility by a Function of Mean and Variance**
Reprinted with permission from Markowitz, H. M. (1979). *Approximating Expected Utility by a Funcion of Mean and Variance.* The American Economic Review, 69(3), 308–317.

**Investment for the Long Run: New Evidence for an Old Rule**
Reprinted with permission from Levy, H. and Markowitz, H. M. (1976). *Investment for the Long Run: New Evidence for an Old Rule.* The Journal of Finance, 31(5), 1273–1286.

**Portfolio Analysis with Factors and Scenarios**
Reprinted with permission from Markowitz, H. M. and Perold, A. F. (1981). *Portfolio Analysis with Factors and Scenarios.* The Journal of Finance, 36(14), 871–877.

**Mean-Variance Versus Direct Utility Maximization**
Reprinted with permission from Kroll, Y., Levy, H. and Markowitz, H. M. (1984). *Mean-Variance Versus Direct Utility Maximization.* The Journal of Finance, 39(1), 47–61.

**EAS-E: An Integrated Approach to Application Development**
Reprinted with permission from Malhotra, A., Markowitz, H. M. and Pazel, D. P. (1983). *EAS-E: An Integrated Approach to Application Development.* ACM Transactions and Database Systems, 8(4), 515–542.

**The ER and EAS Formalisms for System Modeling, and the EAS-E Language**
This article was published in Proceedings of the 2nd International Conference on Entity-Relationship Approach to Modeling and Analysis, Markowitz, H. M., Malhotra, A. and Pazel, D. P., *The ER and EAS Formalisms for System Modeling, and the EAS-E Language*, 29–47, Copyright Elsevier (1983).

**Sparsity and Piecewise Linearity in Large Portfolio Optimization Problem**
This article was published in Sparse Matrices and Their Uses, Markowitz, H. M. and Perold, A. F., *Sparsity and Piecewise Linearity in Large Portfolio Optimization Problems*, 89–108, Copyright Elsevier (1981).

**The System Architecture of EAS-E: An Integrated Programming and Data Base Language**

Reprinted with permission from Markowitz, H. M. (1983). *The System Architecture of EAS-E: An Integrated Programming and Data Base Language.* IBM Systems Journal, 22(3), 187–198.

**The Two Beta Trap**

Reprinted with permission from Markowitz, H. M. (1984). *The Two Beta Trap.* The Journal of Portfolio Management, 11(1), 12–20.

**The Value of a Blank Check**

Reprinted with permission from Markowitz, H. M., Reid, D. W. and Tew, B. V. (1994). *The Value of a Blank Check.* The Journal of Portfolio Management, 21, 82–91.

**Samuelson and Investment for the Long Run**

By permission of Oxford University Press. Chp. "Samuelson and Investment for the Long Run" by Harry Markowitz from "Samuelsonian Economics and the Twenty-First Century" edited by Szenberg, Ramrattan, & Gottesman (2006).

**Chapter 6**

**Individual versus Institutional Investing**

Reproduced with the permission of the Academy of Financial Services and Financial Services Review. Reprinted with permission from Markowitz, H. M. (1991). *Individual versus Institutional Investing.* Financial Service Review, 1(1), 1–9.

**Foundations of Portfolio Theory**

Reprinted with permission from Markowitz, H. M. (1991). *Foundations of Portfolio Theory.* The Journal of Finance, 46(2), 469–477.

**Normative Portfolio Analysis: Past, Present, and Future**

This article was published in Journal of Economics and Business, Markowitz, H. M., *Normative Portfolio Analysis: Past, Present, and Future*, 99–103, Copyright Elsevier (1990).

**Risk Adjustment**

*Risk Adjustment*, Harry Markowitz. Copyright © (1990) by Journal of Accounting, Auditing and Finance. Reproduced with permission of Greenwood Publishing Group, Inc., Westport, CT.

**Investment Rules, Margin, and Market Volatility**

Reprinted with permission from Kim, G. and Markowitz, H. M. (1989). *Investment Rules, Margin, and Market Volatility.* The Journal of Portfolio Management, 16(1), 45–52.

**Data Mining Corrections**

Reprinted with permission from Markowitz, H. M. and Xu, G. L. (1994). *Data Mining Corrections.* The Journal of Portfolio Management, 21(1), 60–69,

**Fast Computation of Mean-Variance Efficient Sets Using Historical Covariances**

Reprinted with permission from Markowitz, H. M., Todd, P., Xu, G. and Yamene, Y. (1992). *Fast Computation of Mean-Variance Efficient Sets Using Historical Covariances.* Journal of Financial Engineering, 1(2), 117–132.

**Computation of Mean-Semivariance Efficient Sets by the Critical Line Algorithm**

Reprinted with kind permission of Springer Science and Business Media. Markowitz, H. M., Todd, P., Xu, G. and Yamene, Y. (1993). *Computation of Mean-Semivariance Efficient Sets by the Critical Line Algorithm.* Annals of Operations Research, 45, 307–317.

**Chapter 7**

**Resampled Frontiers versus Diffuse Bayes: An Experiment**

Reprinted with permission from Markowitz, H. M. and Usmen, N. (2003). *Resampled Frontiers versus Diffuse Bayes: An Experiment.* Journal of Investment Management, 1(4), 9–25.

**DeFinetti Scoops Markowitz**

Reprinted with permission from Markowitz, H. M. (2006). *DeFinetti Scoops Markowitz.* Journal of Investment Management, 4(3), 1–27.

**Efficient Portfolios: Sparse Marices, and Entities: A Retrospective**

Reprinted with permission, Markowitz, H. M., *Efficient Portolios: Sparse Matrices, and Entities: A Retrospective*, Operations Research, 50(1), 2002, 154–160. Copyright 2008, the Institute for Operations Research and the Management Sciences, 7240 Parkway Drive, Suite 310, Hanover, MD 21076 USA.

**Financial Market Simulation**

Reprinted with permission from Jacobs, B. I., Levy, K. N. and Markowitz, H. M. (2004). *Financial Market Simulation.* The Journal of Portfolio Management, 31, 1–10.

**Portfolio Optimization with Factors, Scenarios, and Realistic Short Positions**

Reprinted with permission, Jacobs, B. I., Levy, K. N. and Markowitz, H. M., *Portfolio Optimization with Factors, Scenarios, and Realistic Short Positions,* Operations Research, 53(4), 2005, 586–599. Copyright 2008, the Institute for Operations Research and the Management Sciences, 7240 Parkway Drive, Suite 310, Hanover, MD 21076 USA.

**CAPM Investors Do Not Get Paid for Bearing Risks**

Reprinted with permission from Markowitz, H. M. (forthcoming 2008). *CAPM Investors Do Not Get Paid for Bearing Risks.* The Journal of Portfolio Management.

**The Likelihood of Various Stock Market Return Distributions, Part 1: Principles of Inference**

Reprinted with kind permission of Springer Science and Business Media. Markowitz, H. M. and Usmen, N. (1996). *The Likelihood of Various Stock Market Return Distributions, Part 1: Principles of Inference.* Journal of Risk and Uncertainty, 13: 207–219.

**The Likelihood of Various Stock Market Return Distributions, Part 2: Empirical Results**

Reprinted with kind permission of Springer Science and Business Media. Markowitz, H. M. and Usmen, N. (1996). *The Likelihood of Various Stock Market Return Distributions, Part 2: Empirical Results.* Journal of Risk and Uncertainty, 13: 221–247.

**On Socks, Ties and Extended Outcomes**

Reprinted with kind permission of Springer Science and Business Media. Markowitz, H. M. (1997). *On Stocks, Ties and Extended Outcomes.* Economics and Environmental Risk and Uncertainty, 219–226.

**Single-Period Mean-Variance Analysis in a Changing World**

Reproduced and republished from Markowitz, H. M. and van Dijk, E. L. (2003). *Single-Period Mean-Variance Analysis in a Changing World.* Financial Analysts Journal, with permission from CFA Institute. All rights reserved.

**Market Efficiency: A Theoretical Distinction and So What?**

Reproduced and republished from Markowitz, H. M. (2005). *Market Efficiency: A Theoretical Distinction and So What.* Financial Analysts Journal, with permission from CFA Institute. All rights reserved.

## Chapter 1

# Overview

### Comment

The following chapters each include a Comments section describing the historical background of the articles contained therein. The present chapter consists of one article which supplies the historical background for a number of the articles in the following chapters so that we do not have to describe them there.

### References


Markowitz, H. M. (1993). *Trains of Thought*. The American Economist Journal of The International Honor Society in Economics, Vol. 37, No. 1, Spring, pp. 3–9.

# TRAINS OF THOUGHT

**by Harry M. Markowitz***

My essay will be concerned principally with some philosophical views I have held for much of my life. After recounting the sources (for me) and the nature of these views, I will conclude with some brief personal reflections.

These philosophical views are on a few related topics. My views on any one topic did not spring instantly to mind, but were the results of a train of thought to which I would return many times over weeks, months and years. It was also important to me that the train of thought on one topic did not contradict that on another.

## *What Do We Know?*

The first topic to occupy me, among those reviewed here, concerned what do we know and how do we know it. Until I was thirteen or fourteen I read comic books and "The Shadow" mystery magazines. then I read (I cannot remember why) Darwin's *Origin of Species*. I was especially fascinated with how Darwin marshalled his facts, argued his case and considered possible objections. Subsequently I read popular accounts of physics and astronomy, from the high school library, and original accounts by philosophers, purchased from wonderful big, old, musty used book stores then in downtown Chicago.

The philosopher who impressed me most, who became "my" philosopher, was David Hume. He argued that even though we release a ball a thousand times and each time it falls to the floor, we are not thereby provided proof with certainty that the ball will drop when released a thousand-and-first time. On reflection, one modification to Hume's views seemed necessary. Hume spoke in terms of cause and effect. I release the ball, then it drops to the floor. I eat a substance with a particular appearance and then I feel nourished. I see or do A and then B occurs. At least it always has; but there is no necessary proof that it will. The reason that Hume's view needs modification—really amplification—is that science does not merely catalog cause and effect. Rather it develops theories, what I would now refer to as models, sometimes mistakenly thought to be inevitable universal laws.

Consider the ball once more. What do we mean that it will fall down if I release it? Which way is down if I stand in Australia? Or in space a thousand miles from the earth? The "universal truth" which was observed over and over—as any eighteenth or nineteenth century physicist would tell you—is that the ball attracts, and is attracted to, each other object by a force which is proportional to the product of their masses and inversely proportional to the square of their distance. But this universal law of gravity did not hold universally. In particular it failed to accurately explain the path of the planet Mercury. Einstein's general theory of relativity presents a quite different model of phenomena which the older Newtonian model failed to explain as well as phenomena which the latter has succeeded in explaining. But, as Hume tells us, the fact that the theory of relativity had succeeded in all instances in the past does not prove that it will continue to do so in the future; and, as Einstein himself said, in this case we would have to seek a new theory. For a statement of this view, amply illustrated, see Einstein and Infeld, *The Evolution of Physics*.

Some readers may feel that Hume's views may be true in principle, but of little applicability to Economics or Finance. But if the reader has attended Economics seminars for a few years, he or she can probably supply examples of empirical economic relationships which held in one decade and not in the next; or which held for the preceding twenty years, but not last year.

Better still, the reader should try the exercise which Descartes undertakes in his first meditation, to distinguish between what is known and

*Marvin Speiser Distinguished Professor of Finance and Economics, and the 1990 winner of the Nobel Memorial Prize in economic science, Baruch College. Consultant, Daiwa Securities Trust Co.



The American Economist Journal of The International Honor Society in Economics (1993), Vol. 37, No.1 , Spring, pp. 3-9.

what is conjecture; in the present instance, however, seeking this distinction in financial or economic matters. The reader might reflect on the fact that much of our information on economic and financial matters come from newspaper, radio and television accounts. But we know from our experience as teachers or students that even good, college level students have difficulty in relating more than 80 or 90 percent of any material correctly. This, in itself, is a source of a 10 or 20 percent error rate. In addition, we frequently learn from later accounts of events that earlier accounts were falsified. Even if all participants in events tried to inform reporters accurately, we know that participants and witnesses see and believe different things; and who knows who is correct.

Thus our primary facts delivered to us by newspapers, radio and television are the output of a process full of noise. In addition, we know that different people receiving essentially the same primary facts can fit these into radically different belief structures. Again, who knows who is correct?

Further, databases have errors, programs have bugs; most facts are brought to mind from our memories, and you know how faulty memories can be. Sometimes we dream, and then anything can happen. Perhaps now is a dream. I do not assert that everything we believe is wrong; rather, that much we take as fact is only hypothesis.

### *Probability, Utility and Quadratic Approximations*

When I was a student member of the Cowles Commission at the University of Chicago, Karl Brunner and I worked through parts of von Neumann and Morgenstern's *Theory of Games and Economic Behavior*, including the appendix on expected utility. At first I was skeptical of expected utility as a maximum for rational behavior in the face of risk. But a conversation with Herman Rubin when he visited the Cowles Commission and a reading of Marschak's article on expected utility convinced me that this was a plausible maxim. Not long afterward I was convinced in a course by Leonard J. Savage, that one should act under uncertainty as if one assigned probability beliefs to events for which there are no objective probabilities, and should update probability beliefs according to Bayes rule. At first, I considered questions of expected utility and probability beliefs in the context of economic action in the face of risk and uncertainty. After reading F.P. Ramsey's pioneering essay, and further reflecting on Savage's arguments, I decided that the subject was the older one of "what do we know and how do we know it?" As explained above, I previously concluded that models of the world are never known with certainty. But we are more willing to give up some hypotheses than others. I agreed with Ramsey and Savage that degrees of belief should be formalized in terms of the actions of a rational decision maker, i.e., a decision maker who is not omniscient, but makes no mistakes in logic or arithmetic.[1]

Another train of thought began while reading John Burr Williams' *Theory of Investment Value* as background for my Ph.D. dissertation at the University of Chicago. Williams' asserted that the value of a stock should be the present value of its future dividends. But since the future is uncertain, I interpreted this to be the expected value of future dividends. But if one is concerned only with some expected value for each security, one must be concerned only with expected value for the portfolio as a whole. In this case, the investor would place all his funds in a single security—that with maximum expected return; or he or she would be indifferent between any combination of securities, all with maximum expected return, if there were two or more which tied for maximum. In no case would the investor prefer a diversified portfolio to all undiversified portfolios. But common sense, as well as prior examination of Wiesenberger's *Investment Companies* showed that diversification was a common and sensible aspect of investment practice. The reason, obviously, was to reduce risk. Thus the investor was, and should be, concerned with risk and return on the portfolio as a whole.

As a student in the Economics Department it was natural to think of Pareto optimality. More specifically, as a student of Tjalling Koopman's course on Activity Analysis, it was natural to distinguish between efficient and inefficient risk-return combinations; and to draw, for the first time, what is now referred to as the efficient frontier, then with expected return on the horizontal axis. Standard deviation, or variance, came to mind as the measure of risk. I did not know, off hand, the formula for the variance of a linear combination of random variables. This was supplied by a copy of Uspensky's *Introduction to Mathematical Probability* on the library shelf. I was delighted to see that portfolio variance depended on the covariances between securities as well as the variances of the securities held in the portfolio.

I left the University of Chicago for the RAND Corporation in 1951, having completed all but dissertation. At the invitation of James Tobin I spent the 1954–55 academic year at the Cowles Foundation at Yale, on leave from RAND, writing a book that would be published in 1959 as Cowles Foundation Monograph 16, *Portfolio Selection: Efficient Diversification of Investments*. Much of the time during this period was spent writing drafts of chapters explaining the elements of mean—variance analysis. A parallel activity involved attempting to reconcile mean—variance analysis and expected utility theory. Rather than consider the mean and variance of the present value of future dividend, as I first thought after reading J.B. Williams, I now considered a many period game, and assumed that securities were perfectly liquid. Under certain assumptions, each period the rational investor maximizes the expected value of a single period utility function which depends only on end of period wealth, as explained by R. Bellman. It seemed natural then to approximate this single period utility function by a quadratic, and approximate expected utility by the function of mean and variance which results from taking the expected value of the quadratic. This approach was illustrated by a few examples in my 1959 book, and more extensively by Young and Trent (1969), Levy and Markowitz (1979), and others. For most utility functions reported in the literature, and for probability distributions like historical returns on portfolios, the quadratic approximation does quite well.

It is important to distinguish between the assumption that the investor has a quadratic utility function, and the use of quadratic approximation to a given utility function. For example, Levy and Markowitz show that the Arrow and Pratt objection to a quadratic utility function does not apply to an investor who uses a quadratic approximation to a given utility function. In particular, the latter, quadratic approximation maximizer, has exactly the same risk aversion in the small, in the sense of Pratt, as does the expected utility maximizer whose utility function is approximated.

Markowitz 1959 also notes that under other assumptions the single period utility function may depend on state variables in addition to end of period wealth, and that maximizing the expected value of a quadratic approximation to this function of several variables leads to a mean—variance calculation. However, such quadratic approximations to utility functions of several variables were not explored in Markowitz 1959.

### *Simulation and Systems Descriptions*

In the 1950s Alan S. Manne and I at the RAND Corporation, and others at RAND, UCLA and elsewhere, tried our hand at building industry-wide and multi-industry "activity analysis" models. The first thought was to build models like Leontief's input-output model, except allowing for alternate methods of producing the output of any one industry. Examination of the inverse of a large input-output matrix revealed anomalies that would not be cured by alternate activities, nor by better data. What was required was a more radical departure from the Leontief format; namely, a model in which aggregates of production equipment and aggregates of producer and consumer products were the building blocks of the analysis, as opposed to the Leontief model whose building blocks are "industry capacities" and "interindustry flows". Our reason for departing from the Leontief model, and the results of our collective work, are presented in Cowles Foundation Monograph 18, A.S. Manne and H.M. Markowitz et al., *Studies in Process Analysis: Economy—Wide Production Capabilities*.

Various people provided industry models for this "process analysis" effort. For example, Alan Manne provided a petroleum industry model; Tibor Fabian, a blast furnace, iron and steel industry model; Alan J. Rowe (then at UCLA) and I developed a metal working industries model; etc. As a by-product of this work, I became interested in manufacturing planning and scheduling in the metalworking industries.[2] I soon agreed with those who argued that typical realistic manufacturing planning problems were too complex for analytic solution, or for optimizing algorithms such as linear programming. Simulation techniques were needed for advanced analysis, i.e., to give greater insight than provided by the static analysis of the day. One of the things I did at RAND, after returning form leave at Yale, was to supervise the programming of the computer simulation portion of a large man/machine logistics system simulation. This experience reinforced for me the potential usefulness of simulation techniques and illustrated the difficulty of programming detailed simulation models. These two points had already been illustrated by a previous simulation that had been programmed for me, and large and small simulations programmed by others at RAND. (Programming at the time was done in assembler. FORTRAN was about to make its appearance.)

Not long afterwards I resigned from RAND to accept a tempting offer at the General Electric Computer Department. Soon after I moved from the Computer Department to General Electric's Manufacturing Services where my friend and colleague Alan Rowe was developing a "general purpose" job shop model. It took two or three years for Rowe and one or two programmers to complete the model. Then, when one applied it to a factory other than the one for which it was developed, it turned out to be less "general purpose" than had been hoped. My own theory at the time was to seek "flexibility" rather than generality. This flexibility was to be achieved by building a simulator out of "reusable" FORTRAN modules. The first such General Electric Manufacturing Simulator (GEMS) was built in nine months. This shorter time to program was probably due to the use of FORTRAN rather than assembler language as used in Rowe's job shop simulator. As it turned out, my flexible subroutines were not all that flexible, except for some that performed basic actions such as that of creating or destroying some entity in the simulation (such as a job in the job shop) or inserting an entity into a collection of entities, such as the queue awaiting some resource.

I decided that these basic actions would be more conveniently placed at a programmers disposal by making them part of a programming language rather than leaving them as subroutines as in GEMS. I decided that I would like to develop such a programming language at a place whose mission and environment was like that I had known at RAND. I let my interests to be known to a small number of organizations and, in the end, returned to RAND. Bernard (Bernie) Hausner was assigned to me to implement the new language. Later, Herb Karr was hired as a consultant to write a programming manual. Bernie, Herb and I spent many hours together designing the language which we called SIMSCRIPT (now referred to as SIMSCRIPT I).

Our objective in designing SIMSCRIPT was, insofar as we could, to allow the user to describe the system to be simulated, as distinguished from having the user describe the actions which the computer must take to accomplish the simulation. The status of the system to be simulated was described in terms of *entities* of various *entity types*. Each individual entity was characterized by the *values* of its *attributes*, the *sets* to which it *belonged* and the members of the *sets* it owned. Status changed at points in time called *events*. Subsequently, Ed Russell introduced the notion of a *process* into SIMSCRIPT, which he borrowed from SIMULA, a later simulation programming language. During an event or process, status changes as entities are created or destroyed, attribute values are changed and entities gain and lose set memberships. The SYSTEM as a whole is an entity which can have attributes and own sets. Compound (Cartesian product) entities are also represented.

The SIMSCRIPT programming languages (including the original SIMSCRIPT I and following I.5, II and II.5 versions) have been applied to a wide variety of fields such as manufacturing simulation, from which it evolved, logistics analysis at RAND, and other applications such as to computer systems design, war games, transportation problems and the effects of trading systems on stock price behavior. SIMSCRIPT II.5 continues to have a large number of simulation application users.

The Entity, Attribute and Set (EAS) view of system description has also proven useful for other than simulation programming. For example, the SIMSCRIPT I.5 and SIMSCRIPT II Translators were themselves written in SIMSCRIPT, based on an EAS description of the entities encountered in the translation process.

The SIMSCRIPT II Translator was "bootstrapped" from a SIMSCRIPT I description of the translation process, then recompiled in terms of a SIMSCRIPT II description of its own compilation.

When SIMSCRIPT II was designed in the mid 1960s it was planned that it should be a database as well as a simulation language. A database would consist of the EAS description of the entities of the world represented within the database. In other words, the thought was that not only could entities be represented within a simulation, but also "real" entities could be represented within a database in EAS terms. Because of miscellaneous events, not related to the applicability of the EAS worldview, an implementation of the EAS view of database management, bootstrapped from the SIMSCRIPT II translator, was not completed until the work of Malhotra, Markowitz, and Pazel. We argue that the performance of the EAS-E system, including its use in internal IBM applications, prove the technical success of the approach. However, we were not able to persuade IBM to support EAS-E as a product. In part at least this was because IBM had just announced its support for the relational database methodology, after many years of supporting the hierarchical view of IMS. IBM seemed unlikely to be persuaded to change again in the short run. In the long run I was elsewhere; i.e., after building EAS-E at IBM and seeing that it would be a very long process to sell it internally, I was delighted to accept the offer of the Marvin Speiser Chair at Baruch College where I am now located.

The EAS concepts of system description are described in an article on SIMSCRIPT in the *Encyclopedia of Computer Science and Technology*. This includes a proposal for using this view in managing the entities of a computer operating system as well as those encountered in simulation, compilation and database management. The thought is to provide a uniform method of interacting with the entities encountered in computer systems, whether user defined or defined by the developers or the computer system, whether simulated or real, transient or database, etc. I have not succeeded in persuading the software development profession of the desirability of this approach. On the contrary, "object oriented" programming has emerged as the chief contender for the role which I had hoped for EAS programming. It is not only that I speak of "entities" and they speak of "objects". There is also a difference in the paradigms by which my entities and their objects are manipulated.

Perhaps if I took the time to work with object oriented programming with an open mind on a variety of applications I would conclude that it is at least as good as the EAS approach. As it is, the largest item in my queue of things to do, someday, remains to demonstrate the efficacy of the EAS view of system description.

### *Personal Reflections*

Life has afforded me many pleasurable activities, such as enjoying a fine meal, walking with Mrs. Markowitz on new fallen snow on the path through the woods near our home, flying kites with one or more grandchildren, listening to music, especially J.S. Bach, and the like. But no activity sustains my interest as long as does struggling with some technical or philosophical problem, sometimes alone, sometimes with colleagues. From one point of view these struggles may be classified as "work", since I sometimes get paid for such efforts. From another viewpoint they are play—part of a game like chess or amateur cryptography which I have also enjoyed. Often in my work-game, part of the objective is to produce something that someone will use. It is not only that sometimes someone pays me for such creations. It is also the fact that for a long time I have been primarily concerned with the theory of rational action, especially rational action under uncertainty. One measure of one's success in achieving a useful understanding and techniques for rational action is to have theory and techniques tried, accepted and endure. I have also spent time applying theory and techniques. For example, I currently spend half-time as Director of Research of Daiwa's Global Portfolio Research Department (GPRD) which has money management responsibilities in conjunction with other branches of Daiwa Securities. Previously I was President of Arbitrage Management Company. Such alternating between theory and practice is not uncommon among financial theorists. Sharpe, Rosenberg, Roll, Ross, Black, Vasicek, Leland, and Rubinstein are a few of those who are both theoretician and practitioner. Sometimes they develop the theory of practice, and other times the practice of theory. I find that these two activities reinforce each other.

Some economists report that they entered economics to better mankind's state (e.g., see Szenberg, *Their Life Philosophies: Eminent Economists.*) I have never thought it in my power to much improve the human condition generally. Much of human ill is due to violent aggression, political suppression, ancient hatreds and the like. These are not matters I know how to deal with, either from my training as an economist nor with the decision making techniques I have developed. Together with my wife, I try to be a good neighbor, contribute moderately to charities, try to help my children, grandchildren, students and colleagues when I can be of service, and the like. That done, I feel that I have paid my dues and may indulge myself in life's pleasures, including the struggle with interesting problems and questions of philosophy.

## Notes

1. The question of what it means to "make no mistake in logic and arithmetic" raises further questions which have been extensively explored, without universal agreement, by generations of mathematicians concerned with the foundations of their own discipline. See for example Kleene, *Introduction to Metamathematics*.
2. Another byproduct of the process analysis research was methods for solving large, sparse systems of equations i.e., large systems with relatively few non-zero elements. See Markowitz, H., "The Elimination Form of the Inverse and Its Application to Linear Programming", *Management Science*, 1957. Variants of these methods are now part of large, production linear programming codes.

## References

Arrow, K. (1965), *Aspects of the Theory of Risk Bearing*, Helsinki.

Bellman, R. E. (1957), *Dynamic Programming*, Princeton University Press, Princeton.

Darwin, C. R. (1968), J.W. Burrow, editor, *The Origin of Species by Means of Natural Selection*, Penguin Books.

Descartes, R. (1961), 2nd ed., *Meditations on First Philosophy*, Liberal Arts Press, New York.

Einstein, A. & L. Infeld, (1938), *The Evolution of Physics; the Growth of Ideas from Early Concepts to Relativity and Quanta*, Simon & Schuster, New York.

Hume, D. (1927), *Enquiries Concerning the Human Understanding & Concerning the Principles of Morals*, reprinted from the posthumous edition of 1777, L.A. Selby-Bigge, editor, 2nd ed., The Clarendon Press, Oxford.

Kleene, S. C. (1950), *Introduction to Mathematics*, Van Nostrand, Princeton.

Koopmans, T.C., (1951), "Analysis of Production as an Efficient Combination of Activities", in T. C. Koopmans, (ed.) (1971), *Activity of Production and Allocation*, 7th ed., Yale University Press, New Haven.

Levy, H., H. M. Markowitz (1979), "Approximating Expected Utility By a Function Mean and Variance", *American Economic Review*, June.

Malhotra, A., H. M. Markowitz, D. P. Pazel (1983), "EAS-E: An Integrated Approach to Application Development", *ACM Transactions on Database Systems*, Vol. 8, No. 4, December.

Manne, A. S., H. M. Markowitz, et al, (1963), *Studies in Process Analysis: Economy-wide Production Capabilities*, John Wiley and Sons.

Markowitz, H. M. (1952), "Portfolio Selection", *The Journal of Finance*, Vol. 7, No. 1, March.

Markowitz, H. M. (1957), "The Elimination Form of the Inverse and Its Application to Linear Programming", *Management Science*.

Markowitz, H. M. (1959), *Portfolio Selection: Efficient Diversification of Investments*, John Wiley and Sons, Basil Blackwell, 1991.

Markowitz, H. M. (1979), "SIMSCRIPT", *Encyclopedia of Computer Science and Technology*, Vol. 13, Belzer, J., A. G. Holzman, A. Kent, editors, Marcel Dekker, Inc.

Markowitz, H. M., A. Malhotra, D. P. Pazel (1984), "The EAS-E Application Development System: Principles and Language Summary", *Communications of the ACM*, Vol. 27, No. 8, August.

Marschak, J. (1950), "Rational Behavior, Uncertain Prospects, and Measurable Utility", *Econometrica*, Vol. 18, April.

Pazel, D. P., A. Malhotra, H. M. Markowitz (1983), "The System Architecture of EAS-E: An Integrated Programming and Data Base Language", *IBM Systems Journal*, Vol. 22, No. 3.

Pratt, J. W. (1964), "Risk Aversion in the Small and in the Large", *Econometrica*, January.

Ramsey, F. P. (1931), *The Foundations of Mathematics and Other Logical Essays*, Harcourt Brace and Company, New York.

Savage, L. J. (1954), *The Foundations of Statistics*, Wiley, 2nd edition, Dover, 1972, New York.

Szenberg, Michael, editor (1992), *Eminent Economists, Their Life Philosophies*, Cambridge University Press, Cambridge.

Uspensky, J. V. (1937), *Introduction to Mathematical Probability*, McGraw-Hill, New York.

Von Neumann, J. and O. Morgenstern (1944), *Theory of Games and Economic Behavior*, 3rd ed., Princeton University Press, 1953, Princeton.

A. Wiesenberger and Company, *Investment Companies*, New York, annual editions.

Williams, J. B. (1938), *The Theory of Investment Value*, Harvard University Press, Cambridge, Massachusetts.

Young, W. E. and R. H. Trent (1969), "Geometric Mean Approximation of Individual Security and Portfolio Performance", *Journal of Financial Quantitative Analysis*, June.

## Chapter 2

# 1952

### Comments

The first two articles of this chapter concern portfolio theory. The genesis of portfolio theory is described in the *Trains of Thought* article of Chapter One and will not be described further here.

The *Utility Analysis of Choices Involving Risk* by Friedman and Savage was an assignment to Markowitz by Friedman when I took Friedman's course in approximately 1949. The young Markowitz found some difficulties with the assigned paper. The paper which appears here details these difficulties. At first this paper attracted some but not much attention, and what attention it did attract was short lived. Then Kahneman and Tversky (1979) published their paper on prospect theory. This refers to the current article in a footnote, because the idea of measuring utility as a deviation from current wealth is used by Kahneman and Tversky as well as by me.

Prospect theory differs from the contents of the *Utility of Wealth* in two respects: First, the inflection point at current wealth in prospect theory is convex to the left and concave to the right as opposed to the Markowitz hypothesis that the inflection point is concave to the left and convex to the right. Markowitz also considers the inflection point to be at "customary wealth" rather than current wealth, but this is a minor detail we can overlook in this discussion.

The second way in which the hypotheses of prospect theory differ from the hypotheses in the *Utility of Wealth* is that, rather than use actual probabilities, prospect theory uses weights which are functions of probabilities. Again, for the purposes of comparing these two hypotheses, this is a detail which may be ignored. Incidentally, the *Utility of Wealth* article refers to convex regions as concave and concave regions as convex. This is because Friedman and Savage refer to convex regions as concave from above and concave regions as convex from regions from above. Later they just refer to them as convex and concave. I followed their terminology.

At first, the reference to the *Utility of Wealth* by Kahneman and Tversky was little noticed. But, retrospectively, prospect theorists, a.k.a. behavioral finance professionals, noted the grandfatherly relationship between the *Utility of Wealth* article and the use of this material in prospect theory. In particular, the three-volume handbook by Shefrin (2001) reprints a copy of the *Utility of Wealth* as the first article in its historical review volume. In this volume Shefrin describes the nature and role of the *Utility of Wealth* as follows:

> In 1952 two remarkable articles were published about the behavioral basis upon which individual investors form portfolios. One article was authored by Markowitz and the other by Roy. The concepts that Markowitz and Roy introduced inspired later work by both psychologists and economists.

Markowitz (1952) truly qualifies as a behavioral work, with its focus on how people actually behave. Markowitz addresses a classic question posed by Friedman and Savage (1948): why do people simultaneously purchase insurance and lottery tickets? Friedman and Savage proposed a solution to this question, a solution that Markowitz criticized on behavioral grounds. In arguing against the Friedman-Savage solution, Markowitz described the results of how people behave, citing an experiment conducted by Mosteller and Nogee (1951) about how people bet. Reliance on experimental data fell out of fashion for a while, and still engenders some controversy among financial economists.

Looked at in hindsight, Markowitz showed amazing insight. The theory he proposes as an alternative to Friedman and Savage contains basic elements that were later developed much more fully. His discussion about the difference between present wealth and customary wealth gave rise to the coding of gains and losses relative to a reference point. He recognized that losses loom larger than gains. He proposed a utility function with three inflection points to capture the idea that attitude or risk varied with the situation being faced. In this respect he emphasized the importance of whether a gamble is framed in terms of gains or losses, as well as whether the stakes are small or large. His discussion touches on aspiration points, the preference for positive skewness, and a property Thaler and Johnson (1991) subsequently called the 'house money effect'.

The ideas introduced in Markowitz (1952) were later developed in prospect theory, a framework proposed by psychologists Kahneman and Tversky (1979). Prospect theory combines the insights of Markowitz with those of Allais ([1952] 1979). It draws on Markowitz for the concepts of framing, gains, losses, reference points, and a utility function with concave and convex segments. It draws on Allais for its treatment of probabilities.

## References

Markowitz, H. M. (1952). *Portfolio Selection*. The Journal of Finance, 7(1), March, pp. 77–91.

Markowitz, H. M. (1999). The *Early History of Portfolio Theory: 1600–1960*. Financial Analysts Journal, July/August, pp. 5–16.

Markowitz, H. M. (1952). *The Utility of Wealth*. The Journal of Political Economy, Vol. 60, pp. 151–158.

Reprinted from *Journal of Finance*, March 1952

# PORTFOLIO SELECTION*

HARRY MARKOWITZ
*The Rand Corporation*

THE PROCESS OF SELECTING a portfolio may be divided into two stages. The first stage starts with observation and experience and ends with beliefs about the future performances of available securities. The second stage starts with the relevant beliefs about future performances and ends with the choice of portfolio. This paper is concerned with the second stage. We first consider the rule that the investor does (or should) maximize discounted expected, or anticipated, returns. This rule is rejected both as a hypothesis to explain, and as a maximum to guide investment behavior. We next consider the rule that the investor does (or should) consider expected return a desirable thing *and* variance of return an undesirable thing. This rule has many sound points, both as a maxim for, and hypothesis about, investment behavior. We illustrate geometrically relations between beliefs and choice of portfolio according to the "expected returns—variance of returns" rule.

One type of rule concerning choice of portfolio is that the investor does (or should) maximize the discounted (or capitalized) value of future returns.[1] Since the future is not known with certainty, it must be "expected" or "anticipated" returns which we discount. Variations of this type of rule can be suggested. Following Hicks, we could let "anticipated" returns include an allowance for risk.[2] Or, we could let the rate at which we capitalize the returns from particular securities vary with risk.

The hypothesis (or maxim) that the investor does (or should) maximize discounted return must be rejected. If we ignore market imperfections the foregoing rule never implies that there is a diversified portfolio which is preferable to all non-diversified portfolios. Diversification is both observed and sensible; a rule of behavior which does not imply the superiority of diversification must be rejected both as a hypothesis and as a maxim.

* This paper is based on work done by the author while at the Cowles Commission for Research in Economics and with the financial assistance of the Social Science Research Council. It will be reprinted as Cowles Commission Paper, New Series, No. 60.

1. See, for example, J. B. Williams, *The Theory of Investment Value* (Cambridge, Mass.: Harvard University Press, 1938), pp. 55–75.

2. J. R. Hicks, *Value and Capital* (New York: Oxford University Press, 1939), p. 126. Hicks applies the rule to a firm rather than a portfolio.

The foregoing rule fails to imply diversification no matter how the anticipated returns are formed; whether the same or different discount rates are used for different securities; no matter how these discount rates are decided upon or how they vary over time.[3] The hypothesis implies that the investor places all his funds in the security with the greatest discounted value. If two or more securities have the same value, then any of these or any combination of these is as good as any other.

We can see this analytically: suppose there are $N$ securities; let $r_{it}$ be the anticipated return (however decided upon) at time $t$ per dollar invested in security $i$; let $d_{it}$ be the rate at which the return on the $i^{th}$ security at time $t$ is discounted back to the present; let $X_i$ be the relative amount invested in security $i$. We exclude short sales, thus $X_i \geqslant 0$ for all $i$. Then the discounted anticipated return of the portfolio is

$$R = \sum_{t=1}^{\infty} \sum_{i=1}^{N} d_{it} r_{it} X$$

$$= \sum_{i=1}^{N} X_i \left( \sum_{t=1}^{\infty} d_{it} r_{it} \right)$$

$R_i = \sum_{t=1}^{\infty} d_{it} r_{it}$ is the discounted return of the $i^{th}$ security, therefore

$R = \Sigma X_i R_i$ where $R_i$ is independent of $X_i$. Since $X_i \geqslant 0$ for all $i$ and $\Sigma X_i = 1$, $R$ is a weighted average of $R_i$ with the $X_i$ as non-negative weights. To maximize $R$, we let $X_i = 1$ for $i$ with maximum $R_i$. If several $R a_a$, $a = 1, \ldots, K$ are maximum then any allocation with

$$\sum_{a=1}^{K} X a_a = 1$$

maximizes $R$. In no case is a diversified portfolio preferred to all non-diversified portfolios.[4]

It will be convenient at this point to consider a static model. Instead of speaking of the time series of returns from the $i^{th}$ security $(r_{i1}, r_{i2}, \ldots, r_{it}, \ldots)$ we will speak of "the flow of returns" $(r_i)$ from the $i^{th}$ security. The flow of returns from the portfolio as a whole is

3. The results depend on the assumption that the anticipated returns and discount rates are independent of the particular investor's portfolio.

4. If short sales were allowed, an infinite amount of money would be placed in the security with highest $r$.

$R = \Sigma X_i r_i$. As in the dynamic case if the investor wished to maximize "anticipated" return from the portfolio he would place all his funds in that security with maximum anticipated returns.

There is a rule which implies both that the investor should diversify and that he should maximize expected return. The rule states that the investor does (or should) diversify his funds among all those securities which give maximum expected return. The law of large numbers will insure that the actual yield of the portfolio will be almost the same as the expected yield.[5] This rule is a special case of the expected returns—variance of returns rule (to be presented below). It assumes that there is a portfolio which gives both maximum expected return and minimum variance, and it commends this portfolio to the investor.

This presumption, that the law of large numbers applies to a portfolio of securities, cannot be accepted. The returns from securities are too intercorrelated. Diversification cannot eliminate all variance.

The portfolio with maximum expected return is not necessarily the one with minimum variance. There is a rate at which the investor can gain expected return by taking on variance, or reduce variance by giving up expected return.

We saw that the expected returns or anticipated returns rule is inadequate. Let us now consider the expected returns—variance of returns (*E-V*) rule. It will be necessary to first present a few elementary concepts and results of mathematical statistics. We will then show some implications of the *E-V* rule. After this we will discuss its plausibility.

In our presentation we try to avoid complicated mathematical statements and proofs. As a consequence a price is paid in terms of rigor and generality. The chief limitations from this source are (1) we do not derive our results analytically for the $n$-security case; instead, we present them geometrically for the 3 and 4 security cases; (2) we assume static probability beliefs. In a general presentation we must recognize that the probability distribution of yields of the various securities is a function of time. The writer intends to present, in the future, the general, mathematical treatment which removes these limitations.

We will need the following elementary concepts and results of mathematical statistics:

Let $Y$ be a random variable, i.e., a variable whose value is decided by chance. Suppose, for simplicity of exposition, that $Y$ can take on a finite number of values $y_1, y_2, \ldots, y_N$. Let the probability that $Y =$

5. Williams, *op. cit.*, pp. 68, 69.

$y_1$, be $p_1$; that $Y = y_2$ be $p_2$ etc. The expected value (or mean) of $Y$ is defined to be

$$E = p_1y_1 + p_2y_2 + \ldots + p_Ny_N$$

The variance of $Y$ is defined to be

$$V = p_1(y_1 - E)^2 + p_2(y_2 - E)^2 + \ldots + p_N(y_N - E)^2 .$$

$V$ is the average squared deviation of $Y$ from its expected value. $V$ is a commonly used measure of dispersion. Other measures of dispersion, closely related to $V$ are the standard deviation, $\sigma = \sqrt{V}$ and the coefficient of variation, $\sigma/E$.

Suppose we have a number of random variables: $R_1, \ldots, R_n$. If $R$ is a weighted sum (linear combination) of the $R_i$

$$R = a_1R_1 + a_2R_2 + \ldots + a_nR_n$$

then $R$ is also a random variable. (For example $R_1$, may be the number which turns up on one die; $R_2$, that of another die, and $R$ the sum of these numbers. In this case $n = 2$, $a_1 = a_2 = 1$).

It will be important for us to know how the expected value and variance of the weighted sum ($R$) are related to the probability distribution of the $R_1, \ldots, R_n$. We state these relations below; we refer the reader to any standard text for proof.[6]

The expected value of a weighted sum is the weighted sum of the expected values. I.e., $E(R) = a_1E(R_1) + a_2E(R_2) + \ldots + a_nE(R_n)$ The variance of a weighted sum is not as simple. To express it we must define "covariance." The covariance of $R_1$ and $R_2$ is

$$\sigma_{12} = E\{[R_1 - E(R_1)][R_2 - E(R_2)]\}$$

i.e., the expected value of [(the deviation of $R_1$ from its mean) times (the deviation of $R_2$ from its mean)]. In general we define the covariance between $R_i$ and $R_j$ as

$$\sigma_{ij} = E\{[R_i - E(R_i)][R_j - E(R_j)]\}$$

$\sigma_{ij}$ may be expressed in terms of the familiar correlation coefficient ($\rho_{ij}$). The covariance between $R_i$ and $R_j$ is equal to [(their correlation) times (the standard deviation of $R_i$) times (the standard deviation of $R_j$)]:

$$\sigma_{ij} = \rho_{ij}\sigma_i\sigma_j$$

6. E.g., J. V. Uspensky, *Introduction to Mathematical Probability* (New York: McGraw-Hill, 1937), chapter 9, pp. 161–81.

The variance of a weighted sum is

$$V(R) = \sum_{i=1}^{N} a_i^2 V(X_i) + 2\sum_{i=1}^{N} \sum_{i>1}^{N} a_i a_j \sigma_{ij}$$

If we use the fact that the variance of $R_i$ is $\sigma_{ii}$ then

$$V(R) = \sum_{i=1}^{N} \sum_{j=1}^{N} a_i a_j \sigma_{ij}$$

Let $R_i$ be the return on the $i^{th}$ security. Let $\mu_i$ be the expected value of $R_i$; $\sigma_{ij}$, be the covariance between $R_i$ and $R_j$ (thus $\sigma_{ii}$ is the variance of $R_i$). Let $X_i$ be the percentage of the investor's assets which are allocated to the $i^{th}$ security. The yield $(R)$ on the portfolio as a whole is

$$R = \sum R_i X_i$$

The $R_i$ (and consequently $R$) are considered to be random variables.[7] The $X_i$ are not random variables, but are fixed by the investor. Since the $X_i$ are percentages we have $\Sigma X_i = 1$. In our analysis we will exclude negative values of the $X_i$ (i.e., short sales); therefore $X_i \geqslant 0$ for all $i$.

The return $(R)$ on the portfolio as a whole is a weighted sum of random variables (where the investor can choose the weights). From our discussion of such weighted sums we see that the expected return $E$ from the portfolio as a whole is

$$E = \sum_{i=1}^{N} X_i \mu_i$$

and the variance is

$$V = \sum_{i=1}^{N} \sum_{j=1}^{N} \sigma_{ij} X_i X$$

7. I.e., we assume that the investor does (and should) act as if he had probability beliefs concerning these variables. In general we would expect that the investor could tell us, for any two events (A and B), whether he personally considered A more likely than B, B more likely than A, or both equally likely. If the investor were consistent in his opinions on such matters, he would possess a system of probability beliefs. We cannot expect the investor to be consistent in every detail. We can, however, expect his probability beliefs to be roughly consistent on important matters that have been carefully considered. We should also expect that he will base his actions upon these probability beliefs—even though they be in part subjective.

This paper does not consider the difficult question of how investors do (or should) form their probability beliefs.

For fixed probability beliefs ($\mu_i$, $\sigma_{ij}$) the investor has a choice of various combinations of $E$ and $V$ depending on his choice of portfolio $X_1, \ldots, X_N$. Suppose that the set of all obtainable ($E$, $V$) combinations were as in Figure 1. The $E$-$V$ rule states that the investor would (or should) want to select one of those portfolios which give rise to the ($E$, $V$) combinations indicated as efficient in the figure; i.e., those with minimum $V$ for given $E$ or more and maximum $E$ for given $V$ or less.

There are techniques by which we can compute the set of efficient portfolios and efficient ($E$, $V$) combinations associated with given $\mu_i$

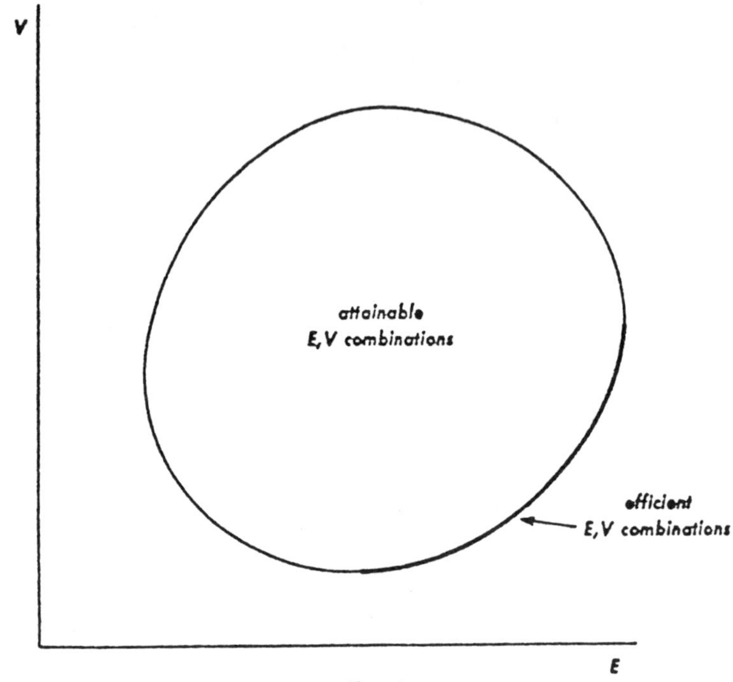


Fig. 1

and $\sigma_{ij}$. We will not present these techniques here. We will, however, illustrate geometrically the nature of the efficient surfaces for cases in which $N$ (the number of available securities) is small.

The calculation of efficient surfaces might possibly be of practical use. Perhaps there are ways, by combining statistical techniques and the judgment of experts, to form reasonable probability beliefs ($\mu_i$, $\sigma_{ij}$). We could use these beliefs to compute the attainable efficient combinations of ($E$, $V$). The investor, being informed of what ($E$, $V$) combinations were attainable, could state which he desired. We could then find the portfolio which gave this desired combination.

Two conditions—at least—must be satisfied before it would be practical to use efficient surfaces in the manner described above. First, the investor must desire to act according to the $E$-$V$ maxim. Second, we must be able to arrive at reasonable $\mu_i$ and $\sigma_{ij}$. We will return to these matters later.

Let us consider the case of three securities. In the three security case our model reduces to

$$1)\qquad E=\sum_{i=1}^{3} X_i\mu_i$$

$$2)\qquad V=\sum_{i=1}^{3}\sum_{j=1}^{3} X_iX_j\sigma_{ij}$$

$$3)\qquad \sum_{i=1}^{3} X_i=1$$

$$4)\qquad X_i\geqslant 0 \qquad \text{for} \qquad i=1,2,3\ .$$

From (3) we get

$$3')\qquad X_3=1-X_1-X_2$$

If we substitute (3′) in equation (1) and (2) we get $E$ and $V$ as functions of $X_1$ and $X_2$. For example we find

$$1')\qquad E=\mu_3+X_1(\mu_1-\mu_3)+X_2(\mu_2-\mu_3)$$

The exact formulas are not too important here (that of $V$ is given below).[8] We can simply write

$$a)\qquad E=E(X_1,X_2)$$

$$b)\qquad V=V(X_1,X_2)$$

$$c)\qquad X_1\geqslant 0,\ X_2\geqslant 0,\ 1-X_1-X_2\geqslant 0$$

By using relations (*a*), (*b*), (*c*), we can work with two dimensional geometry.

The attainable set of portfolios consists of all portfolios which satisfy constraints (*c*) and (3′) (or equivalently (3) and (4)). The attainable combinations of $X_1$, $X_2$ are represented by the triangle $\overline{abc}$ in Figure 2. Any point to the left of the $X_2$ axis is not attainable because it violates the condition that $X_1 \geqslant 0$. Any point below the $X_1$ axis is not attainable because it violates the condition that $X_2 \geqslant 0$. Any

8. $V = X_1^2(\sigma_{11}-2\sigma_{13}+\sigma_{33}) + X_2^2(\sigma_{22}-2\sigma_{23}+\sigma_{33}) + 2X_1X_2(\sigma_{12}-\sigma_{13}-\sigma_{23}+\sigma_{33}) + 2X_1(\sigma_{13}-\sigma_{33}) + 2X_2(\sigma_{23}-\sigma_{33}) + \sigma_{33}$

point above the line $(1 - X_1 - X_2 = 0)$ is not attainable because it violates the condition that $X_3 = 1 - X_1 - X_2 \geqslant 0$.

We define an *isomean* curve to be the set of all points (portfolios) with a given expected return. Similarly an *isovariance* line is defined to be the set of all points (portfolios) with a given variance of return.

An examination of the formulae for $E$ and $V$ tells us the shapes of the isomean and isovariance curves. Specifically they tell us that typically[9] the isomean curves are a system of parallel straight lines; the isovariance curves are a system of concentric ellipses (see Fig. 2). For example, if $\mu_2 \neq \mu_3$ equation 1′ can be written in the familiar form $X_2 = a + bX_1$; specifically (1)

$$X_2 = \frac{E - \mu_3}{\mu_2 - \mu_3} - \frac{\mu_1 - \mu_3}{\mu_2 - \mu_3} X_1 .$$

Thus the slope of the isomean line associated with $E = E_0$ is $-(\mu_1 - \mu_3)/(\mu_2 - \mu_3)$ its intercept is $(E_0 - \mu_3)/(\mu_2 - \mu_3)$. If we change $E$ we change the intercept but not the slope of the isomean line. This confirms the contention that the isomean lines form a system of parallel lines.

Similarly, by a somewhat less simple application of analytic geometry, we can confirm the contention that the isovariance lines form a family of concentric ellipses. The "center" of the system is the point which minimizes $V$. We will label this point $X$. Its expected return and variance we will label $E$ and $V$. Variance increases as you move away from $X$. More precisely, if one isovariance curve, $C_1$, lies closer to $X$ than another, $C_2$, then $C_1$ is associated with a smaller variance than $C_2$.

With the aid of the foregoing geometric apparatus let us seek the efficient sets.

$X$, the center of the system of isovariance ellipses, may fall either inside or outside the attainable set. Figure 4 illustrates a case in which $X$ falls inside the attainable set. In this case: $X$ is efficient. For no other portfolio has a $V$ as low as $X$; therefore no portfolio can have either smaller $V$ (with the same or greater $E$) or greater $E$ with the same or smaller $V$. No point (portfolio) with expected return $E$ less than $E$ is efficient. For we have $E > E$ and $V < V$.

Consider all points with a given expected return $E$; i.e., all points on the isomean line associated with $E$. The point of the isomean line at which $V$ takes on its least value is the point at which the isomean line

9. The isomean "curves" are as described above except when $\mu_1 = \mu_2 = \mu_3$. In the latter case all portfolios have the same expected return and the investor chooses the one with minimum variance.

As to the assumptions implicit in our description of the isovariance curves see footnote 12.

is tangent to an isovariance curve. We call this point $\hat{X}(E)$. If we let $E$ vary, $\hat{X}(E)$ traces out a curve.

Algebraic considerations (which we omit here) show us that this curve is a straight line. We will call it the critical line $l$. The critical line passes through $X$ for this point minimizes $V$ for all points with $E(X_1, X_2) = E$. As we go along $l$ in either direction from $X$, $V$ increases. The segment of the critical line from $X$ to the point where the critical line crosses

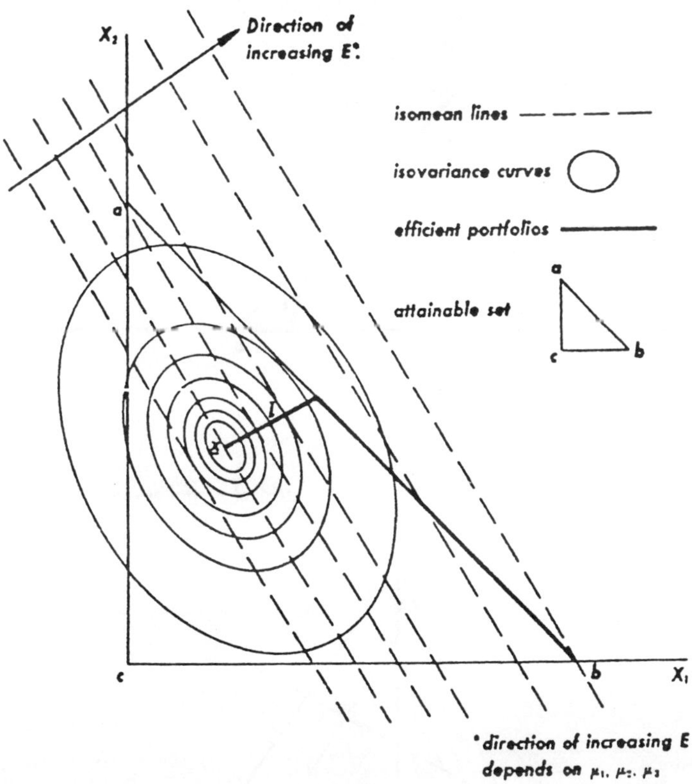


FIG. 2

the boundary of the attainable set is part of the efficient set. The rest of the efficient set is (in the case illustrated) the segment of the $\overline{ab}$ line from $d$ to $b$. $b$ is the point of maximum attainable $E$. In Figure 3, $X$ lies outside the admissible area but the critical line cuts the admissible area. The efficient line begins at the attainable point with minimum variance (in this case on the $\overline{ab}$ line). It moves toward $b$ until it intersects the critical line, moves along the critical line until it intersects a boundary and finally moves along the boundary to $b$. The reader may

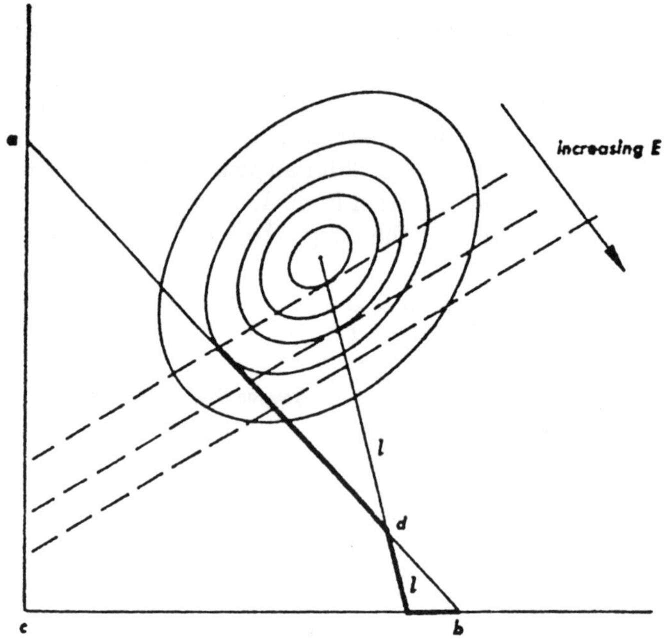
a
increasing E
l
d
l
c
b

FIG. 3

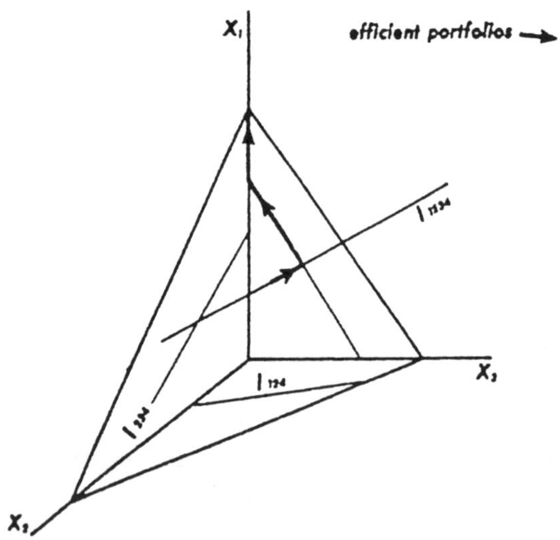
X1
efficient portfolios
l1234
X3
l124
l234
X2

FIG. 4

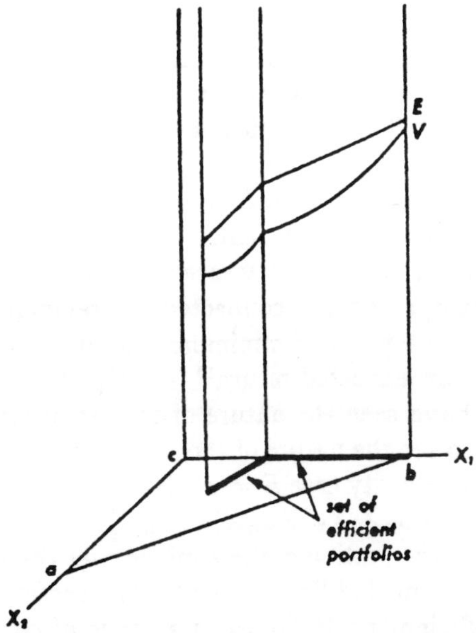
E
V
c
$X_1$
b
set of
efficient
portfolios
a
$X_2$

Fig. 5

V
efficient
E,V combinations
E

Fig. 6

wish to construct and examine the following other cases: (1) $X$ lies outside the attainable set and the critical line does not cut the attainable set. In this case there is a security which does not enter into any efficient portfolio. (2) Two securities have the same $\mu_i$. In this case the isomean lines are parallel to a boundary line. It may happen that the efficient portfolio with maximum $E$ is a diversified portfolio. (3) A case wherein only one portfolio is efficient.

The efficient set in the 4 security case is, as in the 3 security and also the $N$ security case, a series of connected line segments. At one end of the efficient set is the point of minimum variance; at the other end is a point of maximum expected return[10] (see Fig. 4).

Now that we have seen the nature of the set of efficient portfolios, it is not difficult to see the nature of the set of efficient $(E, V)$ combinations. In the three security case $E = a_0 + a_1X_1 + a_2X_2$ is a plane; $V = b_0 + b_1X_1 + b_2X_2 + b_{12}X_1X_2 + b_{11}X_1^2 + b_{22}X_2^2$ is a paraboloid.[11] As shown in Figure 5, the section of the $E$-plane over the efficient portfolio set is a series of connected line segments. The section of the $V$-paraboloid over the efficient portfolio set is a series of connected parabola segments. If we plotted $V$ against $E$ for efficient portfolios we would again get a series of connected parabola segments (see Fig. 6). This result obtains for any number of securities.

Various reasons recommend the use of the expected return-variance of return rule, both as a hypothesis to explain well-established investment behavior and as a maxim to guide one's own action. The rule serves better, we will see, as an explanation of, and guide to, "investment" as distinguished from "speculative" behavior.

10. Just as we used the equation $\sum_{i=1}^{4} X_i = 1$ to reduce the dimensionality in the three security case, we can use it to represent the four security case in 3 dimensional space. Eliminating $X_4$ we get $E = E(X_1, X_2, X_3)$, $V = V(X_1, X_2, X_3)$. The attainable set is represented, in three-space, by the tetrahedron with vertices (0, 0, 0), (0, 0, 1), (0, 1, 0), (1, 0, 0), representing portfolios with, respectively, $X_4 = 1$, $X_3 = 1$, $X_2 = 1$, $X_1 = 1$.

Let $s_{123}$ be the subspace consisting of all points with $X_4 = 0$. Similarly we can define $s_{a1, \ldots, aa}$ to be the subspace consisting of all points with $X_i = 0$, $i \neq a_1, \ldots, a_a$. For each subspace $s_{a1, \ldots, aa}$ we can define a *critical line* $l_{a_1, \ldots aa}$. This line is the locus of points $P$ where $P$ minimizes $V$ for all points in $s_{a1, \ldots, aa}$ with the same $E$ as $P$. If a point is in $s_{a1, \ldots, aa}$ and is efficient it must be on $l_{a_1, \ldots, aa}$. The efficient set may be traced out by starting at the point of minimum available variance, moving continuously along various $l_{a_1, \ldots, aa}$ according to definite rules, ending in a point which gives maximum $E$. As in the two dimensional case the point with minimum available variance may be in the interior of the available set or on one of its boundaries. Typically we proceed along a given critical line until either this line intersects one of a larger subspace or meets a boundary (and simultaneously the critical line of a lower dimensional subspace). In either of these cases the efficient line turns and continues along the new line. The efficient line terminates when a point with maximum $E$ is reached.

11. See footnote 8.

Earlier we rejected the expected returns rule on the grounds that it never implied the superiority of diversification. The expected return-variance of return rule, on the other hand, implies diversification for a wide range of $\mu_i$, $\sigma_{ij}$. This does not mean that the $E$-$V$ rule never implies the superiority of an undiversified portfolio. It is conceivable that one security might have an extremely higher yield and lower variance than all other securities; so much so that one particular undiversified portfolio would give maximum $E$ and minimum $V$. But for a large, presumably representative range of $\mu_i$, $\sigma_{ij}$ the $E$-$V$ rule leads to efficient portfolios almost all of which are diversified.

Not only does the $E$-$V$ hypothesis imply diversification, it implies the "right kind" of diversification for the "right reason." The adequacy of diversification is not thought by investors to depend solely on the number of different securities held. A portfolio with sixty different railway securities, for example, would not be as well diversified as the same size portfolio with some railroad, some public utility, mining, various sort of manufacturing, etc. The reason is that it is generally more likely for firms within the same industry to do poorly at the same time than for firms in dissimilar industries.

Similarly in trying to make variance small it is not enough to invest in many securities. It is necessary to avoid investing in securities with high covariances among themselves. We should diversify across industries because firms in different industries, especially industries with different economic characteristics, have lower covariances than firms within an industry.

The concepts "yield" and "risk" appear frequently in financial writings. Usually if the term "yield" were replaced by "expected yield" or "expected return," and "risk" by "variance of return," little change of apparent meaning would result.

Variance is a well-known measure of dispersion about the expected. If instead of variance the investor was concerned with standard error, $\sigma = \sqrt{V}$, or with the coefficient of dispersion, $\sigma/E$, his choice would still lie in the set of efficient portfolios.

Suppose an investor diversifies between two portfolios (i.e., if he puts some of his money in one portfolio, the rest of his money in the other. An example of diversifying among portfolios is the buying of the shares of two different investment companies). If the two original portfolios have *equal* variance then typically[12] the variance of the resulting (compound) portfolio will be less than the variance of either original port-

12. In no case will variance be increased. The only case in which variance will not be decreased is if the return from both portfolios are perfectly correlated. To draw the isovariance curves as ellipses it is both necessary and sufficient to assume that no two distinct portfolios have perfectly correlated returns.

folio. This is illustrated by Figure 7. To interpret Figure 7 we note that a portfolio ($P$) which is built out of two portfolios $P' = (X_1', X_2')$ and $P'' = (X_1'', X_2'')$ is of the form $P = \lambda P' + (1-\lambda)P'' = (\lambda X_1' + (1-\lambda)X_1'', \lambda X_2' + (1-\lambda)X_2'')$. $P$ is on the straight line connecting $P'$ and $P''$.

The $E$-$V$ principle is more plausible as a rule for investment behavior as distinguished from speculative behavior. The third moment[13] $M_3$ of

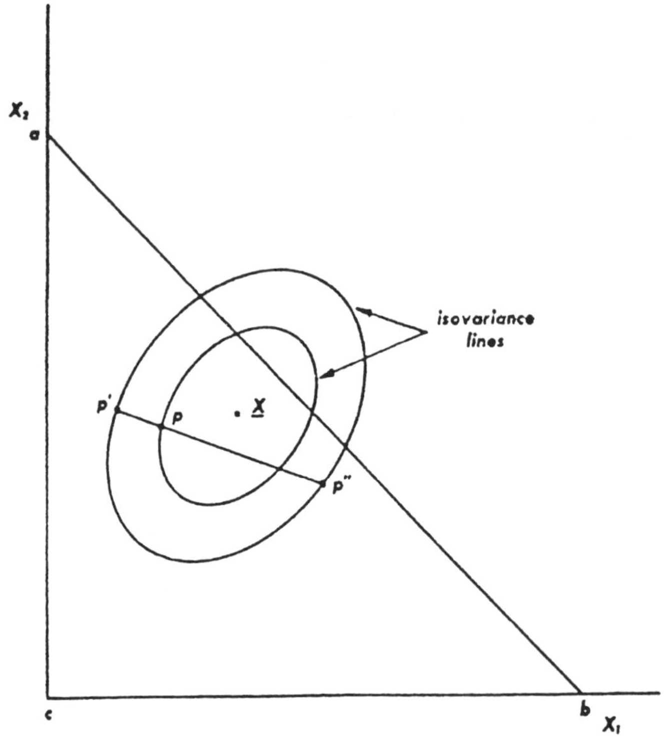


FIG. 7

the probability distribution of returns from the portfolio may be connected with a propensity to gamble. For example if the investor maximizes utility ($U$) which depends on $E$ and $V(U = U(E, V), \partial U/\partial E > 0, \partial U/\partial E < 0)$ he will never accept an actuarially fair[14] bet. But if

13. If $R$ is a random variable that takes on a finite number of values $r_1, \ldots, r_n$ with probabilities $p_1, \ldots, p_n$ respectively, and expected value $E$, then $M_3 = \sum_{i=1}^{n} p_i(r_i - E)^3$

14. One in which the amount gained by winning the bet times the probability of winning is equal to the amount lost by losing the bet, times the probability of losing.

$U = U(E, V, M_3)$ and if $\partial U/\partial M_3 \neq 0$ then there are some fair bets which would be accepted.

Perhaps—for a great variety of investing institutions which consider yield to be a good thing; risk, a bad thing; gambling, to be avoided—$E$, $V$ efficiency is reasonable as a working hypothesis and a working maxim.

Two uses of the $E$-$V$ principle suggest themselves. We might use it in theoretical analyses or we might use it in the actual selection of portfolios.

In theoretical analyses we might inquire, for example, about the various effects of a change in the beliefs generally held about a firm, or a general change in preference as to expected return versus variance of return, or a change in the supply of a security. In our analyses the $X_i$ might represent individual securities or they might represent aggregates such as, say, bonds, stocks and real estate.[15]

To use the $E$-$V$ rule in the selection of securities we must have procedures for finding reasonable $\mu_i$ and $\sigma_{ij}$. These procedures, I believe, should combine statistical techniques and the judgment of practical men. My feeling is that the statistical computations should be used to arrive at a tentative set of $\mu_i$ and $\sigma_{ij}$. Judgment should then be used in increasing or decreasing some of these $\mu_i$ and $\sigma_{ij}$ on the basis of factors or nuances not taken into account by the formal computations. Using this revised set of $\mu_i$ and $\sigma_{ij}$, the set of efficient $E$, $V$ combinations could be computed, the investor could select the combination he preferred, and the portfolio which gave rise to this $E$, $V$ combination could be found.

One suggestion as to tentative $\mu_i$, $\sigma_{ij}$ is to use the observed $\mu_i$, $\sigma_{ij}$ for some period of the past. I believe that better methods, which take into account more information, can be found. I believe that what is needed is essentially a "probabilistic" reformulation of security analysis. I will not pursue this subject here, for this is "another story." It is a story of which I have read only the first page of the first chapter.

In this paper we have considered the second stage in the process of selecting a portfolio. This stage starts with the relevant beliefs about the securities involved and ends with the selection of a portfolio. We have not considered the first stage: the formation of the relevant beliefs on the basis of observation.

15. Care must be used in using and interpreting relations among aggregates. We cannot deal here with the problems and pitfalls of aggregation.

PERSPECTIVES

# The Early History of Portfolio Theory: 1600–1960

Harry M. Markowitz

Diversification of investments was a well-established practice long before I published my paper on portfolio selection in 1952. For example, A. Wiesenberger's annual reports in *Investment Companies* prior to 1952 (beginning 1941) showed that these firms held large numbers of securities. They were neither the first to provide diversification for their customers (they were modeled on the investment trusts of Scotland and England, which began in the middle of the 19th century), nor was diversification new then. In the *Merchant of Venice*, Shakespeare has the merchant Antonio say:

> My ventures are not in one bottom trusted,
> Nor to one place; nor is my whole estate
> Upon the fortune of this present year;
> Therefore, my merchandise makes me not sad.
> Act I, Scene 1

Clearly, Shakespeare not only knew about diversification but, at an intuitive level, understood covariance.

What was lacking prior to 1952 was an adequate *theory* of investment that covered the effects of diversification when risks are correlated, distinguished between efficient and inefficient portfolios, and analyzed risk–return trade-offs on the portfolio as a whole. This article traces the development of portfolio theory in the 1950s (including the contributions of A.D. Roy, James Tobin, and me) and compares it with theory prior to 1950 (including the contributions of J.R. Hicks, J. Marschak, J.B. Williams, and D.H. Leavens).

## Portfolio Theory: 1952

On the basis of Markowitz (1952), I am often called the father of modern portfolio theory (MPT), but Roy (1952) can claim an equal share of this honor. This section summarizes the contributions of both.

My 1952 article on portfolio selection proposed expected (mean) return, $E$, and variance of return, $V$, of the portfolio as a whole as criteria for portfolio selection, both as a possible hypothesis about actual behavior and as a maxim for how investors ought to act. The article assumed that "beliefs" or projections about securities follow the same probability rules that random variables obey. From this assumption, it follows that (1) the expected return on the portfolio is a weighted average of the expected returns on individual securities and (2) the variance of return on the portfolio is a particular function of the variances of, and the covariances between, securities and their weights in the portfolio.

Markowitz (1952) distinguished between efficient and inefficient portfolios. Subsequently, someone aptly coined the phrase "efficient frontier" for what I referred to as the "set of efficient mean–variance combinations." I had proposed that means, variances, and covariances of securities be estimated by a combination of statistical analysis and security analyst judgment. From these estimates, the set of efficient mean–variance combinations can be derived and presented to the investor for choice of the desired risk–return combination. I used geometrical analyses of three- and four-security examples to illustrate properties of efficient sets, assuming nonnegative investments subject to a budget constraint. In particular, I showed in the 1952 article that the set of efficient portfolios is piecewise linear (made up of connected straight lines) and the set of efficient mean–variance combinations is piecewise parabolic.

Roy also proposed making choices on the basis of mean and variance of the portfolio as a whole. Specifically, he proposed choosing the portfolio that maximizes portfolio $(E - d)/\sigma$, where $d$ is a fixed (for the analysis) disastrous return and $\sigma$ is standard deviation of return. Roy's formula for the variance of the portfolio, like the one I presented, included the covariances of returns among securities. The chief differences between the Roy analysis and my analysis were that (1) mine required nonnegative investments whereas Roy's allowed the amount invested in any security to be positive or negative and (2) I proposed allowing the investor to choose a desired portfolio from the efficient frontier whereas Roy recommended choice of a specific portfolio.

*Harry M. Markowitz is president of Harry Markowitz Company.*

Comparing the two articles, one might wonder why I got a Nobel Prize for mine and Roy did not for his. Perhaps the reason had to do with the differences described in the preceding paragraph, but the more likely reason was visibility to the Nobel Committee in 1990. Roy's 1952 article was his first and last article in finance. He made this one tremendous contribution and then disappeared from the field, whereas I wrote two books (Markowitz 1959; Markowitz 1987) and an assortment of articles in the field. Thus, by 1990, I was still active and Roy may have vanished from the Nobel Committee's radar screen.

**Problems with Markowitz (1952).** I am tempted to include a disclaimer when I send requested copies of Markowitz (1952) that warns the reader that the 1952 piece should be considered only a historical document—not a reflection of my current views about portfolio theory. There are at least four reasons for such a warning. The first two are two technical errors described in this section. A third is that, although the article noted that the same portfolios that minimize standard deviation for given $E$ also minimize variance for given $E$, it failed to point out that standard deviation (rather than variance) is the intuitively meaningful measure of dispersion. For example, "Tchebychev's inequality" says that 75 percent of any probability distribution lies between the mean and ±2 *standard deviations*—not two variances. Finally, the most serious differences between Markowitz (1952) and the views I now hold concern questions about "why mean and variance?" and "mean and variance of what?". The views expressed in Markowitz (1952) were held by me very briefly. Those expressed in Markowitz (1959) have been held by me virtually unchanged since about 1955. I will discuss these views in the section on Markowitz (1959).

As for the technical errors: First, it has been known since Markowitz (1956) that variance, $V$, is a strictly convex function of expected return among efficient $EV$ combinations. Markowitz (1952) explained, correctly, that the curve is piecewise parabolic. Figure 6 of Markowitz (1952) showed two such parabola segments meeting at a point. The problem with the figure is that its parabolas meet in such a way that the resulting curve is not convex. This cannot happen. Second, Figure 1 in Markowitz (1952) was supposed to portray the set of all feasible $EV$ combinations. In particular, it showed the "inefficient border," with *maximum* $V$ for a given $E$, as a concave function. This is also an error. Since my 1956 article, we know that the curve relating maximum $V$ for a given $E$ is neither concave nor convex (see Markowitz 1987, Chapter 10, for a description of possibilities).

**The "General" Portfolio Selection Problem.** For the case in which one and only one feasible portfolio minimizes variance among portfolios with any given feasible expected return, Markowitz (1952) illustrated that the set of efficient portfolios is piecewise linear. It may be traced out by starting with the unique point (portfolio) with minimum feasible variance, moving in a straight line from this point, then perhaps, after some distance, moving along a different straight line, and so on, until the efficient portfolio with maximum expected return is reached.[1] Note that we are not discussing here the shape of efficient mean–variance combinations or the shape of efficient mean–standard deviation combinations. Rather, we are discussing the shape of the set of efficient portfolios in "portfolio space."

The set of portfolios described in the preceding paragraph is *not* a piecewise linear *approximation* to the problem; rather, the exact solution is itself piecewise linear. The points (portfolios) at which the successive linear pieces meet are called "corner portfolios" because the efficient set turns a corner and heads in a new direction at each such point. The starting and ending points (with, respectively, minimum variance and maximum mean) are also called corner portfolios.

Markowitz (1952) did not present the formulas for the straight lines that make up the set of efficient portfolios. These formulas were supplied in Markowitz (1956), but Markowitz (1956) solved a much more general problem than discussed in Markowitz (1952). A portfolio in Markowitz (1952) was considered feasible ("legitimate") if it satisfied one equation (the budget constraint) and its values (investments) were not negative. Markowitz (1956), however, solved the (single-period mean–variance) portfolio selection problem for a wide variety of possible feasible sets, including the Markowitz (1952) and Roy feasible sets as special cases.

Specifically, Markowitz (1956) allowed the portfolio analyst to designate none, some, or all variables to be subject to nonnegativity constraints (as in Markowitz 1952) and the remaining variables to not be thus constrained (as in Roy). In addition to (or instead of) the budget constraint, the portfolio analyst could specify zero, one, or more linear equality constraints (sums or weighted sums of variables required to equal some constant) and/or linear inequality constraints (sums or weighted sums of variables required to be no greater or no less than some constant). A portfolio analyst can set down a system of constraints of these kinds such that no portfolio can meet all constraints. In this case, we say that the model is "infeasible." Otherwise, it is a "feasible model."

 

In addition to permitting any system of constraints, Markowitz (1956) made an assumption[2] that assured that if the model was feasible, then (as in Markowitz 1952) there was a unique feasible portfolio that minimized variance among portfolios with any given feasible *E*.

Markowitz (1956) showed that the set of efficient portfolios is piecewise linear in the general model, as in the special case of Markowitz (1952). Depending on the constraints imposed by the portfolio analyst, one of the linear pieces of the efficient set could extend without end in the direction of increasing *E*, as in the case of the Roy model. (Note that if the analysis contains 1,000 securities, the lines we are discussing here are straight lines in 1,000-dimensional "portfolio space." These lines may be hard to visualize and impossible to draw, but they are not hard to work with algebraically.)

Markowitz (1956) presented a computing procedure, the "critical line algorithm," that computes each corner portfolio in turn and the efficient line segment between them, perhaps ending with an efficient line "segment" on which feasible *E* increases without end. The formulas for the efficient line segments are all of the same pattern. Along a given "critical line," some of the variables that are required to be nonnegative are said to be OUT and are set to zero; the others are said to be IN and are free to vary. Variables not constrained to be nonnegative are always IN. On the critical line, some inequalities are called SLACK and are ignored; the others are BINDING and are treated (in the formula for the particular critical line) as if they were equalities. With its particular combination of BINDING constraints and IN variables, the formula for the critical line is the same as if the problem were to minimize *V* for various *E* subject to only equality constraints. In effect, OUT variables and SLACK constraints are deleted from the problem.

At each step, the algorithm uses the formula for the current critical line for easy determination of the next corner portfolio. The next critical line, which the current critical line meets at the corner, has the same IN variables and BINDING constraints as the current line except for *one* of the following—one variable moves from OUT to IN or moves from IN to OUT or one constraint moves from BINDING to SLACK or from SLACK to BINDING. This similarity between successive critical lines greatly facilitates the solution of one line when given the solution of the preceding critical line.[3]

Merton (1972) said, "The characteristics of the mean–variance efficient portfolio frontier have been discussed at length in the literature. However, for more than three assets, the general approach has been to display qualitative results in terms of graphs" (p. 1851). I assume that at the time, Merton had not read Markowitz (1956) or Appendix A of Markowitz (1959).

## Markowitz Portfolio Theory circa 1959

Markowitz (1959) was primarily written during the 1955–56 academic year while I was at the Cowles Foundation for Research in Economics at Yale at the invitation of Tobin. At the time, Tobin was already working on what was to become Tobin (1958), which is discussed in the next section.

I had left the University of Chicago for the RAND Corporation in 1951; my coursework was finished, but my dissertation (on portfolio theory) was still to be written. My RAND work had nothing to do with portfolio theory. So, my stay at the Cowles Foundation on leave from RAND provided an extended period when I could work exclusively on, as well as write about, portfolio theory. The following subsections summarize the principal ways in which my views on portfolio theory evolved during this period, as expressed in Markowitz (1959).

**A Still More General Mean–Variance Analysis.** The central focus of Markowitz (1959) was to explain portfolio theory to a reader who lacked advanced mathematics. The first four chapters introduced and illustrated mean–variance analysis, defined the concepts of mean, variance, and covariance, and derived the formulas for the mean and variance of a portfolio. Chapter 7 defined mean–variance efficiency and presented a geometric analysis of efficient sets, much like Markowitz (1952) but without the two errors noted previously. Chapter 8 introduced the reader to some matrix notation and illustrated the critical line algorithm in terms of a numerical example.

The proof that the critical line algorithm produces the desired result was presented in Appendix A of Markowitz (1959). Here, the result was more general than that in Markowitz (1956). The result in Markowitz (1956) made an assumption sufficient to assure that a unique feasible portfolio would minimize variance for any given *E*. Markowitz (1959) made no such assumption; rather, it demonstrated that the critical line algorithm will work for *any* covariance matrix. The reason it works is as follows: Recall that the equations for a critical line depend on which variables are IN and which are OUT. Appendix A showed that each IN set encountered in tracing out the efficient frontier is such that the associated equations for the critical line are solvable.[4]

**Models of Covariance.** Markowitz (1959, pp. 96–101) argued that analysis of a large portfolio consisting of many different assets has too many covariances for a security analysis team to carefully consider them individually, but such a team can carefully consider and estimate the parameters of a model of covariance. This point was illustrated in terms of what is now called a single-index or one-factor (linear) model. The 1959 discussion briefly noted the possibility of a more complex model—perhaps involving multiple indexes, nonlinear relationships, or distributions that vary through time.

Markowitz (1959) presented no empirical analysis of the ability of particular models to represent the real covariance matrix (as in Sharpe 1963, Cohen and Pogue 1967, Elton and Gruber 1973, or Rosenberg 1974), and I did not yet realize how a (linear) factor model could be used to simplify the computation of critical lines, as would be done in Sharpe (1963) and in Cohen and Pogue.

**The Law of the Average Covariance.** Chapter 5 of Markowitz (1959) considered, among other things, what happens to the variance of an equally weighted portfolio as the number of investments increases. It showed that the existence of correlated returns has major implications for the efficacy of diversification. With uncorrelated returns, portfolio risk approaches zero as diversification increases. With correlated returns, even with unlimited diversification, risk can remain substantial. Specifically, as the number of stocks increases, the variance of an equally weighted portfolio approaches the "average covariance" (i.e., portfolio variance approaches the number you get by adding up all covariances and then dividing by the number of them). I now refer to this as the "law of the average covariance."

For example, if all securities had the same variance $V_S$ and every pair of securities (other than the security with itself) had the same correlation coefficient ρ, the average covariance would be $\rho V_S$ and portfolio variance would approach $\rho V_S$; therefore, portfolio standard deviation would approach $\sqrt{\rho}\sigma_s$. If the correlation coefficient that all pairs shared was, for example, 0.25, then the standard deviation of the portfolio would approach 0.5 times the standard deviation of a single security. In this case, investing in an unlimited number of securities would result in a portfolio whose standard deviation was 50 percent as great as that of a completely undiversified portfolio. Clearly, there is a qualitative difference in the efficacy of diversification depending on whether one assumes correlated or uncorrelated returns.

**Semideviation.** Semivariance is defined like variance (as an expected squared deviation from something) except that it counts only deviations below some value. This value may be the mean of the distribution or some fixed value, such as zero return. Semideviation is the square root of semivariance. Chapter 9 of Markowitz (1959) defined semivariance and presented a three-security geometric analysis illustrating how the critical line algorithm can be modified to trace out mean–semideviation-efficient sets. Appendix A presented the formal description of this modification for any number of securities and a proof that it works.

**Mean and Variance of What? Why Mean and Variance?** The basic ideas of Markowitz (1952) came to me sometime in 1950 while I was reading Williams (1938) in the Business School library at the University of Chicago. I was considering applying mathematical or econometric techniques to the stock market for my Ph.D. dissertation for the Economics Department. I had not taken any finance courses, nor did I own any securities, but I had recently read Graham and Dodd (1934), had examined Wiesenberger (circa 1950), and was now reading Williams.

Williams asserted that the value of a stock is the expected present value of its future dividends. My thought process went as follows: If an investor is only interested in some kind of expected value for securities, he/she must be only interested in that expected value for the portfolio, but the maximization of an expected value of a portfolio (subject to a budget constraint in nonnegative investments) does not imply the desirability of diversification. Diversification makes sense as well as being common practice. What was missing from the analysis, I thought, was a measure of risk. Standard deviation or variance came to mind. On examining the formula for the variance of a weighted sum of random variables (found in Uspensky 1937 on the library shelf), I was elated to see the way covariances entered. Clearly, effective diversification required avoiding securities with high covariance. Dealing with two quantities—mean and variance—and being an economics student, I naturally drew a trade-off curve. Being, more specifically, a student of T.C. Koopmans (see Koopmans 1951), I labeled dominated *EV* combinations "inefficient" and undominated ones "efficient."

The Markowitz (1952) position on the questions used as the heading for this subsection differed little from my initial thoughts while reading Williams. Markowitz (1952) started by rejecting the rule that the "investor does (or should) maximize the discounted . . . [expected] value of future returns," both as a hypothesis about actual behavior and as a maxim for recommended behavior, because it "fails to imply diversification no matter how the anticipated returns are formed." Before presenting the mean–variance rule, Markowitz (1952) said:

> It will be convenient at this point to consider a static model. Instead of speaking of the time series of returns on the *i*th security ($r_{i,1}, r_{i,2}, \ldots, r_{i,t}, \ldots$) we will speak of "the flow of returns" ($r_i$) from the *i*th security. The flow of returns from the portfolio as a whole is $R = \sum X_i r_i$. (pp. 45–46)

The flow of returns concept is not heard from after this point. Shortly, Markowitz (1952) introduced "elementary concepts and results of mathematical statistics," including the mean and variance of a sum of random variables. "The return ($R$) on the portfolio as a whole is a weighted sum of random variables (where the investor can choose the weights)." From this point forward, Markowitz (1952) was primarily concerned with how to choose the weights $X_i$ so that portfolios will be mean–variance efficient.

Markowitz (1952) stated that its "chief limitations" are that "(1) we do not derive our results analytically for the $n$-security case; . . . (2) we assume static probability beliefs." This work expressed the intention of removing these limitations in the future. Markowitz (1956) and Appendix A of Markowitz (1959) addressed the first issue, and Chapter 13 of Markowitz (1959) addressed the second issue.

Chapters 10–12 of Markowitz (1959) reviewed the theory of rational decision making under risk and uncertainty. Chapter 10 was concerned with rational decision making in a single period with known odds; Chapter 11 reviewed many-period optimizing behavior (again, with known odds); Chapter 12 considered single- or many-period rational behavior when the odds might be unknown. The introduction in Chapter 10 emphasized that the theory reviewed there applies to an idealized rational decision maker with limited information but unlimited computing powers and is not necessarily a hypothesis about actual human behavior. This position contrasts with Markowitz (1952), which offered the mean–variance rule both as a hypothesis about actual behavior and as a maxim for recommended behavior.

Chapter 13 applied the theory of rational behavior—which was developed by John von Neumann and Oskar Morgenstern (1944), Leonard J. Savage (1954), Richard Bellman (1957), and others, and was reviewed in Chapters 10 through 12—to the problem of how to invest. It began with a many-period consumption–investment game and made enough assumptions to assure that the dynamic programming solution to the game as a whole would consist of maximizing a sequence of single-period "derived" utility functions that depended only on end-of-period wealth. Chapter 13 then asked whether knowledge of the mean and variance of a return distribution allows one to estimate fairly well the distribution's expected utility. The analysis here did *not* assume either normally distributed returns or a quadratic utility function (as in Tobin 1958). It did consider the robustness of quadratic approximations to utility functions. In other words, if you know the mean and variance of a distribution, can you approximate its expected utility? See also Levy and Markowitz (1979). Furthermore, Chapter 13 considered what kinds of approximations to expected utility are implied by other measures of risk.

The last six pages of the chapter sketched how one could or might ("could" in the easy cases, "might" in the hard cases) incorporate into a formal portfolio analysis considerations such as (1) consumer durables, (2) nonportfolio sources of income, (3) changing probability distributions, (4) illiquidities, and (5) taxes. As compared with later analyses, the Chapter 13 consumption–investment game was in discrete time rather than continuous time (as in Merton 1969), did not reflect the discovery of myopic utility functions (as did Mossin 1968 and Samuelson 1969), and did not consider the behavior of a market populated by consumers/investors playing this game. Its objective was to provide a theoretical foundation for portfolio analysis as a practical way to approximately maximize the derived utility function of a rational investor.

## Tobin (1958)

Tobin was concerned with the demand for money as distinguished from other "monetary assets." Monetary assets, including cash, were defined by Tobin as "marketable, fixed in money value, free of default risk." Tobin stated:

> Liquidity preference theory takes as given the choices determining how much wealth is to be invested in monetary assets and concerns itself with the allocation of these amounts among cash and alternative monetary assets. (p. 66)

Tobin assumed that the investor seeks a mean–variance-efficient combination of monetary assets.

He justified the use of expected return and standard deviation as criteria on either of two bases: Utility functions are quadratic, or probability distributions are from some two-parameter family of return distributions.

Much of Tobin's article analyzed the demand for money when "consols"[5] are the only other monetary asset. The next-to-last section of the article was on "multiple alternatives to cash." Here, Tobin presented his seminal result now known as the Tobin Separation Theorem. Tobin assumed a portfolio selection model with $n$ risky assets and one riskless asset, cash. Because these assets were monetary assets, the risk was market risk, not default risk. Holdings had to be nonnegative. Borrowing was not permitted. Implicitly, Tobin assumed that the covariance matrix for risky assets is nonsingular (or he could have made the slightly more general assumption of Markowitz 1956). Tobin showed that these premises imply that for a given set of means, variances, and covariances among efficient portfolios containing any cash at all, the proportions among risky stocks are always the same:

> . . . the proportionate composition of the non-cash assets is independent of their aggregate share of the investment balance. This fact makes it possible to describe the investor's decisions as if there were a single non-cash asset, a composite formed by combining the multitude of actual non-cash assets in fixed proportions. (p. 84)

The primary purpose of Tobin's analysis was to provide an improved theory of the holding of cash. He concluded that the preceding analysis

> ... is a logically more satisfactory foundation for liquidity preference than the Keynesian theory.... Moreover, it has the empirical advantage of explaining diversification—the same individual holds both cash and "consols"—while the Keynesian theory implies that each investor will hold only one asset. (p. 85)

At a meeting with Tobin in attendance, I once referred to his 1958 article as the first capital asset pricing model (CAPM). Tobin declined the honor. It is beyond the scope of this article, which has a 1960 cutoff, to detail the contributions of William Sharpe (1964), John Lintner (1965), Jan Mossin (1966), and others in the development of capital asset pricing models. A comparison of the assumptions and conclusions of Tobin with those of Sharpe may, however, help locate Tobin in the development of today's financial economics.

Tobin contrasted his interest to mine as follows:

> Markowitz's main interest is prescription of rules of rational behavior for investors: the main concern of this paper is the implications for economic theory, mainly comparative statics, that can be derived from assuming that investors do in fact follow such rules. (p. 85, Note 1)

To this extent, at least, the focus of Sharpe (1964) is the same as that of Tobin. Tobin and Sharpe are also similar in postulating a model with $n$ risky and one riskless security. The principal differences between the two are (1) a difference in assumption between their mathematical models and (2) the economic phenomena concerning which the respective models are asserted.

As for assumptions, Tobin assumed that one can invest (i.e., lend) at the risk-free rate. Sharpe assumed that the investor can either borrow or lend at the same rate. (Tobin usually assumed that the rate is zero, but he noted that this assumption is not essential.) This, perhaps seemingly small, difference between the two models makes for a substantial difference in their conclusions. First, if investors can borrow or lend all they want at the risk-free rate (and the covariance matrix among the $n$ risky stocks is nonsingular), then *all* efficient portfolios consist of one particular combination of risky assets, perhaps plus borrowing or lending. The implication is that, in equilibrium, the market portfolio (plus borrowing or lending) is the *only* efficient portfolio. In the Tobin model, in contrast, if investors have heterogeneous risk tolerances—so some hold cash and others do not—the market portfolio can be quite inefficient, even when all investors have the same beliefs and all hold mean–variance-efficient portfolios (see Markowitz 1987, Chapter 12).

Probably the most remarkable conclusion Sharpe drew from his premises was that in CAPM equilibrium, the expected return of each security is linearly related to its beta and only its beta. This conclusion is not necessarily true in the Tobin model (see Markowitz 1987, Chapter 12).

The second major difference between the two works is that Sharpe postulated that his model applied to all securities, indeed all "capital assets," whereas Tobin postulated only that his model applied to "monetary assets." In fact, Tobin expressed doubts that cash should be considered risk free:

> It is for this reason that the present analysis has been deliberately limited . . . to choices among monetary assets. Among these assets cash is relatively riskless, even though in the wider context of portfolio selection, the risk of changes in purchasing power, which all monetary assets share, may be relevant to many investors.

Between them, Tobin's assumptions were more cautious; Sharpe's revolutionized financial economics.

 

## Hicks (1935, 1962)

The Hicks (1962) article on liquidity included the following paragraph:

> It would obviously be convenient if we could take just one measure of "certainty"; the measure which would suggest itself, when thinking on these lines, is the standard deviation. The chooser would then be supposed to be making his choice between different total outcomes on the basis of mean value (or "expectation") and standard deviation only. A quite simple theory can be built up on that basis, and it yields few conclusions that do not make perfectly good sense. It may indeed be regarded as a straightforward generalisation of Keynesian Liquidity Preference. We would be interpreting Liquidity Preference as a willingness to sacrifice something in terms of mean value in order to diminish the expected variance (of the whole portfolio). Instead of looking simply at the single choice between money and bonds, we could introduce many sorts of securities and show the distribution between them determined on the same principle. It all works out very nicely, being indeed no more than a formalisation of an approach with which economists have been familiar since 1936 (or perhaps I may say 1935). [A footnote to the last sentence of this paragraph explained as follows:] Referring to my article, "A Suggestion for Simplifying the Theory of Money." *Economica* (February 1935). (p. 792)

The formalization was spelled out in a mathematical appendix to Hicks (1962) titled "The Pure Theory of Portfolio Investment" and in a footnote on p. 796 that presents an $E\sigma$ – efficient set diagram.

The appendix presented a mathematical model that is almost exactly the Tobin model with no reference to Tobin. The difference between the Hicks and Tobin models is that Hicks assumed that all correlations are zero whereas Tobin permitted any nonsingular covariance matrix. Specifically, Hicks presented the general formula for portfolio variance written in terms of correlations, rather than covariances, and then stated:

> It can, I believe, be shown that the main properties which I hope to demonstrate, remain valid whatever the $r$'s; but I shall not attempt to offer a general proof in this place. I shall simplify by assuming that the prospects of the various investments are uncorrelated ($r_{jk} = 0$ when $k \neq j$): an assumption with which, in any case, it is natural to begin.

In the discussion that followed, Hicks (1962) derived the Tobin conclusion that among portfolios that include cash, there is a linear relationship between portfolio mean and standard deviation and that the proportions among risky assets remain constant along this linear portion of the efficient frontier. In other words, Hicks presented what we call the Tobin Separation Theorem.

Hicks also analyzed the efficient frontier beyond the point where the holding of cash goes to zero. In particular, he noted that as we go out along the frontier in the direction of increasing risk and return, securities leave the efficient portfolio and do not return. (This last point is not necessarily true with correlated returns.[6])

Returning to the portion of the frontier that contains cash, if the Hicks (1962) results are, in fact, a formalization of those in Hicks (1935)—in the sense of transcribing into mathematics results that were previously described verbally—then the Tobin Separation Theorem should properly be called the Hicks or Hicks–Tobin Separation Theorem. Let us examine Hicks (1935) to see if it did anticipate Tobin as described in the appendix to Hicks (1962).

Within Hicks (1935), the topic of Section V is closest to that of "The Pure Theory of Portfolio Investment" in the appendix of Hicks (1962). Preceding sections of Hicks (1935) discussed, among other things, the need for an improved theory of money and the desirability of building a theory of money along the same lines as the existing theory of value. They also discussed, among other things, the relationship between the Hicks (1935) analysis and that of Keynes as well as the existence of "frictions," such as "the cost of transferring assets from one form to another." In Section IV, Hicks (1935) introduced risk into his analysis. Specifically, he noted, "The risk-factor comes into our problem in two ways: First, as affecting the expected period of investment, and second, as affecting the expected net yield of investment" (p. 7). In a statement applicable to both sources of risk, Hicks continued:

> Where risk is present, the *particular* expectation of a riskless situation is replaced by a band of possibilities, each of which is considered more or less probable. It is convenient to represent these probabilities to oneself, in statistical fashion, by a mean value, and some appropriate measure of dispersion. (No single measure will be wholly satisfactory, but here this difficulty may be overlooked.) (p. 8)

Hicks (1935) never designated standard deviation or any other specific measure as the measure he meant when speaking of risk. After discussing uncertainty of the period of the investment, he concluded Section IV thus:

> To turn now to the other uncertainty—uncertainty of the yield of investment. Here again we have a penumbra.... Indeed, without assuming this to be the normal case, it would be impossible to explain some of the most obvious of the observed facts of the capital market. (p. 8)

The theory of investment that Hicks (1935) presented in Section V may be summarized as follows:

> It is one of the peculiarities of risk that the total risk incurred when more than one risky investment is undertaken does not bear any simple relation to the risk involved in each of the particular investments taken separately. In most cases, the "law of large numbers" comes into play (quite how, cannot be discussed here)....
>
> Now, in a world where cost of investment was negligible, everyone would be able to take considerable advantage of this sort of risk reduction. By dividing up his capital into small portions, and spreading his risks, he would be able to insure himself against any large total risk on the whole amount. But in actuality, the cost of investment, making it definitely unprofitable to invest less than a certain minimum amount in any particular direction, closes the possibility of risk reduction along these lines to all those who do not possess the command over considerable quantities of capital....
>
> By investing only a proportion of total assets in risky enterprises, and investing the remainder in ways which are considered more safe, it will be possible for the individual to adjust his whole risk situation to that which he most prefers, more closely than he could do by investing in any single enterprise. (pp. 9–10)

Hicks (1935) was a forerunner of Tobin in seeking to explain the demand for money as a consequence of the investor's desire for low risk as well as high return. Beyond that, there is little similarity between the two authors. Hicks (1935), unlike Tobin or the appendix to Hicks (1962), did not designate standard deviation or any other specific measure of dispersion as representing risk for the analysis; therefore, he could not show a formula relating risk on the portfolio to risk on individual assets. Hicks (1935) did not distinguish between efficient and inefficient portfolios, contained no drawing of an efficient frontier, and had no hint of any kind of theorem to the effect that all efficient portfolios that include cash have the same proportions among risky assets.

Thus, there is no reason why the theorem that currently bears Tobin's name should include any other name.

## Marschak (1938)

Kenneth Arrow (1991) said of Marschak (1938):

> Jacob Marschak ... made some efforts to construct an ordinal theory of choice under uncertainty. He assumed a preference ordering in the space of parameters of probability distributions—in the simplest case, the space of the mean and the variance.... From this formulation to the analysis of portfolio selection in general is the shortest of steps, but one not taken by Marschak. (p. 14)

G.M. Constantinides and A.G. Malliaris (1995) described the role of Marschak (1938) as follows.

> The asset allocation decision was not adequately addressed by neoclassical economists. ... The methodology of deterministic calculus is adequate for the decision of maximizing a consumer's utility subject to a budget constraint. Portfolio selection involves making a decision under uncertainty. The probabilistic notions of expected return and risk become very important. Neoclassical economists did not have such a methodology available to them.... An early and important attempt to do that was made by Marschak (1938) who expressed preferences for investments by indifference curves in the mean–variance space. (pp. 1–2)

An account of Marschak is, therefore, mandatory in a history of portfolio theory through 1960, if for no other reason than that these scholars judged it to be important. On the other hand, I know of one authority who apparently did not think the article to be important for the development of portfolio theory. My thesis supervisor was Marschak himself, and he never mentioned Marschak (1938). When I expressed interest in applying mathematical or econometric techniques to the stock market, Marschak told me of Alfred Cowles own interest in financial applications, resulting, for example, in Cowles 1939 work.[7] Then, Marschak sent me to Marshall Ketchum in the Business School at the University of Chicago for a reading list in finance. This list included Williams (1938) and, as I described, led to the day in the library when my version of portfolio theory was born. Marschak kept track of my work, read my dissertation, but never mentioned his 1938 article.

So, which authority is correct concerning the place of Marschak in the development of portfolio theory? Like Hicks, Marschak sought to achieve a better theory of money by integrating it with the General Theory of Prices. In the introductory section of the article, Marschak explained that

> to treat monetary problems and indeed, more generally, problems of investment with the tools of a properly generalized Economic Theory ... requires, first, an extension of the concept of human *tastes*: by taking into account not only men's aversion for waiting but also their desire for safety, and other traits of behaviour not present in the world of perfect certainty as postulated in the classical static economics. Second, the *production conditions*, assumed hereto to be objectively given, become, more realistically, mere subjective expectations of the investors—and *all* individ-

 

> uals are investors (in any but a timeless economy) just as all market transactions are investments. The problem is: to explain the objective quantities of goods and claims held at any point of time, and the objective market prices at which they are exchanged, given the subjective tastes and expectations of the individuals at this point of time. (p. 312)

In the next five sections, Marschak presented the equations of the standard economic analysis of production, consumption, and price formation. Section 7 dealt with choice when outcomes are random. No new equations were introduced in this section. Rather, Marschak used the prior equations with new meanings:

> We may, then, use the previous formal setup if we reinterpret the notation: *x*, *y* . . . shall mean, not future yields, but parameters (e.g., moments and joint moments) of the joint-frequency distribution of future yields. Thus, *x* may be interpreted as the mathematical expectation of first year's meat consumption, *y* may be its standard deviation, *z* may be the correlation coefficient between meat and salt consumption in a given year, *t* may be the third moment of milk consumption in second year, etc. (p. 320)

Marschak noted that people usually like high mean and low standard deviation; also, "they like meat consumption to be accompanied by salt consumption" (i.e., *z* as well as *x* in the preceding quotation "are positive utilities" as opposed to standard deviation, *y*, which is "a disutility"). He noted that people "like 'long odds' (i.e., high positive skewness of yields." However, it "is sufficiently realistic . . . to confine ourselves, for each yield, to two parameters only: the mathematical expectation ('lucrativity') and the coefficient of variation ('risk')."

So, is Marschak's article a forerunner of portfolio theory or not? Yes and no. It is not a step (say, beyond Hicks 1935) toward *portfolio* theory because it does not consider *portfolios*. The means, standard deviations, and correlations of the analysis, including the means (and so on) of end products consumed, appear directly in the utility and transformation functions with no analysis of how they combine to form moments of the investor's portfolio as a whole. On the other hand, Marschak's 1938 work is a landmark on the road to a theory of markets whose participants act under risk and uncertainty, as later developed in Tobin and the CAPMs. It is the farthest advance of economics under risk and uncertainty prior to the publication of von Neumann and Morgenstern (1944).

## Williams (1938)

The episode reported previously in which I discovered the rudiments of portfolio theory while reading Williams occurred in my reading early parts of the book. Later in the book, Williams observed that the future dividends of a stock or the interest and principal of a bond may be uncertain. He said that, in this case, probabilities should be assigned to various possible values of the security and the mean of these values used as *the* value of the security. Finally, he assured readers that by investing in sufficiently many securities, risk can be virtually eliminated. In particular, in the section titled "Uncertainly and the Premium for Risk" (starting on p. 67 in the chapter on "Evaluation by the Rule of Present Worth"), he used as an example an investor appraising a risky 20-year bond "bearing a 4 per cent coupon and selling at 40 to yield 12 per cent to maturity, even though the pure interest seems to be only 4 per cent." His remarks apply to any investor who "cannot tell for sure" what the present worth is of the dividends or of the interest and principal to be received:

> Whenever the value of a security is uncertain and has to be expressed in terms of probability, the correct value to choose is the mean value. . . . The customary way to find the value of a risky security has always been to add a "premium for risk" to the pure interest rate, and then use the sum as the interest rate for discounting future receipts. In the case of the bond under discussion, which at 40 would yield 12 per cent to maturity, the "premium for risk" is 8 per cent when the pure interest rate is 4 per cent.
>
> Strictly speaking, however, there is no risk in buying the bond in question if its price is right. Given adequate diversification, gains on such purchases will offset losses, and a return at the pure interest rate will be obtained. Thus the *net risk* turns out to be nil. To say that a "premium for risk" is needed is really an elliptical way of saying that payment of the full face value of interest and principal is not to be expected on the average.

In my 1952 article, I said that Williams's prescription has the investor

> diversify his funds among all those securities which give maximum expected return. The law of large numbers will insure that the actual yield of the portfolio will be almost the same as the expected yield. This rule is a special case of the expected returns–variance of returns rule. . . . It assumes that there is a portfolio which gives both maximum expected return and minimum variance, and it commends this portfolio to the investor.
>
> This presumption, that the law of large numbers applies to a portfolio of securities, cannot be accepted. The returns from securities are too intercorrelated. Diversification cannot eliminate all variance.

That is still my view. It should be noted, however, that Williams's "dividend discount model" remains one of the standard ways to estimate the security means needed for a mean–variance analyses (see Farrell 1985).

## Leavens (1945)

Lawrence Klein called my attention to an article on the diversification of investments by Leavens, a former member of the Cowles Commission. Leavens (1945) said:

> An examination of some fifty books and articles on investment that have appeared during the last quarter of a century shows that most of them refer to the desirability of diversification. The majority, however, discuss it in general terms and do not clearly indicate why it is desirable.

Leavens illustrated the benefits of diversification on the assumption that risks are independent. However, the last paragraph of Leavens cautioned:

> The assumption, mentioned earlier, that each security is acted upon by independent causes, is important, although it cannot always be fully met in practice. Diversification among companies in one industry cannot protect against unfavorable factors that may affect the whole industry; additional diversification among industries is needed for that purpose. Nor can diversification among industries protect against cyclical factors that may depress all industries at the same time.

Thus, Leavens understood intuitively, as did Shakespeare 350 years earlier, that some kind of model of covariance is at work and that it is relevant to the investment process. But he did not incorporate it into his formal analysis.

Leavens did not provide us with his reading list of "some fifty books and articles." This omission is fortunate because I am probably not prepared to read them all and the reader is surely not ready to read accounts of them. Let us assume, until some more conscientious student of this literature informs us otherwise, that Leavens was correct that the majority discussed diversification in general terms and did "not clearly indicate why it is desirable." Let us further assume that the financial analysts who did indicate why it is desirable did not include covariance in their formal analyses and had not developed the notion of an efficient frontier. Thus, we conclude our survey with Leavens as representative of finance theory's analysis of risk as of 1945 and, presumably, until Roy and Markowitz in the 1950s.

## The End of the Beginning

One day in 1960, having said what I had to say about portfolio theory in my 1959 book, I was sitting in my office at the RAND Corporation in Santa Monica, California, working on something quite different, when a young man presented himself at my door, introduced himself as Bill Sharpe, and said that he also was employed at RAND and was working toward a Ph.D. degree at UCLA. He was looking for a thesis topic. His professor, Fred Weston, had reminded Sharpe of my 1952 article, which they had covered in class, and suggested that he ask me for suggestions on a thesis topic. We talked about the need for models of covariance. This conversation started Sharpe out on the first of his (ultimately many) lines of research, which resulted in Sharpe (1963).

For all we know, the day Sharpe introduced himself to me at RAND could have been exactly 10 years after the day I read Williams. On that day in 1960, there was *no* talk about the possibility of using portfolio theory to revolutionize the theory of financial markets, as done in Sharpe (1964), nor was there any inkling of the flood of discoveries and applications, many by Sharpe himself, that were to occur in investment theory and financial economics during the next four decades.

 

## Notes

1. Given the assumptions of Markowitz (1952), if more than one portfolio has maximum feasible $E$, only one of these portfolios will be efficient, namely, the one with the smallest $V$. This one will be reached by the "tracing out" process described.
2. The assumption was that $V$ is strictly convex over the set of feasible portfolios. This assumption is weaker than requiring the covariance matrix to be nonsingular.
3. In the text, I am discussing the shape of efficient sets in portfolio space. As observed in Markowitz (1952), the set of efficient $EV$ combinations is piecewise parabolic, with each line segment in portfolio space corresponding to a parabolic segment in $EV$ space. As discussed previously, Markowitz (1956) understood that successive parabolas meet in such a way that efficient $V$ as a function of $E$ is strictly convex. Markowitz (1956) noted that typically there is no kink where two successive efficient parabola segments meet: The slope of the one parabola equals that of the other at the corner portfolio where they meet. Markowitz (1956) did, however, note the possibility of a kink in the efficient $EV$ set if a certain condition occurred, but the 1956 work did not provide a numerical example of a problem containing such a kink. For numerical examples of problems with kinks in the efficient $EV$ set, see Dybvig (1984) and Chapter 10 of Markowitz (1987).
4. The equations in Markowitz (1956) also depended on which inequalities were BINDING. Markowitz (1959) wrote inequalities as equalities, without loss of generality, by introducing "slack variables" as in linear programming. The critical line algorithm works even if the constraint matrix, **A**, as well as the covariance matrix, **C**, is rank deficient. The critical line algorithm begins with George Dantzig's (1963) simplex algorithm to maximize $E$ or determine that $E$ is unbounded. The simplex algorithm introduces "dummy slacks," some of which remain in the critical line algorithm if **A** is rank deficient (see Markowitz 1987, Chapters 8 and 9). Historically, not only did I have great teachers at the University of Chicago, including Jacob Marschak, T.C. Koopmans, Milton Friedman, and L.J. Savage, but I was especially fortunate to have Dantzig as a mentor when I worked at RAND.
5. Government bonds in Great Britain, originally issued in 1751, that (similarly to an annuity) pay perpetual interest and have no date of maturity.
6. See Chapter 11 in Markowitz (1987) for a three-security example of risky assets in which an asset leaves and later reenters the efficient portfolio. Add cash to the analysis in such a way that the coming and going of the security happens above the tangency of the line from $(0, r_0)$ to the frontier. Perhaps, if you wish, add a constant to all expected returns, including $r_0$, to assure that $r_0 \geq 0$.
7. The Cowles Commission for Research in Economics, endowed by Alfred Cowles, was affiliated with the University of Chicago at the time. Marschak was formerly its director. I was a student member.

## References


Arrow, Kenneth. 1991. "Cowles in the History of Economic Thought." *Cowles Fiftieth Anniversary*. The Cowles Foundation for Research in Economics at Yale University:1–24.

Bellman, Richard. 1957. *Dynamic Programming*. Princeton, NJ: Princeton University Press.

Cohen, K.J., and J.A. Pogue. 1967. "An Empirical Evaluation of Alternative Portfolio Selection Models." *Journal of Business*, vol. 40, no. 2 (April):166–193.

Constantinides, G.M., and A.G. Malliaris. 1995. In *Handbooks in OR & MS*. Edited by R. Jarrow, et al. Vol. 9. Amsterdam, Netherlands: Elsevier Science B.V.

Cowles, A., 3rd, and Associates. 1938. *Common-Stock Indexes, 1871–1937*. Bloomington, IA: Principia Press.

Dantzig, G.B. 1963. *Linear Programming and Extensions*. Princeton, NJ: Princeton University Press.

Dybvig, Philip H. 1984. "Short Sales Restrictions and Kinks of the Mean Variance Frontier." *Journal of Finance*, vol. 39, no. 1 (March):239–244.

Elton, E.J., and M.J. Gruber. 1973. "Estimating the Dependence Structure of Share Prices." *Journal of Finance*, vol. 28, no. 5 (December):1203–32.

Farrell, J.L., Jr. 1985. "The Dividend Discount Model: A Primer." *Financial Analysts Journal*, vol. 41, no. 6 (November/December):16–19, 22–25.

Graham, Benjamin, and David L. Dodd. 1934. *Security Analysis*. New York: McGraw-Hill.

Hicks, J.R. 1935. "A Suggestion for Simplifying the Theory of Money." *Economica* (February):1–19.

———. 1962. "Liquidity." *Economic Journal*, vol. 72 (December):787–802.

Koopmans, T.C. 1951. "Analysis of Production as an Efficient Combination of Activities." In *Activity of Production and Allocation*, 7th ed. Edited by T.C. Koopmans. 1971. New Haven, CT: Yale University Press.

Leavens, D.H. 1945. "Diversification of Investments." *Trusts and Estates*, vol. 80 (May):469–473.

Levy, H., and H.M. Markowitz. 1979. "Approximating Expected Utility by a Function of Mean and Variance." *American Economic Review*, vol. 69, no. 3 (June):308–317.

Lintner, John. 1965. "The Valuation of Risk Assets and the Selection of Risky Investments in Stock Portfolios and Capital Budgets." *Review of Economics and Statistics*, vol. 47, no. 1 (February):13–37.

Markowitz, Harry M. 1952. "Portfolio Selection." *Journal of Finance*, vol. 7, no. 1 (March):77–91.

———. 1956. "The Optimization of a Quadratic Function Subject to Linear Constraints." *Naval Research Logistics Quarterly*, vol. 3:111–133.

———. 1959. *Portfolio Selection: Efficient Diversification of Investments*. New York: John Wiley & Sons.

———. 1987. *Mean–Variance Analysis in Portfolio Choice and Capital Markets*. Oxford, U.K.: Basil Blackwell.

Marschak, J. 1938. "Money and the Theory of Assets." *Econometrica*, vol. 6:311–325.

Merton, R.C. 1969. "Lifetime Portfolio Selection under Uncertainty: The Continuous-Time Case." *Review of Economics and Statistics*, vol. 51, no. 3 (August):247–259.

———. 1972. "An Analytic Derivation of the Efficient Portfolio Frontier." *Journal of Financial and Quantitative Analysis*, vol. 7, no. 4 (September):1851–72.

Mossin, J. 1966. "Equilibrium in a Capital Asset Market." *Econometrica*, vol. 35, no. 4 (October):768–783.

———. 1968. "Optimal Multiperiod Portfolio Policies." *Journal of Business*, vol. 41, no. 2 (April):215–229.

Rosenberg, Barr. 1974. "Extra-Market Components of Covariance in Security Returns." *Journal of Financial and Quantitative Analysis*, vol. 9, no. 2 (March):263–273.

Roy, A.D. 1952. "Safety First and the Holding of Assets." *Econometrica*, vol. 20, no. 3 (July):431–449.

Samuelson, P.A. 1969. "Lifetime Portfolio Selection by Dynamic Stochastic Programming." *Review of Economics and Statistics*, vol. 51, no. 3 (August):239–246.

Savage, Leonard J. 1954. *The Foundations of Statistics*. New York: John Wiley & Sons

Sharpe, William F. 1963. "A Simplified Model for Portfolio Analysis." *Management Science*, vol. 9, no. 2 (January):277–293.

———. 1964. "Capital Asset Prices: A Theory of Market Equilibrium under Conditions of Risk." *Journal of Finance*, vol. 19, no. 3 (September):425–442.

Tobin, James. 1958. "Liquidity Preference as Behavior towards Risk." *Review of Economic Studies*, vol. 25, no. 1 (February):65–86.

Uspensky, J.V. 1937. *Introduction to Mathematical Probability*. New York: McGraw-Hill.

von Neumann, John, and Oskar Morgenstern. 1944. *Theory of Games and Economic Behavior*. 3rd ed. 1967. Princeton, NJ: Princeton University Press.

Wiesenberger, A., and Company. *Investment Companies*. New York, annual editions since 1941.

Williams, J.B. 1938. *The Theory of Investment Value*. Cambridge, MA: Harvard University Press.

THE JOURNAL OF POLITICAL ECONOMY, APRIL 1952, pp 152-158

# THE UTILITY OF WEALTH[1, 2]

HARRY MARKOWITZ

The RAND Corporation

1.1. Friedman and Savage[3] have explained the existence of insurance and lotteries by the following joint hypothesis:

(1) Each individual (or consumer unit) acts as if he (a) ascribed (real) numbers (called utility) to every level of wealth[4] and

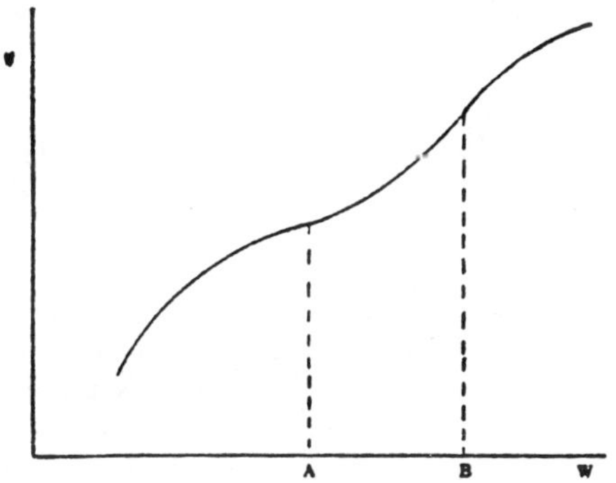


FIG. 1

(b) acts in the face of known odds so as to maximize expected utility.

(2) The utility function is as illustrated in Figure 1. We may assume it to be a continuous curve with at least first and second derivatives.[5] Let $U$ be utility and $W$ be wealth. Below some point $A$, $(\partial^2 U)/(\partial W^2) < 0$; between $A$ and $B$, $(\partial^2 U)/(\partial W^2) > 0$; above $B$, $(\partial^2 U)/(\partial W^2) < 0$.

To tell geometrically whether or not an individual would prefer $W_0$ with certainty or a "fair"[6] chance of rising to $W_1$ or falling to

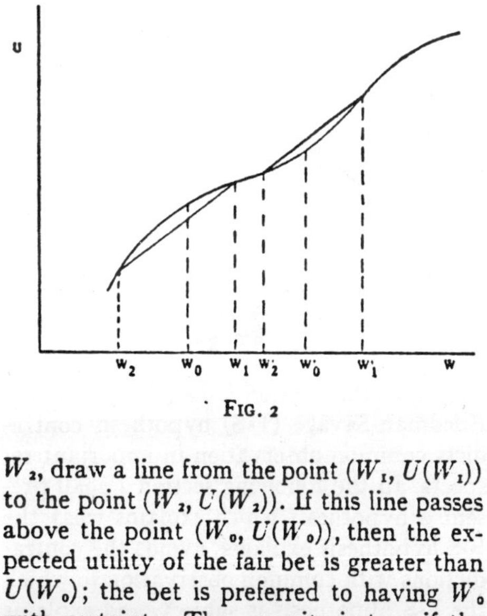


FIG. 2

$W_2$, draw a line from the point $(W_1, U(W_1))$ to the point $(W_2, U(W_2))$. If this line passes above the point $(W_0, U(W_0))$, then the expected utility of the fair bet is greater than $U(W_0)$; the bet is preferred to having $W_0$ with certainty. The opposite is true if the line $\overline{(W_1, U(W_1)), (W_2, U(W_2))}$ passes below the point $(W_0, U(W_0))$. In Figure 2, $W_0$ is preferred to a fair chance of rising to $W_1$ or

[1] This paper will be reprinted as Cowles Commission Paper, New Series, No. 57.

[2] I have benefited by conversations with M. Friedman, C. Hildreth, E. Malinvaud, L. J. Savage, and others. While the present paper takes issue with the article of Friedman and Savage, quoted in n. 3, I take it as axiomatic that the Friedman-Savage article has been one of the major contributions to the theory of behavior in the face of risk. The present paper leads only to a small modification of the Friedman-Savage analysis. This modification, however, materially increases the extent to which commonly observed behavior is implied by the analysis.

[3] M. Friedman and L. J. Savage, "The Utility Analysis of Choices Involving Risk," *Journal of Political Economy*, LVI (August, 1948), 279–304.

[4] I wish to avoid delicate questions of whether the relevant utility function is the "utility of money" or the "utility of income." I shall assume that income is discounted by some interest rate, and I shall speak of the "utility of wealth."

[5] The existence of derivatives is not essential to the hypothesis. What is essential is that the curve be convex below $A$ and above $B$; concave between $A$ and $B$. The discussion would be essentially unaffected if these more general assumptions were made.

[6] A fair bet is defined as one with expected gain or loss of wealth equal to zero. In particular if $a$ is the probability of $W_1$ and $(1 - a)$ is that of $W_2$, then $aW_1 + (1 - a)W_2 = W_0$.

falling to $W_2$. The chance of rising to $W_1'$ or falling to $W_2'$ is preferred to having $W_0'$ with certainty. The first example may be thought of as an insurance situation. A person with wealth $W_1$ would prefer to be sure of $W_0$ than to take a chance of falling to $W_2$. The second example may be thought of as a lottery situation. The person with wealth $W_0'$ pays $(W_0' - W_2)$ for a lottery ticket in the hope of winning $(W_1' - W_2')$. Even if the insurance and the lottery were slightly "unfair,"[7] the insurance would have been taken and the lottery ticket bought.

Thus the Friedman-Savage hypothesis explains both the buying of insurance and the buying of lottery tickets.

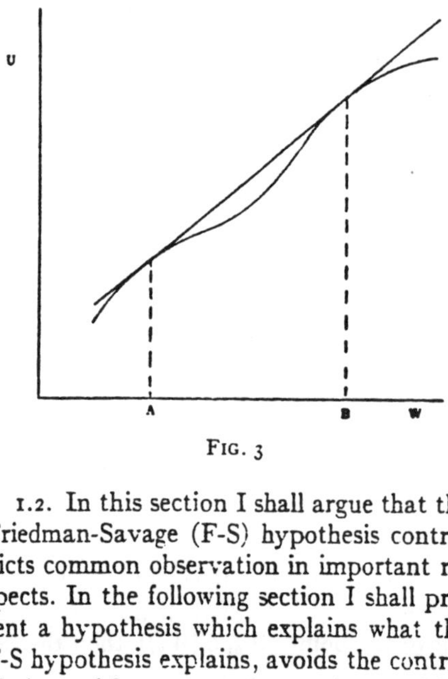


FIG. 3

1.2. In this section I shall argue that the Friedman-Savage (F-S) hypothesis contradicts common observation in important respects. In the following section I shall present a hypothesis which explains what the F-S hypothesis explains, avoids the contradictions with common observation to which the F-S hypothesis is subject, and explains still other phenomena concerning behavior under uncertainty.

In Figure 3 a line $l$ has been drawn tangent to the curve at two points.[8] A person with wealth less than $C$ is presumably "poor"; a person with wealth greater than $D$ is presumably well to do. Friedman and Savage go so far as to suggest that these may represent different social classes. The amount $(D - C)$ is the size of the optimal lottery prize (i.e., the size of prize which it is most profitable for lottery managers to offer). Those poorer than $C$ will never take a fair bet. Those richer than $D$ will never take a fair bet. Those with wealth between $C$ and $D$ will take some fair bets.

We shall now look more closely at the hypothesized behavior of persons with various levels of wealth. We shall see that for some points on the $W$ axis the F-S hypothesis implies behavior which not only is not observed but would generally be considered peculiar if it were. At other points on the curve the hypothesis implies less peculiar, but still questionable, behavior. At only one region of the curve does the F-S hypothesis imply behavior which is commonly observed. This in itself may suggest how the analysis should be modified.

Consider two men with wealth equal to $C + \frac{1}{2}(D - C)$ (i.e., two men who are midway between $C$ and $D$). There is nothing which these men would prefer, in the way of a fair bet, rather than one in which the loser would fall to $C$ and the winner would rise to $D$. The amount bet would be $(D - C)/2$—half the size of the optimal lottery prize. At the flip of a coin the loser would become poor; the winner, rich. Not only would such a fair bet be acceptable to them but none would please them more.

We do not observe persons of middle income taking large symmetric bets. We expect people to be repelled by such bets. If such a bet were made, it would certainly be considered unusual and probably irrational.

Consider a person with wealth slightly less than $D$. This person is "almost rich." The bet which this person would like most, according to the F-S hypothesis, is one which if won would raise him to $D$, if lost would lower him to $C$. He would be willing to take a small chance of a large loss for a large chance of a small gain. He would not insure against a loss of wealth to $C$. On the

[7] I.e., even if $aW_1 + (1 - a)W_2 > W_0$, $a'W_1' + (1 - a')W_2' < W_0'$. For limits on the amount of unfairness which an individual would accept see Friedman and Savage, *op. cit.*, p. 291.

[8] *Ibid.*, p. 300, n. 35.

contrary he would be anxious to underwrite insurance. He would even be willing to extend insurance at an expected loss to himself!

Again such behavior is not observed. On the contrary we find that the insurance business is done by companies of such great wealth that they can diversify to the point of almost eliminating risk. In general, it seems to me that circumstances in which a moderately wealthy person is willing to risk a large fraction of his wealth at actuarially unfair odds will arise very rarely. Yet such a willingness is implied by a utility function like that of Figure 3 for a rather large range of wealth.

Another implication of the utility function of Figure 3 is worth noting briefly. A person with wealth less than *C* or more than *D* will never take any fair bet (and, a fortiori, never an unfair bet). This seems peculiar, since even poor people, apparently as much as others, buy sweepstakes tickets, play the horses, and participate in other forms of gambling. Rich people play roulette and the stock market. We might rationalize this behavior by ascribing it to the "fun of participation" or to inside information. But people gamble even when there can be no inside information; and, as to the joy of participation, if people like participation but do not like taking chances, why do they not always play with stage money? It is desirable (at least according to Occam's razor) to have an alternative utility analysis which can help to explain chance-taking among the rich and the poor as well as to avoid the less defensible implications of the F-S hypothesis.

Another level of wealth of interest corresponds to the first inflection point on the F-S curve. We shall find that the implications of the F-S hypothesis are quite plausible for this level of wealth. I shall not discuss these implicatons at this point, for the analysis is essentially the same as that of the modified hypothesis to be presented below.

2.1. I shall introduce this modified hypothesis by means of a set of questions and answers. I have asked these questions informally of many people and have typically received the answers indicated. But these "surveys" have been too unsystematic to serve as evidence; I present these questions and typical answers only as a heuristic introduction. After this hypothesis is introduced, I shall compare its ability and that of the F-S hypothesis to explain well-established phenomena. The hypothesis as a whole is presented on page 155.

Suppose a stranger offered to give you either 10 cents or else one chance in ten of getting \$1 (and nine chances in ten of getting nothing). If the situation were quite impersonal and you knew the odds were as stated, which would you prefer?

(1) 10 cents with certainty or one chance in ten of getting \$1?

Similarly which would you prefer (why not circle your choice?):

(2) \$1 with certainty or one chance in ten of getting \$10?

(3) \$10 with certainty or one chance in ten of getting \$100?

(4) \$100 with certainty or one chance in ten of getting \$1,000?

(5) \$1,000 with certainty or one chance in ten of getting \$10,000?

(6) \$1,000,000 with certainty or one chance in ten of getting \$10,000,000?

Suppose that you owed the stranger 10 cents, would you prefer to pay the

(7) 10 cents or take one chance in ten of owing \$1?

Similarly would you prefer to owe

(8) \$1 or take one chance in ten of owing \$10?

(9) \$10 or take one chance in ten of owing \$100?

(10) \$100 or take one chance in ten of owing \$1,000?

(11) \$1,000,000 or take one chance in ten of owing \$10,000,000?

The typical answers (of my middle-income acquaintances) to these questions are as follows: most prefer to take a chance on \$1 rather than get 10 cents for sure; take a chance on \$10 rather than get \$1 for sure. Preferences begin to differ on the choice between \$10 for sure or one chance in ten of \$100. Those who prefer the \$10 for sure in situation (3) also prefer \$100 for sure in situation (4); while some who would take a chance in situation (3) prefer the \$100 for sure in situation (4). By situation (6) everyone prefers the \$1,000,000 for sure rather than one chance in ten of \$10,000,000.

All this may be explained by assuming that the utility function for levels of wealth above present wealth is first concave and then convex (Fig. 4).

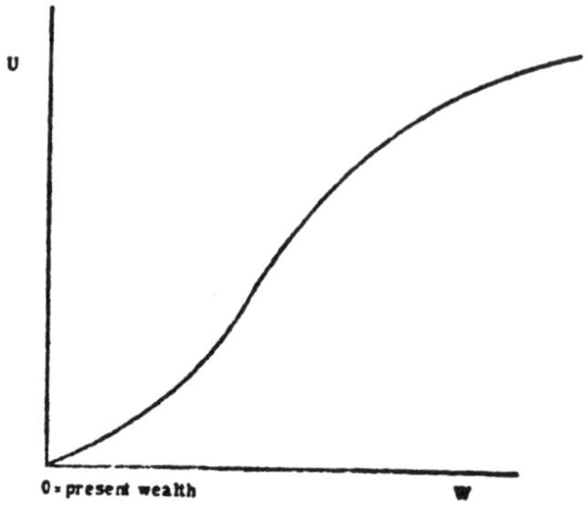


FIG. 4

Let us continue our heuristic introduction. People have generally indicated a preference for owing 10 cents for sure rather than one chance in ten of owing \$1; owing \$1 for sure rather than taking one chance in ten of owing \$10; \$10 for sure rather than one in ten of \$100. There comes a point, however, where the individual is willing to take a chance. In situation (11), for example, the individual generally will prefer one chance in ten of owing \$10,000,000 rather than owing \$1,000,000 for sure. All this may be explained by assuming that the utility function going from present wealth downward is first convex and then concave. Thus we have a curve as in Figure 5, with three inflection points. The middle inflection point is at present wealth. The function is concave immediately above present wealth; convex, immediately below.

How would choices in situations (1)–(11) differ if the chooser were rather rich? My guess is that he would take a chance on getting the \$10 rather than take \$1 for sure; take a chance on \$100 rather than take \$10 for sure; perhaps take a chance on \$1,000 rather than take \$100 for sure. But the point would come when he too would become cautious. For example, he would prefer \$1,000,000 rather than one chance in ten of \$10,000,000. In other words, he would act essentially the same, in situations (1)–(6), as someone with more moderate wealth, except that his third inflection point would be farther from the origin. Similarly we hypothesize that in situations (7)–(11) he would act as if his first inflection point also were farther from the origin.

Conversely, if the chooser were rather poor, I should expect him to act as if his first and third inflection points were closer to the origin.

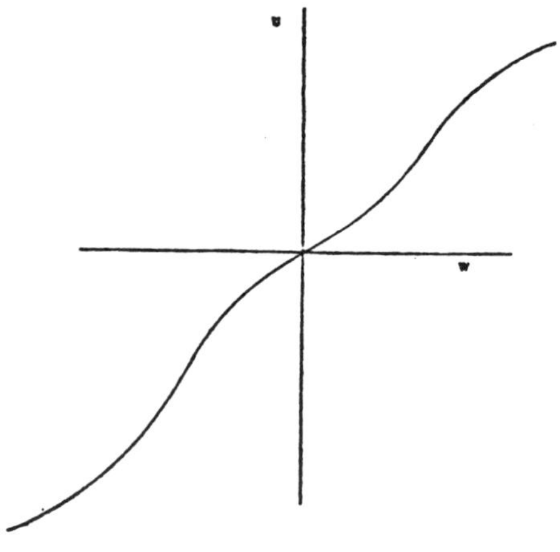


FIG. 5

Generally people avoid symmetric bets. This suggests that the curve falls faster to the left of the origin than it rises to the right of the origin. (I.e., $U(X) > |U(-X)|$, $X > 0$.)

To avoid the famous St. Petersburg Paradox, or its generalization by Cramer, I assume that the utility function is bounded from above. For analogous reasons I assume it to be bounded from below.

So far I have assumed that the second inflection corresponds to present wealth. There are reasons for believing that this is not always the case. For example, suppose that our hypothetical stranger, rather than offering *to* give you \$$X$ or a chance of \$$Y$, had instead first given you the \$$X$ and then had offered you a fair bet which if lost would cost you $-\$X$ and if won would net you $\$(Y - X)$. These two situations are essentially the same, and it is plausible to expect the chooser to act in the same manner in both situations. But this will not always be the implication of our hypotheses if we insist that the second inflection point always corresponds to present wealth. We can resolve this dilemma by assuming that in the case of recent windfall gains or losses the second inflection point may, temporarily, deviate from present wealth. The level of wealth which corresponds to the second inflection point will be called "customary wealth." Unless I specify otherwise, I shall assume that there have been no recent windfall gains or losses, and that present wealth and "customary wealth" are equal. Where the two differ, I shall let "customary wealth" (i.e., the second inflection point) remain at the origin of the graph. Later I will present evidence to support my contentions concerning the second inflection point and justify the definition of "customary wealth."

To summarize my hypothesis: the utility function has three inflection points. The middle inflection point is defined to be at the "customary" level of wealth. Except in cases of recent windfall gains and losses, customary wealth equals present wealth. The first inflection point is below, the third inflection point is above, customary wealth. The distance between the inflection points is a nondecreasing function of wealth.[9] The curve is monotonically increasing but bounded; it is first concave, then convex, then concave, and finally convex. We may also assume that $|U(-X)| > U(X)$, $X > 0$ (where $X = 0$ is customary wealth). A curve which is consistent with our specifications is given in Figure 5.

2.2. An examination of Figure 5 will show that the above hypothesis is consistent with the existence of both "fair" (or slightly "unfair") insurance and "fair" (or slightly "unfair") lotteries. The same individual will buy insurance and lottery tickets. He will take large chances of a small loss for a small chance for a large gain.

The hypothesis implies that his behavior will be essentially the same whether he is poor or rich—except the meaning of "large" and "small" will be different. In particular there are no levels of wealth where people prefer large symmetric bets to any other fair bet or desire to become one-man insurance companies, even at an expected loss.

Thus we see that the hypothesis is consistent with both insurance and lotteries, as was the F-S hypothesis. We also see that the hypothesis avoids the contradictions with common observations to which the F-S hypothesis was subject.

2.3. I shall now apply the modified hypothesis to other phenomena. I shall only consider situations wherein there are objective odds. This is because we are concerned with a hypothesis about the utility function and do not want to get involved in questions concerning subjective probability beliefs. It may be hoped, however, that a utility function which is successful in explaining behavior in the face of known odds (risk) will also prove useful in the explanation of behavior under uncertainty.

It is a common observation that, in card games, dice games, and the like, people play more conservatively when losing moderately, more liberally when winning moderately. Anyone who wishes evidence of this is referred to an experiment of Mosteller and Nogee.[10] Participants in the experiment were asked to write instructions as to how their money should be bet by others. The instructions consisted of indicating what bets

[9] It may also be a function of other things. There is reason to believe, for example, that the distance between inflection points is typically greater for bachelors than for married men.

[10] "An Experimental Measurement of Utility," *Journal of Political Economy*, LIX (1951), 389. The above evidence would be more conclusive if it represented a greater range of income levels.

should be accepted when offered and "further (written) instructions." The "further instructions" are revealing; for example, "A-II—Play till you drop to 75 cents then *stop!!*"; "B-V—If you get low, play only very good odds"; "C-I—If you are ahead, you may play the four 4's for as low as $3"; "C-III—If player finds that he is winning, he shall go ahead and bet at his discretion"; "C-IV—If his winnings exceed $2.50, he may play any and every hand as he so desires, but, if his amount should drop below 60 cents, he should use discretion in regard to the odds and hands that come up." No one gave instructions to play more liberally when losing than when winning. The tendency to play liberally when winning, conservatively when losing, can be explained in two different ways. These two explanations apply to somewhat different situations.

A bet which a person makes during a series of games ("plays" in the von Neumann sense) cannot be explained without reference to the gains and losses which have occurred before and the possibilities open afterward. What is important is the outcome for the whole series of games: the winnings or losings for "the evening" as a whole. Suppose "the evening" consists of a series of independent games (say matching pennies); suppose that the probability (frequency) distribution of wins and losses for a particular game is symmetric about zero. Suppose that at each particular game the player has a choice of betting liberally or conservatively (i.e., he can influence the dispersion of the wins and losses). If he bet with equal liberality at each game, regardless of previous wins or losses, then the frequency distribution of final wins and losses (for the evening as a whole) would be symmetric. The effect of playing conservatively when losing, liberally when winning, is to make the frequency distribution of final outcomes skewed to the right. Such skewness is implied as desirable (in a large neighborhood of customary income) by our utility function. In sum, our utility function implies the desirability of some positive skewness of the final outcome frequency distribution, which in turn implies the desirability of playing conservatively when losing moderately and playing liberally when winning moderately.

This implication holds true whatever be the level of customary wealth of the individual. In the F-S analysis a person with wealth equal to $D$ in Figure 3 would play liberally when losing, conservatively when winning, so as to attain negative skewness of the frequency distribution. This, I should say, is another one of those peculiar implications which flow from the F-S analysis.

Now let us consider the effect of wins or losses on the liberality of betting when we do not have the strategic considerations which were central in the previous discussion. For example, suppose that the "evening" is over. The question arises as to whether or not the game should be continued into the morning (i.e., whether or not a new series of games should be initiated). There is also a question of whether or not the stakes should be higher or lower. We abstract from fatigue or loss of interest in the game.

How do the evening's wins or losses affect the individual's preferences on these questions? Since his gain or loss is a "windfall," the individual is moved from the middle inflection point (presumably by the amount of the gain or loss).

A person who broke even would, by hypothesis, have the same preferences as at the beginning of the evening.

A person who had won moderately would (by definition of "moderate") be between the second and third inflection point. The moderate winner would wish to continue the game and increase the stakes.

A person who had won very much would (by the definition of "very much") be to the right of the third inflection point. He would wish to play for lower stakes or not to play at all. In the vernacular, the heavy winner would have made his "killing" and would wish to "quit while winning."

The moderate loser, between the first and second inflection points, would wish to play for lower stakes or not to play at all.

A person who lost extremely heavily (to the left of the first inflection point) would wish to continue the game (somewhat in desperation).

We see above the use of the distinction between customary and present wealth. In the explanation use was made of both the assumption that (*a*) before windfall gains or losses the second inflection point (customary income) is at present income and (*b*) immediately after such gains or losses customary income and present income are not the same.

To have an exact hypothesis—the sort one finds in physics—we should have to specify two things: (*a*) the conditions under which customary wealth is not equal to present wealth (i.e., the conditions referred to as recent windfall gains or losses) and (*b*) the value of customary wealth (i.e., the position of the second inflection point) when customary wealth is not equal to present wealth. It would be very convenient if I had a rule which in every actual situation told whether or not there had been a recent windfall gain or loss. It would be convenient if I had a formula from which customary wealth could be calculated when this was not equal to present wealth. But I do not have such a rule and formula. For some clear-cut cases I am willing to assert that there are or are not recent windfall gains or losses: the man who just won or lost at cards; the man who has experienced no change in income for years. I leave it to the reader's intuition to recognize other clear-cut cases. I leave it to future research and reflection to classify the ambiguous, border-line cases. We are even more ignorant of the way customary follows present wealth or how long it takes to catch up.

I have assumed that asymmetric bets are undesirable. This assumption could be dropped or relaxed without much change in the rest of the hypothesis; but I believe this assumption is correct and should be kept. Symmetric bets are avoided when moderate or large amounts are at stake. Sometimes small symmetric bets are observed. How can these be explained? I offer three explanations, one or more of which may apply in any particular situation. First, we saw previously that a symmetric bet may be part of a strategy leading to a positively skewed probability distribution of final outcome for the evening as a whole. Second, for very small symmetric bets the loss in utility from the bet is negligible and is compensated for by the "fun of participation." Third (this reason supplements the second), there is an inflection point at $W = 0$; therefore, the utility function is almost linear in the neighborhood of $W = 0$, and, therefore, there is little loss of utility from small symmetric bets.

3.1 Above I used the concept of "fun of participation." If we admit this—as we must—as one of the determinants of behavior under uncertainty, then we must con-

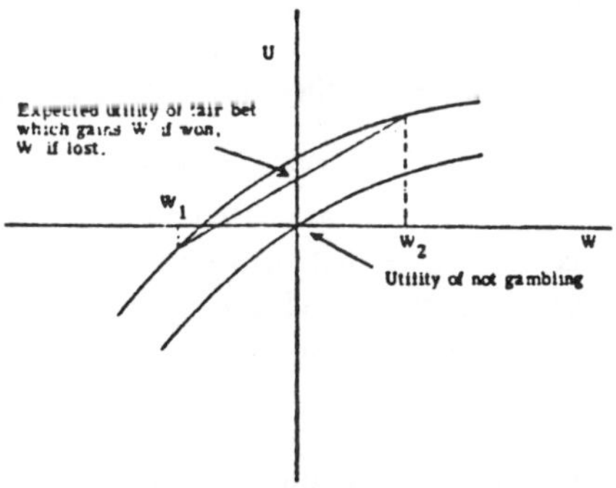


FIG. 6

tend with the following hypothesis: The utility function is everywhere convex; all (fair) chance-taking is due to the "fun of participation." This "classical" hypothesis is simpler than mine and is probably rather popular. If it explained observable behavior as well as my hypothesis, this classical hypothesis would be preferable to mine.

Before examining the hypothesis, we must formulate it more exactly. It seems to say that the utility of a gamble is the expected utility of the outcomes plus the utility of playing the game (the latter utility is independent of the outcome of the game). This can be presented graphically as in Figure 6. One implication of this hypothesis is that, for given (fair) odds, the smaller the amount bet, the higher the expected utility. In particular, when millionaires play poker together, they play for pennies; and no one will buy more than one lottery ticket. This contradicts observation.[11]

One might hypothesize that the utility of the game, to be added to the utility of the outcomes, is a function of the possible loss ($-W_2$) or the difference between gain and loss ($W_1 - W_2$). Neither of these hypotheses explains why people prefer small chances of large gains with large chances of small losses rather than vice versa. Nor do they explain why people play more conservatively when losing than when winning.

In short, the classical hypothesis may be consistent with the existence of chance-taking, but it does not explain the particular chances which are taken. To explain such choices, while maintaining simple hypotheses concerning "fun of participation," we must postulate a utility function as in Figure 5.

4.1. It may be objected that the arguments in this paper are based on flimsy evidence. It is true that many arguments are based on "a priori" evidence. Like most "a priori" evidence, these are presumptions of the writer which he presumes are also held by the reader. Such a priori evidence includes the implausibility of middle-income persons desiring large symmetric bets and the implausibility of the one (moderately rich) man insurance company. Perhaps the only evidence of mine which could, so to speak, "stand up in court" is the testimony of the Mosteller-Nogee experiment. But this does not fully suit our needs, since only a narrow range of wealth positions were sampled. I realize that I have not "demonstrated" "beyond a shadow of a doubt" the "truth" of the hypothesis introduced.[12] I have tried to present, motivate, and, to a certain extent, justify and make plausible a hypothesis which should be kept in mind when explaining phenomena or designing experiments concerning behavior under risk or uncertainty.

[11] The statement that millionaires "ought" to play for pennies is irrelevant. We seek a hypothesis to explain behavior, not a moral principle by which to judge behavior.

[12] Even now we are aware of one class of commonly observed phenomena which seems to be inconsistent with the hypothesis introduced in this paper, as well as the hypotheses which this one was intended to supersede. The existence of multiple lottery prizes with various sized prizes *may* contradict the theory presented. If we are forced to concede that the individual (lottery-ticket buyer) prefers, say, a fair multiple prize lottery to all other fair lotteries, then my hypothesis cannot explain this fact. Nor can any other hypothesis considered in this paper explain a preference for different sized lottery prizes. Nor can any hypothesis which assumes that people maximize expected utility. Even now we must seek hypotheses which explain what our present hypotheses explain, avoid the contradictions with observation to which they are subject, and perhaps explain still other phenomena.

## Chapter 3

# Rand [I] and The Cowles Foundation

### Comments

The article on *The Elimination Form of the Inverse and Its Application to Linear Programming* presents my work on sparse matrices. This was developed as a byproduct of my work on process analysis described later in this chapter. The first sparse matrix code was programmed by William Orchard-Hayes as a demonstration of spares matrix capabilities. Bill usually programmed for George Dantzig, but George loaned Bill to me to demonstrate the notion of sparse matrices which he conjectured might be important in linear programming. It turned out in fact to be very important. Now all large production codes include sparse matrix technology and use a modified version of the "Markowitz rule" described in the article. The Markowitz rule chooses a pivot to be used in a Gaussian elimination. Today the pivot is rejected if it is too close to zero.

The next three articles are on quadratic programming and mean-variance portfolio selection. They concern matters which are described in the *Trains of Thought* article in the first chapter. They will not be discussed further here. The article *DeFinetti Scoops Markowitz,* concerns a matter which was recently brought to my attention by Mark Rubinstein. Its own history is described briefly in the article itself.

The articles *Industry-wide, Multi-industry and Economy-wide Process Analysis and* Studies in Process Analysis*: Economy-wide Production Capabilities*, Chapter Two, *Alternate Methods of Analysis* describe work which Alan Mann and I plus many others pursued when Alan and I were at the Rand Corporation. The objective of this work is described in the first article presented here. The article did not contain a crucial insight which appeared to us later and is documented only in Chapter Two of a subsequent Cowles Foundation book. I have therefore included this chapter.

### References


Markowitz, H. M. (1954). *Industry-wide, Multi-industry and Economy-wide Process Analysis.* "The Structural Interdependence of the Economy", T. Barna, (ed.), John Wiley and Sons.

Manne, A. S. and Markowitz, H. M. (1963). *Alternate Methods of Analysis.* "Studies in Process Analysis: Economy-wide Production Capabilities" Chapter 2. John Wiley and Sons, New York.

Markowitz, H. M. (1957). *The Elimination Form of the Inverse and Its Application to Linear Programming.* Management Science, Vol. 3, pp. 255–269.

Markowitz, H. M. (1956). *The Optimization of a Quadratic Function Subject to Linear Constraints*. Naval Research Logistics Quarterly, Vol. 3, pp. 111–133.
Markowitz, H. M. ( 1994). *The General Mean-variance Portfolio Selection Problem*. Phil. Trans. R. Soc. London, Vol. 347 A, pp. 543–549.

# Chapter 5

# INDUSTRY-WIDE, MULTI-INDUSTRY AND ECONOMY-WIDE PROCESS ANALYSIS

BY

HARRY MARKOWITZ

UNIVERSITY OF PISA - FACULTY OF ECONOMICS

# THE STRUCTURAL INTERDEPENDENCE OF THE ECONOMY

PROCEEDINGS OF AN INTERNATIONAL CONFERENCE
ON INPUT-OUTPUT ANALYSIS

Varenna - 27 June - 10 July 1954

Edited by
TIBOR BARNA

JOHN WILEY & SONS, INC. NEW YORK
A. GIUFFRÈ - EDITORE - MILANO

# INDUSTRY-WIDE, MULTI-INDUSTRY AND ECONOMY-WIDE PROCESS ANALYSIS

## 1. *Introduction*

Process analysis, as the term is used here, refers to the formal analysis of industrial productive processes. It consists of the construction of a mathematical model which reflects the productive processes available to a firm, an industry, a group of industries, or an economy; and the use of this model to estimate, under various conditions, the capabilities of this firm, industry, group of industries, or economy.

Process analysis, thus defined, is not new. The industrial engineer frequently uses mathematical models (although he may not always call them such) to aid in predicting the capabilities of the plant under his responsibility. Economists as well have analyzed actual productive processes on the plant and industry level. [1]

Several persons and organizations in the United States [2] are exploring together the possibility of multi-industry and, eventually, economy-wide process analysis. It is hoped that such an economy-wide analysis can be built by means of a series of subanalyses — each subanalysis having value in itself besides forming a part of the larger whole. By the end of 1954 an analysis of the petroleum industry, and an analysis of machine tool substitution possibilities have been completed [3].

---

[1] Several papers in [1], in particular, discuss the feasibility and desirability of such technological model building, provide illustrative examples of such models, and consider how this and other technological research can contribute to economy-wide models.

[2] The RAND Corporation, Santa Monica, California; Stanford Research Institute, Stanford, California; Management Sciences Research Project, University of California, Los Angeles, California.

[3] Analyses of iron and steel production, fuels and power, and the chemicals industries are under way.

The subanalyses are cast in the linear programming framework. Linear programming provides a powerful computing tool in deciding which set of many possible sets of productive processes will best accomplish some objective. Such a computing tool is especially needed as we move from limited analyses of a few productive processes, to analyses of an industry-wide, multi-industry or economy-wide scope.

It is impossible to set down a-priori rules according to which a process analysis model can be constructed. The person or group who is building a subanalysis must understand, in some detail, the technological relations of the sector, must know what data are obtainable, must know what computing capabilities are available, and from these considerations must attempt to formulate and implement a model which best accomplishes the objectives of the analysis.

In section 2 of this paper an example of a subanalysis is presented to illustrate how a technology can be expressed in linear programming terms; the level of detail we are striving for in our subanalyses; and how a subanalysis may be used. The reader is subjected to this excursion into technology and model building because of the writer's belief that it is impossible to communicate the nature of process analysis by abstract discussion alone. In section 3 the difficulties of, and the aids to, the building of an economy-wide model are discussed. The conclusion of this section is that, in the building of such a model, few difficulties are insurmountable. Section 4 compares an economy-wide process analysis model with other economy-wide models and it discusses what can and cannot be hoped for from an economy-wide process analysis. The moral of this section is that the purpose of an economy-wide process analysis model would be to estimate the over-all capabilities of the economy as a whole. It would not be intended for, or capable of, detailed scheduling of the economy's operations. On the other hand, its job of feasibility testing — estimating what the economy can and cannot do — cannot be adequately accomplished by the simpler models now frequently used for the job.

## 2. *Subanalysis for the petroleum refining industry*

A 109-equation linear programming analysis of the petroleum industry has been developed by Alan Manne of The RAND Corporation. With the immediate objective of constructing a spatial petroleum analysis, Thomas Marschak, also of The RAND Corporation, has been experimenting with aggregate versions of this petroleum model. He now has a 44-equation model which seems able to give roughly the same answers to certain kinds of questions as the larger model. In this section a brief description of the 44-equation model will be presented for illustrative purposes [1]. Since this paper is on the general characteristics of process analysis models rather than on the petroleum analysis in particular, and since the petroleum analyses will be described in detail by their constructors at some later date, the treatment here will be summary and suggestive, rather than complete and definitive.

Figure 1 presents a schematic flow diagram of petroleum refining operations. The numbers in the figure correspond to the numbers in the description below:

1. *Crude oils,* obtained from underground, are heterogeneous mixtures of various hydro-carbons. Crudes from different fields have different compositions. The detailed petroleum analysis distinguishes twenty-five classes of crudes. The aggregate analysis distinguishes three classes of crudes.

2. By means of « fractional distillation » a crude oil is separated into various liquids (cuts) each with a narrower range of boiling temperatures. These cuts include *gasolines* with low boiling temperatures, *kerosene* with a somewhat higher boiling temperature, *oils* with still higher boiling temperetures and greater viscosity; and finally a very viscous *residuum.* All these products of distillation are referred to as « straight run » materials.

3. The straight-run gasolines may be blended into motor fuels as they are, or they may be subject to « reforming ». Reforming changes the structure of the gasoline molecules (technically speaking, it changes many « saturated » molecules into « unsaturated » molecules), thus producing gasolines with better anti-knock qualities.

4. Certain of the products of reforming (such as benzene, toluene, and xylene) are isolated in relatively pure form and sold to the chemicals industry. These chemicals, also produced by the « by-product coke oven », are the basis for a variety of organic chemicals including explosives, detergents, plastics, dies, pharmaceuticals, synthetic rubber and synthetic fibres.

---

1 General descriptions of the 109-equation pretroleum model and the machine tool substitution analysis are available in [2] and [3].

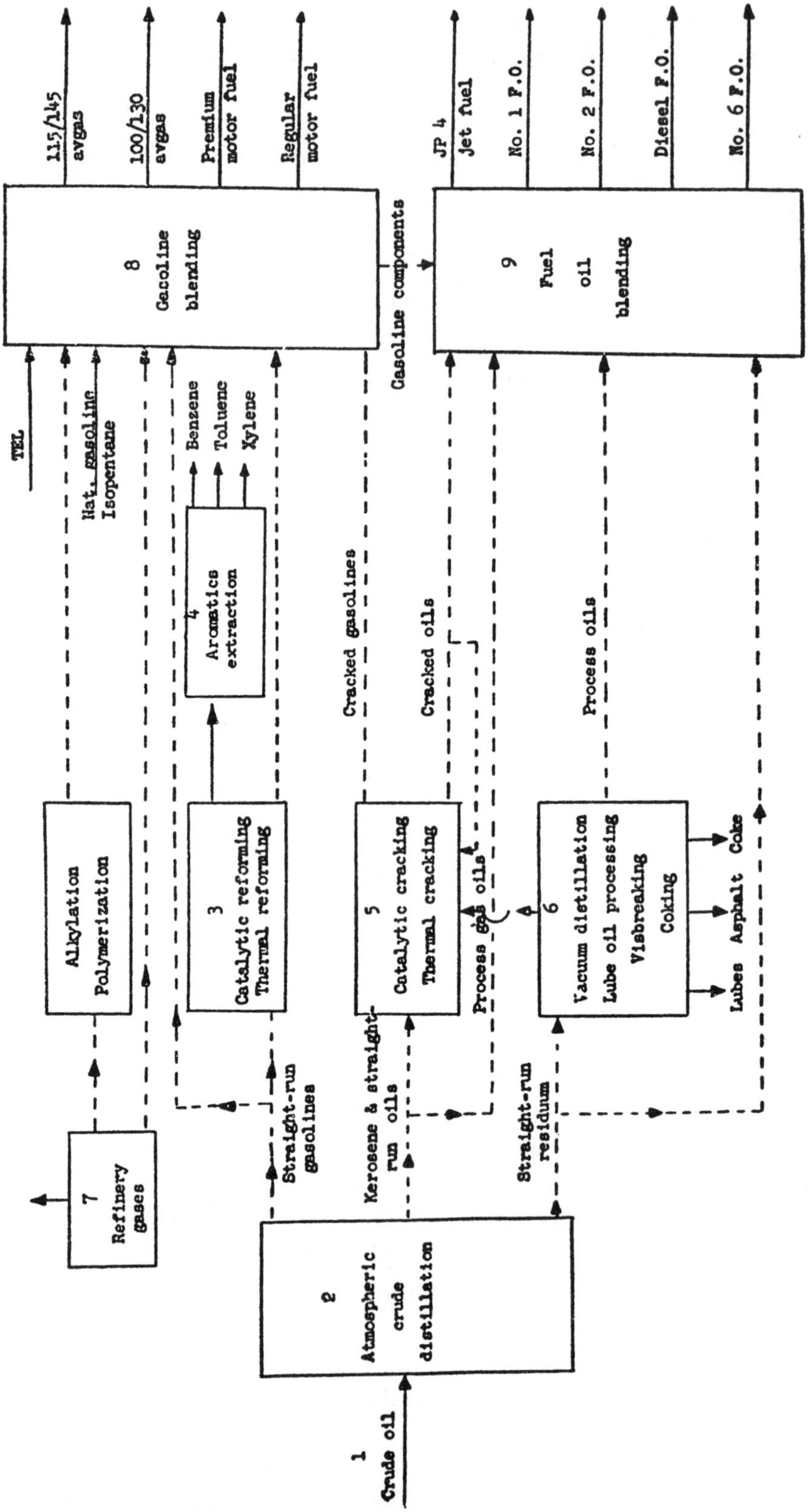


Fig. 1 Schematic flow diagram of petroleum operations

5. Kerosene and heavier straight-run oils may be used directly as end items, may be blended with other materials to form end items, or may be subjected to « *cracking* ». Cracking breaks down many of the larger molecules into smaller gasoline molecules. Oils from the cracking process may be cracked again (recycled) or blended into end items.

6. The « bottoms » or residuum from distillation can be subjected to various processes which result in products such as lubricants, asphalt, coke and « process oils ». The latter may be blended into end items or can be cracked.

7. Refinery « gases » are by-products of most of the previously described refinery operations. These gases are materials that boil under about 100° F, including materials which are liquid under normal temperatures and pressures. These gases can be burned or sold as fuel; sold as chemical raw materials; subjected to *alkylation* or *polymerization* which yield high octane blending stocks; or in the case of one of these « gases », butane, can be blended into motor fuel directly.

8. Straight run, cracked and reformed gasolines of various kinds together with butane, isopentane, natural gasoline and tetraethyl lead can be blended together to form motor fuels of various kinds. In making a particular product, amounts of blending stocks must be chosen so as to meet certain specifications such as octane number (a measure of the antiknock properties of a gasoline). These specifications can usually be met in a great variety of ways.

9. Other end items such as jet fuel, fuel oils and diesel fuel can be made by blending various materials to meet specifications.

Because of the many and devious routes by which crudes can be processed from distillation to end items, it is not a trivial matter to answer questions concerning the capabilities and flexibility of the petroleum industry, even when the pertinent technical coefficients are known. A typical question concerning the capabilities of the industry is: Suppose certain amounts of refining equipment and crude oils are available; suppose we wanted petroleum products in certain proportions, i.e., we wanted a certain product mix among our end items; what is the maximum amount of these end items we can obtain given the available resources and crudes? It is the object of a linear programming analysis to translate such a question into a problem of maximizing a linear function

$$b_1 X_1 + b_2 X_2 + \ldots + b_n X_n$$

subject to linear equations

$$a_{11} X_1 + \ldots + a_{1n} X_n = C_1$$
$$\ldots\ldots\ldots\ldots\ldots\ldots\ldots\ldots$$
$$a_{m1} X_1 + \ldots + a_{mn} X_n = C_m$$

where the variables $X_1 \ldots X_n$ must not be negative. The aggregate petroleum model, in particular, translates the question into a problem with 44 equations and 127 unknows. Once such a linear programming model has been set up it is relatively easy to vary the question, i.e., change the product mix desired and/or the resource availabilities.

The petroleum analysis distinguishes 127 « activities » that could be done in petroleum refining. These are listed along the top the petroleum matrix (Table 1, below). The entries of the table show inputs and outputs per unit of the activity. The amount of the $j$-th activity performed turns out to be the $X_j$ of the linear programming problem referred to above. Illustrative activities are:

Activity 1: The distillation of one unit of crude of type 1. Each unit of this activity uses up one unit of crude 1, uses up one unit of crude distillation capacity, and produces various amounts of straight-run materials as shown in column 1 of the matrix.

Activity 12: The catalytic cracking of straight-run material with a boiling range of 400 to 550° F. One unit of this activity has an input of one unit of catalytic cracking capacity, an input of one unit of straight-run material with the specified boiling range, and outputs of various cracked materials as shown in colum 12.

Activity 32: The adding of 250 to 325° F. straight-run material to the other materials going into motor gasoline. One unit of this activity has an input of one unit of the aforementioned material. It cannot, however, be said to have an output of motor fuel since it does not — by itself — meet all the specifications. The handling of this and other blending activities (activities 19 to 82 [1]) will be explained below.

Activity 83: The « product mix » activity. This « activity » has « inputs » of end items in certain desired proportions. The objective in the linear programming calculation is to maximize the amount of this activity performed.

Activity 84: The « not using » of some of the crude 1 availability. This activity has an input of one unit of crude 1, and no output. There

[1] Of which the last five, which are less interesting, are omitted from Table 1.

are similar « disposal » activities for every item listed along the side of the matrix[1]. Even with an « optimal » allocation of resources it almost always happens that some of these disposal activities are used. We know, on the other hand, that certain disposal activities need never be used, but their inclusion is convenient for computational reasons.

The amount performed of the $j$-th activity is $X_j$. E.g., $X_1$ is the amount per day of crude 1 distilled; $X_{12}$ is the amount per day of 400° - 500° straight-run material cracked; $X_{32}$ is the amount per day of 250° - 325° straight-run material blended into motor gasoline; $X_{83}$ is the amount per day of the product mix obtained; $X_{84}$ is the amount per day of crude 1 available but not distilled.

These variables are subject to various constraints; for example, since the availability of crude 1 = the amount distilled + the amount available but not distilled, we have

$$C_1 = X_1 + X_{84}$$

or, as it appears in the first row of the matrix

$$-C_1 = -X_1 - X_{84}$$

where $C_1$ is the availability of crude 1. Since crude distillation capacity is the amount used up in distilling various crudes plus the amount left idle, we have, according to the fourth row of the matrix,

$$-C_4 = -X_1 - X_2 - X_3 - X_{87}$$

where $C_4$ is the atmospheric distillation capacity. The thirteenth row of the matrix corresponds to the statement that sources and uses of the 550° - 750° straight-run material must be equal; i.e.,

$$0 = .184\,X_1 + .256\,X_2 + .313\,X_3 - X_7 - X_{13} - .421\,X_{57} - .421\,X_{58} - .421\,X_{59} - .029\,X_{65} - .966\,X_{66} - .329\,X_{71} - X_{77} - \alpha_{13}\,X_{83} - X_{96}$$

where $\alpha_{13}\,X_{83}$ is the amount that goes into the product mix as demand for diesel fuel.

Equation 1 to 29 are such crude and equipment availability equations, intermediate material balance equations, and availability of « exogenous » materials equations. Equations 30 to 44

---

[1] But only the first and the last are shown in Table 1.

**PETROLEUM**

| | Items \ Activities | Distillation | | | | | | |
|---|---|---|---|---|---|---|---|---|
| | | Atmos. dist., type 1 crude | Atmos. dist., type 2 crude | Atmos. dist., type 3 crude | Vacuum dist., gas oil oper. | Vacuum dist., lube oil oper. | Ther. cr. of 400-550 SR | Ther. cr. of 550-725 SR |
| | | 1 | 2 | 3 | 4 | 5 | 6 | 7 |
| Crude & Equipment Availabilities | 1. Availability Crude 1 | −1.00 | | | | | | |
| | 2. Availability Crude 2 | | −1.00 | | | | | |
| | 3. Availability Crude 3 | | | −1.00 | | | | |
| | 4. Atmos. crude dist. capacity | −1.00 | −1.00 | −1.00 | | | | |
| | 5. Vacuum distillation capacity | | | | −1.00 | −1.00 | | |
| | 6. Thermal operations capacity | | | | | | −1.00 | −1.00 |
| | 7. Catalytic cracking capacity | | | | | | | |
| | 8. Catalytic reforming capacity | | | | | | | |
| Intermediate Materials | 9. 100-250 Straight run | .138 | .077 | .049 | | | | |
| | 10. 250-325 Straight run | .090 | .065 | .041 | | | | |
| | 11. 325-400 Straight run | .100 | .067 | .042 | | | | |
| | 12. 400-550 Straight run | .174 | .218 | .266 | | | −1.00 | |
| | 13. 550-725 Straight run | .184 | .256 | .313 | | | | −1.00 |
| | 14. 725 + Straight run | .308 | .308 | .283 | −1.00 | −1.00 | | |
| | 15. Vac. distillate 725 + gas oil | | | | .40 | .43 | | |
| | 16. Vac. distillate bottoms | | | | .60 | .22 | | |
| | 17. 100-250 thermally cracked | | | | | | .11 | .10 |
| | 18. 250-325 thermally cracked | | | | | | .057 | .05 |
| | 19. 325-400 thermally cracked | | | | | | .057 | .05 |
| | 20. 400-550 thermally cracked | | | | | | .700 | .042 |
| | 21. 550-725 cracked | | | | | | .021 | .70 |
| | 22. 725 + cracked | | | | | | .006 | .01 |
| | 23. 100-250 catalytically cracked | | | | | | | |
| | 24. 250-325 catalytically cracked | | | | | | | |
| | 25. 325-400 catalytically cracked | | | | | | | |
| | 26. 400-500 catalytically cracked | | | | | | | |
| Exogenous Inputs | 27. Alkylate | | | | | | | |
| | 28. TEL | | | | | | | |
| | 29. Isopentane | | | | | | | |
| End Items Specs. | 30. 115/145 avgas, PN rich | | | | | | | |
| | 31. 100/130 avgas, PN rich | | | | | | | |
| | 32. 100/130 avgas, PN lean | | | | | | | |
| | 33. Regular motor gas, octane no. | | | | | | | |
| | 34. Regular motor gas with 0-1.5 cc/gal of TEL | | | | | | | |
| | 35. Regular motor gas with 1.5-3.0 cc/gal of TEL | | | | | | | |
| | 36. Regular motor gas, RVP | | | | | | | |
| | 37. Jet fuel, % aromatics | | | | | | | |
| End Item Requirements | 38. 115/145 avgas | | | | | | | |
| | 39. 100/130 avgas | | | | | | | |
| | 40. Jet fuel | | | | | | | |
| | 41. Regular motor gas | | | | | | | |
| | 42. Lubes | | | | | .34 | | |
| | 43. # 2 fuel oil | | | | | | | |
| | 44. # 6 fuel oil | | | | | | | |

* For notes see p. 136.

ANALYSIS *

TABLE I

| Conversion → | | | | | | | | | | | ← | |
|---|---|---|---|---|---|---|---|---|---|---|---|---|
| Ther. cr. of 725 + SR | Visbreaking of vac. dist. bottoms | Ther. ref. of 250-325 SR (78% conv.) | Ther. ref. of 325-400 SR (95% conv.) | Cat. cracking of 400-550 SR | Cat. cracking of 550-725 SR | Cat. cracking of vac. dist. 725 + gas oil | Cat. cracking of 400-550 cat. cracked (recycling) | Cat. cracking of 500-725 cat. cracked (recycling) | Cat. ref. of 250-325 SR | Cat. ref. of 325-400 SR | 115/145 Avgas blend (1) | 115/145 Avgas blend (2) |
| 8 | 9 | 10 | 11 | 12 | 13 | 14 | 15 | 16 | 17 | 18 | 19 | 20 |
| -1.00 | -1.00 | -1.00 | -1.00 | | | | | | | | | |
| | | | | -1.00 | -1.00 | -1.00 | -1.00 | -1.00 | | | | |
| | | | | | | | | | -1.00 | -1.00 | | |
| | | -1.00 | | | | | | | -1.00 | | | |
| | | | -1.00 | | | | | | | -1.00 | | |
| | | | | -1.00 | | | | | | | | |
| | | | | | -1.00 | | | | | | | |
| -1.00 | | | | | | | | | | | | |
| | | | | | | -1.00 | | | | | | |
| | -1.00 | | | | | | | | | | | |
| .051 | .030 | | | | | | | | | | | -.625 |
| .025 | .015 | .78 | | | | | | | | | | |
| .026 | .015 | | .95 | | | | | | | | | |
| .057 | | | | | | | | | | | | |
| .126 | | | | | .50 | | | -.25 | | | | |
| .70 | | | | | | .50 | | | | | | |
| | | | | .24 | .24 | .24 | .095 | .095 | | | | |
| | | | | .079 | .079 | .079 | .031 | .031 | .91 | | -.75 | -.375 |
| | | | | .064 | .064 | .064 | .026 | .026 | | .95 | | |
| | | | | .50 | | | -.25 | | | | | |
| | | | | | | | | | | | - 193 | - 193 |
| | | | | | | | | | | | -.25 | |
| | | | | | | | | | | | -15.6 | -29.0 |
| | | | | | | | | | | | 1.00 | 1.00 |
| | -.31 | | | | | | | | | | | |
| | 1.24 | | | | | | | | | | | |

9. — *The structural interdependence of the economy.*

| | Items \ Activities | 115/145 Avgas blend (3) | 115/145 Avgas blend (4) | 115/145 Avgas blend (5) | 115/145 Avgas blend (6) | 100/130 Avgas blend (1) | 100/130 Avgas blend (2) | 100/130 Avgas blend (3) |
|---|---|---|---|---|---|---|---|---|
| | | 21 | 22 | 23 | 24 | 25 | 26 | 27 |
| Crude & Equipment Availabilities | 1. Availability Crude 1 | | | | | | | |
| | 2. Availability Crude 2 | | | | | | | |
| | 3. Availability Crude 3 | | | | | | | |
| | 4. Atmos. crude dist. capacity | | | | | | | |
| | 5. Vacuum distillation capacity | | | | | | | |
| | 6. Thermal operations capacity | | | | | | | |
| | 7. Catalytic cracking capacity | | | | | | | |
| | 8. Catalytic reforming capacity | | | | | | | |
| Intermediate Materials | 9. 100-250 Straight run | | | | | | | |
| | 10. 250-325 Straight run | | | | | | | |
| | 11. 325-400 Straight run | | | | | | | |
| | 12. 400-550 Straight run | | | | | | | |
| | 13. 550-725 Straight run | | | | | | | |
| | 14. 725 + Straight run | | | | | | | |
| | 15. Vac. distillate 725 + gas oil | | | | | | | |
| | 16. Vac. distillate bottoms | | | | | | | |
| | 17. 100-250 thermally cracked | | | −.57 | | | −.625 | |
| | 18. 250-325 thermally cracked | | | | | | | |
| | 19. 325-400 thermally cracked | | | | | | | |
| | 20. 400-550 thermally cracked | | | | | | | |
| | 21. 550-725 cracked | | | | | | | |
| | 22. 725 + cracked | | | | | | | |
| | 23. 100-250 catalytically cracked | −.625 | | | −.57 | | | −.625 |
| | 24. 250-325 catalytically cracked | −.375 | | | | −.75 | −.375 | −.375 |
| | 25. 325-400 catalytically cracked | | | | | | | |
| | 26. 400-500 catalytically cracked | | | | | | | |
| Exogenous Inputs | 27. Alkylate | | −.79 | −.43 | −.43 | | | |
| | 28. TEL | − 193 | − 193 | − 193 | − 193 | − 193 | − 193 | − 193 |
| | 29. Isopentane | | −.21 | | | −.25 | | |
| End Items Specs. | 30. 115/145 avgas, PN rich | −16.1 | .52 | −25.7 | 7.3 | | | |
| | 31. 100/130 avgas, PN rich | | | | | −.6 | −14.0 | −1.1 |
| | 32. 100/130 avgas, PN lean | | | | | −.85 | −5.25 | −8.13 |
| | 33. Regular motor gas, octane no. | | | | | | | |
| | 34. Regular motor gas with 0-1.5 cc/gal of TEL | | | | | | | |
| | 35. Regular motor gas with 1.5-3.0 cc/gal of TEL | | | | | | | |
| | 36. Regular motor gas, RVP | | | | | | | |
| | 37. Jet fuel, % aromatics | | | | | | | |
| End Item Requirements | 38. 115/145 avgas | 1.00 | 1.00 | 1.00 | 1.00 | | | |
| | 39. 100/130 avgas | | | | | 1.00 | 1.00 | 1.00 |
| | 40. Jet fuel | | | | | | | |
| | 41. Regular motor gas | | | | | | | |
| | 42. Lubes | | | | | | | |
| | 43. # 2 fuel oil | | | | | | | |
| | 44. # 6 fuel oil | | | | | | | |

**Table 1 (continued)**

| End Item Blending | | | | | | | | | | | | |
|---|---|---|---|---|---|---|---|---|---|---|---|---|
| Avgas blend (4) | 100/130 Avgas blend (5) | 100/130 Avgas blend (6) | Reg. motor gas — Component 9 | Reg. motor gas — Component 10 | Reg. motor gas — Component 11 | Reg. motor gas — Component 17 | Reg. motor gas — Component 18 | Reg. motor gas — Component 19 | Reg. motor gas — Component 23 | Reg. motor gas — Component 24 | Reg. motor gas — Component 25 | Reg. motor gas — Component 27 |
| 28 | 29 | 30 | 31 | 32 | 33 | 34 | 35 | 36 | 37 | 38 | 39 | 40 |
| | | | −1.00 | | | | | | | | | |
| | | | | −1.00 | | | | | | | | |
| | | | | | −1.00 | | | | | | | |
| | −.57 | | | | | −1.00 | | | | | | |
| | | | | | | | −1.00 | | | | | |
| | | | | | | | | −1.00 | | | | |
| | | −.57 | | | | | | | −1.00 | | | |
| | | | | | | | | | | −1.00 | | |
| | | | | | | | | | | | −1.00 | |
| −.79 | −.43 | −.43 | | | | | | | | | | −1.00 |
| −193 | −193 | −193 | −126 | −126 | −126 | −126 | −126 | −126 | −126 | −126 | −126 | −126 |
| −.21 | | | | | | | | | | | | |
| 15.5 | −10.7 | 7.7 | | | | | | | | | | |
| 23.1 | −.94 | 5.56 | | | | | | | | | | |
| | | | 0.00 | −31.40 | −31.40 | 5.00 | −2.00 | −21.00 | 11.00 | 6.00 | −2.00 | 20.0 |
| | | | 1.00 | 1.00 | 1.00 | 1.00 | 1.00 | 1.00 | 1.00 | 1.00 | 1.00 | 1.00 |
| | | | 1.00 | 1.00 | 1.00 | 1.00 | 1.00 | 1.00 | 1.00 | 1.00 | 1.00 | 1.00 |
| | | | 0.00 | 8.00 | 8.00 | 0.00 | 8.00 | 8.00 | 0.00 | 8.00 | 8.00 | 7.00 |
| 1.00 | 1.00 | 1.00 | | | | | | | | | | |
| | | | 1.00 | 1.00 | 1.00 | 1.00 | 1.00 | 1.00 | 1.00 | 1.00 | 1.00 | 1.00 |

| Items \ Activities | | Reg. motor gas — Component 29 | • Removal • of TEL, 0-1.5 cc/gal. | • Removal • of TEL, 1.5-3.0 cc/gal. | Jet fuel — Component 9 | Jet fuel — Component 10 | Jet fuel — Component 11 | Jet fuel — Component [illegible] |
|---|---|---|---|---|---|---|---|---|
| | | 41 | 42 | 43 | 44 | 45 | 46 | 47 |
| Crude & Equipment Availabilities | 1. Availability Crude 1 | | | | | | | |
| | 2. Availability Crude 2 | | | | | | | |
| | 3. Availability Crude 3 | | | | | | | |
| | 4. Atmos. crude dist. capacity | | | | | | | |
| | 5. Vacuum distillation capacity | | | | | | | |
| | 6. Thermal operations capacity | | | | | | | |
| | 7. Catalytic cracking capacity | | | | | | | |
| | 8. Catalytic reforming capacity | | | | | | | |
| Intermediate Materials | 9. 100-250 Straight run | | | | −1.00 | | | |
| | 10. 250-325 Straight run | | | | | −1.00 | | |
| | 11. 325-400 Straight run | | | | | | −1.00 | |
| | 12. 400-550 Straight run | | | | | | | −1.00 |
| | 13. 550-725 Straight run | | | | | | | |
| | 14. 725 + Straight run | | | | | | | |
| | 15. Vac. distillate 725 + gas oil | | | | | | | |
| | 16. Vac. distillate bottoms | | | | | | | |
| | 17. 100-250 thermally cracked | | | | | | | |
| | 18. 250-325 thermally cracked | | | | | | | |
| | 19. 325-400 thermally cracked | | | | | | | |
| | 20. 400-550 thermally cracked | | | | | | | |
| | 21. 550-725 cracked | | | | | | | |
| | 22. 725 + cracked | | | | | | | |
| | 23. 100-250 catalytically cracked | | | | | | | |
| | 24. 250-325 catalytically cracked | | | | | | | |
| | 25. 325-400 catalytically cracked | | | | | | | |
| | 26. 400-500 catalytically cracked | | | | | | | |
| Exogenous Inputs | 27. Alkylate | | | | | | | |
| | 28. TEL | − 126 | 63 | 63 | | | | |
| | 29. Isopentane | −1.00 | | | | | | |
| End Items Specs. | 30. 115/145 avgas, PN rich | | | | | | | |
| | 31. 100/130 avgas, PN rich | | | | | | | |
| | 32. 100/130 avgas, PN lean | | | | | | | |
| | 33. Regular motor gas, octane no. | 20.0 | −8.00 | −3.00 | | | | |
| | 34. Regular motor gas with 0-1.5 cc/gal of TEL | 1.00 | −1.00 | | | | | |
| | 35. Regular motor gas with 1.5-3.0 cc/gal of TEL | 1.00 | | −1.00 | | | | |
| | 36. Regular motor gas, RVP | −12.0 | | | | | | |
| | 37. Jet fuel, % aromatics | | | | 21.0 | 18.3 | 13.0 | 6.6 |
| End Item Requirements | 38. 115/145 avgas | | | | | | | |
| | 39. 100/130 avgas | | | | | | | |
| | 40. Jet fuel | | | | 1.00 | 1.00 | 1.00 | 1.00 |
| | 41. Regular motor gas | 1.00 | | | | | | |
| | 42. Lubes | | | | | | | |
| | 43. # 2 fuel oil | | | | | | | |
| | 44. # 6 fuel oil | | | | | | | |

**Table 1 (continued)**

| End Item Blending (continued) | | | | | | | | | | | | | |
|---|---|---|---|---|---|---|---|---|---|---|---|---|---|
| | Jet fuel — Component 17 | Jet fuel — Component 18 | Jet fuel — Component 19 | Jet fuel — Component 20 | Jet fuel — Component 23 | Jet fuel — Component 24 | Jet fuel — Component 25 | Jet fuel — Component 26 | # 2 Fuel oil Blend (1) | # 2 Fuel oil Blend (2) | # 2 Fuel oil Blend (3) | # 2 Fuel oil Blend (4) | # 2 Fuel oil Blend (5) |
| | 48 | 49 | 50 | 51 | 52 | 53 | 54 | 55 | 56 | 57 | 58 | 59 | 60 |
| | | | | | | | | | −1.00 | −.579 | | | −.651 |
| | | | | | | | | | | −.421 | −.421 | −.421 | |
| | −1.00 | | | | | | | | | | | | |
| | | −1.00 | | | | | | | | | | | |
| | | | −1.00 | | | | | | | | | | |
| | | | | −1.00 | | | | | | | −.579 | | |
| | | | | | | | | | | | | | −.349 |
| | | | | | −1.00 | | | | | | | | |
| | | | | | | −1.00 | | | | | | | |
| | | | | | | | −1.00 | | | | | | |
| | | | | | | | | −1.00 | | | | −.579 | |
| | 18.5 | 10.0 | 5.0 | −11.0 | 18.5 | −11.0 | −46.0 | −46.0 | | | | | |
| | 1.00 | 1.00 | 1.00 | 1.00 | 1.00 | 1.00 | 1.00 | 1.00 | | | | | |
| | | | | | | | | | 1.00 | 1.00 | 1.00 | 1.00 | 1.00 |

| | Items \ Activities | # 2 Fuel oil Blend (6) | # 2 Fuel oil Blend (7) | # 2 Fuel oil Blend (8) | # 2 Fuel oil Blend (9) | # 6 Fuel oil Blend (1) | # 6 Fuel oil Blend (2) | # 6 Fuel oil Blend (3) |
|---|---|---|---|---|---|---|---|---|
| | | 61 | 62 | 63 | 64 | 65 | 66 | 67 |
| Crude & Equipment Availabilities | 1. Availability Crude 1 | | | | | | | |
| | 2. Availability Crude 2 | | | | | | | |
| | 3. Availability Crude 3 | | | | | | | |
| | 4. Atmos. crude dist. capacity | | | | | | | |
| | 5. Vacuum distillation capacity | | | | | | | |
| | 6. Thermal operations capacity | | | | | | | |
| | 7. Catalytic cracking capacity | | | | | | | |
| | 8. Catalytic reforming capacity | | | | | | | |
| Intermediate Materials | 9. 100-250 Straight run | | | | | | | |
| | 10. 250-325 Straight run | | | | | | | |
| | 11. 325-400 Straight run | | | | | | | |
| | 12. 400-550 Straight run | | | | | | | |
| | 13. 550-725 Straight run | | | | | −.029 | −.966 | |
| | 14. 725 + Straight run | | | | | | | −.979 |
| | 15. Vac. distillate 725 + gas oil | | | | | | | |
| | 16. Vac. distillate bottoms | | | | | | | |
| | 17. 100-250 thermally cracked | | | | | | | |
| | 18. 250-325 thermally cracked | | | | | | | |
| | 19. 325-400 thermally cracked | | | | | | | |
| | 20. 400-550 thermally cracked | −.651 | | | −1.00 | | | |
| | 21. 550-725 cracked | −.349 | −.349 | | | | −.034 | |
| | 22. 725 + cracked | | | | | −.971 | | |
| | 23. 100-250 catalytically cracked | | | | | | | |
| | 24. 250-325 catalytically cracked | | | | | | | |
| | 25. 325-400 catalytically cracked | | | | | | | |
| | 26. 400-500 catalytically cracked | | −.651 | −1.00 | | | | |
| Exogenous Inputs | 27. Alkylate | | | | | | | |
| | 28. TEL | | | | | | | |
| | 29. Isopentane | | | | | | | |
| End Items Specs. | 30. 115/145 avgas, PN rich | | | | | | | |
| | 31. 100/130 avgas, PN rich | | | | | | | |
| | 32. 100/130 avgas, PN lean | | | | | | | |
| | 33. Regular motor gas, octane no. | | | | | | | |
| | 34. Regular motor gas with 0-1.5 cc/gal of TEL | | | | | | | |
| | 35. Regular motor gas with 1.5-3.0 cc/gal of TEL | | | | | | | |
| | 36. Regular motor gas, RVP | | | | | | | |
| | 37. Jet fuel, % aromatics | | | | | | | |
| End Item Requirements | 38. 115/145 avgas | | | | | | | |
| | 39. 100/130 avgas | | | | | | | |
| | 40. Jet fuel | | | | | | | |
| | 41. Regular motor gas | | | | | | | |
| | 42. Lubes | | | | | | | |
| | 43. # 2 fuel oil | 1.00 | 1.00 | 1.00 | 1.00 | | | −.021 |
| | 44. # 6 fuel oil | | | | | 1.00 | 1.00 | 1.00 |

**Table 1** (continued)

| End Item Blending (continued) → | | | | | | | | | | | Excess Availabilities | |
|---|---|---|---|---|---|---|---|---|---|---|---|---|
| # 6 Fuel oil Blend (4) | # 6 Fuel oil Blend (5) | # 6 Fuel oil Blend (6) | # 6 Fuel oil Blend (7) | # 6 Fuel oil Blend (8) | # 6 Fuel oil Blend (9) | # 6 Fuel oil Blend (10) | # 6 Fuel oil Blend (11) | # 6 Fuel oil Blend (12) | # 6 Fuel oil Blend (13) | Product mix | Excess Availibility of Crude 1 | Excess # 6 Fuel oil |
| 68 | 69 | 70 | 71 | 72 | 73 | 74 | 75 | 76 | 77 | 83 | 84 | 127 |
| | | | | | | | | | | | −1.00 | |
| −.014 | | | | | | −.182 | | | | | | |
| | | | | | | | | | | $-\alpha_{12}$ | | |
| | | | −.329 | | | | | | −1.00 | $-\alpha_{13}$ | | |
| −.986 | −.986 | −.986 | | | | | | | | | | |
| | | | | | | | | | | $-\alpha_{16}$ | | |
| | −.014 | | | | | | −.182 | | | | | |
| | | | | −.361 | | | | | | | | |
| | | | −.671 | −.639 | −.737 | −.818 | −.818 | −.818 | | | | |
| | | −.014 | | | | | | −.182 | | | | |
| | | | | | | | | | | $-\alpha_{38}$ | | |
| | | | | | | | | | | $-\alpha_{39}$ | | |
| | | | | | | | | | | $-\alpha_{40}$ | | |
| | | | | | | | | | | $-\alpha_{41}$ | | |
| | | | | | | | | | | $-\alpha_{42}$ | | |
| | | | | | −.263 | | | | | $-\alpha_{43}$ | | |
| 1.00 | 1.00 | 1.00 | 1.00 | 1.00 | 1.00 | 1.00 | 1.00 | 1.00 | 1.00 | $-\alpha_{44}$ | | −1.00 |

## NOTES (Table 1)

The units of measurement for items 1 to 27, 29, and 38 to 44 are millions of barrels per day. Item 28 is measured in thousands of litres per day.

The coefficients of the specification equations are:

Item 30. Rich mixture « performance number » in excess of 145.
Item 31. Rich mixture « performance number » in excess of 130.
Item 32. Lean mixture « performance number » in excess of 100.
See footnote 2, p. 138, for items 33 to 35.
Item 36. 10 minus the vapour pressure (in lbs.) of the blending stock.
Item 37. 25 minus the percent aromatics, by volume.

Cracking and reforming equipment is either « thermal » or « catalytic ». The composition of the product differs depending on which of these is used. Thermal cracking equipment can substitute for thermal reforming equipment with some loss in capacity. In the 109-equation model thermal cracking, thermal reforming, catalytic cracking and catalytic reforming equipment are distinguished. There is also an activity of using thermal cracking equipment for thermal reforming. In the 44-equation model thermal reforming and thermal cracking are aggregated while separate categories are given to catalytic cracking and catalytic reforming.

---

involving blended end items and specifications are not quite so straightforward.

A commodity such as regular grade motor gasoline is not one specific chemical mixture, but any mixture that behaves suitably — that meets certain specifications. In the 44-equation model the assumption is made that jet fuel, for example, is any combination of items 9, 10, 11, 12, 17, 18, 19, 20, 23, 24, 25, and/or 26 such that this combination contains less than 25 per cent aromatics [1] by volume.[2] One way of handling such a specification constraint would be to enumerate various combinations of blending stocks which meet the constraint. For each such way of meeting specifications the model could have an activity which had as inputs these particular amounts of blending stocks, and as output the product (e.g. jet fuel). By the nature of the analysis the model would be permitted to use not

---

[1] An aromatic is a particular kind of hydro-carbon molecule.

[2] The 109-equation model had two other specifications which were omitted from the smaller model on the assumption that this omission would not seriously affect overall results.

only the blends specified but also any convex combination[1] of them. It is sometimes possible to find a small number of such activities which completely characterize the blending possibilities[2]. In other cases this approach only leads to the inclusion in the analysis of part of the possible blending combinations.

Another approach is based on the following considerations: Let $x_k$ be the amount of the $k^{th}$ blending stock used in making jet fuel. The total amount produced is $F = \Sigma x_k$; let $y_k = x_k/F$; be the proportion which $y_k$ is of the total; and let $a_k$ be the percent aromatics in the $k^{th}$ material. Our requirement then is that

$$\Sigma a_k y_k \leq 25$$

or

$$\Sigma a_k y_k - 25 \leq 0$$

Since $\Sigma y_k = 1$ we have

$$\Sigma (a_k - 25) y_k \leq 0$$

Multiplying both sides by $F$ we get

$$\Sigma (a_k - 25) x_k \leq 0$$

or

$$\Sigma (25 - a_k) x_k \geq 0$$

The above inequality can be expressed as an equation as

$$\Sigma (25 - a_k) x_k - e = 0$$

where $e$, like the other variables, must be non-negative.

Thus in the 44-equation model we have an aromatics specification equation (row 37):

$$0 = 21 X_{44} + 18.3 X_{45} + 13 X_{46} + 6.6 X_{47} + 18.5 X_{48} + 10 X_{49} + 5 X_{50} - 11 X_{51} + 18.5 X_{52} - 11 X_{53} - 46 X_{54} - 46 X_{55}$$

---

[1] I.e., if the combination $(y_1, y_2, ..., y_k)$ of $k$ blending stocks meets the specifications and $(y_1^0, y_2^0, ..., y_k^0)$ also meets the specifications, then the model will automatically assume that, e.g., $\frac{1}{2}(y_1, ..., y_k) + \frac{1}{2}(y_1^0, ..., y_k^0) = (\frac{1}{2}y_1 + \frac{1}{2}y_1^0, ..., \frac{1}{2}y_k + \frac{1}{2}y_k^0)$ also meets the specifications. This assumption is always justified since $(\frac{1}{2}y_1 + \frac{1}{2}y_1^0, ..., \frac{1}{2}y_k + \frac{1}{2}y_k^0)$ may be interpreted as having the industry meet one half of its requirements for the product by the combination $(y_1, ..., y_k)$ and the other half by the combination $(y_1^0, ..., y_k^0)$.

[2] I.e., such that every blend which meets the specifications is a convex combinaon of the enumerated blends.

and a jet fuel « quantity requirement » or « sources and uses » equation (row 40):

$$0 = X_{44} + X_{45} + X_{46} + \ldots + X_{55} - \alpha_{40} X_{83} - X_{123}$$

Different end items are handled differently. In the case of the fuel oils various combinations which meet specifications are enumerated and given separate activities. In the case of aviation gasolines pre-blends are formed to meet a vapour pressure specification. The model is required to combine these pre-blends so as to meet octane specifications [1]. The motor gasoline blending equations are somewhat complicated by the fact that octane increases non-linearly as tetraethyl lead (TEL) is added to the blending stocks [2].

---

[1] In the case of aviation gasolines it is assumed that the vapour pressure specification will be met exactly rather than with positive $e$. It is also assumed that the maximum permitted amount of tetraethyl lead, 4.6 ml/gal, will be used.

[2] Tetraethyl lead is a chemical which is added to motor fuel in small amounts to improve its octane rating. Because large amounts are eventually harmful to a motor, upper limits to its use have been set by common practice. These limits, observed by the model, are 4.6 ml/gal for avgas and 3 ml/gal for motor gasoline (ml = millilitre). The model assumes that: as TEL is added to a particular blending stock octane rating increases according to a piecewise linear curve such as abc in the figure below;

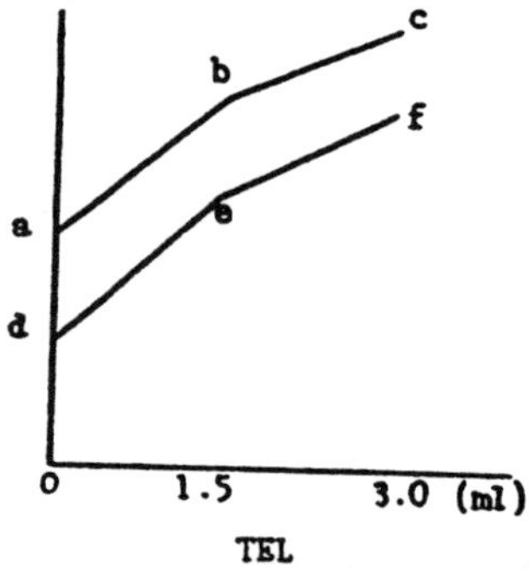


the octane rating versus TEL curves for two different blending stocks are parallel as abc and def in the figure; for a given lead level octane blends linearly. Each of these assumptions is an approximation. The accuracy of or justification for these approximations for the present purposes will not be discussed here. The way these assumptions are incorporated into the model is as follows; for each gasoline blending stock the model contains an activity of adding one unit of this blending stock to the materials going into motor fuel and also allocating (3 ml TEL/gal of blending stock) to the fuel. (The coefficient in the table is 126 ml/barrel which equals 3 ml/gallon). These activities permit only gasoline with 3 ml/gal to be produced by the model. The model therefore contains an activity $r_1$ which, per unit of the activity, reduces the amount of TEL per gallon by 1.5 ml and reduces the octane rating by 3. (This corresponds to moving from $c$ to $b$ in the figure). This activity is not allowed to remove more than

A linear programming model of technology, we see, cannot be constructed mechanically with *a priori* rules. Given the general size limitations of the analysis, an attempt must be made to find a system of equations which best describes the limits imposed by technology.

The particular question we have considered above was the maximizing of a product mix. We could, instead, have had fixed requirements among some items and maximized a product mix among the others; or for each item we could have had a fixed minimum requirement plus a proportional « product mix » requirement; or we could have set values on the end items and maximized the value of the whole. In each of these cases we would be seeking an extreme point in the set of attainable output combinations. This may be illustrated

---

1.5 ml/gal. Another activity $r_2$ also reduces TEL by 1.5 ml/gal, but reduces octane by 8. (This corresponds to moving from $b$ to $a$). In an optimal solution $r_2$ will not be used unless $r_1$ is at its upper limit, since $r_1$ loses less octane per removal of TEL. (Of course, when we speak of « removing » TEL we mean « not putting it there in the first place »). Defining $x_k$, $y_k$ and $F$ as before (but measuring them in barrels), letting $-t$ be the amount of TEL used per barrel (measured in ml), letting $N$ be the octane requirement of the gasoline and $a_k$ be the octane rating of the $k$-th blending stock when it contains 3 ml/gal ( = 126 ml/bbl) of TEL, we have

$$t = -126 + 63\, r_1 + 63\, r_2 \qquad (1)$$

or

$$t = -126\, \Sigma\, y_k + 63\, r_1 + 63\, r_2 \qquad (1')$$

$$N \leq \Sigma\, a_k\, y_k - 3\, r_1 - 8\, r_2 \qquad (2)$$

or

$$\Sigma\, (a_k - N)\, y_k - 3\, r_1 - 8\, r_2 - e_2 = 0 \qquad (2')$$

$$r_1 \leq 1 \qquad (3)$$

or

$$r_1 \leq \Sigma\, y_k \qquad (3')$$

or

$$\Sigma\, y_k - r_1 - e_3 = 0 \qquad (3'')$$

Similarly

$$\Sigma\, y_k - r_2 - e_4 = 0 \qquad (4'')$$

Multiplying (1'), (2'), (3'') and (4'') through by $F$, letting $-t\,F = -T$ be total TEL used in motor gasoline and letting $R_1 = r_1\, F$, $R_2 = r_2\, F$, we get

$$T = -126\, \Sigma\, x_k + 63\, R_1 + 63\, R_2 \qquad [i]$$

$$\Sigma\, (a_k - N)\, x_k - 3\, R_1 - 8\, R_2 - e_2' = 0 \qquad [ii]$$

$$\Sigma\, x_k - R_1 - e_3' = 0 \qquad [iii]$$

$$\Sigma\, x_k - R_2 - e_4' = 0 \qquad [iv]$$

which is essentially what appears in the matrix (plus a vapour pressure specification equation).

graphically in the case of two end items. Suppose, given the basic resources, there exists some production function for the goods $G_1$ and $G_2$. The purpose of computation is to find « interesting » points

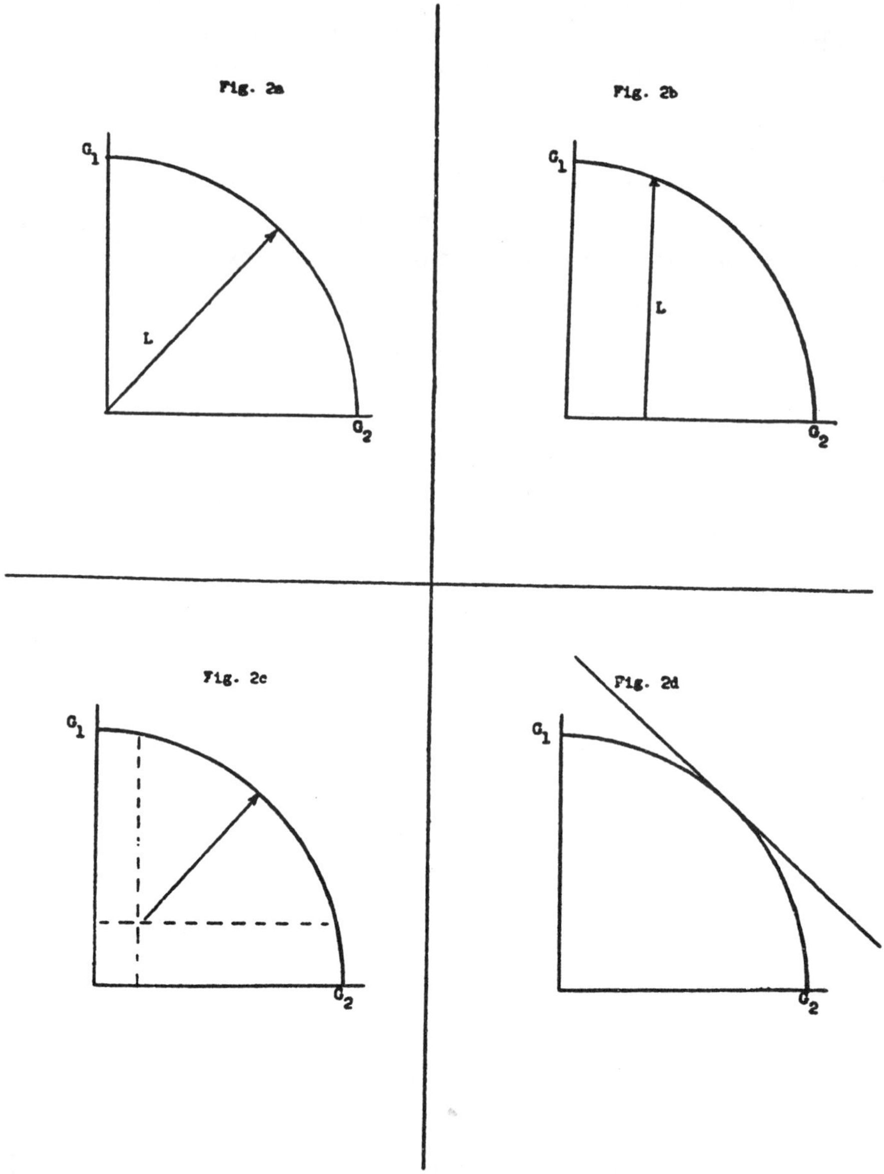


Fig. 2

on this function. If we maximize a product mix we in effect move along a line such as $L$ in Figure *2a* until we meet the production function; if we fix a requirement for one item and maximize the amount attainable of the other, we move along a line such as that in Figure 2b; if we let the requirements for each product have a fixed and a variable part we move along a line such as that in Figure 2c; if we fix values to the two end items and maximize the value of total output, we in effect find a point at which a « utility » or profit line, whose slope is predetermined, is tangent to the production function as in Figure 2d.

Once one such linear programming computation has been run, it is possibile to vary continuously the conditions of the problem and see how the answers are affected. In the petroleum analysis one can see, for example, how the maximum obtainable product mix of some end items varies as the requirements for another set are

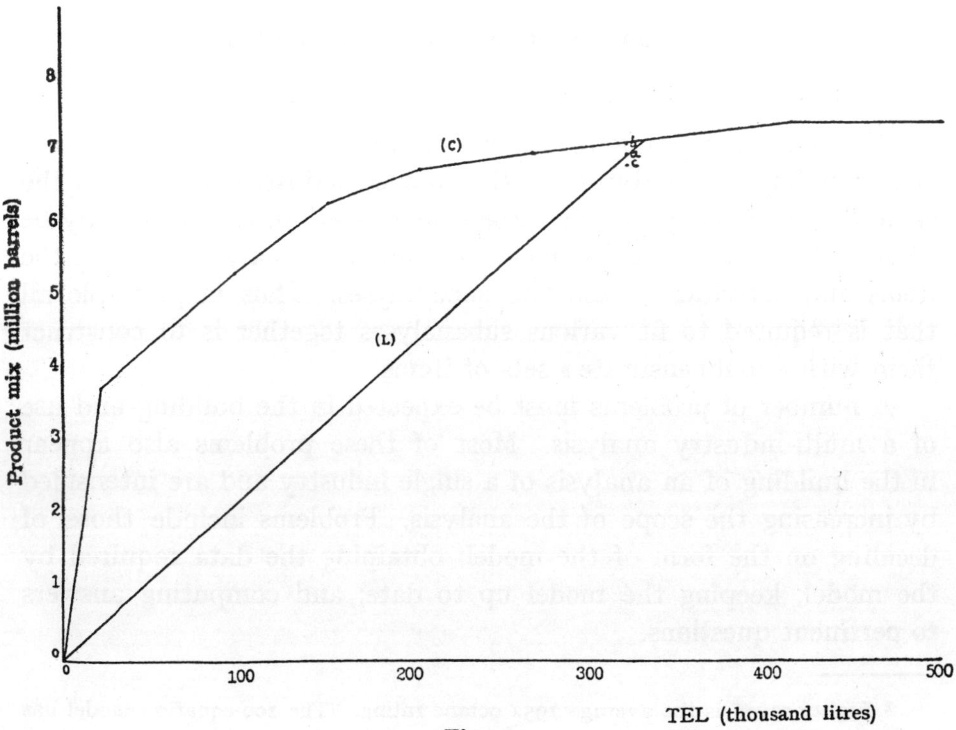


Fig. 3

varied, or how the attainable level of the product mix is affected by varying the availability of one or more crude oil and/or equipment capacity and/or « exogenous » material. In Figure 3, for example, is plotted the 44-equation model's estimates of the maximum amount of product mix attainable for various TEL consumptions, given the 1952 availabilities of crude and equipment capacities and the 1952 set of proportions among end items. This curve (C) is drawn on the « austere » assumption that all gasoline is 85 octane (U.S. regular grade). Figure 3 also shows the actual 1952 U.S. output — TEL consumption combination (point a); the maximum attainable product mix for TEL = 322,000 liters (the 1952 consumption) as estimated by the 109-equation model (point b); the maximum attainable product mix for TEL = 322,000 liters as estimated by the 44-equation model with an 87 octane requirement on motor fuel (point c) [1].

By way of comparison, line *L* shows the maximum obtainable product mix on the assumption that there are fixed requirements for TEL.

## 3. *Problems of economy-wide analysis*

A linear programming model of an industry involves a set of equations corresponding to various « items », and a set of variables corresponding to « activities ». If two such industry studies use the same list of items, insofar as these studies overlap, then they may be combined to form a larger linear programming analysis with all the items and activities of the two subanalyses. Thus in principle all that is required to fit various subanalyses together is to construct them with « commensurate » sets of items.

A number of problems must be expected in the building and use of a multi-industry analysis. Most of these problems also appear in the building of an analysis of a single industry and are intensified by increasing the scope of the analysis. Problems include those of deciding on the form of the model; obtaining the data required by the model; keeping the model up to date; and computing answers to pertinent questions.

---

[1] Eighty-seven is the average 1952 octane rating. (The 109-equation model has two grades of motor fuel, regular and premium). The fact that *c*) in less than *a*) means that the 44-equation model underestimates the maximum amount attainable of the product mix. T. Marschak suggests that this may be chiefly due to the fact that the 44-equation model, unlike the real world and the 109-equation model, does not distinguish between high and low severity reforming. To include this would require about two more rows and two more columns. This modification would raise slightly the attainability curve for 85 octane.

The decision on the *basic concepts, categories and assumptions* of a subanalysis must be based on a number of considerations such as the nature of the technology, the purpose of the analysis, the availability of data, and the capabilities of computing. A mixed team, consisting of persons familiar with the technology and persons familiar with linear programming, seems best able to bring together and weigh these various factors. These matters are not decided upon, once and for all, at the beginning of a project. As a sector is investigated, as the results of trial runs are seen, the « form » of the model changes.

The importance of this decision (as to the basic concepts, categories, and assumptions) should be self-evident to the economist — whose literature is filled with examples in which the form of an economic model pre-determines certain major conclusions, whatever be the parameter values of the model.

An aid in *obtaining information* for a process analysis model is the fact that frequently the industrial engineer has needed and developed essentially the same data. In the analysis of the metal working industries, for examples, we need to know the rates at which various machine tools can perform various basic « tasks ». « Standard data » upon which the industrial engineer bases his estimates are available in abundance both in private and published sources. In the development of coefficients it is not necessary to survey all plants in an industry. Standard performance times — obtained from a variety of sources — seem adequate to indicate whether particular coefficients are fairly stable and to provide an «average» suitable for our purposes. The chief problem encountered in getting many coefficients is that the information is proprietary. This is not an insurmountable difficulty as long as essentially the same information can be obtained from several sources and averaged together for the purposes of the analysis.

Another kind of information, needed in the application of a process analysis model, is inventories of resource (mostly equipment) availabilities. For some industries (steel, petroleum, metal working) we are fortunate enough to have such inventories readily available. It should be noted that an equipment inventory is a basically easier set of information for an establishment to provide than interindustry flow information. Within any plant there is at least one person who must know, for his day-to-day operations, what machinery is present. There is usually no such immediate necessity of knowing the amount of goods sold during the past 12 months to establishments of a particular industry.

Suppose that the optimal solution has been obtained for some linear programming problem but we wish to *add a few new activities* to the model. We do not have to start the problem over, but can usually obtain the new optimal solution in a few iterations. This can be useful in two ways: first, if there is technological change which takes the form of adding new activities, but no new items, such change can be introduced in this way. Second, not every aspect of technology can be explored; not every possible substitution possibility can be considered. In some particular problem it may turn out that a particular aspect of technology deserves further attention, and that substitution possibilities should be included which were missed the first time. Such « new » alternatives can be introduced and a new optimum found without great difficulty.

Usually technological change means that both items and activities must be added to the linear programming analysis. In this case old optimal solutions cannot be « touched up » as easily as in the case of new activities only.

There are two sorts of *computing difficulties* which must be faced in the use of an economy-wide process analysis model: the first is due to the size of such a system, and the second to «non-convexities».

A detailed linear programming model of the economy as a whole must be expected to have literally thousands of equations. At present the largest general-purpose computing code[1] can handle no more than two hundred equations. Computing time rises rapidly as we go from a one-hundred-equation problem (about eight machine hours) to a two-hundred-equation problem (over sixty hours). Problems with more than two hundred equations have been solved, however. Our machine tool substitution analysis, in particular, was used in a linear programming problem with 223 equations, 1500 unknowns and a requirement of between one and two machine hours to solve [4]. The problem had few non-zero coefficients in its matrix and a special code was developed which made use of its special structure.

All our detailed industrial models have, to a great or lesser extent,

---

[1] A code is, in effect, a set of step-by-step instructions which are « given » to a computing machine so that it may « know » how to solve the problem. The code referred to above is for the IBM 701.

special structures and a good many zeros. It is, at this moment, within the capabilities of our machines and know-how to develop codes to handle models of the large size (but « simple » structure) of an economy-wide process analysis model.

Not every technological relationship can be approximated by a linear programming model. It is necessary for the production constraints to be convex [1] — which essentially means that there must be no serious problems of indivisibilities or increasing returns to scale (decreasing costs).

Some problems of increasing returns which appear at a plant level disappear at the industry or economy level. If in the analysis of a plant we found that the solution bought $1\frac{1}{4}$ new lathes, we would be more disturbed than if in an economy-wide analysis we found that we produced $1000\frac{1}{4}$ more lathes. Equipment, such as a distillation tower, which can be built in various capacities frequently shows decreasing costs up to a point. If the optimum size unit is « small » as compared to the total capacity of such units, new capacity can be assumed to be the addition of new, optimal size units, and thus we have « constant » rather than decreasing costs.

Sometimes non-convexities can be avoided by special handling. In our machine tool analysis, for example, we handle the setup time problem [2] by using length of run as part of the characteristics of our basic tasks (e.g., a typical task would be to make a « cyclinder of revolution, small size, precision, large lot »). A requirement for a certain number of aircraft results in certain requirements for such tasks. If the model has one-half of the economy's requirements for a particular task met by one machine and the other half met by another this is interpreted not as a splitting of lots — but as the performance of half such lots on one machine and half on the other [3]. This same side-stepping of non-convexity cannot be used on the small plant level where linear programming formulations have, to the present, been faced with difficulties such as how to avoid having the model set up half a lathe to make some screws and the other half of the same lathe to make the bolts.

---

[1] I.e., if outputs $O_1, \ldots, O_m$ are obtainable with resources $R_1, \ldots, R_n$ and outputs $O_1^0, \ldots, O_m^0$ are obtainable with resources $R_1^0, \ldots, R_n^0$ then outputs $\lambda O_1 + (1 - \lambda) O_1^0, \ldots, \lambda O_m + (1 - \lambda) O_m^0$ must be obtainable with $\lambda R_1 + (1 - \lambda) R_1^0, \ldots, \lambda R_n + (1 - \lambda) R_n^0$.

[2] Cf. [1], pp. 346-359.

[3] See also footnote 1, p. 137.

Often non-convexities can be avoided by asking the right question. For example, in the analysis of an individual petroleum refinery the problem is naturally convex as long as one asks questions such as how to maximize profit with given equipment. This can, of course, be done for various equipment mixes. But if the question is asked, « what is the most profitable refinery which can be constructed with a given initial investment? », the constraint set is no longer convex.

Finally, some non-convexities may not be worth bothering about for the purposes of the model.

But suppose we find ourselves with important non-convexities that cannot be avoided or side-stepped. Then what do we do?

The best solution would be to find a general computing procedure for handling non-convexities. This course, however, is, so to speak, in the laps of the gods. We should not feel that there is no hope in this direction, especially considering a recent « break » in this « convexity barrier ». A famous programming problem with indivisibilities — the travelling salesman problem, which seeks the shortest route touching $n$ points and returning to the point of origin — has been solved by techniques that might possibly generalize to other classes of non-convex problems [5].

In lieu of the above solution, all we can do is to guess values of « bad » variables and use our computing procedure to derive the best values of the « well behaved » variables. This can be done for a range of guesses of the « bad » variables. There is reason to believe that such a procedure might lead to solutions whose « payoffs » were not far from the maximum obtainable, especially if there are not too many « bad variables ». We find that if a person familiar with a technology makes a careful guess he can usually come out within about 5 per cent of the maximum payoff. Frequently the making of the model in itself suggests major changes in practice which are later confirmed by the computation. In the classical case of the travelling salesman problem ($n = 49$) the mathematical technique simply confirmed a result which was already guessed at through the use of a map, a piece of string and some pins.

## 4. *Process analysis and other models of the economy*

Some salient characteristics of an economy-wide process analysis model would be: it would *not* analyze the capabilities of the economy in terms of industries, industry capacities, and interindustry flows (and *a fortiori* not in terms of fixed interindustry flows); it would analyze the capabilities of the economy, rather, in terms of equipment and commodity classes that are much more detailed than those usually found in economic models; it would handle substitution, joint production and decreasing returns, and would attempt to handle increasing returns; for the foreseeable future it would use linear programming as its primary mathematical tool; and it would have as its basic purpose the analysis of the capabilities of the economy, i.e., its job would be that of feasibility testing rather than the actual scheduling of the economy's operations.

It was dissatisfaction with concepts such as « industry capacity » that stimulated work in the field of industry-wide and economy-wide process analysis [6]. Even in industries such as steel and petroleum the concept of capacity is not without difficulties. The value of petroleum products or basic steel forms that can be produced depends on the product mix desired. Single physical measurements such as ignot capacity or crude distillation capacity are not completely adequate as a measure of the entire industry's capacity, since — given the other equipment available — it may not be possible to process all the ignots that could be produced, or all the crude that could be distilled, into desired end items.

In other cases, such as the capacities of metal working industries, the problem becomes even more difficult. Almost without exception the machinery used in making the parts of one metal commodity can be used in making the parts of countless others. During the last war, for example, there were countless instances of machines used for civilian items converted to the making of military items. In peacetime as well the institutions of subcontracting and the used machinery market help to direct machines across industry lines to their most « urgent » needs.

Interindustry flow coefficients can be expected to be unstable under major changes in supply and demand conditions for two reasons. First, an industry, such as defined in the 450-order U.S. matrix, produces a great variety of products. Even if there were no changes in productive processes, interindustry flows could change due to changes in an industry's product mix. Secondly, there are in fact alternative ways of producing the same commodity. Examples of substitution include: electricity can be produced by thermal or hydro means; coal or fuel oil can be used in the thermal production of electricity (many U.S. power stations are set for convenient switch in fuels in case of « shortages » or shifts in relative prices); steel can be produced with various ratios of pig iron and scrap; major raw material for organic chemicals, e.g. benzene, can come from either the petroleum or coke industries (and thus appear as alternate « interindustry » flows); wood, metals and/or plastics can substitute in various uses (e.g., the use of fibreglass instead of steel in sports car bodies); synthetic and natural fibres substitute in the making of textiles; land and fertilizers substitute in agriculture.

This does not mean that the data in an input-output table are not a valuable body of information. It means rather that there is a large gap between having such a picture of what the economy did in a particular year, and being able to evaluate the feasibility of a substantially different set of « final demands ».

A basic presumption of the process analyst is that there are important broad questions (such as the feasibility of alternate government programmes over time) which cannot be answered with broad models. Programmes which are within the bounds of reasonableness, in that they do not require a larger national product than is foreseeable, may nevertheless come across bottlenecks of particular kinds of equipment, skills or materials. Another presumption of the process analyst is that fundamental policy can be better made if the « decision-maker » can obtain rapidly a fairly good idea of what choices are and are not open. This certainly seems better than having to embark on a programme and find out months later that something much less ambitious or much more ambitious or simply something radically different should in fact have been undertaken.

The more we go into commodity, equipment and raw material detail, the more imperative it is to distinguish major substitution possibilities. If we include a particular kind of machinery, skill or material which might be a bottleneck, we must also show the ways in which this resource can be substituted for, if such substitution possibilities exist.

The concept of an economy-wide process analysis model is not synonymous with the concept of an economy-wide linear programming model. For the foreseeable future process analysis uses linear programming as a mathematical tool. If non-linear computing techniques, with the power of linear programming, were to appear they probably would find application in industry-wide and economy-wide process analysis models. Thus not every process analysis need be a linear programming analysis. Conversely, not every economy-wide linear programming model need be a process analysis model. A linear programming model could be built by adding alternate interindustry flows to an input-output table. The model could still retain the concepts of industry capacity and interindustry flow. It might also retain, as its major source of information, observed inter-establishment sales and purchases. It would not be a process analysis model based on the analysis of productive processes. It would be, rather, an « interindustry linear programming model ». As input-output is applied to dynamic questions (where there are the alternatives of stockpiling or capital building), as it is applied to spatial models (where there are alternative of obtaining the same « interindustry flows » from different regions), the use of an interindustry « linear programming model » seems inevitable even if alternative productive processes are not distinguished.

While an industry-wide process analysis model is detailed as compared to other economic models, it nevertheless represents a gross aggregation and simplification of reality. The engineer analyzing the capabilities of an existing refinery can get much more accurate information concerning his particular catalytic cracker than the average yields of an industry-wide petroleum model. The machine-shop scheduler can obtain much more accurate figures for the performances of a turret lathe in his shop than we can for our typical « Turret lathe, 12" to 36" swing ». There are two causes for this gross aggregation. First, our purpose is feasibility testing — the estimation of the capabilities of the industry or economy as a whole — and not detailed allocation and scheduling. Second, even if we wished to do more, we could not. Mathematical models, specially built for individual firms or establishments, can be valuable for detailed production problems; but surely it will be many centuries — if ever — before an economy-wide model schedules for two million machine tools.

Thus an economy-wide process analysis model is not intended to — or capable of — replacing the operating decisions of the industrial engineer or the allocation decisions of the market mechanism. Its purpose is to estimate the overall capabilities of an economy. On the other hand, it is my belief that this job of feasibility testing cannot be adequately handled by simple, overall models. Plans are often seriously upset and delayed because of the lack of particular kinds of machines, skills and materials. How can an analysis anticipate delays unless it considers these machines, skills and materials?.

There is a natural gulf between the industrial engineer and the policy-maker. The industrial engineer knows, within his own domain, technological facts which are important to the policy-maker. The policy-maker, on the other hand, must decide fundamental questions and has little time, and usually little ability, to consider countless technological detail. It would be the purpose of an economy-wide process analysis model to bridge the gulf between the engineer and the policy-maker. It would be designed to bring to bear quickly on the fundamental questions the countless technological detail.

## 5. *A list of works cited.*


[1] W. W. Leontief et alia, *Studies in the Structure of the American Economy* (New York, 1953), Part IV.

[2] A. S. Manne, « A Linear Programming Model of the U. S. Petroleum Refining Industry », *The RAND Corporation Paper* P-563 (3 Sept. 1954).

[3] H. Markowitz, « The Nature and Applications of Process Analysis, » *The RAND Corporation Paper* P.-547 (24 May 1954).

[4] H. Markowitz, « Concepts and Computing procedures for certain $X_{ij}$ Programming Problems », *The RAND Corporation Paper* P-602 (19 Nov. 1954).

[5] G. Dantzig, R. Fulkerson, S. Johnson, « Solution of a Large Scale Traveling Salesman Problem », *Journal of the Operations Research Society of America*, November 1954.

[6] H. Markowitz, « Process Analysis of the Metal Working Industries » *The RAND Corporation Research Memorandum* RM-1085 May 1954).

CHAPTER 2

# ALTERNATE METHODS OF ANALYSIS

*Alan S. Manne and Harry M. Markowitz*

## INTRODUCTION

This chapter discusses three methods of analysis which—like process analysis—seek to predict the ability of the economy to meet desired objectives with limited resources. These three methods—namely, GNP analysis, requirements analysis and input-output analysis—together with process analysis may be grouped under the general heading of feasibility or capabilities analysis. There are some differences in the range of applicability of these methods, but each addresses itself to the following types of question: Is a particular national objective economically feasible? Are there resource limitations which will render the economy incapable of achieving its objective? What are the alternative economic targets that are feasible?

From the viewpoint of the analyst interested in practical policy development, the various methods should be judged in terms of applicability, accuracy, cost, and availability. The policy developer wants to know the types of question to which each method of analysis can be applied. Or, to state the "applicability" consideration conversely, he wants to know which of the available methods bears on his problems. "Accuracy" is concerned with whether or not the answers returned by the analysis are reasonably correct. When a method suggests conclusions counter to initial judgment, is it likely to be a true or a false oracle? If two methods of analysis were applicable to the questions at hand, and both provided equally acceptable levels of accuracy, then choice between them would properly depend on cost and availability. Here cost should include the total costs of preparing for, performing, and interpreting the results of the analysis; while availability may be thought of as the probability that any developmental work needed to make the method applicable to the specific problem at hand can be carried out without undue delay.

For each method, in turn, this chapter briefly discusses its nature and its salient characteristics with respect to applicability, accuracy, cost, and availability. We shall argue that for some problems—and for certain stages of other problems—the simplest method, GNP analysis, serves best. For other problems the more complex methods of requirements analysis, input-output analysis, and process analysis are desirable or essential. In areas where these methods are desirable but unavailable, the gap must be filled by that always available, sometimes accurate method called judgment.

## Gross National Product Analysis

The annual gross national product (GNP) is the money value of goods and services produced by the economy in a year. In contrast to *net* national product, GNP does not subtract depreciation allowances from the value of production. It does, however, exclude the value of intermediate products used within the year in the process of making other goods and services. Thus GNP represents the total money value of the economy's output available to satisfy "final demands" for household consumption, government expenditures, gross fixed capital formation, net inventory changes, and net exports.

Countless questions of detail must be resolved in order to produce an operational definition of GNP: How do we distinguish maintenance expenditures—which are excluded from GNP—from capital expenditures, which are included? Should GNP include the value of the food consumed by the farm families who grew it? Should it include the value of do-it-yourself activities in which people paint, repair, or build their own homes, boats, or furniture? With relative prices and product qualities changing from year to year, how should the value of goods and services be added together to permit reasonably meaningful comparisons of GNP over time? We shall not pursue these questions. Rather, we shall speak of GNP with no more qualifications than we speak of temperature or humidity—implicitly assuming that the various questions have been answered reasonably, and the resulting statistical procedure has been carried out consistently.

A GNP analysis estimates the total gross national product required to meet a proposed economy-wide program, and compares this with the GNP which the economy is likely to have at its disposal. If the proposed use of GNP exceeds the likely supply, something has to give way. Perhaps consumption or investment objectives should be reduced, or perhaps foreign funds should be sought. Or perhaps GNP can be raised through increased labor supply, e.g., increased female participation or longer working hours. In effect, a GNP analysis adds up the bill for proposed economy-wide objectives and compares this with the value of product available to cover this bill. The objectives will have to be reconsidered if the former exceeds the latter.

Although a GNP analysis is simpler to construct than the others, it is not without its difficulties. For example, projections of the supply of GNP usually involve estimates of labor force and productivity. Both of these factors are susceptible to errors of estimate. Nevertheless, in comparison with other methods, GNP analysis is by far the least expensive and most readily available.

Turning to the question of accuracy, we find that GNP analysis is subject to a definite weakness, but can nonetheless serve a useful function. GNP analysis, used by itself, tends to overestimate the capabilities of the economy. It fails to reject programs whose source of infeasibility is the shortage of specific, specialized resources as distinguished from resources in general. Suppose—to take an extreme example—that an agricultural nation attempted in the course of a single year to:

—substantially reduce its output of grain, and
—increase by an equal dollar value its output of steel.

The proposed program does not involve an increased requirement for GNP—yet it is clearly infeasible. Farm land and equipment cannot be converted quickly into steel mills, nor farmers into steel workers.

GNP analysis can serve as a coarse screen, catching proposals whose general demands on resources are out of line in total. Only those proposals which pass through this coarse screen are then subject to the finer screening of the more complex methods of analysis. The more complex methods look for specialized resources which will become bottlenecks if a proposed program is implemented. Not only do they attempt to answer with greater precision the question "Can the economy achieve the specified objectives with limited resources?" but they also address themselves to such questions as "Which resources will become bottlenecks and which will be plentiful?"

Thus, in the division of labor among methods of analysis, the inexpensive coarse screen approach of GNP analysis may have to be supplemented by some finer screen method. The rest of this chapter will discuss two such methods, and the balance of this monograph is concerned with a third.

## Requirements Analysis

A requirements analysis for, say, steel might proceed as follows: $X$ tractors and $Y$ square feet of industrial construction are planned (or expected) for a forthcoming period of time. The average tractor requires $A$ tons of steel; the average square foot of industrial construction requires $B$ tons. Add $A$ times $X$ plus $B$ times $Y$ plus, similarly, any other uses of steel to obtain estimated total requirements. Compare this with the projected steel availability to see if a "steel problem" exists. The analysis may proceed through several levels, reflecting the fact that, e.g., automobile production requires steel, steel requires pig iron, pig iron requires coke, and coke requires coal. For computational convenience the requirements for several resources may be computed simultaneously, thus avoiding the duplicate recalculation of requirements for intermediate goods which contribute ultimately to the requirement for two or more resources.

The inputs to a requirements analysis consist of estimates of resource availabilities, desired levels of final demands, and requirements coefficients estimating the inputs needed per unit of output. From these, the analysis produces a shopping list of ingredients required to support the objectives. Comparisons between requirements and availabilities help to indicate possible trouble areas.

Requirements analysis is applied in practice to problems of various scope. Under different names it is used in manufacturing analysis to estimate expected needs for men, machines, standard materials and purchased parts. The military services use it to determine procurement quantities. At a national level it has been used, during war and peace, for tin, rubber, machine tools, blast furnace capacity, foreign exchange, and countless other potential bottlenecks. The problems of applying requirements analysis at a national level differ somewhat from those of applying it at an intrafirm level. Our concern will be with the former, broader-scope applications.

In an industry-wide steel study based, for example, on *U. S. Census of Manufactures* data, one could distinguish various mill shapes of "carbon steel," "alloy steel except stainless," and "stainless steel." But even "stainless steel" is an aggregate of many individual steels with differing applicabilities in manufacturing. Although the categories of a requirements analysis—e.g., stainless steel—are frequently discussed as if they were homogeneous commodities, in fact they are aggregates. The aggregation system of a requirements analysis in effect assigns specific goods and services to classes, and adds together the members of each class according to some criterion such as weight, volume, piece, or value. The characteristics and data requirements of each specific requirements analysis depend heavily on the aggregation system chosen. The choice of an aggregation system—whether done by design or by passive acceptance of existing categories—is the central, strategic decision in performing a requirements analysis.

Requirements analysis is not alone in its dependence upon aggregation. Aggregation is used with every practical technique of capabilities analysis at a national level. Aggregation problems constitute many, if not all, of the major problems of applying a technique. To a large extent, the differences between one technique and another may be viewed as differences in their approach to aggregation. GNP analysis, for example, represents an extreme form in which all goods and services are added together to form a single money value total. Formally, at least, GNP analysis may be viewed as a requirements analysis in which only one resource, national product, is distinguished. The cost of each good or service, then, is its requirement coefficient for GNP. Thus a GNP analysis is a requirements analysis with an extremely coarse aggregation of resources.

In a similar manner, requirements analysis may be viewed as a special case of process analysis. A process analysis can distinguish alternate ways of producing the same product. Requirements analysis aggregates these alternate productive processes into a representative activity with fixed inputs per unit of output. Thus, if a process analysis model did not distinguish alternate production activities, it would be a requirements analysis, and if it distinguished only one resource (gross national product) it would be a GNP analysis. Input-output analysis, finally, may be viewed as a form of requirements analysis using a somewhat different way of aggregating economic activity. The aggregation principles of input-output and their consequences will be discussed later in this chapter.

Thus each method of analysis has its own ground rules for aggregation. The specific aggregation system chosen within these ground rules determines the specific characteristics and data needs of the particular analysis.

The major cost involved in performing a requirements analysis is that of collecting and organizing data. Requirements analysis does not entail large calculation costs, as may be incurred for process analysis. With the available computing equipment, once the data for a requirements analysis are assembled in suitable form, calculations for even the largest analysis can be performed at a relatively small cost. With respect to the cost and availability of data, requirements analysis stands between GNP and process analysis. It needs a list of specific inputs per unit of output, as distinguished from the single money value figure which is sufficient for GNP analysis. Since requirements analysis does not develop coefficients for alternate productive processes, it can make extensive use of historical inputs and outputs to develop average requirements, avoiding (completely or in great part) the need for engineering data upon which process analysis frequently relies.

The major source of inaccuracy inherent in requirements analysis is its neglect of alternate modes of production. Often, substantially different inputs can be—and, in fact, are—used to produce the same product. Electricity can be produced by water power or, in steam electric plants, from either nuclear fuel, coal, oil, or natural gas; agricultural products can be produced using more or less fertilizer and irrigation; metals can be produced using varying ratios of scrap to ore; the same metalworking tasks can be performed on a variety of machines, and so on. A requirements analysis must attempt to estimate a typical process: e.g., the average use of nuclear fuel vs. coal vs. water power in the production of electricity. But the scarcity of one material relative to another will lead to the use of processes which conserve the one at the expense of the other. Thus, to an important extent, the use of one or another process will depend on the very shortages and surpluses that the analysis seeks to predict.

The manner in which a requirements analysis misestimates the capabilities of an economy, due to its failure to take account of substitution possibilities, depends on the aggregation system used. If two resources (e.g., lathes and milling machines) are aggregated together into the same category (e.g., machine tools) then they are assumed to be perfect substitutes for each other. If they are distinguished as different resource categories, then no substitution is assumed to exist. Thus an extremely coarse classification of resources will tend to overestimate the amount of substitution possible, and hence overestimate the capabilities of the economy. An extremely fine classification, on the other hand, will understate the amount of substitution possible between resources, and hence tend to underestimate the capabilities of the economy. In choosing an aggregation system for a requirements analysis the following dilemma must be faced: In order to anticipate bottlenecks among specific resources a fine classification is needed; but to avoid underestimating substitution possibilities a coarse classification of resources is needed.

Coal is sometimes a substitute for fuel oil, but not in all its applications; a lathe can sometimes substitute for a milling machine, but not always; aluminum and copper are competitors, but only in part of their range of applications. As a consequence, any aggregation system must be a compromise with, rather than a solution to, the dilemma. Any attempt to completely avoid one horn of the dilemma is bound to drive the analysis to the other horn.

In principle one could circumvent this dilemma by the following process of trial and error:

> Choose categories sufficiently fine to identify specific bottlenecks; estimate likely requirements; perform the analysis using these estimates; and inspect the results for bottlenecks.

Then, on the basis of this initial analysis:

> Modify coefficients to reflect processes which substitute plentiful for scarce factors of production; repeat the requirements calculation and again inspect the results; make further adjustments and repeat if necessary.

This is a tedious and time-consuming procedure which must, of necessity, be stopped short of its ultimate end. Process analysis in effect accelerates this procedure by distinguishing alternate processes at the outset, and by letting automatic techniques perform the substitution of one process for another according to overall resource availabilities.

Process analysis and requirements analysis are closely related, as is illustrated by the discussion of the metalworking industries in Part IV of this monograph. Data and procedures are presented in Part IV for a requirements analysis assuming fixed inputs of various kinds of men, machines, and materials per unit output of each metalworking industry. After this, data and procedures for analyzing a certain source of substitution possibilities are presented, plus suggestions concerning the analysis of another source of substitution. The requirements analysis serves as an immediately available technique to which information concerning alternate processes can be added, as appropriate and available.

In itself, without the addition of alternate process information, requirements analysis serves to identify potential trouble areas. In some cases the apparent bottlenecks are not bottlenecks at all. The economy would take care of the shortage naturally, by substituting plentiful for scarce resources. In other cases the bottlenecks are real; the timing and level of objectives should in fact be reconsidered in light of possible infeasibilities. By spotlighting possible trouble areas for further investigation, requirements analysis supplies a valuable service beyond that provided by GNP analysis.

## Input-Output Analysis

A difficult problem of requirements analysis is that of estimating total requirements as distinguished from direct requirements. For example, the production of electricity, say to light homes, requires coal; but the production of coal itself requires electricity, whose production in turn requires more coal; and so on ad infinitum. To make matters worse, the production of both coal and electricity have other requirements whose demands ramify through the economy, further augmenting the total requirements by electricity for coal. In the usual requirements analysis, indirect requirements are, after a point, accounted for by some rule-of-thumb procedure. An example of such a procedure would be to add together direct requirements, second order requirements, and third order requirements of each end item for a particular resource; see what fraction ($X$) of a particular year's use of this resource is thus explained; account for the rest by multiplying ($1/X$) times the sum of first, second, and third order requirements to form estimates of total requirements. Insofar as the fourth + fifth + sixth + · · · order requirement is not proportional to the first + second + third order requirement, the procedure is subject to error. The possible magnitude of this error depends on the extent to which first, second, and third order requirements account for the demands for the resource.

Input-output[1] approaches the problem of estimating total requirements through the use of a complete model of the economy. It classifies business establishments into an exhaustive set of industries and estimates the direct requirements by each industry for each other industry's output. These interindustry demands are arrayed in a square table with industries listed across the top and down the side. With this table (plus the assumption of fixed inputs per unit output) standard mathematical techniques can be used to answer questions such as "How much gross coal production is required to produce an extra one million dollars' worth of electricity, net of all intermediate interindustry requirements?"

As a theoretical matter, the notion of a complete input-output table dates back at least as far as the eighteenth century economist Quesnay. As a practical matter, the construction of an input-output table, in the sense used here, begins with the pioneering work of Wassily Leontief carried out during the 1930's and published in 1941. Subsequently—with electronic computers to trace through the consequences of ever larger systems, and with Leontief's work to demonstrate the feasibility of such an approach—various input-output tables have been built, including a 190-industry model of the United States economy for 1947.

Two forms of interindustry models are generally distinguished. The "closed model" (as first used by Leontief) includes households as an industry with inputs of consumption goods and outputs of labor. The "open model" (Cornfield, Evans, and Hoffenberg, 1947) does not include a household industry but treats demands by households as fixed requirements to be met by the economy. Also treated as fixed are the requirements for other components of "final demand" including government purchases, gross private capital formation, net inventory changes, and net exports. With respect to sectors other than households and other final demands, open and closed models can be identical. For problems of feasibility analysis, the open model is generally the more convenient.

In principle, an input-output analysis could use physical units of measure such as weight, volume, or count. In practice, however, dollar values have

[1] Sometimes referred to as interindustry analysis.

been used almost exclusively. Thus the classic statement of procedure for estimating coefficients for interindustry requirements, expressed here for the closed model, is as follows:

> Each "industry" (including households) is treated as a single accounting entity—comparable to a "country" in official foreign trade statistics—with sales entered on one side of its trading account and purchases on the other. As in the trade between countries the sales of one industry are the purchases of another. Entering the sales and purchase accounts of all the separate industries in one large table we get a comprehensive view of the structure of the national economy as a whole.[2]

From this table of purchases and sales the direct requirement coefficients are calculated by dividing sales from industry $i$ to industry $j$ by the gross output of industry $j$.

The development of data for a large input-output analysis can require tens of thousands of man-hours. The 190-industry analysis of the U. S. economy, for example, required the estimation of thousands of coefficients. Most of these coefficients were not readily found in available statistics but had to be constructed from various sources, sometimes with the aid of rule-of-thumb estimation procedures. Computing costs for tracing out total requirements from the direct requirements, although not negligible, were small compared to the costs of constructing the basic table.

Because of the time required to collect and organize data for a complete interindustry table, "availability" is more of a problem with input-output than with GNP analysis or the usual requirements analysis. The development of a large input-output matrix should be viewed as a major construction project which is not to be rushed to answer some urgent policy question but is to be built carefully to serve many uses through the course of time.

## Input-Output Analysis (Continued)

We shall note two general sources of inaccuracy to which input-output analysis is subject. The first concerns the existence of alternate methods of production. The second concerns the way in which "industry output" and "interindustry flows" are used as basic categories of analysis. Since the former problem area—the existence of alternate methods of production—was discussed previously, it can be dispensed with quickly in the present section. The nature and consequences of inaccuracies introduced through the other source will be discussed in some detail. Despite such inaccuracies in input-output analysis, the table itself—i.e., the basic tabulation of historical interindustry sales and purchases—is a valuable source of data concerning industrial activity. The basic table is tedious and expensive to develop, but so is much of the worthwhile economic data at our disposal.

In a preceding section we discussed difficulties of requirements analysis resulting from its failure to distinguish alternate methods of production. Radi-

[2] Leontief (1951), p. 4. For an introduction to input-output, we also recommend Chenery and Clark (1959).

cally different methods of production exist for many goods and services. The choice of production method depends on relative scarcities of alternate resources, and hence average requirement coefficients depend on the very shortages and surpluses to be predicted. This consideration, already noted for requirements analysis, applies equally to input-output with its assumption of fixed interindustry flows per unit of output.

Input-output is also subject to inaccuracies due to the way in which it makes use of "industry output" and "interindustry flows." For concreteness, we shall illustrate the general nature of these inaccuracies by means of examples drawn from the 190-industry matrix of the United States in 1947. (See Evans and Hoffenberg, 1952.) Difficulties such as those illustrated below have been recognized by a number of analysts who have applied input-output to practical problems. To circumvent these difficulties, various special procedures have been introduced into particular input-output analyses. Our discussion cannot do justice to these various ways of not quite doing input-output. In the examples and generalizations below, we will be dealing essentially with the classical input-output formulation as characterized in the last section. Afterwards, we shall briefly argue our preference for a process analysis approach rather than supplementing input-output with ad hoc procedures.

Suppose that industry I sells to industries X, Y, and Z, and that it purchases from A, B, and C. In tracing out total requirements, the input-output procedure assumes that the proportions among the output of A, B, and C purchased by I to produce output destined for X is the same as those proportions purchased by I to produce output destined for Y or Z. This assumption can cause substantial distortion in estimates of total requirements.[3]

For example, the Non-Ferrous Foundries industry casts both aluminum parts (e.g., for aircraft) and brass parts (e.g., for plumbing fixtures). In tracing out total requirements, input-output analysis assumes that the proportions of aluminum, copper, and zinc in the castings purchased by the Aircraft industries are the same as the proportions purchased by the Plumbing Fixtures and Fittings industry. The Non-Ferrous Foundry industry is treated as if it receives materials destined for different end items, combines them into a homogeneous mixture, and sends this mixture to each purchaser of non-ferrous castings.

The importance of this example depends on three points:

*First: In a case such as the above the input-output procedure introduces substantial inaccuracies in the estimates of indirect requirements.* In 1954,[4] for example, the Aircraft industries actually purchased $49 million worth of aluminum and aluminum-base castings vs. $4 million worth of copper and copper-base castings. The Plumbing Fixtures and Fittings industry, on the other hand, purchased $.5 million of aluminum and aluminum-base castings vs. $14 million worth of copper and copper-base castings. For these two, and

[3] On this point, see also S. B. Noble (1960), especially p. 408.

[4] We use 1954 rather than 1947 figures here since statistics on castings purchased by the Aircraft Equipment n.e.c. industry are more complete for the later year.

for a number of other[5] large purchasers of nonferrous castings, the assumption of equal proportions is untenable.

*Second: This difficulty cannot be avoided by a more detailed industrial classification.* Manufacturing industries are collections of establishments. Interindustry flows are the sums of purchases by establishments in one industry from establishments in another. Since many establishments cast both aluminum and brass, no matter how finely we classify establishments into industries—even if we let each establishment be an industry in itself—brass for plumbing fixtures will appear to end up in aircraft, and aluminum for aircraft will end up in plumbing fixtures.[6]

*Third: The Non-Ferrous Foundry industry is not alone in having this effect on the estimation of indirect requirements.* Similar distortions are caused by any industry which supplies a service performed on a variety of materials on behalf of a variety of consuming industries. Examples include Iron and Steel Forging, Metal Stamping, Metal Coating and Engraving, Machine Shops, and Screw Machine Products.

Input-output analysis is frequently combined with the notion of industry capacity. The input-output analysis predicts gross production required from various industries. By comparing these gross required outputs with the available capacities, potential bottlenecks are identified. This procedure encounters difficulties when industries can, in effect, borrow capacity from each other. Such borrowing of capacity is particularly common among the metalworking industries, which fabricate and assemble metal parts for a great variety of military, household, and industrial durable goods. Skills and equipment needed to perform the tasks of one of these industries typically overlap with those required for other such industries. Many shops regularly or occasionally produce parts destined for commodities of other metalworking industries. It is for such reasons that we find, according to the U. S. input-output table for 1947, that 9 cents' worth of Motor Vehicles, 1.8 cents' worth of Aircraft and 1 cent's worth of Motorcycles and Bicycles were directly "required" to

[5] The following examples present millions of dollars' worth of purchases of aluminum and aluminum-base castings vs. copper and copper-base castings (the data presented in that order) for some 4-digit census industries which consume large amounts of nonferrous castings and show a large discrepancy from the proportionality assumption. High aluminum consumers: Domestic Laundry Equipment (12.7, .1), Electric Appliances (10.1, .1), Metal Doors, Sash and Trim (4.7, .1), Internal Combustion Engines (21.8, 1.5); High copper consumers: Valves and Valve Fittings, except Plumbing (2.2, 22.8), Power Transmission Equipment (1.7, 5.7), Pumps and Compressors (4.4, 13.0).

Source: *U. S. Census of Manufactures*, 1954, Table 1B, pp. 210–238.

[6] This discussion is based upon (*a*) the definition of industry used by the *U. S. Census of Manufactures* and (*b*) the definition of interindustry flow used by Leontief (1951). One modification of these conventional definitions is to segregate the purchases and sales of establishments by product line, thereby departing from the establishment basis of classification. Since nonferrous foundry establishments have labor and equipment which may be used interchangeably for the casting of both brass and aluminum, the use of this product line classification would raise problems of the sort discussed immediately below for the metalworking industries.

make $1 worth of Locomotives; that 7 cents' worth of Motor Vehicles and at least 1 cent each of Aircraft, Ships, and Railroad Equipment were required to make Motorcycles and Bicycles.

Whether or not it is combined with the notion of industry capacity, the input-output procedure is inadequate here for at least two reasons. First, it fails to analyze the capabilities of one such industry to supplement another. Second, it assumes that since 1.8 cents' worth of Aircraft is required directly for Locomotives, a proportionate share of everything that went into Aircraft should also be incorporated into Locomotives. The effect is similar in nature, though less in amount, to the mixing of flows that occurred through intermediate industries such as forges, foundries, and machine shops.

The final difficulty to be discussed occurs when two or more joint products result from the same process. The way this can affect the analysis of economic capabilities may be illustrated by the case of coke production.

The Coke and Products industry produces coke (mostly for blast furnaces) as its main product, and basic organics (for the chemical industries) as a by-product. Suppose there were a fall in the demand for steel. This would reduce the demand by Steel for Blast Furnace output; reduce the demand by Blast Furnaces for Coke output; and thus, according to an input-output matrix based upon purchases and sales, release Coke industry capacity for use by the Organic Chemicals industry. But this implication is opposite in direction from what may be expected in fact. Since basic organics are a by-product of coke production, the reduced production of coke would *reduce* the by-products available for the chemical industries. The chemical industry would either have to use alternate sources of raw materials or reduce its production. Thus, according to a purchase-and-sales analysis, additional "Coke Oven Capacity" would be made available by the fall in steel production, whereas in fact the flow of organics from coke ovens to the chemical industry would be reduced.[7]

To a certain extent, difficulties such as the above can be circumvented without giving up the appearance of an input-output table. The problem of alternate methods of production, for example, can be handled by trial-and-error procedures similar to those described in connection with requirements analysis. The problem discussed in connection with the foundry, forge, stamping, and machine shop industries can be handled by treating the primary metal purchases of such industries as if they were direct purchases by the end item producer.

In some cases, it would be extremely difficult to characterize accurately an aspect of technology within an input-output framework. The sharing of capacity between metalworking industries, for example, could be handled by means of "conversion coefficients" which showed the extent to which the capacity of one industry could be converted to another. Such coefficients would still fail to characterize properly the possibilities for reducing output

[7] An important instance of joint products will arise in multiperiod models of economic development. By investing in durable capital equipment, we obtain a sequence of joint products: capacity available for use during more than one time period.

in one set of industries to supply equipment and labor needed in another set. A more satisfactory approach is to explain the sharing of capacities in terms of the kinds of transferable resources used by these industries.

To many, the attractiveness of the input-output approach is that it permits the construction of a complete model of the economy wihout requiring an understanding of countless technological relationships. After $N$ industries have been chosen, $N^2$ coefficients can be delegated to a data collection team. Data may not be immediately available, but at least the team has a well defined objective: "Find or estimate the amount sold from industry $i$ to industry $j$ during the specified year for each $i$ and $j$."

We have argued above that various supplementary procedures must be used if the implications of such an analysis are not to be completely unreasonable. A serious difficulty with the input-output approach is that it provides no systematic way for seeking out those aspects of technology which require such special handling. Frequently input-output matrices are constructed and used without regard to such pitfalls. Sometimes these pitfalls are revealed through obviously absurd implications of the analysis. Other times pitfalls are found when someone looking at technology asks "What would happen if these technological relationships were forced into the input-output form?" There is no guarantee, however, that such ad hoc finding and patching of difficulties will not leave equally serious problems undetected.

## Summary

Gross national product analysis serves as a coarse screen to reject grossly infeasible programs. It does not detect programs whose infeasibility is due to excessive demands for particular specialized resources.

Requirements analysis compares the demands and supplies of specialized resources. Its chief drawback is its failure to account for alternate modes of production. Despite this difficulty, it can serve a valuable function in pointing out possible trouble areas.

Input-output is a form of requirements analysis which addresses itself particularly to the question of estimating total requirements—both direct and indirect. Input-output analysis fails to account for alternate methods of production. Additional difficulties in its use for capabilities analysis arise from the way in which it uses interindustry sales and purchases as the basic source of data.

Process analysis may be viewed as a generalization of requirements analysis which allows alternate modes of production to be distinguished wherever these are deemed important. Cost, availability, accuracy, and applicability characteristics of process analysis will be discussed in the next chapter.

## References

Cornfield, J., W. D. Evans, and M. Hoffenberg, 1947, "Full Employment Patterns, 1950," *Monthly Labor Review*, LXIV, No. 2, pp. 163–190 (February) and LXIV, No. 3, pp. 420–432 (March).

Chenery, H. B., and P. G. Clark, 1959, *Interindustry Economics,* John Wiley and Sons, New York.

Evans, W. D., and M. Hoffenberg, 1952, "The Interindustry Relations Study for 1947," *Review of Economics and Statistics,* XXXIV, No. 2, pp. 97–142.

Leontief, W. W., 1951, *The Structure of the American Economy, 1919–1939,* second edition, Oxford University Press, New York.

Noble, S. B., 1960, "Some Flow Models of Production Constraints," *Naval Research Logistics Quarterly,* VII, No. 4, pp. 401–419 (December).

"Summary Statistics," *U. S. Census of Manufactures,* 1954, Vol. I, Department of Commerce, Washington, D. C.

# THE ELIMINATION FORM OF THE INVERSE AND ITS APPLICATION TO LINEAR PROGRAMMING*

HARRY M. MARKOWITZ[1]

*The RAND Corporation*

## Introduction

It is common for matrices in industrial applications of linear programming to have a large proportion of zero coefficients. While every item (raw material, intermediate material, end item, equipment item) in, say, a petroleum refinery may be indirectly related to every other, any particular process uses few of these. Thus the matrix describing petroleum technology has a small percentage of non-zeros. If spacial or temporal distinctions are introduced into the model the percentage of non-zeros generally falls further.

The present paper discusses a form of inverse which is especially convenient to obtain and use for matrices with a high percentage of zeros. The application of this form of inverse in linear programming is also discussed.

## The Elimination Form of Inverse

The inverse ($A^{-1}$) of a matrix ($A$) is valuable when a number of sets of equations $AX = b$ are to be solved using different $b$'s and the same $A$. $A^{-1}$, like any matrix, may be expressed as the product of other matrices

$$A^{-1} = M_m M_{m-1} \cdots M_1$$

in an infinite number of ways. E.g. $(2) = (\frac{1}{2})\ (4) = (\frac{1}{8})\ (16)$ etc. If we have such $M_1, \cdots, M_m$ we can solve $X = A^{-1}b$ in a series of steps:

$$X^{(1)} = M_1 b$$
$$X^{(2)} = M_2 X^{(1)}$$
$$\vdots$$
$$X = M_m X^{(m-1)}$$

The expression $M_m \cdots M_1$ is referred to as a "product form" of inverse. In some problems there may be $M_i$ which are easier to obtain and apply than $A^{-1}$ itself.

This paper discusses a particular product form of inverse which is closely related to the Gaussian elimination method of solving a set of simultaneous equations. This "elimination form of the inverse," as we shall call it, is especially valuable when $A$ has a large number of zero coefficients.

* Received October, 1956.

[1] The writer is indebted to George Dantzig and Alan Manne for valuable suggestions.



MANAGEMENT SCIENCE VOL. 3 1957 pgs 255-269

The elimination form of inverse can be illustrated in terms of the solution of three equations in three unknowns:

(1) $$a_{11}X_1 + a_{12}X_2 + a_{13}X_3 = r_1$$

(2) $$a_{21}X_1 + a_{22}X_2 + a_{23}X_3 = r_2$$

(3) $$a_{31}X_1 + a_{32}X_2 + a_{33}X_3 = r_3$$

For the moment we will let the $k^{\text{th}}$ diagonal element be the $k^{\text{th}}$ pivotal element. From equation 1) we get the first equation of our back solution

(B1) $$X_1 = \frac{r_1}{a_{11}} - \frac{a_{12}}{a_{11}} X_2 - \frac{a_{13}}{a_{11}} X_3$$

We eliminate $X_1$ from equations 2) and 3) by adding $\left(-\frac{a_{i1}}{a_{11}}\right)$ times the first equation to the $i^{\text{th}}$ equation, thus obtaining

(2′) $$b_{22}X_2 + b_{23}X_3 = r_2^*$$

(3′) $$b_{32}X_2 + b_{33}X_3 = r_3^*$$

where

$$r_2^* = r_2 - \left(\frac{a_{21}}{a_{11}}\right) r_1$$

$$r_3^* = r_3 - \left(\frac{a_{31}}{a_{11}}\right) r_1$$

Similarly we get

(B2) $$X_2 = \frac{1}{b_{22}} r_2^* - \frac{b_{23}}{b_{22}} X_3$$

and

$$c_{33}X_3 = r_3^{**}$$

where

$$r_3^{**} = r_3^* - \frac{b_{32}}{b_{22}} r_2^*.$$

Finally

(B3) $$X_3 = \frac{1}{c_{33}} r_3^{**}$$

(B3) gives us $X_3$ ; $X_3$ and B2) give $X_2$ ; $X_2$, $X_3$ and B1) give $X_1$.

Consider the transformations which occurred to our original right hand side. First we formed

$$\begin{pmatrix} r_1 \\ r_2^* \\ r_3^* \end{pmatrix} = \begin{bmatrix} 1 & 0 & 0 \\ -\frac{a_{21}}{a_{11}} & 1 & 0 \\ -\frac{a_{31}}{a_{11}} & 0 & 1 \end{bmatrix} \begin{pmatrix} r_1 \\ r_2 \\ r_3 \end{pmatrix}$$

then

$$\begin{pmatrix} r_1 \\ r_2^* \\ r_3^{**} \end{pmatrix} = \begin{pmatrix} 1 & 0 & 0 \\ 0 & 1 & 0 \\ 0 & -\frac{b_{32}}{b_{22}} & 1 \end{pmatrix} \begin{pmatrix} r_1 \\ r_2^* \\ r_3^* \end{pmatrix}$$

then

$$\begin{pmatrix} r_1 \\ r_2^* \\ X_3 \end{pmatrix} = \begin{pmatrix} 1 & 0 & 0 \\ 0 & 1 & 0 \\ 0 & 0 & \frac{1}{c_{33}} \end{pmatrix} \begin{pmatrix} r_1 \\ r_2^* \\ r_3^{**} \end{pmatrix}$$

then

$$\begin{pmatrix} r_1 \\ X_2 \\ X_3 \end{pmatrix} = \begin{pmatrix} 1 & 0 & 0 \\ 0 & \frac{1}{b_{22}} & -\frac{b_{23}}{b_{22}} \\ 0 & 0 & 1 \end{pmatrix} \begin{pmatrix} r_1 \\ r_2^* \\ X_3 \end{pmatrix}$$

and finally

$$\begin{pmatrix} X_1 \\ X_2 \\ X_3 \end{pmatrix} = \begin{pmatrix} \frac{1}{a_{11}} & -\frac{a_{12}}{a_{11}} & \frac{a_{13}}{a_{11}} \\ 0 & 1 & 0 \\ 0 & 0 & 1 \end{pmatrix} \begin{pmatrix} r_1 \\ X_2 \\ X_3 \end{pmatrix}$$

Thus

$$\begin{pmatrix} X_1 \\ X_2 \\ X_3 \end{pmatrix} = \begin{pmatrix} \frac{1}{a_{11}} & -\frac{a_{12}}{a_{11}} & \frac{a_{13}}{a_{11}} \\ 0 & 1 & 0 \\ 0 & 0 & 1 \end{pmatrix} \begin{pmatrix} 1 & 0 & 0 \\ 0 & \frac{1}{b_{22}} & -\frac{b_{32}}{b_{23}} \\ 0 & 0 & 1 \end{pmatrix} \begin{pmatrix} 1 & 0 & 0 \\ 0 & 1 & 0 \\ 0 & 0 & \frac{1}{c_{33}} \end{pmatrix}$$

$$\cdot \begin{pmatrix} 1 & 0 & 0 \\ 0 & 1 & 0 \\ 0 & -\frac{b_{32}}{b_{22}} & 1 \end{pmatrix} \begin{pmatrix} 1 & 0 & 0 \\ -\frac{a_{21}}{a_{11}} & 1 & 0 \\ -\frac{a_{31}}{a_{11}} & 0 & 1 \end{pmatrix} \begin{pmatrix} r_1 \\ r_2 \\ r_3 \end{pmatrix}$$

Since the $a_{ij}b_{ij}c_{ij}$ do not depend on the $r_i$ we have

$$A^{-1} = \begin{pmatrix} \frac{1}{a_{11}} & -\frac{a_{12}}{a_{11}} & -\frac{a_{13}}{a_{11}} \\ 0 & 1 & 0 \\ 0 & 0 & 1 \end{pmatrix} \cdots \begin{pmatrix} 1 & 0 & 0 \\ -\frac{a_{21}}{a_{11}} & 1 & 0 \\ -\frac{a_{31}}{a_{11}} & 0 & 1 \end{pmatrix}$$

Similarly if $A$ is any $mXm$ non-singular matrix we have

$$A^{-1} = B_1 B_2 \cdots B_m R_{m-1} \cdots R_1$$

where $R_k$ is a matrix of the form

$$\begin{pmatrix}
1 & & & & 0 & & & \\
 & 1 & & & \cdot & & & \\
 & & \cdot & & \cdot & & 0 & \\
 & & & \cdot & \cdot & & & \\
 & & & & \cdot \; 0 & & & \\
 & & & & 1 & & & \\
 & & & & & 1 & & \\
 & & & & -\frac{v_{k+1,k}}{v_{kk}} & 1 & & \\
 & & & & \cdot & & \cdot & \\
 & 0 & & & \cdot & & \cdot & \\
 & & & & \cdot & & & \cdot \\
 & & & & -\frac{v_{mk}}{v_{kk}} & & & 1
\end{pmatrix}$$

and $B_k$ is of the form

$$\begin{pmatrix}
1 & & & & & & & & \\
 & \cdot & & & & & & & \\
 & & \cdot & & & & 0 & & \\
 & & & \cdot & & & & & \\
 & & & & 1 & & & & \\
 & & & & & \frac{1}{v_{kk}} & -\frac{v_{kk+1}}{v_{kk}} & \cdots & -\frac{v_{km}}{v_{kk}} \\
 & & & & & & 1 & & \\
 & & & & & & & \cdot & \\
 & 0 & & & & & & \cdot & \\
 & & & & & & & & \cdot \\
 & & & & & & & & 1
\end{pmatrix}$$

Although this elimination form of inverse consists of $2m - 1$ matrices it contains only $n^2$ numbers which are not known a-priori to be 0 or 1. It requires the same number of multiplications and additions to apply the elimination form as it does to apply the conventional $A^{-1}$. The arithmetic operations required to obtain the $B_1 \cdots B_m R_{m-1} \cdots R_1$ and to apply them once are exactly the same as those required to solve a set of linear equations by Gaussian elimination.

Suppose that the $k^{\text{th}}$ pivot is not $v_{kk}^{k}$ but $v_{i_0 j_0}^{k}$ (where $v_{ij}^{k}$ is the value of the parameter of $i^{\text{th}}$ equation $j^{\text{th}}$ variable at the $k^{\text{th}}$ step). This step subjects the right-hand side to a transformation of the form

$$(r)^{\text{new}} = \begin{pmatrix} 1 & & & \gamma_1 & & & \\ & \cdot & & \cdot & & & \\ & & \cdot & \cdot & & & \\ & & & \cdot\,\cdot & & & \\ & & & 1 & & & \\ & & & \cdot\,\cdot & & & \\ & & & \cdot & \cdot & & \\ & & & \cdot & & \cdot & \\ & & & \gamma_m & & & 1 \end{pmatrix} (r)^{\text{old}}$$

At least $k - 1$ of the $\gamma_i$ are zero. The $\gamma$ vector is in the $i_0^{\text{th}}$ column of the matrix. The $k^{\text{th}}$ step also gives rise to a back solution of the form

$$X_{j_0} = \frac{1}{v_{i_0 j_0}^{k}} r_{i_0}^{k} + \sum \eta_j X_j$$

where at least $m - k$ of the $\eta_j$ equal zero.
The elimination form of the inverse in this case is

$$A^{-1} = B_1 \cdots B_m P R_{m-1} \cdots R_1$$

where

$$B_k = \begin{pmatrix} 1 & & & & & & \\ & \cdot & & & & & \\ & & \cdot & & & & \\ & & & \cdot & & & \\ \eta_1 & \cdots & & \frac{1}{v_{i_0 j_0}^{k}} & \cdots & & \eta_m \\ & & & 1 & & & \\ & & & & \cdot & & \\ & & & & & \cdot & \\ & & & & & & \cdot \\ & & & & & & 1 \end{pmatrix}$$

$$R_k = \begin{pmatrix} 1 & & \gamma_1 & & \\ & \cdot & \cdot & & \\ & & \cdot\ \cdot & & \\ & & \cdot\ \cdot & & \\ & & 1 & & \\ & & \cdot\ \cdot & & \\ & & \cdot & \cdot & \\ & & \cdot & & \cdot \\ & & \gamma_m & & 1 \end{pmatrix}$$

and $P$ is permutation matrix such that if $v_{ij}^k$ is a pivotal element $P$ makes the old $i^{\text{th}}$ component of $(r)$ become the new $j^{\text{th}}$ component of $r$.

That $A^{-1} = B_1 \cdots P \cdots R_1$, i.e., that $AX = r$ implies $X = B_1 \cdots P \cdots R_1 r$ may be seen as follows. The transformations $R_{m-1} \cdots R_1 r$ give the right hand sides of the back solutions in the original equation order. We can imagine re-ordering the equations so that new first back solution is the one of the form

$$X_1 = \frac{1}{v_{i1}^k} r_i^k + \sum \eta X$$

and similarly the new $k^{\text{th}}$ back solution is the one in which $X_k$ was eliminated. This rearrangement changes the order of the $r_i$ exactly as does $P$. The last back solution is of the form

$$X_{j_m} = \frac{1}{v} r_{i_m}$$

Since $r_{i_m}$ is now the $j_m th$ component of $(r)$, $B_m$ (as described above) will transform $(r)$ into a vector with $X_{j_m}$ in the $j_m$th position. The next back solution is of the form

$$X_{j_{m-1}} = \frac{1}{v} r_{i_{m-1}} + \eta X_{j_m}$$

Since, thanks to $B_m$ and $P$, $r_{i_{m-1}}$ is the $j_{m-1}$th component and $X_{j_m}$ is the $j_m$th component of $(r)$, $B_{m-1}$ transforms $(r)$ into a vector with $X_{j_{m-1}}$ in the $j_{m-1}$th position as well as $X_{j_m}$ in the $j_m$th position. And so on.

In recording $B_k$ or $R_k$ it is not necessary to write out the entire matrix. In the case of an $R_k$ it is only necessary to record the $i$ eliminated and the non-zero $\gamma_i$. In the case of a $B_k$ it only is necessary to record the $j$ eliminated, the non-zero $n_j$ and 1/pivotal element.

If the matrix $A$ has a large number of zero $\alpha_{ij}$, the elimination form of inverse may have appreciably less than $n^2$ non zero $\gamma s$ and $\eta s$. This may be so even though the usual $A^{-1}$ has no zeros.

The number of non-zero $\eta$ and $\gamma$ in an elimination form of inverse may depend on which pivotal elements are used. Suppose the *'s below represent the non-zero elements of a 5 x 5 matrix.

```
* * * * *
* * * *
* * *
* *
*
```

If $a_{11}$ is the first pivotal element the non-zeros at step two are (barring accidental zeros) as follows

```
* * * *
* * * *
* * * *
* * * *
```

But if $a_{15}$ or $a_{51}$ is the pivotal element the pattern of non-zeros is

```
* * * *
* * *
* *
*
```

A table indicating zero and non-zero coefficients is a valuable aid in the choice of pivotal elements. From such a table an *agenda* (i.e., a complete set of pivotal elements) can be chosen before computation begins. This separation of the choice of the agenda and actual elimination is convenient both in hand and machine computation. There are, however, two dangers attached to deciding on an agenda beforehand. Some pivotal element may accidentally become zero, in which case the agenda cannot be followed to the end. Or some pivotal element may turn out to be so small that its use would adversely affect the accuracy of the results. One solution to these difficulties is to have some test of the acceptability of a pivotal element; form the agenda beforehand and follow it as long as each pivotal element meets the test. If a pivotal element fails the test, a new agenda can be worked out for the remaining equations and variables.

An example agenda is presented in Table 1. The $X$'s represent the original non-zero elements of the matrix. The $M$'s represent coefficients which began as zeros but ended as non-zeros. The numbers $k = 1, \cdots, 43$ in the matrix indicate the $k^{\text{th}}$ pivotal element. The number $(\rho_i)$ at the right of each row indicates the number of elements of that row which were not already eliminated when the row was eliminated. The number $\sigma_j$ at the bottom of each column indicates the number of elements of that column which were not already eliminated when the column was eliminated. One of the by-products of making an agenda beforehand is foreknowledge of all the variables which will appear in any equation and all the equations in which a variable will ever appear.

TABLE 1
*Non-Zero Entries and Agenda*

| j | 1 | 2 | 3 | 4 | 5 | 6 | 7 | 8 | 9 | 10 | 11 | 12 | 13 | 14 | 15 | 16 | 17 | 18 | 19 | 20 | 21 | 22 | 23 | 24 | 25 | 26 | 27 | 28 | 29 | 30 | 31 | 32 | 33 | 34 | 35 | 36 | 37 | 38 | 39 | 40 | 41 | 42 | 43 | No. in Row when Eliminated |
|---|---|---|---|---|---|---|---|---|---|---|---|---|---|---|---|---|---|---|---|---|---|---|---|---|---|---|---|---|---|---|---|---|---|---|---|---|---|---|---|---|---|---|---|---|
| i | | | | | | | | | | | | | | | | | | | | | | | | | | | | | | | | | | | | | | | | | | | | |
| 1 | | | | | | | | | | | | | | | | | | | | | | | | | | | | | | | | | | | | | | | | 1 | | | | 1 |
| 2 | | | | | | | | | | | | | | | | | | | | | | | | | | | | | 4 | | | | | | | | | | | | X | | | 1 |
| 3 | | | | | | | | | | | | | | | | | | | | | | | | | | | | | | | | | | | | | | | | | | 2 | | 1 |
| 4 | | | | | | | | | | | | | | | | | | | | | | | | | | | | | | | | | | | | | | | | X | 3 | X | | 1 |
| 5 | 5 | | | | | | | | | | | | | | | | | | | | | | | | | | | | | | | | | | | | | | | | | | | 1 |
| 6 | | X | 10 | | | | | | | | | | | | | | | | | | | | | | | | | | | | | | | | | | | | X | | | | | 1 |
| 7 | | | | 36 | X | X | X | | | | | | | | | | | | | | | | | | | | | | | | | | | | | | | | | | | | M | 3 |
| 8 | | | | | | | | 6 | | | | | | | | | | | | | | | | | | | | | | | | | | | | | | | | | | | | 1 |
| 9 | | | | | | | | | | | | | | 7 | | | | | | | | | | | | | | | | | | | | | | | | | | X | X | X | | 1 |
| 10 | | 8 | | | | | | | | | | | | | | | | | | | | | | | | | | | | | | | | | | | | | | X | X | X | | 1 |
| 11 | | | X | | | | | X | | | | | | | | | | | | | | | 24 | | | | X | | | | | | | | | | | | | X | X | X | | 2 |
| 12 | | | | X | | | | | | | | | | | | | | | | | | | | 34 | | | | | | | | | | | | | | | | X | X | X | X | 3 |
| 13 | | | | | 35 | | | | | | | | | | | | | | | | | | | | | | | | | | | | | | | | | | | X | X | X | X | 2 |
| 14 | X | | | | | | | | | | | | | | | | | | | | | | | | | | | | | | | | | | | | | | 9 | X | X | X | | 1 |
| 15 | X | | | | | 11 | | | | | | | | | | | | | | | | | | | | | | | | | | | | | | | | | | | | | | 1 |
| 16 | X | | | | | | | | | | | | | | | | | | | | | | | | | | | | | 15 | | | | | | | | | | | | | X | 2 |
| 17 | | | | | | | | | | | | | | | 12 | | | | | | | | | | | | | | | | | | | | | | | | X | | | | | 1 |
| 18 | | X | | | | | | | | | | | | | | 13 | | | | | | | | | | | | | | | | | | | | | | | X | | | | | 1 |
| 19 | | | X | | | | | | | | | | | | | | X | | | | | | | | | | | 26 | | | | | | | | | | | X | | | | | 2 |
| 20 | | | | | | | | | | | | | | | | | | | | | | | | | 14 | | | | | | | | | | | | | | X | | | | | 1 |
| 21 | | | | | X | | X | | | | | | | | | | | | | | | | | X | X | X | | | | | 16 | | | | | | | | X | | | | | 4 |
| 22 | | | | | | X | | | | | | | | | | | | | | | | | | | | | 25 | X | | | | | | | | | | | X | | | | | 2 |
| 23 | | | | X | X | X | X | | | X | | X | | | | | | 28 | | | | | | | | | | | | | | | | | | | | | | | | | | 6 |

| | | | | | | | | | | | | | | | | | | | | | | | | | | | | | | | | | | | | | | | | | | | | |
|---|---|---|---|---|---|---|---|---|---|---|---|---|---|---|---|---|---|---|---|---|---|---|---|---|---|---|---|---|---|---|---|---|---|---|---|---|---|---|---|---|---|---|---|---|
| 24 | | | | X | X | X | X | | | | X | X | | | | | | | 29 | | | | | | | | | | | | | | | | | | | | | | | | | 6 |
| 25 | | | | X | X | X | X | X | | | | | | | | | | | | 30 | | | | | | | | | | | | | | | | | | | | | | | | 4 |
| 26 | | | | X | | | | | | | | | | | | | | | | | | | | | | 33 | | | | | | | | | | | | | | | | | | 2 |
| 27 | | | | | | | | | X | X | | | X | | | | | | | | 31 | | | | | | | | | | | | | | | | | | | | | | | 4 |
| 28 | | | | | | | | | X | X | X | X | X | X | X | X | X | X | X | X | X | X | | | | | | | | | | 17 | | | | | | | | | | | | 11 |
| 29 | | | | | | | | | X | | X | | X | | | | | | | | | 32 | | | | | | | | | | | | | | | | | | | | | | 3 |
| 30 | | | | | | | | | 38 | X | | | | | | | | | | | | | | | | | | | | | | | | | | | | | | | | | | 2 |
| 31 | | | | | | | | | | | 41 | X | X | | | | | | | | | | | | | | | | | | | | | | | | | | | | | | | 3 |
| 32 | | | | | | | | | | | X | 42 | X | | | | | | | | | | | | | | | | | | | | | | | | | | | | | | | 2 |
| 33 | | | | M | M | | M | | M | 40M | M | M | M | | X | X | X | X | X | X | X | X | | | | | | | | | | | | | | | | | | | | | X | 4 |
| 34 | | | | | | | | | | | | | | X | X | X | X | X | X | X | X | X | | | | | | | | | | | 18 | | | | | | | | | | | 7 |
| 35 | | | | | | | | | | | | | | X | X | X | X | X | X | X | X | X | | | | | | | | | | | | 19 | | | | | | | | | | 7 |
| 36 | | | | | | | | | | | | | | | | X | X | | X | X | X | X | | | | | | | | | | | | | 20 | | | | | | | | | 6 |
| 37 | | | | | | | | | | | | | | | | | | | | | | | X | | | | | | | | | | | | | 21 | | | | | | | | 2 |
| 38 | | | | | | | | | X | X | | | | | | | | | | | | | | | | | | | | | | | | | | | | | | | | | 39 | 2 |
| 39 | | | | | | | | | | M | X | X | 43 | | | | | | | | | | | | | | | | | | | | | | | | | | | | | | X | 1 |
| 40 | | | | | | | | | | | | | | | | | | | | | | | X | | | | | | | | | | | | | | 22 | | | | | | | 2 |
| 41 | | | | | | | | | | | | | | X | X | X | 27 | X | X | X | X | X | | | | | | | | | | | | | | | | | | | | | X | 7 |
| 42 | | | | M | | | 37M | | | | | | | | | | | | | | | | | X | X | X | | | | | | | | | | | | | | | | | X | 2 |
| 43 | | | | | | | | | | | | | | | | | | | | | | | | | | | X | X | | | | | | | | | | 23 | | | | | X | 4 |
| | | | | | | | | | | | | | | | | | | | | | | | | | | | | | | | | | | | | | | | | | | | | |
| No. in Column when Eliminated | 4 | 3 | 3 | 3 | 3 | 6 | 2 | 3 | 3 | 2 | 3 | 2 | 1 | 5 | 6 | 7 | 2 | 2 | 2 | 2 | 2 | 2 | 1 | 2 | 3 | 2 | 1 | 1 | 1 | 1 | 1 | 1 | 1 | 1 | 1 | 1 | 1 | 1 | 8 | 8 | 8 | 8 | 3 | |

The matrix which Table 1 represents was the optimum basis of a linear programming problem involving a 43-equation model of petroleum refining. This matrix has 197 non-zero elements. As compared with a possible $(43)^2 = 1849$, the number of non-zero elements in the elimination form of inverse is

$$\sum \rho_i + \sum (\sigma_j - 1) = \sum \rho_i + \sum \sigma_j - 43 = 201.$$

To derive this inverse requires $\sum_{(i,j \text{ pivotal})} \rho_i \sigma_j = 247$ multiplications or divisions and somewhat less additions or subtractions.

It would be desirable to choose an agenda so as to minimize the number of zeros which become non-zero. Generally it is harder to find such an "optimum" agenda than to invert a matrix. An alternative is to choose at each step the pivot which minimizes the number of zeros which become non-zero at that step. A still simpler alternative, which seems adequate generally, is to choose the pivot which minimizes the number of coefficients modified at each step (excluding those which are eliminated at the particular step). This is equivalent to choosing the non-zero element with minimum $(\rho_i - 1)(\sigma_j - 1)$. William Orchard-Hays has coded the "Johniac" for the elimination form using the minimum $(\rho_i - 1)(\sigma_j - 1)$ selection principle. The small percentage of zeros which became non-zeros in our example is typical of cases run thus far (with approximately ten per cent non-zeros in the original matrix) with the Orchard-Hays code.

Figure 1 illustrates matrices with the following properties:

a) all diagonal elements $a_{kk}$ are non-zero

b) if an element $a_{ik}$ or $a_{kj}$ in the $k^{\text{th}}$ column or row is non-zero, then all elements between $a_{ik}$ (or $a_{kj}$) and $a_{kk}$ are non-zero. It is trivial to find an optimum agenda for such matrices. If the $k^{\text{th}}$ diagonal element is used as the $k^{\text{th}}$ pivot, no zero coefficient will become non-zero. If a matrix is almost of the above form except that a few zeros are mixed in with the non-zeros, then using the $k^{\text{th}}$ diagonal element as the $k^{\text{th}}$ pivot may cause the "misplaced" zeros (and only these) to become non-zero.

FIG. 1

## Application to Linear Programming

The simplex method for solving linear programming problems has a number of variants. A recent version[2] requires the solution of two sets of equations. The first set of equations

$$p'A = \gamma' \quad \text{or} \quad A'p = \gamma$$

[2] G. B. Dantzig, Alex Orden, & Philip Wolfe, "The Generalized Simplex Method," RAND P-392-1, 4 August 1953.

yields prices $p$ which are used to select a variable $X_s$, currently equal to zero, to be increased until some "basis" variable becomes zero. The second set of equations is

$$Ae = P_s$$

where $P_s$ is a column vector associated with the variable $X_s$. $e$ is used to determine which basis variable $X_r$ first becomes zero as $X_s$ is increased. The matrix $A$ at one iteration differs from that at the next only in that the column vector $P_r$, (which we will assume to be the $r^{\text{th}}$ column of $A$,) is replaced by the column vector $P_s$. I.e.,

$$A^{(k)} = (P_{j_1} \cdots P_r \cdots P_{j_m})$$

becomes

$$A^{(k+1)} = (P_{j_1} \cdots P_s \cdots P_{j_m})$$

Or

$$A^{(k+1)} = (P_{j_1} \cdots P_r \cdots P_{j_m}) \begin{pmatrix} 1 & & & e_1 & & & \\ & 1 & & \cdot & & & \\ & & \cdot & \cdot & & & \\ & & & \cdot\ \cdot & & & \\ & & & e_r & & & \\ & & & \cdot\ \cdot & & & \\ & & & \cdot & \cdot & & \\ & & & \cdot & & \cdot & \\ & & & e_m & & & 1 \end{pmatrix}$$

$$\text{Letting } E = \begin{pmatrix} 1 & & & e_1 & & & \\ & 1 & & \cdot & & & \\ & & \cdot & \cdot & & & \\ & & & \cdot\ \cdot & & & \\ & & & \cdot & & & \\ & & & \cdot\ \cdot & & & \\ & & & \cdot & \cdot & & \\ & & & \cdot & & \cdot & \\ & & & e_m & & & 1 \end{pmatrix}$$

we have

$$A^{(k+1)} = A^k E$$

The inverse of $A^{(k+1)}$ is

$$A^{(k+1)^{-1}} = E^{-1} A^{(K)^{-1}}$$

where

$$E^{-1} = \begin{pmatrix} 1 & & & -\frac{e_1}{e_r} & & & \\ & \cdot & & \cdot & & & \\ & & \cdot & \cdot & & & \\ & & & \cdot \;\cdot & & & \\ & & & 1 \;\cdot & & & \\ & & & \frac{1}{e_r} & & & \\ & & & \cdot \; 1 & & & \\ & & & \cdot & \cdot & & \\ & & & \cdot & & \cdot & \\ & & & -\frac{e_m}{e_r} & & & 1 \end{pmatrix}$$

In an early variant of the simplex method the new inverse $A^{(k+1)^{-1}}$ was obtained at each step by multiplying out $E^{(k)^{-1}} A^{(k)^{-1}}$.[3] In a more recent version the $E^{(k)^{-1}}$ are carried along and used as product form of inverse.[4] If the first iteration starts with the identity matrix then

$$A^{(k)^{-1}} = E^{(k-1)^{-1}} \cdot E^{(k-2)^{-1}} \cdots E^{(1)^{-1}}.$$

If the product form of inverse is used, as $k$ increases computing time required per iteration also increases. A point is eventually reached when it is desirable to reinvert the current basis $A^{(k_0)}$ and let

$$A^{(k_0+k)^{-1}} = E^{(k_0+k-1)^{-1}} \cdots E^{(k_0)^{-1}} A^{(k_0)^{-1}}$$

At this point the elimination form of inverse can be of value, especially if $A$ has a large number of zeros, since this form requires less time to obtain and apply.

[3] See George B. Dantzig, "Maximization of a Linear Function of Variables Subject to Linear Inequalities," pp. 339–347, and Robert Dorfman, "Application of the Simplex Method to a Game Theory Problem," p. 358 in *Activity Analysis of Production and Allocation*, T. C. Koopmans, Ed., New York, 1951.

[4] See George B. Dantzig, "The Product Form for the Inverse in the Simplex Method," in *Mathematical Tables and Other Aids to Computation*, VIII, No. 46, April, 1954.

Reinverting $A$ is only part of the operations involved in solving a linear programming problem. We therefore cannot expect to obtain, by the use of the elimination form, the same percentage reduction in computing time for the whole linear programming problem as we obtain for the reinversion of $A$. When a more convenient form of inverse is available it may be desirable to reinvert more frequently. To see the effect of a more convenient form of inverse on the frequency of reinversion and the time required to solve a linear programming problem, we must explore the question of optimum reinversion policy.

We will first derive some neat results under several restrictive assumptions. Afterwards we will show computing procedures for obtaining an optimum reinversion policy under more general assumptions.

It has been observed, with stop watch as well as theory, that, with the RAND linear programming code for the IBM 701, the computing time required per iteration increases linearly with the number of iterations. Let us assume, for the moment, that (a) the problem starts with a first basis to be inverted; (b) the time required for this first inversion is the same as that for any subsequent reinversion; (c) computation time per iteration is a linear function of the number of iterations since the last (re)inversion. Hence the time ($t$) since the start of the last reinversion is a quadratic $t = \alpha + \beta I + \gamma I^2$ where $\alpha$, $\beta$, $\gamma > 0$. Let us further assume, for the moment, that the number of iterations $\bar{I}$ required to solve the problem is known beforehand.

Suppose it were decided that there would be $n$ inversions ($n - 1$ reinversions). Let $\Delta I_i$ = the number of iterations between the $i^{\text{th}}$ and $i + 1^{\text{th}}$ inversion (for $i = 1, \cdots, n - 1$). Let $\Delta I_n$ = the number of iterations from the $n^{\text{th}}$ inversion to the end of the problem. Total time ($T$) required is

$$T = \sum_{i=1}^{n} \alpha + \beta \Delta I_i + \gamma (\Delta I_i)^2$$

where

$$\sum_{i=1}^{n} \Delta I_i = \bar{I}$$

The optimum solution must satisfy the Lagrangian equations

$$\frac{\partial \sum_i (\alpha + \beta \Delta I_i + \gamma (\Delta I_i)^2) - \lambda(\Delta I_i)}{\partial \Delta I_{i_0}} = 0$$

$$\therefore\ \beta + 2\gamma \Delta I_i - \lambda = 0 \text{ for all } i$$

$$\therefore\ \Delta I_i \text{ is the same for all } i$$

We can therefore rewrite the expression for $T$ as

$$T = n(\alpha + \beta I + \gamma I^2)$$

$$\text{where } I = \Delta I_i = \frac{\bar{I}}{n}.$$

Or

$$T = \alpha n + \beta \bar{I} + \frac{\gamma \bar{I}^2}{n}$$

$$\frac{dT}{dn} = \alpha - \frac{\gamma \bar{I}^2}{n^2}$$

$$\frac{d^2T}{dn^2} = 2\frac{\gamma \bar{I}^2}{n^3}$$

Since $\frac{d^2T}{dn^2} > 0$ for all $n > 0$, and since $T \to \infty$ as $n \to 0$, any $n > 0$ with $\frac{dT}{dn} = 0$ gives a minimum value of $T$ for all $n > 0$. If such an $n$ is non-integral the best integral value is either that immediately above $\hat{n}$, or that immediately below $\hat{n}$, or both

When

$$\frac{dT}{dn} = 0$$

$$\hat{n} = \sqrt{\frac{\gamma}{\alpha}}\,\bar{I}$$

Let us assume that $n$ is integral. Then the optimum

$$\hat{I} = \hat{\Delta} I = \sqrt{\frac{\alpha}{\gamma}}$$

$$\hat{T} = \alpha \sqrt{\frac{\gamma}{\alpha}}\;\bar{I} + \beta \bar{I} + \gamma \sqrt{\frac{\alpha}{\gamma}}\;\bar{I} = (\beta + 2\sqrt{\alpha\gamma})I$$

The last expression can be used for estimating the time to be saved by using a more convenient form of inverse. Thus—given our various assumptions—if a new method of inversion could produce an inverse in one-fourth the time ($\alpha$) and because of its compactness it permitted the first subsequent iteration to be done in one-half the time ($\beta$, roughly), the whole linear programming problem could be done in one-half the time.

Let us now suppose that $\bar{I}$ is not known but has an a-priori probability distribution (derived presumably from past linear programming problems). We may as well also drop the quadratic assumption on $t$. We define

$$\alpha_{ij} = \text{the expected value of } q_{ij}$$

where

$$q_{ij} = \begin{cases} 0 \text{ when } \bar{I} < i \\ \text{The (expected) time required to (re)invert the matrix at iteration } i \\ \quad \text{and iterate through } j-1 \text{ without reinverting—if } \bar{I} \geqq j. \\ \text{The expected time from the beginning of the reinversion at } i \text{ to the} \\ \quad \text{end of the problem if } i \leqq I < j. \end{cases}$$

Suppose the points of reinversion are before iterations $I_1, \cdots, I_K$ (since reinversion points can be chosen with $I_K$ so large that there is a zero probability that this iteration will occur, there is no loss of generality in assuming a fixed $K$). Expected time, to be minimized, is

$$E = \alpha_{1I_1} + \alpha_{I_1I_2} + \cdots + \alpha_{I_{K-1}I_K}$$

The optimal value of $\hat{I}_1$ of $I_1$ could be calculated if $\hat{I}_2$ were known. It is given by the function $\hat{I}_1(I_2)$ which minimizes $\alpha_{1I_1} + \alpha_{I_1I_2}$ for various values of $I_2$ and which can be readily computed.

Define

$$\alpha_{I_2/I_1} = \alpha_{1,\hat{I}_1(I_2)} + \alpha_{\hat{I}_1(I_2),I_2}$$

We now only need to minimize

$$E = \alpha_{I_2/I_1} + \alpha_{I_2I_3} + \cdots + \alpha_{I_{K-1}I_K}$$

We repeat the process until we have

$$E = \alpha_{I_K/I_1,\cdots I_{K-1}}$$

from which we get $I_K$ and work back through

$$\hat{I}_{K-1}(\hat{I}_K), \qquad \hat{I}_{K-2}(\hat{I}_{K-1}), \qquad \text{etc.}$$

# THE OPTIMIZATION OF A QUADRATIC FUNCTION SUBJECT TO LINEAR CONSTRAINTS

Harry Markowitz[1]

The author discusses a computational technique applicable to the determination of the set of "efficient points" for quadratic programming problems.

## 1. QUADRATIC PROBLEMS

Suppose that variables $X_1, \dots, X_N$ are to be chosen subject to linear constraints:

(1) $$\Sigma a_{ij} X_j = b_i, \quad i = 1, \dots, m_1$$

(2) $$\Sigma a_{ij} X_j \geq b_i, \quad i = m_1 + 1, \dots, m$$

(3) $$X_j \geq 0 \quad , \quad j = 1, \dots, N_1$$

where $0 \leq m_1 \leq m$, $0 \leq N_1 \leq N$ and the matrix $(a_{ij})$ $i = 1, \dots, m_1$ has rank $m_1$ (otherwise the system is inconsistent or has at least one redundant equation). The payoff is a linear function $R = \Sigma r_j X_j$ whose coefficients $r_j$ are not known at the time the $X_j$ are chosen. The $r_j$, rather, are random variables with expected values $\mu_j$ and covariances $\sigma_{jk}$ (including variances $\sigma_{jj} = \sigma_j^2$). The expected value of R is

(4) $$E = \Sigma \mu_j X_j .$$

The variance of R is

(5) $$V = \Sigma\Sigma \sigma_{jk} X_j X_k .$$

Suppose further that some decision-maker likes expected payoff (E) and dislikes variance of payoff (V). Our problem is to compute for the decision-maker (a) the "efficient combinations" of E and V, i.e., those attainable (E, V) combinations which give minimum V for given E and maximum E for given V (Figure 1); and (b) the points in the X space associated with the efficient E, V combinations, i.e., the set of efficient X's.

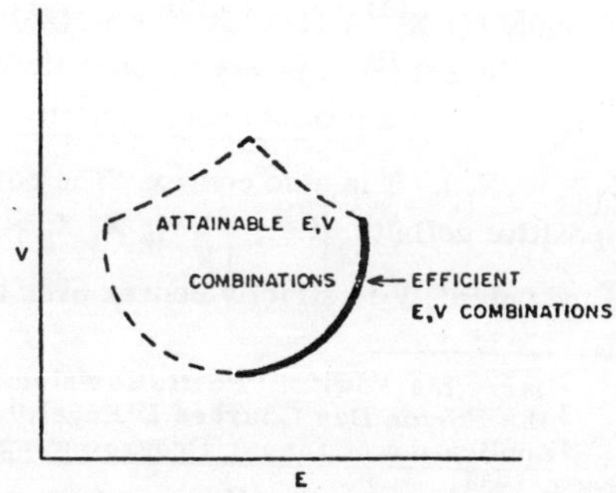


Figure 1

[1] The writer has particularly benefited from discussions with Kenneth Arrow on the subject matter of this paper.

A computing technique is presented in this paper for generating the above efficient sets. An adaptation of this technique can be used for problems of maximizing or minimizing quadratic forms (with the "right" properties) subject to linear constraints.

The practical problem which first suggested the above computing problem was that of selecting a portfolio of securities.[2] Here $X_j$ is the amount invested in the $j^{th}$ security; the $\mu_j$ and $\sigma_{jk}$ are the expected returns and covariances of returns from the various securities. In the simplest case the constraint set is $\Sigma X_j = 1$, $X_j \geq 0$. A problem of very similar structure, analyzed independently by H. S. Houthakker,[3] is that of finding the expenditure on various goods as a function of income for an individual whose utility function is of the form $u = \Sigma a_j X_j + \Sigma\Sigma a_{ij} X_i X_j$. A problem of maximizing a monopolist's quadratic profit function subject to linear constraints is presented by Robert Dorfman.[4] Another problem of this general character is that of maximizing a quadratic likelihood function where there is a priori information concerning the values of parameters to be estimated. Now that reasonably convenient computing procedures exist for such quadratic problems we may be permitted the hope that still other classes of interesting questions can be reduced to this form.

This paper will discuss only minimization problems involving the quadratic form $\Sigma\Sigma \sigma_{ij} X_i X_j$ whose matrix $(\sigma_{ij})$ is positive semi-definite. The reader should have no difficulty in extending the results to minimization problems involving $\Sigma \rho_j X_j + \Sigma\Sigma \sigma_{ij} X_i X_j$ where $(\sigma_{ij})$ is positive semi-definite or maximization problems where $(\sigma_{ij})$ is negative semi-definite.

## 2. ASSUMPTIONS

According to customary usage we say:

(a) A set of points (S) (in Euclidean n-space) is convex if $X^{(1)} \in S$ and $X^{(2)} \in S$ imply $\lambda X^{(1)} + (1-\lambda) X^{(2)} \in S$, for any $0 \leq \lambda \leq 1$.

(b) A set is closed if $X_1, \ldots, X_t, \ldots \longrightarrow y$ and $X_1, \ldots, X_t, \ldots \in S$ imply $y \in S$.

(c) A function $f(X)$ is convex over a set S if $X^{(1)} \in S$, $X^{(2)} \in S$ and $[\lambda X^{(1)} + (1-\lambda) X^{(2)}] \in S$ imply $f(\lambda X^{(1)} + (1-\lambda) X^{(2)}) \leq \lambda f(X^{(1)}) + (1-\lambda) f(X^{(2)})$ for all $0 \leq \lambda \leq 1$.

(d) A function is strictly convex over a set S if $X^{(1)} \in S$, $X^{(2)} \in S$ and $[\lambda X^{(1)} + (1-\lambda) X^{(2)}] \in S$ imply $f(\lambda X^{(1)} + (1-\lambda) X^{(2)}) < \lambda f(X^{(1)}) + (1-\lambda) f(X^{(2)})$ for all $0 < \lambda < 1$.

The set $(\tilde{S})$ of points which satisfy constraints (1), (2), and (3) is a closed, convex set. Variance (V) is a positive semi-definite quadratic form, i.e., $\sum_j \sum_k \sigma_{jk} X_j X_k \geq 0$ for all $(X_1, \ldots, X_N)$. It is also convex. The covariance matrix $(\sigma_{jk})$ is non-singular if, and only if, V is positive definite $\left\{\text{i.e., } \sum_j \sum_k \sigma_{jk} X_j X_k > 0, \text{ if } (X_1, \ldots, X_N) \neq (0, \ldots, 0)\right\}$, which in turn is true if, and only if, V is strictly convex over the set of all X.[5]

---

[2]Harry Markowitz, "Portfolio Selection," Journal of Finance, 1952.

[3]"La Forme Des Courbes D'Engel," Cahiers du Séminaire d'Econometrie, 1953.

[4]Application of Linear Programming to the Theory of the Firm, University of California Press, 1951.

[5]That $V(X) \geq 0$, for all X, is due to the fact that V is the expected value of a square, $E(r-Er)^2$, and therefore cannot be negative. That $|\sigma_{ij}| \neq 0$ if, and only if, V is positive definite is a corollary of material found, e.g., in Birkhoff and MacLane, A Survey of Modern Algebra, Chapter IX, particularly section 8, pp. 243-247. The implications of positive definiteness and semi-definiteness for convexity may be seen as follows: Let $C = (\sigma_{ij})$. Let X and Y be column vectors; X' and Y' be row vectors. C is symmetric so that $C = C'$ and $X'CY = Y'CX$ for any X,Y. We wish to see the implications of

(1) $X'CX \geq 0$

(2) $X'CX = 0 \Longleftrightarrow X = 0$

(Continued)

We will assume[19] that $\widetilde{S}$ is not vacuous. We will also assume that V is strictly convex over the set of X's which satisfy the equations

$$\Sigma a_{ij} X_j = b_i, \quad i = 1, \ldots, m_1.$$

This assures[6] us that V takes on a unique minimum over $\widetilde{S}$ and that if $E = E^0$ is attainable in $\widetilde{S}$, then V takes on unique minimum over the set

$$\widetilde{S} \cap \{X \mid \Sigma \mu_j X_j \geq E_0\}.$$

If a function is convex over a set S, it is convex over any subset of S; therefore $|\sigma_{ij}| \neq 0$ implies that V is strictly convex over $\{X \mid \Sigma a_{ij} X_j = b_i, i = 1, \ldots, m_1\}$. This is not a necessary condition, however. Necessary and sufficient conditions on A and $(\sigma_{ij}) = C$ are discussed in the footnote.[7]

---

5(Continued)
for the difference

$$D = [\lambda X'CX + (1-\lambda) Y'CY] - [(\lambda X' + (1-\lambda)Y') C(\lambda X + (1-\lambda)Y)].$$

Expanding the second term and subtracting we get

$$D = \lambda(1-\lambda) \cdot [X'CX - 2X'CY + Y'CY]$$
$$= \lambda(1-\lambda) \cdot [(X'-Y') C (X-Y)].$$

Assumption (1) implies $D \geq 0$ for all X,Y. Assumptions (1) and (2) imply $D > 0$ if $X \neq Y$. Conversely, if D is positive for all $X \neq Y$, letting $Y = 0$ we find $X'CX > 0$ for all $X \neq 0$.

6Since equations (1) have rank $m_1$, $m_1$ variables and the $m_1$ equations could be eliminated (as in footnote 7) to leave a system with $N-m_1$ variables and $(m-m_1) + N_1$ inequalities. V is strictly convex in these $N-m_1$ variables and therefore the associated quadratic is positive definite. Let Y be any point in the space of the $N-m_1$ variables satisfying the $(m-m_1) + N_1$ inequalities. The points which satisfy these inequalities and have $V \leq V(Y)$ form a compact, convex set. Since V is continuous, it attains its minimum at least once on this set. Since it is strictly convex, it attains its minimum only once. The same argument applies if the constraint

$$\Sigma \mu_j X_j \geq E^0$$

is added to the $(m-m_1) + N_1$ inequalities.

7Suppose $m_1 \geq 1$. Since $\begin{pmatrix} a_{11} \cdots a_{1N} \\ \vdots \quad \vdots \\ a_{m_1 N} \cdots a_{m_1 N} \end{pmatrix}$ has rank $m_1$ we can write, after perhaps relabeling variables,

$$\begin{pmatrix} a_{11} \cdots a_{1m_1} \\ \vdots \quad \vdots \\ a_{m_1 1} \cdots a_{m_1 m_1} \end{pmatrix} \begin{pmatrix} X_1 \\ \vdots \\ X_{m_1} \end{pmatrix} + \begin{pmatrix} a_{1m_1+1} \cdots a_{1N} \\ \vdots \quad \vdots \\ a_{m_1 m_1+1} \cdots a_{m_1 N} \end{pmatrix} \begin{pmatrix} X_{m_1+1} \\ \vdots \\ X_N \end{pmatrix} = \begin{pmatrix} b_1 \\ \vdots \\ b_{m_1} \end{pmatrix}$$

or $A^{(1)} X^{(1)} + A^{(2)} X^{(2)} = b$ where $A^{(1)}$ is non-singular. We thus have $X^{(1)} = (A^{(1)})^{-1} b - (A^{(1)})^{-1} A^{(2)} X^{(2)}$. We can express V in terms of $X^{(2)}$ by substitution, i.e.,

$$X = \begin{pmatrix} X^{(1)} \\ X^{(2)} \end{pmatrix} = \begin{pmatrix} A^{(1)-1} b \\ 0 \end{pmatrix} - \begin{pmatrix} A^{(1)-1} A^{(2)} \\ I \end{pmatrix} X^{(2)}$$

$$V = X'CX = (X^{(1)'} \; X^{(2)'}) \, C \begin{pmatrix} X^{(1)} \\ X^{(2)} \end{pmatrix}$$

(Continued)

## 3. THE CRITICAL LINE $\bar{\ell}$

Let us first note the answer to a simpler problem than that of finding all (E, V) efficient points. Suppose we wished to minimize V subject to the equations

$$\Sigma a_{ij} X_j = b_i , \quad i = 1,\ldots,m_1$$

without regard to the inequalities (2) and (3). A necessary condition for a minimum is that $X_1,\ldots,X_N$ be a solution to the Lagrangian equations,

$$\frac{\partial\,(\Sigma\Sigma\, \sigma_{ij} X_i X_j - 2\,\Sigma\lambda_i\, \Sigma_j\, a_{ij} X_j)}{\partial X_k} = 0 , \quad k = 1,\ldots,N \tag{6}$$

as well as $\Sigma a_{ij} X_j = b_i$, $i = 1,\ldots,m_1$, i.e.,

$$\sum_{j=1}^{N} \sigma_{kj} X_j + \Sigma(-\lambda_i)\, \alpha_{ik} = 0 , \quad k = 1\ldots,N \tag{7}$$

$$\Sigma a_{ij} X_j = b_i , \quad i = 1,\ldots,m_1 . \tag{8}$$

Given the assumption that V is strictly convex over $\{X \mid \Sigma a_{ij} X_j = b_i ,\ i = 1,\ldots,m_1\}$ it follows that

$$\begin{pmatrix} \sigma_{11} & \cdots & \sigma_{1N} & a_{11} & \cdots & a_{m_1 1} \\ \vdots & & \vdots & \vdots & & \vdots \\ \sigma_{N1} & \cdots & \sigma_{NN} & a_{1N} & \cdots & a_{m_1 N} \\ a_{11} & \cdots & a_{1N} & 0 & \cdots & 0 \\ \vdots & & & \vdots & & \vdots \\ a_{m_1 1} & \cdots & a_{m_1 N} & 0 & \cdots & 0 \end{pmatrix}$$

---

[7](Continued)

$$= [(b'\, A^{(1)'-1}, 0) - (X^{(2)'} A^{(2)'} A^{(1)'-1}, X^{2'})]\, C \left[\begin{pmatrix} A^{(1)-1} b \\ 0 \end{pmatrix} - \begin{pmatrix} A^{(1)-1} A^{(2)} X^{(2)} \\ X^{(2)} \end{pmatrix}\right]$$

$$= (b'\, A^{(1)'-1}, 0)\, C \begin{pmatrix} A^{(1)-1} b \\ 0 \end{pmatrix} - 2X^{(2)} (A^{(2)'} A^{(1)'-1}, I)\, C \begin{pmatrix} A^{(1)-1} b \\ 0 \end{pmatrix}$$

$$+ X^{(2)'} \left[ (A^{(2)'} A^{(1)-1}, I)\, C \begin{pmatrix} A^{(1)-1} A^{(2)} \\ I \end{pmatrix} \right] X^{(2)}$$

V is strictly convex for all $X^{(2)}$ if, and only if, the last (i.e., the quadratic) term is strictly convex. This is so if, and only if,

$$(A^{(2)'} A^{(1)-1'}, I)\, C \begin{pmatrix} A^{(1)-1} A^{(2)} \\ I \end{pmatrix}$$

is non-singular.

is non-singular.[8] Since a strictly convex V takes on a unique minimum on a convex set, the unique solution to (7) and (8) is this minimum.

Next consider the problem of minimizing V subject not only to (1) but also to the constraint that $E = E_0$, i.e.,

$$\sum_{j=1}^{N} \mu_j X_j = E_0 \,. \tag{9}$$

We must distinguish two cases:

Case 1: The row vector $(\mu_1, \ldots, \mu_N)$ can be expressed as a linear combination of the $(a_{i1}, \ldots, a_{iN})$, i.e., there exists $\alpha_1, \ldots, \alpha_{m_1}$ such that

$$(\alpha_1, \ldots, \alpha_{m_1}) \begin{pmatrix} a_{11} & \cdots & a_{1N} \\ \vdots & & \vdots \\ a_{m_1 1} & \cdots & a_{m_1 N} \end{pmatrix} = (\mu_1, \ldots, \mu_N) \,. \tag{10}$$

Case 2: There does not exist such a linear combination.

As is shown below,[9] in Case 1 only one value of E, say $E = E^*$, is attainable. Therefore if we require $E \neq E^*$ no solution can be found; if we require $E = E^*$, equations (7) and (8) give the minimum. In Case 2 the matrix

$$\begin{pmatrix} \sigma_{11} & \cdots & 1n\, a_{11} & \cdots & a_{m_1 N} & \mu_1 \\ \vdots & & \vdots & & \vdots & \vdots \\ \sigma_{N1} & \cdots & \sigma_{NN}\, a_{1N} & \cdots & a_{m_1 N} & \mu_N \\ a_{11} & \cdots & a_{m_1 1} & O \cdots & O & O \\ \vdots & & \vdots & \vdots & \vdots & \vdots \\ a_{m_1} & \cdots & a_{m_1 N} & O \cdots & O & O \\ \mu_1 & \cdots & \mu_N & O \cdots & O & O \end{pmatrix}$$

---

[8] If $m_1 = 0$ the statement reduces to one proved in footnote 5. Suppose $m_1 \geq 1$. If $\begin{pmatrix} C & A' \\ A & O \end{pmatrix}$ is singular there is a vector $\begin{pmatrix} Y \\ -\lambda \end{pmatrix} \neq \begin{pmatrix} O \\ O \end{pmatrix}$ such that $\begin{pmatrix} C & A' \\ A & O \end{pmatrix} \begin{pmatrix} Y \\ -\lambda \end{pmatrix} = \begin{pmatrix} O \\ O \end{pmatrix}$ i.e., $\begin{pmatrix} CY \\ AY \end{pmatrix} = \begin{pmatrix} A'\lambda \\ O \end{pmatrix}$. Since the rank of A is $m_1$ there is no $\lambda \neq O$ such that $A'\lambda = O$, therefore $Y \neq O$ in $\begin{pmatrix} Y \\ -\lambda \end{pmatrix}$ above. $V(Y) = Y'CY = Y'A'\lambda = (AY)'\lambda = O$. Let X be any point in $S = \{X \mid AX = b\}$ where

$X' = (X_1, \ldots, X_N)$

$b' = (b_1, \ldots, b_{m_1})$

$A(X + Y) = AX + O = b$; therefore $X + Y$ is in S

$V(X) = X'CX$

$V(X + Y) = X'CX + 2X'CY = X'CX + 2X'A'\lambda = X'CX + 2b'\lambda$

$V(1/2X + 1/2(X + Y)) = V(X + 1/2\,Y)$

$= X'CX + X'CY$

$= X'CX + b'\lambda$

$= 1/2\, V(X) + 1/2\, V(X + Y)$,

thus contradicting strict convexity.

[9] If $\alpha'A = \mu'$ and $AX = b$ then $E = \mu'X = \alpha'AX = \alpha'b$.

is non-singular,[10] and therefore the equations

(11) $$\Sigma\sigma_{jk} X_k + \Sigma(-\lambda_1) a_{ij} - \lambda_E \mu_j = 0 \;,\; j = 1, \ldots, N$$

(12) $$\Sigma a_{ij} X_j = b_i \;,\; i = 1, \ldots, m_1$$

(13) $$\Sigma \mu_j X_j = E^0$$

have a unique solution which gives minimum V for the specified $E^O$. If we let $E^O$ go from $-\infty$ to $+\infty$, the solution to (11), (12), and (13) traces out a line in the $(X,\lambda)$ space. This line may also be described as the solution to the following $N + m_1$ equations in $N + m_1 + 1$ unknowns:

(14) $$\sum_{j=1}^{N} \sigma_{jk} X_k - \sum_{i=1}^{m_1} \lambda_i a_{ij} - \lambda_E \mu_j = 0 \;,\; j = 1, \ldots, N$$

(15) $$\Sigma a_{ij} X_j = b_i \;,\; i = 1, \ldots, m_1$$

or

(16) $$\Sigma \sigma_{jk} X_k + \Sigma(-\lambda_j) a_{ij} = \lambda_E \mu_j \;,\; j = 1, \ldots, N$$

(17) $$\Sigma \alpha_{ij} X_j = b_i \;,\; i = 1, \ldots, m_1$$

$$\text{for } -\infty \le \lambda_E \le +\infty .$$

Since the matrix of equations (16) and (17) is non-singular, given our assumption of strict convexity, they have a solution for every value of $\lambda_E$ whether we are in Case 1 or Case 2 above. In Case 1 the values of $X_1, \ldots, X_N$ do not change (only the values of the $\lambda$'s change) as $\lambda_E$ goes from $-\infty$ to $+\infty$.[11] In Case 2 the X's as well as the $\lambda$'s change. In Case 2 we can define $\hat{V}(E)$ to be minimum V as a function of $E \cdot 2\lambda_E = d\hat{V}/dE \cdot \hat{V}(E)$ must be strictly convex; therefore, E increases with $\lambda_E$. In Section 11 we show that $\hat{V}(E)$ is a parabola.

## 4. CRITICAL LINES $\ell(\mathcal{S},\mathcal{I})$

The set of points $(X,\lambda)$ which satisfy (16) and (17) will be referred to as the <u>critical line</u> $\bar{\ell}$ associated with the subspace

$$\bar{S} = \{X \mid \Sigma a_{ij} X_j = b_i \text{ for } i = 1, \ldots, m_1\}.$$

Critical lines will also be associated with certain other subspaces.

---

[10] Same proof as in footnote 8, using the fact that $\begin{pmatrix} A \\ \mu' \end{pmatrix}$ has rank $m_1 + 1$.

[11] For if $\mu = A'\tilde{\lambda}$ and $\begin{pmatrix} C & A' \\ A & O \end{pmatrix} \begin{pmatrix} X^0 \\ \lambda^0 \end{pmatrix} = R + \begin{pmatrix} \mu \\ O \end{pmatrix} \lambda_E^0$ then $\begin{pmatrix} C & A' \\ A & O \end{pmatrix} \begin{pmatrix} X^0 \\ \lambda^0 + \tilde{\lambda}\theta \end{pmatrix}$ $= R + \begin{pmatrix} \mu \\ O \end{pmatrix} \left(\lambda_E^0 + \theta\right).$

Let $X_{j_1}, \ldots, X_{j_J}$ be a subset of variables. Let

$$\Sigma a_{ij} X_j = b_i \,, \quad i = i_1, \ldots, i_I$$

be a subset of the constraints (1) and (2) with the inequalities replaced by equalities when $i > m_1$. Let $\mathcal{I}$ be the ordered set of indices $(i_1, \ldots, i_I)$; let $\mathcal{J}$ be the ordered set $(j_1, \ldots, j_J)$. We will be particularly interested in $\mathcal{I}$ and $\mathcal{J}$ of the form

(18) $$\mathcal{I} = \{1, 2, \ldots, m_1, i_{m_1+1}, \ldots, i_I\} \quad \text{where } I \geq m_1$$

(19) $$\mathcal{J} = \{j_1, \ldots, j_L, N_1 + 1, \ldots, N\} \quad \text{where } 0 \leq L \leq N_1 .$$

For any indices $\mathcal{I}$ and $\mathcal{J}$ satisfying (18) and (19) we define the submatrix

(20) $$A_{\mathcal{I}\mathcal{J}} = \begin{pmatrix} a_{i_1 j_1} & \cdots & a_{i_1 j_J} \\ \vdots & & \vdots \\ a_{i_I j_1} & \cdots & a_{i_I j_J} \end{pmatrix} .$$

We similarly define subvectors $X_{\mathcal{J}}$, $\lambda_{\mathcal{I}}$ and

(21) $$B_{\mathcal{I}} = \begin{pmatrix} b_{i_1} \\ \vdots \\ b_{i_I} \end{pmatrix} ;$$

also submatrices

(22) $$C_{\mathcal{J}\mathcal{J}} = \begin{pmatrix} \sigma_{j_1 j_1} & \cdots & \sigma_{j_1 j_J} \\ \vdots & & \vdots \\ \sigma_{j_J j_1} & \cdots & \sigma_{j_J j_J} \end{pmatrix}$$

(23) $$M_{\mathcal{I}\mathcal{J}} = \begin{pmatrix} C_{\mathcal{J}\mathcal{J}} & A'_{\mathcal{I}\mathcal{J}} \\ A_{\mathcal{I}\mathcal{J}} & O \end{pmatrix} .$$

If $I = m_1 = 0$, $\mathcal{I}$ is empty. In this case it will sometimes be convenient to think of $A_{\mathcal{I}\mathcal{J}}$ as having no rows and $J$ columns. To every $(\mathcal{I}, \mathcal{J})$ satisfying (18) and (19) we associate a subspace

$$S(\mathcal{I}, \mathcal{J}) = \{X \mid X_j = 0 \text{ for } j \notin \mathcal{J},\ A_{\mathcal{I}\mathcal{J}} X_{\mathcal{J}} = B_{\mathcal{I}}\}.$$

If $A_{\mathcal{I}\mathcal{J}}$ has no rows this reduces to $S(\mathcal{J}) = \{X \mid X_j = 0 \text{ for } j \notin \mathcal{J}\}$. Since $\mathcal{I}$ and $\mathcal{J}$ satisfy (18) and (19), $S(\mathcal{I}, \mathcal{J}) \subset \overline{S}$. Since V is strictly convex over $\overline{S}$, it is strictly convex over $S(\mathcal{I}, \mathcal{J})$.

$A_{\mathcal{I}\mathcal{J}}$ has a rank of I or less. If $A_{\mathcal{I}\mathcal{J}}$ has rank I then the matrix

(24) $$M_{\mathcal{I}\mathcal{J}} = \begin{pmatrix} C_{\mathcal{J}\mathcal{J}} & A'_{\mathcal{I}\mathcal{J}} \\ A_{\mathcal{I}\mathcal{J}} & O \end{pmatrix}$$

is non-singular. (This is a special case of the proposition proved in footnote 8.) If $A_{\mathcal{I}\mathcal{J}}$ has rank less than I, $M_{\mathcal{I}\mathcal{J}}$ is singular, for its last I rows are not independent. For every $(\mathcal{I}, \mathcal{J})$ satisfying (18) and (19) whose $A_{\mathcal{I}\mathcal{J}}$ has rank equal to the number of its rows, we define the <u>critical line</u> $\ell_{\mathcal{I}\mathcal{J}}$ to be the set of points $(X_1, \ldots, X_N, \lambda_1, \ldots, \lambda_m)$ which satisfy

(25) $$\begin{aligned} X_j &= 0 \quad \text{for } j \notin \mathcal{J} \\ \lambda_i &= 0 \quad \text{for } i \notin \mathcal{I} \end{aligned}$$

and

$$\begin{pmatrix} X_{\mathcal{J}} \\ -\lambda_{\mathcal{I}} \end{pmatrix} = M^{-1}_{\mathcal{I}\mathcal{J}} \begin{pmatrix} O \\ B_{\mathcal{I}} \end{pmatrix} + M^{-1}_{\mathcal{I}\mathcal{J}} \begin{pmatrix} \mu_{\mathcal{J}} \\ O \end{pmatrix} \lambda_E .$$

Equations (25) may be written in the form

(26) $$X_j = \alpha_{Xj} + \beta_{Xj} \lambda_E$$

(27) $$\lambda_i = \alpha_{\lambda i} + \beta_{\lambda i} \lambda_E \qquad -\infty < \lambda_E < \infty .$$

Equations (26) by themselves are the projection of $\ell(\mathcal{I}, \mathcal{J})$ onto the X-space. As with $\overline{S}$ and $\overline{\ell}$ we have two cases:

(1) Only one value of E is obtainable in $S(\mathcal{I}, \mathcal{J})$ and the X-projection is a point.

(2) All values of E are obtainable and the X-projection is a line. This line is the set of X's in $S(\mathcal{I}, \mathcal{J})$ which give minimum V for some E. Let

(28) $$\varepsilon_i = \Sigma a_{ij} X_j - b_i \ , \quad i = 1, \ldots, m$$

(29) $$\eta_j = 1/2 \, \frac{\partial V - 2 \sum_{i=1}^{m} \lambda_i \Sigma a_{ij} X_j - 2\lambda_E \Sigma \mu_j}{\partial X_j}$$

$$= \sum_{j=1}^{N} \sigma_{jk} X_k + \sum_i (-\lambda_i) a_{ij} - \mu_j \lambda_E .$$

Constraints (1) and (2) state that

$$\xi_i = 0 \quad \text{for } i = 1,\dots,m_1$$

$$\xi_i \geq 0 \quad \text{for } i = m_1+1,\dots,m.$$

Along any critical line we have

(30)
$$\begin{aligned} X_j &= 0 \quad \text{for } j \notin \mathcal{J}, \\ \eta_j &= 0 \quad \text{for } j \in \mathcal{J}, \\ \xi_i &= 0 \quad \text{for } i \in \mathcal{I}, \\ \lambda_i &= 0 \quad \text{for } i \notin \mathcal{I}. \end{aligned}$$

Also, from (25), letting $m^{st}$ be the $(s,t)^{th}$ element of $M^{-1}_{\mathcal{I}\mathcal{J}}$, we have

(31)
$$\begin{aligned} X_{j_s} &= \sum_{h=1}^{I} m^{s,h+J} b_{i_h} + \left(\sum_{h=1}^{J} m^{sh} \mu_{j_h}\right) \lambda_E \\ &= \alpha_{Xj_s} + \beta_{Xj_s} \lambda_E \qquad \text{for } s - 1,\dots,J. \end{aligned}$$

(32)
$$\begin{aligned} \lambda_{i_s} &= -\sum_{h=1}^{I} m^{s+J,h+J} b_{i_h} - \left(\sum_{h=1}^{J} m^{s+J,h} \mu_{j_h}\right) \lambda_E \\ &= \alpha_{\lambda i_s} + \beta_{\lambda i_s} \lambda_E \qquad \text{for } s = 1,\dots,I. \end{aligned}$$

From (28) and (29) we have.

(33)
$$\begin{aligned} \xi_i &= \sum_{h=1}^{J} a_{ij_h} \alpha_{Xj_h} - b_i + \left(\sum_{h=1}^{J} a_{ij_h} \beta_{Xj_h}\right) \lambda_E \\ &= \alpha_{\xi_i} + \beta_{\xi_i} \lambda_E ; \end{aligned}$$

(34)
$$\begin{aligned} \eta_j &= \left(\sum_{h=1}^{J} \sigma_{jj_h} \alpha_{Xj_h} - \Sigma a_{i_h j} \alpha_{\lambda i_h}\right) \\ &\quad + \left(\sum_{h=1}^{J} \sigma_{jj_h} \beta_{Xj_h} - \sum_{h=1}^{I} a_{i_h j} \beta_{\lambda i_h} - \mu_j\right) \lambda_E \\ &= \alpha_{\eta j} + \beta_{\eta j} \lambda_E . \end{aligned}$$

A corollary of the results of an important paper by Kuhn and Tucker[12] is that a sufficient condition for a point X to give minimum V for a set

$$\tilde{S} \cap \{X \mid \Sigma \mu_j X_j \geq E_0\}$$

for some $E_O$ is that

$$(35)\qquad \begin{aligned} X_j &\geq 0 &&\text{for } j \leq N_1 \text{ and in } \mathcal{J}, \\ \eta_j &\geq 0 &&\text{for } j \notin \mathcal{J}, \\ \xi_i &\geq 0 &&\text{for } i \notin \mathcal{I}, \\ \lambda_i &\geq 0 &&\text{for } i > m_1 \text{ and in } \mathcal{I}, \end{aligned}$$

and $\lambda_E \geq 0$. If $\lambda_E > 0$, the constraint $E \geq E_0$ is effective; if $E_0$ were increased, an equally low value of V could not be obtained. If $\lambda_E = 0$, the point gives minimum V in $\tilde{S}$. In either case the point is efficient.

It will be convenient at times to employ the following relabeling of variables:

$$(36)\qquad \begin{aligned} v_k &= X_k &&\text{for } k = 1, \ldots, N_1, \\ v_k &= \eta_{k-N_1} &&\text{for } k = N_1 + 1, \ldots, 2N_1, \\ v_k &= \xi_{m_1+k-2N_1} &&\text{for } k = 2N_1 + 1, \ldots, 2N_1 + m - m_1, \\ v_k &= \lambda_{k+2m_1-m-2N_1} &&\text{for } k = 2N_1 + m - m_1 + 1, \ldots, 2N_1 + 2m - 2m_1. \end{aligned}$$

Also

$$(37)\qquad K = 2N_1 + 2m - 2m_1$$

and

$$(38)\qquad \mathcal{K} = \{\text{the set of } k \text{ which identify the variables in equation 30)}\}.$$

Thus on any critical line we have

$$(39)\qquad v_k = 0 \quad \text{for } k \in \mathcal{K}$$

and a point X is efficient if it is a projection of a point on a critical line with

$$(40)\qquad v_k \geq 0 \quad \text{for } k \notin \mathcal{K},$$

or

$$(41)\qquad v_k \geq 0 \quad \text{for } k = 1, \ldots, K.$$

[12] H. W. Kuhn and A. W. Tucker, "Nonlinear Programming," in Proceedings of the Second Berkeley Symposium on Mathematical Statistics and Probability.

## 5. INTERSECTIONS OF CRITICAL LINES; NON-DEGENERACY CONDITIONS

In the computing procedure of the next section we move along a critical line until it intersects a plane $v_k = 0$, $k = 1, \ldots, K$. Then either one row and the corresponding column is added to M, or one row and the corresponding column is deleted from M. This raises two questions: (1) under what conditions will the matrix obtained by such additions or deletions be non-singular, and (2) how should the new inverse be obtained? The latter question will not be discussed except to note that the possession of the old inverse is of great value in obtaining the new one.[13]

Concerning the former question, suppose $M_{\mathfrak{I}\mathfrak{J}}$, with $\mathfrak{I}\mathfrak{J}$ satisfying (18) and (19), is non-singular and thus defines a critical line $\ell$. Suppose $\ell$ intersects (but is not contained in) the plane $v_k = 0$, $1 \le k \le K$. We distinguish four cases, depending on whether v corresponds to an X, an $\eta$, a $\lambda$, or a $\xi$:

1. The deletion of a variable. Suppose $\ell$ intersects a plane $X_j = 0$, $j = 1, \ldots, N_1$. Suppose that j is deleted from the set $\mathfrak{J}$ leaving $\mathfrak{J}^*$. Is $M_{\mathfrak{I}\mathfrak{J}^*}$ non-singular? We may suppose without loss of generality that $j = j_1$ and that $A_{\mathfrak{I}\mathfrak{J}}$ may therefore be written

$$A_{\mathfrak{I}\mathfrak{J}} = (\alpha A_{\mathfrak{I}\mathfrak{J}^*}) \tag{42}$$

where $\alpha$ is the column to be deleted. The matrix $\begin{pmatrix} \alpha & A_{\mathfrak{I}\mathfrak{J}^*} \\ 1 & 0 \ldots 0 \end{pmatrix}$ has either rank, I or I + 1. If it has rank I, then

$$\begin{pmatrix} C_{\mathfrak{I}\mathfrak{J}} & A'_{\mathfrak{I}\mathfrak{J}} & \begin{matrix}\mu_j \\ \vdots \\ \mu_{j_J}\end{matrix} \\ A_{\mathfrak{I}\mathfrak{J}} & O & \begin{matrix} 0 \\ \vdots \\ 0 \end{matrix} \\ 1\ 0 \ldots 0 & 0 \ldots 0 & 0 \end{pmatrix} = \widetilde{M} \tag{43}$$

is singular. In this case the equations

$$\widetilde{M} \begin{pmatrix} X_{\mathfrak{J}} \\ \lambda_{\mathfrak{I}} \\ \lambda_E \end{pmatrix} = \begin{pmatrix} 0 \\ \vdots \\ 0 \\ b_{i_1} \\ \vdots \\ b_{i_I} \\ 0 \end{pmatrix} \tag{44}$$

have either no solution or an infinity of solutions. Thus if (44) has one solution, i.e., if $\ell$ intersects $X_{j_1} = 0$ (but is not contained in it), $\widetilde{M}$ is non-singular and the rank of $\begin{pmatrix} \alpha & A_{\mathfrak{I}\mathfrak{J}^*} \\ 1 & 0\ 0 \end{pmatrix}$ is I + 1;

[13] This involves procedures similar to those used in removing a variable from a regression analysis or modifying a basis in linear programming; e.g., see R. A. Fisher, Statistical Methods for Research Workers, p. 164, 10th ed., and R. Dorfman, Activity Analysis of Production and Allocation, p. 358.

hence the rank of $\begin{pmatrix} A_{\mathscr{I}\mathscr{J}^*} \\ 0\dots0 \end{pmatrix}$ is at least I. But the rank of $\begin{pmatrix} A_{\mathscr{I}\mathscr{J}^*} \\ 0\dots0 \end{pmatrix}$ = the rank of $A_{\mathscr{I}\mathscr{J}^*}$. Therefore, the rank of $A_{\mathscr{I}\mathscr{J}^*}$ is I and $M_{\mathscr{I}\mathscr{J}^*}$ is non-singular.

2. The deletion of a constraint. Suppose $\ell$ intersects (but is not contained in) $\lambda_i = 0$ for $i > m_1$. We may assume $i = i_I$ and that

$$A_{\mathscr{I}\mathscr{J}} = \begin{pmatrix} A_{\mathscr{I}^*\mathscr{J}} \\ \alpha' \end{pmatrix}. \tag{45}$$

$A_{\mathscr{I}\mathscr{J}}$ has rank I, $A_{\mathscr{I}^*\mathscr{J}}$ has rank I-1; therefore $M_{\mathscr{I}^*\mathscr{J}}$ is non-singular.

3. Addition of a variable. Continuing the conventions used above, if $A_{\mathscr{I}\mathscr{J}}$ has rank I, so has $A_{\mathscr{I}\mathscr{J}^*} = (A_{\mathscr{I}\mathscr{J}}\ \alpha)$. Therefore $M_{\mathscr{I}\mathscr{J}^*}$ is non-singular.

4. Addition of a constraint. If $\ell$ intersects but is not contained in the plane $\xi_i = 0$, $i > m_1$ then

$$\begin{pmatrix} C_{\mathscr{J}\mathscr{J}} & A_{\mathscr{I}\mathscr{J}} & \mu_{j_1} \\ & & \vdots \\ & & \mu_{j_J} \\ A_{\mathscr{I}\mathscr{J}} & 0 & 0 \\ \alpha' & 0 & 0 \end{pmatrix} = \widetilde{M} \tag{46}$$

(where $\alpha'$ is the row of coefficients of the new constraint) is non-singular. Therefore $\begin{pmatrix} A_{\mathscr{I}\mathscr{J}} \\ \alpha \end{pmatrix} = A_{\mathscr{I}^*\mathscr{J}}$ has rank I + 1 and $M_{\mathscr{I}^*\mathscr{J}}$ is non-singular.

The tracing out of the efficient set is simplified if certain "accidents" do not occur. These accidents are described in the following "non-degeneracy" conditions. The next section of this paper presents a computing procedure for deriving the set of efficient points when all non-degeneracy conditions hold. In Sections 7-10 these conditions are relaxed.

CONDITION 1. On no critical line do we have

$$v_k = \alpha_{vk} + \beta_{vk}\lambda_E = 0 \quad \text{for } k \notin \mathcal{K}.$$

CONDITION 2. On any given critical line $\ell$ we do not have

$$\frac{-\alpha_{vk_1}}{\beta_{vk_1}} = \frac{-\alpha_{vk_2}}{\beta_{vk_2}} \quad \text{for any } k_1 \neq k_2 \text{ with } \beta_{vk_1} \neq 0,\ \beta_{vk_2} \neq 0\,.$$

CONDITION 3. E is bounded from above in $\widetilde{S}$.

We will let $L_E$ stand for "the linear programming problem of maximizing E subject to constraints (1), (2), and (3)."

**CONDITION 4.** $L_E$ has a unique non-degenerate solution.

Condition 4 implies condition 3.

## 6. THE ALGORITHM UNDER CONDITIONS 1 THROUGH 4

We now assume that conditions 1 through 4 are satisfied. Condition 4 implies that the optimum solution to $L_E$ has:[14]

(a) Exactly m variables $X_j$ and $\xi_i$ are not at their lower extreme;

(b) $A_{\mathcal{I}(1)\mathcal{J}(1)}$ (where $\mathcal{I}^{(1)}$ includes all i with $\xi_i = 0$ and $\mathcal{J}^{(1)}$ includes all j with $X_j$ not at its lower limit) has rank equal to the number of its rows;

(c) There exists "prices" $p_i$ and "profitabilities" $\delta_j$ such that

$$p_i > 0 \quad \text{if } \xi_i = 0 \quad \text{for } i = m_1+1, \ldots, m, \tag{47}$$

$$p_i = 0 \quad \text{if } \xi_i > 0 \quad \text{for } i = m_1+1, \ldots, m, \tag{48}$$

$$\delta_j = \sum_i a_{ij} p_i + \mu_j, \tag{49}$$

$$\delta_j = 0 \quad \text{for } j = N_1+1, \ldots, N \text{ and for } X_j > 0 \quad j \le N_1, \tag{50}$$

$$\delta_j < 0 \quad \text{for } X_j = 0 \quad j \le N_1. \tag{51}$$

The matrix

$$M_{(1)} = \begin{pmatrix} C_{\mathcal{J}(1)\mathcal{J}(1)} & A'_{\mathcal{I}(1)\mathcal{J}(1)} \\ A_{\mathcal{I}(1)\mathcal{J}(1)} & O \end{pmatrix} \tag{52}$$

is non-singular and thus defines a critical line $\ell^{(1)}$ along which

$$M_{(1)} \begin{pmatrix} X_{\mathcal{J}(1)} \\ -\lambda_{\mathcal{I}(1)} \end{pmatrix} = \begin{pmatrix} O \\ B_{\mathcal{I}(1)} \end{pmatrix} + \begin{pmatrix} \mu_{\mathcal{J}(1)} \\ O \end{pmatrix} \lambda_E. \tag{53}$$

Since $-A'_{\mathcal{I}(1)\mathcal{J}(1)} \cdot p_{\mathcal{I}(1)} = \mu_{\mathcal{J}(1)}$, if $X^0_{\mathcal{J}(1)}\ \lambda^0_{\mathcal{I}(1)}$ satisfy (53) for $\lambda_E = 0$, then

$$M_1 \begin{pmatrix} X^0_{\mathcal{J}(1)} \\ -\lambda^0_{\mathcal{I}(1)} - p_{\mathcal{I}(1)} \lambda_E \end{pmatrix} = \begin{pmatrix} O \\ B_{\mathcal{I}(1)} \end{pmatrix} + \begin{pmatrix} \mu_{\mathcal{J}(1)} \\ O \end{pmatrix} \lambda_E \tag{54}$$

for all $\lambda_E$. Thus $\ell^{(1)}$ has

---

[14] The following are corollaries of the basis and pricing theorems of linear programming. See, e.g., George B. Dantzig, Alex Orden, Philip Wolfe, "The Generalized Simplex Method for Minimizing a Linear Form under Linear Inequality Restraints," Pacific Journal of Mathematics, Vol. 5, No. 2, June 1955.

(55)
$$X_{\mathcal{J}(1)} = X^0_{\mathcal{J}(1)}$$
$$\lambda_{\mathcal{I}(1)} = \lambda^0_{\mathcal{I}(1)} + p_{\mathcal{I}(1)}\lambda_E$$

for all $\lambda_E$. From (47) it follows that for sufficiently large $\lambda_E$

$$\lambda_i > 0 \quad \text{for } i > m_1 \text{ and in } \mathcal{I}.$$

(56)
$$\begin{aligned}\eta_j &= \Sigma\sigma_{jh}X^0_h - (\Sigma a_{ij}\lambda_i + \mu_j\lambda_E)\\ &= \Sigma\sigma_{jh}X^0_h - \Sigma a_{ij}\lambda^0_i - (\Sigma a_{ij}p_i + \mu_j)\lambda_E.\end{aligned}$$

Equation (51) implies that for sufficiently large $\lambda_E, \eta_j > 0$ for $j \notin \mathcal{J}$. Thus for sufficiently large $\lambda_E, \ell^{(1)}$ satisfies inequalities (40).

Let $\lambda^{(1)}_E$ be the largest value of $\lambda_E$ at which $\ell^{(1)}$ intersects a plane $\eta_j = 0$ for $j \notin \mathcal{J}$ or $\lambda_i = 0$ for $i \in \mathcal{I}$. (The X and $\xi$ do not vary along $\ell^{(1)}$.) If $\lambda^{(1)}_E \le 0$ then $X^0$ gives minimum V as well as maximum E. Suppose $\lambda^{(1)}_E > 0$. Non-degeneracy condition 2 implies $\ell^{(1)}$ intersects only one plane $\eta_j = 0$ or $\lambda_i = 0$ at $\lambda^{(1)}_E$. In the former case we add j to $\mathcal{J}$; in the latter case we delete i from $\mathcal{I}$, to form $\mathcal{I}^{(2)}$, $\mathcal{J}^{(2)}$. The new matrix $M_{(2)} = M_{\mathcal{I}(2)\mathcal{J}(2)}$ is non-singular and defines a critical line $\ell^{(2)}$. Suppose for a moment that it was $\eta_{j_0} = 0$ which $\ell^{(1)}$ intersected at $\lambda^{(1)}_E$. On $\ell^{(2)}$ we have at $\lambda_E = \lambda^{(1)}_E$:

(57)
$$\begin{aligned}&\lambda_i > 0 \quad \text{for } i > m_1 \text{ and } \in \mathcal{I}^{(2)},\\ &\eta_i > 0 \quad \text{for } j \notin \mathcal{J}^{(2)},\\ &\xi_i > 0 \quad \text{for } i \notin \mathcal{I},\\ &\text{and } \begin{cases} X_{j_0} = 0 \\ X_j > 0 \quad \text{for all other } j \le N_1 \text{ and } \in \mathcal{J}^{(2)}.\end{cases}\end{aligned}$$

As always $X_{j_0} = a + b\lambda_E$ along $\ell^{(2)}$. Non-degeneracy condition 1 assures $b \neq 0$. If $b < 0$ the projection of $\ell^{(2)}$ would be efficient for $\lambda^* \ge \lambda_E \ge \lambda^{(1)}_E$ where $\lambda^* > \lambda^{(1)}_E$. This is impossible.[15] Therefore $b > 0$ and $\ell^{(2)}$ is efficient for $\lambda^{(1)}_E \ge \lambda_E \ge \lambda^{(2)}_E$ where $\lambda^{(1)}_E > \lambda^{(2)}_E$. Similar remarks would apply if $\ell^{(1)}$ first intersected $\lambda_{i_0} = 0$ and $i_0$ was deleted from $\mathcal{I}$.

[15]Since $b \neq 0$ the X-projection of the critical line is a line rather than a point. Along this line E increases with $\lambda_E$. If $\lambda_E > \lambda^{(1)}_E$ were feasible, then $E > E^{(1)} = \max E$ would be obtainable, which is impossible.

$\lambda_E^{(2)}$ is the highest value of $\lambda_E < \lambda_E^{(1)}$ at which $\ell^{(2)}$ intersects a plane $v_k = 0$ for $k = 1, \ldots, K$. If this is an $\eta_j$ we again add a j to $\mathcal{J}$. If it is a $\lambda_i$ we delete i from $\mathcal{I}$; if a $\xi_i$, we add i to $\mathcal{I}$; if an $X_j$, we delete j from $\mathcal{J}$. We form $M_{(3)}$ and $\ell_{(3)}$ accordingly and find $\lambda^{(3)} < \lambda^{(2)}$. This process is repeated until $\lambda_E = 0$ is reached. At each step (s) $M_{(s)}$ is non-singular and if $v_{k_s}$ is the new variable ($\eta$, X, $\lambda$, or $\xi$) which is no longer constrained to be zero we have at $\lambda^{(s-1)}$.

$$v_k > 0 \quad \text{for } k \neq k_s \text{ and } \notin \mathcal{K}, \tag{58}$$

$$v_{k_s} = 0 .$$

By condition 1, $b_{vk_s} \neq 0$ along $\ell^{(s)}$. We argue below[16] that we cannot have $b_{vk_s} < 0$. So $b_{vk_s} > 0$ and $\ell^{(s)}$ is efficient for $\lambda_E^{(s-1)} \geq \lambda_E \geq \lambda_E^{(s)}$ where $\lambda_E^{(s-1)} > \lambda_E^{(s)}$. Since there are only a finite number of critical lines, and each can satisfy inequalities (40) for only one segment, $\lambda_E = 0$ is reached in a finite number of steps.

## 7. THE ALGORITHM UNDER CONDITIONS 3 AND 4

Let us now drop non-degeneracy conditions 1 and 2 but still assume conditions 3 and 4. We will use techniques analogous to the degeneracy-avoiding techniques of linear programming.[17]

For every number $\varepsilon$ we define a new problem $P(\varepsilon)$ as follows:

$$\text{minimize } V(\varepsilon) = \Sigma\Sigma\sigma_{ij} X_i X_j + \Sigma \varepsilon^j X_j$$

subject to

$$\Sigma a_{ij} X_j = b_i + \varepsilon^{N+i} \quad , \quad i = 1, \ldots, m_1 , \tag{59}$$

$$\Sigma a_{ij} X_j \geq b_i + \varepsilon^{N+i} \quad , \quad i = m_1 + 1, \ldots, m , \tag{60}$$

$$X_j \geq 0 \quad , \quad j = 1, \ldots, N_1 . \tag{61}$$

---

[16] If $v_{k_s}$ is an $X_j$ or $\xi_i$, $b_{vk_s} < 0$ implies that there are two distinct points which minimize V for some $E > E^{(s-1)}$, which is impossible. This argument also applies if $v_k$ is a $\lambda_i$ or $\eta_j$ unless the X-projection of the new critical line is a point. In the latter case we note (from the Kuhn and Tucker conditions) that an efficient point gives minimum $Q(\lambda_E) = V - \lambda_E E$ subject to (1), (2), (3). For fixed $\lambda_E$, $Q(\lambda_E)$ has a unique minimum. If $v_k < 0$ then two distinct points give minimum $Q(\lambda_E)$ for some $\lambda_E > \lambda_E^{(s-1)}$.

[17] In linear programming these techniques are generally not needed in practice. In quadratic programming arbitrary selection of $v_k = 0$ with $b_{vk} < 0$ to go into $\mathcal{K}$ may (or may not) prove adequate. In any case, the degeneracy-handling techniques are available if needed. See George Dantzig, "Application of the Simplex Method to a Transportation Problem," Activity Analysis of Production and Allocation, Tjalling C. Koopmans, ed.; A. Charnes, "Optimality and Degeneracy in Linear Programming," Econometrica, Vol. 20, No. 2, April, 1952, p. 160; and Dantzig, Orden, and Wolfe, Op. Cit.

For sufficiently small $\varepsilon$ the unique, optimal basis of $L_E$ is feasible and, since it still satisfies the pricing conditions, is optimal.

As we will see shortly for sufficiently small $\varepsilon$, $P(\varepsilon)$ satisfies non-degeneracy conditions (1) and (2). We will also see that for a sufficiently small $\varepsilon^*$, the sequence of indices $(\mathcal{I}\,\mathcal{J})^s$ associated with the critical lines $\ell^{(s)}$, until $\lambda_E = 0$ is reached, is the same for all $P(\varepsilon)$ for $\varepsilon^* \geq \varepsilon > 0$. If we change indices $(\mathcal{I}, \mathcal{J})$ in the same sequence as $P(\varepsilon)$ for small $\varepsilon$, if we let $\lambda_E$ decrease along any critical line when it can without violating $v_k \geq 0$, until we reach $\lambda_E = 0$, then:

(a) We will pass through a finite number of index sets each associated with a non-singular $M_{\mathcal{I}\mathcal{J}}$, before we reach $\lambda_E = 0$.

(b) Since $v_k \geq 0$ is maintained we have the desired solution to the original problem.

Along any critical line of $P(\varepsilon)$ we have

$$\begin{pmatrix} X_{\mathcal{J}} \\ -\lambda_{\mathcal{I}} \end{pmatrix} = M_{\mathcal{I}\mathcal{J}}^{-1}\begin{pmatrix} O \\ B_{\mathcal{I}} \end{pmatrix} + M_{\mathcal{I}\mathcal{J}}^{-1}\begin{pmatrix} \mu_{\mathcal{J}} \\ O \end{pmatrix}\lambda_E + M_{\mathcal{I}\mathcal{J}}^{-1}\begin{pmatrix} \varepsilon^{j_1} \\ \vdots \\ \varepsilon^{N+i_I} \end{pmatrix} \tag{62}$$

or

$$X_{j_s} = \alpha_{Xj_s} + \beta_{Xj_s}\lambda_E + \sum_{h=1}^{I+J} m^{sh}\,\varepsilon^{f(h)} \tag{63}$$

where $f(1) = j_1, f(2) = j_2, \ldots, f(I+J) = N + i_I$

or

$$X_{j_s} = \alpha_{Xj_s} + \beta_{Xj_s}\lambda_E + p_{Xj_s}(\varepsilon)\,. \tag{64}$$

Similarly

$$\lambda_{i_s} = \alpha_{\lambda i_s} + \beta_{\lambda i_s}\lambda_E - \sum_{h=1}^{I+J} m^{J+s,h}\,\varepsilon^{f(h)}$$

$$= \alpha_{\lambda i_s} + \beta_{\lambda i_s}\lambda_E + p_{\lambda i_s}(\varepsilon)\,.$$

$$\xi_i = \alpha_{\xi_i} + \beta_{\xi_i}\lambda_E + \sum_{s=1}^{J} a_{ij_s} p_{Xj_s}(\varepsilon) + \varepsilon^{N+i} = \alpha_{\xi_i} + \beta_{\xi_i}\lambda_E + p_{\xi_i}(\varepsilon)\,. \tag{65}$$

$$\eta_j = \alpha_{\eta_j} + \beta_{\eta_j}\lambda_E + \sum_{s=1}^{J}\sigma_{jj_s} p_{Xj_s}(\varepsilon) - \sum_{s=1}^{I} a_{i_s j} p_{\lambda i_s}(\varepsilon) + \varepsilon^{j} \tag{66}$$

$$= \alpha_{\eta_j} + \beta_{\eta_j}\lambda_E + p_{\eta_j}(\varepsilon)\,.$$

Consider the polynomials:

(67) $$p_{Xj}(\varepsilon) \quad \text{for} \quad j \in \mathcal{J},$$

(68) $$p_{\lambda i}(\varepsilon) \quad \text{for} \quad i \in \mathcal{I},$$

(69) $$p_{\xi i}(\varepsilon) \quad \text{for} \quad i \notin \mathcal{I},$$

(70) $$p_{\eta j}(\varepsilon) \quad \text{for} \quad j \notin \mathcal{J}.$$

None of the polynomials listed above have all zero coefficients, and no two have proportional coefficients. For each polynomial of (69) and (70) has a term with a coefficient of 1 which every other polynomial has with a coefficient of zero. This leaves only the possibilities that some polynomial of (67) or (68) has all zero coefficients or two of these polynomials have proportional coefficients. Both these possibilities imply that $M^{-1}$ is singular and therefore are impossible.

Since $p_{v_k}(\varepsilon)$ has only a finite number of roots, for $\varepsilon$ sufficiently small

$$p_{v_k}(\varepsilon) \neq 0 \qquad \text{for } k \notin \mathcal{K}.$$

Thus

(71) $$v_k = \alpha_{vk} + \beta_{vk}\lambda_E + p_{vk}(\varepsilon) \qquad -\infty < \lambda_E < \infty$$

cannot be identically zero for $k \notin \mathcal{K}$. The critical line intersects the plane $v_{k_1} = 0$ at

(72) $$\lambda_E' = \frac{-\alpha_{vk_1}}{\beta_{vk_1}} - \frac{p_{vk_1}(\varepsilon)}{\beta_{vk_1}}$$

and the plane

$$v_{k_2} = 0 \text{ at}$$

(73) $$\lambda_E'' = \frac{-\alpha_{vk_2}}{\beta_{vk_2}} - \frac{p_{vk_2}(\varepsilon)}{\beta_{vk_2}}.$$

If, say,

(74) $$\frac{-\alpha_{vk_1}}{\beta_{vk_1}} > \frac{-\alpha_{vk_2}}{\beta_{vk_1}}$$

then for sufficiently small $\varepsilon$

$$\lambda'_E > \lambda''_E .$$

On the other hand, since

$$p_{vk_1}(\varepsilon) - p_{vk_2}(\varepsilon) = 0 \quad k_1, k_2 \notin K$$

has a finite number of solutions, for sufficiently small $\varepsilon$

$$\lambda'_E \neq \lambda''_E$$

even if

(75)
$$\frac{-\alpha_{vk_1}}{\beta_{vk_1}} = \frac{-\alpha_{vk_2}}{\beta_{vk_2}} .$$

As $\varepsilon \longrightarrow 0$ the smallest power of $\varepsilon$ dominates; i.e., if, say, $\frac{-p_{vk_1}(\varepsilon)}{\beta_{vk_1}}$ has an algebraically larger coefficient for the first power of $\varepsilon$, then $\lambda' > \lambda''$ as $\varepsilon \longrightarrow 0$. If both have the same coefficient of $\varepsilon$, then it is the coefficients of $\varepsilon^2$ that decide. And so on.

Since there are a finite number of critical lines and a finite number of planes $v_k = 0$, there is a single $\varepsilon^*$ such that for $\varepsilon^* \geq \varepsilon > 0$.

$P(\varepsilon)$ satisfies non-degeneracy conditions (1) and (2); and the order of the index sets $(\mathcal{I}\,\mathcal{J})^S$ is the same for all such $\varepsilon$.

The $m^{st}$ are needed for other purposes and are thus available for resolving degeneracy problems. The other coefficients of $p_{v_k}(\varepsilon)$ can be computed when needed.

## 8. THE ALGORITHM WHEN $L_E$ IS DEGENERATE BUT UNIQUE

Suppose that the solution to $L_E$ is degenerate in that one or more of the basis variables $X_j$ or $\xi_i$ is "accidentally" zero, but is unique in that $\delta_j < 0$ for all $X_j$ not in the basis and $p_i > 0$ for all $\xi_i$ not in the basis.

The constraints of $L_E$ may be written as a system of equations including the $\xi_i$ as variables:

(76)
$$B \begin{pmatrix} X \\ \xi \end{pmatrix} = b .$$

If $\widetilde{B}$ is the submatrix of optimal basis vectors and if $X_{\mathcal{J}}$ and $\xi_{\bar{\mathcal{I}}}$ are the optimal basis variables, then the optimal solution is given by

(77)
$$\begin{pmatrix} X^0_{\mathcal{J}} \\ \xi^0_{\bar{\mathcal{I}}} \end{pmatrix} = \widetilde{B}^{-1} b$$

while all other variables are zero. After we solve $L_E$ we may modify it, forming $L_E(\varepsilon)$ as follows:

$$(78) \qquad B \begin{pmatrix} X \\ \xi \end{pmatrix} = b + \widetilde{B} \begin{pmatrix} \varepsilon \\ \varepsilon \\ \vdots \\ \varepsilon \end{pmatrix} \quad \text{for } \varepsilon > 0$$

$$= b + \varepsilon \begin{pmatrix} r_1 \\ \vdots \\ r_m \end{pmatrix}$$

where $r_i$ is the sum of the $i^{th}$ row of $\widetilde{B}$.

Then

$$(79) \qquad \begin{pmatrix} X_{\mathcal{I}}(\varepsilon) \\ \\ \xi_{\bar{\mathcal{I}}}(\varepsilon) \end{pmatrix} = \widetilde{B}^{-1} b + \begin{pmatrix} \varepsilon \\ \vdots \\ \varepsilon \end{pmatrix}.$$

Thus the original optimal basis is still feasible and therefore uniquely optimal (since it still satisfies the pricing relationships). Also for $\varepsilon > 0$

$$X_j(\varepsilon) > 0 \qquad \text{for } j \in \mathcal{I},$$

$$\xi_i(\varepsilon) > 0 \qquad \text{for } i \in \bar{\mathcal{I}},$$

and

$$\begin{pmatrix} X_{\mathcal{I}}(\varepsilon) \\ \xi_{\bar{\mathcal{I}}}(\varepsilon) \end{pmatrix} \longrightarrow \begin{pmatrix} X^0_{\mathcal{I}} \\ \xi^0_{\bar{\mathcal{I}}} \end{pmatrix}$$

as $\varepsilon \longrightarrow 0$.

The procedures of the last section which apply when $L_E$ has a unique and non-degenerate solution apply with essentially no modification if $L_E$ has a unique but possibly degenerate solution, if we let $P(\varepsilon)$ be

$$\min V = \Sigma\Sigma\, \sigma_{ij} X_i X_j + \Sigma\, \varepsilon^{j+1} X_j$$

subject to

$$(80) \qquad \begin{aligned} \Sigma a_{ij} X_j &= b_i + r_i \varepsilon + \varepsilon^{N+i+1} && \text{for } i = 1, \ldots, m_1, \\ \Sigma a_{ij} X_j &\geqq b_i + r_i \varepsilon + \varepsilon^{N+i+1} && \text{for } i = m_1+1, \ldots, m. \end{aligned}$$

The solution to $L_E(\varepsilon)$ is non-degenerate for sufficiently small $\varepsilon$. Along any critical line we now have

(81)
$$X_{j_s} = \alpha_{Xj_s} + \beta_{Xj_s} \lambda_E + \varepsilon\, p_{Xj_s}(\varepsilon) + \left(\sum_{h=1}^{I} m^{s,h+J} r_{i_h}\right)\varepsilon$$
$$= \alpha_{Xj_s} + \beta_{Xj_s} \lambda_E + q_{Xj_s}(\varepsilon);$$

(82)
$$\lambda_{i_s} = \alpha_{\lambda i_s} + \beta_{\lambda i_s} \lambda_E + \varepsilon\, p_{\lambda i_s}(\varepsilon) - \left(\sum_{h=1}^{I} m^{s+J,h+J} r_{i_h}\right)\varepsilon$$
$$= \alpha_{\lambda i_s} + \beta_{\lambda i_s} \lambda_E + q_{\lambda i_s}(\varepsilon);$$

(83)
$$\xi_i = \alpha_{\xi i} + \beta_{\xi i} \lambda_E + \sum_{s=1}^{J} a_{ij_s} q_{Xj_s}(\varepsilon) + \varepsilon^{N+i+1}$$
$$= \alpha_{\xi i} + \beta_{\xi i} \lambda_E + q_{\xi i}(\varepsilon);$$

(84)
$$\eta_j = \alpha_{\eta j} + \beta_{\eta j} \lambda_E + \sum_{s=1}^{J} \sigma_{jj_s} q_{Xj_s}(\varepsilon) - \sum_{s=1}^{I} a_{i_s j} q_{\lambda i_s}(\varepsilon) + \varepsilon^{j+1}$$
$$= \alpha_{\eta j} + \beta_{\eta j} \lambda_E + q_{\eta j}(\varepsilon)$$

where the $p_v(\varepsilon)$ are as defined in (63) through (66). Since no $p_v(\varepsilon)$ can have zero coefficients and no two can have proportional coefficients, the same is true of the $q_v(\varepsilon)$.

## 9. THE ALGORITHM WHEN $L_E$ IS NOT UNIQUE

A non-degenerate optimal solution to $L_E$ is unique if, and only if,

(85)
$$\delta_j < 0 \text{ for } X_j \text{ not in the basis}$$
$$p_i > 0 \text{ for } \xi_i \text{ not in the basis.}$$

If $L_E$ has a degenerate solution and (85) does not hold, then either the solution is not unique or else only the optimal basis is not unique. If $L_E$ does not have a unique solution we must find the point $\overline{X}$ which gives minimum V for $E = \overline{E} = \max E$. If only the optimal basis of $L_E$ is not unique we still must decide on the $\mathcal{I}\mathcal{J}$ of our first critical line. Both these problems will be resolved in the same manner. Our procedure may be considered as a special case of either approach 1 or approach 4 for minimizing a quadratic subject to linear constraints described in Section 12.

Let us create a new linear programming problem $L_F(\varepsilon)$ by adding a constraint to and modifying the form to be maximized in $L_E(\varepsilon)$. The equation we add is

(86)
$$\sum_{j=1}^{N} \mu_j X_j - \xi_E = \overline{E} + \left(\sum_{j\in\mathcal{J}} \mu_j - 1\right)\varepsilon .$$

If we add $\xi_E$ to the optimum basis variables of $L_E(\varepsilon)$ we have a feasible basis corresponding to a solution with

$$
\begin{aligned}
X_j &= X_j^0 + \varepsilon && j \varepsilon \mathcal{J}, \\
\xi_i &= \xi_i^0 + \varepsilon && i \varepsilon \bar{\mathcal{I}}, \\
\xi_E &= \varepsilon .
\end{aligned}
\tag{87}
$$

Next let us replace the objective function $E = \Sigma \mu X$ with a new one

$$F = \Sigma \nu_j X_j \tag{88}$$

such that the solution in (87) is the unique optimum of $L_F(\varepsilon)$. This may be done easily by assigning any values $p_i > 0$ to $i \not\varepsilon \bar{\mathcal{I}}$, $p_i = 0$ for $i \varepsilon \bar{\mathcal{I}}$ as well as $p_E = 0$. Then choose any set for $\nu_j$ so that $\delta_j = 0$ for $j \varepsilon \mathcal{J}$ and $\delta_j < 0$ for $j \not\varepsilon \mathcal{J}$. Since $L_F(\varepsilon)$ has a unique non-degenerate solution we may use methods already described to trace out the set of points which give minimum $V(\varepsilon)$ for given $F$ until $\lambda_F = 0$. If only a few bases are feasible for $L_F(\varepsilon)$, i.e., if not too many bases are optimal for $L_E(\varepsilon)$, $\lambda_F = 0$ will be reached quickly. At $\lambda_F = 0$ either $\lambda_E = 0$ or $\lambda_E > 0$. In the former case we have arrived at a point with minimum $V$ and maximum $E$. In the latter case we have $\bar{X}$ and are ready to trace out the set of efficient X's. From this point on we let $\lambda_F = 0$, i.e., we ignore $F$ completely. Since at $\lambda_F = 0$, $v_k > 0$ for all $k \not\varepsilon \mathcal{K}$ we may reduce $\lambda_E$ until we intersect a plane $v_k = 0$ and continue as in Section 8.

## 10. THE ALGORITHM, WHEN CONDITION 3 DOES NOT HOLD

If $E$ is unbounded procedure 4 of Section 12 can be used to find the point $\underline{X}$ with minimum $V$. The efficient set can then be traced out in the direction of increasing $\lambda_E$. Since there are only a finite number of critical lines and each critical line is efficient at most once the efficient set is traced out in a finite number of steps.

## 11. THE SET OF EFFICIENT E, V COMBINATIONS

Once the set of efficient X's is found the set of efficient $E, V$ combinations can be obtained easily. The critical line of a subspace in which more than one value of $E$ is obtainable may be expressed as the solution to

$$\Sigma \sigma_{jk} X_k + \Sigma(-\lambda_i) a_{ij} + (-\lambda_E) \mu_j = 0, \qquad j \varepsilon \mathcal{J}, \tag{89}$$

$$\Sigma a_{ij} X_j = b_i, \qquad i \varepsilon \mathcal{I}, \tag{90}$$

$$\Sigma \mu_j X_j = E \tag{91}$$

for $-\infty < E < +\infty$.

If we let $N^{-1}$ be the inverse of the matrix $N$ in (89), (90), (91) we have

(92)
$$\begin{pmatrix} X \\ -\lambda \end{pmatrix} = N^{-1} \begin{pmatrix} 0 \\ b \\ E \end{pmatrix}$$

(93)
$$V = (X', -\lambda') \begin{pmatrix} C & O \\ O & O \end{pmatrix} \begin{pmatrix} X \\ -\lambda \end{pmatrix} = (0', b', E) N^{-1} C N^{-1} \begin{pmatrix} 0 \\ b \\ E \end{pmatrix}$$

from which it follows that along any such critical line V and E are related by a formula of the form

(94)
$$V = a + bE + CE^2 .$$

Thus the set of efficient E, V combinations is piecewise parabolic. We know, or can easily get, the values of E and dV/dE at the end points of each of the pieces. We can also evaluate V at $\bar{X}$.[18] Knowing V at one value of E and $dV/dE = b + 2CE$ at two values of E we can solve for the a, b, and c in (94) for the segment from $\bar{E}$ to $\bar{E} - \varepsilon_1$. Having a, b, and c we can evaluate V at $\bar{E} - \varepsilon_1$ by means of (94). This provides us with the value of V at one value of E on the segment which is efficient from $E - \varepsilon_1$ to $E - \varepsilon_2$. This, combined with the values of dV/dE at two values of E, allows us to obtain the a, b, c of (94) for this next segment—and so on until we trace out the set of E, V combinations.

## 12. MINIMIZING A QUADRATIC

One of the "by-products" of the calculation of efficient sets is the point at which V is a minimum, i.e., where $\lambda_E = 0$. The computing procedures described in Sections 6 through 10 are analogous to the simplex method of linear programming (as contrasted with the "gradient methods" that have been suggested for both linear and non-linear programming). Both the procedure described in the preceding section—considered as a way of getting to min V—and the simplex method require a finite number of iterations, each iteration typically taking a "jump" to a new point which is superior to the old. Each iteration makes use of the inverse of a matrix which is a "slight" modification of the matrix of the previous iteration. The success of the simplex method in linear programming suggests that it may be desirable to use a variant of the "critical line" method in the quadratic case.

Our problem then is to minimize a quadratic

$$V = \Sigma\Sigma\, \sigma_{ij} X_i X_j$$

subject to constraints (1), (2) and (3). We wish to translate this into a problem of tracing out an efficient set. This may be done in several ways.

1. An arbitrary set of $\mu_j$ can be selected and the efficient set traced out until $\lambda_E = 0$. The $\mu_j$ should be selected so that the "artificial" E has a unique maximum.

2. An equality, say,

$$\Sigma a_{1j} X_j = b_1$$

---

[18] If $\bar{X}$ does not exist we can evaluate V at $\underline{X}$ and use the same process "in reverse."

can be eliminated from (1). E can be defined as

$$E = \Sigma a_{1j} X_j$$

and the critical set traced out until $E = b_1$. If $E = b_1$ is reached before $\lambda_E = 0$ the computing procedures of the last section must be continued into the region of $\lambda_E < 0$. While the points thus generated will not be efficient—for they do not give max E for given V—they do give min V for given E. In particular, they will arrive at the point of min V for

$$E = \Sigma a_{1j} X_j = b_1 .$$

3. An inequality, say,

$$\Sigma a_{mj} X_j \geq b_m$$

can be eliminated from (2). E can be defined as

$$E = \Sigma a_{mj} X_j .$$

The efficient set is traced out until either $E = b_m$ or else $\lambda_E = 0$. If the former happens first the constraint is effective; if the latter happens first the constraint is ineffective. In either case, the point associated with the first of these to occur gives min $\dot{V}$ subject to (1), (2), and (3).

4. An initial guess $X_1^0, \ldots, X_N^0$ which satisfies (1), (2), and (3) can be made and $\mu_j$ defined so that, given these $\mu_j$, $X^0$ is efficient. The efficient set can then be traced out until $\lambda_E = 0$. To choose $\mu_j$ so that $X^0$ is efficient, choose arbitrary positive values of $\lambda_i$ ($i \in \mathcal{A}$) and $\lambda_E$. Then choose $\mu_j$ so that

$$\eta_j = 0 \qquad \text{for } X_j \text{ not at its lower bound,}$$

$$\eta_j > 0 \qquad \text{for } X_j \text{ at its lower bound.}$$

If $X^0$ is in the same subspace as the optimal solution, the latter is reached in one iteration.

[19]It has been found recently that the strict convexity assumption can be relaxed. The procedures described in this paper apply, without modification, to the homogeneous quadratic whenever $(\sigma_{ij})$ is positive semi-definite. The non-homogeneous quadratic requires a slight modification of procedure for the general semi-definite $(\sigma_{ij})$. (Footnote added in proof.)

* * *

# The general mean-variance portfolio selection problem

By Harry M. Markowitz
*1010 Turquoise Street, Suite 245, San Diego, California 92109, U.S.A.*

This paper states the 'general mean-variance portfolio analysis problem' and its solution, and briefly discusses its use in practice.

## 1. The problem

We consider $n$ securities whose returns $r' = (r_1, \ldots, r_n)$ during the forthcoming period have expected values $\mu' = (\mu_1, \ldots, \mu_n)$ and a covariance matrix $C = (\sigma_{ij})$. An investor is to select a portfolio $X' = (X_1, \ldots, X_n)$. The return $R = r'X$ on the portfolio has expected value and variance, respectively,

$$E = \mu' X, \quad V = X'CX. \tag{1a, b}$$

The portfolio is to chosen subject to constraints

$$AX = b, \quad X \geqslant 0, \tag{1c, d}$$

where $A$ is $m \times n$ and $b$ is $m \times 1$. Thus, non-negative $X_i$ are to be chosen subject to $m \geqslant 1$ linear inequalities.

A portfolio is *feasible* if it satisfies (1*c*) and (1*d*). An $EV$ combination is feasible if it is the $E$ and $V$ of a feasible portfolio. A feasible $EV$ combination $(E_0, V_0)$ is *inefficient* if there is another feasible $EV$ combination $(E_1, V_1)$ such that either

$$\text{(i)}\ E_1 > E_0 \ \text{ and } \ V_1 \leqslant V_0$$

or

$$\text{(ii)}\ V_1 < V_0 \ \text{ and } \ E_1 \geqslant E_0.$$

A feasible $EV$ combination is efficient if it is not inefficient. A feasible portfolio is efficient or inefficient in accordance with its $EV$ combination.

It is not sufficient to require only condition (i) or condition (ii) in the definition of inefficiency, nor to define an efficient $EV$ combination as one which maximizes $E$ for given $V$ and minimizes $V$ for given $E$. Examples can be constructed of feasible $EV$ combinations which meet the latter two requirements but are nevertheless inefficient as previously defined.

Since $C$ is a covariance matrix it is positive semi-definite. We do not require it to be positive definite because $C$ is singular in some important applications. We will see an example below. If $C$ is singular, there may be more than one efficient portfolio which has a given efficient $EV$ combination. We define a 'complete, non-redundant' set of efficient portfolios as one which contains one and only one efficient portfolio for each efficient $EV$ combination.

The portfolio analysis problem is to determine

1. whether constraints (1*c*) and (1*d*) are feasible and, if they are, calculate
2. the set of all efficient $EV$ combinations and
3. a complete, non-redundant set of efficient portfolios.

*Phil. Trans. R. Soc. Lond.* A (1994) **347**, 543–549
*Printed in Great Britain*

Examples exist of portfolio selection problems with feasible portfolios but no efficient portfolios. This is possible (but not necessary) when $E$ is unbounded and $C$ singular. Excluding this case, every feasible portfolio selection problem has a piecewise parabolic set of efficient $EV$ combinations, with a finite number of pieces. For every such problem there exists a piecewise linear complete, non-redundant set of efficient portfolios. For these assertions to apply with the generality stated, we define 'piecewise' to include a single 'piece' (line or parabolic segment) or only a point. One piece (line segment and parabolic segment) may be unbounded in one direction.

We will consider the computation of these efficient sets after we further consider problem definition and application.

## 2. Application

It might seem that we could gain some generality by allowing linear inequalities ($\geqslant$, $\leqslant$) in (1$c$) or allowing some or all variables to be negative. This is not the case. Specifically, given any mean-variance portfolio analysis problem whose constraints permit some or all variables to be negative and permits some or all linear constraints in (1$c$) to be (weak) inequalities, there is a problem in the standard form of (1$c$) and (1$d$) that has the same answer. That is, it has the same set of efficient $EV$ combinations (even the same set of feasible $EV$ combinations) and, given the complete, non-redundant set of efficient portfolios for the equivalent problem it is easy to determine this for the original problem. (See chapter 2 of Markowitz (1987) for details.)

Thus, as far as applications are concerned, we think of the general mean-variance portfolio selection problem as one of finding mean-variance efficient portfolios in variables that may or may not be required to be non-negative, which are subject to one or more (actually, zero or more) linear equalities or weak inequalities. (Also, certain nonlinear constraints can be approximated, as in linear programming.)

Examples of such linear constraints are the budget constraint, a 'turnover constraint' which limits the amount by which the new portfolio may differ from the previous one, and upper or lower bounds on the amount invested in a security or an industry. A special case is an 'exogenous asset' whose amount is fixed in the portfolio. The exogenous asset may, for example, be a random source of income other than the return on securities. For a given $E$, the variance minimizing portfolio as a whole, including the exogenous asset, depends on the covariances between the exogenous asset and the securities whose amounts are to be selected.

The exogenous asset may be a state variable for which the investor wishes to seek or avoid correlation of portfolio return. The single period mean-variance analysis should be thought of as an approximation to the single period derived utility maximization which is optimal within a many-period investment game. If the derived utility function depends on state variables other than end-of-period wealth, the mean variance approximation may use exogenous assets. (See chapter 3 of Markowitz (1987), starting from the section 'Why mean and variance'.)

We noted that one can convert a model with inequality constraints to an equivalent one with equality constraints. This involves introducing slack variables, as in linear programming. Since slack variables have zero variance, their presence makes $C$ singular. Singular $C$ is no problem for the 'critical line' algorithm described in the next section.

## 3. Computation

Every feasible general portfolio selection problem has a solution of the following nature. Each piece of the piecewise linear set of efficient portfolios has a set $\mathrm{IN} \subset \{1, \ldots, n\}$ of 'in' securities. The others are 'out'. Obtain $A_{\mathrm{IN}}$ and $\mu_{\mathrm{IN}}$ by setting $a_{ki} = 0$ and $\mu_i = 0$ if $i \in \mathrm{OUT}$. Obtain $C_{\mathrm{IN}}$ by setting $\sigma_{ij} = \delta_{ij}$ ($= 1$ if $i = j$, otherwise $= 0$) if either $i$ or $j$ is out.

If

$$M_{\mathrm{IN}} = \begin{pmatrix} C_{\mathrm{IN}} & A'_{\mathrm{IN}} \\ A_{\mathrm{IN}} & 0 \end{pmatrix} \qquad (2a)$$

is non-singular, then the solution to

$$M_{\mathrm{IN}} = \begin{pmatrix} X \\ \lambda \end{pmatrix} = \begin{pmatrix} 0 \\ b \end{pmatrix} + \begin{pmatrix} \mu_{\mathrm{IN}} \\ 0 \end{pmatrix} \lambda_E, \qquad (2b)$$

i.e.

$$\begin{pmatrix} X \\ \lambda \end{pmatrix} = M_{\mathrm{IN}}^{-1} \begin{pmatrix} 0 \\ b \end{pmatrix} + M_{\mathrm{IN}}^{-1} \begin{pmatrix} \mu_{\mathrm{IN}} \\ 0 \end{pmatrix} \lambda_E = \alpha_{\mathrm{IN}} + \beta_{\mathrm{IN}} \lambda_E, \qquad (2c)$$

is referred to as a 'critical line'.

Typically, most critical lines contain no efficient portfolios. If an efficient point exists on a critical line, then there is an interval of efficient portfolios, namely, the solution to (2) for

$$\lambda_E \in [\lambda_{\mathrm{LOW}}, \infty) \quad \text{or} \quad \lambda_E \in [\lambda_{\mathrm{LOW}}, \lambda_{\mathrm{HI}}]. \qquad (3a, b)$$

This is referred to as the 'efficient segment' of the critical line.

In particular, there exists a complete, non-redundant set of efficient portfolios which consists of the efficient segments of $K \geqslant 1$ critical lines. These critical lines are efficient for

$$\lambda_E \in [\lambda^1_{\mathrm{LOW}}, \infty), \quad \lambda_E \in [\lambda^2_{\mathrm{LOW}}, \lambda^2_{\mathrm{HI}}], \ldots, \lambda_E \in [0, \lambda^K_{\mathrm{HI}}],$$

where $\lambda^{k+1}_{\mathrm{HI}} = \lambda^k_{\mathrm{LOW}}$ and $\lambda^k_{\mathrm{HI}} \geqslant \lambda^k_{\mathrm{LOW}}$.

$\mathrm{IN}^{k+1}$ differs from $\mathrm{IN}^k$ by the addition or deletion of one member. In the relatively easy to explain 'non-degenerate' case, to be defined below, $\lambda^k_{\mathrm{HI}} > \lambda^k_{\mathrm{LOW}}$ for $k = 1, \ldots, K$.

Define $\eta' = (\eta_1, \ldots, \eta_n)$ by

$$\eta = (C_{\mathrm{IN}}, A'_{\mathrm{IN}}, -\mu_{\mathrm{IN}}) \begin{pmatrix} X \\ \lambda \\ \lambda_E \end{pmatrix}. \qquad (4)$$

$\eta_i$ is the partial derivative of a lagrangian with respect to $X_i$. Substituting (2$c$) into (4) we obtain $\eta$ as a linear function of $\lambda_E$

$$\eta = \gamma_{\mathrm{IN}} + \delta_{\mathrm{IN}} \lambda_E. \qquad (5)$$

A sufficient condition for a point on a critical line to be an efficient portfolio is that

$$X \geqslant 0, \quad \eta \geqslant 0, \quad \lambda_E > 0. \qquad (6a\text{–}c)$$

A portfolio satisfying (6$a$), (6$b$) and $\lambda_E = 0$ may or may not be efficient. We return to this point below.

The critical line algorithm computes a complete, non-redundant set of efficient portfolios. The computation is simplest when the problem is non-degenerate (still to be defined) and feasible $E$ is bounded above. Part of the non-degeneracy assumptions (to be relaxed below) are that ($a$) there is a unique feasible portfolio which maximizes $E$, and ($b$) the problem of maximizing $E$ subject to (1$c$) and (1$d$) is non-degenerate in the sense defined in linear programming. In this case, the unique optimum solution has exactly $m$ variables – the 'basis variables' – with $X_i > 0$. These basis variables are the first IN set. With $\text{IN}_1$ thus defined, $M_{\text{IN}}$ is non-singular, and $M_{\text{IN}}^{-1}$ is easy to compute. The critical line

$$\begin{pmatrix} X \\ \lambda \end{pmatrix} = \alpha + \beta\lambda_E, \quad \eta = \gamma + \delta\lambda_E, \tag{7$a$, $b$}$$

satisfies (6$a$, $b$) and $\lambda_E \geqslant 0$ for all

$$\lambda_E \in [\lambda_{\text{LOW}}^1, \infty),$$

where $\lambda_{\text{LOW}}$ is the largest $\lambda_E$ below which one of the three conditions becomes false; that is, it is the first (i.e. largest) $\lambda_E$ at which $\eta_i \downarrow 0$ for some OUT $i$ or $\lambda_E \downarrow 0$. With subsequent IN sets we also have the possibility that $X_i \downarrow 0$ for some IN $i$; but with $\text{IN}_1$, i.e. on the first critical line, $X$ is constant and equal to the $E$ maximizing portfolio. Whereas (6$a$, $b$) and $\lambda_E = 0$ do not in general assure an efficient portfolio, if $\lambda_{\text{LOW}}^1 = 0$ then the portfolio with maximum $E$ is also the efficient portfolio with minimum $V$, and is thus the entire complete non-degenerate set of efficient portfolios. We next consider the case of $\lambda_{\text{LOW}}^1 > 0$.

The remaining non-degeneracy assumption is that, at each iteration $k$, a unique $i$ determines $\lambda_{\text{LOW}}^k > 0$. In particular, in the first iteration a unique $\eta_{i*} \downarrow 0$ first. $\text{IN}_2$ is the same as $\text{IN}_1$ except for the addition of $i*$. The new $M_{\text{IN}}$ associated with $\text{IN}_2$ is guaranteed to be non-singular.

The fine print in this guarantee says that the statement would be true if we performed calculations with unlimited precision. This could actually be done by storing the numerator and denominator of all results as unlimited, but always finite, integers; since only rational operations are performed. In practice, floating point arithmetic is used, round-off errors occur, and the aforementioned and other 'guarantees' are no longer certain. As a matter of fact, the computation succeeds, including satisfying conditions which assure optimality, most of the time even for $n \geqslant 1000$. When it fails, some rescaling or a practically equivalent restatement of the problem usually, perhaps always, succeeds. The fact that the algorithm usually works does not diminish the desirability of understanding the round-off error problem much better than we do. From this point on we ignore the round-off error problem and, in effect, assume that computations are performed with unlimited precision.

Given our current non-degeneracy assumptions, $X_{i*}$ will increase as $\lambda_E$ is reduced below $\lambda_{\text{LOW}}^1$. Conditions (6$a$, $b$) and $\lambda_E \geqslant 0$ will hold for

$$\lambda_E \in [\lambda_{\text{LOW}}^2, \lambda_{\text{HI}}^2],$$

where $\lambda_{\text{HI}}^2 = \lambda_{\text{LOW}}^1$, $\lambda_{\text{LOW}}^2 < \lambda_{\text{HI}}^2$ and $\lambda_{\text{LOW}}^2 \geqslant 0$. $\lambda_{\text{LOW}}^2$ is the largest value of $\lambda_E < \lambda_{\text{HI}}^2$ below which one of the three conditions will be violated. If $\lambda_{\text{LOW}}^2 = 0$ then the point on the second critical line at $\lambda_E = 0$ is the efficient portfolio with minimum variance. (When $C$ is singular, there may be other feasible portfolios with the same $V$ but lower $E$.)

In case $\lambda_{\text{LOW}}^2 > 0$, one of our above-stated non-degeneracy assumptions assures us that one and only one $i$ IN will satisfy

$$\alpha_i + \beta_i \lambda_{\text{LOW}}^2 = 0, \tag{8a}$$

or $i$ OUT will satisfy

$$\gamma_i + \delta_i \lambda_{\text{LOW}}^2 = 0. \tag{8b}$$

($X_i = 0$ for $i$ OUT and $\eta_i = 0$ for $i$ IN at all points on the critical line.)

$\text{IN}_3$ differs from $\text{IN}_2$ by the deletion of the $i^*$ that satisfies (8$a$) or the addition of the $i^*$ that satisfies (8$b$). In either case, the new $M_{\text{IN}}$ will be non-singular (even if $C$ is singular). In case (8$a$), $\eta_i$ will increase as $\lambda_E$ is reduced below $\lambda_{\text{LOW}}^2$; in case (8$b$), $X_i$ will increase as $\lambda_E$ decreases.

Our discussion of $\text{IN}_2$, its corresponding critical line and their relation to $\text{IN}_3$ and its critical line, illustrates the general case. The same relations hold for $\text{IN}_k$ and between $\text{IN}_k$ and $\text{IN}_{k+1}$. The computation stops when $\lambda_E = 0$ is reached. This must happen in a finite number of steps, since the same IN set cannot appear twice.

## 4. Degeneracy and other problems

Now we tie up loose ends. First, we note that certain difficulties are handled for us when we use George Dantzig's simplex algorithm for solving the linear programming problem of maximizing $E$ subject to (1$c$) and (1$d$). (See Dantzig (1963) for details concerning the simplex algorithm, cycling in linear programming, etc.) If the model is unfeasible, this is determined by 'phase I' of the simplex calculation, and we are so advised. If the rank of $A$ is less than $m$, phase I provides an equivalent model which is not rank deficient.

Phase II informs us if feasible $E$ is unbounded or if the $E$ maximizing solution is not unique or is degenerate in the sense that some variable in the basis has $X_i = 0$. It also informs us as to which non-basic activities (columns of $A$) have 'zero profitability', i.e. have the partial derivative of the appropriate lagrangian equal to zero.

It has been shown that degenerate linear programming problems can cycle; that is, a sequence of iterations can occur in which variables enter and leave the basis but do not change value, and in which a given basis repeats itself. If the same (non-random) rule is followed to decide which non-basic variable with positive profit is to go into the basis, and which basic variable with zero value is to go out, then the once repeating basis will repeat infinitely often.

The problem of modifying the simplex algorithm so it is guaranteed not to cycle can be solved along the following lines. In principle, for sufficiently small positive $\epsilon$, adding certain powers of $\epsilon$ to each $b_i$ will produce a still feasible linear program which is not degenerate and has almost the same answer as the original linear program. The answer to this perturbed problem approaches the original answer as $\epsilon \downarrow 0$. It is not necessary to actually add these powers of $\epsilon$, since one can calculate the sequence of bases which would occur for any sufficiently small positive $\epsilon$.

In practice, the simplex algorithm never (or perhaps hardly ever) cycles, except in problems especially constructed to show that cycling is possible. In other words, if one ignores the existence of cycling and arbitrarily breaks ties for which variable goes in or out of the basis, then usually no problem is encountered.

It is not known whether the critical line algorithm will cycle if two or more $\eta_i$ and/or $X_i$ go to zero simultaneously and the critical line algorithm chooses one to enter or leave the IN set by an arbitrary rule. To my knowledge, in practice such cycling has never happened. However, in case it ever does, one can build a version of the critical line algorithm that is guaranteed not to cycle, by, in effect, adding suitable powers of $\epsilon$ to the $b_i$ and $\mu_i$. The sequence of IN sets is the same for all sufficiently small positive $\epsilon$. As in linear programming, it is not necessary to actually add these powers of $\epsilon$ to figure out the sequence of IN sets. A solution to the original problem can be determined from the IN sets of the perturbed problem.

A problem with unbounded $E$ can be reduced to one with bounded $E$ by adding the constraint

$$\mu' X \leqslant E_0. \tag{9}$$

The sequence of IN sets is the same for all sufficiently large $E_0$. It is not necessary to actually add constraint (9) to determine this sequence of IN sets. From them, the solution to the original problem can be inferred. The solution includes one unbounded piece (line segment in portfolio space, parabolic segment in $EV$ space) which is efficient for

$$\lambda_E \in [\lambda^1_{\text{LOW}}, \infty).$$

This completes our outline of the solution to the general portfolio problem for all possible inputs.

## References

Dantzig, G. B. 1963 *Linear programming and extensions.* Princeton University Press.

Markowitz, H. M. 1956 *The optimization of a quadratic function subject to linear constraints. Nav. Res. Logistics Q.* vol. III.

Markowitz, H. M. 1987 *Mean-variance analysis in portfolio choice and capital markets.* Oxford: Basil Blackwell.

*Discussion*

R. LACEY (*Derivative Investment Advisers Ltd, U.K.*). How far can transaction costs analysis, described by Professor M. H. A. Davis and Dr P. Wilmott (this Volume), be incorporated into the portfolio selection method optimization module?

H. M. MARKOWITZ. Transaction costs can be incorporated exactly if they are proportional to change in position; see Markowitz (1987). For an approximate solution when costs are linear but not proportional, see Perold (1984).

J. PLYMEN (*Ruislip, U.K.*). Consider the investment scene in the early 1960s when the Markowitz principles were developed. Investment statistics were rudimentary, with long term share indices confined to prices without any dividend record. Computers were too slow and expensive for any elaborate analysis. Mean-variance analysis with its crude one factor input was the only scientific technique available. Although investment inputs were developed using fundamental multifactor betas, this was only a small improvement.

United Kingdom actuaries adopted a different approach. In 1962 they developed the *Financial Times* Actuaries 500 share index. Next they set up equity market models that compare individual share performance with that of the index, obtaining a relative price ranking. (Models by Weaver & Hall, Hempsted and Clarkson have been published in actuarial journals.) With this pricing ability, portfolios are monitored at regular intervals, selling dear shares for cheap ones. This continual programme reduces risk and improves performance.

Mean-variance analysis based on more sophisticated models for performance and using semi-variance rather than mean-variance may have practical value. Mathematicians interested in finance could concentrate on: (i) actuarial market models; (ii) analytical techniques for various forms of derivatives; (iii) mathematical use of semi-variance rather than variance.

M. A. H. Dempster (*University of Essex, U.K.*). The type of index tracking portfolio management policy advocated by Mr Plymen can lead to inefficiencies with respect to any attitude-to-risk criterion, including mean-variance. As pointed out by Hodges (this Volume) the appropriate criterion for portfolio management depends on whose preferences – fund managers or ultimate beneficiaries – it embodies. In any event Professor Markowitz's recent practical portfolio experience with sophisticated mean-semi-variance methods is impressive.

R. G. Tompkins (*Kleinwort Benson Investment Management, U.K.*). In applying a mean-variance framework to emerging markets we find ridiculous results. That is, these markets have extraordinarily high historical returns and extremely low risk. This is counter-intuitive as we know how risky these markets are. We have found data series distributions to be extremely skewed and lepto Kurtic. It would be useful to expand the mean-variance framework in portfolio management to include the third and fourth moments. Perhaps this approach could be applied to the inclusion of contingent claims (such as options) into estimating optional portfolios.

## *Additional references*

Clarkson, R. S. & Plymen, J. 1988 Improving the performance of equity portfolios. *J. Inst. Actuaries* **115**.

Perold, A. 1984 Large-scale portfolio optimization. *Man. Sci.* **30**, 1143–1160.

## Chapter 4

# Rand [II] and CACI

### Comments

The articles in this chapter deal with one or another aspect of SIMSCRIPT. The actual SIMSCRIPT I and SIMSCRIPT II programming manuals are of course not reproduced here. The closest to a definitive description of what SIMSCRIPT means in my mind is in the article SIMSCRIPT. The purpose of the other articles is stated within each.

The SIMSCRIPT article presents a comprehensive view of my conception of SIMSCRIPT II. Leonardo DaVinci's one great regret was that he was never able to complete a giant statue of a horse. This statue took a considerable amount of bronze. The copper in that bronze was of tremendous value. The prince who had promised the copper needed it back to finance a war. My own great regret is that, apparently, I will never see a SIMSCRIPT as visualized in SIMSCRIPT II.

The basic concepts which appeared ultimately in the design of SIMSCRIPT II were conceived while I was at General Electric Manufacturing Services. These concepts were used in a system of routines referred to as GEMS, General Electric Manufacturing Simulator. I left RAND toward the end of the 1950s, attracted by an offer I could not refuse from the GE computer department. The GE computer department proved to be not an interesting place to work. In particular, the management did not understand the role of software or languages in the computer business. To them, computers were just another electrical appliance. When I left the computer department a very talented person who stayed on said, "If I could only make a computer department with all the people that left the computer department, then I would have a real computer department."

After about nine months with the computer department I moved to Manufacturing Services at the invitation of a friend Alan J. Rowe. Alan and I had been partners in the building of models of the metalworking industries for the Process Analysis project. Alan and I had agreed that, in practice, one should use simulation analysis to help the planning, scheduling and dispatching of manufacturing jobshops. (A jobshop is a manufacturing facility which works on individual jobs, one at a time, rather than a flow shop in which work flows down a line.)

I had some previous experience with simulation in the RAND Corporation's first logistics project, LP-1. My conclusion was that a simulator could be very useful, but was very difficult to build. I had a theory that one could make simulation more user friendly by building "reusable subroutines" using the new Fortran subroutine facilities.

Alan had moved to General Electric and had built a large scale simulation model for a GE jobshop. It took him two or three years in which he supervised an excellent programmer who wrote the simulator in assembly language. After I arrived at Manufacturing Services I was offered the opportunity to build a simulator for the Small Transformer Department's jobshop. This was programmed in FORTRAN by Mort Allen of GE. We taught the GEMS philosophy at internal GE courses and got the opportunity to test out the flexibility of our reusable subroutines when we were asked to modify GEMS to accommodate another GE department.

The programming was done by someone in the other department, therefore someone remote from me. It turned out that he was not a very good programmer. Over the phone I kept asking for progress: what had been finished, what were we working on, and so on. I was assured that everything was going well. When nothing worked on schedule Mort and I went up to find out what was going on. It turned out that of all the programming that needed to be done had been written, but the programmer had never thought through how the loops closed. Mort and I put in an intensive effort to make the thing work, and perhaps would have made it work in another month or two, but the manager of manufacturing decided to call the project off. This was surely a correct decision, not because the model was delayed, but because there was no one in his shop with which we had a reliable relationship as we had with the manufacturing engineers of the Small Transformer Department.

Thus the flexible subroutines turned out to be not so flexible after all. But I noticed that what was reusable were basic operations on Entities, Attribute and Sets. In particular, routines which created and destroyed entities, set or read-in attribute values, or linked members of sets into lists were quite usable. This suggested that a language, perhaps like FORTRAN, but including operations on Entities, Attributes and Sets would be of value in building simulations.

I did not stay at GE to build what became SIMSCRIPT, because the Manufacturing Services Department would have considered it proprietary and I wanted the result to be in the public domain. I sought out a "nice RAND-like environment" and ended up returning to RAND.

I had the tremendous good luck to have Bernie Hausner assigned to me as a programmer. Herb Karr, an entrepreneurial sort, called me many months later and asked if I had any work for him. I asked him if he was willing to write a manual. He said he was ready to do anything. (When I was at the GE computer department Herb was at the GE Technical Military Planning Operation (TMPO). When I was at Manufacturing Services, Herb had moved to Planning Research

Corporation (PRC). When I needed a writer for the book, Herb had conveniently (for me) been fired from PRC).

The word "SIMSCRIPT" came from a brainstorming session by Herb and me. We tried various roots that suggested simulation and other roots that represented a written language, and came up with SIMSCRIPT.

While the general concepts of Entities, Attributes, and Sets were suggested by GEMS, the fine details of the SIMSCRIPT language were worked out jointly by Bernie, Herb and me. Once programming had begun and the manual was on its way, the three of us would meet for two or three hours two or three times a week. During this period I began to think about SIMSCRIPT II, especially the features that we did not incorporate into SIMSCRIPT [I] because of implementation difficulties. In particular, I had heard that the Jovial programming language was written in Jovial, and thought that the SIMSCRIPT language could be written in SIMSCRIPT using an Entity, Attribute and Set view of the compilation process. SIMSCRIPT II was eventually written in this manner, bootstrapping from SIMSCRIPT [I].

At some point in time, after SIMSCRIPT [I] and its programming manual were complete, Herb finally persuaded me that we should go into business together. I said, "I have nothing else to do, and it can't be harder than building a compiler." In fact it was harder than building a compiler. At first, to stay off of overhead, I continued to consult at RAND part-time guiding the building of SIMSCRIPT II and the Jobshop Simulation Program Generator, JSPG.

SIMSCRIPT I.5 was a SIMSCRIPT [I] built for IBM's new 360 operating system. It used the SIMSCRIPT II technology to facilitate construction and avoid programming limitations that were imposed by our initial implementation of SIMSCRIPT [I] as a preprocessor into FORTRAN. This became the workhorse of CACI until I was fired in 1968. (Both SIMSCRIPT [I] and SIMSCRIPT II were placed in the public domain, as originally planned when I left GE for RAND.)

Herb and I initially made decisions together. Then we disagreed on the pricing of a product "Qwik Qwery". Then we disagreed on how we should proceed when we disagree. The matter was finally decided when Herb, with 47.5% of the stock and Jim Berkson with 5% of the stock fired me, with 47.5% of the stock, on March 15, 1968 — the Ides of March!

It was conceived that SIMSCRIPT II would be implemented and documented in seven levels, as described in the SIMSCRIPT article. Level Six was to contain the data base storage and retrieval facilities of SIMSCRIPT II. RAND, under Phil Kiviat (who had been recruited to write the SIMSCRIPT II programming manual) and Richard Villanueva, (who replaced Bernie Hausner once Bernie programmed SIMSCRIPT II to the point where SIMSCRIPT II could program

SIMSCRIPT II), decided not to implement Level Six. Kiviat went into business on his own trying to make SIMSCRIPT II into a commercial success for himself. Phil's product was called SIMSCRIPT Plus. When Phil failed commercially Herb and my old outfit, CACI, bought him out and renamed the product SIMSCRIPT II.5.

My first action, after being fired on a Friday, was to call Jack Little at Planning Resource Corporation. Jack was one of the programmers who worked with me when LP-1 was built. I told him that Herb had fired me he said, "Crazy man! Come on over." Subsequently, I consulted for Planning Research Corporation for several months when the problem was posed to me that PRC's own Information Management System (IMS) was obsolete and inflexible. I explained the basic concepts of Entities, Attributes and Sets as a method of storage and retrieval. We did not attempt to implement a SIMSCRIPT including Level Six. Rather I proposed adding Entity, Attribute and Set subroutines to PL/I. I said that I would be out of the office for two months and charge them for one month of consulting time. When I returned I wanted two good programmers for six months to pull out the bugs from my program. In fact it required seven months.

I got my affairs in order and then flew to Hawaii. Walking in one direction in the Honolulu Airport I met Bernie Hausner walking in the opposite direction. Bernie said, "I understand Herb fired you." I said, "How did you know that?" Bernie said that Jack Little told him.

After a couple of weeks I left the Hawaiian Islands and flew south to Tahiti. I found Tahiti unattractive because the entire coastline was private property. I then flew to Bora Bora and found it wonderful. (I gather it is still wonderful, but now tremendously expensive.) I asked how to use the snorkel equipment in the room. I was told, "Put it on and go snorkel." The Bora Bora Resort had a buffet breakfast. I would overeat and not be able to go into the water for an hour or so. I would program for that hour and then I snorkeled for an hour. I overate at lunch, programmed for another hour, snorkeled for an hour and then I probably took a nap. I overate for dinner and then it was dark. So I programmed in the evening. I finished the program, which we refer to as $SIMSCRIPT_{PDQ}$, since it was finished pretty quickly.

As I said, it took a month longer than predicted to finish $SIMSCRIPT_{PDQ}$. It was used for many years. My lasting contact at PRC said that $SIMSCRIPT_{PDQ}$ worked, except that if it crashed at the wrong time it was very difficult to bring up again. He said that they had figured out the problem, but could not tell me the solution. By the time I joined IBM Research in 1974 I had figured out what was wrong and what to do about it, even before reading the standard solutions to the crash-proofing problem.

I joined IBM with the intention of adding Level Six to the public domain version of SIMSCRIPT II. I stripped out the Level Five capabilities — namely the simulation capabilities — to avoid any chance that CACI would sue IBM. They would not have had a leg to stand on, but IBM didn't know SIMSCRIPT from a hole-in-the-ground and would have dropped the project rather than to fight the case.

It took me a couple of years to convince IBM to give me two colleagues from the IBM research staff. A friend, Burt Grad, who had been well-placed in IBM but was leaving with no hard feelings on either side, said my problem was that no one in IBM was interested in SIMSCRIPT. We had to choose a new name. We decided that a good name for a language that was based on Entities, Attributes, Sets and Events was EAS-E.

My colleagues, Ashok Malhotra, Don Pazel and I, completed the Level Six additions to the public domain SIMSCRIPT II in about two years. Don programmed the compiler. Ashok wrote the manual and programmed the "Custodian" which took care of simultaneous requests for data. In particular, the Custodian worried about locking, dead-locking and crash-proofing. When EAS-E was finished we presented it to IBM Research. John Gilvey, head of Central Scientific Services, volunteered to have EAS-E used on his jobshop within IBM Research. This built scientific equipment to the specific requests of the Research Department's scientists. EAS-E turned out to be as satisfactory as conceived. John Gilvey and a couple of members of his staff have remained friends with all the EAS-E team members ever since.

Unfortunately, I was not as successful at selling the idea to IBM as we were in implementing it. EAS-E became available shortly after IBM's System R, including SQL. IBM was not about to consider redoing their data base facilities (recently converted from their IMS) to EAS-E. When this became clear, I left IBM at the invitation of Joel Siegel at Baruch College. Ashok and Don stayed on for many years. Don is still with IBM.

SIMSCRIPT is not dead. The reason SIMSCRIPT is not well known is that CACI has decided, with good economic sense from their point of view, to continue to lease a small number of SIMSCRIPT II.5 copies at a very high price to some very large SIMSCRIPT users. (Herb Karr died many years ago.) They recently decided to produce a SIMSCRIPT III. (I consulted for this project.) SIMSCRIPT III will be an Object Oriented Programming (OOP) language. OOP is the product of a simulation language called SIMULA which came out about six years after the publication of the SIMSCRIPT [I] manual, and approximately at the same time of the publication of the SIMSCRIPT II manual. OOP recognizes Entities, called objects, and Attributes, called properties. Object Oriented Programming languages, like C++ have no convenient way of handling sets. If you want to manipulate sets in C++ it is recommended that you use Microsoft's Foundation Classes. These provide three different ways of filing an entity into a set, depending on how the set is stored. All three ways are very awkward to program as compared to SIMSCRIPT's command, "File job into queue (machine-group)," where queue (machine-group) is pronounced, "Queue of machine group." I consulted for the SIMSCRIPT III project.

I told Ana Marjanski, who headed the SIMSCRIPT III project, that SIMSCRIPT already has entities, attributes plus sets. She explained that the clients want object oriented programming and the clients will get object oriented programming. Steve Rice, who consulted for Ana in this matter, surveyed object oriented programming and advised us as to what features we should add to conform with the general consensus of various object oriented programming languages. We also added modularity, because Bernie Hausner, Herb Karr and I never thought SIMSCRIPT would be dealing with million lines of code including five hundred thousand lines of Preamble, the definitional part of SIMSCRIPT II. Steve Bailey programmed SIMSCRIPT III in a timely manner, much to my surprise and delight. One thing that facilitated the programming was a little advice provided by me. Steve Rice said, "We should never cut into the SIMSCRIPT II.5 code. It is much too complicated." It included the original SIMSCRIPT II which Bernie and I had developed, plus layers of additional coding which CACI programmers had added, sometimes with very little insight as to how the compiler worked. I insisted that we should use the facilities which SIMSCRIPT II had provided, including the Level Seven SIMSCRIPT II language writing language which is described in the SIMSCRIPT Encyclopedia article. I reminded them precisely how this language writing language worked, and how to go about using it. They were amazed that I still remembered these matters after almost forty years. Of course I remember how SIMSCRIPT II works. I will not forget that until I forget the name of my first born son who is even older than SIMSCRIPT II. The SIMSCRIPT III implementation is documented in Rice, Marjanski, Markowitz and Bailey (2005).

After Harry Markowitz Company setup shop, for real, on the West Coast in 1993, and before CACI began the implementation of SIMSCRIPT III, I formed a nonprofit corporation called Rational Decision Making Research Institute (RDMRI). This company has as its only function the maintenance of the EAS-E.org website. This contains the history of SIMSCRIPT, references to IBM's EAS-E, and my own implementation of the set manipulation routines required for very large ranked sets. These were borrowed from the literature, see Malhotra, Markowitz and Pazel (1983). The EAS-E website is still in existence.

SIMSCRIPT II and III have SIMSCRIPT II's Level One through Level Five. IBM's EAS-E had SIMSCRIPT II Level Six, but I imagine that the programs we wrote implementing Level Six are long gone from IBM's archives. My routines on EAS-E.org are merely subroutines, like the PL/1 subroutines I wrote for PRC, but with more sophisticated storage and retrieval capabilities. As I said at the opening of this discussion, my one big regret is that I have never seen the seven levels of SIMSCRIPT II all implemented simultaneously using SIMSCRIPT II's language writing language.

Let me emphasize that SIMSCRIPT should not be conceived as a programming language. Rather it should be considered a way of viewing the world which, in fact, has been implemented in several different ways.

I have this one advantage over Leonardo DaVinci, I can hope that someday someone will put the seven levels of SIMSCRIPT together. Possibly someday someone will build DaVinci's horse according to DaVinci's specifications. But DaVinci wanted to do this himself. I will be delighted if anyone ever puts together the seven levels of SIMSCRIPT.

## References

Markowitz, H. M. (1966). *Simulating with Simscript.* Management Science, Vol. 12, No. 10, June, pp. B396–B405.

Ginsberg, A., Markowitz, H. M. and Oldfather, P. (1965). *Programming by Questionnaire.* The Rand Corporation, Memorandum RM-4460-PR, April, pp. 1–42.

Markowitz, H. M. (1979). SIMSCRIPT. In *Encyclopedia of Computer Science and Technology*, Vol. 13, J. Belzer, A.G. Holzman and A. Kent, (eds.). Marcel Dekker, Inc.

Markowitz, H. M. (1981). *Barriers to the Practical Use of Simulation Analysis.* 1981 Winter Simulation Conference Proceedings, T. I Oren, C. M. Delfosse, C. M. Shub, (eds.), Vol. 1, pp. 3–9.

MANAGEMENT SCIENCE
Vol. 12, No. 10, June, 1966
*Printed in U.S.A.*

# SIMULATING WITH SIMSCRIPT*†

HARRY M. MARKOWITZ

*The RAND Corporation, Santa Monica, California*

**The SIMSCRIPT programming system is especially designed to facilitate the writing of simulation programs. Digital simulations generally consist of a numerical description of "status", which is modified at various points in simulated time called "events". SIMSCRIPT simulations consist primarily of a collection of "event routines" written by the user describing how different kinds of events in a particular simulated world affect current status and cause future events. Status is described in terms of various "entities", "attributes", and "sets" as specified by the user.**

Simulation is presently being used as an analysis and management tool in numerous fields such as manufacturing, logistics, economics, transportation, and military operations. Unfortunately, the development of simulation programs, using conventional programming techniques, can be extremely time consuming. The SIMSCRIPT programming system, on the other hand, is especially designed to facilitate the writing of simulation programs.

For the industrial engineer or operations research analyst, the SIMSCRIPT programming language serves as a convenient notation for formulating simulation models For the programmer, it reduces programming time severalfold as compared to simulations written in FORTRAN and permits relatively easy program modification and expansion. If the analyst and programmer are not the same person, SIMSCRIPT greatly simplifies the problem of communication since the model and the computer program are written in a notation readily understood by both.

There are three aspects of SIMSCRIPT which enable it to reduce the programming time required for simulation. These are its world-view of the model to be simulated; its method of communicating to the computer the world to be simulated; and some features which are useful for programming in general, and thus for simulation programming in particular. In this paper, we will concentrate on the first two aspects—the SIMSCRIPT world-view, and its basic approach to simulation programming.

* Received May 1965.

† Any views expressed in this paper are those of the author. They should not be interpreted as reflecting the views of The RAND Corporation or the official opinion or policy of any of its governmental or private research sponsors. Papers are reproduced by The RAND Corporation as a courtesy to members of its staff.

The SIMSCRIPT language [1] described in this paper was developed at The RAND Corporation to advance the "simulation art" generally, and to facilitate the writing of Air Force logistics simulators in particular.

The present paper was extracted from a RAND memorandum. [2]

This paper was presented at the meeting of the 16th Annual Industrial Engineering Institute of the University of California.

## Simscript's World-View

SIMSCRIPT requires that the world to be simulated be structured in terms of the concepts listed in Fig. 1.

STATUS
- **Entities**
  - **Temporary**
  - **Permanent**
- **Attributes**
- **Sets**

EVENTS
- **Exogenous**
- **Endogenous**

**Fig. 1—Basic Concepts in SIMSCRIPT World-View**

As of any moment of time, the simulated world has a status characterized in terms of how many of which type of *entities* exist; what are the current values of their *attributes*; what *sets* the various entities belong to, and who are the members of the sets which they own. For the sake of programming efficiency, we make a distinction between temporary and permanent entities. Temporary entities can be created and destroyed (can appear and disappear) in the course of a simulation run. Permanent entities do not come and go, but stay with a run from start to finish.

We can illustrate the foregoing in terms of a simple job shop example. Within a shop, labor-classes and machine-groups might be treated as permanent entities; jobs as temporary entities. Respective attributes of these would be the number of idle machines in each machine-group, the number of each kind of personnel, and the due date of each job. A set might be a collection of jobs waiting for processing by a machine from a particular machine-group.

Part of the SIMSCRIPT world-view is an *event*—a point in time when status changes. *Exogenous* events are caused from outside of the simulation process. To continue the job shop example, an exogenous event might be the arrival of a new job. *Endogenous* events are caused by prior occurrences inside the simulation, e.g. the completion of processing one stage of a job. Thus, as illustrated in Fig. 2, during the course of the simulation, the exogenous events (vertical arrows)

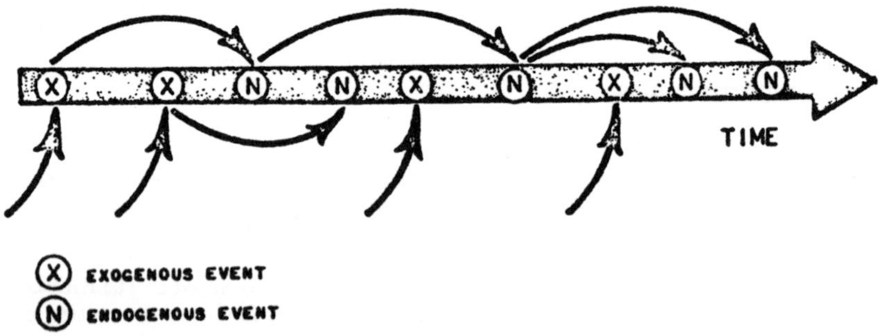


**Fig. 2—Schematic Representation of Exogenous and Endogenous Events through Time**

occur at predetermined times, perhaps causing one or more subsequent endogenous events which, in turn, may cause still other endogenous events.

The SIMSCRIPT supplied timing routine automatically orders all events in time sequence so that the most imminent event occurs next. Simulated time in the model is advanced from event to event rather than at fixed intervals of time.

We will not formally define the basic concepts of entity, attribute, set and event, but will rely mainly on their meaning in common English usage. The precise meaning of these terms, as far as SIMSCRIPT is concerned, is determined by the way they are used. Hence, we cannot understand what we call the SIMSCRIPT viewpoint until we see how a world to be simulated, conceived in these terms, is communicated to the computer.

Let us suppose, then, that an analyst familiar with the precise way we use our basic concepts can, in fact, conceive of a world to be simulated as containing entities of various kinds with their attributes and sets. He further conceives of this world as changing when certain types of events occur. How does he "tell it to SIMSCRIPT?"

## The Simscript Method of Communication

In order to describe *status*, the analyst must fill out a Definition Form as illustrated in Fig. 3. (We will get a closer look at the various panels of this form in subsequent figures.) On this form he must list, by name, each type of entity, each attribute and each type of set distinguished in his simulated world. In addition, the user of SIMSCRIPT must provide a small amount of additional information such as the number of words of computer storage needed to store the values of the attributes of any temporary entity, and also where in the entity record the user would like each of the attributes to be stored.

The first panel of the definition form (Fig. 4) informs SIMSCRIPT of the temporary entities and their attributes. According to the example presented in Fig. 4, this particular simulation contains a type of temporary attribute called a JOB. A four-word record is used to store the current values of its attributes. Thus, if the programmer writes an event routine that says "CREATE JOB," four consecutive words of memory, not otherwise occupied, will be found and subsequently used to store the attributes of the particular job just created. The form also indicates that a temporary attribute called RECT (receipt time) is to be stored in the third word of a JOB record, in the second half of the word.

The second panel of the definition form (Fig. 5) is used to inform SIMSCRIPT about permanent entities and their attributes. Thus according to the Fig. 5 example, the system contains a permanent attribute called MG (short for machine group), there is an attribute called FREE (which represents the number of free machines in the machine group), and there is a random variable called FACTR associated with each machine group.

Figure 6 shows the third panel, which is used for sets. The example has a set called QUE; the "X" in Col. 58 indicates that it is a "ranked" set (rather than a first-in-first-out or last-in-first-out set). Cols. 59 through 63 specify that the

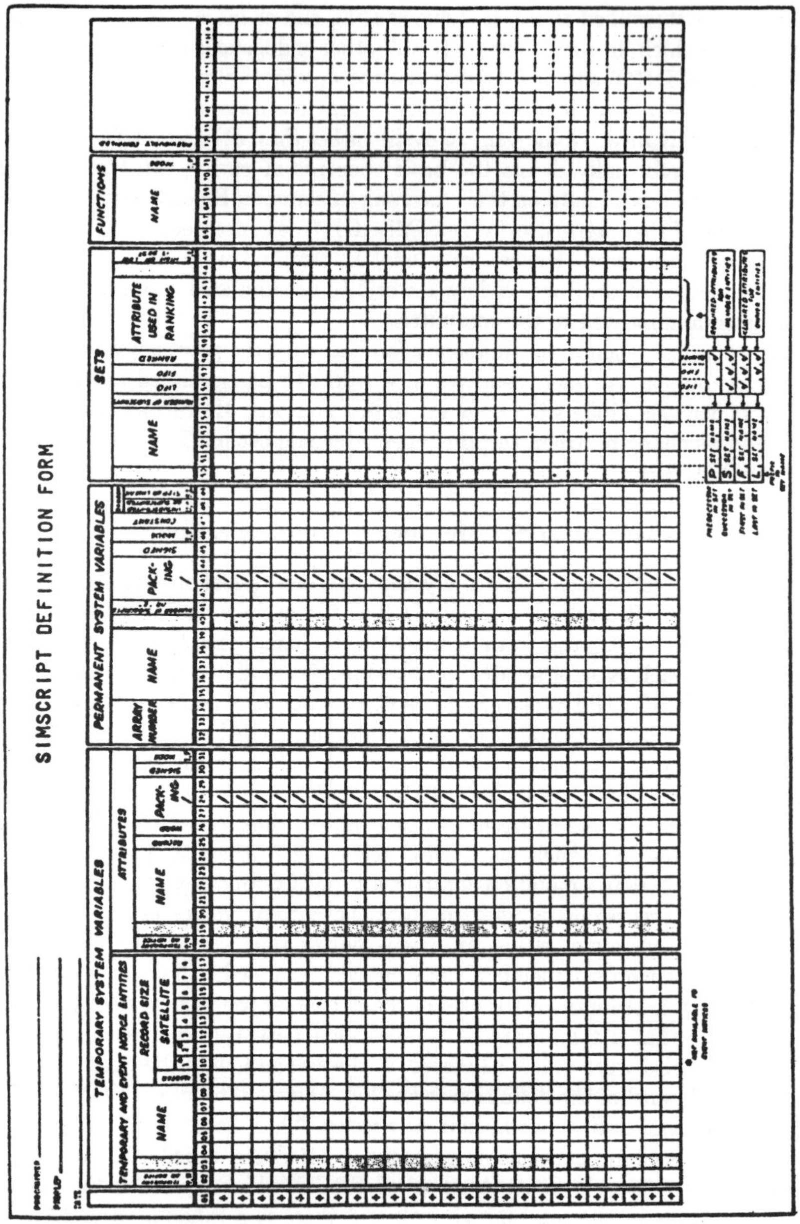


Fig. 3—SIMSCRIPT Definition Form (reduced)

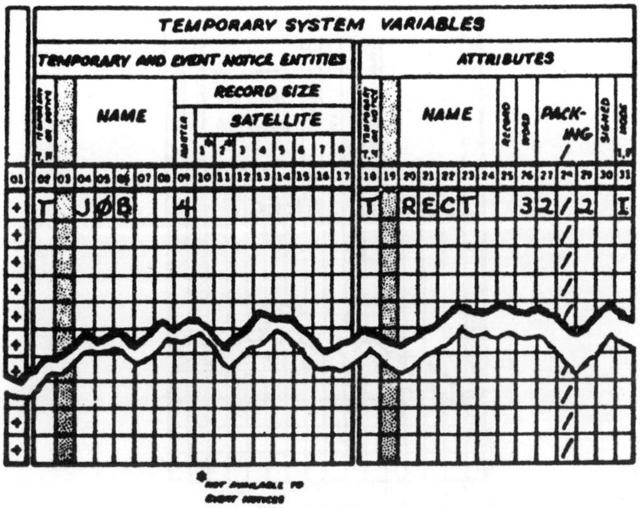

FIG. 4—Example of Temporary Entities and Their Attributes Using First Panel of SIMSCRIPT Definition Form.

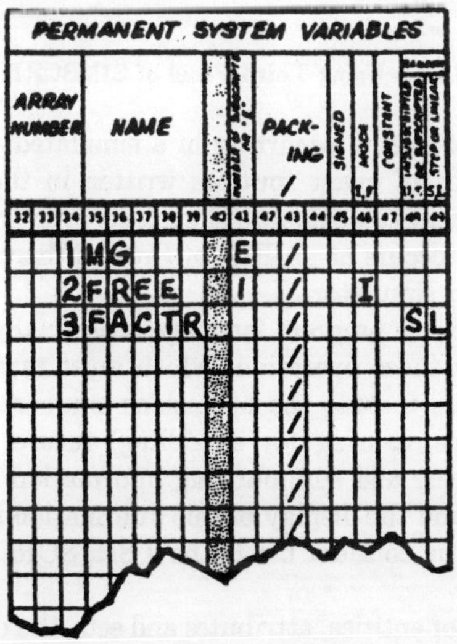

FIG. 5—Example of Permanent Entities and Their Attributes Using Second Panel of SIMSCRIPT Definition Form.

members of this set are to be ranked according to the attribute RECT. The L in column 65 indicates that the *low* values of RECT are at the front end of the set.

Thus the names of the various types of entities, attributes and sets to be distinguished in the specific simulation are indicated on the definition form, together with information needed by SIMSCRIPT to process these properly.

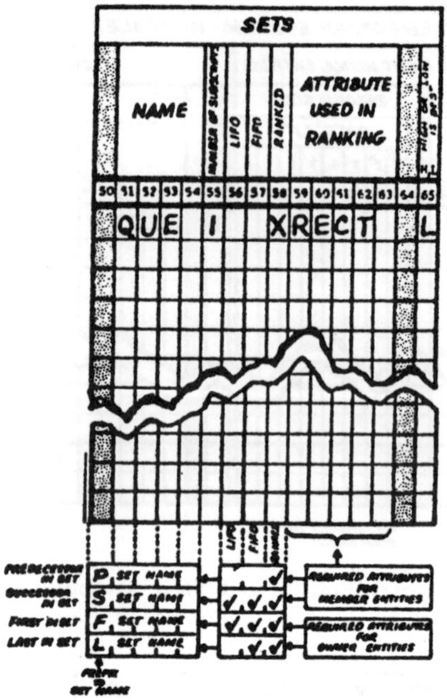


FIG. 6—Example of Sets Using Third Panel of SIMSCRIPT Definition Form

The various types of events occurring in a simulated world are described to SIMSCRIPT by means of event routines written in the SIMSCRIPT source language. Figure 7 presents an example of such an event routine. This particular routine describes what occurs at an End of Process (here abbreviated EPROC) in one simple job shop simulator.

The SIMSCRIPT source program language is especially designed to allow the programmer to specify those operations which must typically be accomplished in event routines. These include the operations enumerated in Fig. 8, namely, changing current status, causing (or cancelling) future events, processing decision rules, accumulating and summarizing information on how well the simulated system is doing, and the display of this information in a form intelligible to the human being. Let us consider briefly how SIMSCRIPT specifies each such action.

Since status consists of entities, attributes and sets, the only ways it can change are if an entity is created or destroyed, a new value is read or computed for some attribute, or some entity gains or loses set membership. Actions of this sort are specified by the commands of CREATE, DESTROY, READ, LET, FILE, REMOVE, illustrated in Fig. 9. Using these commands, the programmer tells SIMSCRIPT how status is to change when a particular kind of event occurs. Similarly, with the aid of the CAUSE and CANCEL statements, the programmer can specify how the occurrence of one event causes some subsequent event or cancels a previously scheduled event that has not yet occurred.

```
  STATEMENT  Continuation                STATEMENT
1 2   NUMBER 5 6 7                                                  72
             ENDOGENOUS EVENT EPROC
             STORE ORDRP(EPROC) IN ORDER
             STORE MGPRC(EPROC) IN MG
             DESTROY EPROC
C            - DISPOSITION OF THE ORDER -
             IF ROUT(ORDER) IS EMPTY, GO TO 10
             CALL ARRVL(ORDER)
             GO TO 20
        10   LET CUMCT = CUMCT + TIME - DATE(ORDER)
             LET NORDR = NORDR + 1.0
             DESTROY ORDER
C            - DISPOSITION OF THE MACHINE -
        20   IF QUE(MG) IS EMPTY, GO TO 30
             REMOVE FIRST ORDER FROM QUE(MG)
             CALL ALLOC(MG, ORDER)
             ACCUMULATE NINQ(MG) INTO CUMQ(MG) SINCE TMQ(MG),
           X     POST NINQ(MG) - 1.0
             RETURN
        30   LET NOAVL(MG) = NOAVL(MG) + 1
             RETURN
             END
```

**Fig. 7—Endogenous Event Routine Describing the End Process for an Order at a Machine Group.**

**a) Change Status**
**b) Cause (or Cancel) Future Events**
**c) Process Decision Rules**
**d) Accumulate and Summarize Information**
**e) Display Results**

**Fig. 8—Types of Operations Performed in Event Routines**

```
CREATE JØB
DESTRØY JØB
LET RECT(JØB) = TIME
READ DUE(JØB)
FILE JØB IN QUE(MG)
REMØVE JØB FRØM QUE(MG)
REMØVE FIRST JØB FRØM QUE(MG)
```

**Fig. 9—Examples of Commands that Change Status**

**As we use it here, the phrase "decision rule" denotes any tests or calculations performed to determine how status will change or what events should be caused. To facilitate such decision calculations, SIMSCRIPT has a complement of**

arithmetic and control statements (some illustrated in Fig. 10) somewhat similar to those contained in other recent programming languages. In addition, SIMSCRIPT has "FIND MIN," "FIND MAX," and "FIND FIRST" commands, illustrated in Fig. 11, which are particularly suited to perform search-type operations frequently found in complex simulations. The case illustrated, for example, instructs the computer by a single "FIND MIN" statement to do the following:

> to search over the set of machine groups (MG) served by a particular labor class (here SRVD is a SET duly defined on the definition form and LC is a labor class determined earlier in the routine). In this search, machine groups with no FREE machines are excluded, as are machine groups with no jobs in queue. Among all machine groups in the set with a free machine and something in queue, the one with lowest (best) priority is chosen. The variable MMG is set equal to the minimizing machine group, the variable MPRI is set equal to the minimum priority value. If, as can happen, there is no machine group that meets the conditions set forth in the FIND statement, the computer is instructed to go to Statement 50. Otherwise, it proceeds with the following command.

```
GØ TØ (10, 20, 25), X(I)
IF (A(B(I))GR(5), GØ TØ 20
DØ TØ 55, FØR EACH JØB ØF QUE(MG), WITH (RECT(JØB)) LS (TIME-LEAD)
IF QUE(I) IS EMPTY, RETURN
```

FIG. 10—Examples of the SIMSCRIPT Versions of Conventional Types of Control Commands.

```
FIND MPRI = MIN ØF PRI(MG), FØR EACH MG ØF
SRVD(LC), WITH(FREE(MG))GR(0), WITH (FQUE(MG))
GR(0), WHERE MMG, IF NØNE, GØ TØ 50
```

FIG. 11—Example Use of FIND MIN Command

Similarly, many fairly complex decision rules can frequently be described with a single FIND MIN, FIND MAX or FIND FIRST command.

Accumulating information over time and summarizing it as of a particular point in simulated time is made easier by the use of ACCUMULATE and COMPUTE STATEMENTS, shown in Fig. 12. The accumulate statement is used for taking an integral under a curve drawn over time; the use of the COMPUTE statement should be apparent from the example.

Finally, the display of information is specified by means of the Report Generator. As illustrated in Fig. 13, the programmer specifies the form, content and row or column repetition desired on a report generator layout sheet. The left-hand side of this sheet is key-punched and then the right-hand side. From the resulting deck, SIMSCRIPT produces a report routine which can be called as required by any event routine or other subroutine of the simulation program.

In sum, the analyst must first conceive of the world to be simulated as having

```
ACC FREE(MG) INTØ CFREE(MG) SINCE
  TFREE(MG), ADD 1.0
COMPUTE MX, SX = MEAN, STD-DEV ØF
  X(I), FØR I = (1)(N)
```

FIG. 12—Examples of Accumulate and Compute Statements

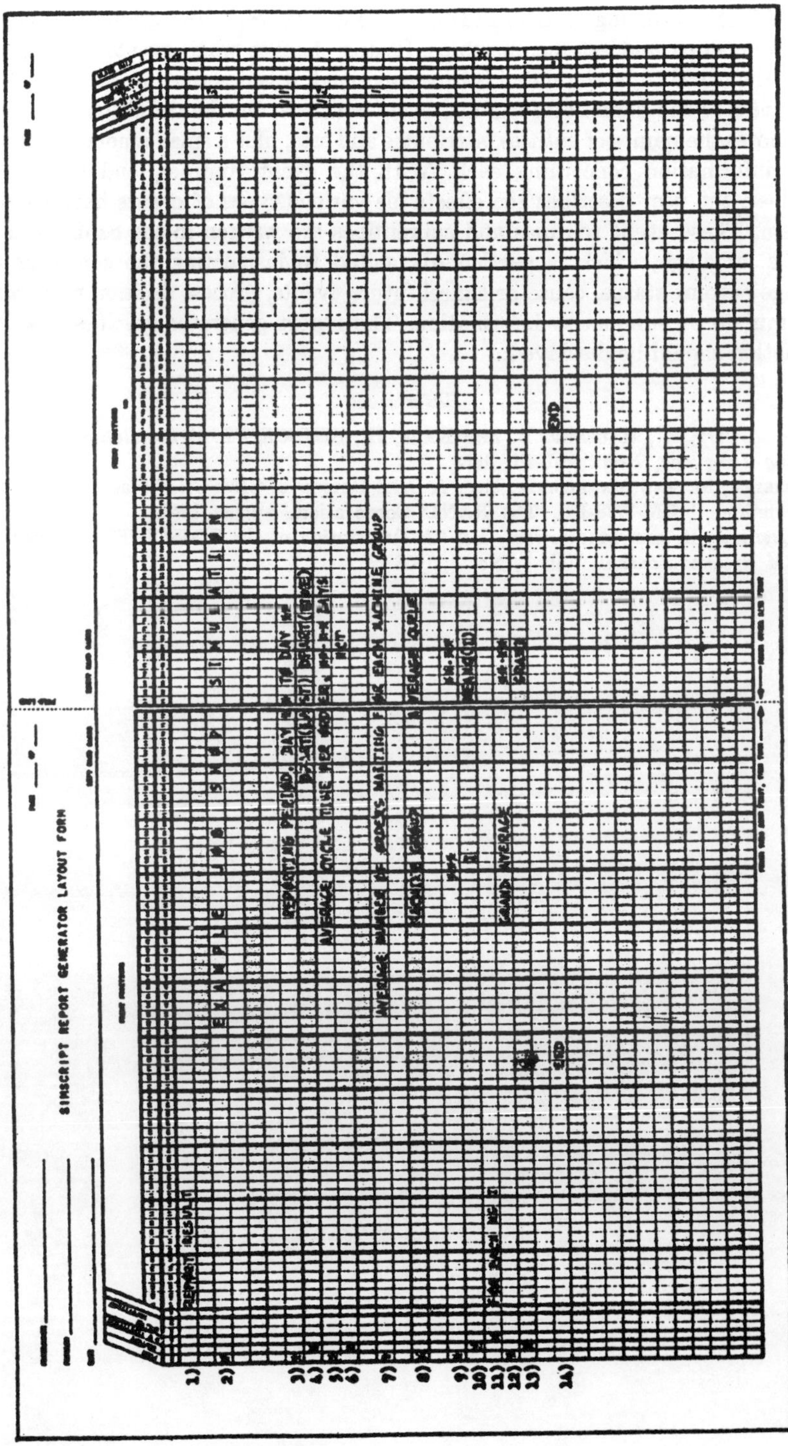


FIG. 13—SIMSCRIPT Report Generator Layout Form (reduced)

a STATUS consisting of ENTITIES of various types, with various attributes, set ownerships and set memberships. Status changes when events take place. Once the world to be simulated is thus conceived, it is described to SIMSCRIPT by means of the definition form and event routines. On the definition form, the user notes the names of entities, attributes and sets, plus a small amount of pertinent information concerning each. With the event routines, and their subroutines, the user describes the effects his various types of events have on the system. These event routines and subroutines are written in the SIMSCRIPT source language, which is particularly suited to instructing the computer to change current status, cause or cancel future events, process decision rules, and accumulate or summarize information. The report generator specifies how information should be displayed.

## References

1. Markowitz, H., Hausner, B. and Karr, H., *SIMSCRIPT: A Simulation Programming Language*, Prentice-Hall, 1963.
2. Geisler, M. and Markowitz, H., "A Brief Review of SIMSCRIPT as a Simulating Technique," RM-3778-PR, The RAND Corporation, August 1963.
3. Dimsdale, B. and Markowitz, H., "A Description of the SIMSCRIPT Language," *IBM Systems Journal*, Vol. 3, No. 1, 1964.

MEMORANDUM
RM-4460-PR
APRIL 1965

# PROGRAMMING BY QUESTIONNAIRE

Allen S. Ginsberg, Harry M. Markowitz and Paula M. Oldfather

This research is sponsored by the United States Air Force under Project RAND—Contract No. AF 49(638)-700 monitored by the Directorate of Development Plans, Deputy Chief of Staff, Research and Development, Hq USAF. Views or conclusions contained in this Memorandum should not be interpreted as representing the official opinion or policy of the United States Air Force.



The RAND Corporation
1700 MAIN ST. • SANTA MONICA • CALIFORNIA • 90406

## PREFACE

RAND has played an active role in the use of digital computers for simulation studies. In particular, the Logistics Department has constructed some large simulations of portions of the Air Force Logistics System. From this work, it has become clear that great amounts of effort and time are required to prepare the necessary computer programs. As an outgrowth of this work, a technique for reducing this effort and program preparation time has been developed. This technique, "Programming by Questionnaire" (or the Program Generation concept), allows a user to obtain a simulation program by filling out an English language questionnaire. This approach has sufficient generality that it promises application to areas of computer programming other than simulation.

This Memorandum describes the questionnaire technique, compares it to existing techniques, and discusses potential applications. The workings of the technique are described in terms of the Job Shop Simulation Program Generator, an example developed to test the feasibility and desirability of the concept.

Programming by questionnaire should be of interest to all those concerned with developing major computer programs. Within the Air Force, particular interest may be found at Headquarters USAF and AFLC because of their interests in simulation, and at ESD because of their general concern with computer usage.

## CONTENTS

PREFACE ........................................................ iii

SUMMARY ........................................................ v

Section

I. INTRODUCTION ........................................................ 1

II. THE MECHANICS OF PROGRAM GENERATION ........................ 3

Using a Program Generator ........................ 3

The Questionnaire ........................ 4

The Statement List ........................ 6

Decision Tables ........................ 8

The Editor ........................ 11

III. OBSERVATIONS AND DISCUSSION ........................ 13

Other Techniques ........................ 13

Extensions of the Program Generation Concept......... 15

Changing the Generator and Modifying Generated Programs........................................16

Variants of the Questionnaire ........................ 18

Difficulties Yet Unsolved ........................ 18

Conclusion ........................ 19

Appendix

A. QUESTIONNAIRE ........................................ 20

B. SAMPLE STATEMENT LIST AND DECISION TABLES ........... 38

## I. INTRODUCTION

The time and effort required to produce a computer program have always been one of the major impediments to the practical application of computers. In the field of simulation, for example, the time required to build a model is critical since the program must be readied, the analysis run, and the conclusions drawn in time to affect a time-dependent decision. Considerable progress has been made in reducing programming time and cost by the development of advanced programming languages such as FORTRAN, COBOL,* SIMSCRIPT,** etc. The main objective of the techniques discussed here is to further reduce the effort required to produce large computer programs within specified areas. The purpose of this Memorandum is twofold: to explain the technique, and to discuss its merits, problems, and future implications.

The Questionnaire approach presents the analyst with a set of options centered around a basic model or technique. By choosing from these options, expressed in English, the user specifies all the information necessary to construct his desired computer program. A "Generator" program, informed of all the chosen options, then constructs the desired program. We characterize this approach as "Programming by Questionnaire," or the Program Generation concept.

The user of a Program Generator specifies the characteristics of the desired program by answering a questionnaire consisting of a set of multiple-choice questions. Submitting these answers to a computer, along with a Program Generator, results in the generation of a program whose logic is described by the options chosen on the Questionnaire. The user need not be aware of the internal parts of a Program Generator -- the Editor, a Statement List, and a set of Decision Tables.

In Sec. II, the workings of the Program Generation technique are explained. Wherever the case for ease of explanation is served, we will refer to a specific example, the Job Shop Simulation Program Generator

---

*COBOL-61 Extended, Report to Conference on Data Systems Language, U. S. Department of Defense, 1962.

**H. M. Markowitz, B. Hausner, H. W. Karr, SIMSCRIPT: A Simulation Programming Language, Prentice-Hall, Englewood Clifs, N. J., April 1963.

(JSSPG), which the authors developed to demonstrate the feasibility of the Program Generator approach and to illustrate its mechanisms. Since the JSSPG is primarily a demonstration vehicle, it does not have some features that are likely to be found in real life shops, such as lot splitting, lap phasing, and alternate routings. The JSSPG will be documented fully in a future Memorandum. The only information presented here regarding the JSSPG is the complete Questionnaire.

The Program Generation concept is not limited to job shops or to simulations, but has potential application to other areas of computer usage. Sec. III will discuss this further. We have delayed our general comments and discussion of the Program Generation concept until Sec. III in order that the reader have an understanding of the workings of the technique.

## II. THE MECHANICS OF PROGRAM GENERATION

### USING A PROGRAM GENERATOR

The user of a Program Generator need concern himself only with the Questionnaire and the input data he specifies for executing the generated program. It is not necessary for him to know anything of the inner workings of the Generator. He need only understand what it is he wishes to model, answer the Questionnaire accordingly, and then, once the program is generated, supply the input data as indicated by the Generator.

For example, a user interested in job shop models specifies which of the statements on the Questionnaire are applicable to the particular model he wants to generate. These statements -- about 140 of them in our JSSPG -- concern the nature of arrivals, routing and process times of jobs, decision rules for job dispatch and resource selection, the existence of shift changes, absenteeism, etc. The answers to these questions, when given to the computer, which contains the Statement List and Decision Tables of the JSSPG and the Editor, result in a job shop simulation computer program. Also generated is a description of the numerical data which the user must supply to the generated program to quantify the characteristics of the shop.

A large number of different job shops can be generated in this manner. While there are not a full $2^{140}$ different models, since not every possible combination of answers is permissible, the number of possible models is at least $2^{30}$, or over one billion models. The work required to develop the JSSPG, however, was comparable to the effort required to build a few large models. We avoided the combinatorial problem by bringing together a number of devices, which, taken with the Questionnaire, form the Program Generator. The devices are:

(a) A "Statement List" which is a set of computer commands and any other statements which may be needed in the construction of any of the many programs.

(b) A set of "Decision Tables" that specify which commands are to be included in the generated program as a function of the answers to the Questionnaire.

(c) An "Editor" that processes the Questionnaire, Statement List, and Decision Tables, thus building the desired program and providing a list of the input data the user must supply in order to use the program.

We will describe each of these components in the following paragraphs.

THE QUESTIONNAIRE

The Questionnaire is the key element of the Program Generator. It completely defines the scope and structure of all models that can be constructed. The Questionnaire for the JSSPG is shown in Appendix A starting on p. 21. It consists of two parts: the questions and an answer sheet. The user responds to the questions by marking the answer sheet, which is keyed to the questions. The answers describe the basic structure of the desired model. At the same time, the user selects the various probability distributions (which describe the form of the statistical behavior of various parts of the model) from another section on the same answer sheet, as directed by instructions on the question sheet. Punched cards prepared directly from this sheet form the input to the computer run which generates the specified computer program.

For ease of understanding, the JSSPG Questionnaire is divided into several sections. Section A, Resource Description, asks the user to specify some of the principal characteristics of the shop itself, such as the shift configuration, the resource availability on each shift, and whether one or two types of resources (i.e., men and/or machines) are used to process the jobs. Section B requires specification of the characteristics of jobs to be processed in the shop, such as frequency of arrival at the shop, routing through the shop, quantity, and process times. Job characteristics required for use in decision rules and analysis are chosen in Section C (e.g., ratio of estimated processing time to actual processing time, due date, value, and cost). Section D presents the options for decision rules such as priority rules for dispatching jobs, and rules for selecting combinations of resources to process jobs. Section E provides a choice of the frequency and types of analysis of results. Section F, which lists the types of probability distributions that may be assigned to certain variables (e.g., arrival frequency, processing time, and quantities), is included for clarification purposes only, since the instructions in Sections A to E have already directed the user to fill in Section F on the answer sheet. In Section G, the user is asked to supply a small amount of numerical data such as the number of different types of resources, and the number of shifts per day.

The listing of probability distributions in Section F illustrates an important feature of the Program Generator. If the Questionnaire does not provide the particular distribution a user wishes, he has the option of obtaining any distribution by writing his own subprogram. For example, even though the Questionnaire does not provide for the quantity of items per lot to be distributed log normally, the user can achieve a log normal quantity distribution by providing his own subprogram.

One important characteristic of the Program Generator concept becomes evident upon reading the Questionnaire; it asks the user to specify only the basic structure of the model he desires. It asks for only a small amount of the numerical data necessary to completely describe the model. The user supplies the remainder of the data when he is ready to execute the program that the Generator constructs. As the data required will vary, depending on which options are selected on the Questionnaire, the Generator specifies exactly what data is needed for execution of any particular program (see "The Statement List," p. 6 for details).

An example of this characteristic of the Questionnaire is the description of shifts. The user can specify, for instance, that he wants every day in his simulation to be the same, each day having more than one shift. While the number of shifts per day are specified on the Questionnaire, their starting times and resource disposition are not, but rather are part of the input data that the user is asked to supply when executing the program. Thus, the user may change the number of resources available on each shift or their position within a day from one run to another by changing the input data without having to return to the Questionnaire.

Another example is the arrival of jobs at the shop. The user chooses only the mechanism for determining the time between job arrivals such as "each type of job has its own probability distribution, each type of job having the <u>same form</u> of distribution" (e.g., exponential, uniform, etc.). If he chooses the option stating that the different types of jobs have different forms of the distribution, at execution time, he must supply data showing the distribution form associated with each type of job. In either case, the parameters of the distributions are supplied as data when executing the program.

The only numerical data the user must supply prior to construction of a program by the Generator are the dimensions of certain "permanent entities" as defined in the SIMSCRIPT language.* In the JSSPG, this amounts to specifying the number of "primary resources" (machine groups, for instance), and if applicable, the number of "secondary resources" (e.g., men), the number of different types of jobs, and other information regarding shifts. In order to change these characteristics of the model, it is not necessary for the user to reconstruct the entire program, but only to change Section G of the Questionnaire where these values are listed. He then obtains only the SIMSCRIPT "Initialization"** portion of the program, rather than an entire program.

## THE STATEMENT LIST

The Statement List is the "menu" from which the generated program is constructed. It consists of all the commands required to build all of the programs that can be described on the Questionnaire. The list also contains any other information the programming system requires, such as definitions of variables and initialization data required by the SIMSCRIPT language. All generated programs contain some subset of the Statement List. A sample routine from the JSSPG is shown starting on p. 39 of Appendix B, accompanied by its corresponding Decision Tables.

As suggested above, the Statement List can be written in any computer language from direct machine language to higher languages such as COBOL and SIMSCRIPT. The effort involved in building a Program Generator is, to a large extent, a function of the length and complexity of the Statement List. Therefore, it is easier to build a Generator using higher languages because of their simplicity, flexibility, and the need for fewer commands to accomplish a given task. The language in the resulting generated program will, of course, be the same one used in the Statement List (i.e., if Fortran is used in the Statement List, a Fortran program will be generated).

---

*Markowitz, Hausner, Karr, SIMSCRIPT, p. 3.

**Ibid., p. 115.

The commands in the Statement List look exactly as they would in any other program, except for their "identification numbers." If a command has no identification number, it is considered part of the preceding command that has an identification number. Thus, whole groups of commands can be called from the Statement List by simply selecting one identification number. A Statement List command may also differ from a normal command since it may be only part of a complete command. For example, when constructing a Statement List in the SIMSCRIPT language, it is sometimes found that a given phrase (such as FIND FIRST) may be controlled by a number of different control phrases (such as FOR EACH JØB), depending on which options the user has chosen on the Questionnaire. Rather than repeat the phrase common to all the options it is convenient to give this portion of the total command its own identification number as well as numbering each of the modifiers. The Editor then combines the appropriate parts into a single command.

In constructing the Statement List, we found it convenient to organize it as we would the most complex of the generated programs. Thus, the Statement List for the JSSPG was divided into the "Exogenous Events," "Endogenous Events,"* and subroutines necessary to write a SIMSCRIPT version of complex job shop simulation. Each of these routines was written separately, an effort being made to include the proper logic and commands to satisfy all feasible combinations of answers to the Questionnaire.

In addition to commands, the Statement List contains other information necessary to run the generated program. In the SIMSCRIPT language this includes the Definition Cards (which specify the variables used in the program) and the Initialization Cards (which specify the array sizes in memory).** Also, the user must know what data he is required to supply in order to use the generated program. All three of these components, the Definition and Initialization Cards and the Input Data Requirements, are lines in the Statement List, just as are the commands, and are selected, based on the user's answers to the Questionnaire, in the same manner as the commands.

---

*Markowitz, Hausner, Karr, SIMSCRIPT, p. 66.

**Ibid., pp. 103, 115.

## DECISION TABLES*

The Decision Tables are the means used by the Editor to decide, based on the answers to the Questionnaire, which lines of the Statement List to include in the generated program. In building the Statement List, the programmer has in his mind the various combinations of commands necessary to satisfy the options chosen on the Questionnaire. The Decision Tables are merely a formal statement of these relationships. They give the Editor a set of rules, in a standardized manner, so it can make the proper choices.

Figures 1a and 1b illustrate the use of Decision Tables in the Program Generator scheme. Figure 1a shows the names usually attached to the four quadrants of a Decision Table. The entries in the condition stub are the numbers given to the questions on the Questionnaire, and the entries in the action stub are the identification numbers given to the lines in the Statement List. On the right-hand side of the vertical double line are the conditions and actions. The condition entries are the answers to the questions listed in the condition stub.

Each column in the conditions section represents a single combination of feasible answers to the questions. For example, in the hypothetical decision table in Fig. 1b, the three questions, Q1, Q2, and Q3, may be answered, in combination, any one of four ways. (There are only four combinations because the nature of the questions precludes the other four in this example.) The first column of three "N's" would apply if the user had not chosen any of the three questions, or, in effect, had answered them all "No." The second column indicates a "Yes" to Q1, a "No" to Q3, and indifference to Q2. The third column illustrates a peculiarity of the Decision Tables as applied in this manner; for the purpose of simplifying the tables, we found it convenient to make the order of the columns meaningful, in the sense that the first column found to match the answers on the Questionnaire, when scanning from left to right in the table, will be the column that applies. Thus, the third column appears to imply indifference to Q3; it actually will be chosen only if all three answers are "Yes" (i.e., Y, Y, Y) since the combination Y, Y, N, will cause the second column to be chosen.

*Burton Grad, "Tabular Form in Decision Logic," Datamation, July 1961.

| CONDITION STUB (Question Nos.) | CONDITIONS (Answers to Questions) |
|---|---|
| ACTION STUB (Statement Nos.) | ACTIONS (Selected Statements) |

Fig. 1a -- Decision Table Terminology

| | | | | | |
|---|---|---|---|---|---|
| Q1 | N | Y | Y | N | |
| Q2 | N | | Y | N | |
| Q3 | N | N | | Y | |
| 16 | | X | X | X | |
| 105 | | X | X | | |
| 114 | | | X | | |
| + 536 | | X | X | X | |

Fig. 1b -- Sample Decision Table

When a column is found to match the answers on the Questionnaire, the actions to be taken are found in the same column in the action portion of the table. For example, if only Q1 is selected on the Questionnaire, then the action taken by the Editor is to include the statements numbered 16, 105, and 536 in the generated program.

Two other features of the use of the Decision Tables are illustrated by Fig. 1b. The "+" in front of the number 536 tells the Editor that this statement is not a regular command but is, instead, a special line of information such as a definition or initialization card (in this case, a definition card), or any other information needed in addition to the regular program commands. The other feature is illustrated by the last column in the table. This column will be reached only if none of the feasible combinations of answers to the questions has been realized, which means the user has made an error in filling out the Questionnaire. The Editor, upon finding both condition and action portions blank, would then write an error message, telling the user where he had erred.

The decision as to how extensive a single Decision Table should be arose a number of times during the construction of the JSSPG. Conceivably each Decision Table could develop, say, one complete routine from the Statement List. For routines of nontrivial length, however, this is a bad practice for two reasons. The complexity in constructing a large table (many entries in the condition stub) goes up much more rapidly than does the size of the table; and the Editor will usually execute its function more rapidly with smaller Decision Tables. This latter feature arises because the execution time will be largely a function of the number of comparisons that must be made in the conditions portion of the table. Thus, the smaller the volume of this portion of the tables, the faster the Editor will operate. Experience has shown that breaking one large table into a number of smaller ones, where convenient, will usually reduce the over-all volume of comparisons, even though some of the question numbers must be repeated in the condition stub portion of the smaller tables. In the JSSPG, we frequently found it economical to let a Decision Table control the selection of only one or two lines of the Statement List. An illustration of a set of Decision Tables used in the JSSPG is shown on pp. 40-42 of Appendix B.

## THE EDITOR

We have been speaking of the Editor's function in general terms. Specifically, the Editor is the computer program that translates the user's responses to the Questionnaire into the desired computer program. The current version of the Editor is written in SIMSCRIPT, though any language could have been used. It treats as input data the three parts of the Generator: answers to the Questionnaire, the Statement List, and the Decision Tables. Even though the content of these three inputs will be different for Program Generators other than the JSSPG, their formats and logic of construction will be the same. Thus, the current version of the Editor can be used for other Program Generators in any area of application, if the language used for the Statement List is SIMSCRIPT, and if the principal components are constructed in the manner outlined herein.

The functions of the Editor are multi-fold:

(1) To translate the answers to the Questionnaire into a computer program.

(2) To supply all control cards and dictionary information necessary to execute the above generated program.

(3) To provide a list of the names, meanings and dimensions of the variables whose values must be supplied to the generated program by the user prior to execution.

(4) To check the answers to the Questionnaire for completeness and consistency.

In addition to a printed listing of the program, the Editor supplies a corresponding deck of cards containing all the required control cards. This deck may be resubmitted, as is, for a SIMSCRIPT compilation. If the user also wishes to execute the program, he need only place the required input data at the back of the deck provided.

Execution time and memory space requirements for the Generator itself are moderate. The generation of relatively large programs has been accomplished in less than three minutes on an IBM 7044 computer. Memory space is conserved by reading in the Statement List and the corresponding Decision Tables in independent blocks. Independent blocks are formed by associating a set of Decision Tables with a portion of the Statement List. The blocks are made independent by assuming that the entries in the action stubs of the Decision Tables call

**only for statement numbers contained in the associated portion of the Statement List.**

## III. OBSERVATIONS AND DISCUSSION

In this section we discuss some of the attributes of the Program Generation concept. Included is a comparison to other methods aimed at the same objective, suggestions for application to other areas, comments on building generators, and on modifying both generators and generated programs. We will also note some of the problems of the technique as well as some of its favorable features.

### OTHER TECHNIQUES

The idea of automatic preparation of computer programs is not new. Other approaches, aimed at reducing programming costs, have been suggested. We will discuss a few of these approaches that have received the greatest attention, pointing out their shortcomings relative to the Program Generator approach.

In one of these techniques, which we shall call the "Modular" approach, the user specifies and builds the desired program by selecting and fitting together a number of pre-programmed or "canned" subroutines. This approach has proven difficult to implement in most situations because of the difficulties in making the various subroutines logically compatible and/or because of the very large library of subroutines necessary to make the method work. If the set of options presented to a user is relatively small, such as in a sort program generator,* the modular approach may prove feasible. Because the Program Generator uses decision tables and compiles programs from individual commands, rather than from larger modules, it alleviates most of the above difficulties, allowing the user a much wider range of options.

The "Generalized Model" approach uses one large program containing all possible options; the options to be executed in a given run are specified by parameter assignment in the input data.** The principal

---

*See, for example, IBM 7090 Generalized Sorting System, IBM Systems Library, File No. 7090-33.

**One illustration: The Job Shop Simulator, Mathematics and Applications Department, Data Systems Division, IBM.

difficulty with this method is the inefficient use of computer time and memory space. In order to decide which options to perform, the program must continually interrogate the parameters the user specified, using the result of the interrogation to decide which parts of the program are to be performed. If the program contains a large number of options, the time spent in deciding which ones to execute will be a significant portion of the total running time. The Program Generator does not suffer from this difficulty since it goes through the process of choosing the options only once, at the time the program is generated.

The inefficient use of computer memory space in the general-type program occurs because all the options and the logic necessary to choose from amongst them must reside in the computer memory during the program's execution. Therefore, memory size places a limit on the number of options that can be made available. Since a Program Generator constructs only the code necessary to execute the chosen options, the generality of the approach is usually not limited by available memory space.

Program generators are not new. An example of a previous generator is the GQS.* As a consequence of how the user specifies his desired model and of the method used to generate the program, the range of options that can be offered in any one such compiler is very limited, as compared to the Questionnaire method.

In summary, the features of Programming by Questionnaire which distinguish it from other methods of program generation are:

(1) An English language questionnaire, requiring no special knowledge of either detailed programming languages or format specifications.

(2) Generation of computer programs which are as efficient as possible (given the limitations of both the language used for the Statement List and the person building the generator) in terms of computer memory space and computer time.

(3) The ability to provide an almost unlimited range of options in any one generator.

---

*George W. Armerding, General Single-Server Queue-Simulation Compiler, Interim Technical Report No. 15, Fundamental Investigations in Methods of Operations Research, M.I.T., July 1960.

## EXTENSIONS OF THE PROGRAM GENERATION CONCEPT

As mentioned previously, the motivation for developing the Generator technique came from our work in the area of simulation. Now that the concept has been developed into something workable, it is clear that the technique is not necessarily limited to simulations. It is not immediately evident, however, about "where" (i.e., which problem areas) or "when" to apply the Generator technique (i.e., as opposed to the use of other techniques or the usual practice of writing the program for the specific application).

An answer to the "where" question lies in the definition of the term "problem area." It is obvious from the description of the workings of the Generator that no one questionnaire can cover all possible computer programs. As in the case of the JSSPG, it must be limited to some fairly specific area. This problem area is characterized by a basic structure or model with many possible variations. In the area of job shop simulations, for example, the basic model we envisioned was a facility with a fixed set of resources, with jobs arriving at the shop at some interval, and with each job having a set of instructions attached to it telling which resources are to process the job at each of its stages. The alternatives superimposed on this basic model were such things as choices of shift configuration, method of job arrivals, job parameters (type, quantity, routing process times, etc.), queue disciplines, and resource selection rules. Even though these and other alternatives may cause various job shop models to look quite different, they all have a basic structure in common, where the commonality might be expressed as a "queueing network," or as a "task-resource system." If it is possible to model an area in this manner, then the Generator concept is likely to be applicable. Examples of some of these areas are given later in this section.

In order to answer the question about "when" to apply the proposed Generator technique as opposed to other alternatives, two aspects to consider are the frequency of usage and the range of options available. If a generator in a particular area, such as the JSSPG, is to be utilized in only one or two applications, clearly it is more economical to write the desired programs as needed. On the other hand, more frequent utilization, say twenty-five or more applications, points toward automatic generation techniques. It is very difficult to locate the cross-over point even if the usage and costs could be accurately projected, since the existence of a generator may have some non-measurable consequences. For example, the user who has a generator at his disposal is likely to try experiments that he otherwise might forego for reasons of lack of time or funds. Thus, the decision about whether or not to supply some form of automatic programming is now, and is likely to remain, a largely subjective evaluation. Likewise, the decision regarding which of the available generation techniques to apply is largely qualitative, since it depends to a large extent on the range of options to be presented to the user. As pointed out earlier, the Questionnaire technique is particularly suitable when the number of options is large. The generalized or modular techniques may be easier to apply when the range of options is small, but by the nature of their mechanisms will almost always result in programs less efficient in use of computer space and time.

From the nature of the Questionnaire and the method used to construct programs, it seems that the Program Generation concept could be profitably applied to many problem areas. In the general field of simulation, there are a number of areas which appear to have both the wide user interest and the wide range of options necessary to justify the construction of a Program Generator. Some of these areas are urban transportation, communication networks, inventory systems, and data processing facilities. Many of the logistics systems of the Air Force could also be modeled in this manner. Outside the area of simulation, one possible application is the area of data reduction. Many organizations store in machine-readable form large volumes of data that they wish to analyze and report in diverse ways. A generator which builds special purpose data retrieval and/or data reduction programs might be feasible and useful. Certain everyday data processing jobs, such as inventory control, payrolls, etc., also appear to be attractive candidates.

CHANGING THE GENERATOR AND MODIFYING GENERATED PROGRAMS

In order to apply the Generator concept to any new problem area, the Questionnaire, the Decision Tables and the Statement List must be written for that specific area. The Editor, as it now exists, need not be rewritten, since it treats the above three elements as data. As long as these three elements of the Generator are constructed as described herein, the Editor will properly perform its function of building a program from the Questionnaire answers.

Constructing a new generator is a fairly difficult task. The builder must know enough about the problem area not only to construct the basic model, but also to anticipate enough of the important options to make the resulting generator useful. At this point in the development of the technique, we cannot recommend the best way to go about building a new generator. In the case of the JSSPG, we first wrote the Questionnaire and from this, constructed the Statement List and Decision Tables. Because the authors had extensive experience in job shop simulations, it is probably not fair to say this is the best procedure, even though it worked very well. It has been pointed out that a reasonable alternative to this procedure is to first build a very large program containing many of the desired options and distill from this the components of a generator.

The usefulness of the Generator concept can also be extended by adding to a given Questionnaire or modifying generated programs. If it becomes clear that some number of desired options have been omitted from a Questionnaire for any particular problem area, it is possible to add this option. The appropriate changes must be made to the three variable elements of the Generator. An alternative to adding to the Generator is to modify the generated program. That is, in cases where the generator can provide a program which is almost, but not quite, what the user wants, he can make the necessary changes to the generated program. Again, it is a fairly difficult task to extend a given generator with either method. However, the orderliness introduced by the Decision Tables and the use of an easily understood language, such as SIMSCRIPT, for the commands in the Statement List, make these kinds of changes entirely feasible.

VARIANTS OF THE QUESTIONNAIRE

Although we have not tried them in practice, we have discussed two variations that can be applied to the Questionnaire. The first concerns the form of the questions. In the JSSPG, the questions are "multiple choice" type, though they could have been "fill-in" type. With fill-in type questions, the Questionnaire would be shorter and simpler, but the Editor would be more complex. The other variant is in the form of presenting the Questionnaire. In the JSSPG, the user must follow various directions which instruct him to answer certain questions, to skip over others, etc., as a function of his answers. If the Questionnaire were stored in a computer, it could be presented to the user on a console, such as a typewriter, question by question, in the proper order. As the user supplied the answers, the computer would choose and present the appropriate questions. Upon completion of the Questionnaire, the computer would compile the desired program, and execute it if directed. This kind of on-line dialogue between user and computer can be a very effective experimental tool. No longer is there a complex programming language or a long time span separating the user from answers to certain kinds of problems.

An example of how the on-line dialogue might be a useful technique is found in the design of computer models of complex systems. The ease and rapidity with which the design of a model might be varied or the decision rules changed might encourage the user to perform sensitivity tests with the model that he might otherwise be reluctant to undertake. Particularly in the case of simulation models, this kind of experimentation can be extremely important and revealing.

DIFFICULTIES YET UNSOLVED

Two principal difficulties with the Program Generator are not yet completely solved. The difficulties are in the area of results presentation and debugging of a generator. Because of the very large number of ways to analyze and summarize raw data and to present the results, it is very difficult to supply a complete and flexible set of options for analysis results. In the JSSPG, the user is supplied with a limited set of choices as to what results will be computed and given no choice as to their form of presentation. We simply made what choices we thought were necessary and useful for analysis of the outcomes. The user who is knowledgeable in computer programming can transcend this difficulty by use of an intermediate output of the JSSPG. As each event takes place during the simulation, a message is written on tape containing pertinent information (i.e., type of event, values of changed variables, time, etc.). A special program can be written to analyze this raw data and present the results as the user desires. This does not solve the problem, however, since the user is now required to write a program manually. More thought on how to enhance the analysis capabilities of the Generator concept is necessary.

The Generator is difficult to debug since it is very difficult to assure that the very large number of different programs that it can generate are all error free. In other words, it is hard to have complete assurance that the generator is fully debugged. Our present attempt to overcome this problem consists of careful construction of the Statement List and Decision Tables, and thorough checking of a large number of generated programs. The testing of these programs should assure that all options on the Questionnaire have been tried, at least independently of each other, along with as many of the combinations of options as practicable. Again, this approach is neither easy nor completely satisfying.

## CONCLUSION

The JSSPG will be published in its entirety in the near future. In the meantime, it is hoped that this Memorandum will serve those who wish to contemplate the proposed technique, or offer suggestions and criticisms, or possibly try construction of their own Generator.

We feel that easy-to-use, flexible, and efficient problem-oriented techniques for the preparation of computer programs are important in broadening the scope of application of digital computers. We hope that the technique described herein will be a contribution to this development.

# Appendix A

# QUESTIONNAIRE

## A. RESOURCE DESCRIPTION

IF ONLY THE INITIAL INPUT VALUES ARE TO BE CHANGED (I.E., ONLY A NEW INITIALIZATION DECK IS DESIRED, NOT AN ENTIRE NEW PROGRAM), COMPLETE SECTION G ONLY. OTHERWISE, CONTINUE HERE.

SHIFT CHANGE OPTIONS - ALL DAYS HAVE 24 HOURS. IF EMPTY SHIFTS ARE DESIRED, THEY MUST BE SPECIFIED.

CHOOSE ONE OF A1, A2, A3, OR A4

EVERY DAY IS THE SAME,...

A1. ONLY ONE SHIFT PER DAY, OF 24-HOUR DURATION. (IF A1 IS CHOSEN, FILL IN G2-G10)

A2. MORE THAN ONE SHIFT PER DAY. (IF A2 IS CHOSEN, FILL IN G2-G14)

NOT EVERY DAY IS THE SAME...

A3. EVERY WEEK IS THE SAME. (IF A3 IS CHOSEN, FILL IN G2-G18)

A4. NOT EVERY WEEK IS THE SAME. (IF A4 IS CHOSEN, FILL IN G2-G22)

TYPE OF RESOURCES

CHOOSE ONE OF A5 OR A6

A5. A JOB IS PROCESSED BY ONE UNIT OF ONE RESOURCE AT EACH OF ITS STAGES. (IF A5 IS CHOSEN, FILL IN G23-G26, THEN GO TO A9)

A6. AT EACH OF ITS STAGES A JOB IS PROCESSED BY ONE UNIT OF A PRIMARY RESOURCE AND ONE UNIT OF ANY OF THE SECONDARY RESOURCES WHICH MAY SERVE THAT PRIMARY RESOURCE. (IF A6 IS CHOSEN, FILL IN G23-G30)

IF A6 WAS CHOSEN AND ONE OF A2, A3, OR A4 WAS CHOSEN, CHOOSE ONE OF A7 OR A8

THE NUMBER OF UNITS OF EACH SECONDARY RESOURCE PER SHIFT IS...

A7. NON-RANDOM

A8. RANDOM

IF A2, A3, OR A4 WAS CHOSEN, CHOOSE ONE OF A9 OR A10

THE NUMBER OF UNITS OF EACH PRIMARY RESOURCE PER SHIFT IS...

A9. NON-RANDOM

A10. RANDOM

# B. JOB CHARACTERISTICS

## TYPES OF JOBS

B1. EACH JOB IS ASSIGNED A JOB TYPE (REQUIRED ONLY IF ONE OR MORE OF THE JOB CHARACTERISTICS LISTED BELOW, SUCH AS FREQUENCY OF JOB ARRIVALS, PROCESS TIMES, ETC., DEPEND ON JOB TYPE). IF B1 IS CHOSEN, FILL IN G31-G34.

## EXOGENOUS INPUTS

CHOOSE EITHER B2 OR ONE OF B9-B11.

B2. JOB ARRIVALS ARE DETERMINED EXOGENOUSLY (I.E. EACH ARRIVAL OCCURANCE IS SPECIFIED INDIVIDUALLY AS INPUT DATA).

IF B2 WAS CHOSEN, CHOOSE ONE OF B3-B8. OTHERWISE, GO TO B9.

B3. THE SEQUENCE OF OPERATIONS PERFORMED ON A JOB AND THE CORRESPONDING PROCESS TIMES ARE SPECIFIED EXOGENOUSLY. IF B3 IS CHOSEN, THE DUE-DATE AND/OR DOLLAR VALUE CAN BE MADE TO DEPEND ON THE TYPE OF THE JOB BY SPECIFYING THEM EXOGENOUSLY (I.E., C6 AND/OR C9 MUST BE CHOSEN). IF B3 IS CHOSEN, GO TO C1.

B4. A TYPE AND A QUANTITY ARE SPECIFIED EXOGENOUSLY FOR EACH JOB. (IF B4 IS CHOSEN, CHOOSE ONE OF B12-B15 FOR THE JOB ROUTINGS, AND THEN GO TO B20)

B5. ONLY THE TYPE IS SPECIFIED EXOGENOUSLY FOR EACH JOB. (IF B5 IS CHOSEN, GO TO B12)

B6. ONLY THE ARRIVAL OF THE JOB IS SPECIFIED EXOGENOUSLY. THE TYPE OF JOB IS DETERMINED BY A RANDOM TABLE LOOK-UP. (IF B6 IS CHOSEN, GO TO B12)

B7. THERE IS NO JOB TYPE (I.E., B1 WAS NOT CHOSEN). THE QUANTITY IS SPECIFIED EXOGENOUSLY FOR EACH JOB. (IF B7 IS CHOSEN, CHOOSE EITHER B12 OR B14, AND THEN GO TO B20)

B8. THERE IS NO JOB TYPE (I.E., B1 WAS NOT CHOSEN). ONLY THE ARRIVAL OF THE JOB IS SPECIFIED EXOGENOUSLY. (IF B8 IS CHOSEN, CHOOSE EITHER B12 OR B14, AND THEN GO TO B16)

## FREQUENCY OF JOB ARRIVALS

IF THERE ARE NO JOB TYPES (I.E., B1 WAS NOT CHOSEN), AND JOB ARRIVALS ARE NOT EXOGENOUS (I.E., B2 WAS NOT CHOSEN), CHOOSE ONE OF F1-F6 FOR THE FORM OF ALL INTER-ARRIVAL TIME DISTRIBUTIONS, THEN GO TO B12. OTHERWISE, CHOOSE ONE OF B9-B11.

B9. THERE IS A DISTRIBUTION OF DELAY TIMES BETWEEN JOB ARRIVALS. WHEN AN ARRIVAL OCCURS, THE TYPE OF JOB IS DETERMINED BY A RANDOM TABLE LOOK-UP. (IF B9 IS CHOSEN, ALSO CHOOSE ONE OF F1-F6 FOR THE FORM OF THE INTER-ARRIVAL DISTRIBUTION)

B10. EACH TYPE OF JOB HAS ITS OWN PROBABILITY DISTRIBUTION OF INTER-ARRIVAL TIMES. THESE DISTRIBUTIONS ALL HAVE THE SAME FORM, BUT THEIR PARAMETERS MAY BE DIFFERENT. (IF B10 IS CHOSEN, ALSO CHOOSE ONE OF F1-F6 FOR THE FORM OF ALL INTER-ARRIVAL TIME DISTRIBUTIONS)

B11. EACH TYPE OF JOB HAS ITS OWN PROBABILITY DISTRIBUTION OF INTER-ARRIVAL TIMES. THESE DISTRIBUTIONS MAY HAVE DIFFERENT FORMS AND MAY HAVE DIFFERENT PARAMETERS.

JOB ROUTINGS

CHOOSE ONE OF B12-B15.

THE SEQUENCE OF OPERATIONS PERFORMED ON A JOB (OF A PARTICULAR TYPE) IS...

B12. FIXED (NON-RANDOM), DOES NOT DEPEND ON TYPE*

B13. FIXED (NON-RANDOM), DEPENDS ON TYPE

B14. RANDOM, DOES NOT DEPEND ON TYPE*

B15. RANDOM, DEPENDS ON TYPE

*IF B12 OR B14 IS CHOSEN, THE PROCESS TIMES CANNOT DEPEND ON THE TYPE OF THE JOB.

PROCESS TIMES

IF EACH NEW ORDER IS ASSIGNED A QUANTITY TO BE USED IN PROCESS TIME CALCULATIONS, GO TO B18. IF EACH ORDER IS NOT ASSIGNED A QUANTITY (I.E. PROCESS TIMES APPLY TO THE LOT AS A WHOLE), CONTINUE HERE. IF THERE ARE NO JOB TYPES (I.E., B1 WAS NOT CHOSEN), OR IF THE ROUTING DOES NOT DEPEND ON THE TYPE OF THE JOB (I.E., B13 OR B15 WAS CHOSEN), CHOOSE ONE OF F7-F14 FOR THE FORM OF ALL PROCESS TIME DISTRIBUTIONS. OTHERWISE, CHOOSE ONE OF B16 OR B17.

B16. EACH TYPE OF JOB HAS ITS OWN PROBABILITY DISTRIBUTION OF PROCESS TIMES PER LOT. THESE DISTRIBUTIONS ALL HAVE THE SAME FORM, BUT THEIR PARAMETERS MAY BE DIFFERENT. (IF B16 IS CHOSEN, ALSO CHOOSE ONE OF F7-F14 FOR THE FORM OF ALL PROCESS TIME DISTRIBUTIONS)

B17. EACH TYPE OF JOB HAS ITS OWN PROBABILITY DISTRIBUTION OF PROCESS TIMES PER LOT. THESE DISTRIBUTIONS MAY HAVE DIFFERENT FORMS AND MAY HAVE DIFFERENT PARAMETERS.

IF EACH NEW ORDER IS NOT ASSIGNED A QUANTITY, GO TO B24. IF THERE ARE NO JOB TYPES (I.E., B1 WAS NOT CHOSEN), CHOOSE ONE OF F15-F19 FOR THE FORM OF ALL QUANTITY DISTRIBUTIONS. OTHERWISE, CHOOSE ONE OF B18 OR B19.

B18. EACH TYPE OF JOB HAS ITS OWN PROBABILITY DISTRIBUTION OF QUANTITIES. THESE DISTRIBUTIONS ALL HAVE THE SAME FORM, BUT THEIR PARAMETERS MAY BE DIFFERENT. (IF B18 IS CHOSEN, ALSO CHOOSE ONE OF F15-F19 FOR THE FORM OF ALL QUANTITY DISTRIBUTIONS)

B19. EACH TYPE OF JOB HAS ITS OWN PROBABILITY DISTRIBUTION OF QUANTITIES. THESE DISTRIBUTIONS MAY HAVE DIFFERENT FORMS AND MAY HAVE DIFFERENT PARAMETERS.

IF THERE ARE NO JOB TYPES (I.E., B1 WAS NOT CHOSEN), OR IF THE ROUTING DOES NOT DEPEND ON THE TYPE OF THE JOB (I.E., B12 OR B14 WAS CHOSEN), CHOOSE ONE OF F20-F27 FOR THE FORM OF ALL SET-UP TIME DISTRIBUTIONS. OTHERWISE, CHOOSE ONE OF B20-B21.

B20. EACH TYPE OF JOB HAS ITS OWN PROBABILITY DISTRIBUTION OF SET-UP TIMES PER LOT. THESE DISTRIBUTIONS ALL HAVE THE SAME FORM, BUT THEIR PARAMETERS MAY BE DIFFERENT. (IF B20 IS CHOSEN, ALSO CHOOSE ONE OF F20-F27 FOR THE FORM OF ALL SET-UP TIME DISTRIBUTIONS)

B21. EACH TYPE OF JOB HAS ITS OWN PROBABILITY DISTRIBUTION OF SET-UP TIMES PER LOT. THESE DISTRIBUTIONS MAY HAVE DIFFERENT FORMS AND MAY HAVE DIFFERENT PARAMETERS.

IF THERE ARE NO JOB TYPES (I.E., B1 WAS NOT CHOSEN), OR IF THE ROUTING DOES NOT DEPEND ON THE TYPE OF THE JOB (I.E., B12 OR B14 WAS CHOSEN), CHOOSE ONE OF F28-F35 FOR THE FORM OF ALL PROCESS TIME DISTRIBUTIONS. OTHERWISE, CHOOSE ONE OF B22 OR B23.

B22. EACH TYPE OF JOB HAS ITS OWN PROBABILITY DISTRIBUTION OF PROCESS TIMES PER UNIT. THESE DISTRIBUTIONS ALL HAVE THE SAME FORM, BUT THEIR PARAMETERS MAY BE DIFFERENT. (IF B22 IS CHOSEN, ALSO CHOOSE ONE OF F28-F35 FOR THE FORM OF ALL PROCESS TIME DISTRIBUTIONS)

B23. EACH TYPE OF JOB HAS ITS OWN PROBABILITY DISTRIBUTION OF PROCESS TIMES PER UNIT. THESE DISTRIBUTIONS MAY HAVE DIFFERENT FORMS AND MAY HAVE DIFFERENT PARAMETERS.

IF THE TOTAL PROCESSING TIME OF A JOB AT ANY STAGE IS MULTIPLIED BY A RANDOM FACTOR WHICH IS ASSOCIATED WITH THE PRIMARY RESOURCE REQUIRED AT THAT STAGE, CHOOSE ONE OF B24 OR B25. OTHERWISE, GO TO B26.

B24. EACH PRIMARY RESOURCE HAS ITS OWN PROBABILITY DISTRIBUTION OF THE FACTOR. THESE DISTRIBUTIONS ALL HAVE THE SAME FORM, BUT THEIR PARAMETERS MAY BE DIFFERENT. (IF B24 IS CHOSEN, ALSO CHOOSE ONE OF F36-F39 FOR THE FORM OF ALL DISTRIBUTIONS OF THE FACTOR)

B25. EACH PRIMARY RESOURCE HAS ITS OWN PROBABILITY DISTRIBUTION OF THE FACTOR. THESE DISTRIBUTIONS MAY HAVE DIFFERENT FORMS AND MAY HAVE DIFFERENT PARAMETERS.

B26. IF JOBS ARE PROCESSED BY A PRIMARY AND A SECONDARY RESOURCE (I.E., A6 WAS CHOSEN), THEN CHOOSING THIS OPTION INDICATES THAT EACH COMBINATION OF PRIMARY AND SECONDARY RESOURCES HAS AN EFFICIENCY FACTOR ASSOCIATED WITH IT, THE TOTAL PROCESSING TIME TO BE MULTIPLIED BY THIS FACTOR.

## C. JOB CHARACTERISTICS REQUIRED FOR DECISION RULES AND ANALYSIS

A NUMBER OF THE JOB SELECTION DECISION RULES LISTED BELOW USE PROCESSING TIME IN SELECTING A JOB. IF THE ESTIMATED PROCESSING TIME USED IN THE DECISION RULES IS DIFFERENT FROM THE ACTUAL PROCESSING TIME AS GENERATED ABOVE, SELECT ONE OF C1-C5. OTHERWISE, GO TO C6.

C1. THE FORM OF THE PROBABILITY DISTRIBUTION, AND ITS PARAMETERS, OF THE RATIO OF ESTIMATED TO ACTUAL PROCESSING TIMES IS THE SAME FOR ALL JOBS AND FOR ALL RESOURCES. (IF C1 IS CHOSEN, ALSO CHOOSE ONE OF F40-F44 FOR THE FORM OF THE DISTRIBUTION)

EACH TYPE OF JOB HAS ITS OWN PROBABILITY DISTRIBUTION OF THE RATIO OF ESTIMATED TO ACTUAL PROCESSING TIMES...

C2. THESE DISTRIBUTIONS ALL HAVE THE SAME FORM, BUT THEIR PARAMETERS MAY BE DIFFERENT. (IF C2 IS CHOSEN, ALSO CHOOSE ONE OF F40-F44 FOR THE FORM OF ALL DISTRIBUTIONS)

C3. THESE DISTRIBUTIONS MAY HAVE DIFFERENT FORMS AND MAY HAVE DIFFERENT PARAMETERS.

EACH TYPE OF PRIMARY RESOURCE HAS ITS OWN PROBABILITY DISTRIBUTION OF THE RATIO OF ESTIMATED TO ACTUAL PROCESSING TIMES...

C4. THESE DISTRIBUTIONS ALL HAVE THE SAME FORM, BUT THEIR PARAMETERS MAY BE DIFFERENT. (IF C4 IS CHOSEN, ALSO CHOOSE ONE OF F40-F44 FOR THE FORM OF ALL DISTRIBUTIONS)

C5. THESE DISTRIBUTIONS MAY HAVE DIFFERENT FORMS AND MAY HAVE DIFFERENT PARAMETERS.

THE 'DUE DATE', BY WHICH TIME A JOB SHOULD BE COMPLETED, IS REQUIRED BY SEVERAL OF THE JOB SELECTION DECISION RULES LISTED BELOW (D5, D9, D10). IT IS ALSO USED IN THE COMPUTATION OF JOB LATENESS STATISTICS. IF JOB STATISTICS ARE NOT DESIRED, AND IF THE JOB SELECTION DECISION RULE IS NOT ONE OF THE AFOREMENTIONED, GO TO C9. OTHERWISE, CONTINUE HERE.

C6. THE DUE-DATE IS READ IN EXOGENOUSLY (ONLY AVAILABLE IF B2 WAS CHOSEN). IF C6 IS CHOSEN, GO TO C9.

THE DUE-DATE IS EQUAL TO THE TIME OF ARRIVAL PLUS A RANDOM INCREMENT. IF THERE ARE NO JOB TYPES OR IF THE INCREMENT DOES NOT DEPEND ON THE JOB TYPE, CHOOSE ONE OF F45-F51 FOR THE PROBABILITY DISTRIBUTION OF THE INCREMENT. OTHERWISE, CHOOSE ONE OF C7 OR C8 (NOT ALLOWED IF B3 WAS CHOSEN)

C7. EACH TYPE OF JOB HAS ITS OWN PROBABILITY DISTRIBUTION OF THE INCREMENT. THESE DISTRIBUTIONS ALL HAVE THE SAME FORM, BUT THEIR PARAMETERS MAY BE DIFFERENT. (IF C7 IS CHOSEN, ALSO CHOOSE ONE OF F45-F51 FOR THE FORM OF ALL DISTRIBUTIONS)

C8. EACH TYPE OF JOB HAS ITS OWN PROBABILITY DISTRIBUTION OF THE INCREMENT. THESE DISTRIBUTIONS MAY HAVE DIFFERENT FORMS AND MAY HAVE DIFFERENT PARAMETERS.

THE DOLLAR VALUE ASSIGNED TO A JOB IS REQUIRED BY ONE OF THE JOB SELECTION DECISION RULES LISTED BELOW (D6). IT IS ALSO USED IN THE COMPUTATION OF SOME OF THE ANALYSIS STATISTICS. IF JOB STATISTICS ARE NOT DESIRED, AND IF THE JOB SELECTION DECISION RULE SELECTED IS NOT ONE OF THE AFOREMENTIONED, GO TO D1. OTHERWISE, CONTINUE HERE.

C9. THE DOLLAR VALUE IS READ IN EXOGENOUSLY (ONLY AVAILABLE IF B2 WAS CHOSEN). IF C9 IS CHOSEN, GO TO D1.

IF THERE ARE NO JOB TYPES (I.E., B1 WAS NOT CHOSEN), OR IF THE DOLLAR VALUE DOES NOT DEPEND ON THE TYPE OF THE JOB, CHOOSE ONE OF F52-F57 FOR THE FORM OF ALL DOLLAR VALUE DISTRIBUTIONS. OTHERWISE, CHOOSE ONE OF C10 OR C11 (NOT ALLOWED IF B3 WAS CHOSEN)

C10. EACH TYPE OF JOB HAS ITS OWN PROBABILITY DISTRIBUTION OF DOLLAR VALUE. THESE DISTRIBUTIONS ALL HAVE THE SAME FORM, BUT THEIR PARAMETERS MAY BE DIFFERENT. (IF C10 IS CHOSEN, ALSO CHOOSE ONE OF F52-F57 FOR THE FORM OF ALL DOLLAR VALUE DISTRIBUTIONS)

C11. EACH TYPE OF JOB HAS ITS OWN PROBABILITY DISTRIBUTION OF DOLLAR VALUE. THESE DISTRIBUTIONS MAY HAVE DIFFERENT FORMS AND MAY HAVE DIFFERENT PARAMETERS.

D. DECISION RULES

SELECT ONE OF D1-D12 FOR THE RULE TO BE USED TO SELECT A JOB FROM THE QUEUE OF A PRIMARY RESOURCE

D1. FIRST COME, FIRST SERVED
D2. SHORTEST PROCESSING TIME AT CURRENT STAGE
D3. LONGEST PROCESSING TIME AT CURRENT STAGE
D4. EARLIEST ARRIVAL AT SHOP
D5. EARLIEST DUE DATE OF JOB COMPLETION
D6. LARGEST VALUE
D7. RANDOM SELECTION FROM QUEUE
D8. A FLOATING POINT FUNCTION SUBPROGRAM, CALLED 'PRITY', WILL BE SUPPLIED BY THE USER. 'PRITY' HAS ONE ARGUMENT, THE JOB. SMALLER VALUES OF 'PRITY' HAVE PRECEDENCE.

D9-D15 PERMITTED ONLY IF FIXED ROUTING (I.E., B12 OR B13 WAS CHOSEN), OR EXOGENOUS ROUTING (I.E., B3 WAS CHOSEN).

SMALLEST SLACK WHERE SLACK = FINAL DUE DATE - CURRENT TIME - 'BACKOFF' WHERE...
D9. 'BACKOFF' = (A)(NO. OF REMAINING OPERATIONS) + (B)(SUM OF REMAINING PROCESSING TIMES)
D10. 'BACKOFF' = A + (B)(PROCESS TIME AT CURRENT OPERATION) + C + (D)(PROCESS TIME AT NEXT OPERATION) + E + (F)(PROCESS TIME AT FOLLOWING OPERATION) +...+ Y + (Z)(PROCESS TIME AT LAST OPERATION, WHERE THE SUM IS OVER ALL REMAINING OPERATIONS AND THE A, B,...Y, Z DEPEND ON THE PRIMARY RESOURCE TO BE USED AT THE OPERATION.

LOOK AHEAD TO THE QUEUE OF THE SUBSEQUENT RESOURCE USED BY THE JOB. AMONG THOSE JOBS WHICH HAVE A SUBSEQUENT OPERATION, THE JOB TO BE PROCESSED IS THE ONE WITH THE SMALLEST...
D11. NUMBER OF JOBS ALREADY IN THE SUBSEQUENT QUEUE AND DUE TO ARRIVE AT THIS QUEUE AT THE COMPLETION OF THEIR CURRENT OPERATION
D12. AS IN D11, BUT WITH TOTAL HOURS OF WORK (I.E., BACKLOG) INSTEAD OF NUMBER OF JOBS

IF A LOOK-AHEAD RULE WAS CHOSEN (I.E., D11 OR D12 WAS CHOSEN), CHOOSE ONE OF D13, D14, OR D15

D13. A JOB WHICH WILL NOT BE COMPLETED AT THE CURRENT OPERATION ALWAYS HAS PREFERENCE OVER A JOB WHICH IS TO BE COMPLETED AT THIS CURRENT OPERATION
D14. A JOB WHICH WILL BE COMPLETED AT THE CURRENT OPERATION ALWAYS HAS PREFERENCE OVER A JOB WHICH IS NOT TO BE COMPLETED AT THIS CURRENT OPERATION
D15. A JOB WHICH WILL NOT BE COMPLETED AT THE CURRENT OPERATION WILL HAVE PREFERENCE OVER A JOB WHICH WILL BE COMPLETED AT THIS CURRENT OPERATION IF AND ONLY IF THE NUMBER IN THE QUEUE (OR BACKLOG, AS APPLICABLE) AT THE SUBSEQUENT OPERATION OF THE JOB NOT TO BE COMPLETED IS LESS THAN A SPECIFIED PARAMETER WHICH DEPENDS ON THE RESOURCE OF THE NEXT OPERATION

IF A LOOK-AHEAD RULE WAS CHOSEN (I.E., D11 OR D12 WAS CHOSEN), CHOOSE ONE OF D1-D7 AS THE RULE OF CHOICE FOR JOBS TO BE COMPLETED AT THE CURRENT OPERATION

SELECTION OF SECONDARY RESOURCE TO USE WITH PRIMARY RESOURCE

IF A JOB IS ALWAYS PROCESSED BY ONE UNIT OF ONE RESOURCE AT EACH OF ITS STAGES (I.E., A5 WAS CHOSEN), GO TO SECTION E. OTHERWISE, CHOOSE ONE OF D16-D18.

AMONG THE SECONDARY RESOURCES THAT CAN SERVE THE PARTICULAR PRIMARY RESOURCE IN QUESTION...

D16. CHOOSE THE FIRST WHICH HAS AN AVAILABLE UNIT. (THE ORDER IN WHICH SECONDARIES ARE EXAMINED FOR A GIVEN PRIMARY RESOURCE IS DETERMINED BY THE USER'S INPUT LIST)

D17. CHOOSE THE AVAILABLE SECONDARY RESOURCE WITH THE LARGEST NUMBER OF AVAILABLE UNITS

D18. CHOOSE THE AVAILABLE SECONDARY RESOURCE WITH THE LARGEST VALUE OF A + (B)(NO. OF AVAILABLE UNITS), WHERE A AND B DEPEND ON THE PARTICULAR PRIMARY/SECONDARY RESOURCE COMBINATION

SELECTION OF PRIMARY RESOURCE TO WHICH A NEWLY AVAILABLE SECONDARY RESOURCE IS TO BE ASSIGNED

CHOOSE ONE OF D19-D23

AMONG THE PRIMARY RESOURCES THAT CAN BE SERVED BY THE PARTICULAR SECONDARY RESOURCE IN QUESTION...

D19. CHOOSE THE FIRST WHICH HAS AN AVAILABLE UNIT. (THE ORDER IN WHICH PRIMARIES ARE EXAMINED FOR A GIVEN SECONDARY RESOURCE IS DETERMINED BY THE USER'S INPUT LIST)

D20. CHOOSE THE AVAILABLE PRIMARY RESOURCE WITH THE LARGEST NUMBER OF JOBS IN QUEUE

D21. CHOOSE THE AVAILABLE PRIMARY RESOURCE WITH THE LARGEST BACKLOG HOURS IN QUEUE*

D22. CHOOSE THE AVAILABLE PRIMARY RESOURCE WHOSE FIRST JOB IN QUEUE HAS THE GREATEST PRIORITY

D23. CHOOSE THE AVAILABLE PRIMARY RESOURCE WITH THE LARGEST VALUE OF A + (B)(NO. OF JOBS IN QUEUE) + (C)(BACKLOG) +/- (D)(PRIORITY OF THE FIRST JOB IN QUEUE), WHERE A, B, C, AND D DEPEND ON THE PARTICULAR PRIMARY/SECONDARY RESOURCE COMBINATION. THE + SIGN IS USED IF THE GREATEST PRIORITY IS ASSOCIATED WITH THE HIGHEST PRIORITY NUMBER AND THE - SIGN IF THE GREATEST PRIORITY IS ASSOCIATED WITH THE LOWEST PRIORITY NUMBER.*

*NOT PERMITTED IF RANDOM OR LOOK-AHEAD JOB SELECTION RULE WAS CHOSEN (I.E., D7, D11, OR D12 WAS CHOSEN)

```
ORDER OF DISPOSING OF PRIMARY AND SECONDARY RESOURCES


   CHOOSE ONE OF D24 OR D25

      WHEN A PRIMARY AND A SECONDARY RESOURCE BECOME AVAILABLE
      SIMULTANEOUSLY AT THE COMPLETION OF A JOB...
       D24. THE PRIMARY RESOURCE IS DISPOSED OF FIRST (AS PER D16-D18
            ABOVE) AND THEN THE SECONDARY IS DISPOSED OF IF STILL
            AVAILABLE (AS PER D19-D23 ABOVE)
       D25. SAME AS IN D24, BUT SECONDARY DISPOSED OF FIRST

   CHOOSE ONE OF D26 OR D27

      RESOURCES ARE ASSIGNED AT THE START OF A SHIFT BY...
       D26. DISPOSING OF THE PRIMARY RESOURCES (PER D16-D18 ABOVE)
       D27. DISPOSING OF THE SECONDARY RESOURCES (PER D19-D23 ABOVE)
```

E. ANALYSIS OF RESULTS

FORM OF OUTPUT

CHOOSE ONE OF E1 OR E2

E1. AN ANALYSIS TAPE WITH EXPLANATIONS OF OUTPUT MESSAGES IS DESIRED. THE ANALYSIS PROGRAM WILL BE WRITTEN BY THE USER. (IF E1 IS CHOSEN, SKIP E3-E10)
E2. THE STANDARD ANALYSIS PROGRAM PRODUCED BY THE GENERATOR IS DESIRED

INTERNALLY GENERATED REPORTS

CHOOSE ONE OF E3 OR E4

INTERIM REPORTS...
E3. CAN BE CALLED FOR EXOGENOUSLY (I.E., EACH REQUEST FOR INTERIM REPORTS IS SPECIFIED INDIVIDUALLY AS INPUT DATA)
E4. OCCUR PERIODICALLY, EACH TYPE OF REPORT HAVING ITS OWN PERIOD

CHOOSE ANY NUMBER OF ALTERNATIVES FROM E5-E10 BELOW TO INDICATE WHICH TYPES OF REPORTS ARE DESIRED ON AN INTERIM AND/OR SUMMARY BASIS.

| TYPE OF REPORT | INTERIM | SUMMARY |
|---|---|---|
| RESOURCE UTILIZATION | E5 | E8 |
| QUEUE STATISTICS | E6 | E9 |
| JOB STATISTICS | E7 | E10 |

THE FOLLOWING TWO SECTIONS, F AND G, ARE INCLUDED FOR CLARIFICATION PURPOSES ONLY. THESE SECTIONS SHOULD HAVE BEEN COMPLETED ON THE ANSWER SHEET BY THIS POINT. HOWEVER, READING THIS PORTION OF THE QUESTIONNAIRE WILL SERVE AS A CHECK ON THE ANSWERS PREVIOUSLY GIVEN TO THESE SECTIONS.

## F. PROBABILITY DISTRIBUTIONS

### JOB ARRIVALS

IF JOB ARRIVALS ARE EXOGENOUS (I.E., B2 WAS CHOSEN), OR IF THE DISTRIBUTIONS MAY HAVE DIFFERENT FORMS (I.E., B11 WAS CHOSEN), GO TO F7. OTHERWISE, CHOOSE ONE OF F1-F6.

THE FORM OF THE PROBABILITY DISTRIBUTION OF INTER-ARRIVAL TIMES IS...

F1. UNIFORM
F2. CONSTANT (NON-RANDOM)
F3. RANDOM TABLE LOOK-UP
F4. EXPONENTIAL
F5. LOG NORMAL
F6. A FUNCTION SUBPROGRAM, CALLED 'TSALE', WILL BE SUPPLIED BY THE USER. IF THERE ARE NO JOB TYPES (I.E., B1 WAS NOT CHOSEN), 'TSALE' HAS NO ARGUMENT. IF B9 OR B10 WAS CHOSEN, 'TSALE' HAS ONE ARGUMENT (I.E., THE TYPE OF JOB).

### PROCESS TIMES PER LOT

IF EACH NEW ORDER IS ASSIGNED A QUANTITY TO BE USED IN PROCESS TIME CALCULATIONS, OR IF THE DISTRIBUTIONS OF PROCESS TIME PER LOT MAY HAVE DIFFERENT FORMS (I.E., B17 WAS CHOSEN), GO TO F15. OTHERWISE, CHOOSE ONE OF F7-F14.

THE FORM OF THE PROBABILITY DISTRIBUTION OF PROCESS TIMES PER LOT IS...

F7. UNIFORM
F8. CONSTANT (NON-RANDOM)
F9. RANDOM TABLE LOOK-UP
F10. EXPONENTIAL
F11. LOG NORMAL
F12. WEIBULL
F13. ERLANG
F14. A FUNCTION SUBPROGRAM, CALLED 'PTIME', WILL BE SUPPLIED BY THE USER. 'PTIME' HAS ONE ARGUMENT (I.E., THE JOB).

IF ONE OF F7-F14 WAS CHOSEN, GO TO F36.

QUANTITIES

IF EACH NEW ORDER IS NOT ASSIGNED A QUANTITY, GO TO F36. IF THE DISTRIBUTIONS OF QUANTITIES MAY HAVE DIFFERENT FORMS (I.E., B19 WAS CHOSEN), GO TO F20. OTHERWISE, CHOOSE ONE OF F15-F19.

THE FORM OF THE PROBABILITY DISTRIBUTION OF QUANTITIES IS...
F15. UNIFORM
F16. CONSTANT
F17. RANDOM TABLE LOOK-UP
F18. POISSON
F19. A FUNCTION SUBPROGRAM, CALLED 'PQTY', WILL BE SUPPLIED BY THE USER. IF THERE ARE NO JOB TYPES (I.E., B1 WAS NOT CHOSEN), 'PQTY' HAS NO ARGUMENTS. OTHERWISE, 'PQTY' HAS ONE ARGUMENT (I.E., THE TYPE OF THE JOB).

SET-UP TIMES PER LOT

IF THE DISTRIBUTIONS OF SET-UP TIME PER LOT MAY HAVE DIFFERENT FORMS (I.E., B21 WAS CHOSEN), GO TO F28. OTHERWISE, CHOOSE ONE OF F20-F27.

THE FORM OF THE PROBABILITY DISTRIBUTION OF SET-UP TIMES PER LOT IS...
F20. UNIFORM
F21. CONSTANT (NON-RANDOM)
F22. RANDOM TABLE LOOK-UP
F23. EXPONENTIAL
F24. LOG NORMAL
F25. WEIBULL
F26. ERLANG
F27. A FUNCTION SUBPROGRAM, CALLED 'STIME', WILL BE SUPPLIED BY THE USER. 'STIME' HAS ONE ARGUMENT (I.E., THE JOB).

PROCESS TIMES PER UNIT

IF THE DISTRIBUTIONS OF PROCESS TIME PER UNIT MAY HAVE DIFFERENT FORMS (I.E., B23 WAS CHOSEN), GO TO F36. OTHERWISE, CHOOSE ONE OF F28-F35.

THE FORM OF THE PROBABILITY DISTRIBUTION OF PROCESS TIMES PER UNIT IS...
F28. UNIFORM
F29. CONSTANT (NON-RANDOM)
F30. RANDOM TABLE LOOK-UP
F31. EXPONENTIAL
F32. LOG NORMAL
F33. WEIBULL
F34. ERLANG
F35. A FUNCTION SUBPROGRAM, CALLED 'UTIME', WILL BE SUPPLIED BY THE USER. 'UTIME' HAS ONE ARGUMENT (I.E., THE JOB).

FACTOR

IF B24 WAS CHOSEN, CHOOSE ONE OF F36-F39. OTHERWISE, GO TO F40.

THE FORM OF THE PROBABILITY DISTRIBUTION OF THE FACTOR IS...
F36. UNIFORM
F37. EXPONENTIAL
F38. NORMAL
F39. A FUNCTION SUBPROGRAM, CALLED 'FACTR', WILL BE SUPPLIED BY THE USER. 'FACTR' HAS ONE ARGUMENT, THE PRIMARY RESOURCE.

ESTIMATED TO ACTUAL PROCESSING TIMES

IF EITHER C1, C2, OR C4 WAS CHOSEN, CHOOSE ONE OF F40-F44. OTHERWISE, GO TO F45.

THE FORM OF THE PROBABILITY DISTRIBUTION OF THE RATIO OF ESTIMATED TO ACTUAL PROCESSING TIMES IS...
F40. UNIFORM
F41. CONSTANT
F42. RANDOM TABLE LOOK-UP
F43. LOG NORMAL
F44. A FUNCTION SUBPROGRAM, CALLED 'EVSA', WILL BE SUPPLIED BY THE USER. IF C1 WAS CHOSEN, 'EVSA' HAS NO ARGUMENTS. IF C2 WAS CHOSEN, IT HAS ONE ARGUMENT (I.E., THE TYPE OF THE JOB). IF C4 WAS CHOSEN, IT HAS ONE ARGUMENT (I.E., THE RESOURCE AT THE CURRENT OPERATION).

DUE-DATE INCREMENT

IF THE 'DUE-DATE' IS NOT REQUIRED, GO TO F52. IF THE DUE-DATE IS EXOGENOUS (I.E., C6 WAS CHOSEN), OR IF THE DISTRIBUTIONS OF THE INCREMENT MAY HAVE DIFFERENT FORMS (I.E., C8 WAS CHOSEN), GO TO F52. OTHERWISE, CHOOSE ONE OF F45-F51.

THE FORM OF THE PROBABILITY DISTRIBUTION OF THE DUE-DATE INCREMENT IS...
F45. UNIFORM
F46. CONSTANT (NON-RANDOM)
F47. RANDOM TABLE LOOK-UP
F48. EXPONENTIAL
F49. LOG NORMAL
F50. NORMAL
F51. A FUNCTION SUBPROGRAM, CALLED 'DINC', WILL BE SUPPLIED BY THE USER. IF THERE ARE NO JOB TYPES (I.E., B1 WAS NOT CHOSEN), OR IF THE INCREMENT DOES NOT DEPEND ON JOB TYPE, 'DINC' HAS NO ARGUMENTS. OTHERWISE, 'DINC' HAS ONE ARGUMENT (I.E., THE TYPE OF THE JOB).

DOLLAR VALUE

IF THE DOLLAR VALUE IS NOT REQUIRED, GO TO SECTION G. IF THE DOLLAR VALUE IS EXOGENOUS (I.E., C9 WAS CHOSEN), OR IF THE DISTRIBUTIONS OF DOLLAR VALUE MAY HAVE DIFFERENT FORMS (I.E., C11 WAS CHOSEN), GO TO SECTION G. OTHERWISE, CHOOSE ONE OF F52-F57.

THE FORM OF THE PROBABILITY DISTRIBUTION OF DOLLAR VALUES IS...

F52. UNIFORM
F53. CONSTANT (NON-RANDOM)
F54. RANDOM TABLE LOOK-UP
F55. EXPONENTIAL
F56. LOG NORMAL
F57. A FUNCTION SUBPROGRAM, CALLED 'VALUE', WILL BE SUPPLIED BY THE USER. IF THERE ARE NO JOB TYPES (I.E., B1 WAS NOT CHOSEN), OR IF THE DOLLAR VALUE DOES NOT DEPEND ON JOB TYPE, 'VALUE' HAS NO ARGUMENTS. OTHERWISE, 'VALUE' HAS ONE ARGUMENT (I.E., THE TYPE OF THE JOB).

## G. INITIAL INPUT VALUES

### GENERAL INFORMATION

G1. CHOOSING THIS OPTION INDICATES THAT ONLY THE INITIAL INPUT VALUES ARE TO BE CHANGED (I.E., ONLY A NEW INITIALIZATION DECK WILL BE PROVIDED, NOT AN ENTIRE NEW PROGRAM). THE USER MUST NOW COMPLETE THE REMAINDER OF SECTION G AND SUBMIT THE NEW 'G' INFORMATION WITH THE PREVIOUS 'A-F' CHOICES.

THE FOLLOWING INFORMATION IS REQUIRED FOR ALL JOB SHOPS...

G2- G5. THE USER'S JOB NUMBER
G6-G10. THE USER'S MAN NUMBER (AN ALPHABETIC CHARACTER FOLLOWED BY A FOUR-DIGIT NUMBER)

### SHIFT CHANGE INFORMATION

IF EVERY DAY IS THE SAME WITH ONLY ONE SHIFT PER DAY (I.E., A1 WAS CHOSEN), GO TO G23. OTHERWISE, CONTINUE HERE.

G11-G14. INSERT ON THE QUESTIONNAIRE ANSWER SHEET THE MAXIMUM NUMBER OF SHIFTS IN ANY DAY

IF EVERY DAY IS THE SAME (I.E., A2 WAS CHOSEN), GO TO G23. OTHERWISE, CONTINUE HERE.

G15-G18. INSERT ON THE QUESTIONNAIRE ANSWER SHEET THE NUMBER OF DIFFERENT TYPES OF DAYS

IF EVERY WEEK IS THE SAME (I.E., A3 WAS CHOSEN), GO TO G23. OTHERWISE, CONTINUE HERE.

G19-G22. INSERT ON THE QUESTIONNAIRE ANSWER SHEET THE NUMBER OF DIFFERENT TYPES OF WEEKS

### RESOURCE INFORMATION

THE FOLLOWING INFORMATION IS REQUIRED FOR ALL JOB SHOPS...

G23-G26. THE MAXIMUM NUMBER OF PRIMARY RESOURCES

IF EACH JOB IS PROCESSED BY ONLY ONE RESOURCE AT EACH STAGE (I.E., A5 WAS CHOSEN), GO TO G9. OTHERWISE, CONTINUE HERE.

G27-G30. INSERT ON THE QUESTIONNAIRE ANSWER SHEET THE MAXIMUM NUMBER OF SECONDARY RESOURCES

JOB TYPES

IF EACH JOB IS ASSIGNED A JOB TYPE (I.E., B1 WAS CHOSEN), THE FOLLOWING INFORMATION IS REQUIRED...

G31-G34. THE NUMBER OF DIFFERENT TYPES OF JOBS

**JSSPG QUESTIONNAIRE**

**ANSWER SHEET**

(NOTE: CC = Card Column)

| A. RESOURCE DESCRIPTION | | | | CC |
|---|---|---|---|---|
| SHIFT CHANGE OPTIONS | NO SHIFTS | | | 1 |
| | EVERY DAY THE SAME | | | 2 |
| | EVERY WEEK THE SAME | | | 3 |
| | GENERAL | | | 4 |
| JOB-RESOURCE RELATIONSHIP | ONE RESOURCE | | | 5 |
| | TWO RESOURCES | | | 6 |
| RESOURCE AVAILABILITY PER SHIFT | SECONDARY RESOURCE | INPUT | | 7 |
| | | RANDOM | | 8 |
| | PRIMARY RESOURCE | INPUT | | 9 |
| | | RANDOM | | 10 |

| B. JOB CHARACTERISTICS | | | CC |
|---|---|---|---|
| THERE ARE JOB TYPES | | | 1 |
| JOB ARRIVALS ARE EXOGENOUS | | | 2 |
| OTHER EXOG INPUTS | ROUTING & PROCESS TIME | | 3 |
| | TYPE & QUANTITY | | 4 |
| | TYPE | | 5 |
| | RANDOM TYPES | | 6 |
| | QUANTITY, NO TYPES | | 7 |
| | ARRIVAL ONLY, NO TYPES | | 8 |
| INTER-ARRIVAL TIMES | SAME FORM, RANDOM TYPES | | 9 |
| | SAME FORM | | 10 |
| | DIFFERENT FORMS | | 11 |
| ENDOG ROUTING | FIXED, NOT BY TYPE | | 12 |
| | FIXED, BY TYPE | | 13 |
| | RANDOM, NOT BY TYPE | | 14 |
| | RANDOM, BY TYPE | | 15 |
| PROCESS TIME/LOT | SAME FORM | | 16 |
| | DIFFERENT FORMS | | 17 |
| QUANTITIES | SAME FORM | | 18 |
| | DIFFERENT FORMS | | 19 |
| SET-UP TIME/LOT | SAME FORM | | 20 |
| | DIFFERENT FORMS | | 21 |
| PROCESS TIME/UNIT | SAME FORM | | 22 |
| | DIFFERENT FORMS | | 23 |
| FACTOR/RESOURCE | SAME FORM | | 24 |
| | DIFFERENT FORMS | | 25 |
| PRIMARY-SECONDARY EFFICIENCY FACTOR | | | 26 |

| C. CHARACTERISTICS FOR DECISION RULES | | | CC |
|---|---|---|---|
| ESTIMATED TO ACTUAL PROCESS TIME | ALWAYS THE SAME | | 1 |
| | SAME FORM/TYPE | | 2 |
| | DIFFERENT FORMS/TYPES | | 3 |
| | SAME FORM/RESOURCES | | 4 |
| | DIFFERENT FORMS/RESOURCES | | 5 |
| DUE-DATE INCREMENT | EXOGENOUS | | 6 |
| | SAME FORM | | 7 |
| | DIFFERENT FORMS | | 8 |
| DOLLAR VALUE | EXOGENOUS | | 9 |
| | SAME FORM | | 10 |
| | DIFFERENT FORMS | | 11 |

| D. DECISION RULES | | | | CC |
|---|---|---|---|---|
| GENERAL PRIORITY RULES | | FIRST COME, FIRST SERVED | | 1 |
| | | SHORTEST PROCESS TIME | | 2 |
| | | LONGEST PROCESS TIME | | 3 |
| | | EARLIEST ARRIVAL | | 4 |
| | | EARLIEST DUE-DATE | | 5 |
| | | LARGEST VALUE | | 6 |
| | | RANDOM | | 7 |
| | | USER'S FUNCTION | | 8 |
| FIXED ROUTING ONLY | SLACK RULES | EQUAL WEIGHTS | | 9 |
| | | UNEQUAL WEIGHTS | | 10 |
| | LOOK-AHEAD | NUMBER OF JOBS | | 11 |
| | | HOURS OF WORK | | 12 |
| | | NOT FINAL STAGE FIRST | | 13 |
| | | FINAL STAGE FIRST | | 14 |
| | | NOT FINAL-FINAL-NOT FINAL | | 15 |
| RESOURCE ASSIGNMENT | SECONDARY FOR PRIMARY | FIRST AVAILABLE | | 16 |
| | | MOST AVAILABLE | | 17 |
| | | WEIGHTED SUM | | 18 |
| | PRIMARY FOR SECONDARY | FIRST AVAILABLE | | 19 |
| | | MOST JOBS | | 20 |
| | | MOST HOURS OF WORK | | 21 |
| | | GREATEST PRIORITY | | 22 |
| | | WEIGHTED SUM | | 23 |
| | DISPOSAL-JOB COMPLETION | PRIMARY FIRST | | 24 |
| | | SECONDARY FIRST | | 25 |
| | DISPOSAL-SHIFT | PRIMARY FIRST | | 26 |
| | | SECONDARY FIRST | | 27 |

| E. ANALYSIS | | | CC |
|---|---|---|---|
| FORM OF OUTPUT | OUTPUT TAPE | | 1 |
| | ANALYSIS REPORTS | | 2 |
| INTERIM REPORTS | EXOGENOUS | | 3 |
| | PERIODIC | | 4 |
| INTERIM CONTENT | RESOURCE UTILIZATION | | 5 |
| | QUEUE STATISTICS | | 6 |
| | JOB STATISTICS | | 7 |
| SUMMARY CONTENT | RESOURCE UTILIZATION | | 8 |
| | QUEUE STATISTICS | | 9 |
| | JOB STATISTICS | | 10 |

F. PROBABILITY DISTRIBUTIONS

| NAME OF CHARACTERISTIC | DISTRIBUTION TYPE | CC | |
|---|---|---|---|
| INTER-ARRIVAL TIMES | UNIFORM | 1 | |
| | CONSTANT | 2 | |
| | RANDOM | 3 | |
| | EXPONENTIAL | 4 | |
| | LOG NORMAL | 5 | |
| | USER'S | 6 | |
| PROCESS TIMES/LOT | UNIFORM | 7 | |
| | CONSTANT | 8 | |
| | RANDOM | 9 | |
| | EXPONENTIAL | 10 | |
| | LOG NORMAL | 11 | |
| | WEIBULL | 12 | |
| | ERLANG | 13 | |
| | USER'S | 14 | |
| QUANTITIES | UNIFORM | 15 | |
| | CONSTANT | 16 | |
| | RANDOM | 17 | |
| | POISSON | 18 | |
| | USER'S | 19 | |
| SET-UP TIMES/LOT | UNIFORM | 20 | |
| | CONSTANT | 21 | |
| | RANDOM | 22 | |
| | EXPONENTIAL | 23 | |
| | LOG NORMAL | 24 | |
| | WEIBULL | 25 | |
| | ERLANG | 26 | |
| | USER'S | 27 | |

| NAME OF CHARACTERISTIC | DISTRIBUTION TYPE | CC | |
|---|---|---|---|
| PROCESS TIMES/UNIT | UNIFORM | 28 | |
| | CONSTANT | 29 | |
| | RANDOM | 30 | |
| | EXPONENTIAL | 31 | |
| | LOG NORMAL | 32 | |
| | WEIBULL | 33 | |
| | ERLANG | 34 | |
| | USER'S | 35 | |
| FACTOR/RESOURCE | UNIFORM | 36 | |
| | EXPONENTIAL | 37 | |
| | NORMAL | 38 | |
| | USER'S | 39 | |
| EST. - ACT. PROCESS TIMES | UNIFORM | 40 | |
| | CONSTANT | 41 | |
| | RANDOM | 42 | |
| | LOG NORMAL | 43 | |
| | USER'S | 44 | |
| DUE-DATE INCREMENT | UNIFORM | 45 | |
| | CONSTANT | 46 | |
| | RANDOM | 47 | |
| | EXPONENTIAL | 48 | |
| | LOG NORMAL | 49 | |
| | NORMAL | 50 | |
| | USER'S | 51 | |
| DOLLAR VALUE | UNIFORM | 52 | |
| | CONSTANT | 53 | |
| | RANDOM | 54 | |
| | EXPONENTIAL | 55 | |
| | LOG NORMAL | 56 | |
| | USER'S | 57 | |

G. INITIAL INPUT VALUES

| | | | MAXIMUM NUMBER OF | | | | | |
|---|---|---|---|---|---|---|---|---|
| INITIAL | JOB NO. | MAN NO. | SHIFTS/DAY | DAY TYPES | WEEK TYPES | PRIMARY RESOURCE | SECONDARY RESOURCE | JOB TYPES |
| 1 | 2 3 4 5 | 6 7 8 9 10 | 11 12 13 14 | 15 16 17 18 | 19 20 21 22 | 23 24 25 26 | 27 28 29 30 | 31 32 33 34 CC |
| | | | | | | | | |

Fig. 2 -- Answer Sheet

## Appendix B

## SAMPLE STATEMENT LIST AND DECISION TABLES

```
*IBFTC SALE                                                          202
      ENDOGENOUS EVENT SALE                                            1
      EXOGENOUS EVENT SALE                                             2
      SAVE                                                             3
      CREATE JOB                                                       4
      LET TYPEJ(JOB) = STYPE(SALE)                                     5
      LET TYPEJ(JOB) = IDIST(PTYPE)                                    6
      CAUSE SALE AT TIME + TSALE(TYPEJ(JOB))                           7
      CAUSE SALE AT TIME + TSALE                                       8
      READ TYPEJ(JOB)                                                  9
      READ TYPEJ(JOB), QTY(JOB)                                       10
      READ QTY(JOB)                                                   11
      READ DDATE(JOB)                                                 18
      READ COST(JOB)                                                  19
     X, DDATE(JOB)                                                   101
     X, COST(JOB)                                                    102
      FORMAT (I4                                                     484
      FORMAT (D3.3                                                   485
      FORMAT (D6.2)                                                  490
     X, D3.3                                                         486
     X, I4                                                           487
     X, D6.2                                                         489
     X)                                                              488
      DIMENSION RS(6), PR(6)                                          14
   10 READ RS(I), PR(I), FOR I = (1)(6)
      FORMAT 6(I4,D4.3)
      DO TO 20, FOR I=(1)(6)
      IF (RS(I)) GE (999), GO TO 30
      CREATE STAGE CALLED S
      LET RES(S) = RS(I)
      LET TIMEA(S) = PR(I)
      FILE S IN RTG(JOB)
   20 LOOP
      GO TO 10
X  30 CONTINUE
      LET QTY(JOB) = PQTY                                             12
      LET QTY(JOB) = PQTY(TYPEJ(JOB))                                 13
      LET DDATE(JOB) = TIME + DINC                                   103
      LET DDATE(JOB) = TIME + DINC(TYPEJ(JOB))                       104
      LET COST(JOB) = VALUE                                          105
      LET COST(JOB) = VALUE(TYPEJ(JOB))                              106
      LET ADATE(JOB) = TIME                                          107
      CALL DISPJ(JOB)                                                 15
      RETURN                                                          27
      END                                                             16
```

```
T       SALE

 B    2 N
 B    4     N
 B    5     N
 B    7     N
 C    6     N
 C    9     N

U

    202 X  X  X
      1 X
      2    X  X
      3       X
      4 X  X  X

E


T       SALE

 B    1 N        Y
 B    2     N
 B    4         N
 B    5         N
 B    9     N

U

      5     X
      6         X

E


T       SALE

 B    2 Y
 B   10     N
 B   11     N

U

      7         X
      8     X

E
```

```
T     SALE

 B   2 N
 B   3   Y Y Y Y                         N N N N
 B   4           Y Y Y Y                 N N N N
 B   5                   Y Y Y Y         N N N N
 B   7                           Y Y Y Y N N N N
 C   6   N Y N Y N Y N Y N Y N Y N Y N Y N Y N Y
 C   9   N N Y Y N N Y Y N N Y Y N N Y Y N N Y Y

U

     9                   X X X X
    10           X X X X
    11                           X X X X
    18     X   X                           X   X
    19       X                               X
   101             X   X   X   X   X   X
   102         X     X X     X X     X X       X
   484           X X X X X X X X X X X X
   485     X   X                           X   X
   487           X X X X
   486             X   X   X   X   X   X
   490       X                               X
   489         X     X X     X X     X X       X
   488     X   X X X X X X X X X X X X X   X   X
    14   X X X X

E


T     SALE

 B  18 N N
 B  19 N N
 F  15 N
 F  16 N
 F  17 N
 F  18 N
 F  19 N

U

    12   X
    13     X

E
```

```
T       SALE

 C   7 N N
 C   8 N N
 F  45 N
 F  46 N
 F  47 N
 F  48 N
 F  49 N
 F  50 N
 F  51 N

U

   103    X
   104        X

E


T       SALE

 C  10 N N
 C  11 N N
 F  52 N
 F  53 N
 F  54 N
 F  55 N
 F  56 N
 F  57 N

U

   105    X
   106        X

E


T       SALE

 D   4 N
 E   5 N
 E   8 N

U

   107    X
    15 X X
    27 X X
    16 X X

E
```

# ENCYCLOPEDIA OF COMPUTER SCIENCE AND TECHNOLOGY

EXECUTIVE EDITORS

*Jack Belzer* *Albert G. Holzman* *Allen Kent*

UNIVERSITY OF PITTSBURGH
PITTSBURGH, PENNSYLVANIA

## VOLUME 13

## *Reliability Theory to USSR, Computing in*

MARCEL DEKKER, INC. • NEW YORK and BASEL

# SIMSCRIPT

## HISTORY

The original SIMSCRIPT language, which we shall refer to here as SIMSCRIPT I to distinguish it from its successors, was developed at The RAND Corporation as a language for programming discrete event simulators. SIMSCRIPT II is a general purpose programming language which extends SIMSCRIPT's entity-attribute-set view to other application areas. SIMSCRIPT I.5 and SIMSCRIPT II.5 are advanced versions of SIMSCRIPT I and SIMSCRIPT II.

In SIMSCRIPT I the "status" of a system to be simulated is described in terms of how many of various types of "entities" exist; for each entity what is the value of its "attributes"; what "sets" does it belong to, and who are the members of the sets it owns. Status changes at points in time called "events." Events occur either "exogenously" (caused from the outside) or "endogenously" (caused by prior occurrences of events within the system).

A "definition form" is used to tell SIMSCRIPT I the names of the entities, attributes, and sets of a world to be simulated. The form also offers options concerning the computer storage of these entities, attributes, and sets.

Events are described in "event routines" programmed in the SIMSCRIPT I source programming language. The language includes commands to "change status" by creating or destroying an entity, reading or assigning an attribute value, filing an entity into a set, or removing an entity from a set. SIMSCRIPT I also includes commands to "cause or cancel event" occurrences; commands to "process decision rules" which determine how to change status and what events to cause (these decision processing commands include FORTRAN-like commands such as IF, GO TO, DO, and additional commands such as FIND and COMPUTE); commands to facilitate the "accumulation of historical performance" of the system over time; and a "report generator" facility. The SIMSCRIPT I translator reads SIMSCRIPT I source statements and, for the most part, writes FORTRAN statements, thereby allowing FORTRAN to compile these into machine language.

The detailed language specification was the result of a joint effort by Bernard Hausner, who also programmed the translator, Herbert W. Karr, who wrote the manual, and myself. SIMSCRIPT I was described in a 1962 RAND manual later published by Prentice-Hall [16]. The SIMSCRIPT I translator became available through SHARE at about the same time.

SIMSCRIPT I.5 [11] was developed by Consolidated Analysis Centers, Inc. (CACI) under contract with IBM. IBM's chief requirement was that the new version run under IBSYS for the 7094. At a language level SIMSCRIPT I.5 is essentially the same as SIMSCRIPT I except for the cleaning up of certain awkwardness in the latter. For example, in SIMSCRIPT I somewhat different rules apply to a DO statement which includes a "for each of set" phrase than apply to other DO statements. The SIMSCRIPT I.5 rules for DO statements are simpler and more general, but still compatible with existing SIMSCRIPT I coding.

Internally the SIMSCRIPT I.5 translator is completely different from that of SIMSCRIPT I. It is built on an entity, attribute, and set view of the translation process, a view which was also used in the building of the SIMSCRIPT II translator. In particular, the source program for the SIMSCRIPT I.5 translator is written for the most part in the SIMSCRIPT I.5 language. The translator (as an object program) reads SIMSCRIPT I.5 and writes assembly language rather than FORTRAN statements.

To use the capabilities which SIMSCRIPT added to FORTRAN—such as its set processing capabilities or its report generator—for a program other than a discrete event simulator, one includes a NONSIMULATION card in one's source program. The language was primarily built for simulation, but the facility for using it for "nonsimulation" was also supplied.

SIMSCRIPT II, on the other hand, was designed primarily as a general purpose programming language with a simulation capability included as one of its major features. No source statement is needed to suppress the simulation capability. Rather, this is placed in the object program if an "events list" is encountered by the translator, and first invoked when a START SIMULATION statement is executed. SIMSCRIPT II was designed with the intention that it be documented in a sequence of "levels." In The SIMSCRIPT II Programming Language [12], Level 1 is presented as a complete but simple programming language that can be quickly learned by someone with no prior programming experience. It includes unsubscripted variables; IF, GO TO, LET, ADD, and SUBTRACT statements; a free form (formatless) READ and a PRINT statement.

Level 2 adds subscripted variables, control phrases and DO statements, subroutines and related features, and other features at roughly the FORTRAN level of complexity without certain FORTRAN rigidities. To this Level 3 adds, for example, more general logical expressions, the WITH, UNLESS, WHILE, and UNTIL control phrases, the FIND and COMPUTE statements, and additional input and output statements.

Level 4 presents the entity, attribute, and set view, and the associated definitional and command capabilities for entities stored in main storage (as distinguished from data base entities).

Level 5 adds simulation capabilities including commands to cause and cancel simulated events, and to accumulate system performance over time. Random number generators for various probability distributions are introduced at this level, even though these can be used in nonsimulation programs.

It was planned to have Level 6 add data base entities, attributes, and sets. These would be defined by what would now be called a data base administrator, their existence noted in a data base dictionary, and individual instances would be manipulated by application programs which create, destroy, file, remove, and assign values to data base entities, attributes, and sets just as they manipulate main storage entities.

It was planned to have Level 7 place at the user's disposal the "language writing language" by which SIMSCRIPT I.5 and II define the syntax of statements, and describe the execution of more complex commands in terms of less complex commands.

Partly because RAND was primarily interested in SIMSCRIPT as a simulation language, it implemented only Levels 1 through 5. Level 6, with its data base entities, has never been implemented as a part of SIMSCRIPT II. However, CACI added data base entities to SIMSCRIPT I.5 for the U.S. Department of Commerce's Economic Development Administration [17]. Planning Research Corporation (PRC) added both main storage and data base entities to PL/I for PRC's own internal MIS.

This PL/I based version, first developed in 1968-1969 and since made more efficient, continues to be used extensively within PRC. The SIMSCRIPT II Level 7 "language writing language" was used in the implementation of SIMSCRIPT II itself, but was never made available to users.

A principal objective in the building of SIMSCRIPT II was to further reduce the amount of programming required to describe a system. For example, once the user had decided on the entities, attributes, and sets of an application in SIMSCRIPT I, he would tell it to SIMSCRIPT by filling out a definition form on which he was also required to specify information concerning the storage of attributes. In SIMSCRIPT II the translator decides these storage options unless the user specifies otherwise. In many large and small ways the extensive use of SIMSCRIPT I, particularly in large simulations at The RAND Corporation, pointed out ways by which SIMSCRIPT programming could be simplified.

A second objective in developing SIMSCRIPT II was to increase the efficiency of the executing program. For example, in SIMSCRIPT I the fetching and storing of attributes was accomplished by calling "get" and "put" routines. The user did not write these calls, but they required execute time. In SIMSCRIPT II most of these calls were replaced by inline assembly code.

The initial design of SIMSCRIPT II was developed during 1963-1964 while the author was a consultant to The RAND Corporation. The specifics of the design were filled in and certain features were added as programming of the translator proceeded at RAND. The first year (approximately) of translator programming was by George Benedict; the next couple of years by Bernard Hausner who developed SIMSCRIPT II to the state where it could be rewritten in SIMSCRIPT II and henceforth be its own source language. Hausner also brought aboard Richard Villanueva whose skill and persistence completed the translator. The design was also filled out in the process of manual writing by Philip Kiviat. The SIMSCRIPT II manual, translator, and final specifications were completed by Kiviat and Villanueva after the present author was no longer associated with RAND.

We must distinguish among three versions of SIMSCRIPT II. The first is the language as described in The RAND Corporation produced manual [12]. As far as Levels 1 through 5 are concerned, with one exception noted below, this is the language which I shall refer to as "SIMSCRIPT II."

A second version is the language as implemented in the RAND-produced SIMSCRIPT II translator, and available through SHARE at a nominal cost. This version has most but not all of the features described in the manual. The omissions are described in a document supplied with the SHARE tape [22]. The most serious omissions are the ACCUMULATE and TALLY statements of Level 5. I shall refer to this as the SHARE version of SIMSCRIPT II. I have used this SHARE version extensively, both for illustrative examples and for one large simulation project, and have found it quite serviceable.

The third version is the SIMSCRIPT II.5 language [13] and translator available commercially through CACI. This version contains almost the complete language as specified in the RAND manual. In particular, the SIMSCRIPT II.5 release 8 implementation for the IBM 360/370 [10] omits only some statements which "do not apply to the S/360/370 implementation," and the text mode which is considered "not basic enough to be a language primitive item." I agree that the implementation of the text mode described in the manual is undesirable. Text mode should be like alpha mode, but with a variable number of characters. (Implementation is easily achieved by using sets or an attribute which points to an array.) I will assume that

SIMSCRIPT II (not otherwise qualified) includes the text mode in this manner. SIMSCRIPT II.5 contains additional features not included in SIMSCRIPT II, such as a structured IF, which is more general than the SIMSCRIPT II's IF but no less Englishlike in its construction, and the often very convenient process view [21] as well as the event view of simulation timing.

Clearly, the concepts of entities, attributes, sets, and events did not originate with SIMSCRIPT. I believe, however, that SIMSCRIPT was the first computer language to permit the user to describe a system in terms of the system's own entities, attributes, sets, and events—allowing the user to select among alternate methods for storing and processing these within the computer. Some of the storage methods used, beyond those of FORTRAN, are like methods previously used by the list processing languages [18, 19]. Some details, such as SIMSCRIPT I's method of obtaining one or more records of size 1, 2, 4, or 8 for piecing together temporary entities (later named the "buddy system"), were apparently original with SIMSCRIPT. These and other implementation details changed between SIMSCRIPT I and SIMSCRIPT II. The entity, attribute, and set view of the world remained constant.

## BASIC CONCEPTS

SIMSCRIPT does not allow every possible use of the words entity, attribute, and set which can arise in conversation of mathematics. The SIMSCRIPT restrictions make for simpler language rules and for a more efficient implementation. At first glance the SIMSCRIPT restrictions would seem to limit the convenience and generality of the language. But it turns out that, with a little practice, systems can be described about as easily subject to SIMSCRIPT's restrictions as they could be allowing more general uses of the basic terms. In the present section we illustrate the notions of entity, attribute, and set; state the restrictions on their use; show how more general uses of the terms can be reformulated into the simpler usage; and discuss how changes in status are specified.

In the next section we show how it all looks when put together into a program.

### Entities

The description of a manufacturing system might distinguish entities such as jobs, workers, and machines. The entities of an air transport system might include flights and airports. The entities of a computer system might include jobs, job steps, data requirements, virtual pages, channels, and IO devices.

Entities may include concrete objects such as machines and jobs, or more abstract objects such as steps in the processing of a job. Webster's Unabridged Dictionary (second edition) defines entity as ". . . that which has reality and distinctness of being either in fact or for thought. . . ." Two competent analysts or system designers may define somewhat different entities in the design or analysis of a given system.

In SIMSCRIPT each individual entity is of one and only one entity type. All instances of that type have the same attributes and own the same sets, although in particular instances any or all attributes may be undefined and any or all sets may be empty. The SIMSCRIPT I or II programmer can circumvent this convention by making use of certain details concerning how entities are referenced and how attributes are stored. This is rarely if ever done, and we shall assume throughout this article that an entity is of one and only one type.

In normal conversation we may speak of an entity as being of more than one entity type. For example, an entity of type "objet d'art" with attributes like artist and price is also a material body with attributes like weight and maximum diameter. For some reason, the fact that one entity can be thought of as being of two or more types rarely suggests itself in actual SIMSCRIPT programming applications; but if and when it does it can be handled in the following manner. Define, in the example, an entity-type material body and an entity-type objet d'art. Whenever an objet d'art comes into existence, create also its material body, which can be noted as an attribute of the objet d'art. Then values can be assigned to, for example, WEIGHT(BODY(OBJET.D.ART)). In this manner the theoretical problem has a solution in theory and, if need be, a solution in fact.

SIMSCRIPT refers to the system as a whole as an entity. It allows the system to have attributes and own sets, but not to belong to sets. If there is only one instance of a set called list, then we say that it is owned by the system as a whole. If A and B are attributes of the system and LIST is a set owned by the system we refer to these in source program statements as A, B, and LIST, rather than as, say, A(SYSTEM), B(SYSTEM), LIST(SYSTEM). The system is conceptually no different than any other entity, since we could achieve the same effect by defining an entity-type called system and creating only one individual of this type.

If man is an entity type and machine is an entity type, then a (man, machine) combination may be treated in SIMSCRIPT as an entity type. An instance, say (man.one, machine.two), is referred to as a "compound entity." Similarly, if C1 refers to one city and C2 refers to another, then (C1, C2) is a compound entity; if s is a security and d is a day then (s, d) is a compound entity. In SIMSCRIPT II any number of entities (E1, E2, ..., En) may appear in a compound entity. A compound entity may have attributes, such as the distance between cities or the closing price of (security, day). A compound entity may own sets, such as the set of items in transit between one city and the next or the transactions executed for a given security on a given day. In principle, a compound entity could also belong to sets, such as the set of (city, city) combinations which are directly linked in some network. This last capability was not incorporated in SIMSCRIPT I or II.

Nothing new in principle is added to SIMSCRIPT by its recognition of compound entities. For example, the same effect can be achieved by defining a new entity type whose attributes include entity 1, entity 2, and what would otherwise be thought of as the attributes of (entity 1, entity 2). This approach is used in fact if SIMSCRIPT's present inability to file compound entities into sets, or if certain restrictions on the storage of attributes of compound entities, are of consequence. For example, we may define a link as having city 1 and city 2 as attributes. We can then file links into sets owned by city 1 and city 2.

We must distinguish between the entities represented (in a simulation or by a data base) as compared to the entities used within the computer to keep track of the former. This is a distinction which German shepherds and some programmers have difficulty making. I have in mind an instance when I pointed out to my shepherd a choice morsel of food on the ground. At first the shepherd sniffed at my finger rather than looking in its direction toward the ground. She failed to distinguish between that which I was pointing with and that which I was pointing to. Similarly, it is sometimes difficult to convince some programmers that an entity in SIMSCRIPT really refers to a machine, or a job, or an airport, or a stocked item rather than, say, several consecutive words of storage, or whatever other mechanism is used within the computer to represent the former. The inability to make this distinction may be likened to an artist who only looks at his paints and never at the landscape to be portrayed.

## Attributes

Examples of attributes include the due date of a job, the destination of a flight, or the number of free machines of a machine group. In high school algebra we learned to read f(x) as f of x. Similarly in SIMSCRIPT notation DUE.DATE(JOB) is read as due date of job and

```
LET VALUE(MORTGAGE)=.85*COST(HOUSE(OWNER(CAR)))
```

would be read as let the value of the mortgage equal .85 times the cost of the house of the owner of the car.

In SIMSCRIPT an attribute may be declared to be a "function." Those not so declared will be referred to here as "stored." If population and square miles are stored attributes of city, then values may be assigned to them in statements such as

```
READ POPULATION(CITY)
```

or

```
LET AREA(CITY)=SQUARE.MILES
```

On the other hand, if POP.PER.SQ.MI is a function attribute, then its value cannot be assigned by a LET or a READ statement, but must be computed by a function subprogram with the name POP.PER.SQ.MI whenever the attribute is referenced. We speak of attributes whose values are assigned by LET or READ statements as stored attributes without committing ourselves to one or another specific storage layout.

As of any instant in time, a specific attribute of a specific individual can have at most one value. "Multivalued" attributes must be handled as sets. For example, SPOUSE(PERSON) can be treated as an attribute (at least in monogamous societies) while CHILDREN(PERSON) must be handled as a set. The assumption that attributes are single-valued is one of the basic conventions of SIMSCRIPT. Other basic conventions are described elsewhere in this section. Presumably a different set of conventions could have been used upon which to build an efficiently implementable description language. It should be noted, however, that small changes in basic conventions may have extensive consequences for syntax and implementation. For example, if we were to decide to let attributes be multivalued, we would have to decide whether

```
LET A(E) = 5
```

means that attribute A of the entity E is to have the value 5 added to its existing (perhaps many) values; or alternatively the one or more current values of A are now to be replaced by the single value 5. Or perhaps we would replace the LET

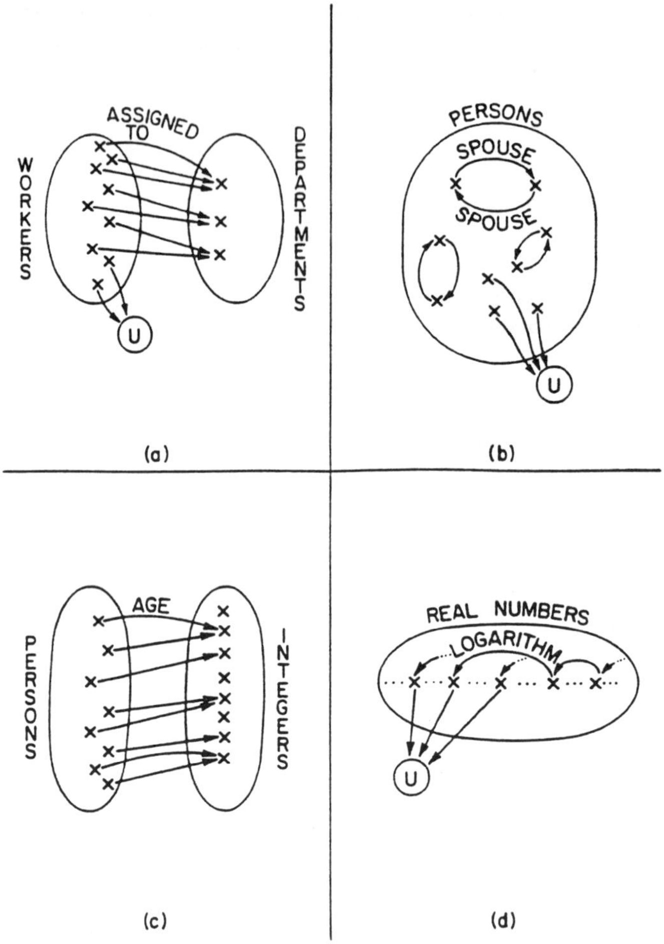


FIG. 1. Attributes.

statement completely by something else. Given the current SIMSCRIPT conventions, this problem does not arise since multivalued attributes are handled as sets.

Figures 1(a) through 1(d) illustrate the nature of attributes. In Fig. 1(a) the X's within the area on the left represent entities of the entity-type worker; those on the right, entities of the entity-type department. The arrow from a worker to a department represents the attribute ASSIGNED. TO. Since an attribute has at most one value, there is at most one arrow emanating from any worker. In the figure we represent a worker with no department (perhaps the general manager and each in his staff) by an arrow to U representing an "undefined" value of the attribute. Several workers may have the same value of the ASSIGNED. TO attribute. Thus several arrows may point to the same department. An attribute therefore may represent a many-one relationship between entities of one type and entities of another type. As a special case only one arrow may point to a given entity, expressing a one-one relationship between individuals of one entity type and some or all individuals of another type. An attribute (like spouse in Fig. 1b) may also point from an entity type to the same entity type.

In Fig. 1(c) an arrow representing the attribute AGE(PERSON) points from a person to an integer. As we discuss later in the section on Entities of Logic and Mathematics, integer may be thought of as an entity type whose individuals include 1, 2, 3, .... More generally, the view taken here is that the value of an attribute is itself always an entity—either an entity of a type defined by the user, such as worker or department, or an entity of a "predefined" type. In some contexts "predefined" entity refers to some entity of logic and mathematics such as an integer, rational number, real number, complex number, array, or integral. Later we show that all of these can be defined in terms of two simple primitive entity types, and in terms of compound entities, attributes, and (finite ordered) sets made up from previously defined types. In some contexts, "predefined" refers more narrowly to those particular types which are recognized by some specific language compiler.

The views concerning attributes expressed here differ somewhat from the views expressed in the SIMSCRIPT manuals. In the manuals we said that attributes have a mode, just as do variables in FORTRAN. Examples of the mode of attributes in SIMSCRIPT II include alpha, text, integer, and real. In the present view we say that these are examples of entities whose types have already been defined for the user. Either the view in the manuals or the view expressed here may serve as a description of the terms entities, attributes, and sets as used in the examples in following sections. One advantage of the view expressed here is that it gives a simpler answer to the question just discussed "what is an attribute value?" It also gives a simpler answer to this question: "If the world consists of entities, attributes, and sets, why do SIMSCRIPT I and II also include arrays?" Consider how the present view would answer the question for a real vector (i.e., a one-dimensional real array). Since an integer is an entity it is allowed to have attributes—including attributes which associate a real number to some or all integers. Suppose A is such an attribute and suppose A happens to be defined only for i = 1, 2, 3, ..., 27. That is, A(1), A(2), A(3), ..., A(27) are real numbers but A(I) is undefined for other integers I. Then A is exactly what we call a vector with (as we say in SIMSCRIPT II) DIM.F(A) = 27. In a similar manner a matrix is an example of an attribute of (integer, integer). Higher dimensional arrays, and arrays of variable dimensionality as used in APL, are also part of the unlimited system of entity types which may be derived in a simple manner from two primitive entity types. Thus the present view of attributes is consistent with such "features" of SIMSCRIPT as various mode and array capabilities, but derives these in a simple systematic way.

Like other programming languages, SIMSCRIPT I and II use variables. For example if W is a recursive local variable in a routine R, then there is one instance of W for each time that R has been called but has not returned control. Thus if A called R called B called R, there would be two instances of W with perhaps two different values. If Y is a "saved" local variable of routine R, then there will be only one instance of Y even though R has been called twice and has not returned from either invocation. The value of a variable is set by a LET or a READ statement as in

```
LET W = 2.5
READ Y
LET P = PLACE(JOB)
```

The value of a variable is either a predefined entity, such as INTEGER or TEXT, or is an entity defined by the user. In these ways a variable is like an attribute. Indeed, a variable may always be viewed as an attribute. A recursive local variable, for example, is an attribute of routine-invocation whereas a saved local variable is an attribute of routine. If this is not immediately obvious, review Fig. 1(a) with routine-invocation instead of worker as the label for the region on the left, and W instead of ASSIGNED.TO as the label on the arrows. W is now a recursive local variable with a department (or U) as its value.

Figure 1(d) is similar to Figs. 1(a) through 1(c) except the arrow points from real number to real number, associating a number with its logarithm. Such a relationship between one of the entities of mathematics and another entity of mathematics is usually called a function. Depending on the tastes of a particular writer and depending on the specific logical or mathematical objects involved, such an association of entities may also be referred to, for example, as a sequence, a mapping, a transformation, or an operator. One difference between the relationships portrayed in 1(d) versus 1(a) through 1(c) is that the arrows in 1(d) do not change with time whereas those in 1(a) through 1(c) change as workers change departments, people get married, and people grow older. The reader may object to speaking of age as a function of man, insisting that a function must not change with time. In this case the reader will please substitute the word attribute whenever I happen to use function as a synonym for it in such situations. The reader may also object to referring to the attributes of an integer, or the attributes of a matrix, or the attributes of a real number. But here I have no good alternative. A word is needed herein that covers the concept portrayed by the arrows in both 1(a) and 1(d). Attribute seems to be the most reasonable word for this purpose as well as the one already used in the SIMSCRIPT languages. We shall therefore speak of rank as an attribute of matrix, logarithm as an attribute of real number, greatest prime factor as an attribute of integer, sum as an attribute of (integer, integer) combination, as well as height, weight, birth date, spouse, and department as attributes of person.

## Sets

Sets are collections of entities. For example, in a manufacturing system we might define a queue to be a collection of jobs waiting for a particular machine group. In an air transport system we might define a stack to be a set of planes waiting for clearance at a particular airport. In a computer system we might define the set of steps associated with a given job, the set of steps now capable of using processor time, or the set of pages associated with a given step or a given job.

In the manufacturing example we would speak of queue as being the name of the set. Each individual instance of a queue would have a machine group as its "owner" entity and would have zero, one, or more jobs as the "member" entities. If the variable MG refers to a particular machine group, then QUEUE(MG), read queue of MG, refers to its queue, i.e., the queue which it owns. Since each airport has its stack, we refer to airport as the owner entity and flight as the member entity. Similarly we may say that each job in a computer system owns a set of steps, and each step owns a set of data requirements.

Like the sets analyzed by Cantor [1] in defining ordinal numbers, SIMSCRIPT sets are ordered. Only finite sets are in fact stored, but one can visualize systems with sets of infinite cardinality with owners and members. For example, consider a system with one or more electrons moving relative to one or more systems of coordinates (e.g., an observer on the earth). We may think of such a system as consisting of electrons, electron-instants (a particular electron at an instant in time), and coordinate systems. Electrons own sets of electron-instants. Each such set is ordered by sequence of occurrence. Each (electron-instant, coordinate-system) combination has attributes like X, Y, Z, and TIME. We shall see later that this system also can be described in terms of entities, attributes, and finite ordered sets.

Since the ordered sets stored by SIMSCRIPT have finite cardinality, they have a first member, a last member, and a finite number of members. Both SIMSCRIPT I and SIMSCRIPT II treat these as attributes of the owner entity. Thus if a machine group owns a queue, then first in queue, denoted by F.QUEUE, is an attribute of machine group. (F.QUEUE, like the other names in the present discussion, is the SIMSCRIPT II generated attribute name. SIMSCRIPT I conventions differ slightly from these.) Unless the programmer explicitly specifies otherwise, if queue is defined as a set, SIMSCRIPT will automatically define first in queue as an attribute and update its value as entities are filed into and removed from queues. The same is true for L.QUEUE (last in queue) and N.QUEUE (number in queue). If the system owns a set called list, then F.LIST, L.LIST, and N.LIST are attributes of the system.

In SIMSCRIPT each member of a set has a predecessor in set and a successor in set as attributes. For example, if a job may belong to a queue, then successor in queue (denoted by S.QUEUE) is an attribute of job. In SIMSCRIPT II the members of a set also have an attribute M.set.name which indicates whether or not the entity is, as of the moment, in a set with this name. Unless the user specifies otherwise, P.set.name, S.set.name, and, in SIMSCRIPT II, M.set.name are updated automatically as entities are filed into and removed from sets during execution.

If the first and successor attributes are not suppressed, a SIMSCRIPT set can be traced out by starting with the first and proceeding to each successor in set until the element with an undefined successor is reached. If the last and predecessor attributes are not suppressed, the set can also be traced out in the reverse direction by starting with the last of set and proceeding to each predecessor. This may be done by explicit reference to the owner and member attributes as in the statement

```
LET JOB = F.QUEUE(MG)
```

or it may be done for the user by means of phrases such as

```
FOR EACH JOB IN QUEUE(MG)
FOR EACH JOB AFTER J2 IN QUEUE(MG)
FOR EACH JOB IN QUEUE(MG) IN REVERSE ORDER
```

Such phrases can be concatenated with other FOR EACH OF set, FOR EACH entity, or FOR variable = phrases, or with phrases such as SIMSCRIPT II's WITH, WHILE, UNTIL, and UNLESS phrases, and used as part of various statements.

The owner and member attributes hold together the sets of SIMSCRIPT and allow them to be traced out by FOR EACH OF set phrases. The types of operations which are allowed, and the way certain operations will be executed, depend on whether the storage of any of these attributes has been suppressed. For example, if a user indicates that the last and predecessor attributes are not to be stored,

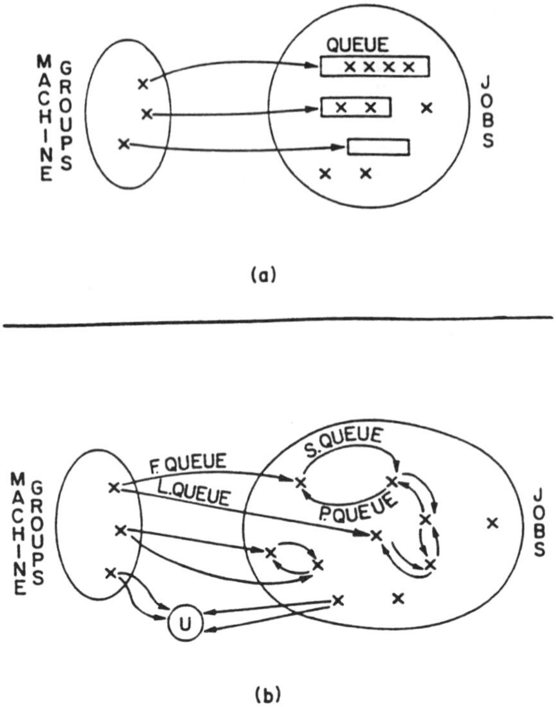


FIG. 2. Sets.

perhaps for space or speed reasons, then SIMSCRIPT will not attempt to update these when a FILE or REMOVE statement is executed, and will give an error message if the phrase FOR EACH JOB IN QUEUE(MG) IN REVERSE ORDER is encountered.

Since successor in set and/or predecessor in set are attributes of the member entity, and since an attribute can have only one value, it follows that a given individual can be in at most one set with a given name. For example, a job can be in only one queue, and can be in the queue at most once at any one time. The job can be in sets with different names such as a queue, a list, and a box, but in only one set at a time with a given name.

Nevertheless, a situation in which there are overlapping sets, such as people who can belong to many clubs and clubs which can have many members, is routinely handled as follows: define a new type of entity which we will call a "membership card." To enroll a new person as a member of a particular club we

```
CREATE A CARD
LET WHO(CARD) = PERSON
LET WHICH(CARD) = CLUB
FILE CARD IN WALLET(PERSON)
FILE CARD IN ROSTER(CLUB)
```

where "who" and "which" are attributes of card, wallet is a set owned by person, and roster is a set owned by club. The card may also have attributes such as the year that the person first enrolled.

This mechanism can also be used when an entity can be in the same set more than once, as when a student can enroll more than once in the set of students who have enrolled for a given course. In this case an enrollment entity may serve the role of the membership card. A particular enrollment belongs to only one set owned by course and one set owned by student. But more than one enrollment for a given student may belong to the set for the course.

Similarly, we may view a word such as CATASTROPHE as a set of character-instances, with each character-instance having a character as an attribute. The character-instance which is the first in set has "C" as the value of its attribute, its successor in the set has "A" as its value, and so on. In this way each member of the set has a unique successor even though the letter A is in one case followed by the letter T and in another S.

Figure 2(a) portrays a set called queue with machine groups as owners and jobs as members. The members of the set are in a rectangle pointed to by the owner. In accord with the SIMSCRIPT conventions the sets are ordered with each machine group owning one such set and each job belonging to at most one such set. Figure 2(b) shows the same information in another way. The owner of the set points, by means of the F.QUEUE attribute, to the first in set; each member of the set points, by means of the S.QUEUE attribute, to the next in set.

Since a given entity may have many attributes, own many sets, and belong to many sets in a given system, any attempt to picture as in Figs. 1(a) and 2(b) all the entity, attribute, and set relationships of a system almost certainly leads to a hopeless mess. A better way to present the same information is to list entity types along with their attributes, their set ownerships, and set memberships—as we shall illustrate in the next section.

## Change of Status

Since status consists of entities, attributes, and sets, the only way that status can change is if an entity comes into the system or goes out of the system, changes an attribute value, or gains or loses set membership. These possibilities would be represented in a figure such as 1(a) or 2(b) by an X appearing on the page, an X disappearing from the page, an arrow in 1(a) pointing to a different X (or U), or an X joining a set or leaving a set in 2(b) (not necessarily from the front or the back of the set). A change in position within the set may be represented by an exit from and a reentry into the set at a different position. SIMSCRIPT source programs indicate that an entity is to come into being (or enter the system) with a CREATE statement, go out of being (or leave the system) with a DESTROY statement, change attribute values with a LET or READ statement, and gain or lose set membership with a FILE or REMOVE statement. Certain variations in the FILE and REMOVE commands are permitted, principally because sets are ordered. For example, we may wish to remove the first, the last, or some specific member of a set; we may wish to file a new member either first, or last, or before or after some existing member. SIMSCRIPT also allows a set to be defined as being ranked according to some attribute of the member entity.

These elemental actions may be performed repeatedly under the control of FOR, UNTIL, WITH, etc., phrases; may be performed selectively as specified by

IF and GO TO statements; may be combined into higher level statements such as COMPUTE and ACCUMULATE; and, in simulation and real-time systems, may occur in exogenously or endogenously triggered events. The examples in the following section illustrate the SIMSCRIPT II syntax and the use of the aforementioned capabilities in actual programs.

## SIMULATION APPLICATIONS

The SIMSCRIPT view of system description was first used for discrete event simulation. We shall illustrate such applications by means of three examples. The first is a simple "job shop" simulator. The other two are extracts from larger simulations. We first present a simple but complete simulator to illustrate how a program looks in its entirety. We then present extracts from large applications to illustrate the generality of the view and the various levels of detail at which it can be used.

### A Job Shop Exercise

Articles on simulation frequently include an example such as customers arriving at a grocery store with several places to pick up items before checking out, or customers arriving at banks where they must queue for service. The simplest of such examples are instances of single server or multiserver queueing models with one queue. More complex examples are instances of networks involving several queues. The latter may be viewed as a special case of a job shop model used extensively, for example, in the simulation of fabrication processes.

It is about as easy to build a "general purpose" job shop simulator as it is to build a specific simulator for a particular hypothetical grocery store. It is also more illustrative of recommended practice. We speak of "general purpose" in quotes because there is a sense in which such a program is general and a sense in which it is specific. It is general in that it reads data to determine how many queues, how many servers for each queue, and certain other matters which are characterized most conveniently by parameters rather than logical relationships. The simulator is specific, on the other hand, in having built into it certain logical relationships (as distinguished from parameters) which determine how job arrivals are to be generated, how jobs are to route, which job is to be processed first when several are waiting, and so on.

Figure 3 contains an exercise which I have used in courses on simulation modeling and programming. It asks the student to produce a simple but "general purpose" job shop simulator with the number of machine groups (therefore the number of queues), the sales rate of different types of jobs, and the routing of various standard types of jobs as input parameters. The exercise also specifies other matters such as the dispatch rules used by the shop and the outputs of a run.

Before writing a simulator in SIMSCRIPT it is important to prepare oneself in two ways: one way is to have as complete a list as possible of the entities, attributes, and sets that are to appear in the simulated system; the other is to list the types of events to occur and a brief description of how each type of event changes status and causes other events. My current practice is to keep the description of entities, attributes, and sets in a computer file, and update this file before making

Each job in the simulated job shop is of one or another "standard type". Jobs of the same standard type have the same routing and process time. For example, perhaps any job with standard type = 1 goes first to machine group 3 and requires there 2.5 hours of processing; then goes (for example) to machine group 17; etc. As of any moment in time the shop may have any number of jobs of standard type = 1. These jobs may be at the same and/or at different steps of their routing. The same is true for any other standard type.

Write a simulation program with the following inputs:

the number of machine groups in the shop;

the number of machines in each machine group;

the number of standard types.

For each standard type

the mean time between sales (order arrivals) for jobs of this type.

For each step in the routing of (any job with) the particular standard type:

the machine group required at this step;

the process time required at this step;

the priority of a job of this type at this step;

the number of jobs of this type at this step in the shop at the start of the simulation.

The model does the following:

For each standard type the model causes sales to occur randomly with an exponential distribution of interarrival times (and a mean interarrival time as read in above).

As a job routes through the shop the model refers back to the standard routing for this type of job to determine machine group and process time at each step of its routing.

Before filing a job into a queue the model sets priority of job equal to the priority associated with its type at the current step.

Queues are ranked by priority and, within priority, by arrival time of job. When a machine needs a job, the job is taken from the front of the queue.

The output of the model should include the minimum, maximum, mean and standard deviation of:

time in shop, for each type of standard job, and for all jobs;

queue size, by machine group;

number of idle machines, by machine group.

FIG. 3. A job shop exercise.

any changes to the program. Figure 4 presents such a description of entities, attributes, and sets for the exercise shop. It shows, for example, that the simulated system will have entities such as MACH.GRP, JOB, and STD.TYPE; that a machine group will have as one of its attributes the number of free machines now

| Entity | Attribute | Owns | Belongs | Comment |
|---|---|---|---|---|
| MACH.GRP | | | | Machine group. |
| | FREE | | | Nr. machines in group now free. |
| | | QUEUE | | JOBs waiting to be processed. |
| JOB | | | | Job to be produced. |
| | JTYPE | | | Standard type of job. |
| | ARR.TM | | | Arrival time in system. |
| | JPRI | | | Priority of this job at this step. |
| | PLACE | | | Current STEP in ROUTING |
| | | | QUEUE | See MACH.GRP. |
| STD.TYPE | | | | A standard type of job. |
| | MT.BS | | | Mean time between sales. |
| | | ROUTING | | STEPs required to produce job of type. |
| STEP | | | | Step in the production of jobs of a type |
| | MG | | | Machine group required at this step. |
| | PTIME | | | Process time required at this step. |
| | SPRI | | | Priority of jobs at this step. |
| | | | ROUTING | See STD.TYPE. |

FIG. 4. Job shop entities, attributes, and sets.

available; that each machine group owns a queue; and that jobs are members of queues.

A solution is presented in Fig. 5. The program assumes that inputs will come in from cards or cardlike lines in a file rather than be entered interactively. The latter can also be done easily with SIMSCRIPT but requires a larger program because of the prompting messages from the computer to the user. Since at present I have a SHARE version translator at my disposal, and this does not have the ACCUMULATE and TALLY statements, I have not debugged the program in Fig. 5. I have debugged a longer SHARE version of the program.

Concerning general style, note that one statement may appear on more than one line, or more than one statement may appear on a single line; that no punctuation marks are required between statements, although one or more periods may optionally appear at the end of any word. Comments begin with a pair of apostrophes and end either with another pair as on Line 41, or at the end of the card. The line numbers are not part of the program but have been added here to facilitate discussion.

The definition of entities, attributes, and sets of the system and other global information is contained between the PREAMBLE statement on Line 1 and the END statement on line 23. For example, the translator is told (please see Line 6) that every machine group has a free (number of machines) attribute and owns a queue. No further specification is needed, since in this example we let SIMSCRIPT II make the memory allocation decisions for us. In a similar manner the attributes, set memberships, and set ownerships of standard types, steps in the routing of a standard type, and jobs are described elsewhere in the PREAMBLE.

The ACCUMULATE statement on line 22 of the program instructs SIMSCRIPT to maintain, for each machine group, a time weighted total of FREE(MACH.GRP) thus:

```
01 PREAMBLE
02 NORMALLY MODE IS INTEGER
03 DEFINE A AS AN ALPHA VARIABLE
04 DEFINE ATIS AS A REAL VARIABLE
05   PERMANENT ENTITIES...
06 EVERY MACH.GRP HAS A FREE AND OWNS A QUEUE
07 EVERY STD.TYPE HAS A MT.BS, A TIS DUMMY AND OWNS A ROUTING
08 DEFINE MT.BS AND TIS AS REAL VARIABLES
09   TEMPORARY ENTITIES...
10 EVERY JOB HAS A JTYPE, A PLACE, A JPRI,AN ARR.TM AND BELONGS TO A QUEUE
11 DEFINE ARR.TM AS A REAL VARIABLE
12 DEFINE QUEUE AS A SET RANKED BY JPRI AND THEN BY ARR.TM
13 EVERY STEP HAS A MG, AN SPRI AND A PTIME AND BELONGS TO A ROUTING
14 DEFINE PTIME AS A REAL VARIABLE
15   EVENT NOTICES...
16 EVERY END.PROC HAS A MPROC AND A JPROC
17 EVERY SALE HAS A STYPE
18 EXTERNAL EVENT IS END.SIM
19 TALLY AMTIS=MEAN, ASTIS=STD.DEV, AMNTIS=MIN, AMXTIS=MAX OF ATIS
20 TALLY MTIS=MEAN, STIS=STD.DEV, MNTIS=MIN, MXTIS=MAX OF TIS
21 ACCUMULATE MQ=MEAN, SQ=STD.DEV, MNQ=MIN, MXQ=MAX OF N.QUEUE
22 ACCUMULATE MFR=MEAN, SFR=STD.DEV, MNFR=MIN, MXFR=MAX OF FREE
23 END

24 MAIN
25 READ N.MACH.GRP   CREATE EVERY MACH.GRP   READ FREE
26 READ N.STD.TYPE   CREATE EVERY STD.TYPE
27   FOR EACH STD.TYPE CALLED S DO THIS...
28     READ MT.BS(S)
29     CAUSE A SALE(S) AT EXPONENTIAL.F(MT.BS(S),1)
30       UNTIL MODE IS ALPHA DO THIS...
31         CREATE A STEP CALLED P      FILE P IN ROUTING(S)
32         READ MG(P), PTIME(P), SPRI(P), N
33         ALSO FOR I=1 TO N, DO THIS...
34           CREATE A JOB    LET JTYPE(JOB)=S    LET PLACE(JOB)=P
35           CALL ALLOC(JOB)
36       LOOP   READ A
37   LOOP
38 START SIMULATION
39 END
```

FIG. 5. Job shop program (continued on next page).

$$S = \sum_{i=1}^{n} (t_i - t_{i-1})\ FREE_i$$

where $t_i$ is the value of time, $FREE_i$ is the old value of FREE(MACH. GRP) the i-th time that FREE(MACH. GRP) changes value, $t_0$ is the last time that this sum was RESET, and $t_n$ is the current time. In the present model the RESET command is not used; hence $t_0 = 0$. The ACCUMULATE statement on Line 22 further instructs SIMSCRIPT to compute and return the time weighted mean of FREE(MACH. GRP)

$$MFR = S/(t_n - t_0)$$

whenever reference is made to the attribute MFR(MACH. GRP), and to compute SFR as the time weighted standard deviation, MNFR as the minimum, and MXFR as the maximum of FREE.

```
40 EVENT SALE(S) SAVING THE EVENT NOTICE
41 CAUSE THE ''NEXT'' SALE IN EXPONENTIAL.F(MT.BS(S),1) DAYS
42 CREATE A JOB
43 LET ARR.TM(JOB)=TIME.V    LET JTYPE(JOB)=S    LET PLACE(JOB)=F.ROUTING(S)
44 CALL ALLOC(JOB)
45 RETURN   END

46 ROUTINE TO ALLOC(J)
47 LET STEP=PLACE(J)    LET MACH.GRP=MG(STEP)
48 IF FREE(MACH.GRP) > 0 SUBTRACT 1 FROM FREE(MACH.GRP)
49   CAUSE AN END.PROC(MACH.GRP,J) IN PTIME(STEP) HOURS
50   RETURN
51 ELSE LET JPRI(J)=SPRI(STEP)    FILE J IN QUEUE(MACH.GRP)
52 RETURN   END

53 EVENT END.PROC(M,J)
54 ADD 1 TO FREE(M)
55 IF S.ROUTING(PLACE(J))=0
56   LET ATIS=TIME.V-ARR.TM(J)
57   LET TIS(JTYPE(J))=ATIS
58   DESTROY JOB CALLED J    GO TO A
59 ELSE   LET PLACE(J)=S.ROUTING(PLACE(J))    CALL ALLOC(J)
60 'A' IF QUEUE(M) IS EMPTY OR FREE(M)=0  RETURN
61   ELSE  REMOVE FIRST JOB FROM QUEUE(M)    SUBTRACT 1 FROM FREE(M)
62     CAUSE AN END.PROC(M,JOB) IN PTIME(PLACE(JOB)) HOURS
63 RETURN   END

64 EVENT END.SIM
65 START NEW PAGE   SKIP 6 LINES    PRINT 1 LINE THUS...
66 TIME IN SHOP STATISTICS, BY TYPE OF JOB
67 SKIP 2 LINES   PRINT 2 LINES THUS...
68   JOB    AVERAGE    STANDARD
69  TYPE     T.I.S.    DEVIATION    MIN     MAX
70 FOR EACH STD.TYPE, PRINT 1 LINE WITH STD.TYPE, MTIS,STIS,MNTIS,MXTIS THUS
71   **        *.**       *.**        *.**    *.**
72 SKIP 1 LINE    PRINT 1 LINE WITH AMTIS,ASTIS,AMNTIS,AMXTIS THUS...
73  ALL        *.**       *.**        *.**    *.**
74 START NEW PAGE   SKIP 3 LINES    PRINT 1 LINE  THUS...
75          QUEUE SIZE AND UTILIZATION STATISTICS BY MACHINE GROUP
76 SKIP 2 LINES   PRINT 2 LINES THUS...
77   MACH   AVERAGE    STANDARD                      AVERAGE  STANDARD
78  GROUP     QUEUE    DEVIATION    MIN     MAX      IDLE   DEVIATION    MIN    MAX
79 FOR EACH MACH.GRP PRINT 1 LINE WITH MACH.GRP, MQ,SQ,MNQ,MXQ,MFR,SFR,
80  MNFR,MXFR THUS...
81   **        * **       *.**        ***     ***     *.**     *.**        ***    ***
82 STOP   END
```

FIG. 5 (continued).

The MAIN routine, starting on Line 24, is the first to receive control during the execution of any run. Its function here, as is usual, is to initialize the simulated system. It can also be used to set-up and initialize a series of simulations, accumulating statistics across these for an overall analysis. In the present case the MAIN routine reads the number of machine groups for the present run (N.MACH.GRP), and creates this number of machine groups with a CREATE EVERY statement. On Lines 5 and 6 of the PREAMBLE, MACH.GRP is defined to be a permanent entity. SIMSCRIPT is thus told that machine groups do not come and go during the course of a run. This implies that the CREATE EVERY statement is applicable, and that individual machine groups are to be identified by their ordinal number, i.e.,

MACH.GRP = 1, or 2, or, ..., N.MACH.GRP. MAIN next reads FREE for each machine group. It reads the number of standard types and creates every standard type. For each standard type it reads the mean time between sales (MT.BS) attribute, CAUSEs the first occurrence of a sale for this standard type (Line 29), and does other initialization concerning standard types such as creating steps (and reading their attributes) for each standard type, and creating a specified number of jobs of the given type at a given step initially in the shop.

Programming note: The DO on Line 30 and the DO on Line 33 both have their ranges closed by the single LOOP statement on Line 36. This happens because Line 33 actually has an ALSO...DO statement which uses the same LOOP as the preceding DO. The ALSO...DO was put into the language to avoid ungrammatical LOOP-LOOP sequences. Also, the word THIS is optional in the DO THIS statement.

The START SIMULATION statement on Line 38 instructs SIMSCRIPT to call upon the timing routine to control the execution of the simulation. The timing routine will repeatedly ask which event is most imminent: one of the endogenous event occurrences now on the calendar or the next (and only) occurrence of the exogenous event END.SIM. SIMSCRIPT II was told of these events on Lines 15 through 18 of the PREAMBLE. Having determined the most imminent event, the timing routine updates its calendar, updates current TIME.V, and calls on the appropriate event routine. At the beginning of our illustrative simulation it will call on the most imminent SALE. As described on Lines 40 through 45, the SALE event reschedules the next sale for the same standard item, reusing the same coming event notice as instructed on Line 41 (and in the SAVING phrase of Line 40). Thus each occurrence of a sale for a particular standard type causes the next occurrence for this type to be placed on the calendar.

On Lines 42-44 the sale event creates a new job, notes its arrival time and type, notes that the current value of PLACE(JOB) is the step which is the first in the routing of the standard type, and calls upon the ALLOCate routine to dispose of the job. When ALLOC returns control to the SALE event, the latter returns control to the timing routine, which again determines which is the most imminent event.

The ALLOC routine notes for its own use the step and the machine group involved. If the number of free machines in the machine group is greater than zero, it decrements the number of free machines and causes an end of process event to occur on behalf of the given machine group and job in PTIME(STEP) hours (Lines 48 and 49). Otherwise, if there are no free machines in the machine group, ALLOC notes the priority of the job at its present step, and files the job into queue (Line 51). When the FILE statement is executed, the job will be put into the appropriate position within the queue, as specified in the DEFINE SET statement in Line 12 of the PREAMBLE.

When an END.PROC becomes the most imminent event, the END.PROC event routine is called with a machine group and a job as its arguments. END.PROC increments the number of free machines of the machine group. If the successor in routing attribute of the job's current step equals zero, i.e., if the job has reached the end of its routing, then END.PROC takes certain actions (Lines 56 and 57) which, together with the TALLY statements on Lines 19 and 20, will update the desired time-in-shop statistics. (Incidentally, TIS is defined as a dummy attribute on Line 8 to instruct SIMSCRIPT II not to store this attribute when assigned, as in Line 57, but only TALLY its totals as specified on Line 20.) Next, END.PROC destroys the completed job, i.e., eliminates it from the simulation.

If the job is not at the end of its routing, END.PROC updates place of job and calls upon ALLOC to dispose of the job (Line 59). In any case, whether the job was destroyed or sent to its next queue, the machine is considered next. If the queue of the machine group is empty or the number of free machines in the machine group is zero (because ALLOC happened to take what Line 54 gave), then control is returned to the timing routine without further action (Line 60). Otherwise the first job in the queue is removed, the number of free machines in machine group is decremented, and an end of process for this machine and job is caused to occur (Lines 61 and 62). The three statements in Lines 61 and 62 in effect start the processing of this job by this machine.

The real action in this simulator takes place between the timing routine, the SALE and END.PROC event routines, and the ALLOCate subroutine. The latter three create and destroy jobs, assign attribute values to jobs and machine groups, file jobs into and remove them from queues, and schedule events to occur. The timing routine keeps asking which is the most imminent event, and transfers to the appropriate event routine after updating TIME.V and its calendar of events.

When the exogenous END.SIM is more imminent than any endogenous event, the END.SIM event is called. This in turn displays the accumulated statistics in the manner pictured in Lines 65 through 81. END.SIM then stops the execution of the program on Line 82.

A SHARE version of the model, with about twice as many statements as in Fig. 5, requires about 16 seconds to translate and assemble on the IBM 370 Model 168. Of this, about 6 seconds are spent for the routines generated by the PREAMBLE. To recompile, END.PROC takes about 2.75 seconds. For shops with short queues the object program will execute about 1500 to 2000 simulated events per CPU second. For shops with very large queues, where queues are ranked rather than LIFO or FIFO, execution performance will degrade unless one uses somewhat fancier programming. CACI reports improved compile and execute times for SIMSCRIPT II.5 although, to my knowledge, no attempt has been made to produce optimized code.

The same basic structure used in the above model can be found in more complex job shop simulations. For example, in many fabrication processes one each of two types of resources—usually a worker and a machine—is required at each step of processing. Suppose for concreteness that each worker is of one or another labor class (LAB.CLASS), that there is a set of machine groups associated with a given labor class, and that any worker in the labor class can run any machine in this set of machine groups. It will be convenient here to assume that, while one labor class can serve many machine groups, a given machine group is served by only one labor class. The more general case is easily represented using the membership card gambit, but raises some allocation problems I would prefer to ignore here. We will assume that an available worker will work on the job with the highest priority, among jobs in the queues of the machine groups which he can serve.

To amend the model to include workers as well as machine groups, the ALLOC routine should check for the availability of a worker as well as that of a machine; i.e., Line 48 should consider

```
IF FREE(MACH.GRP)>0 AND AVAIL(WHO.SERVES(MACH.GRP))>0
```

The END.PROC event must now dispose of the worker as well as the machine. It might do so with the help of a statement such as

```
FOR EACH MACH.GRP IN SERVED.BY(LAB.CLASS),
 WITH FREE(MACH.GRP)>0 AND F.QUEUE(MACH.GRP)>0
 COMPUTE MM=MAX(MACH.GRP) OF JPRI(F.QUEUE(MACH.GRP))
```

This statement will compute and store in MM the value of MACH.GRP which maximizes JPRI (of the job which is first in queue of the machine group) among the specified machine groups. The contingency that no such machine group exists—because none meets the specified conditions—can and should be tested by IF MM=0 .... The documentation file, the PREAMBLE, the MAIN routine, and END.SIM must also be augmented to document, define, initialize, and report on the entity type LAB.CLASS. The basic job shop model should still be apparent, and will represent most of the coding for this particular version.

Other reasonable exercises on modifying a basic job shop model include the following. Make sales an exogenous event and have the routing of each job be part of the data for the specific sale. (See the manual [12] regarding external inputs.) Or, make routings random; or make process times random; or have n shifts of workers per day, where n is an input like 1, 2, or 3; include random absenteeism; or represent each job by a network of steps, rather than a sequence of steps, where the network shows which steps must wait for which others to be completed before the step may queue for resources.

## A Life Insurance Company Corporate Model

Our next example is an extract from a large life insurance company corporate model. SOFASIM (Society of Actuaries Simulator) [23, 25] includes the actuarial, investment, and tax side of a life insurance company. On the actuarial side it sells policies of various types; collects premiums; pays commissions; and pays death benefits, endowments, and surrender benefits. The user may elect to have deaths or sales or lapses computed deterministically or drawn randomly. SOFASIM computes reserves against existing policies as needed for accounting, tax, and decision-making purposes within the model.

On the investment side the model receives coupon payments and processes called and matured bonds. When the company's cash balance becomes too high or too low, it converts cash to bonds by buying or bonds to cash by selling.

The model computes the simulated company's federal income tax according to procedures required of actual life insurance companies. A measure of the complexity of the tax calculation is the fact that the actuarial side of the business—the selling of new insurance, the random or nonrandom determination of deaths, the computation of reserves, and so on—requires 98 lines of SIMSCRIPT coding; the processing of existing investments requires 41 lines of SIMSCRIPT code; decisions concerning the buying and selling of investments require 67 lines of code; while the tax calculation requires 204 lines. These are counts of lines rather than instructions, including comments. The job shop example in Fig. 5 would be counted as 82 lines.

Because of the specific problem for which SOFASIM was built, the simulated company is a stock life insurance company, owned by the stockholders, as compared to a mutual company owned by policyholders. The simulated company decides dividend rates once a year and pays these dividends quarterly.

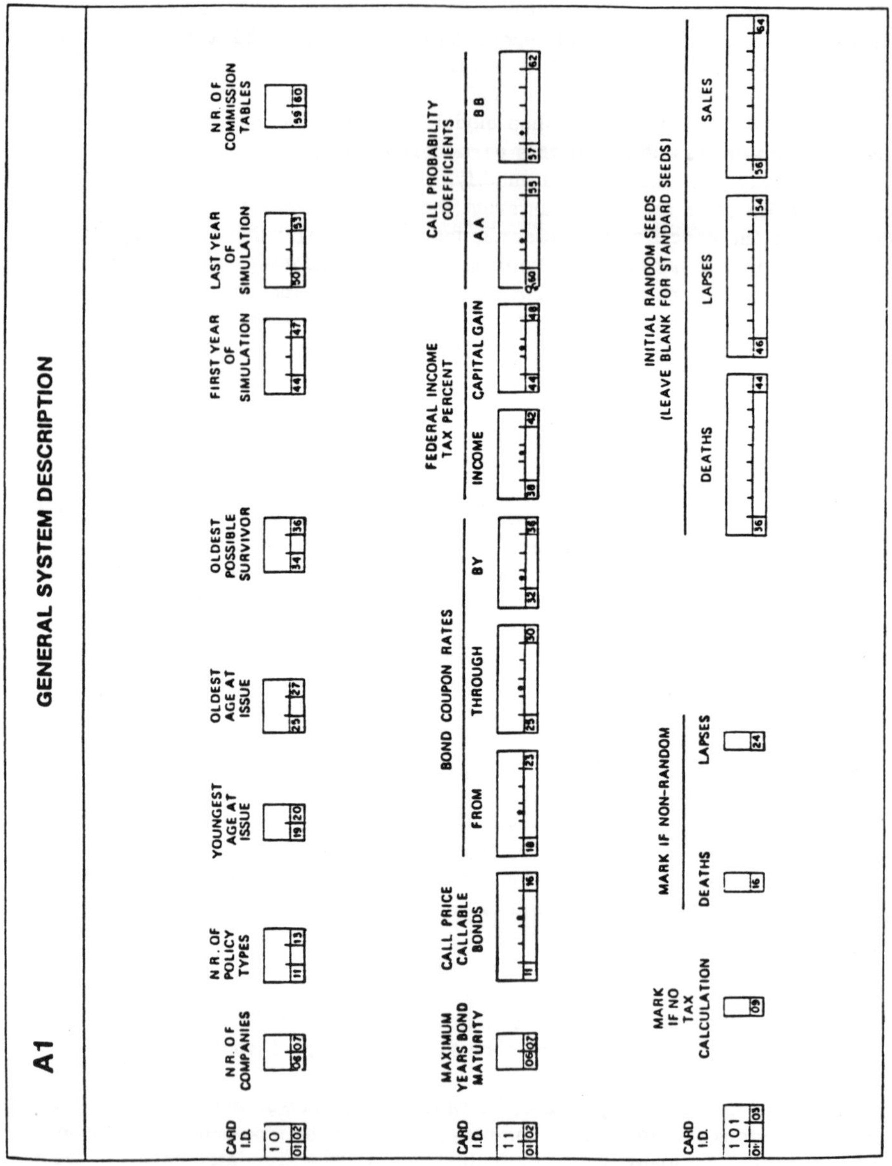


FIG. 6. The SOFASIM A1 form.

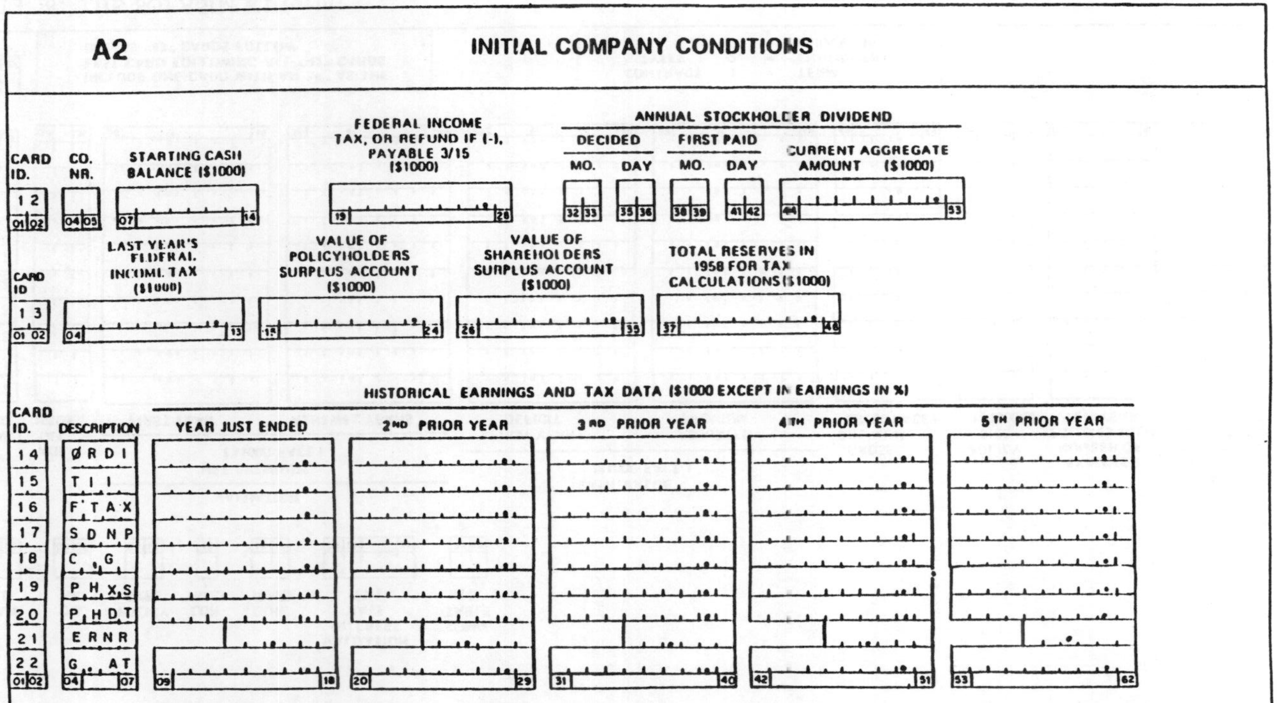


LEGEND: 14 ORDI ORDINARY GAIN OR LOSS (·), BEFORE DEDUCTING SDNPC, NOT YET CARRIED BACK OR FORWARD IN FEDERAL TAX CALCULATION
15 TII. TAXABLE INVESTMENT INCOME
16 FTAX FEDERAL TAX INCLUDING TAX ON CAPITAL GAINS (EXCLUDING TAX ON POLICYHOLDER SURPLUS ACCOUNT WITHDRAWALS)
17 SDNP SPECIAL DEDUCTION FOR NONPARTICIPATING CONTRACTS
18 C.G. CAPITAL GAIN OR LOSS (·) NOT YET CARRIED BACK OR FORWARD
19 PHXS AMOUNT TRANSFERRED FROM POLICYHOLDER SURPLUS ACCOUNT DUE TO MAXIMUM LIMITATIONS
20 PHDT AMOUNT TRANSFERRED FROM POLICYHOLDER SURPLUS ACCOUNT FOR DIVIDEND PURPOSES
21 ERNR CURRENT EARNINGS RATES AS A PERCENT (TO COMPUTE ADJUSTED RESERVE RATE)
22 G.AT GAIN FROM OPERATIONS AFTER TAXES (FOR DIVIDEND POLICY)

FIG. 7. The SOFASIM A2 form.

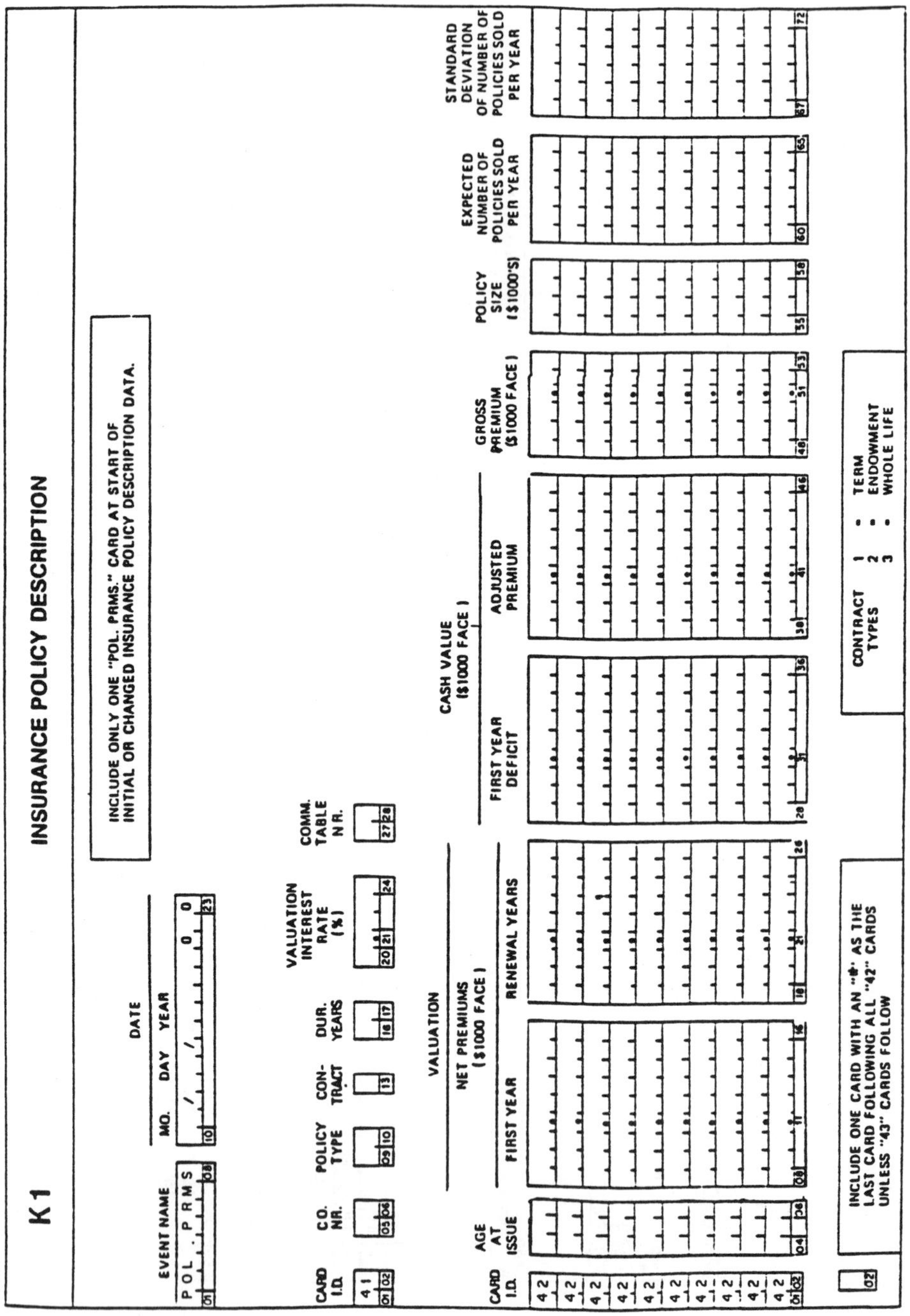


FIG. 8. THE SOFASIM K1 form.

I programmed SOFASIM as a consultant to The Society of Actuaries under the general guidance of Edward A. Lew (then president of the society as well as senior vice president of Metropolitan Life) and under the more specific direction and tutelage of John C. Wooddy (chairman of the Joint Committee on the Theory of Risk, and a senior vice president of North American Reassurance). Dr. Barbara Markowitz helped greatly in debugging. Mrs. Alice Goldstein (member of the Society of Actuaries) succeeded in the difficult job of determining how my original tax computation differed from the tax computation in fact, and then made the program correct in this area.

A realistic SOFASIM run requires a large volume of data. In the first instance this data may be manuscripted on input forms such as those in Figs. 6, 7, and 8. Cards are punched from these forms and read off-line into disk storage. Subsequently, small changes in the data stored on disk are made interactively.

The first form read during system initialization is the A1 form shown in Fig. 6. It presents information such as the number of companies simulated (usually = 1); the first year and the last year of the simulation (e.g., 1978 through 2028); and random seeds if other than standard random seeds are to be used for deaths, lapses, and sales. The A2 form in Fig. 7 shows initial company conditions such as starting cash balance and various tax-related information. The initial status of investments held and the status of initial insurance policies written are described on other forms.

The K1 form in Fig. 8 illustrates one of several SOFASIM forms which are used at the beginning of the simulation to initialize the system, and can be used again any number of times later in the simulation to change system characteristics. The K1 form describes the parameters of life insurance policies issued by a simulated company. The form includes a field which identifies itself as being processed by exogenous event POL.PRMS; a field which indicates the month, day, and year when the specific change is to occur; fields which identify policy type, contract type, and other characteristics of the particular policy; and fields which specify, for example, the expected sales rates for policies of various sizes and various ages of the persons buying the policy.

Other SOFASIM input forms include forms to describe mortality tables, commission tables, company parameters that can change from time to time, interest rates (perhaps as a function of coupon, years to maturity, and whether a bond is callable or not), and parameters describing the investment objectives of the company.

Figure 9 describes some of the SOFASIM entities, attributes, and sets. One entity type is a life insurance company to be simulated; another is a "cohort" of policyholders representing everyone of a given age who bought a policy of a given type and size in a given year. (Individual policyholders are not represented as separate entities.) Other entity types include the tax return and more esoteric entities such as policy type and coupon category.

Figure 10 summarizes most of the events which occur in SOFASIM. It would have been extremely uneconomical to distinguish individual purchases of policies, individual deaths, or individual purchases and sales of bonds. Instead SOFASIM runs on a monthly cycle with the processing of insurance policies, the processing of investments, the computation of operating expenses, and the purchase and sales of investments each happening once a month. Other events such as the quarterly computation of estimated income tax, the paying of dividends, and the resetting of cumulative annual totals happen endogenously but not monthly. Events which change

| *Entity* | *Attributes and Sets* |
|---|---|
| COMPANY | Attributes include cash balance (CASH); total face value of investments (TOT.INV); total policy reserves (TOT.RS). Other company attributes include year to date cumulative totals (such as total premiums collected, and total death benefits paid), cost and lag coefficients (such as expense factors used in computing monthly operating expenses), parameters used in investment policy, parameters used to set stockholders dividend rates, miscellaneous information needed for the annual statement, and thirty attributes involved with the federal income tax calculation. Company has a total of 78 attributes. Company owns a set of cohorts, a set of GAT.MMOs (gain after tax memos) showing past year's gain after taxes, a set of ER.MMOs (earnings memos) showing prior earnings rates, and a set of RETURNs showing past years tax information to be used in loss carry forward and carry back calculations. |
| COMPANY, POL.TYPE, AGE.ATI | A compound entity representing a particular combination of company, policy type and age at issue. Attributes include: contract type e.g., 1 = term insurance, 2 = endowment, 3 = whole life; and duration of premium payment, e.g., equals 20 for a 20 payment life policy. This compound entity owns a set of PSIZEs, where every PSIZE has policy size, expected sales rate and standard deviation of sales rate as attributes. |
| COHORT | Represents a cohort of policyholders who purchased a policy of a given type, in a given year, of a given size, at a given age at issue. Attributes include age at issue, policy size, year of issue and various factors involved in the reserve calculation. |
| CALLCAT, COUPCAT, MATRTY | A particular combination of call category, coupon category and years to maturity. It has one attribute, the current PRICE of a bond in the specified combination of categories. |
| COMPANY, CALLCAT, COUPCAT, MATRTY | A compound entity whose only attribute is the FACE value of bonds held by the particular company, in the particular call and coupon categories, having a particular maturity. |
| RETURN | A prior year's tax return. Attributes include, for example, capital gain or loss not yet carried back or forward. |
| THE SYSTEM | Attributes of the system as a whole include the federal income tax rates on ordinary taxable income and on capital gains; the call price of a callable bond (assumed to be the same for all callable bonds); NO.RL, NO.RM, and NO.TAX which indicate whether lapses and mortality are to be generated randomly, and whether the tax calculation is to be suppressed. |
| Arrays | Commission tables, mortality tables and lapse rate tables are defined as arrays. |

FIG. 9. Some SOFASIM entities.

company policy parameters, sales rates, interest rates, and so on happen exogenously at any time.

Figure 11 contains the PY.DV (pay dividend) event routine. This very simple SOFASIM event routine will serve to illustrate a bit of the SOFASIM coding. The event receives a company (usually the only company in the simulation) as an

| Event Name | Description |
|---|---|
| | *Monthly Events* |
| PR.POL | Processes life insurance policies. For each cohort it computes deaths and lapses, pays death benefits, pays surrender and endowment benefits, collects premiums and processes sales for newly forming cohorts. Computes reserves for the company. |
| PR.INV | Processes existing investments. Collects or pays short term interest rate on the company's positive or negative cash balance. For each category of investment it collects coupon payments and processes investments which mature or are called this month. |
| OP.EXP | Computes current monthly operating expenses. |
| BUY.SL | Perhaps buys or sells investments in order to bring the company's cash balance into a reasonable relationship with its total reserves. |
| | *Other Endogenous Events* |
| DCD.DIV | Decides the stockholder dividend rate for the next twelve months. |
| PY.DIV | Pays dividends at the quarterly rate determined in the DCD.DIV event. |
| TAXCALC | Calculates estimated federal income tax four times a year, and final incomes tax once a year. |
| PAYTAX | Pays the previously calculated estimated or final income tax. |
| REFUND | Receives tax refund if due. |
| RE.RE | Reports and resets. Notes information concerning the year's performance and end of year status to be reported at the end of the simulation. Resets annual cumulative totals. |
| | *Exogenous Events* |
| CO.PRMS | Reads initial or modified values for various company parameters such as parameters used in dividend and investment policies. |
| INTR.RTS | Reads interest rate structure, perhaps as a function of coupon rate, time to maturity, and callability. |
| INV.PROFILE | Reads desired investment profile parameters used by the BUY.SL decision rules. |
| M.OR.L.TABLE | Reads mortality, valuation and lapse tables. |
| COM.TABLE | Reads commission tables. |
| POL.PRMS | Reads policy descriptions as described on form K1, figure 8. Also reads initial policies from another form. |

FIG. 10. SOFASIM events.

argument. The event reschedules itself to happen again in one year. (There are three other event occurrences of PY.DV that happen during the year, each rescheduling themselves in one year. In this manner the event is caused four times a year.) If the quarterly dividend rate (QDV.RT) for the whole company is not positive, then

```
                ''DIVIDEND PAYMENT''
        EVENT PY.DV(C)    SAVING THE EVENT NOTICE
          DEFINE C AS AN INTEGER VARIABLE
          RESCHEDULE THIS
            PY.DV AT DATE.F(MONTH, DAY, YEAR + 1)
          IF QDV.RT(C)<=0.0  RETURN  ELSE...
            SUBTRACT QDV.RT(C) FROM CASH(C)
            ADD QDV.RT(C) TO DIVS(C)
          IF NO.TAX=1  RETURN  ELSE...
            LET A = MIN.F(QDV.RT(C),SH.SA(C))
            SUBTRACT A FROM SH.SA(C)
            LET B = (QDV.RT(C) - A)/(1.0 - INC.TRT)
            IF B > .001 SUBTRACT B FROM PH.SA(C)
              ADD B TO PSA.TR(C)  REGARDLESS...
                RETURN  END
```

FIG. 11. SOFASIM routine to pay dividend.

the event returns to the timing routine without taking action. Otherwise the event subtracts the company's quarterly dividend rate from its cash balance and adds the quarterly dividend rate to the cumulative dividends paid this year. (The four dividend payments paid at a given rate may overlap two calendar years.) The remainder of the routine involves tax matters, perhaps updating the legislatively defined "shareholders surplus account" (SH.SA) and "policyholders surplus account" (PH.SA).

We noted earlier that the basic structure of a job shop can be augmented in various ways. The SOFASIM model, on the other hand, may be viewed as a very elaborate example of a cash flow model. Various actions either increase the cash balance, as with the receipt of premiums and coupon payments, or decrease the cash balance, as with the payments of benefits and dividends. Certain policy actions such as the buying or selling of bonds or the paying of dividends are triggered, at least in part, by the cash balance. A model could have elements of both a job shop and a cash flow, as in a business simulation which included both a detailed production simulation and a financial side. Whether such a simulation would be an economical way to analyze certain problems depends on the specifics of the situation. Be that as it may, the system could be described in terms of its entities, attributes, and sets which change exogenously and endogenously.

## A VM/370 Simulator

Figure 12 presents an extract from the file used to document the entities, attributes, and sets of VSIM, a VM/370 simulator written in the SHARE version of SIMSCRIPT II. VSIM includes a detailed description of certain Control Program (CP) functions, particularly scheduling, dispatching, and paging. The purpose of VSIM is to test alternate decision rules or the settings of certain installation-determined parameters. At present the VSIM simulator itself is operational, but data collection is still in process, to be followed by testing and application.

The entities of VSIM include virtual machines (VMs), pages in the address space of virtual machines, direct access storage devices (DASD) on which pages are kept, and the system as a whole. Attributes include the logon time (ARR.TM)

| Entities | Attributes | Owns | Belongs | Comments |
|---|---|---|---|---|
| VM | | | | Virtual machine. |
| | VM.TNR | | | Type number of machine's user program. |
| | ARR.TM | | | Time virt. mach. entered system. |
| | DSPABLE | | | Dispatchable: 0=>no; 1=>yes. |
| | INTRPD | | | =1=>act interrupted;=0=>not interrupted. |
| | TO.GO | | | Time to go on interrupted action. |
| | QU.NR | | | Queue number. |
| | WSPROJ | | | Nr. pages in projected "working set" |
| | LAST.WSPROJ | | | Previous WSPROJ. |
| | RES.PAGES | | | Nr. of resident pages. |
| | SUMRES.PAGES | | | Sum.over page reads,of resident pages. |
| | NU.RESPAGES | | | Nr. of updates to above tally. |
| | STEALS.QU | | | Nr. times pg wt entered for stolen pg. |
| | EPRIOR | | | Priority of VM in ELG.LIST. |
| | RPRIOR | | | Priority of VM in DSP.LIST. |
| | USR.FCTR | | | Factor used in EPRIOR calculation. |
| | PS.TIME | | | Time spent in problem state. |
| | CP.TIME | | | CP time spent on behalf of this VM. |
| | ETS.PS | | | Time slice will end if PS.TIME->ETS.PS |
| | ETS.SS | | | Time slice will end if CP.TIME->ETS.SS |
| | EQ.PS | | | Quantum will end if PS.TIME->EQ.PS |
| | EQ.SS | | | Quantum will end if CP.TIME->EQ.SS |
| | CUR.LEVEL | | | 1=>USER;2=>UT;3=>CP->US;4=>CP->UT. |
| | CUR.STP.NR | | | Current step number. |
| | CUR.ACT | | | Current action. |
| | CUR.TYPE | | | Current (user,util. or cp) job type. |
| | ST.TIME | | | Time: state entered or last report. |
| | | NBS | | Pages needed but stolen. |
| | | CPRS | | Current page requirements(nonshared). |
| | | SPRS | | Shared page requirements. |
| | | RSQD | | Pages referenced since last qdrop. |
| | | | DSP.LIST | Dispatch list. |
| | | | ELG.LIST | Eligible list. |
| PAGE | | | | |
| | SP.FLAG | | | =0 (not a shared page). |
| | WHOSE | | | Virtual machine of page. |
| | NEEDED.BIT | | | =1=>currently needed;=0 otherwise. |
| | KEEP.BIT | | | =1=>need persists beyond step;=0 otherwise. |
| | LOCK.BIT | | | =1=>Lock page when frame obtained. |
| | LOCKED.BIT | | | =0=>page not locked;=1=>page locked. |
| | MODIFY.BIT | | | =1=>modify page when frame obtained. |
| | MODIFIED.BIT | | | =0=>page not modified;=1=>page modified. |
| | RFRNCD.BIT | | | ->1 by use of page;->0 by STEAL.PG routine.. |
| | TAV.DRUM | | | Time of end trans. to paging drum or disk. |
| | | | FR.LIST | Free list, used by paging mngmnt rules. |
| | | | FL.LIST | Flush list, used by paging mngmnt rules. |
| | | | USR.LIST | User list, used by paging mngmnt rules. |
| | | | NBS | Needed but stolen. |
| | | | RSQD | Referenced since last queue drop. |

FIG. 12. Extract from VSIM status (continued on next page).

of the VM, whether the VM is currently dispatchable, and so on. Sets include the set of nonshared pages which a VM has referred to but has not released, the set of VMs in the system's dispatch list, the set of pages in the system's user list, and so on. Figure 12 does not include all of the attributes and set memberships of the entities listed.

VSIM may be viewed as an elaborate job shop model (indeed a job shop model with jobs of standard types) although the entity, attribute, and set names have been chosen to correspond to VM/370 usage. VSIM differs from a typical manufacturing simulator, just as one detailed realistic manufacturing simulator differs from another.

VSIM coding is illustrated in Fig. 13. VSIM calls the subroutine shown when the simulated CP is required to find a page frame for a page demanded by a VM.

| Entities | Attributes | Owns | Belongs | Comments |
|---|---|---|---|---|
| THE SYSTEM | | | | |
| | CUM.IDLE | | | Cumulative time CPU idle. |
| | CUM.SOH | | | Cum CPU usage for system overhead. |
| | TOT.PROJWS | | | Sum of WSPROJs among VMs in DSP.LIST. |
| | EMPTY.FRAMES | | | Number of page frames without pages. |
| | CPUWORKING.FLAG | | | =0=>CPU idle; =1=>CPU working. |
| | EP.FCTR | | | Factor used in EPRIOR calc. in QDROP. |
| | RP.FCTR | | | Factor used in RPRIOR calc. in QDROP. |
| | DT.CUNP | | | |
| | P1.CUNP | | | PARAMETERS FOR VARIOUS DISTRIBUTIONS. |
| | P2.CUNP | | | |
| | DT.CURE | | | Each parameter name is of the form: |
| | P1.CURE | | | |
| | P2.CURE | | | xx.yyzz |
| | DT.C1DC | | | |
| | P1.C1DC | | | where |
| | P2.C1DC | | | |
| | DT.TTDC | | | xx="DT" means "distr. type" |
| | P1.TTDC | | | ="P1" means "parameter 1" |
| | P2.TTDC | | | ="P2" means "parameter 2" |
| | DT.C2DC | | | |
| | P1.C2DC | | | yy="CU" means "CPU usage" |
| | P2.C2DC | | | ="C1" means "CPU before TT" |
| | DT.C1CD | | | ="TT" means "transmission" |
| | P1.C1CD | | | ="C2" means "CPU after TT" |
| | P2.C1CD | | | |
| | DT.TTCD | | | zz="DC" means "drum to core" |
| | P1.TTCD | | | ="CD" means "core to drum" |
| | P2.TTCD | | | ="IO" means "non-paging IO" |
| | DT.C2CD | | | ="WT" means "long wait" |
| | P1.C2CD | | | |
| | P2.C2CD | | | |
| | DT.C1IO | | | |
| | P1.C1IO | | | |
| | P2.C1IO | | | |
| | DT.C2IO | | | |
| | P1.C2IO | | | |
| | P2.C2IO | | | |
| | DT.C1WT | | | |
| | P1.C1WT | | | |
| | P2.C1WT | | | |
| | DT.C2WT | | | |
| | P1.C2WT | | | |
| | P2.C2WT | | | |
| | LGON.CC | | | CPU time required to logon. |
| | LGOF.CC | | | CPU time required to logoff. |
| | SPFR.CC | | | CPU tm rqd to steal page from FR.LIST. |
| | SPFL.CC | | | CPU tm rqd to steal page from FL.LIST. |
| | SPUS.AA | | | CPU tm to seek to stl pg from USR.LIST. |
| | SPUS.BB | | | Ditto--per page examined in USR.LIST. |
| | | DSP.LIST | | Dispatch list. |
| | | USR.LIST | | User list, used by paging mngr. |
| | | FL.LIST | | Flush list, used by paging mngr. |
| | | FR.LIST | | Free list, used by paging mngr. |

FIG. 12 (continued).

Some background concerning the operation of CP is required to make clear the operation of the routine.

VSIM, like the system it represents, maintains three sets of pages in main storage. One of these, called the free list (FR. LIST), consists of pages whose page frames may be given out without writing their current contents onto a paging drum or disk because the contents have already been written out. The flush list (FL. LIST) consists of page frames whose contents have not been written to the paging DASD, but are to be given out before pages in the user list (USR. LIST). We will not describe here what makes CP move a page from the user list to the flush list. VSIM also keeps a count of the page frames which are not occupied by

```
ROUTINE TO STEAL.PG(REP)
   ''OBTAINS PAGE FRAME.     REP=1=>REPLACEMENT FOR FREE LIST SOUGHT ''
DEFINE PT AND PT2 AS REAL VARIABLES
    ''IF O.K., OBTAIN PAGE FRAME FROM FREE LIST ''
IF FR.LIST IS NOT EMPTY AND REP=0  REMOVE FIRST PAGE FROM FR.LIST
  IF EMPTY.FRAMES+N.FR.LIST<=N.DSP.LIST  CALL STEAL.PG(1) YIELDING PT2
    REGARDLESS...    RETURN.
    '' OR,  OBTAIN PAGE FROM FLUSH LIST  ''
ELSE IF FL.LIST IS NOT EMPTY REMOVE FIRST PAGE FROM FL.LIST   GO TO C
   '' ELSE STEAL PAGE FROM USER LIST ''
ELSE...
'A'  FOR EACH PAGE IN USR.LIST  DO...
  IF LOCKED.BIT(PAGE)=1  GO TO B     ELSE
  IF RFRNCD.BIT(PAGE)=0 GO TO FOUND    ELSE   LET RFRNCD.BIT(PAGE)=0
'B' LOOP    IF SECOND.TIME=0  LET SECOND.TIME=1  GO TO A    ELSE CALL ERR(31)
'FOUND'  REMOVE PAGE FROM USR.LIST
IF SP.FLAG(PAGE)=0 AND NEEDED.BIT(PAGE)=1   FILE PAGE IN NBS(WHOSE(PAGE))
 THEN IF WHOSE(PAGE)=VM    ADD 1 TO SELF.STEALS
    ''PERHAPS TRANSMIT TO DASD. ''
'C' ELSE   IF REP=1 FILE PAGE IN FR.LIST   REGARDLESS
  IF MODIFIED.BIT(PAGE)=0     RETURN    ELSE
  IF STORED.ON(PAGE)>0 LET PGG.DASD=STORED.ON(PAGE)   GO TO D    ELSE
FOR PGG.DASD=1 TO N.PGG.DASD WITH AVAIL.SLOTS(PGG.DASD)>0  GO TO CC
                                                   ELSE CALL ERR(32)
'CC' SUBTRACT 1 FROM AVAIL.SLOTS(PGG.DASD)    LET STORED.ON(PAGE)=PGG.DASD
'D'  ADD 1 TO NR.PG.WRITES
 LET TAV.DASD(PAGE)=TIME +
            WAIT(DT.TTCD(PGG.DASD),P1.TTCD(PGG.DASD),P2.TTCD(PGG.DASD))
 IF REP=0   LET DSPABLE(VM)=0   REGARDLESS   RETURN   END
```

FIG. 13. VSIM routine to steal a page.

any page. Before the STEAL.PG routine is called, VSIM checks to see whether a needed page is on the user list or can be reclaimed from the free or flush lists. If not, a page frame is needed and VSIM calls on STEAL.PG to find one.

STEAL.PG first attempts to obtain a frame whose occupant is in the free list. If such is not available, it takes a frame from the flush list; and if this is not available, it steals a page from the user list. In the last case STEAL.PG, like the system it represents, seeks a page whose reference bit (RFRNC.BIT) equals zero. The reference bit is turned on (set to 1) by 370 hardware whenever a page is referenced. It is turned off (set to zero) by system software. In fact, it is set to zero when each page with referenced bit equal to one is encountered in examining the user list until a page with a zero reference bit is reached. If the end of the user list is reached without encountering a page with reference bit equal to zero, then the user list is scanned once more, now usually encountering one of the pages whose reference bit has been set to zero. In seeking a page from the user list, STEAL.PG is not allowed to take a page frame marked as locked. In case of the presumably rare but conceivable state in which every page frame is locked, VSIM stops with an error message.

If a page is selected from the flush list or the user list, the page may have to be transmitted to a paging drum or disk. This is unnecessary if the page has not been modified during its current stay in main storage. If the page has been modified, its current copy will be written out to the particular slot on the particular DASD device from which its previous (now obsolete) copy was read. If the page has never been assigned a slot, then a slot will be assigned from the first paging device with a slot available.

When a page is removed from the free list (either when a page is reclaimed from the free list or when a page frame is taken for a new demand), if the free list is running low in a certain sense, STEAL.PG is asked to find a page to be moved from the flush or user list to the free list. In this case STEAL.PG is signaled by an argument (rep = 1) that it is not to take a page from the free list, but must take one from the flush list or, failing that, from the user list. STEAL.PG calls itself recursively for this purpose.

The routine in Fig. 13 explains to the computer the same things which the above paragraphs explain to the reader. The routine in the figure has been simplified by deleting from the VSIM version some lines which accumulate certain simulated times, such as the time required to remove a page from the free list. (WAIT, referred to on the next to the last line of the routine, is a function subprogram which takes a distribution type and two parameters as arguments and returns a perhaps random time increment. The reference to TIME rather than TIME.V is not a bug, but further discussion of this point must wait for VSIM documentation, now in queue to be written.)

Examples such as in Fig. 13 suggest, to me at least, that a SIMSCRIPT-like language could be used in the programming of computer systems to CREATE, DESTROY, FILE, REMOVE, and search though sets of entities such as jobs, jobsteps, data requirements, IO requests, and queues for channels. If the translator wrote sufficiently efficient code, its object program could be used in the final product. Short of this, its object code could be used in an initial version of the computer system to test for errors in logic and specifications, while assembly language coders produced hand-honed versions of the components of the system.

## AN ENTITY, ATTRIBUTE, SET, AND EVENT VIEW OF DATA BASE SYSTEMS

Consider the statement

```
FOR EACH MACH.GRP IN SERVED.BY(LAB.CLASS),
  WITH FREE(MACH.GRP)>0 AND F.QUEUE(MACH.GRP)>0
  COMPUTE MM=MAX(MACH.GRP) OF JPRI(F.QUEUE(MACH.GRP))
```

This could be part of a description of the actions to be taken by a simulated factory. It could also be part of the description of the actions to be taken by a computer system in an actual factory in which the status of workers, machines, and queues of jobs is stored in a computer data base. The calculation could appear in a program invoked when a worker reports that he has finished a previously assigned job and now awaits a new assignment.

We saw that the above statement can appear in a SIMSCRIPT II source program describing a simulated factory. Why cannot the same coding appear in the source program for implementing the actual computer-assisted factory? More generally, why cannot the same coding used to describe a system for purposes of simulation also be used to describe a system for purposes of implementation?

One reason why precisely the same code could not be used is that the real factory must keep track of more types of entities than are required for simulation. A data base describing the real factory, for example, would probably distinguish between individual machines as well as machine groups, so that it could keep track of such attributes of machines as date of last scheduled maintenance. It would also keep track of individual workers as well as labor classes, and would probably keep a set of vouchers owned by worker indicating which jobs the worker had been assigned, when he started them, and when he reported their completion.

But these are just more entities, attributes, and sets which can be described in the same manner as the entities, attributes, and sets already included.

Another difference between the coding for the simulated system and that for the real system would be that the description of events would not be the same. In our job shop simulator an end of process was treated as an endogenous event caused within the simulated system. The timing routine kept track of this and other internally scheduled events. In the real system the computer would assign a job to a worker but would not actually cause the completion of this job. Rather the report that the job was completed would come back to it from outside its domain of direct control, and would be treated as if it were an exogenous event.

The computer system nevertheless would have a timing routine and could schedule internally triggered events. For example, at the time a job was assigned to a worker the computer system could schedule an event to check that the job was completed in a reasonable length of time. If the exogenous report that the job was completed occurred before this scheduled time, then the endogenous event would be canceled; otherwise the endogenous event would occur, invoking an event program which would print an inquiry to the worker or his foreman as to the status of the job.

The system, then, would still distinguish between exogenous events triggered from outside its own control and endogenous events scheduled from within. The latter would, in effect, be filed in a set ranked by time and occurrence—as in the managing of endogenous events in a simulation. In the real system the timing routine would consult the real clock to decide whether the next endogenous event occurrence should be invoked yet.

Even if the factory's entity, attribute, and set description were augmented to reflect the detail required for the real system, and even if the event description were altered to reflect the new boundaries between inside and outside, a SIMSCRIPT II Level 5 translator could not compile the real system. One reason is that the entities of the simulated system, such as those in Fig. 5, disappear at the end of program execution. In the real factory the description of the system must persist even when no event program is executing. The function of SIMSCRIPT II Level 6 is to introduce data base entities, i.e., those whose description lasts beyond a single execution and which, if appropriate, can be accessed by many users. It also introduces the scheduling of events in real time.

In many respects the data base entities of Level 6 are like the main storage entities of Level 4; and the real time events of Level 6 are like their counterparts in Level 5. There are, however, some important differences. We shall first sketch the differences and then review the similarities.

In Level 4 the main storage entities (including their attributes and sets) were defined within the program. Data base entities, in contrast, are defined separately from any particular application program. Their existence is noted in a data base dictionary provided to the application programmers, and in an internal dictionary which supplies the same information to the computer. Within a Level 6 program, main storage entities may appear as well as data base entities. The programmer must define the former; the latter are defined for him.

When an object program written by a SIMSCRIPT II compiler with Level 6 capabilities refers to the attributes of a data base entity, it does not access the data base itself directly; rather it calls on a "custodian" program to transmit the attributes of this entity to its main storage area. Similarly it asks the custodian to file an entity into, or remove an entity from, a set rather than take these set processing actions itself. The tasks of the custodian include the satisfying of the demands made upon it by the executing programs, plus a general requirement to protect the security and integrity of the data base.

The way in which the custodian arranges attributes and sets in the data base, and the way that it accomplishes certain basic actions, are not necessarily the same as the way the corresponding matters are handled for main storage entities. For example, filing a main storage (Level 4) entity into a ranked set involves accessing each member of the set in turn until the proper position of the newcomer is located. Since data base entities are typically stored on disks, and the accessing of records within disks is relatively slow as compared to the transferring of contiguous information within a record, it is efficient to store ranked sets differently for data base entities. In the SIMSCRIPT I. 5 and PL/I based implementations discussed in the History section, for example, references to the members of a ranked, data base set are stored so that a number of such references (plus the values of their ranking attributes) are located together in a record to which the owner entity points. Several such records are used when one does not suffice. It is therefore possible to file or remove an entity from a ranked set by accessing a relatively small number of these "set-holder" records rather than accessing the individual set members themselves.

Some methods of storing attributes and sets are more desirable under one set of circumstances; others, under other circumstances. An example of a situation which invites and permits special storage methods is that in which a small amount of the information concerning each of a great many individuals is processed periodically as with the printing of a telephone book. For main storage entities the individual programmer specifies storage options in the preamble of his program. For data base entities the system designer or data base administrator elects among available options in definition statements which are outside of any individual program.

Some differences in the forms of commands appear in Level 6. For example, the FOR EACH OF set phrase at Level 4 (and therefore at Level 6) includes options such as tracing sets in reverse order or tracing sets starting with some specific member in the set. This phrase is augmented at Level 6 by the option to trace out the set IN ANY ORDER. Because of certain details as to DASD accessing, a set traced out in arbitrary sequence may in some cases be traced out more quickly than one that must be traced out in a specified order. The programmer need not have these details in mind when he specifies IN ANY ORDER but only needs to remember that this option may save time if order is not important.

In some cases new options may be added to the wording of certain commands to preserve the self-descriptive nature of the SIMSCRIPT commands. For example, the statement CREATE A WORKER would not be self-descriptive when a new worker is enrolled in the system, since the computer does not create a worker but simply notes that he or she has passed into its domain of cognizance. The create statement can easily be modified to allow the form NOTE A NEW WORKER, or THERE EXISTS A NEW WORKER, where the word NEW like the word A is optional. Such synonyms are already common in SIMSCRIPT II, and are easily added where their desirability justifies the increase in stored dictionary size.

In addition to modifications of existing commands to preserve a self-documenting style or to increase efficiency, three new commands are added at Level 6. A RECORD statement is used for recording changes made to the data base when the programmer wants to override the default treatment as to when changes are only provisional and when they are officially recorded. A LOCK and an UNLOCK statement are used when the programmer wishes to override the default locking of entities. In general, the locking of data base entities is needed to avoid situations in which, say, reservation clerk A is told by the computer that a seat is available, then B is told that a seat is available, then A subtracts one available seat, then B subtracts a seat that is no longer there. User specified locking, as distinguished from default locking, can be used to avoid deadlocks in which program I has entity i but cannot complete without access also to entity j, and Program II has j and cannot complete without i (see Ref. 6). If a deadlock does in fact occur, the custodian must reverse out of the data base the effect of one of the deadlocked programs. This is feasible but time consuming, and in some instances can be avoided by user programmed LOCK statements when default locking could deadlock.

Similarities as well as differences should be noted between Level 6 entities and events and those of Levels 4 and 5. The system designer, data base administrator, or application programmer still thinks of the world to be represented by the data base, like the world represented by a simulator, as consisting of entities of various types with their attributes and set memberships. In designing a system or a new application program, the mind is to be directed in the first instance to that which is to be represented—jobs, queues, machines, men, and so on. The decision as to how to represent these within the computer is a separate and (mostly) subsequent decision. Since the world still (at Level 6) consists of entities, attributes, and sets, the only way that status can change is still if a new entity is created (or its existence noted), an entity is destroyed (or forgotten), it is filed into a set, removed from a set, or a new value of an attribute is assigned or read. These elemental actions may still be done repeatedly under the control of FOR EACH OF set phrases which may be concatenated with other FOR phrases and with WITH, UNTIL, UNLESS, and WHILE phrases, and then used as the control for statements such as READ, COMPUTE, or DO. Events can still be caused and canceled, though the timer now must look at the real clock to decide whether the most imminent endogenous event should occur yet. Statistics concerning attributes can still be ACCUMULATEd over time.

## Comparison with Other Data Base Languages

At present three views are frequently described as the major approaches to data base systems [7]. The first approach is called hierarchical and IMS is cited as the principal example. The second approach is called "network" and the CODASYL committee's DBTG language is cited as the principal example. The third approach is called "relational."

DBTG [3] is in fact an entity, attribute, and set language. The basic concepts of DBTG are the "record," the "field," and the "set." In SIMSCRIPT terms the record is DBTG's way of representing an entity, the field is DBTG's way of storing an attribute, and the DBTG sets (like SIMSCRIPT sets) are ordered collections with owner and member entities.

DBTG is referred to as a network language, presumably because its entity, attribute, and set view can represent a network, in contrast to the hierarchical view which has difficulty here. It is true that the entity, attribute, and set view has no difficulty representing the status of a network. Nor does it have difficulty representing the status of a job shop, or of an accounting system, etc. I therefore feel that the term network is a misnomer.

One attractive feature of DBTG is the automatic maintenance of sets ranked by some attribute of its members. In SIMSCRIPT I and II, if the attribute is to be assigned a new value, the programmer must first remove the entity from the set, change the attribute value, and then file the entity back into the set. In DBTG, if the set is defined appropriately, the member entity will be automatically repositioned if the ranking attribute is changed. This seems to me to be a desirable addition which should become standard in entity, attribute, and set languages.

There are three ways in which the SIMSCRIPT II Levels 1 through 6 language differs from the DBTG language. The first is SIMSCRIPT's greater emphasis on the distinction between (a) that which is to be represented by the data base, and (b) the method by which this is to be represented. For example, DBTG speaks of records where SIMSCRIPT speaks of entities. "A record," in the sense of a sequence of adjacent memory locations on some storage medium, may be used to store the attributes of an entity. It may be desirable, however, to store the attributes of one individual entity on more than one such record—e.g., storing the attributes of various entities in records of a fixed size (to facilitate the handling of dynamic storage requirements) with large entities being stored in several not necessarily adjacent records, while small entities share a single record. Less frequently used attributes of a given entity could be stored on slower, cheaper storage devices than more frequently used attributes. Sometimes a collection of several not necessarily adjacent records representing one individual is referred to as a "logical record" or a "data base record" as opposed to a physical record. Perhaps this is a justifiable terminology. Be that as it may, the SIMSCRIPT view is that the description of a system, and even the programming of its applications, should be pursued first in terms of the entities of the system itself. Decisions concerning the method of storage are a separate and mostly subsequent matter. The separation of the task of system description from that of selecting storage mechanisms simplifies both tasks.

A second difference between DBTG and SIMSCRIPT concerns the instruction repertoire at the disposal of the application programmer. Most basic, perhaps, is the use in SIMSCRIPT of the control phrase. In order for a DBTG programmer to do a FOR EACH OF set WITH ..., he would have to write, in a host language such as COBOL, a call on a DBTG routine which supplies first-in-set. Then he must test to see whether an error condition code is set indicating there is no first-of-set; that is, the set is empty. In the latter case he transfers out of the loop; otherwise the program executes the WITH test which the programmer writes in COBOL referring to the data base attributes which now have been moved into main storage variables. If the WITH test fails, the programmer transfers to the point in the code where a next-in-set is sought; otherwise the program proceeds to the coding controlled by the FOR phrase. Following this is a test (or a transfer to a test) of whether the current member of the set has a succeeding member, and accordingly the succeeding member becomes the current member, or control is allowed to exit beyond the loop.

It is not only that it may take several statements to do the same thing that a FOR and a WITH phrase may do, but it also requires attention by the programmer, and later attention by the reader of the program, to the method of implementation as opposed to the action represented. This makes the program less easy to write and less clear in purpose to the reader.

A third difference between the SIMSCRIPT and DBTG approaches concerns DBTG's "shifting of gears" between the data base and the main storage languages. DBTG has an entity, attribute, and set view of the data base. When the programmer needs the attributes of an entity, or the next entity in a set, he calls for these to be brought into main storage. There they are operated on by a language which is not thought of as an entity, attribute, and set language. When results are to be transferred back to the data base, an interface must again be invoked by the user. In SIMSCRIPT the programmer may freely intermingle references to main storage or data base entities, attributes, and sets, even within a single statement. It is up to the translator to remember which is which. In this way a single command, perhaps compounded from two or more phrases, may reference both main storage and data base entities.

The chief asset of the COBOL/DBTG approach is that the host language is already familiar to a large class of programmers.

IMS began as a hierarchical data base language. It is no longer one in fact, although it still shows traces of its hierarchical past. To explain what it was and what it has become, let us begin by considering the nature of a hierarchy. In a hierarchy there is an entity type (e.g., country) with attributes (such as population and square miles) which belong to no sets, but may own one or more sets, such as a set of cities. Each city belongs only to the set owned by a country. It has attributes and may own sets. The members of the latter sets, say libraries, may in turn only belong to the set owned by city, but may have attributes and may in turn own one or more sets; and so on. Thus a hierarchy may be viewed as a system of entities, attributes, and sets satisfying certain conditions.

Such a hierarchy is referred to in IMS as a data base. Another hierarchy, say with corporations owning departments and departments owning employees, is considered to be a separate data base. When using IMS, a single program may refer to more than one data base. The terminology of IMS does not refer to parts of a hierarchy in terms of entities, attributes, and sets as we have done. Rather it refers to the entity which belongs to no set as a "root segment," and the entities which belong to sets as "dependent segments."

We have described a hierarchy in terms of its entities, attributes, and sets. It is not conversely easy to describe any system of entities, attributes, and sets in terms of one or more hierarchies—the links and nodes of a network being a classic example.

A major feature which has been added to IMS is the "logical relationship." The logical relationship of IMS does precisely this: it allows an attribute of an entity in one hierarchy to refer to an entity elsewhere in the same or another hierarchy. This addition gives IMS a full entity, attribute, and set capability. For example, consider a system whose entities are courses and students, where course is Calculus 1 as distinguished from some particular session of Calculus 1. Suppose we would like to note, for each course, all the students who have taken this course, and for each student all the courses he or she has taken. In SIMSCRIPT we would accomplish this by defining a new entity type, call it an enrollment, which will serve the membership card role in this instance. When a student enrolls in a class we create an enrollment, note class and student as its attributes, and file it into a set belonging to class and another belonging to student. The enrollment itself may have attributes such as date of enrollment and grade achieved. This mechanism will serve even if the student enrolls in a given course more than once.

The hierarchical view, augmented by the logical relationship, can accomplish the same thing in the following manner. A course data base describing the course hierarchy can include enrollment as a dependent segment with a reference to a student in another data base; a student data base describing the student hierarchy can also have enrollment as a dependent segment with the latter dependent segment referring to a course in the course hierarchy. In this manner we can go from a given course to each student in the set of students who have enrolled in the course, or conversely from a given student to each course he has taken. In a similar manner, if we wish to represent the queue of a machine group, we could have a machine group data base with machine groups as the root segment, have another data base with jobs as the root segment, and include in the machine group hierarchy a dependent segment that refers to the jobs of the job data base.

Thus any hierarchy can be described easily in terms of its entities, attributes, and sets; and conversely any system of entities, attributes, and sets can be described in terms of one or more hierarchies plus logical relationships.

The three differences between SIMSCRIPT II and DBTG noted above apply to SIMSCRIPT II and IMS. One difference is concerned with SIMSCRIPT's emphasis on the distinction between the description of the system to be represented and the description of how these entities are to be represented. When I personally look at a job shop or life insurance company, I see various kinds of entities, not various kinds of root segments. A second difference concerns the kinds of commands put at the user's disposal, and the third difference has to do with the shifting of gears as we move from data base to main storage and back again. The paragraphs on this subject for DBTG apply word for word to IMS.

In addition to noting differences between DBTG and IMS on the one hand and SIMSCRIPT II on the other, we should note similarities between them. In each of these data base systems the application program calls on what we refer to as a custodian program (different terminology is used in different systems). The custodian program then must serve functions such as supplying the application program with the information it wants, or modifying the data base in the manner that the application program specifies. Since both DBTG and IMS can represent arbitrary entity, attribute, and set descriptions, their structures may be considered as two examples of how entities, attributes, and sets could be stored. Each elemental action of create, destroy, file, remove, and the reading and assigning of attributes has a realization within DBTG and IMS. The custodian also has functions concerning the security and integrity of the data base. In particular it must be capable of preserving the data base as of some period in time in case of a system crash. In light of the preceding paragraphs, DBTG and IMS may be considered as examples of how integrity and security may be implemented within an entity, attribute, and set system. These mechanisms are applicable whether or not the system designer originally considered the application in terms of what kinds of entities are to be represented or what kinds of root segments are needed; and whether the programmer could write FOR EACH OF set WITH or had to spell this out in terms of get first, test, check, get next, test, and so on.

The "relational" data base view, developed by Codd [4, 5], provides an alternate way of viewing a system to be represented. To illustrate: the relationship father/son may be represented by pairs (father, son). All such pairs, for some reasonable size population, could be listed in a table whose first column is father, whose second column is son, and whose rows therefore are particular father/son pairs. The entire table is referred to as representing "the relationship."

We may define a quantity-used relationship by a triplet

(assembly, component, quantity of component used in assembly)

All such triplets can be represented by a table whose first column indicates assembly, whose second column indicates component, and whose third column indicates the quantity of the latter in the former. The entire table represents the relationship; a line of the table represents a particular triplet which bears this relationship to each other.

In general, a relationship in the relational data base view is an n-tuple whose first element represents an entity of some particular type, whose second entry represents an entity of the same or another type, and so on (where here as elsewhere we include numbers and names as entities). These n-tuples may be conceived of as arranged in a table with n columns and as many rows as there are instances of combinations which bear the particular relationship to each other.

An entity of a system (as a SIMSCRIPT programmer might view the system) may appear in more than one relationship. For example, a part might appear in a quantity-supplied relationship, in a who-supplies relationship, and so on.

The relational view and the SIMSCRIPT view of system status are equally general. To show, first, that the SIMSCRIPT view can easily represent the status of any system that the relational data base view can represent: For each relationship (table) of a relational view, define a SIMSCRIPT entity E such that the first attribute of E is of the same entity type as in the first column of the relational table; the second attribute of E is of the same entity type as in the second column of the table; and so on. Then each individual of type E may represent one row of the table (i.e., one n-tuple) and the entity type E then is equivalent to the relationship. (Incidentally, in SIMSCRIPT courses it is not unusual to occasionally portray an entity type by a table with rows representing entities and columns representing attributes.)

That conversely any system representable in SIMSCRIPT can be represented by by the relational view may be shown as follows. For each entity type in a SIMSCRIPT application define a relationship (i.e., a table) whose columns are the attributes of the entity type. In the case of compound entities define a table whose first column is, for example, worker, whose second column is machine, and the remaining columns are the attributes of the worker, machine combination. We may thus take care of the attributes of a simple or compound type. To represent a set define a relationship whose first column is the owner entity and whose second column is the member entity. The table containing all such (owner, member) pairs contains the same information as the SIMSCRIPT set.

The the two views are equally general is thus shown by showing that SIMSCRIPT's attributes can represent relationships, and relationships can represent attributes and sets. This still leaves open the question as to which is more convenient: the SIMSCRIPT view (<u>not</u> to represent the relational view, but to represent the system directly), or the relational view (<u>not</u> to represent the system by representing the SIMSCRIPT view of the system, but to represent the system directly).

Codd's advice as to the proper use of the relational view is revealing on this point. In Ref. 5 Codd describes the first, the second, and the recommended third normalized form of a relational representation. He starts with a representation which does not meet the conventions laid down in the relational view in that repeating fields appear in single rows of a relationship table. He shows that by describing the system differently the same information can be represented by the simple table described above. This is the first normalized form of the relationship. The second and third normalizations deal with problems such as the following. In general, relationships can be defined so that, e.g., the attributes of vendors appear only in the quantity-purchased relationship. In this case it would be impossible to refer to the attributes of a vendor for whom no purchases appear in the quantity-purchased relationship. Such "anomalies," as Codd calls them, which can occur in general in a relational representation, can be eliminated by transforming such potentially anomalous relationships into third normal form. Codd's example of a system thus transformed has eight relationships representing, respectively, the attributes of a supplier, the attributes of a part, the attributes of a project, the attributes of a (part, project) combination, the attributes of a department, the attributes of an employee, the set of parts actually supplied by each supplier, and the set of parts which a supplier could supply.

As this case illustrates, the third normal form of the relational view uses tables either to represent the attributes of a given entity type or to show the same information as SIMSCRIPT's (owner, member) set representation. While the relational view could be used otherwise, the recommended practice—which Codd says "data base growth tends to force"—is to define relationships which in effect simulate SIMSCRIPT's view of a system. It seems to me that to start with the concept of an arbitrary relationship, and then to avoid the problems which can arise from this view by converting it to an entity, attribute, and set viewpoint, is like starting with a football field, putting a homeplate at the intersection of the goal line and the offsides line, putting a first base along the side line, and so on, calling the resulting game the third normal form of football.

Codd proposes to change status by applying "the powerful operations of relational algebra or the expressions of relational calculus to derive many other relations from those contained in the data base." These operations and expressions process the relationship (i.e., the entity type) as a whole. SIMSCRIPT operations modify individuals, perhaps under the control of FOR phrases. I believe that the latter approach is usually more efficient while being little if any less convenient to the user. The job shop example in Fig. 5 illustrates why. In it, as in many or most systems, the most frequently occurring events modify only one or two individuals of an entity type. Even where many or all individuals are to be referenced or modified, as in the reports in END.SIM, the SIMSCRIPT control phrases can specify this without burdening either the programmer or the readers of the program. This is also the reason that neither SIMSCRIPT I nor SIMSCRIPT II have operations like JOIN or INTERSECT which process sets as a whole.

However, since global operations on entire sets or entity types can be defined in terms of operations on individuals, they could be added to the user's version of the SIMSCRIPT II language by means of the SIMSCRIPT II language writing language discussed in the next section.

## THE SIMSCRIPT LANGUAGE WRITING LANGUAGE

The translators for SIMSCRIPT I.5, SIMSCRIPT II, and SIMSCRIPT II.5 are based on an entity, attribute, and set view of the translation process. Every statement in SIMSCRIPT is viewed as having a set of parts, each part having a set of alternatives. For example, the CREATE statement, as augmented for Level 6, may be viewed as having five parts:

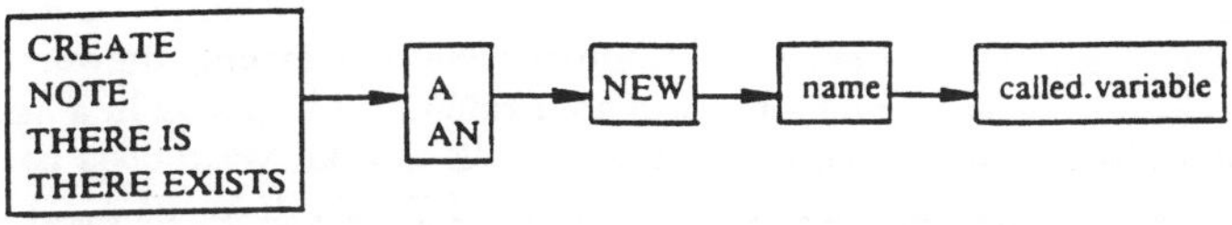


The second of these parts has two alternatives, A or AN. One of the attributes of part is a flag (or Boolean attribute) indicating whether or not at least one instance of the part is required. Another attribute of part is a flag indicating whether or not more than one instance is permitted. In the CREATE statement, for example, an instance of the first part is required, instances of the second and third parts are not required, an instance of the fourth part is required, and the fifth part is not required. No more than one instance of any of the four parts is permitted.

An alternative may specify a literal character string such as CREATE, or may specify a primitive such as integer or name (the latter being a character string not containing blanks or certain special characters such as + - or ,), or an alternative may refer to some other pattern such as called.variable. This other pattern has parts with alternatives, which may in turn refer to other patterns. (This description expresses, in the form that the SIMSCRIPT translator uses, the content of a Backus Naur Form description. See Ref. 15.)

When the SIMSCRIPT translator encounters a source program statement such as

CREATE JOB CALLED J

it "structures" this statement as shown in Fig. 14. Each word in the source statement is represented by a WORD entity (indicated by a W in the figure) filed into a

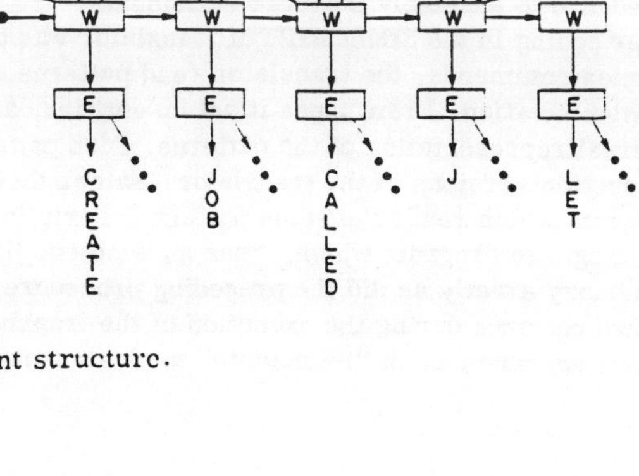


FIG. 14. Statement structure.

set. The WORD representing the word CREATE does not in fact contain the letters C R E . . . ; rather it points to an entity called an ENTRY which is filed into an I-th set, DICT(I), where I is an integer obtained from the letters by hashing. The dictionary is partitioned in this manner to expedite lookup. The letters C R E . . . are, in effect, a TEXT attribute of entry. All information concerning the word CREATE known to the translator is stored either as an attribute of entry or in a set belonging to entry. For example, since the fact that CREATE is a keyword does not prevent its use also as a variable, attribute, or label, SIMSCRIPT II keeps a set of possible uses of this word. The set is owned by ENTRY.

The STRUCT routine attempts to fit a pattern to the incoming stream of words, tracing out in turn any patterns referred to by the statement pattern. In the present example STRUCT will succeed and will leave the entity-attribute-set structure sketched in Fig. 14. It notes the set of parts which directly make up the statement and the parts which in turn make up a given part to any depth required. It notes for the statement as a whole, and each part of the statement, the first WORD and last WORD spanned by the statement or part. Words from the following statement may be read into the set of WORDs but will be ignored once STRUCT determines the largest valid statement it can structure at this point. The arrows emanating from the statement (S), each part (P), each word (W), and each entry (E) in the figure indicate attributes of the respective entities, including attributes due to set ownership and membership. The order of the arrows was chosen to facilitate the drawing of the figure.

Having built this structure, STRUCT returns control, indicating that structuring has been successful. The translator next calls on the routine which knows how to process, in this case, a structured CREATE statement. A translator routine with the responsibility for processing a particular type of statement may proceed in any or all of four ways. (1) As specified by the SIMSCRIPT commands in its source program, it may analyze or modify the entities, attributes, and sets of the structured statement or its dictionary entries. (2) It may call a subroutine, perhaps giving the called routine one or more parts as arguments. (3) It may write assembly code by calling a routine which receives the contents of the assembly language instruction as arguments, rewriting this in the manner expected by the assembler. Or (4) it may "SCRIPT" a more complex statement in terms of one or more less complex SIMSCRIPT statements. These less complex SIMSCRIPT statements are inserted before the next command of the original source program, and are translated in the same manner as any other source statement. The intermediate SCRIPTed statements may be seen as comments along with the original statement in the assembly language listing produced by the translator. Eventually, as any generated statements are translated, and any statements they generate are in turn translated, all is reduced to assembly language statements.

When the earlier coding in the SIMSCRIPT II translator was being used to program more complex commands, the translator read patterns as data described in a pattern description notation. From these it set up entities, attributes, and sets to form an internal representation of the patterns. Such pattern making is not done now in the production versions of the translator. Rather there is a separate pattern-making program which reads the same pattern description notation and writes an assembly language program which, when assembled, link edited, and loaded, arranges memory exactly as did the preceding procedure, but without the time required to make patterns during the execution of the translator. Below we shall refer to the first approach as an "immediate" method of implementation, wherein language extensions are incorporated and used in the same execution of the translator. We shall refer to the second method as a "permanent" method, wherein language extensions cannot be used immediately, but need not be defined again at each execution of the translator since they become a permanent part of the generated translator.

SIMSCRIPT I.5, II, and II.5 also have a notation for describing the SIMSCRIPT statements which will be written by SCRIPT back into the source language stream. This notation may refer to parts of the structure produced by STRUCT. The notation describing statements which may become SCRIPTed are now processed by a separate program which produces a section of assembly code which, when assembled, link edited, and loaded, stores script statements in the form that the SCRIPT routine expects. This section (or CSECT) becomes part of the new translator.

SIMSCRIPT II's current pattern and script description notations were designed for easy implementation, to be used primarily for the building of SIMSCRIPT II itself. If easier-to-use, more self-documenting notation and procedures performing these same functions were implemented, then the original objectives for Level 7 would be completed. An example of the intended use of the language writing language would be the development of commands to facilitate optimization calculations, statistical analyses, or plotting. Such "packages" now usually take the form of subroutine libraries. One objective of Level 7 is to put at the package developer's disposal the ability to provide his consumer with commands as well as library subroutines. In some cases subroutines are simpler for everyone. In other cases well-designed commands provide greater flexibility and a more self-documenting application program.

## Additional Level 7 Needs

The features of SIMSCRIPT II Levels 1 through 5 were in part the result of extensive programming. Some of this was real programming by users of SIMSCRIPT I; the rest was hypothetical programming using proposed features for SIMSCRIPT II. Actual experience with the SIMSCRIPT II language writing language, not available of course when first designed, suggests some hypotheses concerning additional features still needed in the Level 7 area.

Level 7, as implemented for our own use and planned for wider use, facilitates the definition of new commands in terms of old. What now seems needed in addition is the ability to define new modes for attributes, new ways of storing arrays or sets, and new operators for new or old modes. These same facilities may be described in terms of "predefined entity types" as discussed under Basic Concepts and under Entities of Logic and Mathematics. While the solution presented below stays as close as possible to current SIMSCRIPT II terminology, the entity nature of mode, for example, should be apparent.

To illustrate the need, suppose that for some class of users the absence of complex numbers as a SIMSCRIPT II mode is an impediment to the use of the language. A user could program around this deficiency by always defining two real numbers (a real part attribute and an imaginary part attribute) wherever one complex number was wanted, using subroutines to perform complex addition, subtraction, multiplication, and division. Such do-it-yourself incorporation of complex numbers is not objectionable if complex arithmetic is rarely used, but could become burdensome if used frequently. It is desirable therefore for a programmer to be able to explain the nature of complex numbers to his version of the translator or to a version to be used by some class of users. An alternate solution might be to add complex numbers to SIMSCRIPT II for all users. But there seems to be an unlimited number of mode-like objects which might be useful for one class of user or another. For example, another mode that would be highly desirable for certain classes of applications is date, where date has day, month, and year as attributes (just as complex number has real part and imaginary part as attributes). The day, month, and year attributes could either be alpha or integer. Other modes might include height in feet and inches, or location in terms of x-, y-, and z-coordinates, or integers represented by an arbitrarily long set of digits. To meet these and untold other needs, it seems desirable to allow the user or package designer to define new modes for immediate or permanent incorporation into his version of the language.

The definition of new modes requires the definition of new operators upon them. For example, if complex numbers are introduced, then complex addition, and mixed real/complex addition if allowed, must be described to the translator. Users may also desire new operations on existing entity types. For example, a user may wish to use SIMSCRIPT's world view and set operations, and also use some APL-like operations on matrices. Such operations can be described in terms of the elemental SIMSCRIPT actions, but where frequently used it would be convenient to have the translator recognize these or other higher order operators defined for a class of users.

The array of sets, DICT(I), used by the translator to hold entries could be defined as a new method for storing sets and made generally available to users who wished to store and retrieve members of large sets according to a text attribute. A main storage (Level 4) method of storing ranked sets could be defined along the lines described above for storing data base (Level 6) ranked sets. This method of storing sets in main storage should give rise to a smaller number of page faults in a paged computer system. New physical storage devices might suggest new methods of storing data base attributes or sets. A variety of methods has been developed for storing and processing enormous matrices with relatively few nonzero entries (see Ref. 8). Any of these methods might be included in a version of the translator designed for users of such sparse matrices.

The above objectives of allowing new modes, new operators, and new ways of storing sets and arrays could be accomplished with small additions to the SIMSCRIPT II language. For the most part these additions are not new commands, but new alternatives for existing commands. The changes could be implemented within the structure of the existing SIMSCRIPT II translator, although a number of translator routines would have to be modified. A SIMSCRIPT II language incorporating these additions is referred to below as an "augmented" SIMSCRIPT II. Some of the features of the augmented SIMSCRIPT II are analogous to features which appear in other languages, particularly ALGOL, Pascal, PL/I, and SIMULA. We shall not attempt below to point out or analyze individual similarities. We simply note at this point that the individual features of the augmented SIMSCRIPT II were not chosen for their uniqueness, but to extend the SIMSCRIPT II language and translator as little as possible in order to achieve the capabilities noted above.

First we will describe the augmented SIMSCRIPT II procedures for adding new modes, operators, and methods of storage for main storage (Level 4) entities. Afterwards we present these for data base entities.

The adding of new modes, operators, and methods of storage can be implemented in either the immediate method or the permanent method described above. Some statements in augmented SIMSCRIPT II programming are used to define extensions to the language; others are used to make use of these extensions. In the immediate method, both "language extension time" and the usual "compile time" occur during a single execution of the translator. With the permanent method the language extensions are not available immediately, but are a permanent part of the newly generated translator.

The following statements could appear at language extension time in an augmented SIMSCRIPT II.

```
NORMALLY MODE IS INTEGER
NEW MODES...
   EVERY DATE HAS A YEAR, A MONTH, AND A DAY
   DEFINE DATE AS WORD ALIGNED
```

The following statements could refer to the new mode at compile time.

```
TEMPORARY ENTITIES...
   NORMALLY MODE IS DATE
   EVERY PERSON HAS A BIRTH.DATE, A WEDDING.DATE,
   AND A WEIGHT
DEFINE WEIGHT AS A REAL VARIABLE
```

The "new modes" statement is like the temporary entities, permanent entities, and events statements of Levels 4 and 5 in that it presents the translator with information concerning the usage of any entities defined under the given heading. The space requirements for the new mode DATE, and any other new modes, are described exactly as if modes were temporary entities. In the case illustrated the new mode contains three integer attributes: day, month, and year. If desired, these attributes could have been packed into a single word or into space occupying less than a full word, using the existing SIMSCRIPT II packing specifications. In general, a new mode entity may have any number of attributes, own any number of sets, and belong to any number of sets. In particular its own attributes may be previously defined mode entities.

The use of such a new mode is illustrated above in the definition of the temporary entity PERSON. PERSON is declared to have a BIRTH.DATE, WEDDING.DATE, and WEIGHT as attributes whose modes are respectively date, date, and real. The augmented translator will lay out the person record as follows: the first three words of the person record will contain, respectively, the year, month, and day of the person's birthdate. These attributes may either be referred to collectively as BIRTH.DATE(PERSON) or individually as YEAR(BIRTH.DATE(PERSON)). The fourth, fifth, and sixth words of the person record similarly contain the year, month, and day of the person's last wedding date, or contain zeros to indicate no wedding. The seventh word of the record contains the person's weight. Thus the three full words of date, as defined in this example, would be incorporated into the record of a temporary entity in exactly the same way as the two words of the DOUBLE precision mode of SIMSCRIPT II.5. Date could similarly be incorporated into one of the one or more records of a data base entity. We shall consider the attributes and sets of data base entities at greater length below. The storage of the date attributes of a permanent entity would be similar to the present DOUBLE storage in SIMSCRIPT II.5 for permanent or compound entities, except that three words would be allocated for date instead of the two for DOUBLE. In general, the preceding statements apply for the n words of any new mode entity, for n greater than or equal to one. If the entire new mode entity requires less than a full word, it may share space in precisely the same manner in which packed SIMSCRIPT II attributes now share space.

In the SIMSCRIPT II implementation for the IBM 360/370, temporary entities are aligned on double word boundaries, and in SIMSCRIPT II.5 the locations of DOUBLE attributes within records also start on double word boundaries. The define statement would be augmented, as illustrated above, to indicate implementation-dependent information concerning the permitted alignment of a new mode entity.

SIMSCRIPT II already has AFTER CREATING and BEFORE DESTROYING statements which specify routines to be executed after creating or before destroying an entity of a particular type. These statements may be useful for mode entities with complex structures, e.g., which point to one or more arrays or own one or more sets that must be initialized whenever an entity containing an attribute with this mode comes into existence, and which must be purged when such an entity is destroyed. This requires the translator to not only consider whether, for example, person has an after creating or a before destroying specification, but also whether one or more of the mode entities within person has such.

In some cases it is desirable to have one or more attributes of the new mode assigned before the AFTER CREATING routine is executed. This would happen in the natural course of things if the create statement was augmented in such a way that Lines 42 and 43 of Fig. 5 could be written as

```
CREATE A JOB WITH ARR.TM = TIME.V, JTYPE = S,
   AND PLACE = F.ROUTING(S)
```

If a CALLED J phrase were included in the above statement, then the attributes appearing in the WITH clause would be implicitly subscripted by J rather than JOB. An IN sets phrase, similar to the illustrated WITH attributes phrase, should also be permitted.

Operators on new mode entities, or new operators on old modes, could be described in routines which, for the most part, are like existing SIMSCRIPT II function subprograms. Examples of statements that could appear in the augmented operator subprograms are shown in Fig. 15. Line numbers are for our reference

```
 1.  ROUTINE R.DD.ADD(D1,D2) DENOTED BY +
 2.  DEFINE R.DD.ADD AS A DATE FUNCTION
 3.  DEFINE D, D1 AND D2 AS DATE VARIABLES
 4.  RETURN WITH D
 5.  ROUTINE R.DD.ABC(DATE,DATE) DENOTED BY +/*
 6.  ROUTINE TO R.DI.ADD(D,I) DENOTED BY +
 7.  DEFINE R.DI.ADD AS A COMMUTATIVE, DATE FUNCTION
 8.  DEFINE D AS A DATE VARIABLE
 9.  DEFINE I AS AN INTEGER VARIABLE.
10.  ROUTINE TO C.RC(R) CONVERTING REAL TO COMPLEX
11.  FREE.READ ROUTINE FOR DATE
12.  FORMATTED.READ ROUTINE FOR DATE DENOTED BY DR(I,J,K)
```

FIG. 15. Extracts from operator routines.

only. In general an operator appearing between two operands is a function defined on a compound entity. Thus the + in (integer + integer) is a function which the language could have required to be written as +(integer, integer). Lines 1, 2, and 3 of the figure could appear in a function routine in the augmented SIMSCRIPT II language, specifying that the R.DD.ADD function is to be invoked when the translator finds a + standing between two variables (or other subexpressions) whose mode is date. The mode of R.DD.ADD is itself defined to be a date, hence the routine which follows describes a (date, date) operator. The ROUTINE and DEFINE statements in Lines 1, 2, and 3 are the same as those in SIMSCRIPT II except for the DENOTED phrase which specifies an operator name consisting of a special character or sequence of special characters such as + or +/* (as in Line 5). This operator name would be used within expressions just as arithmetic operators are now used. An expression of the form date.one + date.two where date.one and date.two are date expressions would be implemented as a function reference R.DD.ADD(DATE.ONE, DATE.TWO). The RETURN WITH statement, Line 4, is still used to return the result. When the mode entity occupies more than one word, a register points to, rather than contains, the result. Unary operators such as -A, where A is complex, may be treated similarly using operator subprograms with one argument.

The augmentation proposed here not only requires modification of the routines which process expressions and variables, but also modification of the fetch routine which reads the source program and sets up the words (W) of Fig. 14. At present FETCH is prepared to handle arbitrarily long names but not arbitrary combinations of special characters. The combining of special characters into more complex operator names such as ++, +./, */+ in addition to the +, -, *, /, ** is required to permit an arbitrary number of user-defined operators while still recognizing, for example, a+b to be an operand followed by an operator followed by an operand.

Lines 6 through 9, in which a small augmentation of the DEFINE statement appears in Line 7, specifies a routine to be called if a + is encountered between an expression with date mode and an expression with integer mode. Line 10 illustrates the specification of a routine to convert from one mode to another where at least one of the modes is newly defined. The translator will write a call to this routine when the "from" mode appears on the right-hand side and the "to" mode on the left-hand side of = in a LET statement. Such a conversion routine may be called explicitly, if desired, as for example in an operator routine for mixed mode operands. A logical operator, which may appear in a logical expression such as BIRTH.DATE(PERSON) >= GIVEN.DATE, is described by a routine which returns a one or a zero to signal the truth or falsity of the comparison. No such routine is needed for the equality operator (as in DATA.AA = TODAY) for cases in which equality between two attributes with new mode M means equality between each of the components of M.

In addition to operator and conversion routines, new modes require input and output routines if the modes are to appear in READ and WRITE statements. Lines 11 and 12 specify that the following routines perform formatted or free form reads producing an attribute of the mode indicated. The syntax of these statements is patterned after SIMSCRIPT II's LEFT ROUTINE and RIGHT ROUTINE statements for monitored attributes. The formatted read routine statement also includes a phrase to specify the read format descriptor. The syntax for the FORMATTED. WRITE ROUTINE is like that of the formatted read routine. At present I see no easy way of incorporating an arbitrary new mode into a LIST or PRINT statement, as distinguished from a WRITE statement. Of course, any integer, real, or alpha components of a new mode can be LISTed or PRINTed.

The new operator and IO routines are presented to the augmented SIMSCRIPT II translator at language extension time. If a permanent implementation is obtained, the operator routines become part of the SIMSCRIPT II "library" as opposed to becoming part of the translator itself. The current library contains routines to read, write, and convert existing modes, routines to generate random variables, compute various functions, etc. Library routines are loaded into the user's object program only if needed.

Before discussing the defining of new methods for storing sets, we discuss a feature of the proposed augmented SIMSCRIPT II which would be valuable in itself and is assumed in the procedures for new set methods. In SIMSCRIPT I and II, variables and attributes which refer to user defined entities (such as J, S, and P in Fig. 5) are defined as integer variables. They could instead be defined as follows:

```
DEFINE J AS A JOB REFERENCE
```

where JOB is known to be an entity type because of its appearance in an EVERY statement. A global variable JOB would automatically be assumed to be a JOB REFERENCE when JOB is seen to be an entity type, just as JOB, MACH.GRP, and STD.TYPE in Fig. 5 are currently assumed to be global integer variables. For those cases in which a variable or attribute may refer, alternately, to one out of two or more entity types, one could, for example,

```
DEFINE ACT AS A PGG.ACT OR A CALL.ACT
  OR AN ACTION REFERENCE
```

where PGG.ACT, CALL.ACT, and ACTION are entity types. A variable or attribute could also be defined thus

```
DEFINE T AS A TEMPORARY ENTITY REFERENCE
```

or

```
DEFINE T AS A PERMANENT ENTITY REFERENCE
```

but the earlier form specifying the particular type of entity would be the encouraged practice.

One advantage of defining J specifically as a job reference is that the translator can check at compile time to see if an attribute of job is in fact being subscripted by an expression which is, or is allowed to be, a job reference. This feature would be worth implementing if only for the capability it gives the translator to tell the user that he could not have meant PLACE(I) since PLACE is an attribute of job and I is not a job reference.

Another advantage of this feature is the much greater freedom it permits in using the same name for attributes of different entity types. In terms of the entities in Fig. 5, for example, COST can currently be used as an attribute of both JOB and STEP, but only if the user arranges to have COST(JOB) and COST(STEP) located at the same place in their respective temporary entity records; whereas COST could not be used as an attribute of both MACH.GRP and JOB, nor of both

MACH.GRP and STD.TYPE, since a permanent entity currently cannot have an attribute denoted by the same word as that of any other temporary or permanent entity. On the other hand, if the proposed feature is added and the translator is told that J is a JOB reference, it would have no difficulty in writing the correct code at compile time when it encounters COST(J) or COST(MACH.GRP) or COST(exp) where exp is an expression whose type evaluates (according to the types of the operands and the types of the results of operators) an entity type for which COST is defined. The only restrictions on the use of common names for attributes is the fact that if an attribute (A) or a variable (V) can refer to one out of two or more different entity types, and B is the name of an attribute of one or more of these types, there must be no ambiguity concerning how to fetch or store B(V) or B(A(E)).

The storage requirements for the owners and members of a new way of storing a set, say the PARTITIONED method, would be described at language extension time by

```
EVERY PARTITIONED OWNER HAS ...
EVERY PARTITIONED MEMBER HAS ...
```

Partitioned ownership and partitioned membership would be treated like new modes. At compile time memberships, ownerships, and types of sets would be denoted as at present, as in

```
EVERY JOB OWNS A LIST AND A STRING,
  AND MAY BELONG TO A QUEUE AND A BOX
DEFINE BOX AND STRING AS PARTITIONED SETS
```

Because of the feature described above, two different ownership modes—say for set type ABC and set type XYZ—can store FIRST(A(E)) and FIRST(X(F)) in two different ways, where A is a set of type ABC, X a of set of type XYZ, and E and F are owners of A and B, respectively; and similarly for membership modes. If a programmer wishes to refer to, say, first of routing as in Line 43 of Fig. 5, it should not be required that he write F.ROUTING(S) if ROUTING is stored by an original method, or write FIRST(ROUTING(S)) if ROUTING is stored by a method added at some language extension time. It is desirable, therefore, to allow F.ROUTING(S) and FIRST(ROUTING(S)) to have the same effect if queue is organized by a current method. The first form is for compatibility with existing programming, the second to simplify subsequent programming.

FOR EACH OF set phrases would be implemented, as at present, in terms of the first, successor, predecessor, or last of set attributes. Any of these attributes could be stored as specified in the EVERY statement, or could be specified in function subprograms. The WITHOUT phrase could continue to be used at compile time to suppress particular attributes and routines, subject to consistency restraints on combinations that can be suppressed.

At present when SIMSCRIPT II sees that QUEUE is, say, a first-in-first-out set with no attributes or routines suppressed, the translator writes in SCRIPT and then compiles: a routine called A.QUEUE to perform a "file first" action, a routine called B.QUEUE to perform a "file last" action, a routine called C.QUEUE to perform a "file before" action, and so on. The number of arguments in one of these routines depends on the action performed and on the nature of the owner entity. A "file before," for example, refers to one more member type entity than does a "file last." A routine written for a set owned by a compound entity requires more arguments than does a routine performing the same functions for a set owned by the system.

In augmented SIMSCRIPT II there will continue to be routines with unique names to perform various actions for the various user-defined sets. For any particular action, such as a "file first," it may be elected at language extension time to have the action performed in the "standard" way, i.e., as at the present except writing, for example, FIRST(QUEUE( )) for F.QUEUE( ). Or alternatively, the action may be specified, for example, in a

FILE.FIRST ROUTINE FOR PARTITIONED SETS

This routine is presented to augmented SIMSCRIPT II at language extension time. It is not compiled; rather it is stored with the translator, either immediately or permanently, and used as a prototype. Within the routine may appear the words MEMB.E, OWNER.E, SET.N, and perhaps MEMB2.E, as in

LET MEMB.E = FIRST(SET.N(OWNER.E))

When it is determined at compile time that QUEUE is a set of the type in question, the word QUEUE is substituted for every use of the word SET.N in the prototype routine, and the names of the appropriate arguments are substituted for MEMB.E, OWNER.E, and perhaps MEMB2.E. The routine is then compiled and called as needed.

To summarize, as far as main storage entities are concerned, the storage requirements for new modes, including set ownerships and set memberships, are described exactly like the storage requirements of temporary entities, although they can then be used as modes for variables or for temporary or permanent entities. Actions which are taken in the processing of attributes and sets—such as the input and output of attribute values, the filing of members into sets, the conversion from one mode to another, and the processing of logical or arithmetic operators combining two operands of the same or different modes—are described in routines which differ little from current SIMSCRIPT II subprograms. The storage requirements and other matters related to new ways of storing and processing arrays will be discussed in the next section. These various facilities would allow the individual user or package designer considerable flexibility in defining the objects upon which commands operate, in addition to the previously discussed facilities for defining new commands.

In many ways the procedures for defining new modes, etc., for data base entities are like those for main storage entities. There are some differences however. Before discussing these similarities and differences we need more details concerning the division of labor between the data base custodian on the one hand and the object programs written by the translator on the other hand. It will be convenient in this discussion to assume that a data base entity occupies exactly one record. The treatment of multiple record entities is more elaborate but adds little in principle to the present discussion.

The record of a data base entity is laid out like that of a temporary entity. The user's object program (at execute time) cannot access the contents of this record directly but must call on the custodian to BRING it into the object program's main storage. If the entity is modified while in main storage, the executing object program must call on the custodian to STORE the entity. The BRING and STORE commands typically do not appear in the user's source program. Rather they are SCRIPTed into the program when a data base reference variable, e.g., one that appears in a statement such as

DEFINE J AS A JOB REFERENCE

(where JOB is a data base entity) takes on a new value because it is assigned one or because it is used in a FOR EACH OF set phrase. If the new value for J will write over an existing reference to an entity which has been modified, then a STORE will be executed for the old before BRINGing the new.

While the entity resides in main storage, the user's object program may access attributes directly. The program's "understanding" of where particular attributes are stored must correspond to that of the entity's image in the data base. This correspondence is achieved by a step we will refer to as "binding." One object program may be the result of binding source program A with data base E; another may be the result of binding the same source program A with data base F; where both E and F contain the entities, attributes, and sets referred to by A, but may be dissimilar otherwise.

New modes for main storage entities (e.g., "date" or "complex") which do not own or belong to sets and do not require AFTER CREATING or BEFORE DESTROYING routines may be used as new modes for data base entities. The k words of the new mode occupy a place in the n words of the data base record, just as they would in a temporary entity. They are brought in along with the other n - k words by a BRING; are operated on in main storage by the same operator, input-output and conversion routines that operate on them as attributes of Level 4 entities; and are moved out by a STORE along with the rest of the record. Only in the case of these particular main storage modes will we say that the same mode may be used as a data base mode and a main storage mode. In all other cases we insist that a mode be either a main storage or else a data base mode. There is no loss of generality here since we can always define a data base version and a main storage version of essentially the same mode.

Data base entities may have BEFORE and AFTER specifications like main storage entities, including some specifications we have already noted, such as AFTER CREATING, and some we have not had occasion to mention, like BEFORE FILING. In addition, data base entities may have BEFORE or AFTER BRINGING or STORING routines. These last four routines, if needed, plus the AFTER CREATING and BEFORE DESTROYING routines, if needed, should provide for any special set-up or purging requirements for a data base mode. Operators on data base mode entities, operators and conversion routines involving at least one data base mode, and input-output routines involving data base entities are all routines which are loaded with the user's object program and are called exactly like the operators for main storage modes. Once called, the operator routine may perform any augmented SIMSCRIPT II actions, including further data base actions.

The user's executing object program calls on the custodian to take actions on sets, such as the various file and remove actions; or to get the first, next, prior, or last member of a set; or perhaps to get a member with a specific value of the ranking or organizing attribute, as specified by a FOR EACH OF set WITH.... The definer of a new method of data base set storage must write his own such routines, or at least those which are allowed for the particular method of storage.

The actual storage space is still declared as if set ownerships and set memberships were modes. In some cases tricks like the following may be useful: have one of the attributes of the ownership mode refer to a data base entity (say, a "set reference" record) which you BRING when you need to file or remove, and which you then operate on in main storage as if it contained an array or set of data base references.

By these various means a considerable flexibility can be achieved in defining the space and processing requirements for new data base modes and sets. As with main storage modes and sets, the storage requirements may be visualized as those of a temporary entity, and the operators and other processing requirements are specified by routines to do the required jobs. The rules for writing these routines differ little from the current rules for writing routines.

## ENTITIES OF LOGIC AND MATHEMATICS

The SIMSCRIPT manuals speak of attributes as having modes, the modes in SIMSCRIPT II being integer, real, alpha, text, and subprogram. Some attributes defined as having integer mode are further described as being pointers to, or the identification numbers of, user defined entities. The view expressed in the present article is that the value of an attribute is always another entity. This other entity may be of a type which the user defines such as a man, a job, a voucher; or it may be an entity of a type already defined for him such as an integer or other object of logic or mathematics.

I believe that the view expressed here is more consistent than is the original view of attributes with the basic SIMSCRIPT objective of providing a language for describing a system and—separately from this—a choice of options for how to represent the various entities of the system within the computer. If asked to describe the business system of which he is a part, we would understand it if a worker responded "my department is manufacturing." We would be puzzled if he responded "my department is an integer variable which points to manufacturing." Strictly speaking, his department is not manufacturing; rather the name attribute of his department equals m a n u . . . . Thus, in the description of the business system, DEPT(WORKER) equals a department (not equals a pointer to a department) with

NAME(DEPT(WORKER)) = "MANUFACTURING"

Entities such as integers, rational and real numbers are frequently used in the description of systems. In the present section we present a system including these and other entities, all of which can be defined in terms of entities, attributes, and sets starting with a very small list of primitive entities. Thus the various entities of logic and mathematics which are defined already in SIMSCRIPT, or which may be defined using an extended language writing language, are not separate ad hoc creations but are part of a system derivable from a few basics. In this discussion we make no distinction between those entities which are the "entities of logic" as compared to those which are "entities of mathematics." Rather we use the two phrases interchangeably here as including the system of entities derived below.

In SIMSCRIPT I and II if A is an entity type and B is an entity type, then an (A, B) combination may be treated as an entity type. In the present discussion we shall also treat as entity types

(a) a finite, ordered set of A's; and
(b) an attribute associating a B (or an undefined value) with each A.

The former we will denote as an A-set, the latter as an (A->B)-attribute. SIMSCRIPT I and II have managed without these, for example, by ascribing the attributes of any set to the entity which owns the set; and, in the translators, by defining an entity type which represents a user-defined use of a word, such as a word used as an attribute name. The "attributes of the attribute" can be associated with this entity. While such can be done, it is more convenient here to allow the A-set and the (A->B)-attribute as entity types.

Imagine a system in which the only primitive entities are truth-value (an entity type whose only individuals are named "true" and "false") and character (an entity type whose individuals are named A, B, ..., 1, 2, ..., although a much smaller character set would suffice).

Given these two primitive entity types—the truth value and the character—and the above listed ways of forming new entity types from old, we may define: the character-set whose individuals include CAT and FISH; pairs of character-sets, an example of which is (CAT, FISH); and attributes (or functions) from pairs-of-character-sets to character-sets, an example of which is the concatenation function such that CONCATENATION(CAT, FISH) = CATFISH. Such an attribute or function defined from an (entity, entity) combination to an entity of the same type is usually referred to as an operator.

Among character-sets are the entities 1 11 111 .... The concatenation of two individuals among these gives another individual among these. For example, CONCATENATION(111, 1111) = 1111111. This is an example of a positive integer system with concatenation serving the role of addition. The subtraction, multiplication, and division operators may be defined in terms of the addition operator. By allowing the character set 0, and character-sets obtained by concatenating "-" with a positive integer, and by suitably extending the definition of addition, etc., we obtain a general (positive, zero, or negative) integer system. Other systems using other character sets, such as the binary or decimal number systems, may also be used as integer systems. Any of these is isomorphic to any other.

Pick any one such integer system. Henceforth when we speak of an integer we refer to an individual from the system you selected. When we speak of an entity of type integer we refer to an entity of type character-set, but only such character sets as are in your integer system.

A set of A's as the term "set" is most frequently used in mathematics (i.e., an unordered set as defined by Cantor) may be represented by an (A->true)-attribute. For example, a set of integers may be represented by a function or attribute from the integers to truth values, where integers whose attribute value equals "true" are in the set, and those whose value equals false are out of the set. When we have occasion to refer to such sets in the present section, we shall refer to them as $set_C$.

Rational numbers may be represented by (integer, integer) combinations with particular

(rational, rational)->rational

i.e.,

((integer, integer), (integer, integer))->(integer, integer)

attributes for addition, subtraction, multiplication, and division operators.

A real number may be defined as a $set_c$ of rational numbers in the manner of Dedekind. The complex numbers may be defined as pairs of real numbers with particular addition, subtraction, multiplication, and division operators. "Logarithm" is an example of an individual of the entity type (real->real)-attribute. Lesbegue integral (over the entire real line) is an individual of the type

((real->real)-attribute->real)-attribute

A vector of real numbers is an

(integer -> real)-attribute

such that (1)max.def(vector) is an integer attribute of vector, and (2) vector(integer) is defined (and equal to a real number) for every positive integer less than or equal to max.def(vector), and is undefined otherwise. A matrix or other array of fixed dimension may be defined similarly. A general n-dimensional real array, where n can vary as in APL, may be defined as a (vector->real)-attribute defined, at any time, only for integer vectors of a given size n whose i-th component satisfies $1 \leq a_i \leq m_i$, i = 1 to n.

Thus objects such as integers, real numbers, arithmetic operators, integrals, uncountable sets, and arrays of various sorts can be defined in terms of entities, attributes, and (finite ordered) sets, starting with the character set and the truth value as primitive entities. The "proposition" is another type of entity found in logic and mathematics. A proven proposition may be considered as a character-set which is a member of a set of proven propositions. The set of proven propositions is originally stocked with certain character sets called axioms. New propositions are filed into this set only if they can be "proved," i.e., only if there is a set of propositions called "steps" such that each step is the result of applying an inference operator (such as (A, (A implies B)->B)) to statements or pairs of statements which are either in the proven proposition set or are prior steps. See, e.g., Kleene [14].

A system of entity types was developed by Russell [20, 24] to avoid certain paradoxes that can arise if a function is allowed to have itself as an argument or a $set_c$ is allowed to have itself as a member. Some mathematicians have found onerous Russell's requirement to distinguish among the types of various entities of mathematics, and have developed mathematical systems which seek to avoid the paradoxes without introducing types (see Ref. 9). The view taken here is that types are not a nuisance but, on the one hand, a way of warning the computer of the storage and processing requirements of a particular entity, and, on the other hand, to my mind a clarifying taxonomy of the objects with which we deal.

The entities derivable from truth value and character are invaluable in the description of systems even when computer representation is not involved. Perhaps the most outstanding example of this is Euclidian geometry. Three-dimensional Euclidian geometry is a hypothesis about the nature of space. It is not the only possible hypothesis, as non-Euclidian geometries demonstrate. The Euclidian hypothesis can be described in terms of entities such as points and lines, attributes such as intersection(line, line) and line.determined.by(point, point), and sets such as the set of proven propositions. But the description of the system becomes much simpler when the entity type coordinate system is introduced along with the attributes x-coordinate, y-coordinate, and z-coordinate of (coordinate system, point). X, y, and z are not to be seen in the drawings of Euclid. Rather they are entities of the type real number definable in the manner sketched above. More prosaic examples of the use of such entities are found in our daily use of weights and measures.

Above we sketched the derivation of a system of entity types from two primitives, the truth value and the character. A different system can be built upon the entities of the storage media of a particular computer system such as the bit, byte, word, record, and cylinder, and upon the operations of the computer system such as the moving of the i-th word in an array of consecutive words, where the words are in memory and i is in a register. When the entities of logic and mathematics are to be represented within the computer, the entities (including attributes and sets) of the logical system must be represented by the entities (including operators) of the computer system. This representation may be relatively simple as when the mathematical entity integer is represented by a single computer word, or the (integer, integer)-attribute called integer addition is represented by the computer's add operation, or the representation may be more complex as when a higher order stored entity type is described in terms of more primitive modes, which may be described in turn in terms of still more primitive modes, until all are described in terms of computer entities.

The representation of entities in the logical system by entities in the computer system may or may not give rise to a one-to-one correspondence. For example, when integers are represented by an arbitrarily large set of digits, then the correspondence is one-to-one up to the capacity of the computing system to store such information. When integers are represented by a single word of computer memory, then the correspondence is one-to-one within a certain range of integers. Outside this range there is no value of the computer entity to correspond to that of the logical entity. When a rational number is represented by a computer word to be operated upon by floating point arithmetic, then there is no one-to-one correspondence within any nondegenerate range of rationals, since an infinite number of rationals lie between any distinct pair while only a finite number of floatingpoint representations are possible.

A rational number could instead be represented by a pair of integers where these integers are represented by an arbitrary sequence of digits. In this case there is a one-to-one correspondence up to the capacity of the computer. A real number could be represented as a rational number, or it could be represented by an integer part attribute plus a routine which returned the i-th digit of the binary or decimal expansion of the fractional part when given i as an argument, or it could be represented by a routine which returned a true or a false when asked whether a particular rational number was less than the real number represented. In none of these cases is a representation for an arbitrary real number possible, since there are only a countably infinite number of possible computer representations and an uncountable infinity of real numbers.

Since some computer representations of logical entities are not one-to-one, the computer representation R(O) of the logical operator O, when applied to the representations R(E), R(F) of logical entities E and F, may not produce a representation which corresponds to the same operator applied to the logical entities; i.e., we may have $R(O)(R(E), R(F)) \neq R(O(E, F))$. In this case the computer implementation may do the best it can and leave it go at that, hence round off error, or it may signal that the representation has broken down, e.g., when the number to be represented cannot fit in the number of bits provided to represent it.

While the system designer or analyst should think of the system to be represented in terms of its own entities, attributes, and sets in the first instance, he must also keep in mind that eventually the representation will be made in terms of the computer's system of entities. Thus successful SIMSCRIPT programmers have a general understanding of how entities, attributes, and sets are stored within the computer, as well as a knowledge of SIMSCRIPT syntax and an ability to view systems with an entity, attribute, set, and event view.

Suppose that A is an attribute of entity type E. Suppose further that the values of A are of the entity type real matrix. Then if e is an E, then A(e) is an real matrix. If i and j are integers and d=A(e), then d(i, j) is a real number. If we wish to allow expressions on the right-hand side of an assignment statement to be substitutable, in any expression, for variables on the left-hand side, then we must allow A(e)(i, j) to specify the same real number as specified by d(i, j) previously. Similarly more general sequences of subscripts such as F(i, j)(e)(k, l, m) could have meaning. Since statements and phrases do not begin with a left parentheses in the SIMSCRIPT II syntax, this new notation leads to no ambiguity and could be implemented within the present SIMSCRIPT II translator structure. This notation would also be useful in connection with the present SIMSCRIPT II subprogram mode (as described in the handbook [2] or manual [12]).

In the SIMSCRIPT II implementation of arrays there is a word of memory associated with each array, denoted A(*) for the array A, which points to the start of the array. This single word exists whether or not the array has been RESERVED yet, and whatever the dimensionality of the array. Currently this array pointer may be used in assignment statements, as in

```
LET APTR(E) = A(*)
```

or

```
LET A(*) = APTR(E)
```

This can be useful, for example, when an array of information is to be associated with a temporary entity. It should be more useful if we could refer to APTR(E)(I, J) (assuming here that A is a two-dimensional array) as suggested above.

In defining new methods of storing arrays in the augmented SIMSCRIPT II, whose discussion we have postponed until now, the array pointer need not be a single word but may be defined like a new mode. It may be used as an attribute of a permanent, temporary, compound, or data base entity, as well as the mode of an array pointer variable. The routines required to process an array stored by a new method consist of routines to RESERVE, RELEASE, get an element as in LET X = A(I, J), and put an element as in LET A(I, J) = X. The release routine for a new method could begin

```
REL ROUTINE S.FG1(M,N) FOR A SPARSE.FG ARRAY WITH W.SIZE = 1
```

Here SPARSE.FG is a new way of storing arrays; the routine that follows describes the actions to be taken for a two-dimensional array whose dimensions are denoted by M and N. Within the routine, A.PTR denotes the array pointer entity and W.SIZE denotes the size, in words, of the mode of the array; e.g., for a REAL array W.SIZE = 1, whereas for a DATE array W.SIZE = 3. Other REL ROUTINES for a SPARSE.FG organization (with names other than S.FG1) may be defined for dimensionalities other than 2, or with other restrictions on W.SIZE, as in W.SIZE > 1 or W.SIZE = 1/2. The compiler will write a call to the appropriate REL ROUTINE giving it as arguments: the W.SIZE constant, which it knows from the mode of the array; the values of the expressions for M and N; and the array pointer. Similar considerations apply to the routines to reserve, to get, and to put values.

I do not include in the augmented SIMSCRIPT II a provision for defining an n-dimensional array, n variable, since (1) an easy implementation within the current SIMSCRIPT II system is not apparent to me, and (2) the inclusion of such a capability seems redundant for SIMSCRIPT II with its existing ability to define any number of arrays with any (fixed) number of dimensions, as compared to APL, which spares the user from definitional statements generally.

In one sense it is obvious that attributes may be taken as a primitive notion in SIMSCRIPT, and then sets defined in terms of attributes. In particular, the various forms of the file and remove statements, and the "for each of set" phrase, may be described in terms of references to, or modification of, the first, last, successor, and predecessor attributes of owner and member entities. In another sense, however, it is not at all clear that sets are not required as primitives. In writing down commands to be processed by SIMSCRIPT, or even propositions to be stored in a logical data base, characters are filed into sets called words, words are filed into sets called commands or propositions, and these in turn into sets called routines or the set of proven propositions. The ability to distinguish one set of characters from another is usually one of the basic capabilities assumed of the computer or logician. Another basic capability is that of associating a character set with an object, for example the character set S I N E with a particular routine. We have here an example of a set—specifically, a set of characters—and an example of an attribute; namely, the association of this word with that object.

We may define a "kernel" as the capabilities required of a system with which to formally express concepts such as attributes or sets. It is not immediately clear whether or not such a kernel requires both attributes <u>and</u> sets.

Most logical systems assume, implicitly or explicitly, an ability to distinguish one set from another as part of the undefined kernel of capabilities. An exception to this may be found in Chapter 10 of Kleene's <u>Introduction to Metamathematics</u> [14]. Some translation is required to correlate Kleene's system with the view taken here. In particular, ignore Kleene's characterization of his system in terms of "metamathematics as a generalized arithmetic." His 13 zeros, some of which we may denote as &, ¬ , a, |, ... we shall think of, not as zeros of a generalized arithmetic, but the 13 entities of a primitive entity type. Kleene's generalized successor operation is clearly an attribute. Kleene develops a logic, essentially equivalent to that of Hilbert or Goedel, in terms of his 13 individuals of the primitive entity type, his successor attribute, and an ability to define compound entities of either the form (a, b) or (a, b, c). For example "to the variables a, b, c, d, ... of the formal system we correlate, respectively, the entities

a, (|, a), (|, (|, a)), (|, (|, (|, a))), ..."

Of course to explain this system to us Kleene uses sets of characters called words, sets of words called sentences, sets of sentences called paragraphs, and so on;

and to denote a compound entity he uses a set of characters printed side by side on a piece of paper. But if we abstract away the way in which Kleene's primitives are explained to us, or denoted on a printed page, and assume a finite primitive entity type, an attribute, and two simple compound entity capabilities as the kernel capabilities (however these capabilities are stored in a computer or recognized by a logician), then a formal system including integers, propositions, sets of characters, and so on may be derived.

SIMSCRIPT's entity, attribute, and set capabilities, therefore, are an unnecessarily rich set of primitives since a subset of these would suffice. On the other hand the determination of a minimal kernel is mostly an exercise for logicians. In teaching SIMSCRIPT to persons interested in the design and analysis of practical systems, it seems expedient to begin with the full compliment of entity, attribute, and set capabilities as described under Basic Concepts.

## SUMMARY

In SIMSCRIPT the status of a system is described in terms of entities, attributes, and sets. The elemental ways in which status can change are: a new entity comes into existence (or passes from outside to inside the system represented), an entity ceases to exist (or passes outside the system represented), an attribute of a represented entity changes value, or an entity gains or loses set membership. In simulated or real time systems such changes in status occur at points in time called events, which occur exogenously or are scheduled endogenously.

This view of the world was first applied to the description of systems to be simulated. It was later applied to the description of data base supported systems to be implemented, and to the translation process itself. It may be applied to entities of logic and mathematics as well as to entities of the business world.

The successive versions of the SIMSCRIPT language are attempts to allow the user: (1) to conveniently describe a system in this way, and (2) to select among alternatives concerning how the system is to be represented within a computer. The system description and storage options are then automatically processed by a translator and other support programming.

## REFERENCES

1. Cantor, G., Beitrage zur Begrundung der transfiniten Mengenlehre, Math. Ann. 46, 481-512 (1895); 49, 207-246 (1897). English translation by Ph. E. B. Jourdain entitled Contributions to the Founding of the Theory of Transfinite Numbers, Open Court, Chicago and London, 1915; and Dover Publications, New York, 1955.
2. CACI, Inc., The SIMSCRIPT II.5 Reference Handbook, Los Angeles, 1971.
3. CODASYL Data Base Task Group Report, April 1971. Available from Association for Computer Machinery, New York.
4. Codd, E. F., A relational model of data for large shared data banks, Commun. ACM 13, 377-387 (June 1970).
5. Codd, E. F., Normalized Data Base Structure: A Brief Tutorial, IBM Research Report RJ 935, November 1971.

6. Coffman, E. G., Jr., M. J. Elphick, and A. Shoshani, System deadlocks, Comput. Sur. p. 3 (June 1971).
7. Date, C. J., An Introduction to Database Systems, Addison-Wesley, Reading, Massachusetts, 1975.
8. Duff, I. S., A survey of sparse matrix research, Proc. IEEE p. 65 (April 1977).
9. Heijenoort, J. van, From Frege to Godel: A Source Book in Mathematical Logic, 1879-1931, Harvard University Press, Cambridge, Massachusetts, 1967.
10. Johnson, G. D., SIMSCRIPT II.5 User's Manual: S/360-370 Version, Release 8, CACI, Inc., Los Angeles, 1974.
11. Karr, H. W., H. Kleine, and H. M. Markowitz, SIMSCRIPT I.5, CACI 65-INT-1, CACI, Inc., Los Angeles, 1965.
12. Kiviat, P. J., R. Villanueva, and H. M. Markowitz, The SIMSCRIPT II Programming Language, Prentice-Hall, Englewood Cliffs, New Jersey, 1969.
13. Kiviat, P. J., R. Villanueva, and H. M. Markowitz, SIMSCRIPT II.5 Programming Language (E. C. Russell, ed.), CACI, Inc., Los Angeles, 1973.
14. Kleene, S. C., Introduction to Metamathematics, Van Nostrand, New York, 1952.
15. Knuth, D. E., Backus normal form vs. Backus naur form, Commun. ACM 7, 735-736 (1964).
16. Markowitz, H. M., B. Hausner, and H. W. Karr, SIMSCRIPT: A Simulation Programming Language, RAND Corporation RM-3310-PR 1962. Prentice-Hall, Englewood Cliffs, New Jersey, 1963.
17. Markowitz, H. M., Preliminary Software Design for The EDA Information System, CACI-P6-2, CACI, Inc., Los Angeles, 1966.
18. McCarthy, J., et al., LISP 1.5 Programmers Manual, M.I.T. Press, Cambridge, Massachusetts, 1962.
19. Newell, A., Information Processing Language-V Manual, Prentice-Hall, Englewood Cliffs, New Jersey, 1961.
20. Russell, B., Mathematical logic as based on the theory of types, Am. J. Math. 30, 222-262 (1908).
21. Russell, E. C., Simulating with Processes and Resources in SIMSCRIPT II.5, CACI, Inc., Los Angeles, 1975.
22. SHARE Program Library Agency, Program No. 360D-03.2.014, Triangle Universities Computation Center, Research Triangle Park, North Carolina, 1972.
23. Society of Actuaries, SOFASIM Operator's Manual, Chicago, 1977.
24. Whitehead, A. N., and B. Russell, Principia Mathematica, Vol. 1, 1910 (2nd ed., 1925); Vol. 2, 1912 (2nd ed., 1927); Vol. 3, 1913 (2nd ed., 1927), University Press, Cambridge, England.
25. Woddy, J. C., SOFASIM—A computerized model of a stock life insurance company, in Best's Review: Life/Health Insurance Edition, September 1975.


Harry M. Markowitz

1981 Winter Simulation Conference Proceedings Vol. 1
T.I. Ören, C.M. Delfosse, C.M. Shub (Eds.)

# BARRIERS TO THE PRACTICAL USE OF SIMULATION ANALYSIS

Harry M. Markowitz
IBM Thomas J. Watson Research Center,
P.O. Box 218, Yorktown Heights, N.Y. 10598

ABSTRACT

Simulation techniques are used in only a small fraction of instances in which they seem applicable. This paper considers the reasons for such "non-uses." In particular the paper considers simulator programming, the simulation/database interface, and two statistical topics as past, present and future limiting factors in the practical use of simulation techniques.

## 1. INTRODUCTION

In the design and running of businesses--including manufacturing, marketing, finance and the like--simulation techniques are used in only a small fraction of instances in which they seem applicable in principle. True, many simulations are done per year; many simulation papers are given; conferences are held regularly. But it is the exception rather than the rule that businesses use simulation to answer "what if" questions; the exception rather than the rule that simulation is used to think through, for example, the consequences of changes in equipment or procedures. Simulation has rarely become an integral part of business in the way that linear programming has in petroleum refining, the standard actuarial model has in insurance, or double entry bookkeeping has universally in accounting for business receipts and expenditures.

There are various reasons that simulation is not used in particular instances. Reasons for such "non-uses" include the following:

(1) It is not practical (yet?) to simulate a particular situation for reasons that cannot be corrected by better training or software packages. Perhaps, for example, even with the most skillful aggregation and design of experiment, computer execution costs exceed possible benefits.

(2) Simulation would be practical, except that even the most skilful team would take longer to build the model, including data collection and analysis as well as programming and debugging, than the practical situation permits. If it is a situation which arises repeatedly, each instance requires more time to modify the model (again, including data collection and analysis as well as program modification and debugging) than is practical.

(3) Simulation is practical, the concepts and packages are available, but business is wise not to use simulation since skilled simulation analysts are not available.

(4) Simulation is practical; the concepts and packages are available; skilled analysts are also available; the only reason that simulation is not used is that management has not been "sold."

I will discuss limitations of the second type, where simulation is potentially practical but the time required to develop the model, including data collection and analysis as well as programming, is too slow for the problem at hand. Problems in area 3 and 4, including how to train fledgling simulation experts, and how to sell management when simulation is practical in fact, are no less important; but exploring problems of area 2 will be a sufficient task for today.

The opinions expressed here are not based an exhaustive survey, but on my own experiences and reflections. If these comments generate discussion and reflection which in turn effects concept and package development, then perhaps they will be worth the collective manhours, yours and mine, that have been or are about to be expended on them.

## 2. PAST

In the late 1950's and the beginning of the 1960's I believed that the problem was one of simulator programming. Indeed, it could take longer to program a simulation of a factory than it took to build the factory itself. I speak here of detailed simulations in which some portion of the simulated system is described fairly (or quite) literally. If you aggregated sufficiently you could, even then, simulate the whole world in less than the six days it reportedly required to build it.

The programming bottleneck was perceived by many and led to the simulation languages such as GPSS (Efron & Gordon 1964), CSL (Buxton & Laski 1962), GASP (Kiviat 1963) and SIMSCRIPT (Markowitz, et al. 1963) which appeared in the early 60's. I speak of the discrete rather than the continuous languages since the former are the ones usually applicable to what we refer to here as "business" as distinguished from "engineering" problems.

I will not attempt to describe any of these languages except to note, for use later, that the last three listed have an entity, attribute and set view of the simulated world. As of any instant in time the world is viewed as consisting of entities of various types. A given entity is described by the values of its attributes and the sets to which it belongs. SIMSCRIPT sets have owners as well as members; for example, in a jobshop simulator each machine group is said to own (have associated with it) a set of waiting jobs (usually called its queue); in a traffic simulator each stoplight could own a set of waiting cars. An entity, then, is characterized by the values of its attributes, the sets to which it belongs and the members of the sets it owns. An entity can have any number of attributes, own any number of sets and belong to any number of sets.

The simulation languages reduced manyfold the elapsed time and manhours required to program a simulator. In most instances the programming of the simulator is no longer the principal bottleneck in the development and use of detailed business simulation models. Granted, improvements in programming and especially in debugging are always welcome; granted that some large detailed simulations still require many weeks of programming; nevertheless, even in these large-scale cases, the principal bottleneck usually lies elsewhere.

In the 1950s and early 1960s, large, detailed simulation applications would be planned by a team consisting of us modeling and programming specialists plus specialists from the specific substantive area (e.g., manufacturing engineers). The modeling and programming specialists and the substantive experts decided together the contents of the model; the computer types went off and built the programs while the substantive experts arranged for the collection of the data. It took many months to program the model agreed upon, but it took approximately as much elapsed time and more manhours of various sorts to collect the data.

The advent of the simulation languages reduced programming manyfold, but left data collection the largest visible obstacle to the timely development of detailed simulation models. Part of the difficulty with data collection was that needed data was on pieces of paper rather than in computer sensible form. But even when data was on tape or DASD it was difficult to get to it. The substantive experts usually had to work through information systems programmers none of whom had a grasp of their system as a whole. They sometimes remembered the whereabouts of data whose use they programmed; but the uncovering of other data required a painful detective process.

The difference between the time it took for information system programmers to build or modify the database description of a system as compared to the time for a simulation modeler/program to build or modify a detailed simulation of the same system suggested the following theory: Allow business system builders to view their databases as representations of a world (the business and its environment, for example); allow them to describe this world in terms of entities, attributes and sets; and allow them to manipulate this representation with the same commands as are available to the simulation programmers. Perhaps then the programming of business systems would become as (relatively) simple as the programming of their simulation. Further, the structure of the system (specifically, its entity, attribute and set structure) would be as clear as a complex but well documented simulation; and therefore properly educated analysts with special needs, like the need to obtain data for a simulation analysis, would be able to pull out required data themselves much more easily than they could by working through a staff of programmers, none of whom understood the system as a whole.

## 3. PRESENT

Various business reasons delayed the testing of this theory. During the last three years, however, Ashok Malhotra, Donald Pazel and I at IBM Research built a database management system based on the Entity, Attribute and Set view. The name of the system is EAS-E, pronounced "easy" not "ease", and stands for entities, attributes, sets and events; see Malhotra et. al. (1980). The first large-scale application of EAS-E showed a reduction of source program of better than 5 to 1, and an even greater but difficult to quantify superiority in maintenance and evolution as compared to the prior system.

I will describe EAS-E briefly and then return to the topic of the simulation/database interface. But first let me make clear that EAS-E is not an IBM product or part of any IBM plan. It is an experimental language developed at IBM Research and applied only within IBM.

I will illustrate EAS-E in terms of its first application: a rewrite and extension of the Workload Information System of Thomas. J. Watson's Central Scientific Services (CSS). CSS consists of about 90 craftsmen who do glass work, electronics, etc., for Thomas. J. Watson's scientists and engineers. The old Workload Information System, written in PL/I and assembler, was difficult to modify or extend. In the first instance an EAS-E version was built to operate in parallel to the old version, reading the same weekly inputs and generating the same weekly outputs. The EAS-E based system duplicated the functions of the prior system with about one-fifth as much source code. It was then used to program functions--particularly tailored, interactive query and update functions--which would have been difficult to add to the prior system.

Exhibits 1,2 and 3 show programming used to print one of the CSS weekly reports. This looks like SIMSCRIPT II programming (Kiviat, et al. 1969), but remember this code exhibits attributes of database rather than simulated entities. The Exhibits should be almost self explanatory to those who know SIMSCRIPT II, but I will explain a bit for those who do not.

Exhibit 1:

```
ROUTINE TO START_NEW_PAGE_AND_PRINT_HEADING
START NEW PAGE
PRINT 7 LINES WITH PAGE.V, TITLE, SUBTITLE, DATE, AND WEEK THUS...
CSS Information System                                        Page ***
                 Central Scientific Services
    *************************************************
*******************************           ********** **
CSS CSS  DEPT  CHARGE ENTRY COMPL ESTIMT PRCDNG TOTAL TIME CUSTOMER NAME
AREA JOB  NUM   TO  DAY DAY     WEEK TIME RMNING /AREA RESP.

RETURN   END
```

Exhibit 2:

```
PRINT 3 LINES WITH TASK_AREA_NUMBER, JOB_NUMBER, JOB_DEPT_NUMBER,
JOB_PROJ_NUMBER, TASK_ENTRY_DATE, TASK_COMPL_DATE, TASK_ESTIMATED_HOURS,
INHOUSE_HOURS_FOR_WEEK, TASK_INHOUSE_HOURS+TASK_VENDOR_HOURS,
TASK_ESTIMATED_HOURS - TASK_INHOUSE_HOURS - TASK_VENDOR_HOURS,
JOB_CUSTOMER, JOB_DESCRIPTION, TASK_EST_VENDOR_HOURS,
VENDOR_HOURS_FOR_WEEK, TASK_VENDOR_HOURS, TASK_ASSIGNEE_NAME, AND STAR THUS...
*** ***** **** **** *** *** ****** ****** ****** ****** *************
********************************** ****** ****** ******   ************ *
-------------------------------------------------------------------------------
```

Exhibit 3:

```
FOR EVERY GROUP IN CSS_GROUPS, FOR EVERY AREA IN GROUP_AREAS
        WITH AREA_TASKS NOT EMPTY AND AREA_TYPE ¬= VENDOR_TYPE, DO THIS...
   LET SUBTITLE = AREA_NAME
   CALL START_NEW_PAGE_AND_PRINT_HEADING
   FOR EACH TASK IN AREA_TASKS WITH TASK_COMPL_DATE=0, CALL PRINT_TASK_LINES
REPEAT
```

Exhibit 4:

```
FIND THE JOB IN CSS_JOBS WITH JOB_NUMBER = PROPOSED_JOB_NUMBER
  IF ONE IS FOUND...
     CALL REJECT (|JOB ALREADY EXISTS WITH SPECIFIED JOB NUMBER.|)
     RETURN
  ELSE...
```

Exhibit 1 is a routine which starts a new page and prints a heading on it. This routine is called in the printing of several different reports. The START NEW PAGE statement is self-explanatory. The seven lines (including the blank seventh line) following the PRINT 7 LINES... statement are printed during run-time just as they appear in the source program, except that the first variable, PAGE.V (automatically maintained by EAS-E) is printed in place of the first grouping of ***s, the second variable TITLE is printed in place of the second groupings of *s, etc. The PRINT statement in exhibit 2, slightly simplified from its CSS version, prints 3 lines containing the specified variables and expressions in the places indicated in the three form lines (last 3 lines of the exhibit). These lines will print data under the headings of exhibit 1. Most of the data being printed are attributes of database entities of type JOB and TASK.

I mention as background to exhibit 3 that tasks to be performed by CSS are divided into areas. These areas, in turn, are divided into groups. Thus the "Device and Mask Generation" group includes the "Electrochemistry" and "Mask Gen Lab" areas.

The first phrase of the exhibit (FOR EVERY GROUP IN CSS_GROUPS) instructs the compiled EAS-E program to have the reference variable GROUP point in turn to each member of the set called CSS_GROUPS. For each such GROUP, the next two phrases ("FOR... WITH...") instruct the executing program to look at each area in the group's areas that meet certain conditions. For each GROUP and AREA thus generated, the program sets the variable SUBTITLE used in the START_NEW_PAGE_PRINT_HEADING routine, and then calls the latter routine. Next, under the control of a FOR and a WITH phrase, it calls on a subroutine which includes the PRINT 3 LINES... statement of exhibit 2 for tasks in the AREA_TASKS set which have not yet been completed. These six lines thus select groups, areas and tasks for which heading lines and body lines are printed for one of the CSS weekly reports.

The example of a FIND statement reproduced in exhibit 4, including the IF ONE IS (NOT) FOUND statement that usually follows a FIND, finds the job in the set of CSS_JOBS with its job number as specified. EAS-E has a different syntax for the FIND statement than does SIMSCRIPT II, to improve readability. Again, it is a job represented in the database, rather than a simulated job, which is found here.

In addition to the commands illustrated in the exhibits, EAS-E has SIMSCRIPT II-like commands to CREATE (the database representation of) an entity, FILE an entity into a set, REMOVE an entity from a set, do something FOR EACH in a set, define database entity types by telling EAS-E what attributes, ownerships and memberships EVERY entity of the type has, etc.

EAS-E allows the user to manipulate database entities much as SIMSCRIPT II allows him to manipulate simulated entities. But the mechanisms behind the scenes are necessarily different and often more complex. For example:

> (1) One or more users can interrogate or update parts of the database simultaneously. But an individual user's program does not read from or write onto the database directly, since one program might thereby clobber the work being done by another. Rather each communicates with a "Custodian" program which remembers which entities are already being accessed read-only or read-write.
>
> (2) The custodian must assure that if the system crashes at any point in time, either all of the changes or none of the changes of a given program are, in effect, reflected in the database. The programmer can ask to RECORD ALL DATABASE ENTITIES modified thus far; but again the custodian must assure that either all or none of these changes are recorded in case of a crash.
>
> (3) LIFO and FIFO sets are stored as linked lists as in SIMSCRIPT. But it would be very inefficient for large ranked database sets to be stored this way. For example, if a set of cars owned by a department of motor vehicles has a million members ranked by owners name, then on the average one-half million members would have to be accessed to FILE or FIND a car in the set if stored as a linked list. EAS-E actually stores the set by means of a balanced tree of subsets, in such a way that very few accesses are required to FILE or FIND in sets with millions of members.

The above goes on behind the scenes. The user still says CREATE, DESTROY, FILE, REMOVE, FIND, FOR EACH...WITH..., and the like.

Another difference between simulated systems and real business systems is that the simulated system whizzes along through time as fast as the computer can move it. The real system moves at a relatively more leisurely pace. Hence the analyst may wish to browse the current status of parts of the database. Towards this end EAS-E provides the analyst with direct (nonprocedural) abilities to browse the system and, if authorized, to update it directly. Similar facilities might be of use in simulated systems, perhaps to assist debugging, if combined with the ability to stop the flow of simulated time at designated events, times or conditions.

Exhibit 5 is an example of a Browser display. In this case the attributes, ownerships and memberships of one entity is shown on the analyst's screen. The first line of the screen shows the type of the entity plus certain internal identification information which we will ignore here. The region marked (2) shows attributes of the entity. We see, for example, that the JOB in the exhibit has JOB_NUMBER = 80150, JOB_DEPT_NUMBER = 461, etc. The line marked (3) tells us that the JOB owns a set called JOB_TASKS, that the set for this particular job has only one member, and that the set is ordered by job number. The line marked (4) indicates that the job belongs to a set called PROJ_JOBS.

You move through the database with the aid of the PF (program function) keys whose template is shown in exhibit 6. To look at every task in the set JOB_TASKS, for example, place an X as shown at (3), then press PF10 labeled FORWARD on the template. This brings the first (and in this case, only) task of the set. Generally, to move forward to the first or next member of a set press PF10, to move backwards in a set press PF11. Further capabilities include scrolling up or down

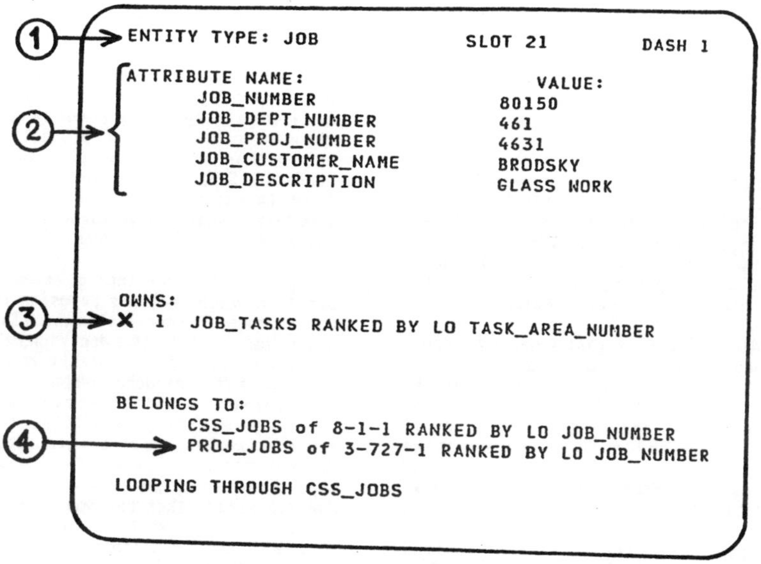


Exhibit 5

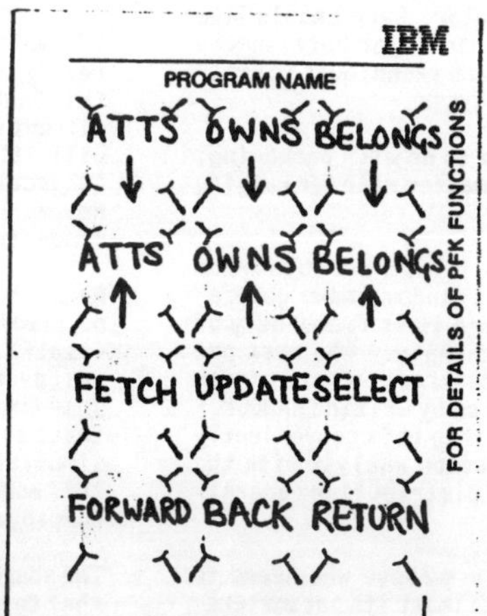


Exhibit 6

the displayed attributes, ownerships or memberships of the entity if there are more than can fit in the space allotted on the screen, specifying selection conditions to be used in looping through a set, and modifying attributes by overwriting their values on the screen.

My colleagues and I believe that the experience of EAS-E thus far will prove typical, and that application development can be reduced manyfold by a language (perhaps EAS-E, perhaps something like it) which uses the same worldview and basic concepts as some simulation language, and allows the application developers to manipulate database entities in the same way that simulation programmers manipulate information about simulated entities. This should help the simulation/database interface in two ways. First, it should be easier for any analyst with special needs to learn as much of the structure of the database as he needs to know, to browse the database and to program routines to pick up desired data. Second, the simulation programmer in particular should be helped by having database facilities which use the same general worldview, some of the same entity, attribute and set structure and perhaps even some of the event coding as will appear in his simulator.

## 4. FUTURE

Assume for the sake of the current argument that the problem of getting to the data stored in databases is "solved in principle." Imagine that such software and the knowledge to use it were widespread. What then would be the principal bottleneck to the application of simulation techniques? I believe this bottleneck would be of a statistical nature.

For example, rather than using past demands for products as exogenous inputs to a model, an analyst may want to generate demands randomly. This allows the analyst to test the effects of shifts of demand by changing parameters of the distribution, but it raises questions as to the form and parameters of the demand distribution. In particular, what form of model and values of parameters would explain past demands reasonably well?

I will not attempt to review the large body of good and relevant work on statistical analysis related to simulation, and beyond that the ocean of literature relevant to statistical inference in general, and hence to simulation in particular. Here I would like to point out what seems to me two major bottlenecks limiting the application of such techniques in practical business simulation analysis.

The first limiting factor has to do with packaging; the second is a philosophical matter with far ranging implications.

The notion of packaging may be illustrated by random number generators. As long as a random number generator is buried in the literature it is frequently of little use to the real-world analyst who must put models together quickly. In contrast, when a random number generator can be invoked by writing RANDOM.F or NORMAL.F or GAMMA.F, etc., then it is conveniently at the disposal of any simulation analyst with the background to understand the distribution generated.

The same applies to the analyst above who needs to select a model of demand and estimate its parameters. Ideally he should be able to specify the model and the data to be used in the analysis in a manner most convenient to himself, and have the rest done for him. Too often the world offers him much less. I realize that I am only offering a gripe rather than a solution. But it is a complaint which I believe many in the field will confirm as identifying a limiting factor. Furthermore, this limiting factor could be greatly alleviated in this age in which compilers can deal with just about any unambiguous interface one cares to devise.

The next problem area concerns a basic philosophical matter. The business application of simulation analysis usually involves choice under uncertainty. No one tells the analyst the form of the demand distributions, much less their parameters. The object of the simulation analysis is to help guide action, even in circumstances when objective statistical analysis cannot conclusively reject all but one plausible hypothesis.

Without trying to untangle how much was already understood by Bayes, Ramsey, Jeffreys, and De Finetti, as cited in Savage (1954), one may say that certainly the most influential, and probably the most important work on action under uncertainty was that of Leonard J. Savage (1954). On the whole, however, the literature on statistical methodology as applied to simulation analysis is surprisingly oblivious to Savage's revolutionary ideas.

Savage provided a convincing, axiomatic defense of a position which includes Bayesian inferences as a consequence. The conclusion which Savage draws from his axioms is that, if a decision maker is to avoid ridiculous behavior (contrary to the axioms), he must act as if he attaches probabilities to possible states of nature, and acts so as to maximize expected utility given these probability beliefs. As information accumulates he alters his probability beliefs according to Bayes' rule. If evidence is strong, in that the power of the test is great in the Neyman-Pearson sense, then persons of widely differing a priori beliefs will be drawn close together in their a posteriori beliefs. If the power of the test is weak, specifically if several hypotheses could well have generated the data, then beliefs among these will shift little.

If you have not studied the axiom system you cannot really make up your mind whether you should accept this "subjective probability" approach. I will not attempt to present the Savage axioms here. An example will illustrate one problem with using so called "objective" statistics (of the "old fashioned" Neyman-Pearson variety) and perhaps encourage the reader to consider seriously the Savage alternative.

Returning to our analyst who needs to develop a model of demand, perhaps the first model that suggests itself is that the demands for a particular product has, over the last k years, been drawn repeatedly from some (unchanging) normal distribution. Suppose he tests this hypothesis and accepts it; then builds a simulation model that assumes normality, and uses this model to determine policy. The analyst and his Company may be in BIG trouble.

The source of this trouble is twofold. First, recall that to accept a hypothesis on the basis of an objective test means merely to fail to reject it. If asked, the same data might also accept (i.e., not reject) all sorts of other plausible hypotheses. Perhaps if the analyst had assumed that the (unchanging) distribution was from the broad family of Pearson distributions, which includes the normal as a special case, and had used a Bayesian calculation given some a priori beliefs, the data might have swung belief away from the normal rather than towards it--even though the evidence was not strong enough for a Neyman-Pearson test to reject the hypothesis at the specified level of significance.

Once the analyst has "accepted" a hypothesis by an objective test the second step toward big trouble is to act as if the hypothesis, e.g. normality, has been proved with certainty. This is done by building it into a simulation model, then using the model to answer policy questions. For example, suppose that the estimate of the mean you give the model is m, and the estimate of the standard deviation is s. Suppose that, in effect, the model is offered a 10,000 to 1 bet that the observed demand will not exceed m + 10s. Since this has virtually zero probability for a normal distribution, the model will presumably accept the bet. But with other distributions, perhaps other distributions which would be accepted by the same data which accepted the normal, a 10s deviation from the mean is far from "virtually zero" (e.g., the Tchebychev inequality assures only that the probability does not exceed .01).

What the subjective probability approach decrees in the above situation is to shift probability beliefs, based on the data, according to Bayes' rule; and evaluate alternative strategies as if nature first drew a model randomly, with probabilities equal to the updated beliefs, then drew a history randomly from this model. In particular it would not assume that a 10s move was virtually impossible unless (roughly speaking) the analyst's a posteriori beliefs virtually ruled out all models that give such an event any perceptible probability.

Suppose that you could be sure (perhaps by construction) that an unchanging distribution had mean zero and unit variance, but all you knew about the form of the distribution was that the normal hypothesis had been "accepted", i.e., not rejected. Would you bet $10,000.00 of your own money to win $1.00. Of course not. This shows that your intuition is more Bayesian than are the statistical procedures commonly used in building a simulator. I believe that if you study the Savage axioms many of you will decide that your instincts are right and the procedures are wrong.

Simulation is sometimes used in situations where objective statistical tests are appropriate, e.g., when trying to determine facts about a given complex random process which cannot be solved analytically. Thus I cannot fault any particular methodological study because it is objective rather than subjective. But taken as a whole the literature on statistics for simulation is quite deficient in Bayesian analysis. Certainly the "well packaged" statistical facilities called for above should include Bayesian as well as "objective" capabilities.

## 5. POSTSCRIPT AND SUMMARY

Imagine a business in the 21st century when planning is more rational than it is today. In particular, top management no longer expects someone to hand it a point estimate of what the future will be, as the start of a process which produces a great plan if the predicted occurs, but can be disastrous if a plausible alternative occurs instead.

This 21st century planning process seeks a strategy which, in effect, is robust against adversity as well as profitable when the likely occurs. When the answer is not obvious and optimization techniques not applicable, simulation will be used to trace out the consequences for business performance of major contingencies such as shifts in demand, in price or in technology. Since a particular business in the year 20XY will usually be much like it was in 20X(Y-1), model modification will consume more resources than new model construction.

For large complex businesses, models of components of the business will be used for more detailed questions, and to test more aggregate versions to be incorporated into broader scope models. The internal database of the business, and available relevant external databases will be used to estimate relationships for models, and to obtain initial conditions and exogenous events for certain runs. In the case of some detailed simulation models, the same coding will run the simulated shop (or other business component) as runs the real shop.

I have not said whether the above business exists at the beginning of the 21st century, 19 years from now, or at the end of the century, 119 years from now. Above I have outlined my views as to the concepts and packages needed to reach this stage. I would be most interested in hearing your views on these matters.

## REFERENCES


Buxton, J.N., & Laski, J.G., Control and Simulation Language, *The Computer Journal,* Vol 5, 1962, pp. 194-199

Efron, R. and Gordon G., A general purpose systems simulator program and examples of its application: Part I - Description of the simulator *IBM Systems Journal* Vol. 3, No. 1, pp. 21-34, (1964).

Kiviat, P.J., *GASP -- A General Activity Simulation Program,* Applied Research Laboratory, U.S.Steel Corp, Monroeville, PA, July 1963.

Kiviat, P.J., Villanueva, R., & Markowitz, H.M., *The SIMSCRIPT II Programming Language,* Prentice Hall; Englewood Cliffs, NJ, 1969.

Malhotra, A., Markowitz, H.M., & Pazel, D.P., *EAS-E: An Integrated Approach to Application Development,* RC 8457, IBM T. J. Watson Research Center, Yorktown Hts., NY 10598, August 29 1980.

Markowitz, H.M., Hausner, B., & Karr, H.W., *SIMSCRIPT: A Simulation Programming Language,* The RAND Corporation RM-3310-PR 1962. Prentice-Hall NJ, 1963.

Savage, Leonard J. *The Foundations of Statistics,* John Wiley and Sons, New York, 1954.

## Chapter 5

# IBM's T. J. Watson Research Center

### Comments

The first five articles of this chapter are concerned with one or another aspect of portfolio theory. The first three articles follow up a notion in my 1959 book. The latter does not justify the use of mean-variance analysis by assuming that probability distributions are normal or that utility functions are quadratic. Rather, it justifies it by the assertion that perhaps a quadratic approximation to an investor's actual utility function provides portfolios which have utility almost as good as the utility-maximizing portfolio. It supports this view with an "empirical study" involving nine securities. The three articles which begin the chapter follow this thought in various ways.

*The Two Beta Trap* article is the ultimate result of a request to lecture to decide a controversy at the time. The issue and my view on the issue are described in the article.

The general portfolio selection model needs "covariances". This can be supplied by a factor model such as Sharpe's (1963) one-factor model and Rosenberg's (1974) many-factor model. The next two articles, by André Perold and me, consider scenarios as well as factors as sources of covariance.

My main activity at IBM's T. J. Watson Research Center was to extend SIMSCRIPT's Entity, Attribute, Set and Event view of system modeling to data base entities as well as the simulated entities of SIMSCRIPT. The resulting Entity, Attribute, Set and Event language was not called SIMSCRIPT III, for internal IBM reasons, but was called EAS-E. The next three articles of this chapter discuss one or another aspect of EAS-E modeling and the EAS-E programming language. There is also an EAS-E programming manual written by Ashok Malhotra, bearing all our names, Malhotra, Markowitz and Pazel.

Paul Samuelson and I have a long standing argument about investment for the long run. The final two articles of this chapter concern that controversy. The second of the two articles is a technical paper by me on the subject. The first of the two articles is a somewhat less technical paper which I wrote for a volume in honor of Paul Samuelson on his 90$^{th}$ birthday, edited by Szenberg, Ramrattan, & Gottesman (2006).

## References

Levy, H. and Markowitz, H. M. (1979). *Approximating Expected Utility by a Function of Mean and Variance.* The American Economic Review, June, Vol. 69, No. 3, pp. 308–317.

Kroll, Y., Levy, H. and Markowitz, H. M. (1984). *Mean-variance Versus Direct Utility Maximization.* The Journal of Finance, Vol. 39. No. 1, pp. 47–61.

Markowitz, H. M., Reid, D. W. and Tew, B. V. (1994). *The Value of a Blank Check.* The Journal of Portfolio Management, Summer, pp. 82–91.

Markowitz, H. M. (1984). *The Two Beta Trap.* Journal of Portfolio Management, Vol. 11, No. 1, Fall, pp. 12–20.

Markowitz, H. M. and Perold, A. F. (1981a). *Portfolio Analysis with Factors and Scenarios.* The Journal of Finance, 36(14), September, pp. 871–877.

Markowitz, H. M. and Perold, A. F. (1981b). *Sparsity and Piecewise Linearity in Large Portfolio Optimization Problems.* In "Sparse Matrices and Their Uses", I. S. Duff (ed.), pp. 89–108. San Diego: Academic Press.

Markowitz, H. M., Malhotra, À. and Pazel, D. (1983). *The ER and EAS Formalisms for System Modeling, and the EAS-E Language.* In "Entity-Relationship Approach to Information Modeling and Analysis", P. P. Chen (ed.), pp. 29–47. (North-Holland) ER Institute: Elsevier Science Publishers B. V.

Malhotra, A., Markowitz, H. M. and Pazel, D. P. (1983). *EAS-E: An Integrated Approach to Application Development.* ACM Transactions on Database Systems, Vol. 8, No. 4, December, pp. 515–542.

Pazel, D. P., Malhotra, A. and Markowitz, H. M. (1983). *The System Architecture of EAS-E: An Integrated Programming and Data Base Language.* IBM Systems Journal, Vol. 22, No. 3, pp. 187–198.

Markowitz, H. M. (2006). *Samuelson and Investment for the Long Run.* In "Samuelsonian Economics and the Twenty-First Century". M. Szenberg, L. Ramrattan, and A. A. Gottesman (eds.), pp. 252–261. Oxford University Press.

Markowitz, H. M. (1976). *Investment for the Long Run: New Evidence for and Old Rule.* The Journal of Finance, Vol. 31, No. 5, pp. 1273–1286.

# Approximating Expected Utility by a Function of Mean and Variance

*By* H. LEVY AND H. M. MARKOWITZ*

Suppose that an investor seeks to maximize the expected value of some utility function $U(R)$, where $R$ is the rate of return this period on his portfolio. Frequently it is more convenient or economical for such an investor to determine the set of mean-variance efficient portfolios than it is to find the portfolio which maximizes $EU(R)$. The central problem considered here is this: would an astute selection from the $E,V$ efficient set yield a portfolio with almost as great an expected utility as the maximum obtainable $EU$?

A number of authors have asserted that the right choice of $E,V$ efficient portfolio will give precisely optimum $EU$ if and only if all distributions are normal or $U$ is quadratic.[1] A frequently implied but unstated corollary is that a well-selected point from the $E,V$ efficient set can be trusted to yield almost maximum expected utility if and only if the invector's utility function is approximately quadratic, or if his a priori beliefs are approximately normal. Since statisticians frequently reject the hypothesis that return distributions are normal, and John Pratt and Kenneth Arrow have each shown us absurd implications of a quadratic utility function, some writers have concluded that mean-variance analysis should be rejected as the criterion for portfolio selection, no matter how economical it is as compared to alternate formal methods of analysis.

Consider, on the other hand, the following evidence to the contrary. Suppose that two investors, let us call them Mr. Bernoulli and Mr. Cramer, have the same probability beliefs about portfolio returns in the forthcoming period; while their utility functions are, respectively,

$$(1)\qquad U(R) = log(1 + R)$$

$$(2)\qquad U(R) = (1 + R)^{1/2}$$

Suppose that Mr. Cramer and Mr. Bernoulli share beliefs about exactly 149 portfolios. In particular suppose that the 149 values of $E$, $V$, $Elog(1 + R)$ and $E(1 + R)^{1/2}$ they share happen to be the same as that of the annual returns observed for 149 mutual funds during the period 1958 through 1967, as reported below. (We are not necessarily recommending unadjusted past data as predictors of the future; rather we are using these observations as one example of "real world" moments.)

Now let us suppose that Mr. Bernoulli, having read William Young and Robert Trent, decides that when he knows the $E$ and the $V$ (or the standard deviation $\sigma$) of a distribution he may guess its expected utility to him by the formula:

$$(3)\qquad Elog(1 + R) \cong (log(1 + E + \sigma) + log(1 + E - \sigma))/2$$

He would find that there is a .995 correlation between the pairs (actual mean log$(1 + R)$, estimated mean log$(1 + R)$) for the 149 such pairs provided by the 149 historical distributions. Furthermore, the regression relation (over the sample of 149) between the actual mean $log(1 + R)$ and the estimate provided by (3) is

$$(4)\qquad \text{actual} = 0.002 + 0.996 \cdot \text{estimated}$$

As it happens, the portfolio which maximized the approximation (3) also maximized the expected value of the true utility (1). If Mr.

*Jerusalem School of Business Administration, Hebrew University, and Thomas J. Watson Research Center, IBM Corporation, respectively. We have benefited greatly from technical assistance and the helpful comments of Y. Kroll, and from discussions with Marshall Sarnat.

[1]Analyses of relationships between $E$, $V$ efficiency, on the one hand, and quadratic utility and/or normal distributions, on the other hand, may be found for example in James Tobin (1958, 1963), Markowitz, Martin Feldstein, Giora Hanoch and Levy (1969, 1970), and John Chipman.

Bernoulli selected among the 149 portfolios on the basis of (3) he would, in this instance, do precisely as well as if he had used the true criteria (1). Finally, as will be shown later, (3) always increases with $E$ and decreases with $\sigma$, thus is always maximized by an $E,V$ efficient portfolio.

Mr. Cramer, seeing the good fortune of Mr. Bernoulli in finding an approximation to his expected utility based on $E$ and $V$ alone, might try the corresponding approximation to his own utility function, namely:

$$(5)\quad EU \simeq (U(1 + E + \sigma) + U(1 + E - \sigma))/2$$

where $U$ is now given by equation (2). Mr. Cramer would be delighted to find that the correlation between predicted and actual for his utility function is .999; the regression relationship is

$$(6)\quad \text{actual} = -.013 + 1.006 \cdot \text{estimated}$$

The portfolio, among the 149, which maximized the approximation (5) also maximized the true expected utility (2); and (2) is always maximized by a portfolio in the $E,V$ efficient set.

Suppose that a third investor, a Mr. $X$, does not know his current utility function—has just not taken the time recently to analyze it as prescribed by John von Neumann and Oskar Morgenstern—but does know that equation (5) provides about as good an approximation to his utility function as it does to those of Mr. Bernoulli and Mr. Cramer. He also knows, from certain properties which he is willing to assume concerning his utility function, that equation (5) is maximized by an $E,V$ efficient portfolio. If Mr. $X$ can carefully pick the $E,V$ efficient portfolio which is best for him, then Mr. $X$, who still does not know his current utility function, has nevertheless selected a portfolio with maximum or almost maximum expected utility.

In this paper we present a class of approximations $f_k(E, V, U(\cdot))$ where $k \geq 0$ is a continuous parameter distinguishing one method of approximation from another. For $k = 1$ we get equation (5); for $k = 0$ we have a method proposed by Markowitz. We shall examine some empirical relationships between $EU$ and $f_k(E,V,U(\cdot))$ for various utility functions, empirical distributions, and values of $k$. We shall explain these empirical results in terms of a simple analysis of the expected difference between a utility function and an approximating function. We shall also consider certain objections to $E$, $V$ analysis, due to Karl Borch, Pratt, and Arrow in light of our empirical results, our analysis of expected difference, and a reconsideration of Pratt's analysis of risk aversion for the kinds of quadratic approximations we use.

## I. A Class of Approximations

Markowitz used two methods to approximate $EU$ by a function $f(E, V)$ depending on $E$ and $V$ only. The first is based on Taylor-series around zero:

$$(7)\quad U = U(0) + U'(0)R + .5U''(0)R^2 \ldots$$

hence

$$(8)\quad EU \simeq U(0) + U'(0)E + .5U''(0)(E^2 + V)$$

The second approximation is based on a Taylor-series around $E$:[2]

$$(9)\quad U = U(E) + U'(E)(R - E) + .5U''(E)(R - E)^2 \ldots$$

hence

$$(10)\quad EU \simeq U(E) + .5U''(E)V$$

In tests with empirical distributions and the logarithmic utility function (by Markowitz, and by Young and Trent) the approximation in (10) performed markedly better than the approximation in (8).

Both approximations involve fitting a quadratic to $U(R)$ based on properties of $U$ (i.e., $U$, $U'$, and $U''$) at only one value of $R$ ($=0$ or $E$, respectively). The present authors conjectured that a better approximation perhaps could be found by fitting the quadratic to three judiciously chosen points

[2]A somewhat different use of this Taylor-series to justify mean-variance analysis is presented by S. C. Tsiang. See also Levy regarding the Tsiang analysis.

on $U(R)$. To produce a mean-variance approximation the three points must themselves be functions of at most $E$, $V$ and the function $U(\ )$. A class of such functions was selected in which the quadratic was fit to the three points

$$(11)\quad (E-k\sigma, U(E-k\sigma)), (E, U(E)), (E+k\sigma, U(E+k\sigma))$$

The quadratic passing through these three points can be written as

$$(12)\quad Q_k(R) = a_k + b_k(R-E) + c_k(R-E)^2$$

To simplify notation we will often write $Q$, $a$, and $b$ for $Q_k$, $a_k$, and $b_k$, the subscript $k$ being understood. Equation (12) implies

$$(13)\quad EQ = a + cV$$

Solving

$$(14)\quad \begin{aligned} U(E-k\sigma) &= a + b((E-k\sigma)-E) + c((E-k\sigma)-E)^2 \\ &= a - bk\sigma + ck^2\sigma^2 \\ U(E) &= a + b0 + c0^2 \\ U(E+k\sigma) &= a + bk\sigma + ck^2\sigma^2 \end{aligned}$$

we find that

$$(15)\quad \begin{aligned} a &= U(E) \\ b &= \frac{U(E+k\sigma) - U(E-k\sigma)}{2k\sigma} \\ c &= \frac{U(E+k\sigma) + U(E-k\sigma) - 2U(E)}{2k^2\sigma^2} \end{aligned}$$

hence

$$(16)\quad f_k(E, V, U(\cdot)) = EQ = U(E) + \frac{U(E+k\sigma) + U(E-k\sigma) - 2U(E)}{2k^2}$$

If we substitute $k = 1$ and simplify we obtain equation (5). If we define the approximation in (10) as $f_0$, and let $k \to 0$ in (16) we find that[3]

$$(17)\quad f_k \to f_0 \text{ as } k \to 0$$

For given $k$ and $U$, (16) clearly depends on only $E$ and $V$. It is not immediately clear that $f$ is always maximized by an $E,V$ efficient portfolio. Each of the utility functions used in our experiments has $U' > 0$, $U'' < 0$, and $U''' \geq 0$ for all rates of return $R > -1.0$. These three properties are sufficient to assure us that $f$ is maximized by an $E,V$ efficient portfolio, provided that $E - k\sigma > -1$ for all portfolios considered.[4]

## II. Analysis of Error Functions

For a given $k$, and for any probability distribution for which the specified moments exist, the difference between $EU$ and $f(E, V, U(\cdot))$ may be written as

$$(18)\quad D_k = EU - EQ_k = Ed_k(R; E, V, U(\cdot))$$

where $Q_k$ is given in (12) and

$$(19)\quad d_k(R) = U(R) - Q_k(R)$$

[3]That $f_k \to f_0$ as $k \to 0$ follows readily if we show that

$$\frac{U(E+k\sigma) + U(E-k\sigma) - 2U(E)}{k^2} \to U''(E)V$$

This may be shown by computing the second derivative of the numerator with respect to $k$, the second derivative of the denominator, and applying L'Hospital's rule.

[4]Differentiating (16) with respect to $\sigma$ we find, since $U'' < 0$, that

$$\frac{\partial f_k}{\partial \sigma} = \frac{U'(E+k\sigma) - U'(E-k\sigma)}{2k} < 0$$

Differentiating (16) with respect to $E$, and substituting $\xi = k\sigma$, we find that

$$(a)\quad \partial f_k/\partial E = U'(E) + [U'(E+\xi) + U'(E-\xi) - 2U'(E)]V/2\xi^2$$

For positive or negative $\eta$ Taylor's theorem implies

$$(b)\quad U'(E+\eta) = U'(E) + U''(E)\eta + .5U'''(E+\theta)\eta^2$$

for some $\theta$ between 0 and $\eta$. Substituting (b) for $U'(E+\xi)$ and $U'(E-\xi)$ in (a), and using $U' > 0$ and $U''' \geq 0$, we find (for some $\theta_1$ and $\theta_2$ between 0 and $\xi$) that $\partial f_k/\partial E = U'(E) + [U'''(E+\theta_1) + U'''(E-\theta_2)]V/4 > 0$.

TABLE 1—$d_k(R) = U(R) - Q_k(R)$ FOR $U = log_e(1 + R)$; $E = .1$; $k = 0$ or 1; AND $\sigma = .15$

| R | U(R) | $Q_0$(R) | U − $Q_0$ | $Q_1$(R) | U − $Q_1$ |
|---|---|---|---|---|---|
| -.70 | -1.20397 | -.89643 | -.30755 | -.90348 | -.30049 |
| -.50 | -.69315 | -.59891 | -.09424 | -.60373 | -.08942 |
| . . . | | | | | |
| -.30 | -.35667 | -.33444 | -.02223 | -.33735 | -.01933 |
| -.25 | -.28768 | -.27349 | -.01419 | -.27596 | -.01172 |
| -.20 | -.22314 | -.21461 | -.00854 | -.21667 | -.00648 |
| -.15 | -.16252 | -.15779 | -.00473 | -.15946 | -.00306 |
| -.10 | -.10536 | -.10304 | -.00232 | -.10433 | -.00103 |
| -.05 | -.05129 | -.05035 | -.00094 | -.05129 | .00000 |
| .00 | .00000 | .00027 | -.00027 | -.00034 | .00034 |
| .05 | .04879 | .04882 | -.00003 | .04853 | .00026 |
| .10 | .09531 | .09531 | -.00000 | .09531 | -.00000 |
| .15 | .13976 | .13973 | .00003 | .14001 | -.00024 |
| .20 | .18232 | .18209 | .00023 | .18262 | -.00030 |
| .25 | .22314 | .22238 | .00077 | .22314 | .00000 |
| .30 | .26236 | .26060 | .00176 | .26158 | .00078 |
| .35 | .30010 | .29676 | .00335 | .29794 | .00217 |
| .40 | .33647 | .33085 | .00563 | .33221 | .00427 |
| .45 | .37156 | .36287 | .00869 | .36439 | .00718 |
| .50 | .40546 | .39283 | .01263 | .39449 | .01098 |
| .55 | .43825 | .42072 | .01753 | .42250 | .01576 |
| .60 | .47000 | .44655 | .02345 | .44842 | .02158 |
| . . . | | | | | |
| 1.00 | .69315 | .57878 | .11437 | .58075 | .11240 |
| 1.50 | .91629 | .55812 | .35817 | .55846 | .35783 |
| 2.00 | 1.09861 | .33086 | .76775 | .32761 | .77100 |
| 2.50 | 1.25276 | -.10301 | 1.35577 | -.11179 | 1.36455 |
| 3.00 | 1.38629 | -.74349 | 2.12978 | -.75975 | 2.14604 |

For example, Table 1 presents $d_k(R)$ for $U = log(1 + R)$, for $k = 0$ and $k = 1$, for $E = .1$, for $\sigma = .15$, and for various values of $R$. Much about the joint distribution of $EU$ and $f_k(E, V, U(\cdot))$ is explained by such tables plus general properties of the distributions involved. Consider the fourth column of Table 1 showing $d_k(R)$ for $k = 0$. Among the 149 mutual fund distributions mentioned earlier, those with $E$ in the neighborhood of .10 all have every year's return between a 30 percent loss and a 60 percent gain for the year. (For example, 64 distributions had $.08 \leq E \leq .12$; all were within the range indicated.) $d_0(R)$ goes from $-.022233$ at $R = -.3$, to $+.023454$ at $R = .6$, with substantially smaller values of $|d_0|$ in between. If we imagine spreading a probability distribution throughout the interval $-.3$ to $+.6$, keeping $E = .1$, there is a limit to how large we can make $|Ed_0|$. In fact, if we assume that the distribution can take on only the values listed in the table ($-.3$ to $+.6$ by steps of .05), then a little linear programming will show us that the worst distribution, the one with the greatest $|E(d)|$, is one with a probability of 3/8 of $R = -.3$, a probability of 5/8 of $R = +.35$, and with $E(d) = -.00649$. None of the 149 historical distributions had this worst possible distribution. Insofar as they were less skewed, positive errors above $E$ tended to cancel negative errors below $E$. Insofar as they clustered closer to the mean than in our worst case, the absolute value of the deviations were smaller.

If we recomputed Table 1 for some other $E$, say $E = .15$, the new table would appear very much like the old. The principal difference in column 4 would be that the smallest values of $|d_0(R)|$ would be centered around the new mean, $R = .15$. An analysis similar to that above would again explain why $|Ed|$ was small for those historical distributions which had $E$ in this new neighborhood.

We should examine the concept of $|Ed|$ being small. A statement to the effect that the difference between $EU$ and $EQ$ is less than some number (like .0065) is, in itself, of absolutely no value either in explaining the correlation between $EU$ and $EQ$, or in judging whether $Q$ is a good approximation to $U$ in practice. For a utility function is only defined up to an arbitrary choice of unit and scale. In particular, if we multiply $U$ by an arbitrary positive constant, obtaining a precisely equivalent utility function, we also multiply by the same constant the approximations (8), (10), and (16), and therefore multiply by the same constant the difference between $EU$ and each of these. Thus we can make $|Ed|$ arbitrarily close to zero by using the utility function $V = bU(R)$ for sufficiently small $b$.

The arbitrary choice of unit and scale, however, does not change certain measurements: the correlation and regression coefficients are unaffected; and the appearance of a comparison between the plot of $U$ against $R$ vs. a plot of $Q$ against $R$ is also unaffected in the following sense. Suppose that we plot $U$ in Table 1 against $R$ for $R = -.3$ to $+.6$. Since $U$ rises from about $-.36$ to about $+.47$ we might allow .1 utility units per inch of vertical scale. If we also plot on the same graph $Q_0(R)$ from the third column of the table, $Q$ would be two-tenths of an inch above $U$ at

$R = -.3$, about two-tenths of an inch below $U$ at $R = .6$; but for much of their lengths the two curves would be virtually indistinguishable as they rose from the lower left to the upper right-hand corner of the page. Suppose now that we change the origin and scale of the utility measure. If we still want the two curves to fill the page we simply relabel the vertical axis leaving the two curves unchanged.

If we define $\sigma_{Ed}$ to be the standard deviation of $Ed$ over a set of distributions, where $Ed$ is the mean value of $d$ for a given distribution, and similarly define $\sigma_f = \sigma_{EQ}$ as the standard deviation of $f$ over the set of distributions, then it can be shown[5] that the correlation $\rho_{EU,f}$ between $EU$ and $f$ over the set of distributions is at least

(20) $$\rho_{EU,f} \geq (1 - \gamma^2)^{1/2}$$

where

(21) $$\gamma = \sigma_{Ed}/\sigma_f$$

Thus if $f$ is ten times as variable as $Ed$ then $\rho_{EU,f}$ is at least $(.99)^{1/2} = .995$.

Column 6 of Table 1 presents $d_1(R)$. This equals zero at $R = E - \sigma$, $R = E$, and $R = E + \sigma$ as planned. Just considering $-.3 \leq R \leq .6$, the $f_1$ approximation is definitely superior to $f_0$ in the range from $R = -.05$ through $-.3$, and from $+.25$ to $.6$. On the other hand, $f_0$ fits better near the mean, i.e., from .00 through .20 among the values listed. The empirical results presented in the following sections indicate, for the distributions and utility functions explored, whether it was more beneficial to approximate a bit better near the mean, as does $f_0$, or to hold up well over wider range as does $f_1$.

Table 1 also shows values of $d_0 = U - Q_0$ and $d_1 = U - Q_1$ for more extreme values of $|R - E|$ than discussed thus far. We see for example that, for an investor with a logarithmic utility function, $f_0$ is likely to be a poor approximation to $EU$ for a distribution with $E = .1$ and with nontrivial probabilities of, say, $R = -.7$ and $R = 1.5$. More generally, the empirical results reported below would be less favorable to mean-variance approximations if we were dealing with much more speculative distributions. The mean-variance analysis is thus more suitable for investor's and opportunity sets in which such extremes have very low probabilities in the portfolios which maximize both $EU$ and $f_k$.

### III. Empirical Results

For annual returns of 149 investment companies, 1958–67,[6] Table 2 shows the correlation between $EU(R)$ and $f_k$ $(E, V, U(\cdot))$ for $k = .01, .1, .6, 1.0$, and 2.0 and for various utility functions. (We also computed correlation coefficients for a few other values of the $a$ and $b$ coefficients in the utility functions, with results which one might expect by interpolating or extrapolating the results reported in Table 2. For example, the exponential utility with $b = 20$ was even more of an exception to the general rule than is the case with $b = 10$ reported here.) A row of Table 2 presents correlation $\rho$ as a function of $k$, for some given utility function. In every case reported in Table 2, with the exception of the exponential utility function with $b = 10$, $\rho$ is a nonincreasing function of $k$; hence $\rho_{.01} \geq \rho_k$ for all $k$. Note $\rho_k$, as a function of $k$, is frequently quite flat between $k = .01$ and 1.0, but drops faster from $k = 1$ to $k = 2$. We did not calculate the $\rho_{0.0}$ correlations but, from continuity considerations, we assume

[5]From (19) and $f = EQ$ it follows that

(a) $$\rho_{EU,f} = cov\,(f, f + Ed)/(\sigma_f \sigma_{f+Ed})$$

A short calculation shows that $cov\,(f, f + Ed) = var\,(f) + \rho_{f,Ed}\sigma_f\sigma_{Ed} = var\,(f)(1 + \gamma\rho_{f,Ed})$, and that $\sigma_{f+Ed} = \sigma_f(1 + \gamma^2 + 2\gamma\rho_{f,Ed})^{1/2}$. Substituting these two formulas into (a) we get

(b) $$\rho_{EU,f} = (1 + \gamma\rho_{f,Ed})/(1 + \gamma^2 + 2\gamma\rho_{f,Ed})^{1/2}$$

For $\gamma < 1$ this is a continuous, positive function of $\rho_{f,Ed}$ in the range $-1 \leq \rho_{f,Ed} \leq +1$. If we differentiate (b) with respect to $\rho_{f,Ed}$, set the resulting expression equal to zero, assume $0 < \gamma < 1$, and solve for $\rho_{f,Ed}$, we find that $\partial\rho_{EU,f}/\partial\rho_{f,Ed} = 0$, only at $\rho_{f,Ed} = -\gamma$. Substituting this into (b) we find that at this stationary point $\rho_{EU,f} = (1 - \gamma^2)^{1/2}$. We may confirm that is a minimum rather than a maximum or inflection point by noting, from (b), that $\rho_{EU,f} = +1 > (1 - \gamma^2)^{1/2}$ for $\rho_{f,Ed} = \pm 1$.

[6]The annual rate of return of the 149 mutual funds are taken from the various annual issues of A. Wiesenberger and Company. All mutual funds whose rates of return are reported in Wiesenberger for the whole period 1958–67 are included in the analysis.

TABLE 2—CORRELATION BETWEEN $EU(R)$ AND $f_k(E, V, U(\cdot))$ FOR ANNUAL RETURNS OF 149 MUTUAL FUNDS, 1958–67

| Utility Function | | k = 0.01 | 0.1 | 0.6 | 1.0 | 2.0 |
|---|---|---|---|---|---|---|
| Log(1 + R) | | 0.997 | 0.997 | 0.997 | 0.995 | 0.983 |
| $(1+R)^a$ | a=0.1 | 0.998 | 0.998 | 0.997 | 0.997 | 0.988 |
| | a=0.3 | 0.999 | 0.999 | 0.999 | 0.998 | 0.995 |
| | a=0.5 | 0.999 | 0.999 | 0.999 | 0.999 | 0.998 |
| | a=0.7 | 0.999 | 0.999 | 0.999 | 0.999 | 0.999 |
| | a=0.9 | 0.999 | 0.999 | 0.999 | 0.999 | 0.999 |
| $-e^{-b(1+R)}$ | b=0.1 | 0.999 | 0.999 | 0.999 | 0.999 | 0.999 |
| | b=0.5 | 0.999 | 0.999 | 0.999 | 0.999 | 0.999 |
| | b=1.0 | 0.997 | 0.997 | 0.997 | 0.996 | 0.995 |
| | b=3.0 | 0.949 | 0.949 | 0.941 | 0.924 | 0.817 |
| | b=5.0 | 0.855 | 0.855 | 0.852 | 0.837 | 0.738 |
| | b=10. | 0.447 | 0.449 | 0.503 | 0.522 | 0.458 |

that they are close to those found for $k = .01$. For most cases considered $\rho_{.01} > .99$.

The correlations for the exponential with $b = 10$ are much lower than those of the other utility functions reported in Table 2. In our 1977 paper we analyze this utility function at a greater length than space permits here, and arrive at two conclusions. The first conclusion is that an investor who had $-e^{-10(1+R)}$ as his utility function would have some very strange preferences among probabilities of return. Reasonably enough, he would not insist on certainty of return. For example, he would prefer (a) a 50–50 chance of a 5 percent gain vs. a 25 percent gain rather than have (b) a 10 percent gain with certainty. On the other hand there is no $R$ which would induce the investor to take (a) a 50-50 chance of zero return (no gain, no loss) vs. a gain of $R$ rather than have (b) a 10 percent return with certainty. Thus a 50-50 chance of breaking even vs. a 100 percent, or 300 percent, or even a 1000 percent return, would be considered less desirable than a 10 percent return with certainty. We believe that few if any investors have preferences anything like these. A second conclusion, more important than the first as far as the present discussion is concerned, is that even if some unusual investor did have the utility function in question, if he looked at his $d_k(R)$ in advance he would be warned of the probable inapplicability of mean-variance analysis. The corresponding version of Table 1 (scaled to have about the same $\sigma_f$ for the two approximations, as (20) suggests) shows $d_k$ to generally be more than an order of magnitude greater for the exponential with $b = 10$ than for the logarithmic utility function.

TABLE 3—CORRELATION BETWEEN $EU(R)$ AND $f_{.01}(E, V, U(\cdot))$ FOR 3 HISTORICAL DISTRIBUTIONS

| Utility Function | | Annual returns on 97 stocks[8] | Monthly returns on 97 stocks[8] | Random portfolios of 5 or 6 stocks[9] |
|---|---|---|---|---|
| Log(1 + R) | | 0.880 | 0.995 | 0.998 |
| $(1+R)^a$ | a=0.1 | 0.895 | 0.996 | 0.998 |
| | a=0.3 | 0.932 | 0.998 | 0.999 |
| | a=0.5 | 0.968 | 0.999 | 0.999 |
| | a=0.7 | 0.991 | 0.999 | 0.999 |
| | a=0.9 | 0.999 | 0.999 | 0.999 |
| $-e^{-b(1+R)}$ | b=0.1 | 0.999 | 0.999 | 0.999 |
| | b=0.5 | 0.961 | 0.999 | 0.999 |
| | b=1.0 | 0.850 | 0.997 | 0.998 |
| | b=3.0 | 0.850 | 0.976 | 0.958 |
| | b=5.0 | 0.863 | 0.961 | 0.919 |
| | b=10. | 0.659 | 0.899 | 0.768 |

Table 3 shows the correlation between $EU$ and $f_{.01}$ for three more sets of historical distributions. While $\rho_k$ was computed for the same values of $k$ reported in Table 2, we confine our attention to $k = .01$ since, almost without exception, $\rho_k$ was a nonincreasing function of $k$. The first column of data in Table 3 shows $\rho_{.01}$ for annual returns on 97 *U.S.* common stocks during the years 1948–68.[7] It is understood, of course, that mean-variance analysis is to be applied to the portfolio as a whole rather than individual investments taken one at a time. Annual returns on individual stocks were used in this example, nevertheless, as an example of historic distributions with greater variability than that found in the portfolios reported in Table 2. As expected, the correlations are clearly poorer for the individual stocks than they are for the mutual fund portfolios. For $U = log(1 + R)$, for example, the correlation is .880 for the annual returns on stocks as

[7]This data base of 97 *U.S.* stocks, available at Hebrew University, had previously been obtained as follows: a sample of 100 stocks was randomly drawn from the CRSP (Center for Research in Security Prices, University of Chicago) tape, subject to the constraint that all had reported rates of return for the whole period 1948–68. Some mechanical problems reduced the usable sample size from 100 to 97. The inclusion only of stocks which had reported rates of return during the whole period may have introduced selection bias into the sample. It might prove worthwhile to experiment with alternate methods of handling the appearance and disappearance of stocks.

compared to .997 for the annual returns on the mutual funds.

Since monthly returns tend to be less variable than annual returns we would expect the correlations to be higher for the former than the latter. The $\rho_{.01}$ for monthly returns on the same 97 stocks are shown in the second column of data in Table 3. For the logarithmic utility function, for example, the correlation is .995 for the monthly returns on individual stocks as compared to .880 for annual returns on the stocks and .997 for annual returns on the mutual funds. On the whole, the $\rho_{.01}$ for the monthly returns on individual stocks are comparable to the annual returns on the mutual funds.

Annual returns on individual stocks (i.e., on completely undiversified portfolios) have perceptibly smaller $\rho_{.01}$ than do the annual returns on the well diversified portfolios of mutual funds. The third column of data in Table 3 presents $\rho_{.01}$ for "slightly diversified" portfolios consisting of a few stocks. Specifically, it shows the correlations between $EU$ and $f_{.01}$ on the annual returns for nineteen portfolios of 5 or 6 stocks each randomly drawn (without replacement) from the 97 U.S. stocks.[8] We see that for the logarithmic utility function $\rho_{.01} \doteq .998$ for the random portfolios of 5 or 6, up from .880 for individual stocks. The $\rho_{.01}$ for the annual returns on the portfolios of 5 or 6 were generally comparable to those for the annual returns on the mutual funds. These results were perhaps the most surprising of the entire analysis. They indicate that, as far as the applicability of mean-variance analysis is concerned, at least for joint distributions like the historical returns on stocks for the period analyzed, a little diversification goes a long way.

In addition to the correlation coefficient, we examine in our 1977 paper other measures of the ability of $f_k$ to serve as a surrogate for $EU$, and conclude that $f_{.01}$ does as well in these comparisons as it does in terms of correlations. For example, we computed the frequency with which any other available portfolio was better than the portfolio which maximized $f_k$. We found, in particular, that in every case with $\rho > .9$ in Tables 2 or 3 the portfolio with maximum $f_{.01}$ was also the portfolio (among the 149 or 97 or 19 considered) with the greatest $EU$. We cannot say with any precision how high a correlation between $f_k$ and $EU$ is high enough. Be that as it may, for many of the utility functions and distributions considered (chosen in advance as representative of utility functions frequently postulated, or distributions clearly "real world") $f_{.01}$ was an almost faultless surrogate for $EU$. Where $f_{.01}$ performed poorly, the user would have been warned in advance by an analysis of expected error.

### IV. Some Objections Reconsidered

Since $\rho_{EU,f} < 1$ it can happen that portfolio $A$ has a higher $f_k$, than portfolio $B$, while portfolio $B$ has a higher $EU$. In fact, given any function $f$ of $E$ and $V$, Borch presents a method for finding distributions $A$ and $B$ such that $f(E_A, \sigma_A) = f(E_B, \sigma_B)$, yet clearly $EU_A > EU_B$ because distribution $A$ stochastically dominates distribution $B$. (In terms of equation (20), this example has $\gamma = \infty$.)

Borch's argument shows that it is hopeless to seek an $f$ that will be perfectly correlated with $EU$ for all collections of probability distributions. The evidence on the preceding pages nevertheless supports the notion that the imperfect approximations $f_k(E, V)$ are frequently good enough in practice to choose almost optimally among realistic distributions of returns; and the $d_k$ function can be used in advance to judge the suitability of $f_k$.

The approximation $f_k$ was obtained by fitting the quadratic (12) to three points. We have seen that for certain utility functions and historical distributions of returns, $f_k$ is highly correlated with $EU$. Pratt and Arrow, on the other hand, have shown that any quadratic utility function had highly undesirable theoretical characteristics. How do we reconcile these two apparently contradictory observations?

[8]We randomly drew 5 stocks to constitute the first portfolio; 5 different stocks to constitute the second portfolio, etc. Since we have 97 stocks in our sample, the eighteenth and nineteenth portfolios include 6 stocks each. Repetition of this experiment with new random variables produced negligible variations in the numbers reported, except for the case of $U = e^{-10(1+R)}$. A median figure is reported in the table for this case.

It is essential here to distinguish between three types of quadratic approximations:

1) Assuming that utility as a function of wealth $V(W)$ remains constant from period to period, a quadratic $q(W)$ is fit to $V(W)$ once and for all. As $W$ changes from time to time, the same $q(W)$ function is used to select portfolios. (Note that $V$ expresses utility as a function of wealth $W$, in contrast to the previously defined $U$ function which expressed utility as a function of rate of return $R$. Note also that an unchanging $V(W)$ implies the existence of some $U(R)$ at each point in time, though not necessarily the same $U(R)$ each time. The converse is not true: the existence of a $U(R)$ at each point in time does not necessarily imply an unchanging $V(W)$.)

2) At the beginning of each time period a $q(W)$ function is fit to $V(W)$ at the point $W$ = current wealth; or equivalently $Q(R)$ is fit to $U(R) = V((1 + R)W_{t-1})$ at $R = 0$. Even if $V(W)$ remains constant through time, the quadratic fit $Q(R)$ is changed as wealth changes.

3) The quadratic fit (of $Q(R)$ to $U(R)$) depends on at least $E$ and perhaps $\sigma$ as well. In this case the fit varies between one distribution and another being evaluated at the same time.

The Pratt and Arrow objections apply to quadratic approximations of type 1.[9] The approximation in (8) is of type 2. Approximations (10) and (16) are of type 3. Each of these three classes of approximations have different risk-aversion properties. We shall confine our attention here to a comparison between the first and third of these. To illustrate the difference between the first and third, we shall compare their "risk premiums" for small risks as defined by Pratt.

Pratt's results may be expressed as follows. Let $Pr_i(R)$ $i = 1, 2, 3, \ldots$ be a sequence of probability distributions such that each has the same mean:

$$\text{(22a)} \quad \int R\,dPr_i(R) = E_0 \qquad i = 1, 2, 3, \ldots$$

and such that their standard deviations approach zero:

$$\text{(22b)} \qquad \lim \sigma_i = 0$$

where the limit, here and elsewhere in this section unless otherwise specified, is taken as $i \to \infty$. The risk premium for the $i$th distribution is defined implicitly by

$$\text{(22c)} \quad U(E_0 - \pi_i) = \int U(R)\,dPr_i(R) \qquad i = 1, 2, 3, \ldots$$

where the right-hand side equals the expected utility of the $i$th distribution.

*The omitted material contains an inconsequential "bug".*

*(Harry Markowitz)*

Pratt reasonably asserts that $\pi/\sigma^2$, and hence $r((1 + E_0)W_{t-1})$ should be a decreasing function of its one argument, $W = (1 + E_0)W_{t-1}$. He notes that for a quadratic (of type 1) $r(W)$ is an increasing function of $W$ and concludes:

> Therefore a quadratic utility cannot be decreasingly risk-averse on any interval whatever. This severely limits the usefulness of quadratic utility, however nice it would be to have expected utility depend only on the mean and variance of the probability distribution. Arguing "in the small" is no help: decreasing risk aversion is a local property as well as a global one. [p. 132]

That Pratt's conclusion is not correct for an approximation of type 3 will be shown in the following way. If an investor maximized some mean-variance approximation $f(E,\sigma)$, such as $f_k$ for fixed nonnegative $k$, then his risk premium would be given implicitly by

$$\text{(24)} \qquad f(E - \pi, 0) = f(E, \sigma)$$

[9]Pratt correctly asserts that his analysis does not require constant $V(W)$ over time. We must distinguish, however, between a $V$ function varying with time (but not depending on $W$ itself) vs. approximations of type 2 or 3 in which the choice of the quadratic function depends on $W$ or even $E$ and $\sigma$. For simplicity, we describe approximations of type 1 in terms of a fixed $V(W)$, rather than try for greater generality here.

Writing $\sigma_R$ and $\pi_R$ to emphasize that we refer to the standard deviation and risk premium of R, as opposed to $\sigma_W$ and $\pi_W$ which will stand for the standard deviation and risk premium of wealth, $W = W_0(1+R)$, Pratt showed that

$$\lim_{\sigma_R \to 0} \frac{\pi_R}{(\sigma_R)^2} = -\tfrac{1}{2}\frac{U''(E_0)}{U'(E_0)} \qquad (1)$$

$$= (\tfrac{1}{2})r^*(W_0(1+E_0))$$

where $r^*$ is "relative risk aversion". Recall that

$$r^*(W) = W \cdot r(W) \qquad (2)$$

where

$$r(W) = -\frac{V''(W)}{V'(W)}$$

is defined as the absolute risk aversion. In his paper on "Risk Aversion ...", Pratt discusses r(W) first and shows that

$$\lim_{\sigma_W \to 0} \frac{\pi_W}{(\sigma_W)^2} = -\tfrac{1}{2}\frac{V''(W)}{V'(W)}$$

$$= (\tfrac{1}{2})r(W)$$

Towards the end of the paper he discusses $r^*$ and shows (1) and (2).

Below we define a class of type 3 approximations including, in particular, $f_k$ for all $k \geq 0$. We consider a sequence of probability distributions $Pr_i(R)$ $i = 1, 2, 3, \ldots$ satisfying (22a) and (22b), and find that

$$(25) \quad \lim \pi_i/\sigma_i^2 = -1/2U''(E_0)/U'(E_0)$$

for all approximations in the class. Thus the risk aversion "in the small" of $f_k$ is precisely the same as that of $U$ itself. We have seen that not all $f_k$ are equally good "in the large." But in the small, they are asymptotically the same as the utility function which they approximate.

We assume that the investor maximizes an approximation $f(E,\sigma)$ which is the expected value of a quadratic of type 3 such that $f = EQ$ in (13) satisfies

$$(26a) \qquad a = U(E)$$

$$(26b) \qquad c \to .5U''(E) \text{ as } \sigma \to 0$$

It follows immediately from (10) that $f_0$ satisfies (26a) and (26b), and from (15) that $f_k$ $k > 0$ satisfies (26a). That $f_k$ also satisfies (26b) follows from L'Hopital's rule applied to the expression for $c$ in (15).

From (13) and (26a), (24), (13) again, and (26b) we infer that

$$\begin{aligned} U(E_0 - \pi) &= f(E_0 - \pi, 0) \\ &= f(E_0, \sigma) \\ &= U(E_0) + c(E_0, \sigma)\sigma^2 \\ &\quad \to U(E_0) \text{ as } \sigma \to 0 \end{aligned}$$

But $U(E_0 - \pi) \to U(E_0)$, and $U' > 0$ throughout, implies

$$(27) \qquad \lim \pi_i = 0$$

Using Taylor's theorem we may write

$$\begin{aligned} (28) \quad f(E_0 - \pi, 0) &= U(E_0 - \pi) \\ &= U(E_0) - U'(E_0)\pi \\ &\quad + .5U''(\xi)\pi^2 \end{aligned}$$

where $\xi$ is between $E_0 - \pi$ and $E_0$. Hence, from (27), $\xi \to E_0$ as $i \to \infty$. Using (28) on the left of (24), (13) on the right of (24), and rearranging terms we get

$$(29) \quad \pi/\sigma^2 = -c(E_0,\sigma)/(U'(E_0) - .5U''(\xi)\pi)$$

for each distribution in the sequence. This together with (26b) and (27) implies (25).

### V. The E,V Investor

Let us return to Mr. $X$ who has not analyzed his utility function recently. Suppose that, when presented with probability distributions of $E,V$ efficient portfolios, he can pick that portfolio which has greater $EU$ than any other $E,V$ efficient portfolio. By definition, his choice will be at least as good as the portfolio which maximizes $f_{.01}$. In addition to the functions of $E$ and $V$ discussed above there may very well be others—not yet explored, or perhaps not yet conjectured—which perform better than does any $f_k$ as a surrogate for $EU$. Mr. $X$'s choice of portfolio will also be at least as good as the best of all of these in each particular situation.

## REFERENCES


**Kenneth Arrow,** *Aspects of the Theory of Risk Bearing*, Helsinki 1965.

**K. Borch,** "A Note on Uncertainty and Indifference Curves," *Rev. Econ. Stud.*, Jan. 1969, *36*, 1–4.

**J. S. Chipman,** "The Ordering of Portfolios in Terms of Mean and Variance," *Rev. Econ. Stud.*, Apr. 1973, *40*, 167–90.

**M. S. Feldstein,** "Mean Variance Analysis in the Theory of Liquidity Perference and Portfolio Selection," *Rev Econ. Stud.*, Jan. 1969, *36*, 5–12.

**G. Hanoch and H. Levy,** "The Efficiency Analysis of Choices Involving Risk," *Rev. Econ. Stud.*, July 1969, *36*, 335–46.

——— **and** ———, "Efficient Portfolio Selection with Quadratic and Cubic Utility," *J. Bus., Univ. Chicago*, Apr. 1970, *43*, 181–89.

**H. Levy,** "The Rationale of the Mean-Standard Deviation Analysis: Comment," *Amer. Econ. Rev.*, June 1974, *64*, 434–41.

——— **and H. Markowitz,** "Mean-Variance Approximations to Expected Utility," T. J. Watson Res. Ctr., IBM, Aug. 1977.

**Harry Markowitz,** *Portfolio Selection: Efficient Diversification of Investments*, New York 1959; New Haven 1970.

**J. W. Pratt,** "Risk Aversion in the Small and in the Large," *Econometrica*, Jan. 1964, *32*, 122–36.

**J. Tobin,** "Liquidity Preference as Behavior Towards Risk," *Rev. Econ. Stud.*, Feb. 1958, *25*, 65–87.

——, "The Theory of Portfolio Selection" in Frank H. Hahn and Frank P. R. Brechling, eds., *Theory of Interest Rates*, New York 1963.

**S. C. Tsiang,** "The Rationale of the Mean Standard Deviation Analysis, Skewness Preference, and the Demand for Money," *Amer. Econ. Rev.*, June 1972, *62*, 354–71.

**John von Neumann and Oskar Morgenstern,** *Theory of Games and Economic Behavior*, 3d ed., Princeton 1953.

**W. E. Young and R. H. Trent,** "Geometric Mean Approximation of Individual Security and Portfolio Performance," *J. Finan. Quant. Anal.*, June 1969, *4*, 179–99.

**A. Wiesenberger and Company,** *Investment Companies*, New York, annual editions.

# Mean-Variance Versus Direct Utility Maximization

YORAM KROLL, HAIM LEVY, and HARRY M. MARKOWITZ*

ABSTRACT

Levy and Markowitz showed, for various utility functions and empirical returns distributions, that the expected utility maximizer could typically do very well if he acted knowing only the mean and variance of each distribution. Levy and Markowitz considered only situations in which the expected utility maximizer chose among a finite number of alternate probability distributions. The present paper examines the same questions for a case with an infinite number of alternate distributions, namely those available from the standard portfolio constraint set.

IT IS FREQUENTLY ASSERTED that mean-variance analysis applies exactly only when distributions are normal or utility functions quadratic, suggesting that it gives almost optimum results only when distributions are approximately normal or utility functions look almost like a parabola. On the other hand, in a recent paper Levy and Markowitz [6] showed empirically that the ordering of portfolios by the mean-variance rule was almost identical to the order obtained by using expected utility for various utility functions and historical distributions of returns.

For example, the authors calculated an "exact" expected utility (of, say, a logarithmic utility function) for each of 149 mutual funds by attributing an equal probability for each year in the sample. Using the same data, the expected utility was approximated by a function of mean and variance $U = f(E, V)$, where $E$ represents the mean and $V$ represents the variance. The exact expected utility and the approximation based only on $E$ and $V$ were found to be highly correlated. The analysis was repeated for various sets of data and various utility functions, and the same results were obtained in almost every case.

Good results were obtained, then, when $EU$ and $f(E, V)$ were compared for a finite number of portfolios, e.g., 149 mutual funds. However, it may be that $f(E, V)$ would do less well when asked to find the best portfolio among the infinite number of possible mixtures of a finite number of securities. In this case, the exact maximizing of expected utility might lead to quite different results than those obtained by using the mean-variance approximation. The aim of the present paper is to compare the expected utility of the optimum portfolio for given utility functions with the expected utility of well-selected portfolios from the mean-

*Kroll and Levy from Jerusalem School of Business Administration, Hebrew University, Israel, and Markowitz from Baruch College, The City University of New York.

variance efficient frontier. Specifically, for various probability distributions and utility functions:

(a) We derive a fine mesh of points along the mean-variance efficient frontier, and calculate the expected utility of all these portfolios. From these we select the portfolio with the highest expected utility.
(b) Using the same set of data of individual stocks, we select the portfolio which maximizes expected utility of the given utility function—not just maximum among mean-variance efficient portfolios, but among all feasible portfolios.

Comparison of the expected utilities obtained in (a) and (b) indicates the possible error in investment decision made using the mean-variance framework.

We also compare the portfolios obtained in (a) and (b). In addition, we compare the expected utility of the optimum portfolio with the expected utilities of portfolios containing equal weights of a few securities with highest means. Finally, we examine the effect of leverage on the approximation of the $E$-$V$ maximization to the direct utility maximization.

The present paper is similar to that of Pulley [9] who also compares mean-variance efficient portfolios which approximately maximize expected utility with portfolios which actually maximize expected utility. There are two principal differences between the results presented here and those of Pulley. The most important difference between Pulley's and the present work is the criteria used to compare the expected utility $EU_M$ achieved by the mean-variance approximation with the actual maximum expected utility $EU_A$. Pulley made an error in choice of criterion which completely invalidates his comparisons. Recall that if $U(R)$ is a utility function then, for any real number $c$, $c + U(R)$ is another utility function with exactly the same ranking of probability distributions. Pulley's ratic criteria $EU_M/EU_A$ is not invariant to the choice of $c$. In fact, it can always be made arbitrarily close to 1.0 (the best possible score) or near to zero, or ever negative by choice of the irrelevant additive constant.

For example, suppose someone tested a great many utility functions including among others, $U(R)$, $U(R) + 1000$, and $U(R) - 5$. Of course, what we call $U(R$ and what we call $U(R) + 1000$ is arbitrary—an accident of history. They ar equivalent utility functions. Suppose that the portfolio which maximizes expecte utility happens to have $EU(R) = 8$ while the mean-variance approximatio happens to have $EU(R) = 4$. This gives us a mediocre Pulley score of $4/8 = 0.$ But the equivalent utility function, $U(R) + 1000$, has the great Pulley score c $1004/1008 = 0.99+$; while the equally equivalent utility function $U(R) - 5$ ha the terrible Pulley score of $-1/3 = -0.33$.

In general, equivalent utility functions can always be made to give arbitrari good or arbitrarily bad Pulley scores. Pulley's results should, therefore, be viewe like the results of an experiment on heat made with a broken thermometer—n just a slightly inaccurate thermometer, but a capricious one that can read freezir or boiling for two bodies with the same temperature.

Another difference between the present article and the Pulley article is th the latter is concerned, as its title states, with "short holding periods." Accor ingly, Pulley's analyses deal with monthly and semiannual holding periods. The analyses in the present paper, on the other hand, use annual holding periods. As both Pulley and Levy-Markowitz explain, the higher the portfolio variance the less likely is a mean-variance approximation to do almost as well as actual expected utility maximization. Thus, our use of annual data poses a greater challenge for mean-variance than do Pulley's monthly and semiannual analyses. We further challenge mean-variance by including an analysis which allows borrowing.

## I. The Problem

A portfolio is mean-variance efficient if it maximizes expected rate of return ($E$) for a given variance ($V$), and minimizes the variance for a given expected return. Let us denote by $X_i$ the proportion of the $i$th asset in the portfolio. Thus, assuming the standard constraint set without borrowing, an efficient portfolio $\underline{X}' = (X_1, X_2, \cdots, X_n)$ solves the following problem:

$$\text{Minimize } \underline{X}' \textstyle\sum \underline{X}$$

subject to

$$X_i \geq 0 \quad i = 1, 2, \cdots, n$$

$$\underline{X}'\underline{M} = E$$

and

$$\underline{X}'1 = 1$$

where $\sum$ is the covariance matrix, $M$ is the vector of mean returns of the $n$ securities, $E$ is the mean return of the portfolio, and 1 represents either the number 1 or a vector of 1's as needed.

For various values of $E$ we obtain various efficient portfolios. For each $E$-$V$ efficient portfolio, one may calculate its expected utility.

$$EU(\textstyle\sum_{i=1}^{n} X_i R_i)$$

where $R_i$ is the return on the $i$th security. In principle, by calculating expected utility for all efficient portfolios, one can select the $E$-$V$ efficient portfolio which maximizes expected utility. The maximum expected utility obtained on the $E$-$V$ efficient set will be denoted by $E^*U(\ )$.

While $E^*U(\ )$ is the solution to the maximization of the expected utility over the set of the $E$-$V$ efficient set, the optimal portfolio is obtained by allowing all possible investment mixes, and not only the $E$-$V$ efficient portfolios. The optimal portfolio is given by solving the following optimization problem:

$$\max_{X} EU(\textstyle\sum_i X_i R_i)$$

subject to

$$X_i \geq 0 \quad i = 1, 2, \cdots, n$$

and

$$\underline{X}'1 = 1$$

The value obtained by this maximization, which we shall refer to as the direct maximization, will be denoted by $EU(\ )$ as distinguished from $E^*U(\ )$.

Since $EU(\ )$ is a general maximization without the constraint that $\underset{\sim}{X}$ be $E$-$V$ efficient, $E^*U(\ ) \leq EU(\ )$. Finding $\underset{\sim}{X}$ which solves for $EU(\ )$ (direct maximization) is not a trivial exercise even by computer. It has the following disadvantages:

(a) It requires a large number of calculations, typically several times the number necessary to trace out the $E$-$V$ efficient frontier. More details on the method of direct maximization are given in the Appendix.
(b) In calculating $EU(\ )$ for a given set of data, one has to repeat all the calculations for each of the various $U$'s which one considers. In contrast, the $E$-$V$ efficient portfolios can be found once for all utility functions leaving the choice of $E^*U(\ )$ from the $E$-$V$ frontier a relatively minor task.

Consider an investment consultant who would like to find his customers' optimal investment strategies. However, he cannot do so without knowledge of the investors' particular utility functions. Moreover, he has many clients who may differ with respect to their preferences. On the other hand, if $E^*U(\ )$ is almost equal to $EU(\ )$, the consultant can overcome the difficulty of not knowing the investors' preferences simply by deriving the $E$-$V$ efficient portfolios and presenting only these alternative porfolios to his customers. Each investor would choose from the $E$-$V$ efficient set a portfolio according to his particular preferences, which may not be explicitly stated.

We must analyze, however, the loss of welfare incurred by using $E^*U(\ )$ as the optimal criterion rather than $EU(\ )$. The remainder of this paper is primarily concerned with the development and measurement of an index of this welfare loss. We shall test the approximation of $E^*U(\ )$ to $EU(\ )$ for certain frequently cited utility functions.

## II. The Quality of the Approximation

One could measure the loss of utility incurred by choosing among $E$-$V$ efficient portfolios by the difference $D = EU(\ ) - E^*U(\ )$; but, $D$ is not invariant to linear transformations of the utility functions. A natural choice instead for an index is:

$$I = \frac{E^*U(\ ) - E_N U(\ )}{EU(\ ) - E_N U(\ )}$$

where $E_N U(\ )$ is the expected utility of a "Naive" portfolio in which $\frac{1}{n}$ is investec in each security, namely:

$$E_N U(\ ) = EU\left(\sum_{i=1}^{N} \frac{1}{n} R_i\right)$$

By definition, $I$ is less than one. It can be negative, but one would not expect th best $E$-$V$ portfolio to be worse than a naive portfolio; thus, in most cases $0 \leq$ $\leq 1$. If $I$ is close to zero, we can conclude that the $E$-$V$ efficiency criterion is nc very promising, since the naive method gives almost the same expected utilit

When the index $I$ is close to one, the approximation is good and the error in using the $E$-$V$ criterion is small.

Another possible index for measuring the welfare loss is:

$$I_R = \frac{E^*U(\ ) - E_R U(\ )}{EU(\ ) - E_R U(\ )}$$

where $E_R U(\ )$ is the expected utility of a portfolio selected at random from a uniform distribution, i.e., with every subset of the constraint set having the same probability of including the selected portfolio as any other subset of equal volume. We repeat the random selection several times and calculate the average expected utility across all random portfolios. In fact, we found this average almost equal to $E_N U(\ )$. Thus, we report only the results for the index $I$.

## III. The Selected Utility Functions

The following utility functions are used in the empirical tests:

$-e^{-(1+R)}$
$(1 + R)^a$, $(a = 0.1, 0.5, 0.9)$
$(2 + R)^a$, $(a = 0.1, 0.5)$
$\ln(i + R)$, $(i = 1, 2)$

where $R$ is defined as the rate of return on investment. All of these functions have $U' > 0$, $U'' < 0$, and $U''' > 0$. In Table I we list the properties of these functions with respect to the absolute and relative risk aversion measures of Arrow [1] and Pratt [8]. Note that in Table I, $W$ corresponds to $R + 1$ above. For example, $(2 + R)^a = (B + W)^a$ where $B = 1$.

## IV. The Data

We selected three mutually exclusive samples of 10, 12, and 20 stocks from the CRSP tape. Since the results are very similar, we report here only the results of the 20-stock sample. It is not that we recommend past history alone as a predictor

**Table I**

Some Properties of the Selected Utility Functions Defined on Wealth, $W$

| Utility | Absolute Risk-Aversion Measure | Proportional Risk-Aversion Measure |
|---|---|---|
| $-e^{-aW}$, $(a > 0)$ | Constant | Increasing |
| $(W + B)^a$, $(0 < a < 1)$ | Decreasing | Increasing for $B > 0$<br>Constant for $B = 0$<br>Decreasing for $B < 0$ |
| $\ln(W + B)$, $(B > 0)$ | Decreasing | Increasing if $B > 0$<br>Constant if $B = 0$<br>Decreasing if $B < 0$ |

of future returns. Rather we use this data as examples of real world security and portfolio moments.

In Table II, we present the means, standard deviations, coefficient of variation, the relative skewness, and the kurtosis of the annual returns of these 20 stocks in the years 1949–1968.

The question has been raised concerning the Levy-Markowitz results whether the ability of $f(E, V)$ to approximate $EU(R)$ was due to normality of the underlying distributions rather than the asserted robustness of the quadratic approximation. For the present data, Table II clearly rejects the notion that the return distributions are normal. First note that only one security is negatively skewed. If these were independent drawings from any symmetric distribution, then the probability would be only $21/2^{20} \approx 0.0002$ that only zero or one sample would be negatively skewed. The significance of this observation is clouded by the lack of independence between securities. Recall that if returns were normal then the coefficient of skewness would be roughly normal with mean 0 and standard deviation $= \sqrt{6/N} = 0.55$, and the deviation of the coefficient of kurtosis from 3 would be roughly normal with mean 0 and standard deviation $= \sqrt{24/N} = 1.1$ (see Kendall and Stuart [5]). In Table II, 6 out of the 20 securities have

**Table II**

The Average Rate of Return, Standard Deviation, Coefficient of Variance, and Relative Skewness of the 20-Stock Sample in the Years 1949–1968

| Stock | Means % | Standard Deviations % | Coefficient of Variation* | Relative Skewness** | Kurtosis*** |
|---|---|---|---|---|---|
| 1. Conelco | 30.67 | 100.24 | 3.27 | 2.97 | 10.17 |
| 2. Texas Gulf | 23.50 | 59.36 | 2.53 | 2.61 | 8.93 |
| 3. Carpenter | 23.25 | 37.19 | 1.60 | 0.71 | 0.705 |
| 4. Cerro | 21.38 | 44.51 | 2.08 | 0.55 | −1.00 |
| 5. Chrysler | 20.49 | 50.31 | 2.45 | 2.31 | 7.32 |
| 6. California Pack. | 20.30 | 24.22 | 1.19 | 0.26 | 0.78 |
| 7. Dana Co. | 20.04 | 30.26 | 1.50 | 1.03 | 1.52 |
| 8. Sterling Drugs | 17.55 | 17.34 | 0.99 | −0.49 | 1.11 |
| 9. Copperweld Steel | 17.53 | 38.27 | 2.18 | 1.17 | 2.57 |
| 10. Crucible Steel | 16.95 | 38.48 | 2.27 | 0.42 | −0.08 |
| 11. Mobil Oil | 16.87 | 23.69 | 1.40 | 0.04 | −0.94 |
| 12. Colt | 15.15 | 43.76 | 2.89 | 0.50 | 0.26 |
| 13. Standard Oil | 14.82 | 19.26 | 1.30 | 0.22 | −0.68 |
| 14. Sucrest Co. | 14.20 | 29.33 | 2.06 | 0.56 | 1.66 |
| 15. Sunray | 14.14 | 24.18 | 1.68 | 0.50 | −0.68 |
| 16. Chemway | 13.84 | 35.65 | 2.58 | 0.84 | 0.18 |
| 17. Continental Can | 13.53 | 15.23 | 1.13 | 0.21 | −0.58 |
| 18. Detroit Steel | 13.45 | 32.44 | 2.41 | 0.37 | −0.56 |
| 19. Spartan | 12.84 | 44.82 | 3.49 | 1.21 | 2.15 |
| 20. City Stores | 9.10 | 24.62 | 2.71 | 1.96 | 4.49 |

* Coefficient of Variation is the standard deviation over the mean.

** Relative skewness is measured by $\mu_3/(\mu_2)^{3/2}$ where $\mu_2$ and $\mu_3$ are the second and third central moments, respectively.

*** Kurtosis is defined here as $(\mu_4/\mu_2^2) - 3$ where $\mu_4$ is the fourth central movement.

observed skewness at least 2 standard deviations from zero, and some of these are 4 or 5 standard deviations from expected under the normal hypothesis. Furthermore, several have observed kurtosis which is 4 to 9 standard deviations from expected. Clearly, not all security returns are normal.

## V. The Empirical Results

Table III presents the optimum portfolios selected by direct maximization for various utility functions. From the table we see that:

> Ignoring the small proportion invested in Chrysler in the case of the negative exponential utility function, only 4 securities out of the available 20 securities appear with positive proportions in the optimal portfolio.

The 4 selected securities are from the group of 6 securities with the highest mean (see Table II).

Conelco, with highest mean, always appears in the optimal portfolio. The securities, Texas Gulf and Carpenter, with second and third highest mean, almost always appear in the optimal portfolio. The fourth highest (Cerro) never appears in the optimal portfolio; and the fifth (Chrysler) almost never appears. California Pack. which is ranked only sixth according to mean is selected in many cases, perhaps due to its low variance. (California Pack. has almost the lowest coefficient of variation.)

In order to compare the above with $E$-$V$ efficient portfolios, we first derived a "mesh" of $E$-$V$ efficient portfolios with mean ranging from 16.5 to 30.3 by steps of 0.2. Table IV presents a selection of portfolios from this set. The table reports for each portfolio, the mean, the standard deviation, and the investment allocation. Note that, like the direct maximization, the $E$-$V$ efficient portfolios are not well-diversified. Out of the 20 securities, only 9 are ever used. Moreover, for much of the efficient frontier only 4–5 stocks account for 100% of the portfolios.

We next calculated the expected utility of each portfolio in our mesh of $E$-$V$ efficient portfolios. Since the investment allocation is given, no maximization is

**Table III**

Optimal Investment Strategies with a Direct Utility Maximization

| Utility Function | California (6)* | Carpenter (3) | Chrysler (5) | Conelco (1) | Texas Gulf (2) | Total | Average Return | Standard Deviation |
|---|---|---|---|---|---|---|---|---|
| $-e^{-X}$ | 44.3 | 34.7 | 0.2 | 5.5 | 15.3 | 100% | 22.4 | 27.3 |
| $X^{0.1}$ | 33.2 | 36.0 | - | 13.6 | 17.2 | 100% | 23.3 | 32.3 |
| $X^{0.5}$ | — | 42.2 | — | 34.4 | 23.4 | 100% | 25.9 | 49.4 |
| $X^{0.9}$ | — | — | — | 97.6 | 2.4 | 100% | 30.5 | 98.6 |
| $\ln(X)$ | 37.9 | 34.8 | — | 11.1 | 16.2 | 100% | 23.1 | 29.4 |
| $\ln(X+1)$ | 3.7 | 46.8 | — | 26.1 | 23.4 | 100% | 25.1 | 43.7 |
| $(X+1)^{0.1}$ | 0.4 | 44.9 | — | 30.3 | 24.4 | 100% | 25.5 | 46.7 |
| $(X+1)^{0.5}$ | — | 33.5 | — | 66.5 | — | 100% | 28.2 | 74.2 |

* Numbers indicate rank according to mean, Table II.

**Table IV**

**Proportions of Stocks Average Return, Standard Deviation of Portfolios on the *E*-*V* Efficient Frontier, and the Implied Risk-Free Rate.***

| Security | | | | | | |
|---|---|---|---|---|---|---|
| California | 9.46 | 13.62 | 16.40 | 21.45 | 25.13 | 30.65 |
| Carpenter | | | | 3.46 | 6.27 | 10.47 |
| Chrysler | | | | 0.46 | 1.00 | 1.81 |
| Continental | 32.84 | 25.41 | 20.46 | 15.28 | 17.35 | 7.97 |
| Dana | 5.85 | 7.96 | 9.36 | 6.91 | 4.55 | 1.02 |
| Mobil Oil | 0.62 | 0.47 | 0.37 | 1.67 | 2.41 | 3.59 |
| Sterling Drugs | 37.32 | 39.25 | 40.53 | 39.87 | 38.84 | 37.29 |
| Sucrest Co. | 10.33 | 8.50 | 7.28 | 4.82 | 3.12 | 0.58 |
| Texas Gulf | 3.58 | 4.79 | 5.60 | 6.09 | 6.30 | 6.62 |
| TOTAL | 100.00 | 100.00 | 100.00 | 100.00 | 100.00 | 100.00 |
| Average Return | 16.5 | 17.10 | 17.50 | 18.10 | 18.50 | 19.10 |
| Standard Deviation | 11.70 | 12.51 | 13.19 | 14.36 | 15.70 | 16.53 |
| Risk-free Rate | 6.66 | 9.47 | 10.50 | 11.19 | 11.51 | 11.90 |

| Security | | | | | | |
|---|---|---|---|---|---|---|
| California | 34.31 | 38.99 | 41.91 | 46.08 | 48.55 | 46.04 |
| Carpenter | 12.73 | 15.45 | 19.70 | 25.86 | 29.53 | 35.07 |
| Chrysler | 2.39 | 3.08 | 2.80 | 2.38 | 2.13 | 0.90 |
| Conelco | | | | 0.24 | 0.67 | 2.62 |
| Continental | 5.18 | 0.21 | | | | |
| Mobil Oil | 2.41 | | | | | |
| Sterling Drugs | 35.69 | 33.65 | 25.66 | 13.56 | 6.00 | |
| Texas Gulf | 7.28 | 8.62 | 9.94 | 11.88 | 13.12 | 15.36 |
| TOTAL | 100.00 | 100.00 | 100.00 | 100.00 | 100.00 | 100.00 |
| Average Return | 19.5 | 20.10 | 20.50 | 21.10 | 21.50 | 22.10 |
| Standard Deviation | 17.46 | 18.91 | 19.97 | 21.77 | 23.08 | 25.23 |
| Risk-free Rate | 12.17 | 12.70 | 13.53 | 14.39 | 14.74 | 16.04 |

| Security | | | | | |
|---|---|---|---|---|---|
| California | 39.37 | 28.50 | 21.29 | 10.39 | 3.15 |
| Carpenter | 38.59 | 43.38 | 46.57 | 51.36 | 54.56 |
| Conelco | 4.97 | 8.65 | 11.11 | 14.80 | 17.26 |
| Texas Gulf | 17.08 | 19.47 | 21.06 | 23.45 | 25.04 |
| TOTAL | 100.00 | 100.00 | 100.00 | 100.00 | 100.00 |
| Expected Return | 22.50 | 23.10 | 23.50 | 24.10 | 24.50 |
| Standard Deviation | 26.96 | 29.97 | 32.20 | 35.80 | 38.34 |
| Risk-free Rate | 16.85 | 17.64 | 18.00 | 18.39 | 18.59 |

| Security | | | | | | |
|---|---|---|---|---|---|---|
| Carpenter | 50.07 | 44.55 | 36.28 | 30.76 | 22.49 | 16.97 |
| Conelco | 24.06 | 29.45 | 37.53 | 42.91 | 50.99 | 56.38 |
| Texas Gulf | 25.87 | 26.00 | 26.19 | 26.32 | 26.52 | 24.65 |
| TOTAL | 100.00 | 100.00 | 100.00 | 100.00 | 100.00 | 100.00 |
| Expected Return | 25.10 | 25.50 | 26.10 | 26.50 | 27.10 | 27.50 |
| Standard Deviation | 42.45 | 45.59 | 50.78 | 54.48 | 60.31 | 64.34 |
| Risk-free Rate | 19.57 | 20.05 | 20.55 | 20.79 | 21.08 | 21.23 |

| Security | | | | | | |
|---|---|---|---|---|---|---|
| Carpenter | 8.70 | 3.18 | | | | |
| Conelco | 64.46 | 69.85 | 78.10 | 83.68 | 92.05 | 97.63 |
| Texas Gulf | 26.84 | 26.97 | 21.90 | 16.32 | 7.95 | 2.37 |
| TOTAL | 100.00 | 100.00 | 100.00 | 100.00 | 100.00 | 100.00 |
| Expected Return | 28.10 | 28.50 | 29.10 | 29.50 | 30.10 | 30.[illegible] |
| Standard Deviation | 70.56 | 74.79 | 81.33 | 85.92 | 93.12 | 98.[illegible] |
| Risk-free Rate | 21.41 | 21.51 | 21.94 | 22.22 | 22.56 | 22.[illegible] |

* This implied risk-free rate is the rate at which a straight line which is tangent to the efficient risky frontier crosses the vertical axis. The risk-free rate obtained here is a technical result which is not necessarily equal to the actual risk-free rate.

For example, for the first $E$-$V$ efficient portfolio in Table IV we calculated a rate of return for each period $t$,

$$R_t = 0.0946R_{1t} + 0.3284R_{2t} + 0.058R_{3t}$$
$$+ 0.00062R_{4t} + 0.3732R_{5t} + 0.1033R_{6t} + 0.0358R_{7t}$$

where $R_{it}$ is the annual rate of return in year $t$ of the $i$th security included in the efficient portfolio. From this we calculate the expected utility

$$EU(\ ) = \frac{1}{20}\Sigma_{t=1}^{20}\ U(R_t).$$

For each given utility function, we repeat this calculation for all efficient portfolios and then select the $E$-$V$ efficient portfolio which yields the highest expected utility, $E^*U(\ )$.

For example, the portfolio in our mesh of $E$-$V$ efficient portfolios with the highest expected utility for $U = \ln(1 + R)$ is that which yields 24.7 percent with a standard deviation of 39.6 percent (see Table V below). There may exist a better $E$-$V$ efficient portfolio in the neighborhood of this mesh point; thus, our results underestimate the $E$-$V$ approximation.

Table V presents the mesh portfolios with highest expected value for various utility functions. A comparison of the investment strategies given in Tables III and V shows that the direct maximization of $EU(\ )$ and maximization of $E^*U(\ )$ often results in quite similar portfolios. Even if the investment allocations differ, the performance index $I$ may be close to 1.0, indicating a small welfare loss from maximizing $E^*U(\ )$ rather than $EU(\ )$.

Table VI shows the derived expected utility as well as the values of the index $I$. The results are quite impressive. The index is equal or close to one in all cases. The lowest index of 0.978 is obtained in the case of $\ln(X)$. This implies that these risk-averse investors would lose almost nothing by selecting the optimum portfolio from an $E$-$V$ efficient set rather than using direct maximization over all feasible portfolios considered.

Figure 1 presents the $E$-$V$ efficient frontier, the location of the naive portfolio (denoted by "N"), the portfolios with a direct maximization of $EU(\ )$ (denoted by "O"), and the portfolios which maximize $EU$ among $E$-$V$ efficient portfolios (denoted by "×"). By definition the direct utility function cannot lie to the left

**Table V**

Optimum $E$-$V$ Portfolios for Various Utility Functions

| Utility Function | California (6) | Carpenter (3) | Conelco (1) | Texas Gulf (2) | Total | Average Return | Standard Deviation |
|---|---|---|---|---|---|---|---|
| $-e^{-X}$ | 39.4 | 38.6 | 5.0 | 17.0 | 100% | 22.5 | 27.0 |
| $X^{0.1}$ | 28.5 | 43.4 | 8.6 | 8.6 | 100% | 23.1 | 30.0 |
| $X^{0.5}$ | — | 41.8 | 32.1 | 26.1 | 100% | 25.7 | 47.3 |
| $X^{0.9}$ | — | — | 97.6 | 2.4 | 100% | 30.5 | 98.1 |
| $\ln(X)$ | 32.9 | 41.8 | 7.4 | 18.7 | 100% | 22.9 | 28.9 |
| $\ln(X + 1)$ | — | 55.6 | 18.7 | 25.7 | 100% | 24.7 | 39.6 |
| $(X + 1)^{0.1}$ | — | 50.1 | 24.0 | 25.9 | 100% | 25.7 | 47.3 |
| $(X + 1)^{0.5}$ | — | 3.2 | 69.8 | 27.0 | 100% | 28.5 | 75.8 |

**Table VI**

Direct and Approximated Expected Utility and the Approximation Index

| Utility Function | The Expected Utility from Direct Maximization $EU(\ )$ | The Highest Utility from an $E$-$V$ Efficient Portfolio $E^*U(\ )$ | Expected Utility of the Naive Portfolio $EU_N(\ )$ | Approximation Index* $I$ |
|---|---|---|---|---|
| $-e^{-X}$ | −0.30382 | −0.30390 | −0.31610 | 0.993 |
| $X^{0.1}$ | 1.01842 | 1.01838 | · 1.01466 | 0.989 |
| $X^{0.5}$ | 1.10465 | 1.10459 | 1.07940 | 0.998 |
| $X^{0.9}$ | 1.24933 | 1.24934 | 1.15482 | 1.000 |
| $\ln(X)$ | 0.18016 | 0.17935 | 0.14387 | 0.978 |
| $\ln(X+1)$ | 0.79575 | 0.79572 | 0.77215 | 0.999 |
| $(X+1)^{0.1}$ | 1.08300 | 1.08300 | 1.08033 | 1.000 |
| $(X+1)^{0.5}$ | 1.49668 | 1.49664 | 1.47308 | 0.998 |

* The Approximation Index is given by:

$$I = \frac{E^*U(\ ) - E_N U(\ )}{EU(\ ) - E_N U(\ )}$$

where $E_N U(\ )$ assumes investment of $\frac{1}{n}$ in each security.

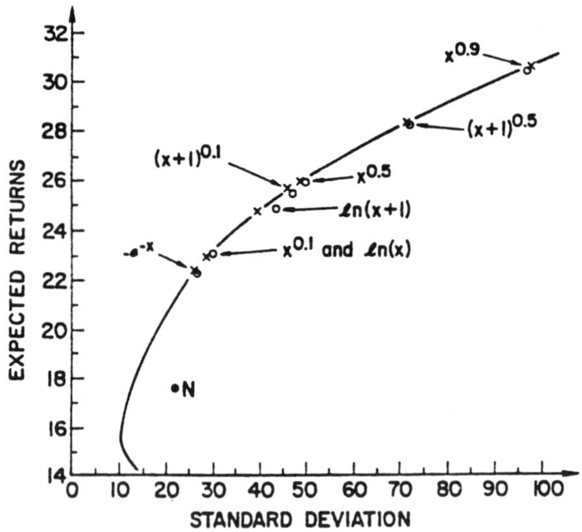


**Figure 1.** $E$ and $\sigma$ of Various Portfolios

of the efficient frontier; but may lie on or to the right of the frontier. In fact, all points "O" lie almost on the efficient set. This comparison is given numerically in Table VII. Again we see that the $E$ and $\sigma$ of both maximization methods are very much the same and that the direct maximization portfolios have $E$'s and $\sigma$'s which are almost on the efficient $E$-$V$ frontier. The optimal portfolios from direct maximization have a standard deviation which is higher only by 3.8–0.0% than the minimum standard deviation which can be obtained on the $E$-$V$ frontier for the same mean. These deviations from the $E$-$V$ frontier are relatively small in comparison to the naive portfolio which has a standard deviation which is 70.5% greater than that of the $E$-$V$ portfolio with the same mean of 17.5%.

Previously we noted that the optimal portfolios usually contain 3 or 4 securities, where these securities tend to have the highest mean return. It is tempting to conjecture, therefore, that holding equal amounts of 2 to 5 securities with the highest mean may yield almost optimum results. Table VIII tests this hypothesis. The performance index ($I$) of the $E$-$V$ criterion is compared with the indexes of

**Table VII**

$E$ and $\sigma$ of Optimal Portfolios According to $E$-$V$ and Direct Maximization Methods, Proximity of Direct Maximization Optimal Portfolios to $E$-$V$ Efficient Frontier

| | E and σ of Optimal Portfolios | | | | | |
|---|---|---|---|---|---|---|
| | $E$-$V$ Maximization | | Direct Maximization | | | |
| Utility Function (1) | $E$ (2) | $\sigma$ (3) | $E$ (4) | $\sigma$ (5) | $(E\text{-}V)^*$ (6) | $\frac{\sigma-\sigma(E\text{-}V)}{\sigma(E\text{-}V)}$ (7) |
| $-e^{-X}$ | 22.5 | 27.0 | 22.4 | 27.3 | 26.5 | 3.0% |
| $X^{0.1}$ | 23.1 | 30.0 | 23.3 | 32.3 | 31.1 | 3.8% |
| $X^{0.5}$ | 25.7 | 47.3 | 25.9 | 49.4 | 49.0 | 0.8% |
| $X^{0.9}$ | 30.5 | 98.1 | 30.5 | 98.6 | 98.1 | 0.5% |
| $\ln(X)$ | 22.9 | 28.5 | 23.5 | 29.9 | 29.9 | 0.0% |
| $\ln(X+1)$ | 24.7 | 35.6 | 25.1 | 43.7 | 42.5 | 2.8% |
| $(X+1)^{0.1}$ | 25.7 | 47.3 | 25.5 | 46.7 | 45.6 | 2.4% |
| $(X+1)^{0.5}$ | 28.5 | 75.8 | 28.2 | 74.2 | 71.6 | 3.6% |
| The naive portfolio of equal proportions | | | 17.5 | 22.5 | 13.2 | 70.5% |

* $\sigma(E\text{-}V)$ denotes the standard deviation of portfolios on the efficient $E$-$V$ frontier that have the same mean as the mean of the optimal portfolios obtained by the direct maximization.

**Table VIII**

The Indexes ($I$) of $E$-$V$ Portfolios and Portfolios of Equal Proportions of 1–5 Stocks with the Highest Means

| Utility Function | $E$-$V$ Portfolio | Portfolio with Equal Proportions of $K$ Stocks with Highest Mean: $K=1$ | $K=2$ | $K=3$ | $K=4$ | $K=5$ |
|---|---|---|---|---|---|---|
| $-e^{-X}$ | 0.993 | −2.375 | −0.321 | 0.400 | 0.405 | 0.512 |
| $X^{0.1}$ | 0.989 | −0.823 | 0.400 | 0.791 | 0.703 | 0.726 |
| $X^{0.5}$ | 0.998 | 0.506 | 0.888 | 0.989 | 0.872 | 0.820 |
| $X^{0.9}$ | 1.000 | 1.000 | 0.798 | 0.716 | 0.623 | 0.555 |
| $\ln(X)$ | 0.978 | −1.178 | 0.234 | 0.693 | 0.615 | 0.659 |
| $\ln(X+1)$ | 0.999 | 0.209 | 0.812 | 0.979 | 0.875 | 0.833 |
| $(X+1)^{0.1}$ | 1.000 | 0.379 | 0.863 | 0.989 | 0.882 | 0.832 |
| $(X+1)^{0.5}$ | 0.998 | 0.913 | 0.955 | 0.940 | 0.827 | 0.755 |

portfolios with equal proportions of securities with the $k$ highest means, for $k = 1, \cdots, 5$.

In most cases the approximation index of the equal proportions portfolio of the highest mean stocks is much lower than the index of the best $E$-$V$ efficient portfolio. In some cases the former index is even negative. In particular, holding only one stock with the highest mean will yield very poor and even negative approximation indexes in almost all cases with the exception of the $X^{0.9}$ function where the investor is almost risk neutral.

Thus, using only the mean, ignoring the variances and covariances, in order to construct equal proportions portfolios of 2 to 5 stocks frequently improves expected utility relative to the naive portfolio of equal proportions of all stocks. However, using both the portfolio mean and variance to select portfolios will usually improve expected utility much more. Additional moments besides the mean and variance will not improve the approximation substantially, as the approximation index of the $E$-$V$ criterion is really very close to one.

## VI. The Effect of Leverage

Leverage increases the risk of the portfolio. If the investor borrows part of the funds invested in the risky portfolio, then the fluctuations of the return on these leveraged portfolios will be proportionately greater. If unlimited leverage is allowed the quadratic approximations described in [6] and [7] can be forced to fail, e.g., since losses approaching 100% will give an expected logarithm approaching $-\infty$ while the quadratic remains finite. In this section, we examine the effect of limited leverage on the approximation index. Specifically we restricted the amount of borrowing to a maximum of 50% of the investment.

Assuming a 10% risk-free rate,[1] we derived a mesh of portfolios on the efficient $E$-$V$ frontier for the $n$ securities plus an $n + 1$st risk-free security allowed to take negative values constrained by 50%. For each utility function, we calculated the expected utility of each of these efficient portfolios, and selected the portfolios which maximized expected utility. In addition, we obtained the true optimum portfolios using exactly the same basic technique as in the unleveraged case but allowing 50% borrowing.

Thus the direct maximization process solved the following problem:

$$\text{Max } EU(\textstyle\sum_{i=1}^{n} X_i R_i + (1 - \sum_{i=1}^{n} X_i) R_F)$$

subject to $\quad X_i \geq 0 \quad i = 1, 2, \cdots, n$

and $\quad X_{n+1} = 1 - \sum_{i=1}^{n} X_i \geq -0.5$

where $X_{n+1}$ is the proportion invested in the risk-free asset.

The index of approximation requires the choice of a "naive" portfolio. In the previous case, without leverage, we simply let $X_i = 1/n$ for each $i$. In the present case we use a leveraged naive portfolio, i.e., the solution to the following:

$$\text{Max } U[(1 - X_{n+1}) \textstyle\sum_{i=1}^{n} R_i/n + X_{n+1} R_F]$$

[1] When we repeated the analysis with a 7% rate, we got similar results.

subject to $-0.5 \le X_{n+1} \le 1$, where $X_{n+1}$ is the proportion invested in the riskless asset with yield $R_F$. Thus the "naive" portfolio was allowed optimum leverage.

Table IX presents the approximation index for the $E$-$V$ maximization. Once again the approximation indexes are high; indeed, about the same as the indexes in the unleveraged case. At least in our empirical sample, then, leverage limited to 50% does not lead to perceptibly worse approximation indexes. Notice that, with the risk-free rate of 10%, the maximum amount of borrowing of 50% was used in all cases. Even the naive portfolio, with a mean return of 17.5%, is leveraged by 50% for all utility functions.

## VII. Conclusions

Levy and Markowitz reported that, for various finite populations of portfolio returns such as the returns on 149 investment company portfolios, the best mean-variance efficient portfolio was frequently the portfolio which maximized expected utility—or at least had a near optimum expected utility. The present paper reports that, for various utility functions and the historical returns on 3 different sets of securities, when a portfolio may be chosen from any of the infinite number of portfolios of the standard constraint set the best mean-variance efficient portfolio has almost maximum obtainable expected utility. This remained true when 50% borrowing was allowed.

The hypothesis was tested, and rejected, that similar excellence could be obtained by investing equally in the $k$ securities with highest expected returns, for $k$ equal to about the number of securities in the optimum portfolio. It was also found that the excellence of the mean-variance result was not due to normality of data. Rather it illustrates the robustness of the quadratic approximation as reviewed by Levy and Markowitz.

**Table IX**

**Direct and Approximated Expected Utility with Borrowing and Lending at Risk-Free Interest* and the Approximation Index**

| Utility Function | Optimum Leverage % | Expected Utility of Direct Maximization $EU(\ )$ | Approximated Expected Utility with $E$-$V$ $E^*U(\ )$ | Expected Utility of Naive Portfolio $E_NU(\ )$ | Approximation Index $I$** |
|---|---|---|---|---|---|
| $-e^{-X}$ | 50 | −0.29633 | −0.29641 | −0.31333 | 0.996 |
| $X^{0.1}$ | 50 | 1.02105 | 1.02100 | 1.01593 | 0.990 |
| $X^{0.5}$ | 50 | 1.12341 | 1.12258 | 1.09110 | 0.973 |
| $X^{0.9}$ | 50 | 1.31880 | 1.31847 | 1.18642 | 0.998 |
| $\ln(X)$ | 50 | 0.20384 | 0.20303 | 0.15370 | 0.983 |
| $\ln(X+1)$ | 50 | 0.81392 | 0.81352 | 0.78345 | 0.987 |
| $(X+1)^{0.1}$ | 50 | 1.08504 | 1.08499 | 1.08162 | 0.986 |
| $(X+1)^{0.5}$ | 50 | 1.51393 | 1.51389 | 1.48374 | 0.999 |

* The risk-free rate is 10% and the amount of leverage is constrained by 50%.

** $I = \dfrac{E^*U(\ ) - E_NU(\ )}{EU(\ ) - E_NU(\ )}$.

## Appendix

### The Direct Maximization Algorithm[2]

*A. The Problem*

There are $n$ risky options and $m$ periods. Let $X_{ij}$ be the return on security $i$ in period $j$. Let $u$ be a utility function with $u' \geq 0$ and $u'' \leq 0$. The maximization problem is:

$$\text{Max}\left\{\frac{1}{m} \sum_{j=1}^{m} u\left(\sum_{i=1}^{n} y_i X_{ij}\right)\right\}$$

subject to $\sum_{i=1}^{n} y_i = 1$.

Let us first solve the optimization problem for the case where short sales are not allowed. Thus, we add the constraints $y_i \geq 0$ for $i = 1, \cdots, n$. In order to switch from the inequality constraints $y_i \geq 0$ to equality constraints, we use the following method. Redefine $y$ by $z_i^2 = y$, and let us denote

$$f(z) \equiv -\sum_{j=1}^{m} u\left(\sum z_i^2 X_{ij}\right)$$

and the constraint will be:

$$g(z) \equiv \sum_{i=1}^{n} z_i^2 - 1 = 0.$$

Let us use an augmented-Lagrangian method,[3] where

$$L(z, \lambda, r) = f(z) + \lambda g(z) + \tfrac{1}{2} r \times g(z)^2$$

and the maximization constrained problem will be:

$$\text{Min } L(z, \lambda, r) \qquad \text{subject to} \quad g(z) = 0.$$

We use Powell's Hestenes Algorithm for finding the solution of this problem[4] which is described below.

*Step 1* Determine $r_0$, $\lambda_0$, $z_0$ for $K = 0$. In our cases we determined $r_0 = 10$, $\lambda_0 = 1$, $z_0^i = \dfrac{1}{n}$, $i = 1, \cdots, n$.

*Step 2* Find $z_K$ such that: $L(z_K, \lambda_K, r_K) = \underset{z}{\text{Min}}(z, \lambda_K, r_K)$. We selected the Fletcher and Powell Method[5] to find this unconstrained minimum. The method will be presented later with more details.

*Step 3* Updating by: $\lambda_{K+1} = \lambda_K + r_K g(z_K)$. (This is the updating formula that was proposed by Hestenes and Powell—see Hestenes [4].)

*Step 4* Updating $r_{K+1} = \tau r_K$ where $\tau = 1.1$. The size of $\tau = 1.1$ was determined by us empirically after a few trials.

*Step 5* Checking the stopping criterion.

$$\| \nabla_Z L(z_K, \lambda_K, r_K) \| + \| g(z_K) \| < \epsilon$$

[2] This section was written, and the application of this algorithm to our problem was carried out, by Ammon Gonan from Technion, Institute of Technology in Haifa, Israel.

[3] See Hestenes [4].

[4] See Hestenes [4].

[5] See Fletcher and Powell [3].

where $\nabla_Z L(,,)$ is the gradient of $L$ with respect to $z$. We selected $\epsilon = 10^{-3}$.[6] If the criterion is not satisfied, define $K = K + 1$ and go back to *Step 2.*

## *B. Details on the Maximization Method of Step 2*

The Fletcher and Powell Method which was selected belongs to the Quasi-Newton algorithms group. The method is in the IMSL library and a description of the method is given in textbooks of nonlinear search methods (see for example, Avriel [2] pp. 322–34).

The F & P Algorithm for Min $f(x)$ is given by the following steps.

$$\text{Denote:}\quad P^K = x^K - x^{K-1}$$

$$\gamma^K = \nabla f(x^K) - \nabla f(x^{K-1})$$

*Step 1* Given $x_0$, $\nabla f(x_0)$, and an arbitrary symmetric $n \times n$ positive definite matrix $H_0$. (For example $H_0 = 1$.) Initially $K = 0$.

*Step 2* Find $\bar{\lambda}$ such that:

$$f(x_K - \bar{\lambda} H_K \nabla f(x_K)) = \underset{\lambda}{\text{Min}}\, f(x_K - \lambda H_K \nabla f(x_K))$$

and define $Y_{K+1} = x_K + \bar{\lambda} H_K \nabla f(x_k)$.

*Step 3* Calculate $\nabla$, $P^{K+1}$, $\gamma^{K+1}$, and define

$$H_{K+1} = H_k + \frac{P^{K+1}(P^{K+1})^T}{(P^{K+1})^T, \gamma^K} - \frac{(H_K \gamma^{K+1})(H_K g^{K+1})T}{(\gamma^{K+1})^T H_K \gamma^{K+1}}$$

$H_K$ is an approximation to the inverse of the Hessian of $t$, i.e., $H_K \approx (\nabla^2_{xx} f(x_K))^{-1}$.

*Step 4* Check a stopping rule otherwise go back to Step 2. The stopping rule is again $\| \nabla f(x_K) \| < \epsilon$.

[6] Stopping rules of $\epsilon = 10^{-6}$ and $\epsilon = 10^{-8}$ were also examined. The results were not meaningfully different from the results obtained by $\epsilon = 10^{-3}$.

## REFERENCES


1. K. J. Arrow. *Aspects of the Theory of Risk Bearing.* Yrjo Johnsson Lectures, Helsinki, 1965.
2. M. Avriel. *Nonlinear Programming.* Englewood Cliffs, NJ: Prentice-Hall, 1976.
3. R. Fletcher and M. J. D. Powell. "A Rapidly Convergent Descent Method for Minimization." *Computer Journal* 6 (1963), 163–68.
4. M. R. Hestenes. "Multiplier and Gradient Methods." *Journal of Optimization Theory and Application* 4 (1969), 303–20.
5. M. G. Kendall and A. Stuart. *The Advanced Theory of Statistics.* Volume 1, 3rd Edition, New York: Hafner, 1969.
6. H. Levy and H. M. Markowitz. "Approximating Expected Utility by a Function of Mean and Variance." *American Economic Review* 69 (1979), 308–17.
7. H. M. Markowitz. *Portfolio Selection: Efficient Diversification of Investments.* New York: John Wiley and Sons, 1959; New Haven: Yale University Press, 1970.
8. J. W. Pratt. "Risk Aversion in the Small and in the Large." *Econometrica* 32 (1964), 122–36.
9. M. Pulley. "A General Mean-Variance Approximation to Expected Utility for Short Holding Periods." *Journal of Financial and Quantitative Analysis* 16 (1981), 361–73.

# The Value of a Blank Check

*A pathological level of risk aversion?*

Harry M. Markowitz, Donald W. Reid, and Bernard V. Tew

In standard microeconomic theory, an individual's preference for one good over another arises as if the individual is incrementally maximizing a utility or satisfaction function. Assuming all goods can be bought and sold, an individual can achieve the utility-maximizing level and combination of goods by maximizing profits or returns or, more generally, by maximizing wealth.

The problem with maximizing returns, however, is that returns are uncertain or risky because each potential profit-making choice or strategy has many potential outcomes. Thus, when choosing a portfolio of securities, investors should maximize the mean or expected value of their utility over all possible outcomes.[1]

In investment practice, the maximum expected utility principle is difficult to implement, even if we assume that utility depends only on portfolio return. One method for approximating expected utility is by using a function of mean and variance of return. Typically, it is far more economical in terms of both estimation and optimization costs to select a mean-variance efficient portfolio than to maximize expected utility. Yet, the best choice of mean-variance efficient portfolios, in general, will provide less than maximum feasible expected utility.

This raises the question of how closely a portfolio chosen from the mean-variance efficient set can approximate the portfolio solution that maximizes an investor's expected utility. If the answer is "very close"

**HARRY M. MARKOWITZ** is president of Harry Markowitz Co., in San Diego (CA 92109), and a consultant at Daiwa Securities Trust Co., in Jersey City (NJ 07302).

**DONALD W. REID** and **BERNARD V. TEW** are special partners and co-directors of quantitative equity research at Weiss, Peck and Greer in Chicago (IL 60606).

in some appropriate sense, then the cost of performing the theoretically correct analysis would exceed the benefit gained as compared to the simpler mean-variance analysis.

A number of researchers have investigated this question of "closeness" of the mean-variance approximation for various mathematical representations of utility. Young and Trent [1969] and Levy and Markowitz [1979] find that a function of only mean and variance provides an extremely accurate approximation to the expected value of utility described by the logarithmic function, $U = E[\ln(1 + R)]$, where R is rate of return. Ederington [1986] reports mixed results for mean-variance approximations of exponential and negative power functions, $U = -\exp[-\alpha W_0(1 + R)]$ and $U = -(1 + R)^{-\beta}$, $\beta \geq 0$, respectively.

With low sensitivity to risk, mean-variance approximations achieve solutions very close to maximum expected utility. Poorer results are obtained for conditions with higher sensitivity to risk. For the exponential function, Simaan [1987] reports conclusions consistent with those of Ederington. Grauer [1986] concludes that the investment policies of the mean-variance criterion and power function utility for similar risk aversion measures are not as similar as commonly believed, especially for high sensitivity to risk as represented by functions such as $U = -(1/50)(1 + R)^{-50}$.

This summary of research reveals that the ability of mean-variance approximations to achieve maximum or near maximum expected utility is excellent for what appear to be low sensitivities to risk, but decreases as the function representing utility reflects a higher aversion to risk. The problem with this conclusion is that no practical insight is given as to how well the functional forms and risk sensitivities depict actual investor behavior. Thus, whether investors should seek risk efficiency beyond mean-variance efficiency remains unclear.

We propose that certain utility functions described in the financial economic literature to evaluate investment policies, and thereby judge the efficacy of mean-variance approximations, are not representative of investor behavior. Pratt [1964] and Arrow [1965] argue that most investors do not have increasing absolute risk aversion as their wealth increases; that is, sensitivity to risking some fixed amount does not increase as wealth increases. This helps to narrow the range of functional forms that should be considered for representing utility.

To add to the Pratt-Arrow criterion, we argue that one should rule out functions that assign implausibly low values to a chance for infinite wealth. As an example, consider the function used by Grauer. An investor with the utility function $U = -(1/50)(1 + R)^{-50}$ would be indifferent between receiving 1) a rate of return of $R_c$ = 0.014 (i.e., 1.4%) with certainty, and 2) a fifty-fifty chance of no return (zero gain, zero loss) or infinite wealth.

One could perhaps imagine a situation in which a return of 1.4% is so critical that an investor would be indifferent between it and a 50% chance of winning inexhaustible wealth. But we doubt if many, if any, pension funds, investment companies, or large private or institutional investors would assign such a low value to a chance for a "blank check."

We check our assessment of "implausibly low" values of a chance for a blank check by surveying clients of a brokerage firm to get their indications of such a value. Then we extend the blank check concepts to utility functions including human capital and show the effect on the level of risk aversion applied to the investment portfolio decision.

## CONCEPTS OF A BLANK CHECK LOTTERY

The value investors place on a chance for infinite wealth provides a helpful method for evaluating the realism of a mathematical representation of investor utility. The value elicited from the investor can be used to infer parameter values for the mathematical function that reflects the risk behavior needed to give rise to such a value. Furthermore, the value placed on the chance for infinite wealth can be related to the Pratt-Arrow measure of risk aversion that has been prevalent in the research literature.

Assume an investor wins a prize — a lottery ticket that provides a fifty-fifty chance of receiving either a blank check (infinite wealth) or no additional wealth. If the investor maximizes the expected utility of a function of return, R, and if the function is bounded above (there is a value that the function's value cannot exceed), there exists a risk-free return ($R_c$) that gives the same expected utility as the blank check lottery. In other words, there exists a certainty equivalent return for which the investor is just induced to surrender the claim to the blank check lottery.

We consider two bounded functions, previously used in the research literature, to provide some insight about mean-variance analysis as a practical method for approximating expected utility of investors. First, consider the exponential function given as

$$U(R) = -\exp[-\alpha W_0(1 + R)] \text{ for } \alpha > 0 \qquad (1)$$

where $W_0$ is the initial value of the portfolio (or other wealth amount on which the rate of return, R, is considered). With some algebraic maneuvering, the certainty equivalent return, $R_c^e$, for the blank check lottery can be solved as

$$R_c^e = \ln(2)/\alpha W_0 \qquad (2)$$

In words, Equation (2) shows the relationships among the investor's value of the blank check lottery in terms of a risk-free rate of return on the initial wealth level, $R_c^e$, the initial level of wealth, $W_0$, and the $\alpha$ parameter that controls the risk sensitivity of the utility function. Therefore, if an investor's initial wealth and the value placed upon the blank check lottery in terms of a risk-free rate of return on this initial wealth are known, then the appropriate value for the $\alpha$ parameter can be solved that allows this particular value of the blank check lottery.

The prevailing measure of risk aversion used in financial and risk theory research is the Pratt-Arrow relative risk aversion coefficient (RRA). The RRA indicates the degree to which investors have an aversion to risk relative to their level of wealth.

The intuition behind the Pratt-Arrow risk aversion coefficient is as follows: An investor's additional utility or desire for an additional unit of wealth decreases as the investor becomes more wealthy. The RRA shows the speed of decline in value of the additional utility obtained for an additional unit of wealth. Thus, relatively large RRAs indicate that obtaining additional wealth becomes less and less important very rapidly; but conversely, losing wealth becomes more and more important very rapidly.

The Pratt-Arrow relative risk aversion coefficient is derived as

$$RRA = -U''/U' \qquad (3)$$

where $U'$ and $U''$ are derivatives with respect to R, evaluated at $R = 0$.[2] Since the RRA is a mathematical derivation, the investor's utility must be expressed as a well-behaved mathematical function. In particular, for the exponential utility function in (1), the RRA can be solved as

$$RRA^e = \alpha W_0 \qquad (4)$$

Equation (4) shows that for the negative exponential function the indicated risk aversion is controlled by the level of wealth and the $\alpha$ parameter. Again with some algebraic manipulations, we can infer the RRA of an investor with an exponential utility function from the value the investor places on the blank check lottery, resulting in

$$RRA^e = \ln(2)/R_c^e \qquad (5)$$

Equation (5) establishes the relationship between the certainty equivalent return of the blank check lottery and the Pratt-Arrow relative risk aversion coefficient. The RRA value established through certainty equivalent return of a blank check lottery can then be compared to values that have been used in research for this functional form. This will allow conclusions about closeness of mean-variance approximations to expected utility solutions for the risk behavior reflected in realistic ranges of risk aversion.

Next, consider the negative power function given by

$$U(R) = -[W_0(1 + R)]^{-\alpha} \qquad (6)$$

for $\alpha > 0$. As before, the value of the blank check lottery in terms of certainty equivalent return can be solved in terms of the utility function, and the RRA can be solved and related to the certainty equivalent return.

Solving for the value of the blank check lottery in terms of the certainty equivalent return for the negative power function results in

$$R_c^p = 2^{(1/\alpha)} - 1 \qquad (7)$$

The RRA can be solved as

$$RRA^p = \alpha + 1 \qquad (8)$$

Solving the relationship between RRA and the certainty equivalent, $R_c$, results in

**EXHIBIT 1**
Risk Aversion Associated with Various $R_c$

| $R_c$ | RRA $U = -\exp(-\alpha w)$ | RRA $U = -W^{-\alpha}$ |
|---|---|---|
| 0.10 | 6.93 | 8.27 |
| 0.25 | 2.77 | 4.11 |
| 0.50 | 1.39 | 2.71 |
| 1.00 | 0.69 | 2.00 |
| 2.00 | 0.34 | 1.63 |

$$RRA^p = 1 + \ln(2)/\ln\left(1 + R_c^p\right) \quad (9)$$

Exhibit 1 lists selected values for the relationship between relative risk aversion (RRA) and certainty equivalent return ($R_c$), and Exhibit 2 graphically illustrates the relationship. As certainty equivalent return increases, relative risk aversion for the exponential function approaches zero from above, while the relative risk aversion for the negative power function approaches one from above. As the certainty equivalent return approaches zero, the RRAs for both functions approach infinity. Note that the RRA for the power utility function ($RRA^p$) is greater than the RRA for the exponential utility function ($RRA^e$) for each certainty equivalent value.

**EXHIBIT 2**
**RRA VERSUS $R_c$ FOR TWO UTILITY FUNCTIONS**

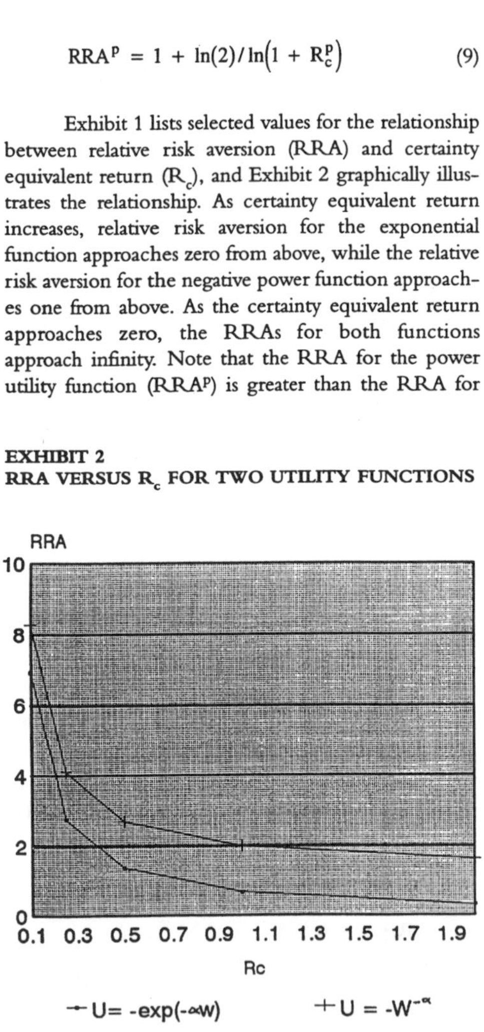


## A SURVEY

It seems to us absurd that many, if any, investors would trade a fifty-fifty chance of unlimited wealth for little more than a 1.4% gain, as in the example from Grauer. Yet it is difficult to judge what larger values of certainty equivalent returns, $R_c$, might be typical.

As a rough guide to plausible certainty equivalent returns for a blank check lottery, we asked investors to place themselves in the situation we have described, and, in effect, to assign their certainty equivalent value to the opportunity for infinite wealth. A local office of a national brokerage firm mailed a brief questionnaire to 125 investors of its choice. Forty-nine investors responded. Of these, forty completed all applicable questions.

The sample includes people in a variety of situations: several students or part-time students with little income or net worth, several retired people over age sixty with net worth ranging from hundreds of thousands to a few millions of dollars, three retired respondents under forty with net worths again in the hundreds of thousands or millions of dollars, and working people in various jobs and professions including a bookkeeper, a store manager, an accountant, an auditor, a teacher, a professor, a psychologist, an engineer, a caterer, and two well-to-do farmers.

Admittedly, our forty respondents do not constitute a large or "scientific" sample. On the other hand, they are forty actual investors, who we presume are disinterested in questions concerning the explicit form of their respective utility functions, and who have taken the time to answer our questions.

We use the certainty equivalent relationships developed for the exponential and power utility functions in the previous section to consider the implications of the values of certainty equivalent returns obtained from our survey. The certainty equivalent returns are considered for several measures of initial wealth ($W_0$):

1. Portfolio worth.
2. Total net worth excluding human capital.
3. Total net worth including human capital.
4. Value of the retirement fund.

**EXHIBIT 3**
Survey $R_c^a$ and $R_c^b$

| Rank | $R_c^a$ | $R_c^b$ |
|---|---|---|
| 1 | 0.13 | 0.08 |
| 2 | 0.24 | 0.08 |
| 3 | 0.25 | 0.09 |
| 4 | 0.26 | 0.13 |
| 5 | 0.40 | 0.15 |
| 6 | 0.42 | 0.17 |
| 7 | 0.48 | 0.20 |
| 8 | 0.48 | 0.20 |
| 9 | 1.00 | 0.24 |
| 10 | 1.25 | 0.29 |
| 11 | 1.52 | 0.36 |
| 12 | 1.67 | 0.63 |
| 13 | 2.00 | 0.80 |
| 14 | 2.00 | 0.92 |
| 15 | 2.99 | 1.00 |
| 16 | 3.00 | 1.00 |
| 17 | 3.38 | 1.18 |
| 18 | 3.57 | 1.29 |
| 19 | 4.00 | 1.43 |
| 20 | 4.00 | 1.43 |
| 21 | 4.08 | 1.50 |
| 22 | 5.00 | 1.67 |
| 23 | 8.33 | 1.92 |
| 24 | 10.53 | 2.00 |
| 25 | 12.50 | 3.33 |
| 26 | 13.33 | 3.57 |
| 27 | 20.00 | 5.00 |
| 28 | 25.00 | 5.71 |
| 29 | 30.30 | 6.67 |
| 30 | 52.63 | 12.50 |
| 31 | 75.00 | 12.50 |
| 32 | 100.00 | 16.67 |
| 33 | 100.00 | 20.69 |
| 34 | 125.00 | 28.57 |
| 35 | 400.00 | 50.00 |
| 36 | 476.19 | 52.63 |
| 37 | 500.00 | 100.00 |
| 38 | 1,176.47 | 160.00 |
| 39 | 2,000.00 | 333.33 |
| 40 | 37,500.00 | 20,000.00 |

The certainty equivalent return using portfolio worth as the wealth measure, $R_c^a$, is calculated as the dollar value of certainty equivalent given by the respondent, divided by the value of the portfolio given by the respondent; these calculated values are listed for all respondents in column (2) of Exhibit 3. Likewise, the certainty equivalent return using total net worth excluding human capital, $R_c^b$, is calculated as the dollar value of the respondent's certainty equivalent return, divided by the respondent's value of total net worth. These values are listed in column (3). The two columns are arranged in order of increasing $R_c^a$ and $R_c^b$, respectively; therefore the two entries in a row do not necessarily correspond to the same investor.

Some summary statistics for $R_c^a$ and $R_c^b$ are presented in Exhibit 4. Exhibit 4 also summarizes statistics for $R_c^d$, the respondent's instructions to the manager of the particular retirement fund as to the dollar value of the certainty equivalent desired for the retirement portfolio, divided by the retirement portfolio value, where these figures were provided (N = 28).

## A PROBLEM WITH EXPONENTIAL UTILITY

Suppose, for the moment, that all forty respondents have exponential utility functions. Also assume that initial wealth, $W_0$, in Equation (1) is measured as total net worth excluding human capital, and R represents return on this value.

According to Exhibit 4, the median value of certainty equivalent return on total net worth excluding human capital, $R_c^b$, is 1.46, or 146%. No respondent has this $R_c^b$, which lies between the twentieth and twenty-first highest response. Nevertheless, we will speak of the median respondent. Given the assumed exponential form of utility, Equation (5) implies that the median respondent has RRA = 0.47.

According to empirical research, some assert that investors *do not* have RRA less than one (Friend and Blume [1975] and Cohn et al. [1975]). Others assert that investors *should not* have RRA less than one because it would impair long-run portfolio asset growth (Markowitz [1976]). Equation (5) says that the RRA for the exponential function is less than one, pro-

**EXHIBIT 4**
Summary Statistics for $R_c^a$, $R_c^b$, and $R_c^d$

| | $R_c^a$ | $R_c^b$ | $R_c^d$ |
|---|---|---|---|
| Maximum | 37,500.00 | 20,000.00 | 953.33 |
| Q3 | 69.41 | 12.50 | 50.00 |
| Median | 4.04 | 1.46 | 4.96 |
| Q1 | 1.32 | 0.30 | 1.16 |
| Minimum | 0.13 | 0.08 | 0.09 |
| N | 40.00 | 40.00 | 28.00 |

vided that the certainty equivalent return is greater than 0.693 (69.3%); i.e., $RRA^e < 1.0$ provided $R_c > \ln(2) = 0.693$. It seemed to us a priori that a 70% increase in net worth is a rather meager alternative to a fifty-fifty chance of infinite wealth. In fact, twenty-eight out of the forty respondents choose a higher certainty equivalent return ($R_c^b$) when total net worth is used as the wealth measure.

An investor with $R_c \geq 70\%$ and $RRA \geq 1.0$ cannot have an exponential utility function for any $\alpha$ and $W_0$. If you seek a form of utility function consistent with these reasonable ranges of certainty equivalent return and RRA, you must reject the exponential utility function.

In terms of certainty equivalent return using portfolio and retirement fund values as the initial wealth measure, $R_c^a$ and $R_c^d$, respectively, an even larger fraction of respondents have certainty equivalents greater than 70%; $R_c^a > 0.70$ for thirty-two out of forty respondents, and $R_c^d > 0.70$ for twenty-six out of twenty-eight respondents. If the latter twenty-six also want $RRA \geq 1.0$ for their retirement fund, they do not want their retirement money invested to maximize the expected value of an exponential utility function.

## THE EFFECTS OF HUMAN CAPITAL

Next suppose that the investor chooses a portfolio to maximize the expected value of a utility function in which human capital is considered:

$$U = U[W_H + W_P(1 + R)] \tag{10}$$

where $W_H$ is the value of human capital, $W_P$ the value of the investor's portfolio, and R, as before, the return on the portfolio. We are able to make some progress in analyzing the relationship between the certainty equivalent value of a blank check lottery ($R_c$) and RRA, when the value of human capital is included in the utility function, without answering the hard questions concerning how to measure the value of human capital ($W_H$).

Since the blank check, should it be won, is delivered immediately, we may treat $W_H$ as constant rather than random in determining the certainty equivalent return, $R_c$. In this case we may view U in Equation (10) as a function of constants and the random variable R.

Risk aversion in terms of portfolio return R ($RRA_P$) can be shown to be

$$RRA_P = RRA_T \tau \tag{11}$$

where $RRA_T = -(W_H + W_P)U''/U'$ is relative risk aversion measured with respect to total wealth, $W_T = W_H + W_P$, and

$$\tau = W_P/(W_H + W_P) \tag{12}$$

is the ratio of portfolio wealth to total wealth.

In words, given the assumption that human capital can be treated as a constant, the relative risk aversion in making portfolio decisions is simply a proportion of the relative risk aversion that applies to total wealth. The proportionality factor is the proportion of portfolio value ($W_P$) to total wealth value ($W_T = W_H + W_P$).

Including a constant human capital value as an asset has several interesting implications for portfolio risk aversion. One broad implication is that no matter the form of the utility function, if the investment portfolio is a small enough proportion of total wealth including human capital, then the RRA applied to the portfolio decision will be less than one.

As a more specific implication, consider the Markowitz [1959, 1976] advice against choosing a mean-variance efficient portfolio with greater mean and variance than those that approximately maximize $E[\ln(1 + R)]$, with $RRA = 1$. With probability one, the "net asset value" of the investor who maximizes $E[\ln(1 + R)]$ will eventually pull ahead and stay ahead of an investor with any distinctly different investment strategy (Brieman [1960, 1961]).

To choose an efficient portfolio with higher arithmetic mean and variance is to get lower growth in the long run and greater volatility in the short run. But this advice cannot lead to an expected utility-maximizing decision when applied only to the portfolio decision if utility is given by Equation (10). Similarly, the problem with the exponential function discussed in the previous section becomes less significant if only the investment portfolio decision is considered.

A second implication of the result of Equation (11) is that it provides an alternate explanation of portfolio investment behavior over time. Samuelson [1988] reminds us that if an investor has constant relative risk aversion for terminal wealth, and neither deposits (beyond an initial deposit) nor withdraws for T periods, his or her ranking of probability distributions of R is the same in every period, no matter how large T. Thus, the portfolio decision in terms of riskiness does not change as T gets smaller over time.

It disturbs Samuelson that the young man with large T is not more speculative than the old man with small T. He shows that such a difference in behavior would result if return R followed a mean-reverting process. While a young man and an old man may have the same total asset value when human capital is considered, for the young man $W_P$ is a smaller fraction of $W_T$. Using Equation (11), this implies that $\tau$ and $RRA_P$ are smaller, and therefore the young man is less risk-averse in his investment portfolio decision.

The relationship between the certainty equivalent return of a blank check lottery ($R_c$) and the RRA applied in the portfolio decision for the utility function including human capital ($RRA_P$) is developed in the appendix. The derivations show that the relationships for the utility functions with human capital included are bounded by the relationships already derived and illustrated in Exhibit 2. For $\tau = 1$ we recover a form of Equation (4), as shown in the upper curve of the figure. For $0 < \tau < 1$, the curve relating $RRA_P$ and $R_c$ lies strictly between that of upper and lower curves; as $\tau \downarrow 0$, the curve approaches the lower curve in the figure, i.e., the exponential utility curve.

## APPLICATION TO EVALUATING MEAN-VARIANCE APPROXIMATIONS

To illustrate how our results on the value of a blank check supplement research on the efficacy of mean-variance approximation, we use Simaan's [1987] analysis of the "optimization premium" to get a sense of how closely mean-variance analysis approximates results from directly maximizing expected utility.

Simaan defines an optimization premium $\Theta$ as follows. Suppose $R_p$ is the random return on the feasible portfolio that maximizes expected utility, E[U(R)]; $R_m$ is the random return on the mean-variance efficient portfolio that provides greater E[U(R)] than any other mean-variance efficient portfolio; then

$$E[U(R_p - \Theta)] = E[U(R_m)] \tag{13}$$

**EXHIBIT 5**
Optimization Premiums

| $RRA = \alpha W_0$ | $\Theta$ |
|---|---|
| 2 | 0.00023 |
| 4 | 0.00073 |
| 6 | 0.00144 |
| 8 | 0.00229 |
| 10 | 0.00323 |
| 15 | 0.00581 |
| 20 | 0.00859 |
| 25 | 0.01147 |
| 30 | 0.01441 |
| 40 | 0.02040 |
| 50 | 0.02646 |
| 100 | 0.05719 |

Source: Simaan [1987, Exhibit 4.2].

In words, if $\Theta = 0.01$, it is worth exactly one cent per dollar of portfolio value, paid out of the portfolio, to have the true expected utility-maximizing portfolio rather than the best of the mean-variance efficient portfolios.

Simaan is able to solve for $\Theta$ under several assumptions: the portfolio is fully invested; utility is exponential; the distribution of returns is of the form

$$r_i = \alpha_i + \beta_i F + u_i \qquad i = 1, ..., n \tag{14}$$

where $\alpha_i$ and $\beta_i$ are constants; the $u_i$ are normally distributed but not necessarily uncorrelated; and F has a Pearson Type III distribution, which includes the gamma distribution as a special case. When F is non-symmetric, typically the best mean-variance portfolio provides less than the maximum expected utility, thus $\Theta > 0$.

To provide a numeric illustration for his analytic solution, Simaan estimates the parameters of the distribution in Equation (14) from monthly returns on ten securities for the period January 1973-December 1982. Exhibit 5 shows $RRA_e$ and $\Theta$ for this case of n = 10. The optimization premium $\Theta$ is typically smaller when a risk-free asset is introduced as an eleventh security.

As $RRA_e$ ranges from 2 to 100, the worth of using an optimum portfolio under Simaan's assumptions, rather than the best mean-variance portfolio, ranges from two-hundredths of a penny to 5.7 cents per dollar. The $\Theta$s for our survey sample, assuming each respondent has an exponential utility function, are obtained by combining information from Exhibits 1, 4, and 5.

The optimization premium is less than 0.1 of a

penny for $RRA_e$ less than about 5; i.e., for $R_c$ greater than about 0.14. This is the case for all but one respondent when $W_0$ is portfolio value, and for all but four respondents when $W_0$ is total wealth excluding human capital.

Thus, for most of our respondents, assuming that they have an exponential utility function and given Simaan's other assumptions, it is worth less than one-tenth of a cent per dollar of portfolio to have returns from the portfolio-maximizing expected utility ($R_p$) rather than returns from the mean-variance portfolio giving the highest expected utility ($R_m$).

Optimization premiums or comparable results are not available for the negative power utility function. These results, however, also apply approximately to the power utility function when human capital is considered and the portfolio value proportion of total wealth value ($\tau$) approaches zero (see Equation (A-5) and the derived Equation (A-9) in the appendix).

## SUMMARY AND CONCLUSION

We define a quantity $R_c$ equal to the return on an investor's wealth that the investor finds exactly as attractive as a fifty-fifty chance of unlimited wealth versus the status quo. We refer to this quantity as the investor's certainty equivalent value of a blank check lottery. To provide a rough range of plausible $R_c$ values, we report those of forty investors who responded to a survey. If we assume that the investor's utility function is an exponential function of wealth, then the investor's relative risk aversion, RRA, can be expressed as a function of $R_c$.

An immediate consequence is this: If the investor has RRA $\geq 1$, as many believe is true and/or desirable, and has $R_c > \ln(2)$, as do most of our respondents, then the investor cannot have an exponential utility function. The relationship between RRA and $R_c$ is also shown assuming that the investor's utility is a negative power function of wealth. The negative power function does not have the problem exhibited by the exponential utility function.

Explicitly including human capital in the utility function, assuming that it is a constant, provides insight into the portfolio decision. First, the relative risk aversion that applies to the portfolio allocation decision ($RRA_P$) becomes a proportion of the relative risk aversion that is considered for total wealth ($RRA_T$); in other words, the risk aversion toward the portfolio decision will be less than the risk aversion applied to total wealth. This helps diminish the problem of using an exponential function when applied only to the portfolio decision, and negates the idea that the investment portfolio decision should be based on an $RRA_P = 1$.

Second, this relationship between $RRA_P$ and $RRA_T$ provides an explanation as to why younger investors with a high proportion of their wealth in the form of human capital are less risk-averse toward their investment portfolios than older investors.

We obtain another important insight from considering human capital value as part of the total wealth in the utility function. If we assume that utility is an exponential or negative power function of total wealth including human capital value, without knowing the specific value of human capital, we can nevertheless put bounds on the relationship between the RRA of the portfolio and the $R_c$ of the portfolio. In the case of the exponential, the inclusion of human capital does not alter the relationship between RRA and $R_c$. In the case of the negative power function, the curve relating RRA to $R_c$ lies between that of the curves for the exponential and for the negative power function when human wealth is excluded.

In the financial economics research literature, various specific utility functions have been used in evaluating the efficacy of mean-variance approximations to expected utility. Results from studies using a good method of approximating expected utility from mean and variance indicate that if you know the mean and variance of a distribution you can estimate its expected utility quite closely except for utility functions we consider to exhibit "pathological risk aversion." Of course, the level of RRA that is "pathological" is a matter of debate.

The evidence we provide indicates that for most investors very little is lost by confining portfolio choice to the mean-variance efficient set. At the same time, we do not assert that we have generally settled the question of loss in utility due to using a mean-variance efficient portfolio; there are many empirical issues to consider. For example, Simaan's analysis uses monthly return, for which mean-variance approximations are more effective than for quarterly or annual holding-period returns (Levy and Markowitz [1979]).

Our discussion nevertheless illustrates how results on plausible certainty equivalents on a blank check lottery ($R_c$) and their implications for plausible RRA and functional forms can supplement research on the efficacy of mean-variance approximations by ruling out or de-emphasizing utility functions that are unrealistic representations of investor preferences.

## APPENDIX

Suppose that the investor chooses a portfolio to maximize the expected value of a utility function in which human capital is considered:

$$U = U[W_H + W_P(1 + R)] \qquad \text{(A-1)}$$

Risk aversion in terms of portfolio return R is given by

$$\begin{aligned} RRA_P &= -(d^2U/dR^2)/(dU/dR) \\ &= -W_P U''/U' \\ &= RRA_T \tau \qquad \text{(A-2)} \end{aligned}$$

where $U'$ and $U''$ are the derivatives of U with respect to $W_T = W_H + W_P$; $RRA_T = -(W_H + W_P)U''/U'$ is relative risk aversion measured with respect to total wealth $W_T$; and

$$\tau = W_P/(W_H + W_P) \qquad \text{(A-3)}$$

is the ratio of portfolio wealth to total wealth.

Next, we derive the relationship between $R_c$ and $RRA_P$, assuming (A-2), for the exponential and negative power utility function. Now $R_c$ satisfies

$$U[W_H + W_{Pc}(1 + R_c)] = 1/2U(W_H + W_P)$$

since $U(\infty) = 0$ for both functions.

In the case of exponential utility we note that

$$U[W_H + W_P(1 + R_c)] = \exp(-\alpha W_H) \times$$

$$\{-\exp[-\alpha W_P(1 + R)]\} \qquad \text{(A-4)}$$

But multiplication of a utility function by a positive constant does not change its preferences among probability distributions; in particular it does not change its RRA or $R_c$. Thus, the relationship between $R_c$ and $RRA_P$ for (A-4) is the same as in Equation (9) in the text and shown as the lower curve in Exhibit 2; that is, the relationship for the exponential utility function given by Equation (2).

Next consider

$$\begin{aligned} U[W_H + W_P(1 + R)] &= [W_H + W_P(1 + R)]^{-\alpha} \\ &= (W_H + W_P)^{-\alpha}\,[(1 - \tau) + \\ &\quad \tau(1 + R)]^{-\alpha} \qquad \text{(A-5)} \end{aligned}$$

This has the same ranking of probability distributions of R as

$$\begin{aligned} U &= [(1 - \tau) + \tau(1 + R)]^{-\alpha} \\ &= (1 + \tau R)^{-\alpha} \qquad \text{(A-6)} \end{aligned}$$

Since with R = 0 in (A-6), U = 1, we have

$$(1 + \tau R_c)^{-\alpha} = 1/2 \qquad \text{(A-7)}$$

A short calculation produces

$$\alpha = \ln(2)/\ln(1 + \tau R_c) \qquad \text{(A-8)}$$

$$\begin{aligned} RRA_P &= \tau(1 + \alpha) \\ &= \tau\left[\frac{\ln(2) + \ln(1 + \tau R_c)}{\ln(1 + \tau R_c)}\right] \qquad \text{(A-9)} \end{aligned}$$

For a given $R_c$, $RRA_P$ is strictly increasing in $\tau$ for $0 < \tau \leq 1$. (From $d(1/\alpha\tau)/d\tau = d[\ln(1 + \tau R_c)/\tau\ln(2)]/d\tau < 0$ for $\tau \in (0, \infty)$, we infer $1/\alpha\tau$ is strictly decreasing,[3] and therefore $\alpha\tau$ and $RRA_P = \alpha\tau + \tau$ are strictly increasing.)

$$\begin{aligned} RRA_P &= \left[\frac{\ln(2) + \tau R_c - 1/2(\tau R_c)^2 + \ldots}{\tau R_c - 1/2(\tau R_c)^2 + \ldots}\right] \\ &= \frac{\ln(2) + \tau R_c + \ldots}{R_c - 1/2\tau R_c^{\,2} + \ldots} \qquad \text{(A-10)} \end{aligned}$$

we see that

$$RRA_P \downarrow \ln(2)/R_c \text{ as } \tau \downarrow 0 \qquad \text{(A-11)}$$

(Compare (A-11) to Equation (9).)

## ENDNOTES

There is an elder author but no senior authors.

[1]This assumes that the investor accepts or should accept basic principles such as the von Neumann and Morgenstern or the Leonard J. Savage axioms for action under risk or uncertainty.

[2]Utility often is represented as a function of end-of-period wealth, V(W). Utility expressed as a function of R, U(R), can therefore be expressed as

$$U(R) = V[(1 + R)W_0]$$

Thus,

$$U''/U' = -W_0 V''/V'$$

which is the Pratt-Arrow risk aversion.

[3]Since $d[\ln(1 + \tau R_c)/\tau\ln(2)]/d\tau$

$$= \left[\frac{\tau R_c}{1 + \tau R_c} - \ln(1 + \tau R_c)\right] \Big/ \tau^2 \ln(2)$$

and since $\tau^2\ln(2) > 0$ and $y = \tau R_c > 0$, the sign of the above is the same as the sign of

$$\Phi = \frac{y}{1 + y} - \ln(1 + y)$$

for y > 0. At y = 0, Φ = 0. For y > 0

$$\frac{d\Phi}{dy} = \frac{1}{(1 + y)^2} - \frac{1}{1 + y} < 0$$

It follows that $\Phi < 0$ for $y > 0$; hence the assertion in the appendix follows.

## REFERENCES


Arrow, K.J. *Aspects of the Theory of Risk Bearing*. Helsinki, 1965.

Brieman, L. "Investment Policies for Expanding Business Optimal in a Long Run Sense." *Naval Research Logistics Quarterly*, 7, 4 (1960), pp. 647-651.

——. "Optimal Gambling Systems for Favorable Games." *Fourth Berkeley Symposium on Probability and Statistics*, 1961, pp. 65-78.

Cohn, R.A., W.G. Lewellen, R.C. Lease, and G.G. Schlarbaum. "Individual Investor Risk Aversion and Investment Portfolio Composition." *Journal of Finance*, 30 (May 1975), pp. 605-620.

Ederington, L.H. "Mean-Variance as an Approximation to Expected Utility Maximization." Working Paper 86-5, School of Business Administration, Washington University, St. Louis, Missouri, 1986.

Friend, I., and M.E. Blume. "The Demand for Risky Assets." *American Economic Review*, 65 (December 1975), pp. 900-922.

Grauer, R.R. "Normality, Solvency, and Portfolio Choice." *Journal of Financial and Quantitative Analysis*, 21 (September 1986), pp. 265-278.

Levy, H., and H.M. Markowitz. "Approximating Expected Utility by a Function of Mean and Variance." *American Economic Review*, 69 (June 1979), pp. 308-317.

Markowitz, H.M. "Investment for the Long Run: New Evidence for an Old Rule." *Journal of Finance*, 31, 5 (December 1976), pp. 1273-1286.

——. *Portfolio Selection: Efficient Diversification of Investments*. New Haven: Yale University Press, 1970.

Pratt, J.W. "Risk Aversion in the Small and in the Large." *Econometrica*, 32 (January 1964), pp. 122-136.

Samuelson, P.A. "Longrun Risk Tolerance when Equity Returns are Mean Regressing: Pseudoparadoxes and Vindication of 'Business Man's Risk.'" James Tobin Colloquium, Yale University, May 6-7, 1988.

Simaan, Yusif. "Portfolio Selection and Capital Asset Pricing for a Class of Non-Spherical Distributions of Asset Returns." Dissertation, Baruch College, The City University of New York, 1987.

Young, W.E., and R.H. Trent. "Geometric Mean Approximation of Individual Security and Portfolio Performance." *Journal of Financial and Quantitative Analysis*, 4 (June 1969), pp. 179-199.

# The "two beta" trap

*It lies in differing but specific assumptions about what beliefs investors do and do not hold.*

*Harry M. Markowitz*

12
FALL 1984

Two distinct meanings of the word "beta" are used in modern financial theory. These meanings are sufficiently alike for people to converse — some with one meaning in mind, some with the other — without realizing that they are talking about two different things. The meanings are sufficiently different, however, that we can validly derive diametrically opposed conclusions, depending on which one we use.

The net result of all this can be an Abbott and Costello skit with portfolio theory rather than baseball as its setting. Take, for example, the apparently shocking exposé, "Is Beta Dead?" [14], which contrasted assertions about beta by Richard Roll and Barr Rosenberg. Speaking on behalf of *Institutional Investor* magazine, the article reported that, "After years of chronicling the rise of modern portfolio theory, we saw that a serious crack had developed in its very foundations — a crack that could eventually bring the entire house that MPT built tumbling down." In fact, the article had simply fallen into the two beta trap.

In what follows here, I review the background and definitions of the two betas, and then tabulate propositions that are true for one concept and false for the other.

The basic distinction between the two betas has often been described before. Nevertheless, some of the contrasting implications of the two are not well known, and are sometimes still a source of confusion. For example, I will show, in connection with Propositions three and four, that the ideal weights for forming an index to compute one kind of beta may be completely different from the ideal weights for forming an index to compute the other. This appears to contradict results reported by Stapleton and Subrahmanyam [12]. The reconciliation of the two results appears in the footnote to the section on Propositions three and four.

Later in this article, I return to the "Is Beta Dead" controversy. I will also point out other instances of confusion (to be found even in technical journals) that arise when properties of one beta are incorrectly ascribed to the other.

#### NORMATIVE AND POSITIVE MEAN-VARIANCE ANALYSIS

The first meaning of beta arose in early attempts to use mean-variance analysis to aid in the management of actual portfolios — that is, in "normative portfolio analysis." The second meaning arose in the assumption in the theory of capital markets that investors in fact use mean-variance analysis — that is, in positive mean-variance theory.

In certain ways, the positive and normative theories have traveled distinct paths. Sometimes the implications of the two are confused. For example, one can read from time to time that MPT asserts that no security or portfolio can have a return, on average, greater than a rate based on its level of riskiness. This is true for the standard positive theory, but not for the normative theory. We must be clear about the normative versus positive theories if we are to be clear about the two betas.

The normative theory, as presented by Markowitz [5], shows how an investor or investing institution can minimize variance for different levels of expected return, subject to various constraints. These constraints consist of zero, one, or more linear equalities or inequalities in variables that may or may not be required to be nonnegative. For example, short positions may or may not be allowed; maximum positions may be imposed on individual securities or groups of securities; constraints may be placed on current income in addition to mean-variance objectives for total return, and nonportfolio income may be introduced as an exogenous asset.[1]

For input, the analysis requires estimates of the means, variances, and covariances of various securities, as well as the constraints to be satisfied. *The analysis does not assume that all investors hold these same beliefs.* Nor does it assume that everyone else, or anyone else for that matter, uses mean-variance analysis. It takes the beliefs of the investor or investment team as given to it, and traces out the mean-variance efficient set. No assumption is made that the market as a whole will be one of the portfolios produced in this efficient set analysis.

The positive theory, developed by Tobin [13], Sharpe [11], and Lintner [4], considered what the market for capital assets would be like *if everyone acted according to mean and variance.* Their theories made specific assumptions concerning investors' beliefs and opportunities in order to derive interesting, quantitative answers to questions raised. These are assumptions that the normative theory permits but does not require. In particular, the Tobin analysis and the Sharpe-Lintner Capital Asset Pricing Model (CAPM) assume that all investors have the same beliefs about individual securities, that they all act according to mean and variance, that all investors can borrow at the same rate at which they can lend (the "risk-free" rate), and that they can borrow any amount they want at this rate.

Tobin, Sharpe, and Lintner knew, as well as you and I do, that investors have different beliefs, that borrowing rates are typically higher than lending rates, and that there are credit restrictions on the amount that one can borrow. They chose these assumptions to have a theory with neat, quantitative implications. They left it to empirical research to see whether the conclusions deduced from these idealized assumptions fit aggregate economic data.

Later positive theorists used other assumptions. For example, the assumption that the investor can borrow and lend any amount desired at the risk-free rate is frequently dropped; in its place, it is commonly assumed (following Black [1]) that the investor may sell short any amount of a stock and use the proceeds of the short for other equity investments. For example, it is assumed that an investor with an equity of $10,000 can short $1 million worth of stock A and invest $1 million plus $10,000 in stock B.

I have argued elsewhere [6] that the more recent assumptions of positive theories can hardly be considered more plausible than the original Tobin, Sharpe, and Lintner assumptions. Be that as it may, the Sharpe-Lintner CAPM is handier here, and I use it below. The distinction between the beta of the positive theory and that of the normative theory would be about the same if we used the Black model [1] instead.

**BETA$_{1959}$**

The inputs to a normative analysis include estimates by the investor or investment team of the expected returns, variance of returns, and either covariance or correlation of returns between each pair of securities. For example, an analysis that allows 200 securities as possible candidates for portfolio selection requires 200 expected returns, 200 variances of return, and 19,900 correlations or covariances. An investment team tracking 200 securities may reasonably be expected to summarize their analyses in terms of 200 means and 200 variances, but it is clearly unreasonable for them to produce 19,900 carefully considered correlation coefficients.

It was clear from the start that we need some kind of model of covariance for the practical application of normative portfolio analysis to large portfolios. Markowitz [5] did little more than point out the problem and suggest some possible models of covariance for further research.

One model proposed that we could explain the correlation among security returns by assuming that the return on the ith security is:

$$r_i = \alpha_i + \beta_i F + u_i, \tag{1}$$

where the expected value of $u_i$ is zero, and $u_i$ is uncorrelated with F and every other $u_j$. Originally, F was denoted by I and described as an "underlying factor, the general prosperity of the market as expressed by some index." I have changed the notation from I to F here, to emphasize that $r_i$ depends on the underlying Factor rather than the Index used to estimate the factor. The index never measures the factor exactly, no matter how many securities are used in the index, provided that every security with $\beta_i \neq 0$ has positive variance of $u_i$. The index I will equal:

$$\begin{aligned} I &= \Sigma w_i r_i \\ &= \Sigma \alpha_i w_i + F(w_i \beta_i) + \Sigma u_i w_i \\ &= A + BF + U, \end{aligned} \tag{2}$$

where $w_i$ is the weight of return $r_i$ in the index, and

$$A = \Sigma \alpha_i w_i,$$

$$B = \Sigma w_i \beta_i, \text{ and}$$

$$U = \Sigma u_i w_i$$

1. Footnotes appear at the end of the article.

is the error in the observation of F. In choosing the $w_i$ it is not necessary to require that A = 0 and B = 1, provided that B ≠ 0, for the same reason that Centigrade and Fahrenheit are both valid scales for measuring temperature. Under the conditions stated, the variance of U is:

$$V_U = \sum_{i=1}^{N} w_i^2 V_{u_i} > 0$$

for any choice of $w_i$ such that B ≠ 0.

Sharpe [10] tested Equation (1) as an explanation of how security returns tend to go up and down together. He concluded that Equation (1) was as complex a model of covariance as seemed to be needed. This conclusion was supported by research of Cohen and Pogue [2]. King [3] found strong evidence for industry factors in addition to the market-wide factor. Rosenberg [9] found other sources of systematic risk beyond market-wide factor and industry factor.

We will refer to the beta coefficient in Equation (1) as $\text{beta}_{1959}$. We will contrast the properties of this beta with that of the beta that arises from the Sharpe-Lintner CAPM.

### $\text{BETA}_{1964}$

We noted that the Sharpe-Lintner CAPM makes various assumptions about the world, including that all investors are mean-variance efficient, have the same beliefs, and can borrow or lend at the same rate. Note, however, one assumption that CAPM does *not* make: The Sharpe-Lintner CAPM does not assume that the covariances among securities satisfy Equation (1). On the contrary, the assumptions it makes concerning covariances are more general. They are consistent with Equation (1) but do not require it. They are also consistent with the existence of industry factors as noted by King, or other sources of systematic risk such as those identified by Rosenberg. Thus, if a beta somehow comes out of the CAPM analysis, it is not because Equation (1) is assumed in advance.

The beta that emerges from the Sharpe-Lintner CAPM is a consequence of the conditions for minimizing variance for a given level of expected returns subject to certain kinds of constraints. Given the constraints of the Sharpe-Lintner CAPM, each investor I and security i must satisfy an equation of the following form:

$$\sum_{i=1}^{N} \sigma_{ij} X_{jI} = k_I(\mu_i - \mu_o),$$

where $\sigma_{ij}$ is the covariance between security returns $r_i$ and $r_j$; $X_{jI}$ is the percent of the value of the portfolio of investor I held in security j; $k_I$ is a constant that may vary among investors but is independent of i; $\mu_i$ is the expected value of $r_i$, and $\mu_o$ is the rate at which any investor is assumed to be able to borrow or lend, i.e., "the risk-free rate."

When these equations are weighted by the worth of the Ith investor and summed over all investors, they yield the equation:

$$\sum_{j=1}^{N} \sigma_{ij} X_j = k(\mu_i - \mu_o) \tag{3}$$

for each security i, where $X_j$ is the percent that security j is of the "market portfolio." But

$$\sum_{j=1}^{N} \sigma_{ij} X_j$$

is the covariance between the random variable $r_i$ and the random variable $M = \Sigma r_j X_j$. That is:

$$\Sigma \sigma_{ij} X_j = \text{cov}(r_i, M).$$

Thus, we may write Equation (3) as:

$$\text{cov}(r_i, M) = k(\mu_i - \mu_o)$$

for i = 1, 2, . . ., N. Dividing both sides by $V_M$, the variance of the market, we derive:

$$\frac{\text{cov}(r_i, M)}{V_M} = \frac{k}{V_M}(\mu_i - \mu_o). \tag{4}$$

Now, the least-squares regression coefficient between any two jointly distributed random variables, such as $r_i$ and M, is given by:

$$\beta_{r_i M} = \frac{\text{cov}(r_i, M)}{V(M)}. \tag{5}$$

Since $c = k/V_M$ does not depend on i, we may write Equation (4) as:

$$\beta_i = c(\mu_i - \mu_o) \tag{6}$$

or[2]

$$\frac{\mu_i - \mu_o}{\beta_i} = (1/c), \tag{6'}$$

where we write $\beta_i$ for $\beta_{r_i M}$.

Equation (5) is true whether or not $r_i$ and M are related as in Equation (1), with M in place of F. Even if Equation (1) does not hold, we can still find $\alpha_i$ and $\beta_i$ to minimize expected $(r_i - \alpha_i - \beta_i R)^2$. The $\beta_i$ thus obtained is given by Equation (5). Therefore, the Sharpe-Lintner CAPM implies Equation (6): "excess return" on the ith security, $\mu_i - \mu_o$, is proportional to the beta of the security — where beta is given by Equation (5). I will refer to it in this article as $\text{beta}_{1964}$.

14 FALL 1984

Note that Equation (6) is an assertion about the expected return of a security and how it relates to the regression of the security return against the market. *Unlike Equation (1), it is not an assertion about how security returns tend to covary.*

Here is one source of confusion between $\text{beta}_{1959}$ and $\text{beta}_{1964}$: William Sharpe had an important role in the development of each. William Sharpe, however, has never been confused on this point. In particular, when he explained $\text{beta}_{1964}$ to me two decades ago, he emphasized that he had derived it without assuming Equation (1).

#### PROPOSITIONS ABOUT BETAS

Table 1 lists various propositions about betas and indicates whether they are true or false for $\text{beta}_{1959}$ and $\text{beta}_{1964}$. The first column presents the proposition, the second indicates whether the proposition is true (T) or false (F) for $\text{beta}_{1959}$, and the third column indicates the same for $\text{beta}_{1964}$. Most of the propositions in Table 1 are true for one of the betas and false for the other.

TABLE 1

PROPOSITIONS ABOUT BETA

| | $\beta_{1959}$ | $\beta_{1964}$ |
|---|---|---|
| 1. The $\beta_i$ of the ith security equals $\text{cov}(r_i, R)/V(R)$ for some random variable R. | T | T |
| 2. R is "observable;" specifically, it may be computed exactly from security returns ($r_i$) and market values ($X_i$). | F | T |
| 3. R is a *value*-weighted average of the $r_i$. | F | T |
| 4. An index I that estimates R should ideally be weighted by $\lambda_A(1/V_{u_i}) + \lambda_B(\beta_i/V_{u_i})$. Unfortunately, the $\beta_i$ and $V_{u_i}$ needed to determine these weights are unobservable. | T | F |
| 5. If ideal weights are not used, then equal weights are "not bad" in computing I; specifically, nonoptimum weights can be compensated for by increased sample size. | T | F |
| 6. Essentially, all that is important in computing I is to have a large number of securities; it is not necessary to have a large fraction of all securities. | T | F |
| 7. The ideally weighted index is an efficient portfolio. | F | T |

**Proposition 1**

As noted above, and repeated as the first entry of Table 1, both $\text{beta}_{1959}$ and $\text{beta}_{1964}$ equal:

$$\beta_i = \text{cov}(r_i, R)/V(R) \tag{7}$$

for some random variable R. In the case of $\text{beta}_{1959}$, R is the F of Equation (1); in the case of $\text{beta}_{1964}$, R is the M in Equation (2).

**Proposition 2**

Also, as noted before, F cannot be observed exactly, no matter how many securities are used to estimate it, provided that no security has a nonzero $\beta_i$ and a zero variance of $u_i$. In contrast, M in Equation (2) is observable, at least in principle, if only we are diligent enough to measure each $X_i$ and $r_i$ in the market. Thus, the assertion that R in Equation (7) is observable is true in principle for $\text{beta}_{1964}$ and false for $\text{beta}_{1959}$.

**Propositions 3 and 4**

One source of confusion about the two betas concerns whether an index estimating R should be "value weighted;" that is, should the $w_i$ used in computing an estimate of R from the $r_i$ equal the $X_i$? We have seen that in the case of $\text{beta}_{1964}$ $R = M = \Sigma X_i r_i$. In this case $w_i = X_i$ = market value weights.

We can make a considerable error in this case if we use equal weights instead of value weights. To illustrate, suppose:

$$V_1 = V_2$$
$$\text{cov}(r_1, r_2) = 0$$
$$\text{cov}(r_1, r_i) = \text{cov}(r_2, r_i)$$
$$\text{for } i = 3, 4, \ldots, N.$$

Suppose also that $X_1$ is much larger than $X_2$. Then:

$$\text{cov}(r_1, M) = X_1V_1 + \sum_{i=3}^{N}\sigma_{1i}X_i$$

will be much larger than:

$$\text{cov}(r_2, M) = X_2V_2 + \sum_{i=3}^{N}\sigma_{1i}X_i,$$

but $\text{cov}(r_1, I)$ will equal $\text{cov}(r_2, I)$ if an index I weights $r_1$ and $r_2$ equally. An incorrect calculation of equilibrium $\mu_i$ will result if we use I instead of M. Also, other things being equal, securities that are highly correlated with $r_1$ will be less desirable than those highly correlated with $r_2$. This will not be implied by the equally weighted I.

The answer is different in the case of $\text{beta}_{1959}$. Ideally, we would like to eliminate the error term U from Equation (2). Our index would be perfect if $V_U = 0$, provided of course $B \neq 0$. Nevertheless, as long as no security with $\beta_i \neq 0$ has $V_{u_i} = 0$, the perfect index cannot be achieved with a finite number of securities. Short of this, it might seem that the best to be wished is for $V_U$ be a minimum. In this case, $w_i$ should equal $1/V_{u_i}$.

But the matter is more complex. Suppose that index $I_a$ has half as much $\sigma_U = \sqrt{V_U}$ as does $I_b$. Suppose also that $I_a$ has a B that is half as great as that of $I_b$, and suppose that both have A = 0. We can hardly say that $I_a$ is a better index than $I_b$, since it has the same A, B and $V_U$ as $I_b/2$, and $I_b/2$ is no better or worse an index than $I_b$. Thus, smaller $V_U$ does not necessarily mean that the index is better.

A plausible alternative is to choose weights so as to:

$$\text{minimize } \frac{\sigma_U}{B} \tag{8}$$

$$\text{subject to } \Sigma w_i = 1.$$

This is equivalent to minimizing $V_U/(B^2V_F)$, or maximizing the fraction of:

$$V_I = B^2V_F + V_U$$

due to variation in F.

More generally, we could seek weights w = $(w_1, w_2, \ldots w_N)$ from among the set of weights that:

$$\begin{aligned} &\text{Minimize } V_U \\ &\text{for various levels of B} \\ &\text{subject to } \Sigma w_i = 1. \end{aligned} \tag{9}$$



We will define the various solutions to Equation (9) as the "efficient weights." Note, in particular, that both the weights that minimize $V_U$ and those that minimize $\sigma_U/B$ are examples of "efficient weights." Note, further, that the problem of finding efficient weights is mathematically the same as that of finding efficient portfolios when nonnegativity is not required. Using the Lagrangian multiplier technique, we find that any set of efficient weights satisfy:

$$w_i = \lambda_A\left(\frac{1}{V_{u_i}}\right) + \lambda_B\left(\frac{\beta_i}{V_{u_i}}\right) \tag{10}$$

$$\text{for } i = 1 \text{ to } N$$

for suitably chosen $\lambda_A$ and $\lambda_B$. The $\lambda$'s themselves depend on the $\beta_i$, the $V_{u_i}$, and the desired efficient combination of B and $V_U$.

Equation (10) raises two questions:

1. The ideal weights for an index $I_{1959}$, to be used in estimating $\beta_{1959}$, depend on unknowns like $V_{u_i}$ and the betas themselves. Does this mean that we cannot estimate $\beta_{1959}$?
2. Equation (10) does not involve the percent of the market represented by each security. It would appear, then, that the ideal weights for an index $I_{1959}$ for estimating $\beta_{1959}$ differ from $(X_1, \ldots, X_N)$, the ideal weights for $I_{1964}$. At least the formulas *look* different. But isn't it possible that Equation (10) is just a different way of always computing $X_i$, arriving at the same answer by a different path?

Concerning question 1, we will see later that, when ideal weights are not known, equal weights are "not bad" for $I_{1959}$. In particular, a larger number of securities in the index can compensate for nonoptimal weights. We will also consider the statistical consistency of estimates of $\beta_{1959}$ based on either an optimal or equally-weighted $I_{1959}$.

The remainder of this section will be concerned with the second question: Are the optimum weights for $I_{1959}$ presented in Equation (10) really different from the $X_i$?

It is possible for both Equation (1) and the assumptions of CAPM to hold. Recall that each permits, but does not require, the other. We shall show that even if both Equation (1) and CAPM are true, the ideal weights for one index need bear no relationship to the ideal weights for the other.

To see this, start by choosing any numbers $\beta_i^a$ and $V_{u_i} > 0$ as the parameters of Equation (1). The set of efficient w follow from this choice. Pick one of the efficient w's, for example, the one that minimizes $\sigma_U/B$. We have now determined the ideal weights for $I_{1959}$. Call this $w^a = (w_1^a, w_2^a, \ldots, w_N^a)$.

Having chosen $w^a$, now arbitrarily choose a different set of weights $w^b$. We will keep $\beta_i^a$ and $V_{u_i}$ as chosen above and choose $(\mu_i - \mu_o)$ so that $w^b$ are the ideal weights for $I_{1964}$. We will thus show that any choice of $w^a$ as the ideal weights for $I_{1959}$ can go with any choice, $w^b$, of ideal weights for $I_{1964}$.

The $\beta_i^a$ and $V_{u_i}$ of Equation (1) determine the covariance matrix $(\sigma_{ij})$. The variances of the $r_i$ (as distinguished from the variance of the $u_i$) are given by $(\beta_i^a)^2V_F + V_{u_i}$, while the covariance between $r_i$ and $r_j$ (for i = j) is $\beta_i^a\beta_j^aV_F$.

Let $X_i = w_i^b$, choose any $\mu_o$, and define $\mu_i$ by:

$$(\mu_i - \mu_o) = \sum_{i=1}^{N} X_i\sigma_{ij}. \tag{11}$$

With the $(\mu_i - \mu_o)$ as thus defined, the $X_i$'s, $\mu_i$'s, and $\sigma_{ij}$ are a CAPM equilibrium (see Equation (3) which is equivalent to Equation (6)). To illustrate the model concretely, postulate shares outstanding and company prospects such that the prices that make the percents of market value equal to $X_i$ also give the assumed $\mu_i$, $\beta_i^a$ and $V_{u_i}$ per dollar invested.

For example, for simplicity, imagine an economy that perpetually repeats itself in that the same firms exist each period; each firm has the same number $(S_i)$ of shares outstanding; the earnings $(e_i)$ of the firm, each period, is drawn from the same probability distribution, though the particular draw will vary from period to period; all earnings are paid as dividends, and the earnings per share each period are:

$$e_i = a_i + b_iF + v_i,$$

where $a_i$ and $b_i$ are constants and $v_i$ is uncorrelated with F and every other $v_j$. Choose equilibrium stock prices ($p_i$) arbitrarily; then choose $a_i$, $b_i$, $S_i$, and the distribution of $v_i$ so that:

$$a_i = \alpha_i p_i,$$

$$b_i = \beta_i p_i, \text{ and}$$

$$X_i = \frac{S_i p_i}{\sum_{i=1}^{N} S_i p_i}.$$

That is:

$$S_i = X_i T/p_i$$

for some arbitrary T, and:

$$v_i = u_i p_i.$$

We have now worked backwards from the answer we wish to obtain to the model that will give us that answer. If we start with particular numerical values of $\beta_i$, $V_u$, $w^b$, and $w^a$; compute $a_i$, $b_i$, $S_i$, and $v_i$ as described above; hide the $w^a$ and $w^b$; announce the $a_i$, $b_i$, $V_{u_i}$, and $S_i$, we can then show that certain $p_i$'s (the ones we chose arbitrarily above) are CAPM equilibrium prices; we constructed the numbers so that Equation (6) would hold with c = 1. We can then obtain the optimum weights for $I_{1959}$ and $I_{1964}$, which turn out to be — no surprise to us — the original $w^a$ and $w^b$.

Whether or not Equation (1) is true in fact — or almost true in some sense, whether CAPM is true or almost true, and if so whether the ideal weights for $I_{1959}$ and $I_{1964}$ are similar or different in fact, are empirical questions beyond the scope of this paper. I have shown that, as far as the logical implications of Equation (1) and the CAPM assumptions are concerned, any possible ideal weights for $I_{1959}$ can exist with any other (similar or different) ideal weights for $I_{1964}$.[3] The nature of these weights is summarized as Propositions 3 and 4 in Table 1.

#### Proposition 5

The fifth proposition in Table 1 asserts that if ideal weights cannot be obtained, equal weights are good enough. In particular, an increase in the number of securities can compensate for nonoptimum weights by increasing the number of securities in the index. We have already seen that this proposition is false for $\text{beta}_{1964}$. It is easily seen to be true for $\text{beta}_{1959}$ under mild restrictions on how fast the $V_{u_i}$ increase as i increases.

For example, if there is an upper bound on $V_{u_i}$, then $V_U$ will approach zero as N approaches infinity. This says that, if large enough, a larger equally-weighted index will be as good as a smaller optimally-weighted index.

Thus far I have equated the "goodness" of $I_{1959}$ with the smallness of $V_U$ (at least for given $B = \Sigma w_i\beta_i$). I have yet to demonstrate any relationship between the estimates $\hat{\beta}_i$, obtained by regressing $r_i$ against $I_{1959}$, and the actual $\beta_i$ of Equation (1). In the remainder of the present section, I will consider the consistency of $\hat{\beta}_i$ for both equally-weighted and optimally-weighted $I_{1959}$. That is, we shall consider $\beta_i^* = \lim \hat{\beta}_i$ — almost always — as T (the number of historical observations) increases, with N (the number of securities in $I_{1959}$) held fixed. We shall see that $\beta_i^* \neq \beta_i$, but that the difference falls as N increases, and for N = 500 the difference is already negligible.

For ($w_1$, $w_2$, . . . $w_N$) equal to either equal or optimal weights, or for any weights at all for that matter, we have (e.g.):

$$\beta_1^* = \frac{\text{cov}(r_1, I)}{V(I)}$$

$$= \frac{\text{cov}(\alpha_1 + \beta_1 F + u_1, A + BF + \Sigma w_i u_i)}{V(A + BF + \Sigma w_i u_i)}.$$

Note that Equation (1) is unchanged if we multiply each $\beta_i$ by $\lambda \neq 0$, and divide F by the same constant $\lambda$. We may choose $\lambda$ so that B = 1. With this choice of scale factor we have:

$$\beta_1^* = \frac{\text{cov}(\alpha_1 + \beta_1 F + u_1, A + F + \Sigma w_i u_i)}{V(A + F + \Sigma w_i u_i)}$$

$$= \frac{\beta_1 V_F + w_1 V_{u_1}}{V_F + V_U}. \quad (12)$$

For a security not in the index, say $\beta_{N+1}$, the second term in the numerator is zero. To roughly compare $\beta_i$ with $\beta_i^*$ for, say, N = 500 and $w_i = 1/N$ for i = 1 to N, let:

$$\theta_1 = \frac{V_{u_1}}{V_F}$$

$$\bar{\theta} = \frac{\frac{1}{N}\Sigma V_{u_i}}{V_F}.$$

Then,

$$\beta_1^* = \frac{\beta_1 V_F + \theta_1 V_F/N}{V_1 + \frac{\frac{1}{N}\Sigma V_{u_i}}{N}}$$

$$= \frac{\beta_1 V_F + \theta_1 V_F/N}{V_F + \bar{\theta}\, V_F/N}$$

$$= \frac{\beta_1 + \theta_1/N}{1 + \bar{\theta}/N}.$$

With $\theta_1 = \bar{\theta} = 2$, for example,

$$\beta_1^* = \frac{\beta_1 + .004}{1 + .004},$$

which should be close enough for practical purposes and is usually negligible as compared to sampling error. Typically, $\theta_i < 10$ (more like 2 or 3) and exceptions should not be included in an equally-weighted index, if possible.

#### Proposition Six

The next proposition asserts that all that is important in designing a good index is to have many securities, as opposed to having a large percentage of the population represented in the index. This proposition is true for $I_{1959}$ and false for $I_{1964}$, as illustrated by two extreme examples.

First, suppose that there are only a few securities in the entire population, and all of them are used in computing the index. Then $I_{1964}$ would, in fact, be M. It would be precisely correct. In the case of $I_{1959}$, on the other hand, the error in the estimate of the underlying factor F would show that there were only n = 6 securities.



$$V_U = \sum_{i=1}^{6} w_i^2 V_{u_i}$$

is the same, whether the six securities are 100% or 1% of the universe. The same applies to the other error term in Equation (12). Thus, a 100% sample of a small universe would give a correct $I_{1964}$, but would be of no more or less value in constructing $I_{1959}$ than an equally small sample with the same $\beta_1$ and $V_{u_i}$ from a large population.

At the other extreme, imagine that the sample is large but its percentage of the total population is small. For example, suppose N = 1,000 out of 100,000 securities. Equation (12) shows that the use of $I_{1959}$ then gives essentially consistent estimates of $\beta_{1959}$; but $I_{1964}$ may give seriously inconsistent estimates of $\beta_{1964}$. First, the covariance with the index of an asset not in the index will tend to be too low. Second, if the index contains more of certain kinds of assets than is characteristic of the entire population, then assets of this sort will tend to have a higher correlation with the index than with the true M, and assets of other sorts will tend to have lower correlations. More precisely, the covariance between return $r_i$ and the market is a weighted average of the $\sigma_{ij}$ (including $V_i = \sigma_{ii}$) weighted by market values. If the index chosen does not have approximately the same average $\sigma_{ij}$ for a given i, the estimate of $\beta_{i,1964}$ will be inconsistent statistically.

For $I_{1959}$, then, large sample size is sufficient. For $I_{1964}$, large sample size may, in itself, be insufficient.

#### Proposition Seven

This proposition asserts that the ideal index is an efficient portfolio. This is true for $I_{1964}$ and false for $I_{1959}$.

One of the conclusions of the Sharpe-Lintner CAPM assumptions is that the market portfolio is efficient. In fact, except for borrowing or lending, the market portfolio is the only combination of risky assets that is efficient in this CAPM. All other efficient portfolios consist of either investment in the market portfolio plus lending at the risk-free rate, or of investment in the market portfolio financed in part by borrowing at the risk-free rate.

But we saw that, in a model for which both Equation (1) and the CAPM assumptions held, the market (and hence the ideal weights for $I_{1964}$) could be any set of positive numbers that sum to one, whatever the ideal weights for $I_{1959}$. Hence, while the ideal weights for $I_{1964}$ make an efficient portfolio, given the CAPM assumptions, those for $I_{1959}$ need not.

#### EPILOGUE AND EDITORIAL

Barr Rosenberg and Richard Roll, the main figures in the "Is Beta Dead?" article, both understand the difference between the two betas discussed here. The following two extracts will show this and will be useful in our subsequent discussion, which depends on who had what to do with which beta.

In "Extra-Market Components of Covariance in Security Returns," Rosenberg [9] presents a sophisticated multifactor model of security returns, including an overall market factor, extra-market factors, and "elements of return that are specific to the individual securities and are therefore assumed to be uncorrelated with one another and with the factors." As the title of the paper states, Rosenberg presents this multifactor model (including the market factor among others) as an explanation of covariance in security returns. He considers the use of this model for mean-variance portfolio optimization, which requires mean returns as inputs. At one point in his exposition, he states "Thus far, the model of [expected] returns has been completely general. Now consider the implications of the capital asset pricing model, which, in its simplest form (see [Sharpe and Lintner]), implies [our Equation (6)]." Rosenberg then shows some implications of adding the CAPM assumptions to his model of covariance.

In "Ambiguity When Performance Is Measured by the Securities Market Line," Roll [8] shows that, under one or another CAPM assumption, a common practice of measuring a portfolio performance (as a deviation from a beta-mean return market line) is nonsense. Toward the end of the paper, after proving his principal point, Roll states that he has "presented some negative aspects of the securities market line as a performance measuring device. Yet the theory of portfolio diversification and the concept of 'beta' . . . as a systematic risk measure constitute a pervasive paradigm. . . . When discussing the securities market line and beta, most people are actually thinking intuitively about a one-factor linear return generating process of the form [our Equation (1)]. . . . [Our F] is not a portfolio. [It] is the *unique* source of common variation in the ensemble of asset returns. This seemingly innocuous distinction is actually critical." Indeed it is!

It must be admitted that financial theorists are sometimes a disputatious crowd. Rosenberg's paper argued against a simpler model of covariance that many before him had thought good enough; Roll argued against a method of performance measurement that others before him thought was an implication of CAPM but was not; Roll elsewhere has, on occasion, disputed empirical work put forth as verifying some CAPM; the present author, admittedly, has engaged in a dispute or two on matters such as choice of criteria for portfolio selection; and so on.

Withal, the big fight between Roll and Rosenberg reported in *Institutional Investor* was a non-event. Rosenberg was arguing about models of covariance, Roll about implications of CAPM for performance measurement. It was a fight between two boxers in different rings, in different arenas, in different cities.

I should say in defense of the *Institutional Investor* article that it was easy to have the impression that the central contributions of Roll and Rosenberg were in conflict. For example, an important issue for Roll in [8] is whether the index I used for performance measurement is an efficient portfolio. (The use of the mean-beta market line measurement is wrong in either case, but the nature of the error is different.) But Rosenberg uses a market index as an estimate of a general factor, without concern as to whether this index is an efficient portfolio. As we have seen, however, if both Equation (1) and CAPM are correct, $I_{1964}$ is an efficient portfolio but $I_{1959}$ need not be; the latter just needs enough securities. The same is also true if Equation (1) is replaced by Rosenberg's more complex model, and the market index is used as an estimate of the market (among other) factors.

Confusion about beta did not start or end with the thought that perhaps it was dead. The following is a common example. Someone wants to test whether CAPM applies in some sector of the market for capital assets. He builds an index for the assets in the sector and then tests to see whether observed average excess returns are proportional to the regression of returns against this index. But the $I_{1964}$ that CAPM says will explain expected returns is the market portfolio. A broad index, not a narrowly defined sector index, is required by CAPM. (The index for the sector could perhaps be a factor in a model of covariance in the manner of King and Rosenberg.)

On the other hand, following from a trivial generalization of my discussion at the end of the section on Propositions 3 and 4, the weights of one index may have absolutely no relationship to the other; and, as reviewed in the first part of the same section, a considerable error can occur, if we use incorrect weights with $I_{1964}$.

But, Is Beta Dead? In fact both are doing fine, and enjoying their respective grandchildren.


REFERENCES

1. Black, F. "Capital Market Equilibrium with Restricted Borrowing." *Journal of Business*, July 1972.
2. Cohen, K. J., and J. A. Pogue. "An Empirical Evaluation of Alternative Portfolio Selection Models." *Journal of Business*, April 1967.
3. King, B. F. "Market and Industry Factors in Stock Price Behavior." *Journal of Business*, Supplement, January 1966.
4. Lintner, J. "Security Prices, Risk, and Maximal Gains from Diversification." *Journal of Finance*, December 1965.
5. Markowitz, H. M. *Portfolio Selection: Efficient Diversification of Investment*. New York: John Wiley & Sons, 1959; New Haven, Conn.: Yale University Press, 1970.
6. ——. "Nonnegative or Not Nonnegative: A Question About CAPMs." *Journal of Finance*, May 1983.
7. Perold, A. F. "Solving Large Portfolio Optimization Problems." Manuscript, Graduate School of Business Administration, Harvard University, 1981.
8. Roll, R. "Ambiguity When Performance Is Measured by the Securities Market Line." *Journal of Finance*, September 1978.
9. Rosenberg, B. "Extra-Market Components of Covariance in Security Returns." *Journal of Financial and Quantitative Analysis*, March 1974.
10. Sharpe, W. F. "A Simplified Model for Portfolio Analysis." *Management Science*, January 1963.
11. ——. "Capital Asset Prices: A Theory of Market Equilibrium under Conditions of Risk." *Journal of Finance*, September 1964.
12. Stapleton, R. C., and M. G. Subrahmanyam. "The Market Model and Capital Asset Pricing Theory: A Note." *Journal of Finance*, December 1983.
13. Tobin, J. "Liquidity Preference as Behavior toward Risk." *Review of Economic Studies*, February 1958.
14. Wallace, A. "Is Beta Dead?" *Institutional Investor*, July 1980.


[1] Most of [5] discusses portfolio analysis subject to the constraints:

$$\sum_{j=1}^{N} a_{ij}X_j = b_i \qquad i = 1 \text{ to } M$$

$$X_j \geq 0 \qquad j = 1 \text{ to } N,$$

but it also shows how to reduce to the above form any system of linear equalities or inequalities in variables that may or may not be required to be nonnegative.

[2] The Sharpe-Lintner assumptions imply $c > 0$; thus, division by c is permitted.

[3] The above results seem to contradict the recent note by Stapleton and Subrahmanyam [12] concerning the market model and CAPM. Except for notation, it would seem that Stapleton and Subrahmanyam assume both CAPM and Equation (1), then reach conclusions different from those presented here. But they make different assumptions concerning the random variables here called F and $u_i$. We assume $u_i$ is uncorrelated with F and every other $u_j$; this is the assumption made in [5] and [10]. It is also the assumption that will be made for you if you use a general portfolio selection code, such as that of Perold [7], and elect the "one factor model" as the particular form of your covariance matrix. One consequence of this assumption is that the market return will rarely equal F. Stapleton and Subrahmanyam assume that (what we denote by) F in (1) is, in fact, the return on the market M. Consistent with this, they do *not* assume that the $u_i$ are uncorrelated. I do not contend that their model is worse than the one presented here — it is just different. We have shown that a model that satisfies the CAPM assumptions and also Equation (1) — including the stated independence assumption — can have any combination of optimal weights for $I_{1959}$ and $I_{1984}$. Stapleton and Subrahmanyam start with different premises and reach different conclusions.

THE JOURNAL OF FINANCE • VOL. XXXVI, NO. 14 • SEPTEMBER 1981

# Portfolio Analysis with Factors and Scenarios

HARRY M. MARKOWITZ and ANDRÉ F. PEROLD


## ABSTRACT

Recently there has been a growing interest in the scenario model of covariance as an alternative to the one-factor or many-factor models. We show how the covariance matrix resulting from the scenario model can easily be made diagonal by adding new variables linearly related to the amounts invested; note the meanings of these new variables; note how portfolio variance divides itself into "within scenario" and "between scenario" variances; and extend the results to models in which scenarios and factors both appear where factor distributions and effects may or may not be scenario sensitive.


MODELING THE COVARIANCES OF intersecurity returns is one of the most important aspects of a portfolio analysis, especially for large numbers of securities. For a universe of 1,000 securities, say, the direct estimation of the roughly half a million covariances is not practicable, let alone the ensuing high cost of computation. The most widely used model today is the multifactor approach, where the source of covariation amongst returns is attributed to a few (say 10) common factors, e.g. [5, 6]. For 1,000 securities, only about 10,000 coefficients have to be estimated, a small fraction of what was originally required. Further, an equally dramatic reduction in the computational effort to compute efficient portfolios can be realized by exploiting the fact that the covariance matrices resulting from these models can be easily diagonalized, e.g. [1, 6].

An alternative approach is the scenario or states of the world model as described, for example, in [2]. This involves selecting, say, 10 future states of the world, assigning probabilities of occurrence and estimating the returns of the securities in each of them. The appeal of this approach is that it frequently seems an ideal vehicle for quantifying one's subjective views of the future. Its drawback, and one of the main reasons why it has not been used for portfolio analyses involving any substantial number of securities, is that with fewer states of the world than securities, the resulting covariance matrix is singular, i.e. riskless portfolios of risky securities can be formed.

In this paper we show how this drawback can be overcome by explicitly modeling the uncertainty remaining once a scenario has been realized. We then show how the mathematical form of these models is precisely the same as in the multifactor case, and thus, upon diagonalization, able to yield the same savings in computation. The question of computation with this general class of models is pursued in detail in [4].

## I. The Model

We begin by partitioning the event space into $S$ mutually exclusive and exhaustive events, to be called scenarios,[1] the $s$th occurring with probability $p_s$. Typically,

[1] When a scenario is the resolution of all relevant future uncertainties, we shall call it a "state of the world."

the scenarios might be obtained from an event tree such as the following:

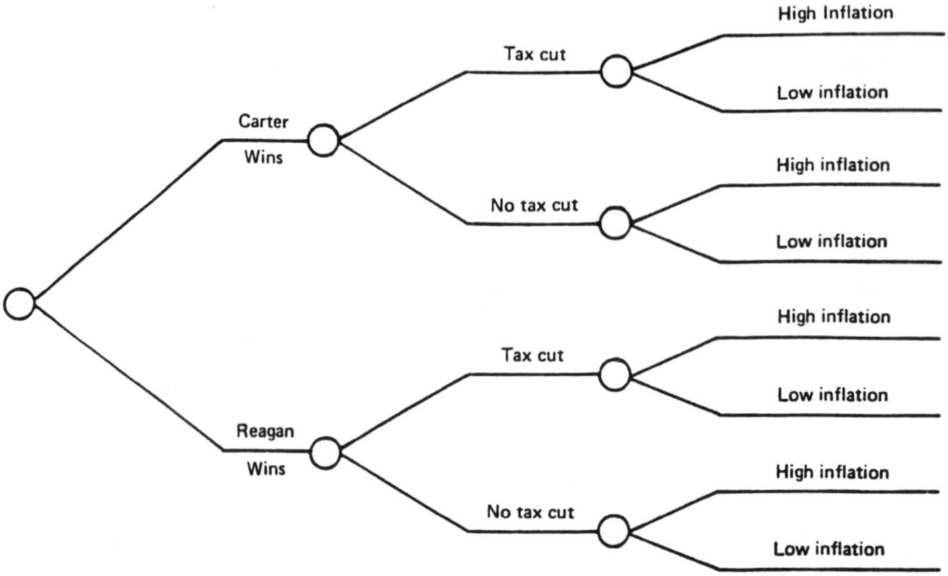


Here, there are $S = 8$ scenarios, and their probabilities are simply the products of the conditional probabilities of the appropriate branches of the tree.

Let there be $N$ securities in the inverse and let $r_i$ be the return on the $i$th security. Then, without loss of generality, and conditional on scenario $s$ occurring, $s = 1, \cdots, S$, $r_i$ may be written as

$$r_i = \mu_{is} + \epsilon_i \tag{1}$$

where $\mu_{is}$ is a constant and $\epsilon_i$ satisfies

$$E(\epsilon_i | \text{scenario} = s) = 0 \tag{2}$$

In the following let $\nu_i$ denote the expected value of $r_i$, $\sigma_i^2$ the variance of $r_i$, and $\sigma_{ij}$ the covariance between $r_i$ and $r_j$. Note that

$$\nu_i = \textstyle\sum_{s=1}^{S} p_s \mu_{is}. \tag{3}$$

The question now is how to model the conditional distributions of the $\epsilon$'s. We consider different assumptions in increasing order of generality.

### *A. The Usual States of the World Model*

In this case (e.g. [2]) it is assumed that sufficiently many states have been chosen so that it is reasonable to regard the $\epsilon$'s as being identically zero. Since the resulting covariance between $r_i$ and $r_j$ is then

$$\sigma_{ij} = \textstyle\sum_{s=1}^{S} p_s (\mu_{is} - \nu_i)(\mu_{js} - \nu_j)$$

the covariance matrix of returns, $C$, has the form

$$C = GPG' \tag{4}$$

where $G$ is an $N \times S$ matrix with $(i, s)$th entry given by

$$g_{is} = \mu_{is} - \nu_i \tag{5}$$

and $P$ is an $S \times S$ diagonal matrix with $s$th diagonal entry $p_s$. While this will follow by specializing the result to be established next, note that $C$ in (4) has exactly the same form as a covariance matrix formed from historical observations, where $\mu_{is}$ can be interpreted as the return on security $i$ in period $s$ and all states are equally likely, i.e. $p_s = 1/S$ for all $s$.

*B. Modeling the $\epsilon$'s as being Conditionally Uncorrelated*

The first improvement over the above model is to choose the scenarios in such a way that the $\epsilon$'s can be assumed to satisfy

$$E(\epsilon_i\epsilon_j|\text{scenario} = s) = 0 \qquad i \neq j \tag{6}$$

and

$$E(\epsilon_i^2|\text{scenario} = s) = \sigma_{is}^2 \tag{7}$$

In other words, while there is some remaining uncertainty in the return of a security once the scenario becomes known, it is uncorrelated with the remaining uncertainty in the return of any other security. Here the covariance matrix has the form

$$C = D + GPG' \tag{8}$$

where $G$ and $P$ are as in Section $A$ and $D$ is an $N \times N$ diagonal matrix with $i$th diagonal entry

$$\textstyle\sum_{s=1}^{S} p_s\sigma_{is}^2$$

To see this, observe that

$$\sigma_{ij} = \textstyle\sum_{s=1}^{S} p_s E\{(r_i - \nu_i)(r_j - \nu_j)|\text{scenario} = s\} \tag{9}$$

From (1) and (5) it follows that, given scenario $s$,

$$r_i - \nu_i = \epsilon_i + g_{is}$$

Thus, given scenario $s$,

$$(r_i - \nu_i)(r_j - \nu_j) = g_{is}g_{js} + \epsilon_i\epsilon_j + \epsilon_i g_{js} + \epsilon_j g_{is}$$

By (2), (6), and (7), the right-hand side of this equation has conditional expectation

$$g_{is}g_{js} \qquad i \neq j$$

and

$$g_{is}^2 + \sigma_{is}^2 \qquad i = j$$

From this and (9), the form of (8) follows immediately. Clearly, when $\sigma_{is}^2 = 0$ for all $i$ and $s$, (4) obtains.

This model is much more realistic than the States of the World Model in $A$, especially when there are only a few scenarios, and presents little additional difficulty as regards implementation. The only addition is that of $\sigma_{is}^2$, the uncertainty in one's best guess at the return of security $i$ in scenario $s$.

### C. Combining Scenarios and Factors

If the zero correlation assumption (6) is unreasonable, the next step might be to model the covariances of the $\epsilon$'s using common factors. In the following, covariances amongst factors will all be assumed to be scenario dependent; however, we shall distinguish between factors whose sensitivities (betas) are scenario independent and factors whose betas are scenario dependent. Let $K$ be the number of factors of the first type and $L$ the number of factors of the second type. The following results specialize to various simpler cases, e.g. when $K$ or $L$ equals zero or the factors are orthogonal.

In place of (6) and (7), and conditional on the occurrence of scenario $s$, the model now becomes

$$\epsilon_i = \sum_{k=1}^{K} \beta_{ik} F_k + \sum_{l=1}^{L} \gamma_{ils} G_l + \delta_i \tag{10}$$

where the factors $F_k$, $G_l$, and disturbances $\delta_i$ all have conditional expectation zero. Further the $\delta$'s are assumed conditionally uncorrelated with each other and the factors. Let

$$E(\delta_i^2|\text{scenario} = s) = \sigma_{is}^2$$
$$E(F_k F_r|\text{scenario} = s) = u_{krs}$$
$$E(G_l G_t|\text{scenario} = s) = v_{lts} \tag{11}$$

and

$$E(F_k G_l|\text{scenario } s) = w_{kls}$$

where $s = 1, \cdots, S$, $i = 1, \cdots, N$, $k, r = 1, \cdots, K$, and $l, t = 1, \cdots, L$.

A manipulation similar to that presented in Section $B$ yields the following expressions for the variances and covariances of the returns:

$$\sigma_i^2 = \sum_{s=1}^{S} p_s\{\sigma_{is}^2 + g_{is}^2 + \sum_{k=1}^{K}\sum_{r=1}^{K} \beta_{ik}\beta_{ir}u_{krs}$$
$$+ \sum_{l=1}^{L}\sum_{t=1}^{L} \gamma_{ils}\gamma_{its}v_{lts} + 2\sum_{k=1}^{K}\sum_{l=1}^{L} \beta_{ik}\gamma_{ils}w_{kls}\} \tag{12}$$

$$\sigma_{ij} = \sum_{s=1}^{S} p_s\{g_{is}g_{js} + \sum_{k=1}^{K}\sum_{r=1}^{K} \beta_{ik}\beta_{jr}u_{krs}$$
$$+ \sum_{l=1}^{L}\sum_{t=1}^{L} \gamma_{ils}\gamma_{jts}v_{lts}$$
$$+ \sum_{k=1}^{K}\sum_{l=1}^{L} (\beta_{ik}\gamma_{jls} + \beta_{jk}\gamma_{ils})w_{kls}\} \tag{13}$$

Interpreting (13), for example, we see that the covariance $\sigma_{ij}$ is derived from the four terms in { }: respectively, these are covariation due to scenarios; covariation due to the $F$ factors; covariation due to the $G$ factors; and covariation due to interrelationships between the $F$'s and $G$'s.

To obtain the form of $C$ in matrix notation as was done in Sections $A$ and $B$, let $B$ be the $N \times K$ matrix of coefficients $(\beta_{ik})$, and $\Gamma_s$ the $N \times L$ matrix of coefficients $(\gamma_{ils})$, $s = 1, \cdots, S$. Let $U$, $V_s$, and $W_s$ respectively be the $K \times K$, $L \times L$ and $K \times L$ matrices of coefficients $(\sum_{s=1}^{S} p_s u_{krs})$, $(v_{lls})$ and $(w_{kls})$, and $D$, $G$, and $P$ be as before. Then, as can be confirmed by multiplying out $HRH'$ and comparing the result with (13), $C$ may be written as

$$C = D + HRH' \tag{14}$$

where $H$ is the $N \times (S + K + SL)$ matrix

$$H = (G, B, \Gamma_1, \cdots, \Gamma_S)$$

and $R$ is the $(S + K + SL) \times (S + K + SL)$ matrix.

$$R = \begin{bmatrix} p_1 & & & & & \\ & \ddots & & & O & \\ & & p_S & & & \\ & & & U & p_1 W_1 \cdots p_S W_S & \\ & O & & p_1 W_1' & p_1 V_1 & O \\ & & & \vdots & \ddots & \\ & & & p_S W_S' & O & p_S V_S \end{bmatrix}$$

The need to consider separately the factors whose sensitivities are scenario independent should now be clear: each such factor contributes one to the dimension of $R$, whereas each factor whose sensitivities are scenario dependent contributes $S$. Some factors, such as "investor preference for small companies," fall naturally into the former category since there the sensitivity of the $i$th return is simply the size of the $i$th company. Others might have sensitivities which vary only slightly from one scenario to the next, so that scenario independence can be imposed with little loss.

The choice of a particular factor model to capture the covariation amongst the $\epsilon$'s will be largely dependent on the degree to which the scenarios themselves explain the covariation amongst returns. In some cases, a single factor such as the market (perhaps with scenario dependent betas) might be deemed sufficient, so limiting the additional estimation requirements to a market beta for each security in each scenario.

In concluding this secion, we remark that (14) is also the well-known form [5] of the conventional multifactor model in which $R$ is the covariance matrix of the factors and $H$ is the matrix of sensitivities.

## II. Diagonalization

We have seen that the covariance matrices resulting from these models all have the form (14) where $D$ is diagonal, $H$ is $N \times M$, and $R$ is $M \times M$. In the scenario model, $M = S$, the number of scenarios; in the multifactor model $M = L$, the number of factors; and in the combined model, $M = S + K + SL$ where $K$ and $L$ are respectively the number of factors with scenario independent and dependent sensitivities. In all cases we assume $M$ to be small relative to $N$, at most perhaps

50 and often more like 10. This is a reasonable assumption in practice since the modeling effort simply becomes too great otherwise.

The variance of a portfolio of holdings $x = (x_1, \cdots, x_N)$ is given by the quadratic form

$$x'Cx$$

Using (14), this is equal to

$$x'Dx + x'HRH'x$$

which becomes

$$x'Dx + y'Ry \tag{15}$$

provided

$$Hx = y \tag{16}$$

Since (16) is linear in the "old" variables $x$ and the "new" variables $y$, it may be introduced as $M$ additional constraint equations in a general portfolio selection model [3]. The resulting quadratic form (15) is then diagonal in all except possibly the $y$ variables.

The only off-diagonal entries (in $R$) are due to nonzero covariances amongst factors. In the combined scenario and factor case $R$ can be left as is and treated as a sparse matrix, i.e. a record kept of nonzero entries only. In the case of a multifactor model where all or most factors are correlated, $R$ can be written as the product

$$R = TT'$$

where $T$ is a triangular matrix. Then (15) becomes fully diagonalized by substituting the identity matrix for $R$, and the product $HT$ for $H$. In Rosenberg's multifactor model [4], where $M = L = 47$, this is not desirable since $H$ is very sparse (i.e. it has a high percentage of zeros), and forming the product $HT$ would make it 100% dense. In this case, however, $R$ itself has the form (14), so that a second level of additional variables and constraints can be defined to achieve further diagonalization.

That these models can be diagonalized in the above manner is highly significant: the cost of computing efficient portfolios can be made to depend directly on the size of the input i.e. $NM$ rather than the a priori quantity $N^2$. See e.g. [4].

## III. Conclusion

This paper has shown (1) how the scenario model can be extended to yield more meaningful estimates of covariance amongst security returns; and (2) how the well-known computational advantages of the multifactor model can also be realized for a scenario, or mixed scenario and factor, model. These developments should greatly enhance the practicability of the scenario approach for large scale portfolio analysis.

**REFERENCES**

1. K. J. Cohen and J. A. Pogue. "An empirical evaluation of alternative portfolio selection models." *Journal of Business* 40 (1967), 166–93.
2. R. J. Hobman. "Setting investment policy in an ERISA environment." *Journal of Portfolio Management* (Fall 1975), pp. 17–21.
3. H. M. Markowitz. *Portfolio Selection, Efficient Diversification of Investments.* Cowles Foundation Monograph 16, Yale University Press, 1959.
4. H. M. Markowitz and A. F. Perold. "Sparsity and piecewise linearity in large portfolio optimization problems." in *Sparse Matrices and Their Uses*, I.S. Duff (ed), Academic Press, 1981.
5. B. Rosenberg. "Extra-market components of covariance in security returns." *Journal of Financial and Quantitative Analysis* 9 (1974), 263–74.
6. W. F. Sharpe. *Portfolio Theory and Capital Markets.* New York: McGraw-Hill, 1970.

# SPARSITY AND PIECEWISE LINEARITY IN LARGE PORTFOLIO OPTIMIZATION PROBLEMS

H.M. Markowitz

*(IBM T.J. Watson Research Center, Yorktown Heights, N.Y.)*

and

A.F. Perold

*(Graduate School of Business Administration, Harvard University, Boston, MA.)*

ABSTRACT

The general mean-variance portfolio selection problem is usually formulated as a parametric quadratic program that has a dense Hessian, viz. the covariance matrix of security returns. In applications with large numbers of securities, the covariances are estimated with the use of return generating models that require the estimation of far fewer coefficients. This paper discusses the exploitation of this structure using the technique of defining additional variables and constraints in order to transform the problem into one that is sparse. Then it is shown how piecewise linear transactions costs and upper and lower bounds can be implemented implicitly and in a unified way. This approach avoids the need to define separate variables for each piece of linearity, as is currently practised.

## 1. INTRODUCTION

The general portfolio optimization problem is the following: Given a universe of n securities (almost all of them risky) find the proportion of wealth to be invested in each so as to yield a portfolio with minimal risk for a given expected return. The problem was first posed as such by Markowitz (1959) and may be stated as the quadratic program

$$\begin{aligned} \text{minimize:}\quad & V = \underline{x}^T C \underline{x} \\ \text{subject to:}\quad & \underline{\mu}^T \underline{x} \geq E \\ & A\underline{x} = \underline{b} \\ & \underline{x} \geq \underline{0} \end{aligned} \tag{1}$$

where C is the n x n positive (semi) definite variance-covariance matrix of the security returns, $\underline{\mu}$ is the n-vector of expected returns, and the pair A,$\underline{b}$ constitutes a set of linear constraints. E and V are the mean and variance respectively of the portfolio's return.

The collection of all portfolios with minimal V for some E and maximal E for some V is called the efficient frontier (Figure 1). Each of these portfolios may also be characterized as being the solution of the quadratic program

$$\begin{aligned} \text{maximize:} \quad & \underline{\mu}^T \underline{x} - \lambda\, \underline{x}^T C \underline{x} \\ \text{subject to:} \quad & A\underline{x} = \underline{b};\ \underline{x} \geq \underline{0} \end{aligned} \tag{2}$$

for some $\lambda \geq 0$, where $\lambda$ can be interpreted as the investor's tradeoff between risk and return. Rather than specifying E (or V) in advance, the investor will either specify $\lambda$ and solve (2), or determine the whole frontier (solve (1) for each value of E) and then select a portfolio by directly comparing the mean-variance pairs. The latter is usually preferred in practice.

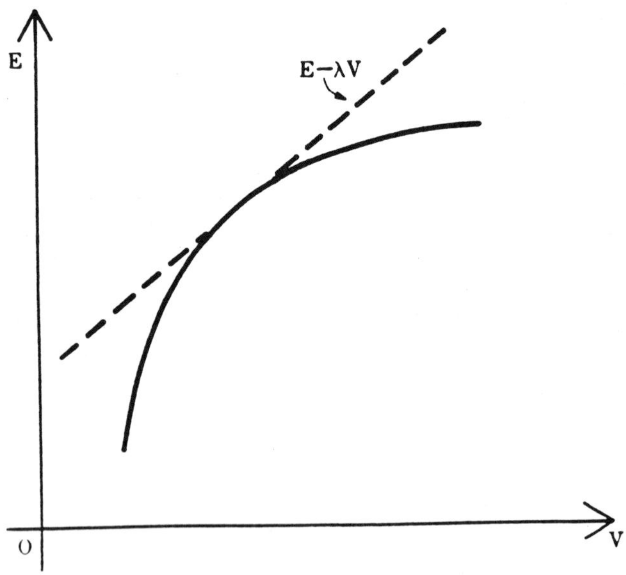

Often the investor already has a portfolio, say $\underline{z}$, and wishes to revise it because expectations of the future have changed. The optimization problem is the same as before except that due to transactions costs the expected return becomes

$$\underline{\mu}^T\underline{x} - \sum_{i=1}^{n} s_i(x_i - z_i)^- - \sum_{i=1}^{n} t_i(x_i - z_i)^+$$

where $s_i$ and $t_i$ respectively are the per unit costs of selling and purchasing stock in the $i^{th}$ security and where $(x)^+$ is the maximum of x and zero and $(x)^-$ is the maximum of -x and zero.

During the last decade, formal portfolio optimization of this kind has gained widespread acceptance by practising investment managers, primarily in the United States. This has created the need to develop techniques for large scale portfolio optimization, perhaps for as many as 1000 securities. Two problems are immediate: with 1000 securities, half a million covariances of returns have to be estimated. Even if this can be accomplished there is the question of economical computation and fast turn around.

To estimate the covariance matrix a number of simplifying models of security returns have been suggested that require the direct estimation of far fewer coefficients. While several optimization procedures exploiting this fact have been suggested, all either place further unnecessary restrictions on the class of problems that can be solved, or perform the calculations in a cumbersome manner.

In this paper we briefly review the models used to estimate covariances, and discuss the shortcomings of existing optimization methods. Then we show that probably the most effective method for solving these problems is an implementation of the original complementary pivot algorithm for quadratic programming due to Markowitz (1956) and Wolfe (1959), together with the idea of "sparsification", namely, reformulating the problem as one that is sparse by defining a few additional variables and constraints. This very simple approach has been used for single factor models (for example IBM (1969), Sharpe (1963)) and for multi-factor models (for example Sharpe (1970), but its greater generality and ease of implementation seem to have gone unnoticed. In the last part of the paper, we show how transactions costs can be easily implemented and extended to the more general convex piecewise linear case without the burdensome and currently practised procedure of splitting each variable into at least two others and so vastly increasing the size of the problem.

## 2. MODELS FOR THE ESTIMATION OF COVARIANCES

The most widely used model of security returns is the multi-factor model. This involves the selection of a few (say 10) common factors that impact the returns of all or a large group of securities simultaneously, and then making the assumption that any remaining variability in the return of a security is specific to that security only, and largely independent of events that impact other securities. The return on security i, $R_i$, may be written as

$$R_i = \mu_i + \beta_{i1}F_1 + \ldots + \beta_{i\ell}F_\ell + \varepsilon_i$$

for some constants $\mu_i$ and $\beta_{ik}$, and random variables $F_k$ and $\varepsilon_i$ all with mean zero, the $\varepsilon_i$ being uncorrelated with each other and the $F_k$. In this case the covariance matrix of the n securities may be written as

$$C = D + GPG^T \qquad (3)$$

where D is a diagonal matrix with $i^{th}$ diagonal entry $\sigma_i^2 = \text{var}(\varepsilon_i)$, G is the $n \times \ell$ matrix of coefficients $\beta_{ik}$, and P is the $\ell \times \ell$ covariance matrix of the factors $(F_k)$. Often the factors are chosen to be uncorrelated, so that P is diagonal. The widely used model of Rosenberg (1974) has P of the same form as C.

A very different approach is a states of the world model as used for example by Hobman (1975) and recently generalized by Markowitz and Perold (1980). Here it is assumed that there are $\ell$ possible future states of the world, the $s^{th}$ occurring with probability $p_s$, $s = 1,\ldots,\ell$. Further, the return on security i in state s, $R_{is}$, is given by

$$R_{is} = \mu_{is} + \varepsilon_i$$

where the $\mu_{is}$ are constants and the $\varepsilon_i$ are random variables satisfying $E(\varepsilon_i|s) = 0$, $E(\varepsilon_i^2|s) = \sigma_{is}^2$, and $E(\varepsilon_i\varepsilon_j|s) = 0$. The

G is the matrix of coefficients $(\mu_{is} - \sum_{t=1}^{\ell} p_t \mu_{it})$ and P is a diagonal matrix with $s^{th}$ diagonal entry $p_s$.

These are perhaps the two most important models of security returns, and it is the form of the covariance matrix in (3) that will be exploited in the remainder of the paper. It is worth noting that this structure is also present in all other models that have been used to date in practice.

## 3. EXISTING OPTIMIZATION METHODS

One of the most widely used methods in current day practice is Von Hohenbalken's algorithm (Von Hohenbalken (1975), Rosenburg and Rudd (1976), and Wilshire Associates (1979)), an iterative method that makes successive approximations of the constraint set using simplices of increasing dimension. It very quickly gives a good approximate solution but is extremely slow in attaining the optimum exactly. Points on the efficient frontier are determined by solving (2) for several (for example, five) discrete values of $\lambda$. A similar method, due to Sharpe (1978), has the advantage of being extremely well suited for use out-of-core, even for an arbitrary covariance matrix, but can handle only upper and lower bounds on the variables, and a budget constraint ($\Sigma x_i = 1$). All other constraints have to be introduced with the use of penalty functions (fixed penalties) and, once again, only one point at a time can be determined on the efficient frontier.

There is essentially only one algorithm for computing the whole efficient frontier, namely the complementary pivot algorithm (Markowitz (1956), Wolfe (1959)) that solves (2) parametrically for all values of $\lambda$. (See section 4 for further elaboration on this point). Existing implementations of this method that take advantage of the covariance matrix structure in (3) can handle only very special cases: the codes of Williamson and Downs (1970) and IBM (1969) require G to be of rank one and can treat only a budget constraint. Pang (1980) requires C to be positive definite[1], and can also not treat general linear constraints. None of these can handle transactions costs without doubling the problem size.

Lastly, we mention the work of Elton et al (1976, 1977a, 1977b) who proposed simple ranking devices that can be used almost by hand. Like those in IBM (1969) or Williamson and

---

[1] With riskless assets and slack variables, C will always have rank less than n.

Downs (1970), however, these apply only to the case when G has rank 1.

## 4. REVIEW OF THE COMPLEMENTARY PIVOT ALGORITHM

Following Markowitz (1959, p. 338), the necessary and sufficient conditions for $\underline{x}$ to solve (1), for fixed E, are that there exist multipliers $\underline{\pi}$, $\underline{\eta}$, $\rho$ which together with $\underline{x}$ satisfy

$$\begin{bmatrix} I & C & A^T \\ O & A & O \end{bmatrix} \begin{bmatrix} \underline{\eta} \\ \underline{x} \\ \underline{\pi} \end{bmatrix} = \begin{bmatrix} \underline{O} \\ \underline{b} \end{bmatrix} + \begin{bmatrix} \underline{\eta} \\ \underline{O} \end{bmatrix} \rho \tag{4}$$

$$\eta_i x_i = 0, \; i = 1,\ldots,n, \quad \rho.(E - \underline{\mu}^T \underline{x}) = 0 \tag{5}$$

$$\underline{x} \geq \underline{O}, \; \underline{\eta} \leq \underline{O}, \; \rho \geq 0. \tag{6}$$

It is shown in Markowitz (1959) that the efficient frontier can be found by varying $\rho$ over the interval $(0,\infty)$ with accompanying $\underline{x}(\rho)$, $\underline{\pi}(\rho)$, $\underline{\eta}(\rho)$ and $E(\rho)$ satisfying (4)-(6). Moreover, any solution to (4)-(6) for a given $\rho$ is the solution of (2) with $\lambda = 1/\rho$; E and V are monotonically increasing as functions of $\rho$.

A typical iteration of the algorithm begins with a given value of $\rho$, say $\rho^O$, and a nonsingular basis

$$B = \begin{bmatrix} I & C_{\alpha\beta} & A^T_{\cdot\alpha} \\ O & C_{\beta\beta} & A^T_{\cdot\beta} \\ O & A_{\cdot\beta} & O \end{bmatrix},$$

a submatrix of the coefficient matrix in (4). The index sets $\alpha$ and $\beta$ are disjoint and $\alpha \cup \beta = \{1,2,\ldots,n\}$. $\underline{x}$, $\underline{\pi}$ and $\rho$ are determined by

$$B \begin{bmatrix} \underline{\eta}_\alpha \\ \underline{x}_\beta \\ \underline{\pi} \end{bmatrix} = \begin{bmatrix} \underline{O} \\ \underline{O} \\ \underline{b} \end{bmatrix} + \begin{bmatrix} \underline{\mu}_\alpha \\ \underline{\mu}_\beta \\ \underline{O} \end{bmatrix} \rho^O \tag{7}$$

$$\underline{x}_\alpha = \underline{O}, \; \underline{\eta}_\beta = \underline{O},$$

and it is assumed that

$$\underline{x}_\beta \geq \underline{0},\ \underline{\eta}_\alpha \leq \underline{0},$$

i.e. relations (4)-(6) are satisfied.

The next step is to vary $\rho$, for example downwards if the procedure began with $\rho = +\infty$, while simultaneously adjusting $\underline{x}_\beta$, $\underline{\pi}$ and $\underline{\eta}_\alpha$ to satisfy (7). At some point, say $\rho = \rho^1 < \rho^0$, the further movement of $\rho$ will cause a component of $\underline{x}_\beta$ or $\underline{\eta}_\alpha$ to violate its sign restriction. If this component is $x_r$, $r\varepsilon\beta$, then $\eta_r$ is introduced into the basis in place of $x_r$, and the new basic solution is determined by the index sets

$$\beta_{new} = \beta_{old} \sim \{r\},\ \alpha_{new} = \alpha_{old} \cup \{r\}.$$

If it is $\eta_s$, $s\varepsilon\alpha$, then $x_s$ is introduced into the basis in place of $\eta_s$, the new basic solution then being determined by

$$\beta_{new} = \beta_{old} \cup \{s\},\ \alpha_{new} = \alpha_{old} \sim \{s\}.$$

It is shown in Markowitz (1959) that this procedure yields a nonsingular basis at each step, that $\underline{x}$ and $\underline{\eta}$ remain feasible, and that $\rho$ indeed varies over the full range $(0,\infty)$; the efficient frontier is given by the line segments between pairs of adjacent "corner" or basic portfolios. Note that this procedure may be used to <u>define</u> the efficient frontier, and therefore is the only algorithm to compute it, modulo getting started.

### 4.1 *Getting started*

There are essentially two methods for getting started. The most economical one is for the case occurring most often in practice, namely that $E=\underline{\mu}^T\underline{x}$ is bounded above on $\{\underline{x}: A\underline{x} = \underline{b},\ \underline{x} \geq \underline{0}\}$. Here it is best to solve the linear program

$$\text{maximize} \quad \underline{\mu}^T\underline{x} : A\underline{x} = \underline{b},\ \underline{x} \geq \underline{0}. \tag{8}$$

If (8) has a unique optimal solution (easily determined by inspection of the resulting vector of "reduced costs") this is also an efficient point, and a starting solution is immediately at hand for some sufficiently large $\rho$. Else, the minimum variance solution must be sought from among the optimal solutions to (8). This can be found using the same procedure as outlined above with the exception that in place of $\underline{\mu}$, an artificial $\bar{\underline{\mu}}$ is used for which the optimal solution obtained in (8) is efficient, and the variable corresponding to strictly positive reduced costs in (8) are deleted temporarily from the problem.

In the event that E is unbounded above, the more expensive procedure must be used of first determining the minimum variance solution with $\underline{\mu}^T \underline{x}$ unrestricted, using an artificial $\bar{\underline{\mu}}$ as indicated above, but now with all variables being considered. If this solution is unique (for example if C is positive definite) it is also efficient, and we have a starting solution with $\rho = 0$. If not, the starting solution must be sought from among those with minimum variance that yields the maximum E. This can be done by solving a linear program, again with certain variables temporarily omitted.

*4.2 Basis changes*

From the above description, it is clear that at each iteration, we require an explicit expression of the basic variables $(\underline{\eta}_\alpha, \underline{x}_\beta, \underline{\pi})$ in terms of $\rho$. By (7) we thus require two vectors $\underline{y}$ and $\underline{z}$ that satisfy

$$B\,\underline{y} = \begin{bmatrix} \underline{0} \\ \underline{0} \\ \underline{b} \end{bmatrix}, \quad B\,\underline{z} = \begin{bmatrix} \underline{\mu}_\alpha \\ \underline{\mu}_\beta \\ \underline{0} \end{bmatrix}, \tag{9}$$

for then

$$\begin{bmatrix} \underline{\eta}_\alpha \\ \underline{x}_\beta \\ \underline{\pi} \end{bmatrix} = \underline{y} + \underline{z}\rho. \tag{10}$$

From (10), the basis change can be determined by means of the usual minimum ratio test.

To perform the next iteration all that is required is the $\underline{y}$ and $\underline{z}$ satisfying (9) with respect to the new basis. Since the new basis differs from the old in a single column, these vectors can be updated by means of an elementary matrix as in the product form of the simplex method (Dantzig (1963)). This requires the solution of the set of equations

$$B\underline{w} = \underline{h} \tag{11}$$

where $\underline{h}$ is the column entering the basis (being the complement of the column leaving the basis). The new $\underline{y}$ and $\underline{z}$ are then given by

$$\underline{y}_{new} = W\underline{y}_{old}, \quad \underline{z}_{new} = W\underline{z}_{old}$$

where

$$W = I + \frac{1}{w_k}(\underline{w} - \underline{e}_k)\,\underline{e}_k^T,$$

k being such that the leaving column corresponds to the $k^{th}$ basic variable, and $\underline{e}_k$ being the $k^{th}$ unit vector.

We are left with the problem of finding a representation of B that allows (11) to be solved in an efficient and stable manner, and that can be easily updated from one iteration to the next.

## 5. EXPLOITING THE STRUCTURE OF C

We can now turn to the implementation of the above algorithm when the covariance matrix has the structure given in (3):

$$C = D + GPG^T.$$

To transform this into a sparse matrix problem notice that by defining new variables $\underline{u}$ and adding the constraints

$$G^T\underline{x} - \underline{u} = \underline{0},$$

the quadratic form in (1) becomes

$$\underline{x}^T C\underline{x} = \underline{x}^T D\underline{x} + \underline{u}^T P\underline{u}.$$

In the case of Rosenberg's model (Rosenberg (1974)), P, too, has the same form:

$$P = E + HFH^T$$

where E is diagonal and F is dense, symmetric positive definite. Proceeding as above, the quadratic form may now be written as

$$\underline{x}^T C\underline{x} = \underline{x}^T D\underline{x} + \underline{u}^T E\underline{u} + \underline{v}^T F\underline{v}$$

where in addition

$$H^T\underline{u} - \underline{v} = \underline{0}.$$

Problem (1) thus becomes

$$
\begin{aligned}
\text{minimize:}\quad & \underline{x}^T D\underline{x} + \underline{u}^T E\underline{u} + \underline{v}^T F\underline{v} \\
\text{subject to:}\quad & \underline{\mu}^T \underline{x} \ge E \\
& G^T \underline{x} - \underline{u} = \underline{0} \\
& H^T \underline{u} - \underline{v} = \underline{0} \\
& A\underline{x} \quad = \underline{b} \\
& \underline{x} \ge \underline{0},
\end{aligned}
\tag{12}
$$

where the first two terms in the minimand are weighted sums of squares. The resulting constraint matrix for the complementary pivot algorithm (cf (4)) is sparse, viz.

$$
\left[\begin{array}{cccc|ccc}
I & D & & & G & & A^T \\
 & & E & & -I & H & \\
 & & & F & & -I & \\
\hline
 & G^T & -I & & & & \\
 & & H^T & -I & & & \\
 & A & & & & &
\end{array}\right]
\tag{13}
$$

In Rosenberg's model G is n x 47, H is 47 x 9, and F is 9 x 9. Furthermore, G is sparse, and H and F are dense. In a states of the world model, or a multi-factor model with uncorrelated factors, G is typically n x 10 and dense, and P (=E) is 10 x 10 and diagonal (that is,no need for F and H). A typical applications run might have n=500, and the constraint matrix A consisting of one or two dense rows (for example the budget constraint). Sometimes sparse sector constraints are also imposed, perhaps as many as 30. Upper and lower bounds are treated implicitly, as will be discussed in section 6.

In short, we have transformed a dense tableau with n + m rows, m being the number of explicit constraints, into one that is sparse with n + m + $\ell$ rows, where $\ell$ is the number of added variables. $\ell$=56 in the case of the Rosenberg model, and will be the number of factors or states of the world when P is not further decomposable.

*5.1 Use with a sparse matrix package*

On a large computer, by far the easiest way to implement the reformulation (12) is to use the sparse matrix routines available as part of any large scale linear programming package. These commonly consist of an LU factorization routine, a routine to solve equations with respect to the LU factors, and a routine to update the LU factors after each change of basis.

Here, each basis has the form

$$B = \left[\begin{array}{cccc:ccc} I & & & & G_{\alpha .} & & A^T_{.\alpha} \\ & D_{\beta\beta} & & & G_{\beta .} & & A^T_{.\beta} \\ & & E & & -I & H & \\ & & & F & & -I & \\ \hdashline & G^T_{\beta .} & -I & & & & \\ & & H & -I & & & \\ & A_{.\beta} & & & & & \end{array}\right] \tag{14}$$

$$\underbrace{\qquad\qquad}_{\text{I}} \qquad \underbrace{\qquad\qquad\qquad\qquad\qquad\qquad}_{\text{II}}$$

Notice that the columns in section II of B always remain there. This is because both the dual variables on the equality constraints and the additional variables ($\underline{u},\underline{v}$) are unrestricted in sign, and can thus never "block". Any basis change is the addition of a column to $\beta$ and the deletion of a unit column, or vice versa.

Whenever the basis change is the case of a unit column replacing a $\beta$ column, the number of non-zeros in B decreases. This means that a factorization method placing as much weight as possible into U will be preferable, since during updating, the (denser) outgoing column in U will cause a net reduction in the total number of non-zeros in L and U. The LAØ5 routines of Reid (1975), for example, have this property.

Besides ease of implementation, the use of general sparse matrix routines like those in Reid (1975) allows one to take full advantage of any sparsity present in G and A. As mentioned earlier, these can range from being completely dense to being rather sparse, depending on the particular constraints and model of covariance. The drawback of such an approach is that it fails to take into account the symmetry present in a large part of B and the fact that much of B (non-zero wise) never changes. We show next how exploiting this can save substantially on in-core storage, but makes implementation more complex.

### 5.2 *Implementation out-of-core*

The need to consider out-of-core implementations has arisen primarily because of the greatly increased use of in-house mini-computers. The following proposal requires a single read-only file in secondary storage, and in-core storage requirements of order at most $(m + \ell + r)^2$ where r (usually very small) is to be defined below.

We assume without loss of generality that F is diagonal[1]. Consider now the solution of (11), viz. $B\underline{w}=\underline{h}$. The most natural way to proceed is to perform Gaussian elimination along the diagonal of B in (14) up to the point where all the $\underline{\eta}_\alpha$, $\underline{x}_\beta$, $\underline{u}$ and $\underline{v}$ variables have been eliminated, or have been moved forward in a principal rearrangement because of zero or unstable diagonal pivots. Let the number of such skipped pivots be r. This leaves a "remaining matrix" in the elimination scheme that has the form

$$V = \begin{bmatrix} NN^T & M \\ M^T & Z \end{bmatrix}$$

and is of order $(m + \ell + r)$, where Z is the diagonal matrix of rejected pivots (usually Z will be zero) and (M,N) is a partition of

$$\begin{bmatrix} G^T_{\beta .} & -I & 0 \\ 0 & H & -I \\ A_{\cdot \beta} & 0 & 0 \end{bmatrix}$$

with its columns suitably scaled. V will be treated in-core as a dense matrix.

Since, at each iteration, V may change in size up or down and is in addition modified by a matrix of rank one, we need to

---

[1] If F is not diagonal then diagonalize it beforehand and modify H appropriately.

choose a factorization for V that can be easily and stably updated, and yet is storage efficient. While the QR (Lawson and Hanson (1974)) and LU factorizations certainly qualify as far as updating is concerned, they require twice as much storage as a factorization exploiting the symmetry of V. A method that does accomplish the desired task is the symmetric indefinite factorization of Bunch and Parlett (1971):

$$V = L\tilde{D}L^T$$

where L is a permuted lower triangular matrix and $\tilde{D}$ has either 2 x 2 or 1 x 1 blocks on the diagonal, each 2 x 2 block having one positive and one negative eigenvalue. This can be updated from one iteration to the next, as indicated in Bunch and Kaufman (1977), with D requiring the work space of at most a five-diagonal matrix. In-core storage requirements for V would thus be limited to L and the five-diagonal matrix.

Note that the elimination prior to operating with V, and the back substitution involved thereafter can all be performed by having access to the diagonal entries of B and the columns of

$$\begin{bmatrix} G & -I & O \\ O & H & -I \\ A & O & O \end{bmatrix}$$

which can be stored on disk in packed form, say in a binary file, and then read repeatedly.

## 6. IMPLICIT TREATMENT OF UPPER AND LOWER BOUNDS AND TRANSACTIONS COSTS

As indicated in Section 1, the problem solved more commonly in practice is the portfolio revision problem where $\underline{x}$ is constrained to lie between upper and lower bounds, and a concave piecewise linear transactions cost is incurred for deviations of $\underline{x}$ from an existing portfolio. We show how both the bounds and transactions costs can be treated in a unified manner by working with the existing variables rather than by defining a separate variable for each piece of linearity.

In the following let the portfolio return be given by the separable function

$$f(\underline{x}) = \sum_{i=1}^{n} f_i(x_i)$$

where each $f_i(x_i)$ is concave and piecewise linear, and takes on the value $-\infty$ whenever $x_i$ lies outside its bounds. For example, suppose that we currently have 10% of our portfolio in security number 3, and are willing to hold as little as 5% and no more than 20% in this security. If the current expected return on the security is 9% and brokerage commissions are incurred at a flat rate of 1% for both buying and selling, then $f_3(x_3)$ is given by

$$\begin{aligned} f_3(x_3) &= -\infty && \text{if} \quad x_3 < .05 \\ &= .09x_3 - .01\,|x_3 - .1| && \text{if} \ .05 \le x_3 \le .20 \\ &= -\infty && \text{if} \quad x_3 > .20 \end{aligned}$$

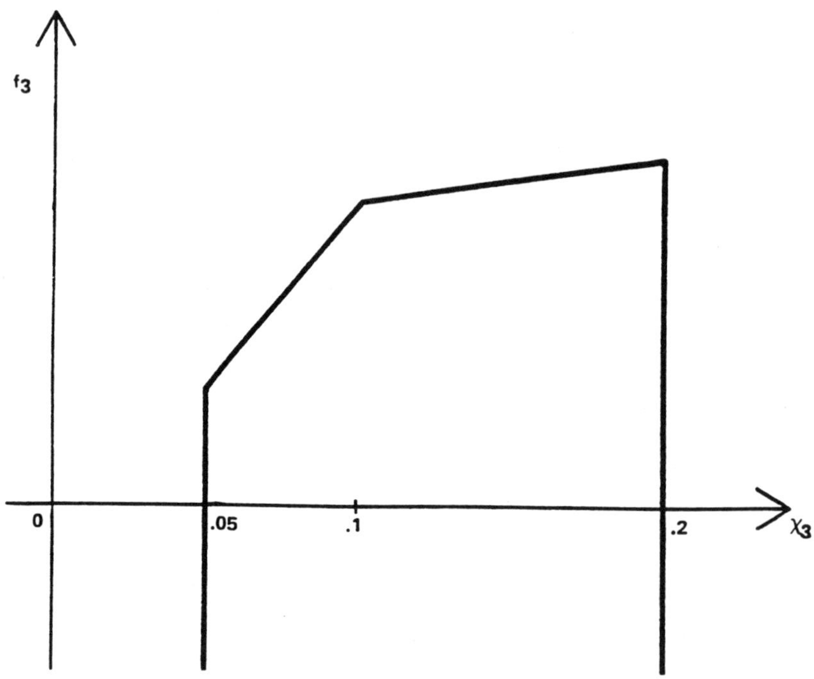


Figure 2

We allow the upper and lower bounds to take infinite values, or be equal.

The original problem (1) then becomes

$$\text{minimize} : \quad V = \underline{x}^T C \underline{x}$$
$$\text{subject to:} \quad f(\underline{x}) \geq E$$
$$A\underline{x} = \underline{b},$$

where now the inequality constraints on $\underline{x}$ have been taken into account by f. The optimality conditions for this problem are that there exist $\underline{x}$ and multipliers $\underline{\xi}$, $\underline{\pi}$, $\rho (\rho \geq 0)$ satisfying

$$C\underline{x} + A^T\underline{\pi} - \underline{\xi} = \underline{0}$$
$$A\underline{x} = \underline{b}$$

and

$$\underline{\xi} \in \rho . \underline{\partial f}(\underline{x}), \ \rho . (E - f(\underline{x})) = 0 \tag{15}$$

where the set $\underline{\partial f}(\underline{x})$ is the subgradient of f at $\underline{x}$. See for example, Rockafellar (1970).

It is easily checked that the earlier conditions (4)-(6) for the case of nonnegativity restrictions and no transactions costs are a special case of the above: For example, when $x_i > 0$, $\partial f_i(x_i) = \{\mu_i\}$ so that $\xi_i$ becomes $\rho\mu_i$. This is equivalent to having, in (4), a right hand side value equal to $\rho\mu_i$, and $\eta_i = 0$. When $x_i$ is at its lower bound, $\partial f(x_i) = [\mu_i, \infty)$ so that $\xi_i$ lies in the interval $[\rho\mu_i, \infty)$. This is equivalent to having a right hand side value equal to $\rho\mu_i$, and $\eta_i \leq 0$. Sargent's extension (Sargent (1978)) to include upper bounds can also be immediately derived from (15).

More generally, it can be shown that the basic solutions to (15) have the following characterization: Each $x_i$ is either out of the basis, $i \epsilon \alpha$, at one of the breakpoints defined by its piecewise linear return function $f_i$, or is in the basis, $i \epsilon \beta$, at some intermediate value. For each $i \epsilon \beta$, $\xi_i$ is out of the basis, and assumes the value $\rho\nu_i$ where $\nu_i$ is the slope of $f_i(x_i)$ on the segment currently containing $x_i$. For each $i \epsilon \alpha$, $\xi_i$ is in the basis and free to vary in the range $[\rho\nu'_i, \rho\nu''_i]$ where $\nu'_i$ and $\nu''_i$ are the slopes of $f_i$ to either side of the breakpoint $x_i$.

Implementation of the general case is surprisingly straightforward if the slopes and breakpoints are stored consecutively in a packed array, and a basis status vector is maintained, indicating at which breakpoint or in which segment each $x_i$ lies. The only other modification required is in the minimum ratio test (determining the leaving column) since the intervals of feasibility for $\xi_i$, namely $[\rho\nu'_i, \rho\nu''_i]$, are now functions of the parameter $\rho$. The change is simply to compute two ratios for $\xi_i$, one involving $\nu'_i$ and the other involving $\nu''_i$.

## 7. CONCLUSIONS AND SUMMARY

This paper has proposed efficient solution techniques for a class of portfolio optimization problems that are being solved routinely by practising investment managers. The first recommendation was to sparsify the covariance matrix by defining a few additional variables and constraints. Any profundity in this suggestion lies perhaps in the fact that it is extremely simple to carry out, and that it appears to be the most effective alternative. Then we showed how to treat bounds and piecewise linear transactions costs implicitly and in a unified manner, without the need to more than double the size of the problem.

Both suggestions have been implemented on the DEC 10 at the Harvard Business School and found to work extremely well. However, direct and meaningful comparisons with other portfolio optimizers have yet to be carried out.

## REFERENCES

Bunch, J.R. and Parlett, B.N. (1971) Direct methods for solving symmetric indefinite systems of linear equations, SIAM J. Numer. Anal., 8, pp. 639-655.

Bunch, J.R. and Kaufman, L. (1977) Indefinite quadratic programming, Technical Report No. 61, Dept. Computing Science, Bell Laboratories, Murray Hill, New Jersey.

Dantzig, G.B. (1963) Linear Programming and Extension, Princeton University Press.

Elton, E.J., Gruber, M.J. and Padberg, M.W. (1976) Simple criteria for optimal portfolio selection, J. Finance, 31, pp. 1341-1357.

Elton, E.J., Gruber, M.J. and Padberg, M.W. (1977a) Simple rules for optimal portfolio selection: the multi-group case, J. Financial and Quant. Anal., 12, pp. 329-344.

Elton, E.J., Gruber, M.J. and Padberg, M.W. (1977b) Simple criteria for optimal portfolio selection with upper bounds, Op. Res., 25, pp. 952-967.

Hobman, R.J. (1975) Setting investment policy in an ERISA environment, J. Portfolio Management, pp. 17-21.

IBM (1969) 1401 Portfolio Selection Program (1401-F1-04X) Program Reference Manual, IBM, New York.

Lawson, C.L. and Hanson, R.J. (1974) Solving Least Squares Problems, Prentice Hall.

Markowitz, H.M. (1956) The optimization of a quadratic function subject to linear constraints, Naval Res. Logist. Quart., 3, pp. 111-133.

Markowitz, H.M. (1959) Portfolio Selection, Efficient Diversification of Investments, Cowles Foundation Monograph 16, Yale University Press.

Markowitz, H.M. and Perold, A.F. (1980) Portfolio analysis with factors and scenarios, Working Paper HBS 80-47 Graduate School of Business Administration, Harvard University.

Pang, J.-S. (1980) A new and efficient algorithm for a class of portfolio selection problems, Ops. Res., 28, pp. 754-767.

Reid, J.K. (1975) Fortran subroutines for handling sparse linear programming bases, Report CSS 20, Computer Science and Systems Division, AERE Harwell, Oxon, England.

Rockafellar, R.T. (1970) Convex Analysis, Princeton University Press.

Rosenberg, B. (1974) Extra-market components of covariance in security returns, J. Financial and Quant. Anal., 9, pp. 263-274.

Rosenberg, B. and Rudd, A. (1976) Portfolio optimization algorithms: a progress report, Working Paper 42, Research Program in Finance, Graduate School of Business Administration, University of California, Berkeley.

Sargent, R.W.H. (1978) An efficient implementation of Lemke's algorithm and its extension to deal with upper and lower bounds, Math. Prog. Study, 7 pp. 36-54.

Sharpe, W.F. (1963) A simplified model for portfolio analysis, Management Sci., 9, pp. 277-293.

Sharpe, W.F. (1970) Portfolio Theory and Capital Markets, McGraw-Hill.

Sharpe, W.F. (1978) An algorithm for portfolio improvement, Research Paper No. 475, Graduate School of Business, Stanford University.

Von Hohenbalken, B. (1975) A finite algorithm to maximize certain pseudo-concave functions on polytopes, Math. Prog., 9, pp. 189-206.

Williamson, J.P. and Downs, D.H. (1970) Manuals for computer programs in finance and investments, Dartmouth College, Hanover, New Hampshire.

Wilshire Associates (1979) PRISM: portfolio review and investment system management, Wilshire Associates, Santa Monica, California.

Wolfe, P.S. (1959) The simplex method for quadratic programming, Econometrica, 27, pp. 382-298.

DISCUSSION

MR. R.E. BORLAND (NPL). Does not the actual use of your model by investment managers have an influence on the market itself?

PEROLD. Very little, if any, since there is so much subjectivity that goes into the estimation of the parameters. Besides, while larger and larger sums of money are being invested on the basis of mean-variance analysis, they do not yet constitute a fraction large enough to have any observable effects.

DR. A.M. ERISMAN (Boeing). One of your main considerations seemed to lie in developing techniques for use on small computers. With all the money that your investment managers have at their disposal why don't they buy or rent bigger machines?

PEROLD. The money is not their own, and it is very much in their interests to charge their clients as small a management fee as possible. Their clients always have the option of investing their money in a so-called "passive" portfolio, like a broadly diversified market index, at little or no cost. The investment managers thus have to outperform this passive portfolio by at least the management fee, in order to stay alive. Since large computers usually serve many different users, each optimization run incurs a substantial marginal cost. On a small computer the marginal cost is usually next to nothing, thus making it much easier to justify repeated sensitivity runs and the like.

DR. J.K. REID (Harwell). I would like to open up the discussion and ask what you currently consider to be the best techniques in linear programming?

PEROLD. Naturally such a broad question as that does not admit to a one-line answer. Many good techniques that will happily solve the "average" LP are part and parcel of a package like IBM's MPSX. However, there are problems like time-period (staircase) linear programs for which no truly satisfactory solution techniques have been found. Such problems are often numerically unstable and require disproportionately many simplex iterations. With respect to these, I would like to reiterate the comment I made after Duff's lecture regarding the benefits or otherwise of taking into account the structure of LP bases. It seems as if straight Markowitz does much better on these problems than do the bump and spike, and block factorization methods. On Dantzig's PILOT energy model, for example, Reid's LAØ5A code produced LU factors containing about 5000 nonzeros on a basis with initially 3500 nonzeros. The bump and spike codes in the MINOS and WHIZZARD packages, and certain block factorization techniques all produced factors with around 10,000 nonzeros. Moreover, these factorizations were all rather ill-conditioned, whereas that produced by LAØ5A was not.

DR. R.C. DANIEL (SCICON). We have found big gains in using supersparsity where we take advantage of similar valued elements in the tableau. Also, we find ourselves requiring to work out-of-core for most realistic problems.

PEROLD. The need to use supersparsity and/or work out-of-core has certainly diminished with the greater availability and decreased cost of large amounts of in-core storage. For small machines and for very large LP's, however, the need will remain. I am always a little sceptical, though, of the value of extremely large LP models (say with hundreds of thousands of variables) since one invariably ends up aggregating the answer. One might be just as well off aggregating the model. Very large LP models arising say by discretizing a continuous time model are probably best solved by a specialized procedure.

DR. H.M. LIDDELL (Queen Mary College). Have you any experience with array processors? We have a feasibility study under way at Queen Mary College although we have not yet found any encouraging results.

PEROLD. My computing environment has never included array processors although workers in other areas may have more common access to them.

(UNKNOWN). The NAG library has only routines for small linear programs. MPSX is not available to me so where else would I obtain routines for the large sparse case?

PEROLD. It all depends on what you want to pay. The Harwell routine LAØ5, written by Reid, is very good but only handles the basis matrices. For users of CDC machines, the APEX system is excellent.

Entity-Relationship Approach to
Information Modeling and Analysis, P.P. Chen (ed.)
Elsevier Science Publishers B.V. (North-Holland)


# The ER and EAS Formalisms for System Modeling, and the EAS-E Language

H.M. Markowitz, A. Malhotra, and D.P. Pazel

IBM Thomas J. Watson Research Center,
P.O. Box 218, Yorktown Heights, N.Y. 10598

This paper reviews the relationships between the ER (Entity-Relationship) and the EAS (Entity-Attribute-and-Set) formalisms; describes the EAS-E implementation of the EAS view as far as we have developed it; considers where EAS-E could go from here, i.e., considers the natural long run goal of an EAS-based application development language; and draws implications of these observations for ER as well as EAS programming.

## 1. INTRODUCTION

EAS-E is an experimental application development system being developed at IBM's T. J. Watson Research Center. It is based on the Entity, Attribute and Set (EAS) formalism for system modeling. We were delighted to see in the call for papers that topics of interest for this conference include "relationships with...EAS-E." In this connection we would like to review the relationships between the ER and EAS views; describe the EAS-E implementation of the EAS view as far as we have developed it; consider where EAS-E could go from here, i.e., consider the natural long run goal of an EAS-based application development language; and draw implications of these observations for ER as well as EAS programming.

## 2. THE EAS WORLDVIEW

The Entity, Attribute and Set formalism for system description was introduced almost simultaneously by three simulation languages--CSL(1), GASP(7), and SIMSCRIPT(10)--in the early 1960s. For each of these languages an entity is some concrete or abstract "thing" represented by the simulation; an attribute is some property or characteristic of the entity; and a set is an ordered collection of entities. In SIMSCRIPT, each instance of a set has one owner entity as well as zero, one or more member entities. For example, the simulation of a factory might include a type of set whose name is QUEUE; machine 1 owns one instance of a QUEUE, machine 2 owns a second instance, etc. Zero, one or more jobs belong to a particular queue.

The elemental acts for changing the status of such a system are to CREATE or DESTROY an entity, FILE an entity into or REMOVE an entity from a set, or assign a value to an attribute. A simulated activity or event may involve many such elemental acts.

The Entity, Attribute and Set view of the simulation languages is closely related to the network view of the database languages, with record corresponding to entity, field corresponding to attribute, and set corresponding to set. In a DBTG database, as in a SIMSCRIPT simulation, sets have owners as well as members; if only a single set with a given name is allowed, it is said to be owned by the system as a whole.

There are minor differences between the EAS concepts as incorporated, e.g., in SIMSCRIPT II(8) and the network concept as incorporated in DBTG(3). For example, SIMSCRIPT allows an entity of a given type to both own and belong to a set with a given name. This indeed is the easiest way to represent a tree structure: if an entity type, say ORGANIZATIONAL_UNIT owns a set called SUBORGANIZATIONs whose members are themselves ORGANIZATIONAL_UNITS (which own SUBORGANIZATIONS, etc.) an organization tree is represented. On the other hand SIMSCRIPT does not offer, but could and surely should offer, the options of mandatory and automatic sets as used by DBTG.

A more important difference, though some would discount this as mere quibbling over choice of words, is the use of entity-attribute as primitives in the one language as compared to record-field in the other. In part the difference is philosophical, the one terminology emphasizing that which is represented as distinguished from how this representation is achieved. The difference also has to do with such "practical" matters as how information is stored. To illustrate, let us briefly consider certain details concerning the SIMSCRIPT and EAS-E implementations.

The entity types in a SIMSCRIPT program, and the "main storage", as distinguished from database, entities of EAS-E, are divided into two classes: (a) those types whose individuals may be created and destroyed individually during the course of an execution; and (b) those types whose individuals are created en masse usually at the beginning of an execution and usually exist throughout the execution (e.g. throughout a simulation run). The former are called "temporary" entities, the latter "permanent" entities (at least they are "permanent" during the course of the run).

In part for reasons of efficiency, the attributes of temporary entities are stored in records whereas those of permanent entities are stored in arrays. Thus all the attributes of one temporary entity are stored in successive words of memory; whereas all the values on one attribute of all instances of some type of permanent entity are stored together in an array. Consequently, information concerning a particular permanent entity is dispersed in many arrays rather than grouped together into one record for the entity.

Both permanent and temporary entities can own sets, both can belong to sets, both have attributes which can be modified or referenced. The programmer FILEs, REMOVEs and assigns values to attributes with the same source program statements in either case. It is up to the software behind the scenes to remember what is stored where, and how to update it.

It is sometimes argued that, even in the case of a permanent entity, though its attributes are scattered they constitute a "logical record". While such terminology may be helpful in some contexts, we feel it normally more descriptive to say that EAS-E represents entities of various types, and it is up to the implementation to decide which are stored in records, which in arrays, and which in perhaps other ways.

## 3. EAS and ER

Chen(2) describes how entities and their relationships can be described by a network, therefore an EAS, model. In particular, an attribute of one entity may "point" to another entity; hence may be used to represent a one-one or many-one relationship. Sets obviously show a one-many relationship between the owner entities and the member entities respectively. An m-n relationship between, e.g., courses and students requires an intermediate entity like an "enrollment". Enrollment belongs to a set owned by course and another owned by student and thereby connects the two. (In EAS-E, entities belonging to database sets automatically point to their owner entities; hence each enrollment would know its student and course without further definition or programming.) Usually--in fact, in every case we have dealt with in either simulation or database applications--the intermediate entity carries other useful information, such as the date and grade of the enrollment.

In considering the representation of entity, attribute and set structures by the ER formalism, one must keep in mind that sets are ordered collections. A set of backorders for a given item, for example, not only remembers who is waiting for the item but also remembers that Mr. Jones is first, Ms. Smith is second, and so on. This sequence information is most easily represented in ER terms by use of an "ordinal path", in the sense of Griffith (5), from first member to last member as well as a 1-n relationship between owner and members.

The ER and the EAS views are similar in that they allow their users to model a (real or imagined) world rather than require him to spell out how the computer is to represent this world. They each provide their users with a simple but general set of modeling concepts. The strength of each view is at the same time its weakness. The fact that the entity, attribute and set view requires the user to declare in advance, in effect, which are 1-1 or many-1 relationships versus 1-many and m-n relationships, allows an EAS implementation to lay out a program to execute more efficiently. But more trouble is usually involved in an EAS than an ER system when database definitions are changed; e.g., the aforementioned efficient program must be recompiled when it is decided that an employee can belong to more than one department, therefore a set must now be used where previously an attribute sufficed.

The ER and EAS views seem to us sufficiently similar that experience with the implementation and use of an EAS-based application development system should be at least suggestive for, and perhaps formally generalizable to, the ER view. Accordingly, in the next section we briefly describe and illustrate the EAS-E implementation of the EAS view for database management; thereafter we consider possible extensions of EAS-E to other areas; and finally consider the corresponding possibilities for the ER view.

## 4. THE EAS-E LANGUAGE

EAS-E (which stands for "entities, attributes, sets and events") includes a procedural language which is now operational, and "direct" or nonprocedural facilities currently under development. We begin by describing the procedural language.

EAS-E is an integrated language, as distinguished from the usual database language embedded in a conceptually unrelated host language. Other salient features of the EAS-E procedural language are its entity, attribute and set view, and its English-like syntax. It also includes a simplified interface with DMS/CMS.

Below we illustrate these features in terms of the first application system developed using EAS-E: a rewrite and extension of the Workload Information System of Thomas. J. Watson's Central Scientific Services (CSS). CSS consists of about 90 craftsmen who do glass work, electronics, etc., for Thomas. J. Watson's scientists and engineers. The old Workload Information System, written in PL/I and assembler, was difficult to modify or extend. In the first instance an EAS-E version was built to operate in parallel with the old version, reading the same weekly inputs and generating the same weekly outputs. The EAS-E based system duplicated the functions of the prior system with about one-fifth as much source code. The new system shows an even greater but difficult to quantify advantage over the old in terms of ease of modification and extension.

Interface With DMS/CMS

EAS-E provides a simplified interface with the Display Management System for CMS (DMS/CMS), an IBM program product that facilitates the specification and manipulation of display screens (6). Panels designed through DMS/CMS can be displayed by EAS-E programs via the DISPLAY statement. This consists of the word DISPLAY, the name of the panel to be displayed, perhaps followed by any or all of the following: a list of variables "given" from the calling program to the panel, a list of variables "yielded" back from the panel to the calling program, instructions as to cursor position, intensity of specified fields, special message at the bottom of the screen, and audible signal.

The DISPLAY statement in exhibit 1 appears in the EAS-E version of the CSS Workload Information System. The statement itself should be as clear as any brief verbal explanation we could add here. The meaning of the statement depends partly on a SUBSTITUTE statement, exhibit 2, which appears in the program PREAMBLE and thus is global to the program's various routines. Because of this SUBSTITUTE statement, whenever the word REPORT_SPEC_DATA appears in the source program, as it does twice in the DISPLAY statement of exhibit 1 and many times elsewhere in the program, a list of 17 variables is substituted in its place. In particular, the DISPLAY statement gives current values of these 17 variables and receives updated values as filled in by the user.

English-like Syntax

In general we have tried to design the EAS-E syntax to be English-like and self-documenting, in part to facilitate review, communication and maintenance. Examples of English-like EAS-E coding are given in the preceding and the following sections illustrating certain aspects of the language. Here we will discuss the PRINT and FIND statements as further examples of EAS-E's English-like syntax.

Exhibit 3 is a routine from the CSS system whose function is, as it says, to start a new page and print a heading on it. This routine is called in the printing of several different reports. The START NEW PAGE statement is self-explanatory. The seven lines (including the blank seventh line) following the PRINT 7 LINES... statement are printed during run-time just as they appear in the source program, except that the first variable, PAGE.V (automatically maintained by EAS-E) is printed in place of the first grouping of ***s, the second variable TITLE is printed in place of the second groupings of *s, etc. Though the routine is without comment, its action should be clear to anyone.

```
Exhibit 1:
DISPLAY FORM GIVEN REPORT_SPEC_DATA YIELDING REPORT_SPEC_DATA
 WITH DATA FIELDS 1,2,3 AND 4 BRIGHT, WITH SIGNAL,
 WITH CURSOR AT DATA FIELD 1, AND WITH COMMENT =
 |AT MOST ONE ON THIS LINE CAN BE SPECIFIED. REENTER OR PRESS PF12 TO QUIT.|
```

```
Exhibit 2:
SUBSTITUTE THESE 4 LINES FOR REPORT_SPEC_DATA
REP_DEPT_XX, REP_PROJ_XX, REP_JOB_XX, REP_WORKER_XX, REP_DETAIL_LINE_1,
REP_DETAIL_LINE_2, REP_DETAIL_LINE_3, REP_DETAIL_LINE_4, REP_ACTIVE_ONLY_XX
REP_COMPL_ONLY_XX, REP_BOTH_XX, REP_CUSTOMERS,
REP_HEADING, REP_ENTRY_FROM, REP_ENTRY_TO, REP_COMPL_FROM, REP_COMPL_TO
```

```
Exhibit 3:
ROUTINE TO START_NEW_PAGE_AND_PRINT_HEADING
START NEW PAGE
PRINT 7 LINES WITH PAGE.V, TITLE, SUBTITLE, DATE, AND WEEK THUS...
CSS Information System                                     Page ***
                     Central Scientific Services
      **************************************************
*****************************            ********** **
CSS CSS  DEPT  CHARGE ENTRY COMPL ESTIMT PRCDNG TOTAL TIME CUSTOMER NAME
AREA JOB  NUM   TO  DAY  DAY    WEEK  TIME RMNING /AREA RESP.

RETURN   END
```

```
Exhibit 4:
PRINT 3 LINES WITH TASK_AREA_NUMBER, JOB_NUMBER, JOB_DEPT_NUMBER,
JOB_PROJ_NUMBER, TASK_ENTRY_DATE, TASK_COMPL_DATE, TASK_ESTIMATED_HOURS,
INHOUSE_HOURS_FOR_WEEK, TASK_INHOUSE_HOURS+TASK_VENDOR_HOURS,
TASK_ESTIMATED_HOURS - TASK_INHOUSE_HOURS - TASK_VENDOR_HOURS,
JOB_CUSTOMER, JOB_DESCRIPTION, TASK_EST_VENDOR_HOURS,
VENDOR_HOURS_FOR_WEEK, TASK_VENDOR_HOURS, TASK_ASSIGNEE_NAME, AND STAR THUS
*** ***** ****  **** *** *** ****** ****** ****** ****** *************
********************************** ****** ****** ******    ************* *
---------------------------------------------------------------------------
```

```
Exhibit 5:
FIND THE JOB IN CSS_JOBS WITH JOB_NUMBER = PROPOSED_JOB_NUMBER
  IF ONE IS FOUND...
     CALL REJECT (|JOB ALREADY EXISTS WITH SPECIFIED JOB NUMBER.|)
     RETURN
  ELSE...
```

```
Exhibit 6:
FOR EVERY GROUP IN CSS_GROUPS, FOR EVERY AREA IN GROUP_AREAS
          WITH AREA_TASKS NOT EMPTY AND AREA_TYPE ¬= VENDOR_TYPE, DO THIS...
   LET SUBTITLE = AREA_NAME
   CALL START_NEW_PAGE_AND_PRINT_HEADING
   FOR EACH TASK IN AREA_TASKS WITH TASK_COMPL_DATE=0, CALL PRINT_TASK_LINES
REPEAT
```

The PRINT statement in exhibit 4, slightly simplified from its CSS version, prints 3 lines containing the specified variables and expressions in the places indicated in the three form lines (last 3 lines of the exhibit). These lines will print data under the headings of exhibit 3. When controlled in a manner illustrated later, the print statements in exhibits 3 and 4 precisely duplicate reports from the old CSS system, but with a shorter and much more self-descriptive source program.

The example of a FIND statement reproduced in exhibit 5, including the IF ONE IS (NOT) FOUND statement that usually follows a FIND, finds the job in the set of CSS_JOBS with its job number as specified. The statement should be self-explanatory to anyone who understands the notion of a set as various database and simulation languages use the term.

Integrated Language

Exhibit 6 illustrates that a substantial amount of EAS-E's power could not be achieved by imbedding a database language in an unrelated host language. As background for this example we mention that tasks to be performed by CSS are divided into areas. These areas, in turn, are divided into groups. Thus the "Device and Mask Generation" group includes the "Electrochemistry" and "Mask Gen Lab" areas.

The first phrase of the exhibit (FOR EVERY GROUP IN CSS_GROUPS) instructs the compiled EAS-E program to have the reference variable GROUP point in turn to each member of the set called CSS_GROUPS. For each such GROUP, the next two phrases ("FOR... WITH...") instruct the executing program to look at each area in the group's areas that meet certain conditions. For each GROUP and AREA thus generated, the program sets the variable SUBTITLE used in the START_NEW_PAGE_PRINT_HEADING routine, and then calls the latter routine. Next, under the control of a FOR and a WITH phrase, it calls on a subroutine which includes the PRINT 3 LINES... statement of exhibit 4 for tasks in the AREA_TASKS set which have not yet been completed. These six lines thus select groups, areas and tasks for which heading lines and body lines are printed for one of the CSS weekly reports.

Now let us consider the coding required to perform the same functions if a database language (DBL) is imbedded in a host language (like PL/I-IMS or COBOL-DBTG) where the host language knows how to loop, call and print whereas the DBL knows how to access the database. Specifically, consider the coding required to perform the function expressed in the phrase "FOR EVERY GROUP IN CSS_GROUPS". In either of the aforementioned host-DBL systems (the coding details differ but the sequence of actions is the same) the following must be coded:

> The DBL is called upon to "get first" GROUP in CSS_GROUPS; host coding is used to test a flag or error-code to see if no member was fetched because the set was empty; in the latter case host coding transfers out of the loop. If the set was not empty, after processing the member the host coding transfers back to a point which calls on the DBL to do a "get next" to obtain the next GROUP in CSS_GROUPS; the result of this operation must also be tested to see if the last-in-set has been processed.

Thus, several lines of host/DBL coding are required to perform the same function, and the function performed is much less clear to the reader of the coding than is the phrase "FOR EVERY GROUP IN CSS_GROUPS."

The FOR EVERY phrase is one example in which greater power (one short phrase serving in place of several statements) and readability can only be achieved by rejecting the customary division of labor between host language and database language. The single statement at the beginning of exhibit 6 (FOR EVERY GROUP IN CSS_GROUPS, ... ) fetches database entities, tests for conditions such as AREA_TYPE not equal to VENDOR_TYPE, and controls looping--thus combining traditionally separated host-vs.-DBL functions in a single statement.

In general, database and core (main storage) variables can appear anywhere they make logical sense in any EAS-E statement-- READ, WRITE, LET, IF, etc. To do otherwise is as restrictive and arbitrary as would be a rule saying that such statements could contain only local variables, not global variables, of a host (algorithmic) language.

Direct Query and Update Facilities

In addition to the procedural language, we have developed a direct browsing facility ("Browser") and plan further direct (nonprocedural) capabilities of a similar nature. Exhibits 7 through 11 illustrate how the user interacts with the database using Browser.

Browser displays individual entities of the database on a screen as in exhibit 7. The first line of the screen shows the type of the entity plus certain internal identification information which we will ignore here (see (9)). The region marked (2) in exhibit 7 shows attributes of the entity. We see, for example, that the JOB in the exhibit has JOB_NUMBER = 80150, JOB_DEPT_NUMBER = 461, etc. The line marked (3) tells us that the JOB owns a set called JOB_TASKS, that the set for this particular job has only one member, and that the set is ordered by job number. The line marked (4) indicates that the job belongs to a set called PROJ_JOBS.

You move through the database with the aid of the PF keys whose template is marked as shown in exhibit 8. To look at every task in the set JOB_TASKS, for example, place an X as shown at (3), then press PF8 labeled EVERY on the template. This brings the first (and in this case, only) task of the set as shown in Exhibit 9. (Pressing the appropriate arrow on the keyboard moves the cursor only to allowed input positions.)

```
(1)->  ENTITY TYPE: JOB                      SLOT 21        DASH 1

       ATTRIBUTE NAME:                           VALUE:
             JOB_NUMBER                        80150
(2)->        JOB_DEPT_NUMBER                   461
             JOB_PROJ_NUMBER                   4631
             JOB_CUSTOMER_NAME                 BRODSKY
             JOB_DESCRIPTION                   GLASS WORK


       OWNS:
(3)->  X  1  JOB_TASKS RANKED BY LO TASK_AREA_NUMBER


       BELONGS TO:
             CSS_JOBS of 8-1-1 RANKED BY LO JOB_NUMBER
(4)--------> PROJ_JOBS of 3-727-1 RANKED BY LO JOB_NUMBER

       LOOPING THROUGH CSS_JOBS
```

Exhibit 7

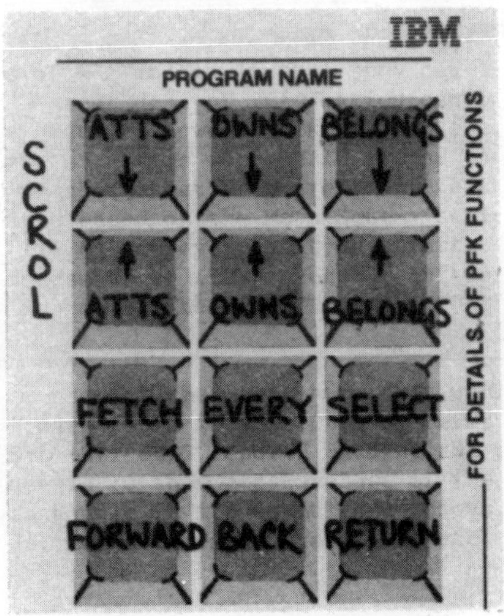


Exhibit 8

```
ENTITY TYPE: TASK                SLOT 141         DASH 1
ATTRIBUTE NAME:                        VALUE:
      TASK_AREA_NUMBER              113
      TASK_ASSIGNEEE_NAME           FISHER
      TASK_DEPT_NUMBER              461
      TASK_JOB_NUMBER               80150
      TASK_PROJ_NUMBER              4631
      TASK_ENTRY_DATE                     41
      TASK_COMPL_DATE                      0
      DATE_TASK_ENTRY_NOT                 41
      TASK_TYPE                            1
      TASK_REAPPROVED_FLAG                 1
                                                        MORE... ← (5)
OWNS:
   0  SUBTASKS FIFO



BELONGS TO:
      SUBTASKS OF 9-129-1
      PR.AR_TASKS OF 11-70-1 RANKED BY LO TASK_JOB_NUMBER
      AREA_TASKS OF 6-18-1 RANKED BY LO TASK_DEPT_NUMBER

LOOPING THROUGH JOB_TASKS
```

Exhibit 9

The "more..." message at (5) in exhibit 9 indicates that there are more attributes of task than can be displayed in the space provided. PF1 scrolls attribute information down; PF4 scrolls attribute information up. PF2,3,5,6 similarly scroll ownership and membership information up and down. PF10 is used to advance to the next entity in a set (or to the next in a selected subset); PF11 is used to move back to the preceding member of the set. If the end of the set is reached, pressing the "forward" button pops the user up to the owner of the set. You can also pop up to the owner by pressing "return", PF12.

If you put an X in front of a set to which an entity belongs, e.g., on the line labelled (4) in exhibit 7, and press "fetch", PF7, Browser brings the owner of the set (in this case, the project shown in exhibit 10). To look at a subset of the 12 jobs in the set PROJ_JOBS of the project in exhibit 10, place an X on the screen as at the line marked (6); then press "select", PF9. A screen like that in exhibit 11 will be presented, but without entries under the headings "relation" or "value". In the exhibit a user is asking to see all jobs in the set with JOB_NUMBER not equal to 80150 and with JOB_CUSTOMER_NAME = BRODSKY.

The browsing session is initiated when the user types BROWSE at his terminal. Browser asks the user to specify the name of a database. After this is entered the authorized user is presented a screen, like those in the exhibits, showing the attributes of, and the sets owned by, the database as a whole. He then proceeds to browse as illustrated above, using PF functions as described on the template in exhibit 8. If he presses "enter" instead of a PF key, he is given the same 12 options again, plus an extended list of options. These allow him, for example, to review the still open paths by which he got to his current screen. In addition, we plan to add the ability to update the database by overwriting the value of an attribute on the screen, and to generate a tabulation and/or summary of specified entities.

## 5. THE REST OF THE ICEBERG

Preceding sections have described EAS-E's current procedural and direct facilities for examining and manipulating database and main-storage entities. But these capabilities are but a small part of what should be encompassed in an integrated EAS application development system. In the present section we cite 3 instances of desirable extensions; in the next section we generalize these instances into an "EAS principle"; and in the final section we consider the corresponding generalization for ER systems.

(I) As noted before, EAS-E's entity, attribute and set view has been used for simulation for two decades, and in database management prior to EAS-E under the name "network" model. The EAS view is also used, in effect, in computer graphics (see, for example, (4)). It is common in the

```
ENTITY TYPE: PROJECT              SLOT 727          DASH 1

ATTRIBUTE NAME:                         VALUE:
      PROJ_NUMBER                   4631
      PROJ_NAME                     INTERFACE PHYSICS
      PROJ_MANGR_NAME               YOUNG
      PROJ_DEPT_NUMBER              461




OWNS:
   8  AREAS_USED_BY_PROJECT RANKED BY LO PR.AR_AREA
(6)→X 12  PROJ_JOBS RANKED BY LO JOB_NUMBER


BELONGS TO:
      LOC_PROJECTS of 10-1-1 RANKED BY LO NUMBER
      DEPT_PROJECTS of 2-35-1 RANKED BY LO PROJ_NUMBER
```

Exhibit 10

ENTITY TYPE: JOB

| ATTRIBUTE NAME: | MODE: | RELATION: | VALUE: |
|---|---|---|---|
| JOB_NUMBER | TEXT | NE | 80150 |
| JOB_PROJ_NUMBER | ALPHA | | |
| JOB_CUSTOMER_NAME | TEXT | = | BRODSKY |
| JOB_DESCRIPTION | TEXT | | |
| O.PROJ_JOBS | IDENTFR | | |
| O.CSS_JOBS | IDENTFR | | |

Exhibit 11

latter field to speak of geometric entities of types such as point, line, polygon, circle, plane and sphere. These entities have attributes such as the x and y coordinates of a point and the radius of a circle. Compound geometric entities are composed (e.g. by union or intersection, or variants thereof) from sets of primitive entities and/or other compound geometric entities. Sometimes these sets are called sets; sometimes they are called lists or some other synonym for set; sometimes an alternate structure, such as the binary tree, is used instead.

Clearly any general purpose programming system that aspires to build a wide variety of application systems should interface with some kind of graphics package. An application development system based on an EAS worldview would most naturally use an EAS view of the relationships among geometric objects. Furthermore, a programming system which aspires to be integrated in the sense already illustrated should allow the user to manipulate geometric entities in the same manner as he manipulates the database objects which he defines.

A graphics package for an EAS system, therefore, should include predefined entity types such as point, line, plane, sphere, etc. as well as compound objects made up of such primitive objects. The entity, attribute and set structure for such geometric objects should be made known to the user in the same manner as are other database objects. If the user knows that an entity of type circle has an attribute called radius he could, if required, create a circle, assign its radius attribute, and file it into the set of geometric objects of a picture using the same commands that he uses for database entities. Thus once the user knows the names of the geometric entity types, their attributes, the sets owned and the sets to which they belong, he has a substantial graphics capability and, for the most part, does not need to learn new commands beyond the already familiar CREATE, DESTROY, FILE, REMOVE, LET, FIND and FOR EACH.

A collection of the aforementioned capabilities--to create, destroy, file, etc.,-- for graphic entities is not sufficient in itself to constitute a convenient graphics package. Two other things are needed. First, certain complex higher order operations must be provided even though they can be synthesized from the elemental operations. One example is the clipping of lines wh[illegible] geometric world and its set of objects is associated with a window. The second requirement is for special input/output facilities to enter and display geometric objects. The latter capability is in part a matter of sensing the attributes of an I/O device such as the X- and Y- coordinates of a light-pen; and in part is a complex operation which may itself be described in terms of device sensing and other elemental EAS operations.

(II) A suitable graphics package for an integrated EAS language, then, should include predefined entity types with their attribute and set structures; the ability for the user to manipulate these geometric objects in the same way as he manipulates the objects which he has defined himself; commonly used complex operations which could be described in terms of elemental EAS operations, but would be a burden to do so for each particular application; and special input/output facilities. But such capabilities (with the possible exception of special I/O facilities) would be at least as valuable in another area.

The EAS-E programming system, like most programming systems is dependent upon an operating system. In general, users of operating systems frequently experience difficulties trying to get them to do their bidding. Some operating systems can be instructed in only a strange job control language. Large manuals are common for explaining available facilities, and job control or EXEC languages. Users report much frustration in trying to get the operating system to do things which it should, but no one around knows how to instruct it to do.

Much of this difficulty is quite unnecessary. Job control systems have quite ordinary EAS structures typically including jobs with one or more tasks to be done, each task owning a set of resource requirements. Resources include peripheral devices with queues of waiting work. Users may also own sets such as virtual readers and virtual printers (or whatever they are called in the particular system). The actions which frustrated users seek from their recalcitrant operating systems are the acts of any job shop: to schedule tasks, perhaps contingent on the outcomes of other tasks; to examine the contents of queues; to insert or remove tasks from various queues, and the like.

This could be done easily if the user was told the names of the various entities, attributes and sets of the particular operating system; was allowed to use the same procedural commands or direct facilities to create, destroy, file, remove, assign attribute values, and find entities as he can in an EAS-based database or simulation language, and the same event scheduling commands as he can in an EAS-based simulation language. The system itself is not particularly complex. The impression of great complexity comes from the fact that the user must use a different, and usually not very powerful language, to direct it. Worse, he often must use different "languages" with unrelated syntax rules to direct different parts of the operating system. (Here "operating system" is broadly defined as encompassing all the entities of the computer system with which the user must interact.)

There are elemental actions which the user could specify which the computer system cannot or should not fulfill. He may ask for information for which he is not authorized. Justifiably or otherwise he may ask to DESTROY THE (real) PRINTER. He may ask to take some elemental action when certain prerequisite conditions (describable in EAS terms) have not been met. In such cases the user should be instructed that he is not authorized to take such an action, or that particular preconditions are not met. At least the system should recognize the request, and responds with the action or an appropriate message.

As at present, frequently used combinations of elemental actions would be prepackaged and invocable by the user. Also, certain descriptive synonyms would be allowed. For example the user would probably prefer to say PRINT and specify a file (n.) rather than say FILE (v.t.) this file in the queue of the printer. It is not necessary to have a great number of such higher order actions involving rarely used cases, since these can be composed from typically few elemental actions when and if necessary.

Thus, as in the case of the graphic entities, the user would be told the attribute and set structure of the entities of the operating system and, where authorized, could manipulate these as if these were his own defined simulation or database entities. In addition he could schedule events, either immediately or with delay, unconditionally or conditionally, as easily as he can do such now in EAS based simulation.

(III) Next consider the entities of text processing. We may think of an "ordinary" file (n.), the kind we build with an editor, as being an entity containing a set of lines, each line containing one text attribute. The file itself has attributes like name and date_last_modified. The lines could be thought of as containing a set of characters, or alternatively it may be thought of as containing a set of words. The word entities are not stored in records or arrays, as with other main storage and database entities, but are stored implicitly like words in a book, by putting at least one blank after each successive word in the set. We may alternately think of the file (document) as owning a set of paragraphs, each paragraph owning a set of words. We may speak of the one type of entity (the paragraph) on one occasion, and the other type (the line) on another.

Typically we do not think in terms of creating individual entities such as characters and words and filing them into their sets. Rather we think of higher order operations or special input/output operations such as the actions of an editor. On the other hand there are frequent occasions when it would be desirable to find and perhaps modify the entity, attribute and set structures of documents using the same commands applicable to other entities. This is illustrated in exhibit 12.

In the first instance assume the routine in exhibit 13 deals with user defined entities including the entity type PROGRAM that owns a set of ROUTINES, each ROUTINE pointing to an entity called its SOURCE program which owns a set of WORDS. The entity WORD has an attribute TEXT which is the actual contents of the word. The program in exhibit 13 searches the set of words of each routine in the set of routines of the program, changing the contents of certain words. It also schedules a recompilation of any routine for which at least one word is changed.

```
Exhibit 12
FIND THE PROGRAM IN MY.PROGRAMS WITH NAME = GIVEN.NAME
IF NONE ... ELSE...
FOR EACH ROUTINE IN ROUTINES ''OF PROGRAM'' DO...
   FOR EACH WORD IN WORDS(SOURCE(ROUTINE))
                    WITH TEXT(WORD) = OLD.WORD, DO...
      LET TEXT(WORD) = NEW.WORD
      LET FLAG = 1
   LOOP
   IF FLAG = 1
      SCHEDULE RECOMPILE(ROUTINE) AT NEXT_NONPRIME_TIME
   REGARDLESS...
LOOP  END
```

Except for the SCHEDULE statement, as long as we assume that PROGRAM, ROUTINE and WORD are user defined entities exhibit 13 only contains concepts and commands already discussed. What we add now is the notion that PROGRAM, ROUTINE and WORD should be automatically defined entities. A user would usually manipulate these implicitly with some editor. The proposal is that the user should also be allowed to manipulate these entities as he would entities which he defined himself. The exhibit illustrates a circumstance in which it would be convenient to do so; e.g., when we wish to loop through sets of such entities selecting those which meet certain conditions, taking one or more actions on those which do. The scheduling of the recompile in the exhibit is an operating system function rather than a text editing function. The exhibit thus also illustrates a circumstance in which it is convenient to mix the two.

## 6. THE EAS PRINCIPLE.

The ultimate goal of an integrated application development system should be to permit the user to process all entities, attributes and sets in a uniform manner whether these are entities which he defines, or entities which have been defined for him such as graphical entities, text entities or the entities of the computer system itself with which he is obliged to interact. The application development system may offer some alternatives for dealing with entities depending on the tastes and needs of the user. For example, it may offer the user a compact rather than an English-like syntax, allowing him to type ∀ rather than FOR EACH, and ∃ rather than CREATE. The purpose of this option would be to accommodate users who do not type very fast. But it would not force the user to change his style of programming, change his keywords, change his syntax rules simply because he has passed from a database entity to a graphic entity to an entity of the operating system.

Problems are encountered when you consider in detail how to implement this general principle. For example:

(1) There are alternate possible EAS structures for implementing a given function; e.g., alternate possible details for the EAS structure of an operating system.

(2) The same word may be desirable as the name of an entity, attribute or set in different contexts; e.g., as a name used in the graphics package and/or in the operating system and/or as a user defined entity.

(3) It is not a trivial matter to implement a proper response for each elemental action which the user may give for any entity relevant to him.

But, as far as we have seen, the problems are all solvable. For example

(1) The fact that there are alternate EAS structures for an operating system does not contradict the assertion that we should tell the user the EAS structure adopted, and respond reasonably (with action or message) to any request for any elemental act on this structure.

(2) The proper interpretation of words that can be used in different contexts can be solved by having a set of current contexts, each context owning a dictionary of words, the set of contexts being ordered by which should be considered first and which next in interpreting a word. (We know, of course, how to arrange and rearrange this set, since we know how to do this for ANY set with which we must interact.) Where the context order is to be overriden for a specific word usage, a qualifier can be added to the word (e.g., G.POINT referring to POINT as used in the Graphics context rather than any other).

(3) The above EAS principle was not proposed because it is easy to implement, but because we believe that it will make application development one or two orders of magnitude easier.

## 7. APPLICABILITY TO ER

We have argued that an integrated EAS language should allow the user to examine and manipulate entities in a uniform manner, whether these are entities of types which he defines himself, or those which are defined for him. In this way the user is spared having to learn new keywords and syntax to do the same old actions to new things.

It seems to us that this same desideratum applies if entities are characterized by their 1-1, 1-m, m-1 and m-n relationships rather than in terms of their attributes and sets. If it were necessary (not just convenient at the user's option, but actually necessary) for the user to switch in and out of the ER view as he moves from one type of entity to another, or express himself in a different manner as he moves from type to type, he would be hindered by the burden of learning two or more modes of expression, and further burdened by the requirement to know when to switch from one to the other. But such seems to us as unnecessary for ER as for EAS. We suggest that "integrated ER" should be as much a focal point for you as integrated EAS is for us.

## 7. REFERENCES


1. Buxton, J.N., & Laski, J.G., Control and Simulation Language., The Computer Journal 5, 1962, pp. 194-199.

2. Chen, P.P-S, The Entity-Relationship Model -- Towards a Unified View of Data., ACM Trans. Database Systems Vol 1, No. 1, March 1976, pp. 9-36.

3. CODASYL Data Base Task Group Report, Available from ACM, New York, NY, April 1971.

4. Giloi, W.K., Interactive Computer Graphics., Prentice-Hall, NJ, 1978.

5. Griffith, R.L., & Harlan, V.G., Theory of IDEA Structures., TR 02.559, IBM Systems Development Division, San Jose, CA, April 1973.

6. IBM Corporation, Virtual Machine/370 Display Management System for CMS: Guide and Reference. Program Number 5748-XXB File No. 5370-39 SC24-5198-0.

7. Kiviat, P.J., GASP -- A General Activity Simulation Program., Applied Research Laboratory, U.S.Steel Corp, Monroeville, PA, July 1963.

8. Kiviat, P.J., Villanueva, R., & Markowitz, H.M., The SIMSCRIPT II Programming Language., Prentice Hall, Englewood Cliffs, NJ, 1969.

9. Malhotra, A, Markowitz, H.M., & Pazel, D.P., EAS-E: An Integrated Approach to Application Development., RC 8457, IBM T. J. Watson Research Center, Yorktown Hts., NY 10598, August 29 1980.

10. Markowitz, H.M., Hausner, B., & Karr, H.W., A Simulation Programming Language., The RAND Corporation RM-3310-PR 1962. Prentice-Hall NJ, 1963.

# EAS-E: An Integrated Approach to Application Development

A. MALHOTRA, H. M. MARKOWITZ, AND D. P. PAZEL
IBM Thomas J. Watson Research Center

*EAS-E* (pronounced EASY) is an experimental programming language integrated with a database management system now running on VM/370 at the IBM Thomas J. Watson Research Center. The EAS-E programming language is built around the entity, attribute, and set (*EAS*) view of application development. It provides a means for translating operations on EAS structures directly into executable code. EAS-E commands have an English-like syntax, and thus EAS-E programs are easy to read and understand. EAS-E programs are also more compact than equivalent programs in other database languages.

The EAS-E database management system allows many users simultaneous access to the database. It supports locking and deadlock detection and is capable of efficiently supporting network databases of various sizes including very large databases, consisting of several millions of entities stored on multiple DASD extents. Also available is a nonprocedural facility that allows a user to browse and update the database without writing programs.

Categories and Subject Descriptors: H.2.3 [**Database Management**]: Languages—*data description languages* (*DDL*), *data manipulation languages* (*DML*); H.2.4 [**Database Management**]: Systems; D.2.6 [**Software Engineering**]: Programming Environments

General Terms: Languages

Additional Keywords and Phrases: Entity relationship model

## 1. INTRODUCTION

*EAS-E* (pronounced EASY) is an experimental programming language integrated with a database management system now running on VM/370 at the IBM Thomas J. Watson Research Center. The programming language is in fact a manifestation of the entity, attribute, and set (*EAS*) philosophy of application development. In this, a conceptual framework in terms of entities, attributes, and sets is used to analyze the application or system to be developed. The entities, attributes, and sets describe the status of the application at a given moment in time. Points in time at which the status changes are called events. (EAS-E is an acronym from Entities, Attributes, Sets, and Events.)

The EAS-E programming language provides commands for manipulating EAS structures. Thus, the power and convenience of the language is based on the

Authors' address: IBM Thomas J. Watson Research Center, P.O. Box 218, Yorktown Heights, NY 10598.

power and convenience of the EAS formalism. We shall see later that real-world relationships can be modeled easily and directly in terms of EAS structures. Changes to these structures can be expressed immediately in terms of EAS-E commands. Thus, the programming language is directly tied into a philosophy for modeling the application domain. The entity, attribute, and set model is discussed in Section 2. The EAS-E programming language is discussed in Section 3.

EAS-E commands have an English-like syntax. This, combined with its direct relation to the EAS model, makes EAS-E programs readable and understandable. Counterintuitively, EAS-E programs are also considerably shorter than equivalent programs in other database languages. This is demonstrated in Sections 3.5 and 3.6.

The EAS status of an application is either changed by procedural programs written in the EAS-E programming language or by *BROWSE*, a nonprocedural facility for examining and updating a database. BROWSE is discussed in Section 4.

Finally, EAS-E programs are efficient. This is partly due to the efficiency of network databases and partly due to our implementation: for example, the manner in which we hold database ranked sets. This is discussed in Section 5.1. Other implementation topics are discussed in Sections 5.2 and 5.3.

The EAS-E database management system allows several users simultaneous access to the database. It supports locking and deadlock detection and is capable of efficiently supporting network databases of various sizes, including very large databases consisting of several millions of entities stored on multiple DASD extents. Section 6 discusses the EAS-E database management system.

## 2. THE EAS MODEL

*Webster's Unabridged Dictionary* (second edition) defines an *entity* as "that which has reality and distinctness of being either in fact or for thought . . . ." In an EAS-E representation, it is usually some "thing" (account, check, job) of the real world that is to be represented in the database. It may also be a less tangible thing, such as a task to be performed, which we find convenient to postulate as an entity to facilitate the description of the status of the application.

The *attributes* of an entity can be considered to be its properties or characteristics. At any instant in time an attribute has at most one value or it may be undefined.

In EAS-E, as in SIMSCRIPT [10] and DBTG [5], a set is an ordered collection of zero, one, or more entities which is owned by some entity. To illustrate, each account owns a set of current transactions, that is, checks and deposits which have occurred since the last monthly statement. Thus, the account for John Smith owns one such set, that for Mary Jones owns another such set, and so on. In general, we say that each account owns its current transaction set. Accounts may also own a set of old transactions, outstanding loans, etc.

An entity may have any number of attributes, own any number of sets, and belong to any number of sets. An entity can both own a set of a given name and belong to a set with the same name.

These general facilities allow more specialized structures, such as trees and networks, to be expressed more or less trivially within the general framework of entities, attributes, and sets. This is discussed in Section 3.1. A more detailed discussion of the power of the EAS formalism can be found in [14].

Thus, along with the EAS-E programming language and the EAS-E database management system, comes a methodology for application development. This states that the application design team should first decide on the Entities, Attributes, and Sets that are needed to describe the status of the system. These should be documented in some standard manner. Once this is done, events can be designed to alter the EAS status as necessary. Since the status is described completely in terms of the entities, attributes, and sets in existence at any point in time, events can change status in only a limited number of ways. They can

create or destroy entities,
assign or change attribute values,
file or remove entities from sets.

The design team can begin by writing the sequence of the above types of actions in English. Typically, a paragraph or two, per event, is sufficient and this translates readily into EAS-E code.

Commands for the above actions have an English-like syntax. For example,

```
CREATE A TASK
DESTROY THE TASK CALLED T2
LET PRIORITY(TASK) = 5
FILE THIS TASK IN JOB.TASKS
REMOVE THE FIRST TASK FROM JOB.TASKS
```

Thus, EAS-E programs, which are collections of such commands, can be read like English prose. Note that the commands must conform to a predefined syntax, they cannot be written in free English, but the syntax has been designed to make their intentions clear and unambiguous.

## 3. THE EAS-E PROGRAMMING LANGUAGE

This section discusses the EAS-E programming language. As mentioned before, the language is designed to facilitate the manipulation of EAS structures (i.e., it provides commands to create and destroy entities, assign attribute values, and file and remove entities from sets). Section 3.1 discusses how the set facilities of EAS-E allow a wide variety of data structures to be defined and manipulated by these commands.

EAS-E is also an integrated language, in that main storage and database entities can be manipulated by the same statements. This is discussed in Section 3.2.

In addition to the commands illustrated in Section 2 that operate on individual entities, EAS-E provides the FOR EACH phrase that can be used to operate on collections of entities: the members of a set or entities of a given type. The FOR EACH phrase is discussed in Section 3.3.1. Sections 3.3.2 and 3.3.3 discuss the FIND statement that can be used to find entities that meet given criteria and the DISPLAY statement that provides full-screen input-output capability.

The EAS-E procedural language contains other commands such as IF, GO TO, PRINT, CALL, and so on. These are not discussed here. The complete syntax is available in [12].

Section 3.4 discusses the facilities for defining entities, attributes, and sets in EAS-E. Sections 3.6 and 3.7 compare EAS-E programs with equivalent programs in PL-I/SQL [1] and PLAIN [24].

### 3.1 Sets as Standard Data Structures

In an experiment conducted by one of the authors [13], subjects were asked to design a program to be implemented in a higher level language that would retrieve information from three files in response to certain specified types of queries. Each of the eight subjects designed a different data structure to hold the information in main storage. If the designs were to be implemented, programs would need to be written to initialize each data structure and manage the pointers for all operations on it.

We argue that the set facilities in EAS-E provide a general purpose data structure that can be used by the programmer without the need to initialize the structures and maintain their pointers.

EAS-E provides main storage and database sets that may be ordered *FIFO* (first in, first out), *LIFO* (last in, first out), or ranked by one or more attribute values. Each set must have an owner and may contain zero or more entities of one or more types.

FILE and REMOVE commands can be used to operate upon these sets. The FIFO or LIFO ordering refers to the default FILE and REMOVE operations. In a FIFO set, the defaults are FILE LAST and REMOVE FIRST. In a LIFO set, they are FILE FIRST and REMOVE FIRST. The programmer can override these defaults by specifying FILE LAST or FILE FIRST or REMOVE LAST or REMOVE FIRST. He can also file before or after a specific member and remove a specific member.

In a ranked set, entities are filed in order of attribute value. Thus, there is no FILE FIRST, FILE LAST, FILE BEFORE, or FILE AFTER for a ranked set.

Data structures that the programmer may require are frequently included in these facilities or can be almost trivially defined in terms of EAS structures. For example, a stack is a LIFO set. To use a stack in a program the programmer has merely to use FILE and REMOVE statements on a set defined as LIFO. Similarly, pipelines and queues are FIFO sets and can be used by writing FILE and REMOVE statements. A tree structure can be created by defining an entity that may be called a NODE that owns a set called LINKS and also belongs to a set called LINKS. The tree is generated by creating NODES and filing them into LINKS of other NODES.

Attributes provide one-to-one and many-to-one relations. The value of an attribute can be considered to always be another entity, either a user defined entity or a system defined entity such as an integer or text string. For example, the DEPT attribute of EMPLOYEE points to a user defined entity of type DEPARTMENT, while the AGE attribute of EMPLOYEE points to an integer. In the same way, a logarithm can be considered to be an attribute of one real number entity that points to another real number entity

Sets provide for 1 to $n$ relations: the owner being related to the members. For example, the EMPLOYEE set of DEPARTMENT relates a DEPARTMENT to its EMPLOYEEs.

In EAS-E an entity can belong to only one set of a given name. To represent $m$ to $n$ relations we need to introduce an intermediate entity. For example, we may want STUDENTs to belong to many courses and COURSEs to own a set of students. Let the intermediate entity be called a REGISTRATION.CARD. Each STUDENT owns a set of REGISTRATION.CARDs and each REGISTRATION.CARD belongs to a set owned by a COURSE. This simple mechanism is further justified by the fact that the intermediary entity often turns out to have attributes in its own right, in this case, a GRADE attribute and perhaps a set called ATTENDANCE.RECORD.

### 3.2 Integrated Language

The original approach to creating a database system was to imbed special statements in an existing general purpose programming language (PL/I, COBOL, or C). The special statements were subroutine calls, as for IMS [7], or statements that were recognized and processed by a preprocessor as in System R [1] and INGRES [23].

This approach failed to provide a satisfactory programming environment mainly because the programming language could not operate directly on database objects. Values of their attributes had to be copied into main storage variables for manipulation and back to the database objects for updating. A better approach is to define database objects as datatypes in the programming language. To work with relational databases, several languages [17, 19, 22, 24, 25] defined relation and cursor (or other means of referring to a tuple) as datatypes. We call such languages *integrated*, as they contain commands to manipulate both main storage and database objects.

Another approach, exemplified by UDL [6] and DAPLEX [21], is to define a front-end to manipulate relational, hierarchical, and network databases in a uniform manner. These proposals imbed mechanisms for manipulating database objects in an existing programming language such as COBOL or PL/I. They define as datatype a means of referring to a database object (cursor, reference variable) and provide facilities for iterating over and selecting from collections of database objects. These proposals are integrated in spirit, but their convenience and usability will depend substantially on the kind of interface that is provided with the host language and the manner in which the general-purpose facilities of the host language mesh with the database manipulation facilities.

EAS-E is an operational integrated network language and could, in principle, be adapted to work with relational or hierarchical database languages. In this sense it resembles UDL and DAPLEX. We feel, however, that the database manipulation facilities and the main storage facilities of the language should stem from a consistent data modeling philosophy and have designed EAS-E to allow the programmer to manipulate database and main storage structures in an entirely equivalent manner.[1] This is illustrated in the following paragraphs. The

[1] PLAIN [25] recognized the need for main storage relations to store intermediate results. Its "markings" are, however, derived from relations in the database, and it does not seem possible to create markings from scratch in main storage.

EAS-E compiler and database custodian both use main storage entities, attributes, and sets extensively. Details of the EAS-E compiler are discussed in [16]. The custodian is discussed in Section 6.

The statement shown below prints the number and name attributes of all employees who belong to a set called workers, owned by a specific department, and whose salary is greater than a specified salary.

```
FOR EACH EMPLOYEE IN WORKERS(DEPARTMENT)
   WITH SALARY > SPECIFIED SALARY
PRINT 1 LINE WITH NUMBER AND NAME THUS ...
******     *********************
```

If EMPLOYEE and DEPARTMENT are main storage entity types, then the variables EMPLOYEE and DEPARTMENT are automatically defined to refer to main storage entities. The FOR EACH assigns the variable EMPLOYEE to point in turn to each of the employee entities that are members of the WORKERS set of DEPARTMENT and whose SALARY attribute has a value greater than the value of the main storage variable SPECIFIED.SALARY. The PRINT statement prints the NUMBER and NAME attributes of these employees on one line as specified by location of the asterisks.

If EMPLOYEE and DEPARTMENT are database, rather than main storage, entity types, then the variables EMPLOYEE and DEPARTMENT are automatically defined to refer to database entities. As we shall see later, if WORKERS is a set ranked by SALARY, the FOR EACH brings into main storage all the EMPLOYEE entities in the WORKERS set of DEPARTMENT whose SALARY attribute has a value greater than the value of the main storage variable SPECIFIED.SALARY. If WORKERS is defined to be a FIFO or LIFO set, or if it is ranked, but not by SALARY, then all the employee entities in WORKERS of DEPARTMENT are brought into main storage and their SALARY attributes are examined. The variable EMPLOYEE is assigned in turn to point to those employees who meet the specified condition on SALARY. The PRINT statement prints the values of the NAME and NUMBER attributes as before.

Thus, EAS-E treats database and main storage entities in an equivalent manner. Database and main storage entity identifiers and attribute names can appear wherever they make sense in EAS-E statements: IF, LET, READ, FOR EACH, CALL, and so forth.

### 3.3 Some Examples of EAS-E Syntax

Section 2 illustrated statements which took actions like create, destroy, and file on individual entities. In this section we discuss how EAS-E controls actions to be taken on selected entities of a type or selected members of a set. We begin with the discussion of the FOR EACH phrase which plays a central role in such control. Later we discuss two statements: the FIND and the DISPLAY.

3.3.1 *The FOR EACH Phrase.* Section 3.2 had an example of a FOR EACH phrase. We now discuss in detail its syntax and function. FOR EACH phrases are examples of control phrases because they control the execution of a statement such as a PRINT or a CREATE. A group of statements can be controlled by attaching the control phrase to a DO. The end of the controlled phrases is indicated by a LOOP or REPEAT.

Section 3.2 introduced the FOR EACH phrase for processing selected members of a set. In fact, the phrase is more general. FOR EACH phrases can also be used to process all entities of a given type. They can be nested together and can be modified by logical phrases.

The phrase that processes entities in a set has two basic forms:[2]

(a) FOR EACH variable OF set
(b1) FOR EACH variable FROM arithmetic expression OF set
(b2) FOR EACH variable AFTER arithmetic expression OF set

Form (a) assigns the named variable to point in turn to the entities that belong to the indicated set. If the set is empty, all of the statements controlled by the FOR statement are bypassed.

Form (b1) does the same task as form (a), except that it starts with the set member identified by the indicated expression. Form (b2) is similar to (b1), but starts with the set member that follows the identified member. If the identified member is not in the set, the program terminates with an error message.

Another form of the FOR EACH phrase can be used to process entities of a particular type. This has the syntax:

(1) FOR EACH entity
(2) FOR EACH entity CALLED variable

Form (1) assigns the variable with the same name as the entity type to point to the entities of that type in turn. Form (2) assigns the named variable to point to the entities of the type in turn.

The words EVERY and ALL can be used instead of EACH, and the words IN, ON, and AT used as synonyms for OF. To step backward through a set, the phrase IN REVERSE ORDER is placed after the set name.

FOR EACH phrases can be nested inside one another. For example, the following statements process all the EMPLOYEEs that belong to the WORKERS set of all the DEPARTMENTs.

```
FOR EACH DEPARTMENT,
  FOR EACH EMPLOYEE IN WORKERS, DO...
```

Logical control phrases can be used to select from among the members of a set or entities of a type. They can also be used to control the termination of the iteration. A logical control phrase contains a logical expression and a *logical*

[2] The notation used to specify the syntax of EAS-E is similar to that used for this purpose by other programming languages. The following are its basic components:
(1) Words in capitals are *keywords*.
(2) Words in lower case are *primitives*, such as "entity" and "set" or syntactic constructions defined elsewhere.
(3) Brackets [ ] and braces { } denote choices. When brackets appear a choice *may* be made. When braces appear a choice *must* be made. Items for selection appear in a vertical list within brackets or braces.

*control operator.* The logical control phrases are

$$\left\{\begin{matrix}\text{WITH}\\ \text{UNLESS}\\ \text{WHILE}\\ \text{UNTIL}\end{matrix}\right\} \quad \text{logical expression}$$

A WITH phrase modifies the sequence of values that pass from a FOR EACH to the statements that it controls. Its logical expression is tested each time a new candidate entity is generated by the FOR EACH, and if the expression is false the entity is skipped. This phrase is useful for screening values before they pass into the statements controlled by the FOR EACH.

WHEN can be used as a synonym for WITH. UNLESS and its synonym EXCEPT WHEN can be used to specify phrases such that the items passing the indicated test are rejected from the loop, rather than accepted.

A WHILE phrase limits the iteration of a FOR EACH statement. Its logical expression is reevaluated each time the FOR EACH generates a new entity. If the logical expression evaluates false, the loop ends. The effect of an UNTIL phrase is complementary to that of the WHILE phrase: The loop continues until the logical expression is false.

WITH, UNLESS, WHILE, and UNTIL phrases can be attached to nested FOR EACH phrases. When this is done, each WITH or UNLESS phrase applies to the FOR EACH statement immediately preceding it, and each WHILE or UNTIL phrase applies to *all preceding* FOR EACH statements. Sequences of WITH, UNLESS, WHILE, and UNTIL phrases can be attached to FOR phrases in any combination. More than one of each type of phrase is permitted.

To find five temporary workers in the production departments who can do welding (perhaps to make them permanent) we would write:

```
FOR EACH DEPARTMENT WITH TYPE(DEPARTMENT) = | PRODUCTION |
  FOR EACH EMPLOYEE IN WORKERS(DEPARTMENT)
    WITH TYPE(EMPLOYEE) = | TEMPORARY |
      AND SKILL(EMPLOYEE) = | WELDING |
    UNTIL NO.FOUND = 5 DO
    LIST NAME(EMPLOYEE) ADD 1 TO NO.FOUND LOOP
```

3.3.2 *The FIND Statement.* The FOR EACH phrase is used to process the entities in a collection that meet given criteria. In contrast, the FIND statement is used to locate one entity, namely the first entity that meets stated criteria.

The search may be carried out in two distinct ways: for entities of a given type or for entities in a given set. Somewhat different syntax applies to these two cases.

The FIND statement consists of the word FIND (or its synonym BRING) followed by one or more selection phrases (*s-phrases*) that specify the variables and the selection criteria. Successive *s*-phrases that follow the FIND or BRING must be separated by commas.

$$\text{FIND-}s\text{-phrase}\left[\left\{\begin{matrix}\text{AND}\\ ,\end{matrix}\right\} s\text{-phrase}\right]\ldots$$

Where any number of *s*-phrases, preceded by commas or "AND"s can be substituted for the ellipses (. . .).

*Entities of a Given Type.* The structure of the *s*-phrase to select from among entities of a given type is

$$\begin{bmatrix}\text{THE}\\ \text{A}\\ \text{AN}\end{bmatrix}\ \text{entity [CALLED variable]}\ \ldots$$

$$\text{[variable = ]}\ \begin{bmatrix}\text{THE}\\ \text{A}\\ \text{AN}\end{bmatrix}\ \text{entity}\ldots$$

This will locate the entity that meets the conditions given by the ellipses (. . .). The conditions can be any combination of logical phrases. The UNTIL and WHILE phrases can, optionally, be preceded by ONLY SEARCHING.

If the "CALLED variable" or "variable =" constructions are used, the named variable will be set to point to the found entity. If not, the variable with the same name as the entity type will be set to point to the found entity.

Consider, for example, a bank that has entities of type ACCOUNT. To find a particular ACCOUNT we may use

FIND THE ACCOUNT WITH ACCT.NO = | 12-345 |

where ACCT.NO is a attribute of ACCOUNT. To set a reference variable called ACCT equal to the found ACCOUNT, we may use

FIND ACCT = THE ACCOUNT WITH ACCT.NO = | 12-345 |
or FIND THE ACCOUNT CALLED ACCT WITH ACCT.NO = | 12-345 |

*Entities in a Set.* To search through entities in a set, the following *s*-phrase construction is used:

$$\text{[ variable = ]}\ \begin{bmatrix}\text{THE}\\ \text{A}\\ \text{AN}\end{bmatrix}\begin{bmatrix}\text{FIRST}\\ \text{LAST}\end{bmatrix}\ \text{variable}\ \begin{bmatrix}\text{FROM} & \ldots\\ \text{AFTER} & \ldots\\ \text{UP TO} & \ldots\\ \text{BEFORE} & \ldots\end{bmatrix}\ \text{IN set}\ldots$$

The FIRST or LAST indicates the direction in which the set is to be searched. The words FROM, AFTER, UP TO, and BEFORE should be followed by a variable or expression that evaluates to a member in a set and specifies where the search should begin and end. As before, any combination of logical phrases can complete the *s*-phrase and be used to specify the selection criteria. For example:

```
FIND THE DEPOSIT IN TRANSACTIONS(ACCOUNT)
  WITH AMOUNT = 100.00
```

*The Compound FIND Statement.* Consider a selection problem on a system that has MACHINE and JOB as entities. Every MACHINE owns a QUEUE of JOBs waiting for it. FREE and DUE.DATE are attributes of MACHINE and

JOB respectively. We want to find a MACHINE which has FREE > 0 and a JOB in its QUEUE with DUE.DATE < TIME.F.[3]

If we write two FIND statements:

```
FIND THE FIRST MACHINE WITH FREE > 0
FIND THE FIRST JOB IN QUEUE(MACHINE) WITH DUE.DATE < TIME.F
```

then the first MACHINE that has FREE > 0 may not have a JOB in its QUEUE with DUE.DATE < TIME.F and we may not get the combination we want. The following compound FIND statement will get the MACHINE that has an appropriate value of FREE and also a JOB in its QUEUE with DUE.DATE < TIME.F.

```
FIND THE FIRST MACHINE WITH FREE > 0 AND
  FIRST JOB IN QUEUE(MACHINE) WITH DUE.DATE < TIME.F
```

*Terminating the FIND Statement.* A pair of special IF statements can follow the FIND statement.

$$\text{IF} \begin{bmatrix} \text{ONE} \\ \text{A CASE} \end{bmatrix} \text{[IS] FOUND} \ldots \text{ELSE}$$

$$\text{IF NONE} \begin{bmatrix} \text{IS FOUND} \\ \text{FOUND} \end{bmatrix} \ldots \text{ELSE}$$

The statements represented by the ellipses are executed if the FIND statement found or failed to find a suitable candidate.

The power of the FIND statement is illustrated in the following program. It finds prime numbers from 1 to 10000 by testing whether a given number is divisible by any of a set of previously generated primes, only searching through primes whose square does not exceed the number in question.

```
PREAMBLE
NORMALLY MODE IS INTEGER
EVERY PRIME HAS A VALUE AND BELONGS TO SOME PRIMES
THE SYSTEM OWNS SOME PRIMES
END

FOR I = 1 TO 10000 DO
  FIND THE PRIME IN PRIMES WITH MOD.F(I,VALUE(PRIME)) = 0
    ONLY SEARCHING WHILE VALUE(PRIME) < = SQRT.F(I)
  IF NONE IS FOUND
    CREATE A PRIME
    LET VALUE(PRIME) = I FILE PRIME IN PRIMES
  REGARDLESS LOOP
END
```

3.3.3 *Full-Screen Input/Output Capability.* EAS-E provides full screen input/output facilities via an interface to the Display Management System for CMS (DMS/CMS). DMS/CMS is an IBM program product [8] that facilitates the specification and manipulation of display screens.

[3] TIME.F is a function in the EAS-E library that provides the current time.

In accordance with DMS/CMS convention we will use the word *screen* to mean what is visible on the face of the terminal at one time. We will use the *panel* to mean the display designed by the user, consisting of specifications regarding what is to be displayed.

DMS/CMS provides facilities for designing panels interactively. Essentially, a panel is specified by displaying an empty screen and locating the constant information (prompt fields) and the variable information (data fields) on it in the position it is to appear in the display. The start of each field is identified by a special character. For prompt fields, this initial character is followed by the text to be displayed. For data fields, the initial character is followed by as many nonblank characters as are required to fill out the field.

Panels designed through DMS/CMS can be displayed by EAS-E programs via the DISPLAY statement. This consists of the word DISPLAY followed by the panel name. In general, this is followed by giving arguments whose values are displayed in the data fields, and yielding arguments whose values are read from them.

The DISPLAY statement for displaying a panel and specifying, giving, and yielding arguments can begin in either of the following two forms:

DISPLAY panel [(argument list)] [YIELDING argument list]

$$\text{DISPLAY panel} \left[\left\{\begin{matrix}\text{GIVEN}\\ \text{GIVING}\\ \text{WITH}\end{matrix}\right\}\text{ argument list}\right] \text{[YIELDING argument list]}$$

this statement may, optionally, be followed by the word WITH, followed by one or more specification phrases that can take the following four forms:

$$\left\{\begin{matrix}\text{INPUT}\\ \text{DATA}\\ \text{TEXT}\\ \text{PROMPT}\end{matrix}\right\}\left\{\begin{matrix}\text{FIELD}\\ \text{FIELDS}\end{matrix}\right\}\text{ integer expression list }\left\{\begin{matrix}\text{BRIGHT}\\ \text{NORMAL}\\ \text{DARK}\end{matrix}\right\}$$

$$\text{CURSOR AT}\left[\begin{matrix}\text{INPUT}\\ \text{DATA}\end{matrix}\right]\text{[FIELD] integer expression}$$

COMMENT = text expression

SIGNAL

If more than one specification phrase follows the WITH, successive phrases must be separated by a comma or an AND. The optional specification phrases correspond to features of the screen display that can be specified differently for each call to the panel. The first type of phrase allows individual fields to be displayed at NORMAL or BRIGHT intensity or to be suppressed by the DARK option.

The second type of specification phrase allows the cursor to be positioned at a particular data field when the panel is displayed. The third type of specification phrase allows a line of text to be displayed as the last line on the screen when the panel is displayed. This is particularly useful when redisplaying panels that have erroneous or inappropriate data entered into them. The offending field is brightened and the comment is used to specify the nature of the error. The last type of the specification phrase sounds the audible alarm when the panel is displayed.

Some examples of DISPLAY statements are

```
DISPLAY OPTIONS WITH SIGNAL AND
  COMMENT = | PRESS APPROPRIATE PF KEY |
DISPLAY REQUIREMENTS YIELDING PRODUCT.NO AND QUANTITY
DISPLAY AREA.INFO(224) YIELDING PASSWORD, WITH INPUT FIELD 2 DARK,
  CURSOR AT FIELD 2 AND SIGNAL
DISPLAY FORM YIELDING NAME, ADDRESS, PHONE AND CUST.CLASS
  WITH TEXT FIELD CLASS BRIGHT AND TEXT FIELD 1 DARK
DISPLAY BOARD GIVEN MOVE YIELDING NEXT MOVE WITH SIGNAL
```

*The Global Variable KEY.V.* EAS-E provides a global variable KEY.V which assumes special significance when a DMS/CMS panel is displayed. In this situation,

$$\text{KEY.V} = \begin{matrix} n & \text{if} & \text{PF key}^4\ n \text{ is depressed} \\ 0 & \text{if} & \text{the ENTER key is depressed} \end{matrix}$$

Thus, this variable provides a convenient method for the user to communicate a selection from a menu to a program. The program displays a screen specifying menu options (to be indicated by pressing PF keys), and follows this by IF statements or computed GO TOs on the value of KEY.V. For example,

```
DISPLAY DINNER.MENU WITH SIGNAL AND
  DATA FIELD SPECIAL BRIGHT
IF KEY.V = 0 GO COFFEE.ONLY ELSE
GO TO OPTION(KEY.V)
```

In all other ways KEY.V can be used as any other global variable. Its value can be set and tested. It can be used to pass arguments to routines, and so forth.

*Error Handling.* DMS panels are often used to provide forms (or templates) in which a user fills in data. The data is then read by a program. If inappropriate data is entered into a field, such as the characters BROWN in an integer field, the EAS-E system flags an error. Data errors can also occur if numbers are too large to be converted to integers or have improper formats.

In case of data errors, the EAS-E system redisplays the panel with the offending data field highlighted and an appropriate error message displayed as a comment. The user is then allowed to change the data on the panel. All yielding variables are reread when the user is finished with the panel.

### 3.4 Entity, Attribute, and Set Definitions

Like all EAS-E statements, the entity, attribute, and set definitions attempt to be English-like and readable. The statements shown in Figure 1 define the entity types DEPARTMENT, STUDENT, and ENROLLMENT.

[4] PF keys refer to Program Function keys on the IBM 3270 family of terminals.

```
EVERY DEPARTMENT HAS A DEPT_NAME, A DABBR AND OWNS SOME STUDENTS
DEFINE DEPT_NAME AS A TEXT VARIABLE
DEFINE DABBR AS AN ALPHA VARIABLE

EVERY STUDENT HAS A ST_NAME, A ST_NUMBER, OWNS SOME ENROLLMENTS
  AND BELONGS TO SOME STUDENTS
DEFINE ST_NAME AND ST_NUMBER AS TEXT VARIABLES

EVERY ENROLLMENT HAS A COURSE_NUMBER, A GRADE AND A SIGN
EVERY ENROLLMENT BELONGS TO AN ENROLLMENTS
DEFINE COURSE_NUMBER AS A TEXT VARIABLE
DEFINE GRADE AS AN INTEGER VARIABLE
DEFINE SIGN AS AN ALPHA VARIABLE
```

Fig. 1. Some entity, attribute, and set definitions.

The definitions consist of attribute names for each entity type and the sets each entity type may own or belong to. Other EAS-E declarations can be used to locate attributes in specific parts of the entity record, give multiple names to attributes, and define sets without the full complement of pointers and operational routines. These declarations will not be discussed here.

The definitions shown in Figure 1 can be used to define main storage entity types or database entity types. If they are used to define main storage entity types, they will appear in the PREAMBLE of the program. If they are to be used to define database entity types, they must be packaged as a special file and sent to a definition processor that processes them and sends the processed definitions to the specified database. The database custodian stores the processed definitions and sets up the structures necessary to store entities of these types.

A program that wishes to work with database entity types, for example the entity types of Figure 1, from a database called SCHOOL, must include the following statement in its preamble.

```
DATABASE ENTITIES INCLUDE DEPARTMENT, STUDENT
  AND ENROLLMENT FROM SCHOOL
```

When the program is compiled, this statement causes the definitions of DEPARTMENT, STUDENT, and ENROLLMENT to be brought from the SCHOOL database and used to generate the executable code. When the program is executed, the presence of this statement causes links to be established with the database SCHOOL and the mechanisms for working with database entities to be included in the program.

Database definitions specify the layouts of the entity types. If the attributes and sets of some database entity types are to be changed, new definitions must be prepared for them, much like Figure 1, processed, and sent to the database. Once this is done, entities of these types can exist in the old layout, the new layout, and in a dual layout. The dual layout is in fact a combination of the old layout and the new layout. To convert entities from old to new layout, they are brought into main storage in dual layout: the old layout plus a blank new layout. Programs can now be written to assign values to the new layout and file it into sets. (The old and new attributes and sets are referred to by prefixing the attribute and set names by O_ and N_ respectively.) The old layout can then be destroyed. When all the entities of a type are converted, the old definition can be purged.

```
 1 PAYRAISE: PROC (XDEPT);
 2   $DCL(XEMPNO, XDEPT, XSAL, XRATING, XRAISE);
 3   $LET C1 BE
 4        SELECT  EMPNO, SAL, RATING
 5        INTO    $XEMPNO, $XSAL, $XRATING
 6        FROM    EMP
 7        WHERE   DEPT = $XDEPT;
 8   $OPEN C1;
 9   DO WHILE ('1' B);
10     $FETCH C1;
11     IF SYR_CODE ¬ = 0 THEN GO TO WRAPUP;
12     /* COMPUTE XRAISE BASED ON
13       XSAL AND XRATING */
14     $UPDATE EMP
15       SET SAL = SAL + $XRAISE
16       WHERE CURRENT OF C1;
17   END;
18 WRAPUP;
19 $CLOSE C1;
20 END PAYRAISE;
```

Fig. 2. An example PL/I program with SQL statements.

### 3.5 A PL-I/SQL Program and an Equivalent EAS-E Program

We now compare a PL/I program working against System R [1, 4] with an equivalent EAS-E program. The PL/I-SQL program shown in Figure 2 has 20 lines. The equivalent EAS-E program shown in Figure 3 has 6 lines.

System R supports relational data structures. Relations can be considered to be tables of entities, with each tuple (row) of a relation being a collection of attributes in a given order. The program shown in Figure 2 is taken from [4]. It updates the salary of some of the employees whose records are contained in a relation called EMP.[5] Each row of EMP refers to a particular employee and stores the EMPNO, DEPT, SAL, RATING and possibly other attributes. Statements that start with "$" are SQL statements.

The first line of the program in Figure 2 contains its name and specifies that it has one argument: XDEPT. The statement on line 2 declares the main storage variables XEMPNO, XDEPT, XSAL, XRATING, and XRAISE as SQL variables. Subsequently these variables, referred to as $XEMPNO, $XDEPT, etc., can be used in SQL statements.

The SQL statement on lines 3 to 7 defines a cursor called C1 and associates it with a set of tuples consisting of the EMPNO, SAL, and RATING attributes of the tuples in the EMP relation that meet the condition DEPT = $XDEPT. The $OPEN statement on line 8 places the cursor before the first tuple and binds the value of $XDEPT. The DO WHILE on line 8 starts a perpetual loop whose scope is terminated by the END statement on line 17. Within this loop the $FETCH delivers the next tuple in the set and copies its EMPNO, SAL, and RATING

[5] This program illustrates the general method of working with database relations via PL-I/SQL. In the more limited case where the new salary can be computed from the attributes of the employee and from PL/I variables by the operators add, subtract, multiply, and divide, the more compact SQL set update facility can be used. This requires about the same amount of coding as the EAS-E program in Figure 3.

```
ROUTINE PAYRAISE (XDEPT)
DEFINE XDEPT AS A REFERENCE VARIABLE
FOR EACH EMPLOYEE IN EMPLOYEES (XDEPT)
  LET SALARY = . . . RETURN
END
```

Fig. 3. An equivalent EAS-E program.

attributes into $XEMPNO, $XSAL, and $XRATING respectively. The raise is then computed and stored in $XRAISE. The $UPDATE statement on line 14 stores the new value of the SAL attribute in the tuple currently pointed to by C1.

Line 11 tests the return code from the $FETCH and jumps out of the loop when the set of tuples is exhausted. The $CLOSE statement deactivates C1 and the program ends.

The function performed by the program in Figure 2 can be performed by the single EAS-E statement:

```
FOR EACH EMPLOYEE WITH DEPT = XDEPT
  LET SALARY = . . .
```

This brings each employee in turn from the database into main storage and tests whether its DEPT attribute equals XDEPT. If this is so, it updates its salary as specified. The database modifications are made permanent when the program ends.

For compilation, the above statement must be preceded by

```
DATABASE ENTITIES INCLUDE EMPLOYEE FROM . . .
```

Bringing all the entities of a given type into main storage and processing those that meet given conditions is exactly equivalent to processing a relation in a relational database. In a network database it is possible to do better. If the selection criterion is one that is commonly used, such as employees within a department, then the EMPLOYEE entities should be filed into sets owned by the DEPARTMENT entities. With this structure, only the EMPLOYEEs in the particular department are brought into main storage.

Figure 3 shows an EAS-E subroutine that updates the salary of all employees in a given department. For compilation, the subroutine (Figure 3) would have to be preceded by

```
DATABASE ENTITIES INCLUDE EMPLOYEE AND DEPARTMENT FROM . . .
```

### 3.6 A PLAIN Program and an Equivalent EAS-E Program

Wasserman [25] includes a PLAIN program called honor_roll that produces a report of student names, identification numbers, and grade point averages for all students with a grade point average of 3.5 or higher, grouped by departmental major, for a specific set of departments. Figure 4 shows an EAS-E program that uses the database structure defined in Figure 1 to produce the same report.

Note that the EAS-E program is a great deal smaller. It is approximately one-third as long as the PLAIN program, and we believe is easier to read and comprehend. Some of the difference in length can be explained by the fact that the EAS-E program does not need to include the database definitions.

```
PREAMBLE
DATABASE ENTITIES INCLUDE DEPARTMENT, STUDENT AND ENROLLMENT
  FROM SCHOOL
DEFINE GPA AND SCORE AS REAL FUNCTIONS
END
FOR EACH DEPARTMENT
  WITH DABBR = "MATH" OR DABBR = "PHYS" OR DABBR = "CHEM"
    OR DABBR = "EECS" OR DABBR = "BIOL" OR DABBR = "HIST"
    OR DABBR = "ENGL" OR DABBR = "ECON"
  PRINT 2 LINES WITH DEPTH_NAME THUS ...
           HONOR STUDENTS IN THE DEPARTMENT OF *******************
GPA      NAME                                   NUMBER
FOR EACH STUDENT IN STUDENTS(DEPARTMENT), DO ...
LET G = GPA
IF G > = 3.5 PRINT 1 LINE WITH G, ST_NAME, AND ST_NUMBER THUS ...
  *.*      ************************************    *******
REGARDLESS LOOP
END
ROUTINE TO "COMPUTE" GPA
FOR EACH ENROLLMENT IN ENROLLMENTS(STUDENT) WITH "A" ≦ GRADE ≦ "F"
  COMPUTE M = MEAN OF SCORE
RETURN WITH M      END
ROUTINE TO "COMPUTE NUMERICAL" SCORE
IF GRADE = "A" LET SCORE = 4.0 ELSE
IF GRADE = "B" LET SCORE = 3.0 ELSE
IF GRADE = "C" LET SCORE = 2.0 ELSE
IF GRADE = "D" LET SCORE = 1.0 REGARDLESS
IF GRADE < 4.0 AND SIGN = "+" ADD .3 TO GRADE   RETURN WITH GRADE
ELSE IF GRADE > 0 AND SIGN = "−" SUBTRACT .3 FROM GRADE
REGARDLESS RETURN WITH GRADE
END
```

Fig. 4. An EAS-E program to produce on honor roll.

A real EAS-E program would either produce the honor roll for all departments or for a department whose abbreviations was read in. In both these cases the three line WITH statement following the FOR EACH DEPARTMENT would not be necessary.

GPA and SCORE are defined on line 4 in Figure 4 as functions that return REAL values. This concept is similar to the "derived relations" of DAPLEX [21]. Functions must be specified as executable routines and provide a means for a program, or set of programs, to tailor the model embodied in the database to their purposes. Unlike DAPLEX, they are not stored in the database and used to provide/enforce different views of the database.

The definition of the GPA function uses the COMPUTE statement. This statement can be used to compute the mean, standard deviation and other statistics of variables that change value within a loop. In EAS-E, comments are delimited by a pair of single quotes (") or the end of the line.

## 4. THE BROWSE FACILITY

The BROWSE facility allows a user to examine and update a database without programming. This facility is invoked from a terminal, and the user can navigate through the database by typing a few characters on the screen and pressing PF keys. The initial screen displays the names of all the entity types in the database. From this screen the user can choose to examine all the entities of a type in turn, examine selected entities of a type, or display a specific entity. An entity display consists of the datatypes and values of its attributes, the names and populations of the sets it owns, and the names and owners of the set it belongs to. When an entity is displayed on the screen, the user can continue to move along the path that brought him there or he can initiate another path by starting to move through all or selected entities of a set owned by the entity or displaying the owner of a set the entity belongs to or one that is the value of an IDENTIFIER attribute.

In addition to the above actions that allow him to examine the database, the user can make changes to the database. To update an attribute value the user overwrites the displayed value and pushes the update key. The same key, with the addition of brief commands typed on the screen, allows him to file and remove entities from sets.

The above actions can all be accomplished by pressing PF keys. If the user presses ENTER instead, he is presented with an extended menu that allows him to create and destroy entities and ask for information regarding the BROWSE commands. The extended menu also allows him to make changes permanent in the database or rescind provisional changes. It can also be used to display the state of BROWSE (i.e., the path by which the user got to the current display).

Thus, BROWSE allows all the basic EAS actions to be performed quickly and conveniently from an interactive display. We find it very useful in debugging and correcting the consequences of bugs, as well as a substitute for small EAS-E programs. It is planned to extend BROWSE to include nonprocedural specification of queries and reports. BROWSE is discussed in more detail in [14].

## 5. SOME IMPLEMENTATION DETAILS

This section discusses some of the details behind the implementation of EAS-E. The database ranked sets discussed in Section 5.1 make the language efficient, while the reference variables discussed in Section 5.2 and the automatic recording discussed in Section 5.3 make the language convenient to use. All of these mechanisms are transparent to the user.

Some aspects of the features discussed in this section are not currently operational. They are identified by footnotes and are in queue for implementation.

### 5.1 Database Ranked Sets

Database FIFO and LIFO sets are maintained by the owner entity pointing to the first (and perhaps the last) member, and each member pointing to its successor (and perhaps its predecessor). Finding a member whose attributes meet given criteria in such a set is very slow as, on the average, half the members in the set have to be brought into main storage before the target member is found.

If such a FIND needs to be done often, then the set should be ranked by the important attributes. Database ranked sets are organized in a special manner. The owner of each ranked set stores in a fixed place the offset to a linear structure that contains information about the set. This structure has some fixed information such as the number of members in the set, and one record for each of the members. The record contains the values of the ranking attributes and the database identification of the number.

To find a member with given values of ranking attributes, the owner is brought into main storage and its ranked set structure is located. A binary search can now be conducted amongst the records to find the member with the required attribute values.

If a database ranked set gets large, the ranked set structure of the owner gets too long to bring conveniently into main storage. To avoid this, the structure is split up and stored in two or more special database entities called subsets. The ranked set structure of the owner now stores the ranking attribute values of the first member record in each subset and the database identification of the subset. As the set gets still larger, this organization is extended into a balanced tree [2, 3] of subsets. The owner points to *a* first level of subsets, the subsets may point to other subsets, and the lowest level of subsets point to members. This is shown in Figure 5. With the attribute values of the members stored in balanced trees, it is necessary to bring only the owner and at most a few subsets into main storage to locate a member with specific ranking attribute values.

An important special case of database ranked sets are the sets owned by the entity that represents a database entity type and containing all the entities of that type as members.[6] These sets make it possible to find efficiently an entity with given attribute values from among entities of that type. If such a set is desired, the names of the ranking attributes have to be specified in the database definitions. It is then automatically updated whenever an entity of that type is created or destroyed or has the value of a ranking attribute changed.

## 5.2 Reference and Identifier Variables

To facilitate working with the database, EAS-E provides two special variable modes or datatypes: Identifier and Reference. Identifier variables store the database identification of entities. For example, if there were a database entity type called PERSON and each PERSON had an attribute called SPOUSE that pointed to *a* another PERSON, then SPOUSE would have to be an identifier attribute.

Reference variables provide a means of giving a main storage name to a database entity. They serve the same function as cursors in PLAIN [24] and UDL [6]. The concept is similar to the "currents" of DBTG [5] and the cursors of System R [1], but is less restrictive. Reference variables can be defined anywhere in the program and are not restricted to refer to entities of a particular type. Each reference variables has, however, an access mode: read.only or read.write, and this determines the manner in which the referenced entity is available in the program. A reference variable of the same name is implicitly defined for every entity type in the DATABASE ENTITIES INCLUDE statement. Thus, in the program of Figure 4, DEPARTMENT, STUDENT, and ENROLLMENT are automatically defined as reference variables with the default access mode: read.only.

[6] This feature is not implemented yet.

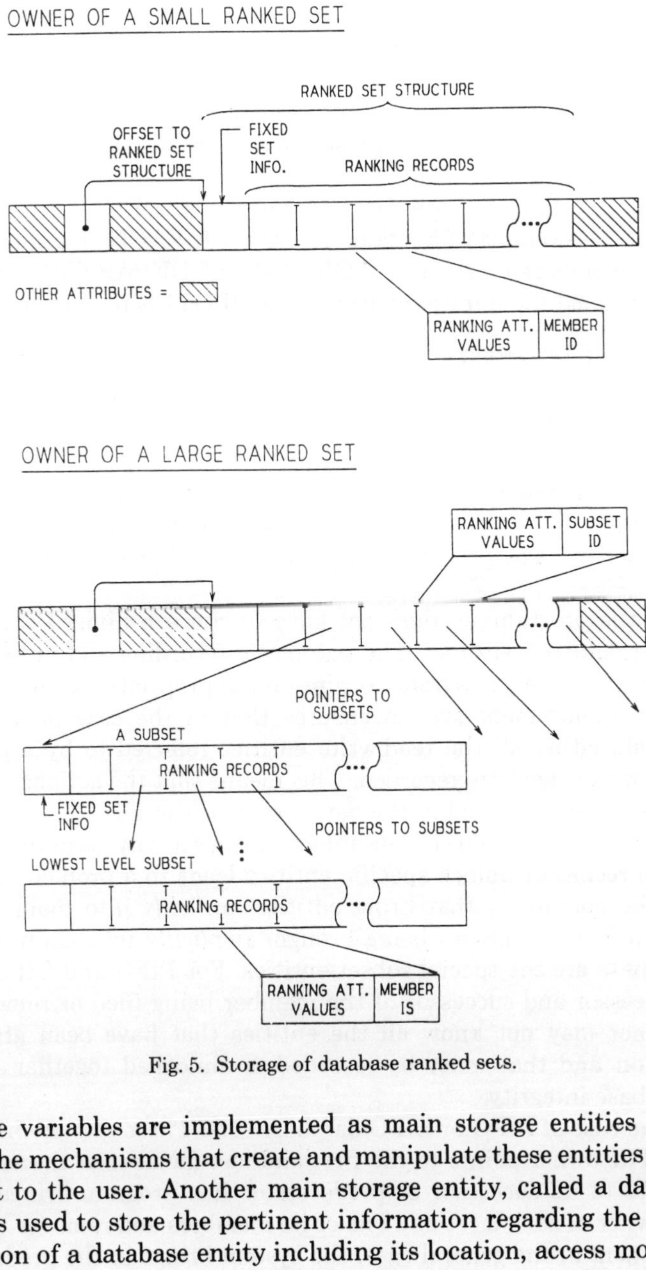


Fig. 5. Storage of database ranked sets.

Reference variables are implemented as main storage entities in the user's program. The mechanisms that create and manipulate these entities are, however, transparent to the user. Another main storage entity, called a database representative, is used to store the pertinent information regarding the main storage manifestation of a database entity including its location, access mode (read.only or read.write), and so on. The database representative is unique to a database entry, while any number of reference variables may refer to a database entity. Each reference variable that refers to a database entity points to its database representative and belongs to a set owned by the database representative.

Let REF1 and REF2 be reference variables. If REF1 refers to a particular database PERSON, then the statement

LET REF2 = SPOUSE(REF1)

would cause the database entity pointed to by SPOUSE(REF1) to be brought into main storage, a database representative created for it, and REF2 set to point to it and filed into a set owned by it. If SPOUSE(REF1) was already in main storage, then it would not be brought in, and REF2 would merely point to its database representative and belong to a set owned by it. The implementation of reference variables and database representatives are discussed in more detail in [16].

Entities are also brought into main storage by the FIND and FOR EACH statements. In the example of Section 3.2, the FOR EACH statement automatically brought members of the set WORKERS of DEPARTMENT into main storage and assigned the iteration variable EMPLOYEE to point to them.

### 5.3 Recording and Unlocking

A typical program that works against a database brings a few entities into main storage, examines their attributes and their sets, and perhaps modifies some of them. It may also create new entities and destroy some old ones. When the program ends, any new or modified entities, and the change in status of the destroyed entities, are automatically recorded in the database. After this, all the entities accessed by the program are unlocked and made available to other programs.

In general, the programmer does not have to concern himself with recording and unlocking database entities. The automatic recording and unlocking at the end of the program is transparent to him and appropriate for most situations. The database management system ensures that in the case of a software or hardware crash, either all the read.write entities referred to by a program are recorded or none of them are recorded. This means that the actions of a program are either completely reflected in the database or not at all.

The programmer can also record or unlock entities at any point in the program. The ability to record or unlock specific entities leads to a problem, however. In addition to the operations that bring entities *explicitly* into main storage, set operations can lead to entities being brought *implicitly* into main storage. For ranked sets these are the special subset entities. For FIFO and LIFO sets these are the predecessor and successor of the member being filed or removed. Thus, the programmer may not know all the entities that have been affected by a database action and that must be recorded or unlocked together in order to preserve database integrity.

Consider the case of FIFO or LIFO database sets. These sets are held together by pointers. The owner points to the first and last member of the set, and each member points to its successor or predecessor. If a program brings into main storage the owner of such a set and one of its members and removes the member from the set then, in the general case, the system brings in the predecessor and successor of the set and changes their pointers appropriately. If only the modified owner and erstwhile member are now recorded, the database could lose integrity if the program or system crashes without recording the successor and predecessor.

To prevent this from happening, EAS-E maintains main storage entities called *equivalence classes.*[7] Each equivalence class owns a set of memos that contain the identification of database entities that must all be recorded or unlocked if any

[7] This feature is not implemented yet. At present only the RECORD ALL is safe.

one of the them is recorded or unlocked. In the above example, as part of the remove operation, the database representatives of the owner, the member, the successor, and the predecessor would be checked to ascertain whether any one of them already belongs to an equivalence class. If one of them does, then memos for the others would be filed into the equivalence class and pointers to the equivalence class stored in their database representatives. If none of the database representatives points to an equivalence class, then one would be created, memos for the four entities filed into it, and the four database representatives set to point to it. Finally, if the database representatives of the four entities pointed to more than one equivalence class, the classes would be merged into a single class.

If the set was a ranked set, one or more subsets may have been brought into main storage without the programmer's knowledge. The remove operation would affect some of the subsets and perhaps the owner. Thus, memos for the affected entities should be filed into an equivalence class.

With this mechanism the programmer need specify records and unlocks for only those entities he has worked with. The related entities are picked up from the equivalence classes and recorded and unlocked automatically to preserve database integrity.

## 6. THE EAS-E DATABASE MANAGEMENT SYSTEM

In the current VM/370 implementation, every EAS-E database is controlled by a *custodian.* The custodian is a program that resides in an independent virtual machine. Only the custodian can read or write the database directly. User programs run in their own virtual machines and access the database by making requests to the custodian. Several user programs can work against the database concurrently.

Database entities have unique identifiers and appear to the user program as contiguous pieces of storage. The smallest individually addressable unit of DASD storage is called a physical block. Physical blocks are of fixed size, say 1024 words, and are divided into segments. The custodian also keeps track of logical blocks. Each logical block is written on a physical block. Entities are stored on full logical blocks and/or a partial logical block consisting of an integral number of segments within the block, the segments being linked together by pointers. The custodian maintains a catalog that associates the entity identifier with the first logical block and segment number and an index that associates logical blocks with physical blocks.

The size of a database can be increased by obtaining additional physical storage and introducing it as a new extent to the custodian. This is a simple procedure, and it allows a database to be increased incrementally to an arbitrary size.

The status of the database system consists of the status of the entities it contains along with the system control status. The system control status consists of the entity identifier to logical block and segment catalog, the logical block to physical block index, the lists of logical blocks with 1, 2, . . . , *n* available segments, and the list of free physical blocks.

The custodian's responsibilities include

(1) Retrieving and recording entities.

(2) Providing an access control mechanism that maintains locks on entities and queues of users waiting for entities not currently available.
(3) In the event of a main storage (or soft) crash during the recording process, ensuring that either all or none of the entities which a user program wishes to record are in fact recorded.
(4) Providing a mechanism for recovering from software crashes.
(5) Providing a mechanism to recover from physical damage to the storage device (hard crash).

The mechanisms that implement these features are discussed below. The status of the database management system is maintained in terms of entities, attributes, and sets in the custodian's main storage.

### 6.1 Concurrency Control

User programs can request database entities from the custodian with one of two types of locks: read.only or read.write. An entity that has been given out read.only can be given out read.only to other users. Read.write requests for such an entity must wait until all read.only users have released the entity. An entity that has been given out read.write cannot be given read.write or read.only to another user until it is released.

Main storage entities in the custodian, called slots, maintain the status of database entities. Each of these entities remembers, for one database entity, whether or not it has been built in the custodian and what its lock status is. For entities given out read.only, it maintains a list of *readers.* For entities given out read.write, it stores the identity of the *writer.* Users waiting for the entity are filed into a set of *requestors.*

When a request for an entity comes in, the custodian checks whether the entity exists in its main storage or is being built. If not, the relevant physical blocks are read, and the entity is built and its location stored in the slot.

If the entity exists in the custodian's main storage, it checks if it is available. If it is available (i.e., it is not issued to anybody or it is issued read.only and requested read.only), it is transmitted to the requestor. If the entity is not available (i.e., it is issued read.write or it is issued read.only and the request is read.write), the user is filed last into the set of requestors for the entity.

Every time a user requests an entity that is not available, the custodian checks whether a deadlock has been created. If a deadlock has been created, the user who made the last request is terminated and all the entities locked by him are unlocked, while the rest of the users execute as usual.

The user issues a record or an unlock by sending a list of entities or entity identifiers to the custodian along with a transaction code. In case of a record, the entities to be recorded are written on DASD. The database status is then modified to reflect the recorded and destroyed entities (see Section 6.2).

The user has the ability to specify whether he wants to maintain locks on the recorded entities. If he does, the record is complete. If not, and in the case of an unlock, the entities are released.

When an entity issued read.write is released, its set of requestors is examined. If the first user on the set requests the entity read.write, the entity is issued to him and he is removed from the set. If the first user on the set requests the entity read.only, the entity is issued to him and he is removed from the set. The first user on the set is examined again. If he too wants the entity read.only, it is issued to him and he is removed from the set. This continues until the set ends or a read.write request is encountered.

When a user releases a read.only entity he is removed from the set of readers. If this set is empty and the set of requestors for the entity is not empty, then the first user in the set must want the entity read.write. The entity is sent to him and he is removed from the set.

## 6.2 Protecting Against Software Crashes

When a user program ends normally or issues a RECORD, either all entities issued read.write or specified entities and members of their equivalence classes are sent to the custodian. In case of a software crash, the custodian must ensure that either the actions of a program are completely reflected in the database or not at all (i.e., either all the entities sent to the custodian are recorded or none of them are recorded). This is accomplished as follows: In general, the process of recording a few entities and updating the control status consists of (a) finding some free physical blocks, (b) writing entity records on DASD, and (c) writing the system control status on DASD to reflect the new entity records. The process is so arranged that if the system crashes before the new control status is completely written, then the old status is valid, and the physical blocks on which the modifications were written remain in the set of free physical blocks.

The traditional approach is to maintain the control status so that it is written correctly on DASD after every RECORD. Lorie [11] describes such a system, in which relevant blocks of the control status are brought into main storage as necessary. After writing the entities, all the status blocks that have been modified are written out on DASD. The central idea behind the EAS-E solution is to maintain the control status in main storage. After a RECORD, the status is updated in main storage, and the *changes* are written onto DASD as a sequential file called the TO.DO file. The complete system control status is written out periodically on DASD, say once a day. After a software crash the control status is read into main storage from the physical blocks where it was last written out. The TO.DO file is then read, and the control status is updated in main storage to include the changes subsequent to the last writing of the complete control status.

Consider in detail now the problem of saving the status after some entities have been modified. In EAS-E, the custodian writes each entity on the available segments of one or more logical blocks in main memory. The logical blocks are then written out on available physical blocks and the index modified in main storage to point to the new physical blocks. A modified entity may stay the same size, or its size may decrease or increase. If the size decreases or remains the same, it can be accommodated in the same data block. If its size increases it may still be possible to accommodate it in the same data block, provided it has enough free space. If the modified entity can be accommodated in the old data block, it starts at the same place as the original entity, and the catalog does not have to be updated. Otherwise, the catalog is modified in main storage to associate the entity identifier with new logical block and segment numbers. After this is done for all the entities, the changes in control status of the database, in the form of TO.DO memos, are written on the TO.DO file.

If the modified entity was written on the old data block, then only one TO.DO memo is needed to record the writing of the data block onto another physical block. If the modified entity had to be written on another data block and the catalog block modified to point to the new data block, then three TO.DO memos are required: one for catalog block, one for the new data block, and one to record the freeing up of the old data block. A TO.DO memo needs two words in our current implementation, and physical blocks are 1024 words in length. A typical interactive business program, such as that used by a bank teller or a clerk at the Department of Motor Vehicles, looks at a few entities in a large database, modifies some of these entities and stops. Thus, most of the time, the TO.DO memos resulting from the actions of one program will fit on a single TO.DO block. Occasionally they will go over a block boundary and require two TO.DO blocks. Thus, one or two DASD accesses are required to record the action of a program, with the average being close to one.

Compare this to a system which saves the correct status on DASD after every set of modifications. The appropriate comparison is the case where the time interval between two saves is very small. According to Lorie [11], this involves "updating the master, a single (index) block, its copy and one or two bit map blocks"—in other words, 5 or 6 accesses as compared with 1 or 2 accesses for EAS-E. In addition to this, EAS-E also has some advantage in reading in the status. In Lorie's system this has to be done, for a portion of the database, every time a program wants to modify some records. In EAS-E this is done once for every time the system is brought up. Thus, it takes a little longer to bring up the system, but there is no need to read-in a part of the status every time a program wants to modify some entities. Due to these advantages the EAS-E system has a much lower "overhead" in recording a batch of entities.

EAS-E also has to do some extra work in writing the system status, erasing the TO.DO file and the old status. There is no corresponding job in a system such as Lorie's because the relevant part of the status is written after every RECORD. As we have mentioned earlier, this sequence of system actions can be done at a time when the load is low, such as in the middle of the night, and without any interruption in system availability. If the computer is taken down periodically for maintenance, the new status can be written and the TO.DO file and old status erased when the computer is brought up again.

The advantage of maintaining the current system status in main memory is that every time some entities are modified the status on DASD does not have to be brought up-to-date. Instead, the status is updated in main storage and a few memos written on DASD. If the computer goes down and has to be brought up again, these memos can be used to update the status in main memory after it is read in from DASD. Since the memos are much smaller than the portions of DASD that would have to be updated, this presents a great saving in the accesses required to save the status after recording a batch of entities.

If the custodian were to operate in a real machine, then the decreased DASD writes would translate directly into savings in elapsed time for the users of the custodian. If the custodian operates in a virtual environment then, in view of its special status, it should be provided with locked pages of real storage. In this situation the approach of maintaining status in main storage would also be directly advantageous.

If the custodian runs in a paged virtual environment, then the status maintained in main storage would be written and read from the paging device. In this case the advantage of maintaining status in main storage would be diluted to the difference in speeds between DASD and the paging device. In a "fair share" environment [9] the EAS-E approach is guaranteed to do at least as well as an approach that keeps all its status information on DASD.

Maintaining system status completely in main storage or completely on DASD are the two ends of a continuum. The system status for a custodian supporting a large database will occupy a great deal of main storage. If enough main storage is not available, the status can be divided into two parts, one part residing on DASD and updated much as in the Lorie system, and the other residing in main storage and updated by the TO.DO mechanism. In such a case the most frequently updated parts of the status should be maintained in main storage to take advantage of the TO.DO mechanism.

The technique of updating status in main memory by means of the TO.DO file has other advantages as well. Consider a database system implemented for a bank. This will contain an entity called the BANK, one of whose attributes will be CASH on hand. Every time a deposit is made or a cheque cashed, the value of CASH changes. If it were desired to maintain a correct cash value on a continuous basis then, in a regular database management system, the BANK would have to be recorded after every transaction. In the EAS-E system the BANK can be made part of the system status. This means that changes in its attribute values can be written on the TO.DO file. In this way, writing the physical blocks necessary to record the BANK, and other specifically nominated entities, is avoided at the small cost of adding a few memos to the TO.DO file.[8]

Thus, the TO.DO file provides a facility that can be used to make the recording of frequently modified entities more economical. This is particularly important as such entities can cause bottlenecks in database management systems.

### 6.3 Protecting Against Physical Damage

The EAS-E database management system uses a variation of the dump and journal technique to protect against physical damage to the data storage devices. Essentially, this consists of copying the database periodically (dumping) and keeping a journal of the changes made between dumps. In case of physical damage, the database can be reconstructed by starting from the last dump and applying the journal to it.

The problem with the dump and journal technique is that dumping a large database takes a long time and, typically, the system is inaccessible during this time. Severance and Lohmann [20] provide an illustrative example where a complete dump takes six hours.

A superior alternative is differential dumping [18, 26] in which sections of the database are examined sequentially, and the blocks that have changed are

[8] Adding attributes of user defined entities to the custodian status is not implemented yet.

dumped. This is logical dumping, as opposed to physical dumping in which whole tracks or cylinders of the database are copied. Logical dumping is much slower than physical dumping. Also, some mechanism has to be provided to keep track of exactly which blocks have been altered.

EAS-E uses a process that may be called continuous fractional dumping. Periodically (i.e., every $T$ seconds), a track is read from the database into main memory. Each block on the track is immediately affixed with the current sequence number. No writing is permitted between reading the track and affixing the sequence stamps. After these are in place, the track may be dumped at any convenient time. Some of the blocks on the track may already be in main memory, and their contents may have been modified. When these blocks are written on DASD they will have sequence numbers later than the number on the track and will represent a later version.

With continuous fractional dumping the entire database is dumped in rotation with a cycle time that may be arranged to be any convenient period, say a week. Very little CPU time is required, and if the dumping is carried out under lightly loaded conditions, there will be little interference with user processing.

Along with the dump, a journal is maintained by writing out on tape, or on a spare disk pack, every physical block that is written on the database. Each physical block carries along with it a ten word trailer that contains the sequence stamp, the physical block number, the logical block number, the type of block (data or index), and some other information that allows the status to be completely reconstructed from an examination of the database. Status information and TO.DO information, thus, need not be recorded on the journal. There is one exception to this, however. If attribute values of specially nominated entities are kept on the TO.DO file, rather than being recorded every time, this information must be included in the journal.

The fractional dump and the journal can be used to reconstruct the database after a hard crash. When the crash occurs, a copy of the database will be available on the dump in which different parts are current as of different times. Let us assume there are five fractions on one disk pack that are current as of $S1$, $S2$, $S3$, $S4$, and $S5$ ($S1 < S2 < S3 < S4 < S5$). A journal will also be available starting at or before $S1$ and going to or past $S5$. To reconstruct the database, a fresh disk pack is obtained and formatted to replace the disk pack that was destroyed. The relevant blocks are now copied from the dump onto this disk pack. The journal is now used to correct this information and bring it up to date. To do this, the journal is, in effect, read backwards. Blocks found with sequence numbers greater than $S5$ are updated and their block numbers filed in a set of correct physical blocks. After this is done, the numbers of the blocks that are correct on the dump disk for $S5$, and are not already in the correct block set, are added to it. Reading the journal backwards is continued and now the records found with sequence numbers greater than $S4$ are updated and their block numbers added to the correct block set. When the journal reaches $S4$, the numbers of the blocks on the dump disk that are correct at $S4$, and are not already on the correct block list, are added to it. This process continues until all the physical blocks are on the correct block list. This may or may not occur before the journal reaches $S1$.

The time period $T$ can be set to vary with the system load and, in fact, the mechanism can be controlled in many different ways. For example, it can be shut off during the first shift and activated with a fixed $T$ during the second and third shifts, until a specified fraction of the database has been dumped.

## 7. SUMMARY AND STATUS

This paper has discussed the EAS formalism for application modeling and the EAS-E programming language. We have shown that EAS-E programs are more compact than equivalent programs in other languages and discussed some of the features (unified language for database and main storage entities, control phrases, the FIND statement, and the full-screen I/O capability) that make it a powerful application and system development language.

EAS-E also provides a nonprocedural interface that allows a user to browse and update a database without programming.

The existing VM/370 implementation will let practically any number of users interact with a database of practically any size. DASD extents can be added as the database's total space requirements grow. *Ranked* sets (i.e., sets ordered by one or more attributes of the member entities) are stored as balanced trees, so that entities can be filed or found quickly in sets of any size from few to millions of members.

EAS-E has been running for some time at Yorktown. It has been subjected to one field test: a rewrite and extension of the Workload Information System of Thomas J. Watson's Central Scientific Services (CSS). CSS consists of about 100 craftsmen who do model shop, glass work, electronics, etc., for Thomas J. Watson's scientists and engineers. The old Workload Information System, written in PL/I and assembler, was difficult to modify or extend. The EAS-E version duplicated the function of the old system: it read the same weekly inputs and generated the same outputs. It achieved this with about one-fifth as much source code as the old system. It also showed an even greater, but difficult to quantify, advantage over the old system in terms of ease of modification and extension. The application has now been extended to also accept inputs and provide outputs online rather than batch.

The CSS Workload Information System provided a good test of EAS-E's utility and power. More field tests are planned to provide further proof. Several enhancements are also planned, including an improved authorization package (the current facilities only allow a user to be authorized either read.only or read.write), subsidiary entities that exist only within their owner and move and change with him (such as accounts of a transaction and lines of a file), and an extension to the locking mechanism that would allow a user to read a copy of (part of) the database at a particular point in time while other users continued to read and update it in the normal manner.[9]

## REFERENCES

1. ASTRAHAN, M.M., ET AL. System R: a relational approach to database management. *ACM Trans. Database Syst. 1*, 2 (June 1976), 97–137.
2. BAYER, R., AND MCCRIEGHT, E. Organization and maintenance of large ordered indexes. *Acta Inf. 1*, 3 (1972), 173–189.

[9] This feature was suggested by Walt Daniels.

3. Bayer, R., and Unterauer, K. Prefix B-trees. *ACM Trans. Database Syst. 2*, 1 (March 1977), 11–26.
4. Blasgen, M.W. System R: an experimental relational database management system. Lecture Notes, IBM Research, San Jose, Calif., May 19, 1979.
5. *CODASYL Data Base Task Group Report.* Available from ACM, New York, April 1971.
6. Date, C.J. An introduction to the Unified Database Language (UDL). TR 03.106, IBM Santa Teresa Laboratory, San Jose, Calif., May 1980.
7. *IBM Corporation Information Management System/360, General Information Manual.* Prog Prod 5734-XX6, GH20-0765 06061.
8. *IBM Corporation Virtual Machine/370 Display Management System for CMS: Guide and Reference.* Program No. 5748-XXB, File No. 5370-39 SC24-5198-0.
9. *IBM Corporation VM/370 Resource Management Programming RPQ PO-9006 Programmer and System Logic Guide.* Program No. 5799-ARQ LY20-1996-0.
10. Kiviat, P.J., Villanueva, R., and Markowitz, H.M. *The SIMSCRIPT II Programming Language.* Prentice Hall, Englewood Cliffs, N.J., 1969.
11. Lorie, R.A. Physical integrity in a large segmented database. *ACM Trans. Database Syst. 2*, 1 (March 1977), 91–104.
12. Malhotra, A., Markowitz, H.M., and Pazel, D.P. The EAS-E programming language. RC 8935, IBM T.J. Watson Research Center, Yorktown Heights, N.Y. 10598.
13. Malhotra, A., Thomas, J.C., Carroll, J.M., and Miller, L.A. Cognitive processes in design. *Int. J. Man-Mach. Stud. 12*, 2 (Feb. 1980), 119–140.
14. Markowitz, H.M. Simscript. In *Encyclopedia of Computer Science and Technology*, J. Belzer, A.G. Holzman, and A. Kent, Eds., Marcel Dekker, New York, 1979, pp. 79–136. Also RC 6811, IBM T.J. Watson Research Center, Yorktown Heights, N.Y. 10598.
15. Markowitz, H.M., Malhotra, A., and Pazel, D.P. The ER and EAS formalisms for system modeling and the EAS-E language. In *Proc. 2nd Int. Conf. Entity-Relationship Approach* (Washington, D.C., Oct. 12–14, 1981) pp. 29–48. Also RC 8802, IBM. T.J. Watson Research Center, Yorktown Heights, N.Y. 10598.
16. Pazel, D.P., Malhotra, A., and Markowitz, H.M. The system architecture of EAS-E: an integrated programming/database language. *IBM Syst. J. 22*, 3, pp. 188–198. Also RC 9085, IBM T.J. Watson Research Center, Yorktown Heights, N.Y. 10598.
17. Rowe, L. A., and Shoens, K.A. Data abstraction, views and updates in RIGEL. In *Proc. ACM 1979 SIGMOD Conf.* (Boston, Mass., 1970) ACM, New York, pp. 71–81.
18. Sayani, H.H. Restart and recovery in a transaction-oriented information processing system. In *Proc. ACM SIGMOD Workshop on Data Description, Access and Control* (May 1974), ACM, New York, pp. 351–366.
19. Schmidt, J.W. Some high level constructs for data of type relation. *ACM Trans. Database Syst. 2*, 3 (Sept. 1977), 247–261.
20. Severance, D.G., and Lohman, G.M. Differential files: their application to the maintenance of large databases. *ACM Trans. Database Syst. 1*, 3 (Sept. 1976), 256–267.
21. Shipman, D.W. The functional data model and the data language DAPLEX. *ACM Trans. Database Syst. 6*, 1 (March 1981), 140–173.
22. Shopiro, J.W. Theseus—a programming language for relational databases. *ACM Trans. Database Syst. 4*, 4 (Dec. 1979), 493–517.
23. Stonebraker, M.R., Wong, E., Kreps, P., and Held, G.D. The design and implementation of INGRES. *ACM Trans. Database Syst. 1*, 3 (Sept. 1976), 189–222.
24. van de Riet, R.P., Wasserman, A.I., Kerstein, M.L., and De Jonge, W. High-level programming features for improving the efficiency of a relational database system. *ACM Trans. Database Syst. 6*, 3 (Sept. 1981), 464–485.
25. Wasserman, A.I., et al. The data management facilities of PLAIN. *ACM SIGPLAN Notices 16*, 5 (May 1981), 59–80.
26. Yourdon, E. *Design of On-line Computer Systems.* Prentice-Hall, Englewood Cliffs, N.J., 1972, pp. 340–353, 515–542.

Received November 1980; revised October 1981; accepted February 1983

*Reprinted from*

IBM

**Systems Journal**

Vol. 22, No. 3, 1983

# The system architecture of EAS-E: An integrated programming and data base language

by D. P. Pazel
A. Malhotra
H. M. Markowitz

# The system architecture of EAS-E: An integrated programming and data base language

by D. P. Pazel
A. Malhotra
H. M. Markowitz

***EAS-E is an application development system based on an entity-attribute-set view of system description. It consists of a procedural language for manipulating data base and main storage entities, and direct (nonprocedural) facilities for interrogating and updating data base entities. The EAS-E software itself was implemented with the entity-attribute-set view. This paper reviews some of the EAS-E features and considers some of its implementation details. This paper is both an introduction to the EAS-E software architecture and an example of the usefulness of the entity-attribute-set view.***

EAS-E (pronounced "easy") is an application-developed system based on an entity-attribute-set view of system description. EAS-E allows the application developer to manipulate higher-level data structures in main storage and in the data base with equal facility. In particular, it allows him to work with entities, attributes, and sets in main storage and in the data base as easily as he works with main storage variables in conventional programming languages.

The application designer conceives the application in terms of the *entities* (objects and things) that must be remembered, their *attributes* (properties), and the *sets* (order collections of entities) that they own or belong to. The application can then be implemented in EAS-E, which directly supports operations on entities, attributes, and sets.

EAS-E consists of a procedural language for manipulating data base and main storage entities, and direct (nonprocedural) facilities for interrogating and updating the data base entities. The virtues of the EAS-E entity-attribute-set view with respect to building application systems are presented in References 1 and 2. Reference 1 also compares programs written in EAS-E with programs written in other programming systems. In the present paper we discuss the EAS-E software and the advantages of the entity-attribute-set view in building system software.

Simple EAS-E commands written by an application programmer, e.g., to CREATE a data base entity, FILE it into a set, or FIND it later, can result in quite complex actions. The consequences of such actions, although transparent to the user, are to perform such actions as communicating information to and from the data base, efficiently manipulating very

**Very little is required to go from the conceptual model to the application program.**

large sets, and ensuring the integrity of the data base. Thus the programmer communicates his intentions in a compact, readable form and the EAS-E software translates those intentions into the detailed computer actions required to execute them efficiently.

In the first part of this paper, the general character of the EAS-E language is explored. A simple example is first modeled in the EAS-E philosophy or world view. This same example is then presented as a simple main storage program written in the EAS-E language. Finally, the example is seen again as a data base application written in EAS-E. Very little is required to go from the conceptual model to the application program. The paper then shows the data base query and modification aspects of EAS-E. The concluding section shows the modeling of the execution environment of the EAS-E application system using the very philosophy on which EAS-E is based. Using this, structure and actions required by the EAS-E execution environment become clear. Thus, despite the complexity of these actions, the EAS-E software was implemented in a relatively short time by the authors with a relatively small amount of source code. We believe that the large amount of function produced per unit of resource expended was due in large part to the use of the entity-attribute-set view in the building of the EAS-E software itself. In summary, the following are highlights of EAS-E:

- It is an integrated data base and programming language that simplifies the specification of complex data base actions.
- It incorporates a powerful modeling philosophy.
- EAS-E is easy to use.

## The EAS-E world view

The primitive concepts that comprise the world view behind our integrated programming and data base language are Entity, Attribute, Set, and Event (thus EAS-E). The power of these concepts has been used previously, for example, in two widely used simulation languages, SIMSCRIPT[3,4] and GASP.[5,6]

An *entity* may be thought of as a distinct structure or object. If one were to model a governmental system, for example, one might consider states and cities as *entity types*. When we refer to an entity type, we refer to a class of similar entities. For example, in a model containing 99 cities, CITY is the entity type of which there are 99 instances or entities. STATE is another entity type with one or more instances.

An *attribute* is a property or characteristic of an entity. Each attribute takes on values from some domain, e.g., real numbers, integers, alphanumeric strings, and pointers to user-defined entities. An attribute has at most one value at any time, or it may be undefined. For example, the entity type CITY may have CITY_NAME and MAYOR as attributes, and the entity type STATE may have STATE_NAME as an attribute.

A *set* is a collection of zero, one, or more entities of one or more entity types; a set is owned by an entity. For example, in the system with entity types CITY and STATE, we may introduce the set called CITY_SET owned by entities of type STATE, with member entities of type CITY. In the representation of the system, each STATE owns a CITY_SET consisting of some number of CITY entities.

These entity, attribute, and set facilities can be used to implement more complex structures such as stacks, pipelines, trees, and bills of material in a straightforward manner, as discussed in Reference 3. There are no limits on the number of sets an entity type may own or belong to. Also, there is flexibility in allowing entities of various types to belong to the same set or to own the same set. With this flexibility, the application modeler has full power in modeling structures as complex as he could possibly wish.

A visual aid for expressing the relationships among entities, attributes, and sets is illustrated in Figure

**Figure 1 An entity-attribute-set (EAS) description of a system**

| ENTITY | ATTRIBUTE | OWNS | BELONGS |
|---|---|---|---|
| STATE | STATE_NAME | CITY_SET | |
| CITY | CITY_NAME<br>MAYOR | | CITY_SET |

1. The entity types are listed with their attributes, the sets they own, and the sets to which they belong. The information in Figure 1 is referred to as the entity-attribute-set (EAS) description of a system.

In working with an EAS structure, there are five basic actions out of which higher-level actions may be built. One may CREATE an entity, that is, make an instance of an entity type. One may assign values to attributes of entities. Entities may be FILEd in or REMOVEd from sets. Finally, entities may be DESTROYed; that is, an instance of an entity of some type may be annihilated. To add a new city to the system, for example, one CREATEs a CITY entity, assigns values to its CITY_NAME and MAYOR, and FILEs it into the CITY_SET of STATE.

## The programming and data base language

EAS-E has been implemented on VM/370 at the IBM Thomas J. Watson Research Center, where it supports several applications. In this section, we first present EAS-E as a programming language. All basic operations are defined for private work spaces in main storage. We then present EAS-E as a data base language in which the basic operations extend naturally to data base concepts. Finally, we discuss the facilities that EAS-E provides to change data base definitions and modify existing entities to conform to new definitions. In the following sections, we discuss the implementation of these facilities.

**EAS-E as a programming language.** EAS-E includes an English-like procedural programming language that embodies many such standard language features as various data types, input/output, and control statements. EAS-E also features simple methods of defining and manipulating main storage EAS structures. Such structures are defined by means of the EVERY and DEFINE statements. The information in Figure 1, for example, is written as follows:

```
EVERY STATE HAS A STATE_NAME, AND OWNS A CITY_SET
EVERY CITY HAS A CITY_NAME, A MAYOR,
        AND BELONGS TO A CITY_SET
DEFINE STATE_NAME, CITY_NAME, AND MAYOR AS TEXT VARIABLES
```

Sets in EAS-E may be ordered in any one of the following three ways: (1) First In First Out (FIFO), (2) Last In First Out (LIFO) and (3) RANKED, that is, sorted according to the values of one or more attributes of the member entities. In main storage, these three types of sets are implemented as linked lists. In the given example, CITY_SET may be defined as FIFO or LIFO. Alternatively, CITY_SET can be ranked by CITY_NAME to provide an alphabetical ordering of the cities, specified as follows:

```
DEFINE CITY_SET AS A SET RANKED BY CITY_NAME
```

Entity and set definitions are contained in the initial section of an EAS-E program, called the PREAMBLE. This is followed by the executable source code. As mentioned earlier in this paper, EAS-E has many of the standard features of programming languages. We shall not discuss these here; instead we concentrate on the statements that manipulate EAS structures special to EAS-E. Statements corresponding to the five basic actions listed in the previous section are the following: CREATE an entity, DESTROY an entity, assign values to attributes with LET or READ statements, FILE entities into sets, and REMOVE entities from sets. The following example illustrates how these statements may be combined to add a city to a state:

```
CREATE A CITY
LET CITY_NAME=|GREENVILLE|
LET MAYOR=|JEAN GREEN|
FILE CITY IN CITY_SET(STATE)
```

EAS-E also provides a simple syntax for finding specific entities that are based on given attribute values. For example, if one wants to find the CITY with a CITY_NAME of GREENVILLE in the CITY_SET owned by STATE, and if found remove it from the set and delete it from the system, he writes the following procedure:

```
FIND THE CITY IN CITY_SET(STATE) WITH CITY_NAME=|GREENVILLE|
IF FOUND
  REMOVE CITY FROM CITY_SET(STATE)
  DESTROY CITY
ELSE
```

**EAS-E as a data base language.** In the development of an EAS-E data base application, the user has a particular overview. Entity, attribute, and set structures are stored in the data base. The data base resides in a separate virtual machine and is overseen by a *custodian* program. The custodian manages the data and can respond to simultaneous requests from several virtual machines that are running EAS-E programs.

To define data base entity types, the user prepares a file of definitions similar to that of a PREAMBLE and communicates them to the data base. Then the user writes a series of programs to work with the data base, manipulating the previously defined structures. These programs are then compiled and executed.

**To write a program that works with the data base, the user must specify which entity types are to be maniplated.**

If the entities and sets of the preceding section were to define a new data base system called GOVERNMENT, the user would begin by transmitting the following file of definitions to the custodian:

```
DATA BASE DEFINITIONS FOR DATA BASE GOVERNMENT

EVERY STATE HAS A STATE_NAME AND OWNS A CITY_SET
DEFINE STATE_NAME AS A TEXT VARIABLE

EVERY CITY HAS A CITY_NAME, A MAYOR,
        AND BELONGS TO A CITY_SET
DEFINE CITY_NAME, MAYOR AS TEXT VARIABLES

DEFINE CITY_SET AS A SET RANKED BY CITY_NAME

END
```

After these definitions have been stored in the data base, the user is ready to work with them, creating, destroying, filing, and removing entities, and setting attribute values. This can be done either with the EAS-E procedural language discussed in this paper, or with the direct (nonprocedural) facilities. Procedural commands for doing this are essentially the same as those for main storage entities, but with a few differences.

To write a program that works with the data base, the user must specify which entity types are to be manipulated. This is done in the PREAMBLE of the program with the DATA BASE ENTITIES INCLUDE statement. For example, to work with the entity types STATE and CITY, the user writes the following statement:

```
DATA BASE ENTITIES INCLUDE STATE AND CITY FROM GOVERNMENT
```

Attributes and sets of STATE and CITY need not be specified in the PREAMBLE. The compiler obtains this information from the definitions stored in the data base.

The user works with data base entities much as though working with main storage entities. For example, to loop over all data base entities of a given type, such as CITY, he may write the following statement:

```
FOR EACH CITY, DO...
```

With that, the executing program acquires each CITY entity from the data base, one at a time. Alternatively, the FIND statement can be used to find a particular entity of a given type whose attributes satisfy certain criteria. This and some of the previous concepts are used in the following example of a program to add the city GREENVILLE to the state OHIO:

```
PREAMBLE
NORMALLY ACCESS IS READ.WRITE
DATA BASE ENTITIES INCLUDE STATE AND CITY FROM GOVERNMENT
END
FIND THE STATE WITH STATE_NAME=|OHIO|
CREATE A CITY
LET CITY_NAME=|GREENVILLE|
LET MAYOR=|JEAN GREEN|
FILE CITY IN CITY_SET(STATE)

END
```

Unlike main storage entities, data base entities are addressed by variables of data type REFERENCE. Reference variables may be declared anywhere in a program with an access mode of read.write or read.only. A reference variable with the prevailing access mode is automatically provided for each entity type specified in the DATA BASE ENTITIES INCLUDE statement.

Each data base entity has a unique data base *identification number.* In the current implementation, the identification number consists of three integer values, namely a *type* number that identifies the entity type, a *slot* number that serializes the instance of this type, and a *dash* number that indicates the number of times this slot has been occupied. Identification numbers are stored in variables of type IDENTIFIER. They are helpful in accessing entities explicitly. For example, if ID is an identifier variable and REF is a reference variable, the assignment

**Simple queries can be written as small EAS-E programs.**

```
LET REF=ID
```

results in bringing from the data base the entity whose identification number is given in ID. The entity is brought read.only or read.write, depending on the access mode of the reference variable.

When a program ends normally (i.e., not by crashing) the changes made to the data base are committed (made permanent). In addition, changes made up to a particular point may be committed by execution of the following statement:

```
RECORD ALL DATA BASE ENTITIES
```

If the user wishes to undo the data base manipulations prior to a RECORD, the UNLOCK ALL DATA BASE ENTITIES statement may be used. In case of system crashes, either the entire contents of an implicit or explicit record will be reflected by the data base or none of its contents will be reflected.

**Queries in EAS-E.** Simple queries can be written as small EAS-E programs. For example, the following program prints the names and mayors of all the cities in New York:

```
PREAMBLE
DATA BASE ENTITIES INCLUDE STATE AND CITY FROM GOVERNMENT
END
FIND THE STATE WITH STATE_NAME=|NEW YORK|
FOR EACH CITY IN CITY_SET(STATE)
PRINT 1 LINE THUS...
             CITY                                  MAYOR
PRINT 1 LINE WITH CITY_NAME(CITY) AND MAYOR(CITY) THUS...
      **********************          ************************


END
```

Here the PRINT statement prints the line(s) following it exactly as specified, except that the asterisks are replaced by variable names.

Such a program is easy to write and allows a common query to be run over and over again. A more sophisticated version of the example would parameterize the program on the state name and perhaps the information to be displayed. In fact, EAS-E users tend to write query generators tailored to the queries they need most frequently.

For unusual queries and for looking through the data base, BROWSER, the full-screen nonprocedural facility mentioned earlier, is the most convenient. BROWSER allows one to move through the data base mostly by pressing program function keys on the terminal; it also allows the specification of simple reports with headings, totals, etc. BROWSER is described in References 2 and 7.

**Modifying data base definitions.** At any time after the GOVERNMENT data base is first defined, the definitions stored in it may be modified by statements such as those in the following program:

```
DATA BASE DEFINITIONS FOR DATA BASE GOVERNMENT

MODIFIED DEFINITION
EVERY STATE HAS A STATE_NAME, OWNS A CITY_SET
            AND A COUNTY_SET
DEFINE STATE_NAME AS A TEXT VARIABLE

NEW DEFINITIONS
EVERY COUNTY HAS A COUNTY_NAME, A COUNTY_SEAT,
      AND BELONGS TO A COUNTY_SET
DEFINE COUNTY_NAME, COUNTY_SEAT AS TEXT VARIABLES
DEFINE COUNTY_SET AS A SET RANKED BY COUNTY_NAME

END
```

Since the definition for CITY has not changed, it need not be resubmitted.

At this point, there are both an old and a new definition for STATE. Existing entities of type STATE are in the format defined by the old definition. In transforming the individual entities to the format defined by the new definition, they pass through another format called a dual format. The dual format is a combination of the old and the new formats; i.e., it consists of two distinct entities, one in the old and the other in the new format. Thus each individual STATE can be in one of the following three formats: (1) old format, (2) dual format, or (3) new format. Entities are converted from old to dual format when they are first pointed to by variables of data type DUAL REFERENCE. This consists of the entity in the old format followed by an empty new format. One can now copy the common attributes and sets from the old into the new format with the following statement:

```
MOVE THE COMMON ATTRIBUTES AND SETS OF STATE
```

One may then fill in the other attributes of the entity in the new format and populate the new sets that it owns. In doing so, one can refer to attributes and sets of the old or new entity by prefixing their names with O_ and N_, respectively. When no longer needed, the old version of the entity may be destroyed, putting the entity into new format. When all the entities of a given type are in the new format, the custodian permits the old definition to be purged.

## EAS-E implementation structures

Both the EAS-E compiler and data base custodian have been designed and built using the EAS philosophy. This has reduced several-fold the time required to implement the language and data base management system. The EAS view also pervades the background environment for an executing program. This environment is the principal topic of the remainder of this paper.

**Data base entity structure.** Data base entities are maintained on permanent storage and managed by the custodian. When a program requests an entity, the custodian fetches the entity from permanent storage and sends it to the program's main storage. At this point, the data base entity consists of a contiguous piece of storage with attributes laid out sequentially. The compiler uses the definition of the entity type to compute the offsets of these attributes. The offsets occur in fixed positions for all attributes of a given type. In the case of such variable-length attributes as TEXT, which may have arbitrary length, the offset points to a field that contains a relative displacement value. Figure 2 illustrates the way in which this displacement indicates where in the entity structure the text and its length are embedded.

**Figure 2 A data base entity with a text attribute**

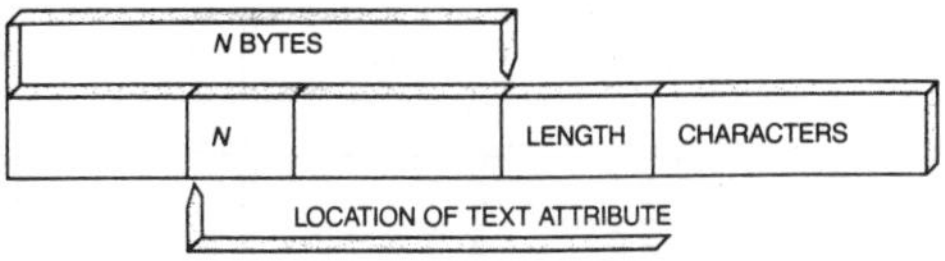


Another consideration in the layout of a data base entity concerns ranked set structures. Data base ranked sets are implemented in a manner quite different from main storage ranked sets. In order to provide rapid access to members in large sets, some information about the ranking attributes of the member entities is embedded in the entity structure of the owner.[1] Since the amount of ranking information is variable, the displacement of the ranking information, like the displacement of TEXT attributes, is stored in a position indicated by a fixed offset.

To summarize, a data base entity starts with fixed-length attributes and the displacements of variable-length attributes stored in positions that are constant for the entity type. This is followed by the variable-length attributes.

**Fundamental system structures.** To enable a user to work with data base entities in a manner equivalent to main storage entities, EAS-E provides mechanisms that (1) bring data base entities into main storage; (2) keep track of what has been brought into main storage and where it is located; and (3) record information in the data base. The structures used for this are now described.

Every EAS-E executing program is automatically provided with a main storage entity called the PROGRAM. Attributes assigned to the PROGRAM are in essence global variables. Sets owned by the PROGRAM are also global or *universal* sets.

A *data base representative*, or DB.REP, is a type of main storage entity used to keep track of data base entities brought into main storage. Each data base entity brought into the environment of an executing EAS-E program has a unique DB.REP. Essential attributes of DB.REP are the pointer to the main storage representation of the entity (DB.CORE), the access mode (DB.ACC), and the unique identifica-

**Figure 3 An EAS description of EAS-E system structures**

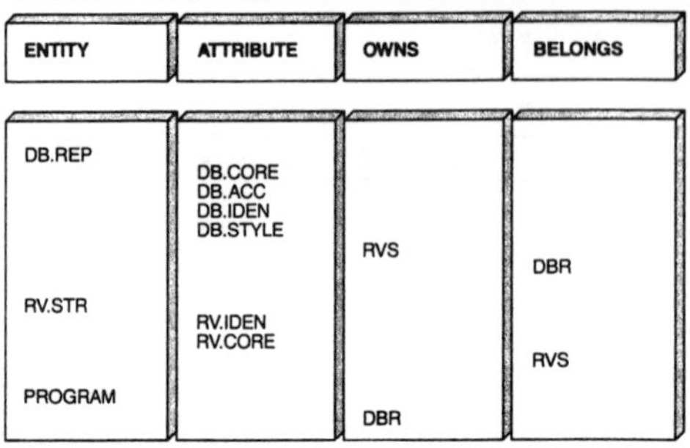


tion number for that entity (DB.IDEN). The PROGRAM owns a FIFO set called DBR to which the DB.REPs belong.

As mentioned earlier, reference variables are used to refer to data base entities. In fact, the reference variable does more than merely point to the main storage location of a data base entity. It encapsulates all the information about the existence of the data base entity in main storage. Thus each reference variable is a pointer to a main storage entity called a *reference variable structure* or RV.STR, which has as attributes the identification number (RV.IDEN), and a pointer to the main storage representation of the data base entity (RV.CORE). Also, since more than one reference variable can point to a data base entity, the RV.STR belongs to a set called RVS which is owned by the DB.REP entity.

The EAS structure of these relationships is shown in Figure 3. The entity-attribute-set names shown are not the actual names used in our implementation. The names used here provide readability. Our implementation is designed to conform to conventions that distinguish system-defined and user-defined names; it is discussed in Reference 8. Later in this paper, we show that DB.REP and RV.STR have additional attributes. These may be ignored for the moment until we discuss the structures needed for the modification of data base definitions.

As an example of how these structures are used at execution time, consider the actions associated with the statement CREATE A CITY. First, a request is made to the data base custodian for a unique identification number for the new entity. Using information provided in the object code, the executing program builds an empty main storage version of the entity. A DB.REP is created for the entity, its attributes are filled in, and it is filed into the DBR set. Next, an RV.STR is created, its attributes are filled in, and it is filed into the RVS set of the DB.REP. The reference variable CITY is now made to point to the RV.STR.

As another example, consider the actions associated with a statement that brings an entity from the data base, for example, by setting a reference variable to an identifier variable. The PROGRAM's DBR set is searched for an entity with that identification number. If the entity is not found there, it is requested from the custodian. Then a DB.REP is created, its attributes filled in, and it is filed in DBR. If the entity is found in the DBR set, its access mode is checked and upgraded from read.only to read.write if necessary. If the access mode is upgraded, notification is sent to the custodian. Finally, the attributes of the RV.STR to which the reference variable will point are filled in, and the RV.STR is filed into the RVS set of the DB.REP.

**Data base set organization and manipulation.** As previously mentioned, the EAS-E language provides three kinds of set organizations—First In First Out (FIFO), Last In First Out (LIFO), and RANKED organizations. LIFO and FIFO data base sets are held together as linked lists (like main storage sets). First- and last-member attributes are automatically defined for the owner entity, and successor and predecessor attributes are defined for any member entity. These member attributes point to the first and last members of the set and the successor and predecessor in the set of the particular member, respectively. The owner entity may also keep a membership count attribute, and the member entity may have an attribute pointing to the owner. For FIFO and LIFO sets, the difference between main storage sets and data base sets is that set pointer attributes are IDENTIFIER variables.

Ranked data base sets, on the other hand, are based on an entirely different organization. If ranked data base sets were implemented as linked lists, a search for a member with given attribute values would proceed by considering each member in turn until the desired one was found. Since accessing many members of a large data base set can be very time-consuming, such an implementation is unreasonable. To overcome this problem, ranked data base sets are organized as balanced trees.[9] A balanced tree has a root node associated with the owner entity from which a tree of subnodes extends, with the leaf nodes pointing to the member entities. Information is maintained at each node that relates to the value ranges of the ranking attributes on the offspring nodes. The fanout at each node is quite large and is based upon keeping the size of the node entity close to one page. The implementation of ranked data base sets is discussed in more detail in Reference 1.

The implementation of ranked data base sets requires that the structure of data base entities be somewhat different from main storage entities, where ranked sets are implemented as linked lists. For example, the top node of a ranked set is included in the structure of the owner. In order for the system routines to locate that information, EAS-E generates an attribute in the entity that contains a displacement to the ranked set information relative to the beginning of the owner. Also, the tree nodes require the existence in the data base of an automatically defined entity type. The EAS-E system manipulates these entities according to the needs of the balanced trees. The extra entity attributes and the data base node entities are transparent to the user.

**Looping through data base sets.** The compilation of a loop statement in EAS-E, such as

```
FOR EACH CITY IN CITY_SET(STATE)
```

results in a simple loop control when the set is a linked list. More precisely, the *first* or *last* pointer of the owner is used to initiate the loop, and the *successor* or *predecessor* pointers of the members are used to continue the loop until the last member is processed. If selection clauses such as WITH CITY_NAME=|GREENVILLE| are appended to the previously looping statement, a series of tests (IF statements) are generated within the loop proper. In this case, although the domain of the loop appears smaller, the full set must still be searched. That is, each member of the set must be brought into main storage and tested.

The ranked set organization just described is designed to help locate entities that meet given conditions, without bringing in each member of the set. To do this, the compiler must identify and extract the bounds given on the loop constraints in the source program and arrange to pass these restrictions at execution time to a set-scanning mechanism.

To identify the selection criteria, the compiler scans the selection clauses on a loop statement for constraints of the following form:

```
< RANKING ATTRIBUTE > < LOGICAL RELATION > < EXPRESSION >
```

Here EXPRESSION is an arithmetic expression that does not contain any ranking attributes. We call such a constraint a P-CONSTRAINT. For each ranked-set loop, the compiler examines the appended constraints and seeks out the first *n* P-

**The problem of passing the selection information at execution time to the loop-searching mechanism has been addressed by designing an EAS structure to contain that information.**

CONSTRAINTs that are related to the ranking attributes in order and are linked by "and"s, for maximum *n*; that is,

```
< P-CONSTRAINT > < "AND" > < P-CONSTRAINT > < "AND" > ......
```

Here the P-CONSTRAINTS are ordered by the ranking attribute. These P-CONSTRAINTS are then translated into object code that generates a selection structure that imposes these limits on the domain of the loop. The selection structure is discussed more fully later in this paper. Constraints that do not fit into this pattern are translated into a set of conditional clauses in the main body of the loop, as usual. Thus, members are selected in accordance with the P-CONSTRAINTS that embody the specifications on the ranking attributes; if necessary, they are brought into main storage and tested for other criteria.

The problem of passing the selection information at execution time to the loop-searching mechanism has been addressed by designing an EAS structure to contain that information. Calls are then generated in the loop structure that fill in those structures and

**Figure 4 EAS view of selection structures**

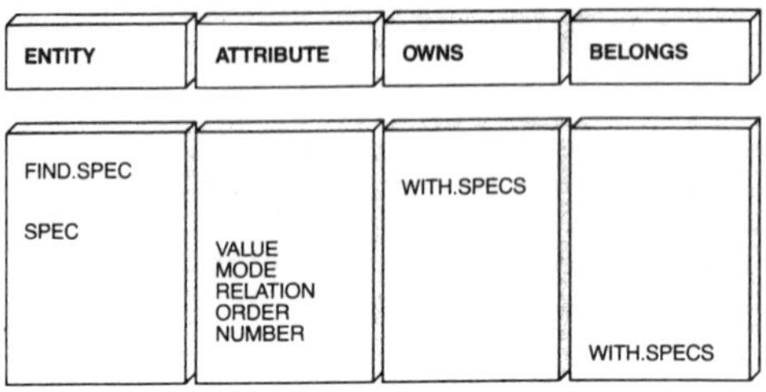


eventually pass them to the loop mechanism at execution time. An EAS view of a selection structure is shown in Figure 4.

The selection structure consists of an owner node (FIND.SPEC) that owns a set of SPECS, each of whose members contain all the information about a P-CONSTRAINT. VALUE is the execution-time value of the EXPRESSION portion of the P-CONSTRAINT. MODE indicates the data type of the ranking attribute being constrained. RELATION specifies how the ranking attribute relates to the VALUE, e.g., =, <, >, etc. ORDER indicates whether ranking on that attribute is by high or low value. NUMBER is the ordinal number of the ranking attribute, that is, first, second, third, and so forth. This structure is used by the looping mechanism to search the balanced tree of the set. Selection structures are also used in FILE and REMOVE operations to find where a new member should go and to find the member that is to be removed. The creation, filling in, utilization, and clean-up of these structures is completely transparent to the user.

**RECORD and UNLOCK.** The execution of the statement RECORD ALL DATA BASE ENTITIES results in making all creates, destroys, and other alterations of data base entities permanent to the data base. A RECORD may be issued at any time during program execution and may be specified with a HOLD option, in which case all read.write entities that have been accessed remain accessed upon completion of the RECORD. This saves communication overhead if the entities are used again. If HOLD is not specified, all read.write entities not pointed to by a reference variable (i.e., not being used) are released. This saves main storage space in the program. The following paragraph shows how the DB.REP and RV.STR structures are used in the RECORD operation.

The identification number of each read.write entity, as represented by DB.REPs in the DBR set, is written onto a communication buffer. If the entity has been destroyed, this fact is noted. Otherwise, the entity is rebuilt as a contiguous piece of storage, with text attribute and ranked set information appended to the entity and copied into the communication buffer. This buffer is transmitted to the custodian, who commits these changes to the data base. When control is returned from the custodian, the DBR set must be scanned once more to find the DB.REPs of entities recorded, to free the main storage version of those entities, to null out and remove RV.STRs from RVS sets, and finally to destroy their DB.REPs.

The effect of executing an UNLOCK statement is to release all data base entities. With this statement, any changes made in an executing program since the last RECORD are rescinded. This is, in effect, an undoing of data base actions before they are committed. The actions for this statement for each DB.REP in the DBR set are as follows: the main storage for the data base entity is released, the RV.STRs are nulled and removed from RVS sets, and the DB.REP is destroyed. The custodian is told that all entities this user has accessed are to be released.

**Looping and auto-unlock.** EAS-E provides the capability of looping over all data base entities of a particular type, as in executing FOR EACH CITY . . . . This is accomplished through a get-next operation, based on a sequential catalog of entities of a given type maintained by the custodian. To obtain the first entity, the get-next operation transmits a special identification to the custodian. On successive passes, get-next communicates the identification number of the previously obtained entity to the custodian and receives the next entity from the custodian, or it receives an indication of loop termination.

In looping over a large number of entities—as in FOR EACH CITY . . .—main storage can all be used up unless unneeded entities are released. To relieve the programmer from having to do this, EAS-E unlocks read-only entities that are not being used—that is, they have no reference variables pointing to them—whenever there are 30 such entities.

**Modifying data base definitions.** As previously noted, sometimes both an old and a new definition may exist for any entity type. An individual entity of a type may then be in one of three formats—old, new, or dual. Because of this, the DB.REP and RV.STR are more complex than shown in Figure 3. Figure 5 provides a more complete description of EAS-E modification structures. (DB.CORE and RV.CORE in Figure 3 are referred to as DB.NEW.VER and RV.NEW.VER, respectively.) Since an entity can be in old, new, or dual format, each DB.REP and RV.STR has a pointer to the old version of the entity in main storage (DB.OLD.VER or RV.OLD.VER) and a pointer to the new version of the entity in main storage (DB.NEW.VER or RV.NEW.VER). Also, DB.STYLE indicates the current format of the entity.

When an entity that has two definitions is requested from the custodian, the following actions are performed. The custodian provides a code to the executing program indicating the format of the entity received from the data base. If the entity is in old format and is being referenced by a DUAL REFERENCE variable, it is put into dual format with the creation of a blank new version of the entity and by setting the DB.REP and RV.STR main storage pointers to the respective versions. If the entity is already in dual format, the data received from the custodian consist of the concatenation of the two versions of the entity. The system code then sets the DB.REP and RV.STR pointers to the respective entities. If the entity is in new format, then only DB.NEW.VER and RV.NEW.VER point to the received data.

**EAS-E has been designed to accommodate data bases of arbitrary size, from very small to very large.**

Depending on the format of the entity, different ways of packaging it at RECORD time are selected. In particular, if the entity is in dual format, both versions of the entity must be reassembled separately in a manner specified in the section on RECORD and UNLOCK. Then both entities are moved to contiguous storage in the communication buffer with the old version preceding the new version.

**Figure 5 Description of EAS-E modification structures**

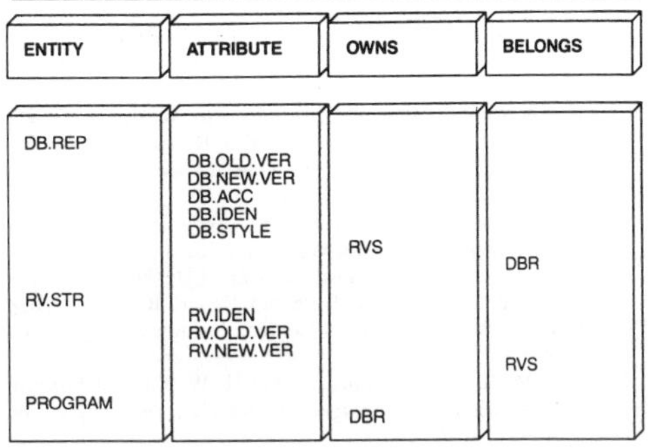


## Concluding remarks

This paper has presented a brief overview of the EAS-E modeling philosophy or world view and the EAS-E programming language. This approach to application development focuses sharply on clarity in data structures (i.e., entities-attributes-sets) and actions (i.e., events). We have shown that by using a programming and data base language based upon these elements, there is little effort in going from a conceptual model of an application to the application program itself. EAS-E has been designed to accommodate data bases of arbitrary size, from very small to very large.

The EAS-E modeling philosophy has been used throughout the design and implementation of the EAS-E application development system. The EAS-E compiler, custodian, and library routines are all written in the EAS-E language.

Areas for further research include such capabilities as the following: building of data base utility routines to be used by host languages, communication with several EAS-E data bases simultaneously, and single-user EAS-E data bases. In the latter case, the data would exist as a user's private data base, as opposed to residing in a separate service machine.

## Cited references

1. A. Malhotra, H. M. Markowitz, and D. P. Pazel, *EAS-E: An Integrated Approach to Application Development*, Research

Report RC 8457, IBM Thomas J. Watson Research Center, Yorktown Heights, NY 10598; also submitted to the *ACM Transactions on Database Systems*.
2. H. M. Markowitz, A. Malhotra, and D. P. Pazel, "The ER and EAS formalisms for system modeling, and the EAS-E language," *Proceedings of the Second International Conference on Entity-Relationship Approach*, Washington, DC, October 12–14, 1981, pp. 29–48; also, Research Report RC 8802, IBM Thomas J. Watson Research Center, Yorktown Heights, NY 10598.
3. H. M. Markowitz, "SIMSCRIPT," *Encyclopedia of Computer Science and Technology*, Vol. 13, J. Belzer, A. G. Holtzman, and A. Kent, Editors, Marcel Dekker, New York (1979), pp. 79–136; also Research Report RC 6811, IBM Thomas J. Watson Research Center, Yorktown Heights, NY 10598.
4. H. M. Markowitz, B. Hausner, and H. W. Karr, *A Simulation Programming Language*, Prentice-Hall, Inc., Englewood Cliffs, NJ (1963).
5. P. J. Kiviat, *GASP—A General Activity Simulation Program*, U. S. Steel Corporation, Applied Research Laboratory, Monroeville, PA (July 1963).
6. A. A. B. Pritsker, "GASP," *Encyclopedia of Computer Science and Technology*, Vol. 8, J. Belzer, A. G. Holtzman, and A. Kent, Editors, Marcel Dekker, New York (1977), pp. 408–430.
7. H. M. Markowitz, A. Malhotra, and D. P. Pazel, *The EAS-E Application Development Summary: Principles and Language Summary*, Research Report RC 9910, IBM Thomas J. Watson Research Center, Yorktown Heights, NY 10598.
8. A. Malhotra, H. M. Markowitz, and D. P. Pazel, *The EAS-E Programming Language*, Research Report RC 8935, IBM Thomas J. Watson Research Center, Yorktown Heights, NY 10598.
9. R. Bayer and K. Unterauer, "Prefix B-trees," *ACM Transactions on Database Systems* **2**, No. 1, 97–137 (March 1977).

## General references

M. M. Astrahan, M. W. Blasgen, D. D. Chamberlin, K. P. Eswaran, J. N. Gray, P. P. Griffiths, W. F. King, R. A. Lorie, P. R. McJones, J. W. Mehl, G. R. Putzolu, I. L. Traiger, B. W. Wade, and V. Watson, "System R: A relational approach to database management," *ACM Transactions on Database Systems* **2**, No. 1, 97–137 (June 1976).

*CODASYL Data Base Task Group Report*, available from the Association for Computing Machinery, Inc., 1133 Avenue of the Americas, New York, NY 10036.

C. J. Date, *An Introduction to Database Systems, Third Edition*, Addison-Wesley Publishing Company, Inc., Reading, MA (1981).

*Information Management System/360, General Information Manual: Program Product 5734-XX6*, GH20-0765 06061, IBM Corporation; available through IBM branch offices.

**Ashok Malhotra** *IBM Research Division, Thomas J. Watson Research Center, P.O. Box 218, Yorktown Heights, New York 10598.* Dr. Malhotra has been a research staff member in the Computer Sciences Department at the Research Center since 1975. His current research in improved application development facilities is a manifestation of a general interest in making computers more accessible to the nonspecialist, especially the manager. Dr. Malhotra received his Ph.D. from M.I.T. in management; he has several years experience as a management consultant. He is editor of the *Program Development Quarterly*, an internal IBM newsletter devoted to advances in program and application development technology.

**Harry M. Markowitz** *IBM Research Division, Thomas J. Watson Research Center, P.O. Box 218, Yorktown Heights, New York 10598.* Dr. Markowitz has been a research staff member in the Computer Sciences Department at the Thomas J. Watson Research Center since he joined IBM in 1974. During this period his principal interest has been the design and development of the EAS-E application development system. Previously, while at the RAND Corporation (1952–1959 and 1960–63), he developed techniques for inverting very large but sparse matrices which are now widely used. Also at RAND he designed and supervised the development of the SIMSCRIPT simulation programming language. Dr. Markowitz received his Ph.D. in economics from the University of Chicago. In his Ph.D. dissertation he developed the "portfolio theory," which is now regularly taught in finance departments of business schools.

**Donald P. Pazel** *IBM Research Division, Thomas J. Watson Research Center, P.O. Box 218, Yorktown Heights, New York 10598.* Mr. Pazel is a research staff member in the Computer Sciences Department at the Thomas J. Watson Research Center. He joined IBM in 1973 at Morris Plains, New Jersey, where he worked on the Safeguard project for the Federal Systems Division. Since joining the Research Center staff in 1975, Mr. Pazel has worked in the areas of operating systems, compilers, and data bases. Mr. Pazel graduated maxima cum laude from LaSalle College, Philadelphia, in 1972 with a B.A. in mathematics, and received an M.S. degree in mathematics from the University of Virginia in 1973. He is a member of the Mathematics Association of America and the Association for Computing Machinery.

Reprint Order No. G321-5190.

# Samuelson and Investment for the Long Run

**Harry M. Markowitz**

When I was a student in the Economics Department of the University of Chicago, Karl Brunner and I diligently read through Paul Samuelson's *Foundations of Economics*. Karl and I found a bug in the book and wrote Professor Samuelson concerning it. Samuelson replied that several other people had called his attention to this bug, but we were the first non-Asians to do so. Years later, I was surprised and delighted to see Samuelson cite my work and write about portfolio theory, albeit sometimes critically as well as creatively.

Through the years Paul and I have had one ongoing debate on the following topic. If an investor invests for the "long run" should she or he choose each period the portfolio which maximizes the expected logarithm of 1+ return for that period? I say yes; Paul says no. Our written works on the subject include Samuelson (1963, 1969, 1979) and Markowitz (1959, 1976). We also debated the matter at a meeting in Vale, Colorado many years ago. To this day both of us feel that their view has been completely vindicated. But, I must admit, Samuelson's 1979 article titled "Why We Should Not Make Mean Log of Wealth Big Though Years To Act Are Long" is a particularly remarkable expository achievement. As he explains in the last paragraph, "No need to say more. I've made my point. And, save for the last word, have done so in prose of but one syllable". It is hard not to feel intimidated in a debate with an opponent who is a combination of Albert Einstein and Dr. Seuss.

In the present essay I note the primary positions of the two sides and give one example that illustrates their differences. I chose an example which most simply supports my side of the debate. I believe that any other example will, upon close examination, also support my side but not necessarily as directly. I make no attempt to provide arguments on the other side since Paul, despite or because of his 90 years, is perfectly capable of doing so.

**Background**

The expected log criteria was proposed by Kelly (1956) and embraced by Latane (1957, 1959). Markowitz (1959) accepts the idea that the expected logarithm (of one plus return) of a portfolio is its rate of growth in the long run. Markowitz concludes that the cautious investor should not choose a mean-variance combination from the mean-variance efficient frontier with higher arithmetic mean (and therefore higher variance) than the mean-variance combination which approximately maximizes expected log, or, equivalently, geometric mean return. A portfolio higher on the frontier subjects the investor to more volatility in the short run and no greater return *in the long run*. The cautions investor, however, may choose a mean-variance combination lower on the frontier, giving up return in the long run for greater stability of return.

Breiman (1960, 1961) supplied a strong law of large numbers argument supporting the expected log rule. Samuelson (1963, 1969, 1979) provides an expected utility argument which contradicts the expected log rule. Markowitz (1976) provides an alternate expected utility argument which supports the expected log rule.

**The Expected Log Rule in General and Particular**

Throughout this paper I consider an investor who starts with an initial wealth $W_0$ and allocates resources, without transaction costs, at discrete times 0, 1, 2 … separated by a day, month, year or millisecond. The return on a portfolio P during time interval t--between time point t-1 and t--is denoted $r_t^P$. In general, as of time t-1, the probability distribution of $r_t^P$ may depend on the values of state variables as of time t-1 and may be jointly distributed with the values of state variables as of t. The max E log rule says that, whatever this dependence on and joint distribution with state variables, as of time t-1 choose the portfolio P which maximizes current, single-period

$$E \log\ (1+r_t^P) \tag{1}$$

where E is the expected value operator.

The issues which separate the Kelly, Latané, Brieman and Markowitz arguments *for* and the Samuelson arguments *against* are already present, and can be discussed more simply, in the special case wherein the returns $r_t^P$ on a given portfolio are i.i.d. (independent and identically distributed) and the investor must rebalance to the same portfolio P (therefore the same probability distribution of $r_t$) at every point in time. We shall deal only with this special case in this paper. See Markowitz (1976) for the more general case.

**First Argument *For* Max E log**

If an investor repeatedly draws from a rate of return distribution without adding or withdrawing funds beyond the initial $W_0$, then at time T the investor's wealth is

$$W^T = W_0 \prod_{t=1}^{T} (1+r_t^P) \tag{2}$$

where $r_t^P$ here represents the rate of return actually achieved on the portfolio at time *t*. The rate of return, $g^P$, achieved during the entire history from 0 to $T$ satisfies

$$\begin{aligned} (1+g^P) &= (W_T^P / W_0)^{1/T} \\ &= \left( \prod_{t=1}^{T} (1+r_t^P) \right)^{1/T} \end{aligned} \tag{3}$$

$g^P$ is the rate of return which, if earned each period, would grow wealth from $W_0$ to $W_T$ in $T$ periods. Thus, wealth at time $T$ is a strictly increasing function of

$$\log(1+g^P)=(1/T)\left(\sum_{t=1}^{T}\log(1+r_t^P)\right) \tag{4}$$

The assumption that $r_t^P$ is i.i.d. implies that $\log(1+r_t^P)$ is i.i.d. If $r_t^P=-1$ is possible then log $(1+r_t^P)$ is an "extended real" random variable defined on $[-\infty,\infty)$. If $\log(1+r_t^P)$ has expected value

$$E\log(1+r_t^P)=\mu\in[\infty,\infty] \tag{5}$$

then the strong law of large numbers says that--with probability 1--

$$\lim_{T\to\infty}\sum_{t=1}^{T}\left(\log(1+r_t^P)\right)/T\to\mu. \tag{6}$$

In particular, if μ is finite then, with probability 1, for every positive $\varepsilon$ there is a $T_0$ such that the difference between μ and the sample average of $\log(1+r_t^P)$ is less than $\varepsilon$, for all T greater than $T_0$:

$$\forall\varepsilon>0\ \exists T_0 \quad \text{such that } \forall T>\mathrm{T}_0$$

$$\left|\sum_{t=1}^{T}\log(1+r_t^P)/T-\mu\right|<\varepsilon \tag{7}$$

$T_0$ is random, and will typically vary from one randomly drawn sequence to the next (independently drawn) random sequence. If expected $\log(1+r_t^P)=\infty\ (or-\infty)$ then, with probability 1, for every $b$>0 there is a random time $T_0$ such that for all time thereafter average $\log(1+r_t^P)$ is greater than $b$ (respectively, less than $-b$):

$$\forall b > 0 \; \exists T_0 \text{ such that } \forall T > T_0$$

$$\sum_{t=1}^{T} \log(1 + r_t^P) / T > b \tag{8}$$

(<-b, respectively)

One of the principal justifications for identifying max $E\log(1+r)$ with investment for the long run follows from the above. If $r_1^P, r_2^P, \ldots$ and $r_1^Q, r_2^Q$ are two rate of return sequences, each i.i.d. but $r_t^P$ may be correlated with $r_t^Q$, and the first has a higher $E\log(1+r)$ than the second, i.e.,

$$\begin{aligned} \mu_P &= E\log(1+r_t^P) \\ &> E\log(1+r_t^Q) \\ &= \mu_Q \end{aligned} \tag{9}$$

then (3),(4) and (7) or (8) imply that--with probability 1--there is a time $T_0$ such that $W_T^P$ exceeds $W_T^Q$ ever after

$$\exists T_0 \; \forall T > T_0 \; W_P^T > W_Q^T. \tag{10}$$

Since, with probability one, there comes a time such that forever after the wealth of the investor who rebalances to portfolio P exceeds that of the investor who rebalances to portfolio Q, surely one can say that P does better than Q in *the long run*. This does not necessarily imply that any particular investor, with a finite life and imminent consumption needs, should invest in P rather than Q. But it seems an unobjectionable use of language to summarize relationship (10) by saying that portfolio P does better than portfolio Q "in the long run".

**Argument *Against* Max E log**

Consider an investor who invests $W_0$ at time 0, and lets this investment "ride", without additional investments or withdrawals until some fixed, distant time T. At time T the investor, or his or her heirs, will "cash in" the investment. The investor must decide whether the trustees of this investment are to be instructed to rebalance each period to portfolio P or Q. We continue to assume that successive returns to a given portfolio are i.i.d., although the simultaneous returns $r_t^P, r_t^Q$ may be correlated. Suppose that the investor seeks to maximize expected utility of final wealth, where the utility function is the form

$$U = \operatorname{sgn}(\alpha) W_T^\alpha \tag{11}$$

for some $\alpha \neq 0$. Since returns to a given portfolio are i.i.d., expected utility equals

$$\begin{aligned} EU &= \operatorname{sgn}(\alpha)\, E\left(\prod_{t=1}^{T}(1+r_t^P)\right)^\alpha \\ &= \operatorname{sgn}(\alpha)\left(E(1+r^P)^\alpha\right)^T \end{aligned} \tag{12}$$

Thus, the expected utility maximizing portfolio is whichever has greater $E(1+r)^\alpha$. This is not necessarily the one with greater $E\log(1+r)$.

Samuelson (1969) and Mossin (1968) show a much stronger result than shown above (in which it is *assumed* that the investor rebalances to the *same* portfolio each period). Even if the investor may switch portfolios, e.g., choose one when there is a long way to go and another when the end is imminent, the optimum strategy for the utility function in (11) is to stay with the same portfolio from beginning to end, whether "time is long" or not.

Thus, no matter how distant the goal, the optimum strategy is not the max E log rule.

**Example**

Consider two portfolios P and Q. P provides 6 percent per year with certainty. Q provides, each year, a fifty-fifty chance of 200 percent gain or 100 percent loss. The expected return and expected $\log(1+\text{return})$ of P are 0.06 and $\log_e(1.06) = 0.058$, respectively. The expected return and expected log of Q are

$\frac{1}{2}(2.00) + \frac{1}{2}(-1.00) = 0.50$ (fifty percent) and $\frac{1}{2}\log(3.00) + \frac{1}{2}\log(0.0) = -\infty$.

An investor who followed the max E log rule would prefer P. For any fixed investment horizon T, the investor who maximized expected utility of form (11) with $\alpha = 1$, *i.e.*, an investor who maximized expected terminal wealth, would prefer Q.

The arguments for and against the max E log rule can be illustrated with this example. Imagine that the return on Q are determined by a flip of a fair coin, heads being favorable. If the coin is flipped repeatedly, with probability one eventually a tail will be tossed. From that time on $0 = W_T^Q < W_T^P = (1.06)^T$. Thus, in the particular case, as in general, with probability 1 there comes a time when the max E log strategy pulls ahead and stays ahead of the alternate strategy, forever.

On the other hand, pick some point in time, such as T=100. At that time P provides $(1.06)^T$ with certainty. Q provides nothing if a tail has appeared in the first 100 tosses. If not, $W_T^Q = 3^T$. Since this has probability $(\frac{1}{2})^T$ expected wealth (equals expected utility here) is

$$EW_T^Q = (\tfrac{1}{2})^T 3^T = (1.50)^T > (1.06)^T = W_T^P \tag{13}$$

Thus, in the particular case as in general, the portfolio which maximizes $EU$ for T=1 also maximizes $EU$ for arbitrarily large $T$ fixed in advance.

**Another Argument *For* Max E log**

Markowitz (1976) argues that an assertion that something is best (or not best) in the long run should be an asymptotic statement that some policy or strategy does (or does not) approach optimality as $T \to \infty$. The Samuelson argument against Max E log is presented in terms of a (long) game of fixed length. Since this fixed length is arbitrarily long, the Samuelson argument can be transformed into an asymptotic argument as follows. Imagine a sequence of games $G_1$ $G_2$, $G_3$, …, $G_{100}$, … . The second game $G_2$ is "just like" the first except it is two periods long, T = 2, rather than one period long T = 1. The third game $G_3$ is just like the first two except that it is three periods long, T = 3, and so on.

In general, the notion that game $G_T$ is "just like" game $G_{T-1}$, only longer, would require that the same opportunities be available in the first T – 1 moves of game $G_T$ as were available in all of $G_{T-1}$. For the simple example in the last section, it implies that the same two probability distributions, P and Q, be used T times instead of T-1. Let $EU_T^P$ and $EU_T^Q$ represent the expected utility of the T-period game, obtained by repeatedly investing in distribution P or Q respectively. Samuelson's complaint about identifying P as the better investment for the long run is that it is *not* true that

$$\lim_{T\to\infty} EU_T^P \geq \lim_{T\to\infty} EU_T^Q$$

even though P has greater $E\log(1+r)$ on each spin of the wheel.

One way in which the Samuelson games stay the same as T varies is that each is scored by the expected value of *the same function* of *final wealth*. We could instead score the games by the same function of rate of return g defined in (3). In the example of the last section, $P$ always supplies a rate of return of 0.06. The rate of return supplied by $Q$ is

$$q^Q = \begin{cases} -1.0 \text{ with probability } 1-(\frac{1}{2})^T \\ \phantom{-}2.0 \text{ with probability } \quad (\frac{1}{2})^T \end{cases}$$

Let $f$ be any strictly increasing function of $g$. Let us define a sequence of games, $H_1, H_2, \ldots, H_{100} \ldots$ which are just like the Samuelson games except that they are each scored by expected value of the same function $V = f(g)$. Then

$$\begin{aligned} EV_T^P &= f(.06) \to f(.06) \\ EV_T^Q &= (1-(\tfrac{1}{2})^T)\, f(-1.0) \\ &\quad +(\tfrac{1}{2})^T\ f(2.0) \\ &\to f(-1.0) \end{aligned}$$

Thus, indeed,

$$EV_T^P > EV_T^Q \text{ as } T \to \infty.$$

If we score each game by the same function of g (rather than the same function of $W_T$) then the max E log rule *is* asymptotically optimal.

Suppose we wish to compare the performances of two investment strategies for varying horizons: e.g., for a five year period, a ten year period, …, a fifty year period, etc. How should we decide whether increasing time is more favorable to one than the other? No matter how long or short the horizon there is *some* chance that one will do better, and *some* chance the other will do better. The question is how to "add up" these various possibilities. One way---the constant utility of final wealth way---assumes that the trade-offs should be the same between making a dollar grow to \$1.10 versus \$1.20 versus \$1.30 after 50 years as after 5 years. The other way—constant utility of rate of growth—assumes that the trade-offs should be the same between achieving a 3 percent, 6 percent and 9 percent rate of growth during the 5 or 50 years. For a fixed T, any utility of final wealth $U(W_t)$ can be expressed as a utility of growth $f(g) = U(W_0(1+g)^T)$. But, as our example illustrates, assuming that *U* remains the same versus assuming *f* remains the same as *T* increases, has very different implications for the asymptotic optimality of the max Elog rule.

**Summary**

One argument in favor of the Elog rule is that (under broad assumptions) eventually the wealth of the investor who follows the rule will become greater than, and stay greater forever than, an investor who follows a distinctly different strategy. Samuelson's argument against the rule is that if the investor seeks to maximize the expected value of a certain kind of function of final wealth, for a long game of fixed length, then maximizing Elog is *not* the optimal strategy. Indeed, if we let the length of the game increase, the utility supplied by the max Elog strategy does not even approach that supplied by the optimum strategy. This assumes that utility of final wealth remains the same as game length varies. On the other hand, if we assume that it is the utility of rate-of-growth-achieved, rather than utility of final wealth, that remains the same as length of game varies, then the Elog rule *is* asymptotically optimal.

As Keynes said, "In the long run we are all dead". Even if you buy the notion, for either reason, that the max Elog rule is asymptotically optimal for the investor who lets her, his or its money ride, it may not be optimal for the individual or institution with fixed or random cash flow needs. Perhaps this is a sufficient caveat to attach to the observation that the cautious investor should not select a mean-variance efficient portfolio higher on the frontier than the point which approximately maximizes expected log (1+return); for a point higher on the frontier subjects the investor to greater volatility in the short fun and, almost surely, no greater rate-of-growth in the long run.

**References**

Breiman, Leo (1960), *Investment Policies for Expanding Businesses Optimal in a Long Run Sense*. Naval Research Logistics Quarterly, Vol. 7, No. 4, pp. 647-651.

_____(1961), *Optimal Gambling Systems for Favorable Games* Fourth Berkeley Symposium on Probability and Statistics, I, pp. 65-78.

Kelly, J.L., Jr. (1956) *A New Interpretation of Information Rate*, Bell System Technical Journal, pp. 917-926.

Latane, H. A. (1957) *Rational Decision Making in Portfolio Management*, Ph.D. dissertation, University of North Carolina.

_____(1957) *Criteria for Choice Among Risky Ventures,* Journal of Political Economy, April.

Markowitz, H. M. (1959) Portfolio Selection: Efficient Diversification of Investments, John Wiley and Sons, New York, 1959, Yale University Press, 1972.

_____(1976) *Investment for the Long Run: New Evidence for an Old Rule*, The Journal of Finance, Vol. XXXI, No. 5, December. pp.1273-1286.

Mossin, Jan (1968) *Optimal Multiperiod Portfolio Policies* Journal of Business, Vol. 41, No. 2, pp. 215-229.

Samuelson, P. A. (1963) *Risk and Uncertainty: A Fallacy of Large Numbers*, Scientia, 6th Series, 57th year, April-May.

_____(1969) *Lifetime Portfolio Selection by Dynamic Stochastic Programming*, Review of Economics and Statistics, August.

_____(1979) *Why We Should Not Make Mean Log of Wealth Big Through Years To Act Are Long.*

*Reprinted from*

# *The Journal of* FINANCE

VOL. XXXI DECEMBER 1976 No. 5

# INVESTMENT FOR THE LONG RUN: NEW EVIDENCE FOR AN OLD RULE

HARRY M. MARKOWITZ*

## I. BACKGROUND

"INVESTMENT FOR THE LONG RUN," as defined by Kelly [7], Latané [8] [9], Markowitz [10], and Breiman [1] [2], is concerned with a hypothetical investor who neither consumes nor deposits new cash into his portfolio, but reinvests his portfolio each period to achieve maximum growth of wealth over the indefinitely long run. (The hypothetical investor is assumed to be not subject to taxes, commissions, illiquidities and indivisibilities.) In the long run, thus defined, a penny invested at 6.01% is better—eventually becomes and stays greater—than a million dollars invested at 6%.

When returns are random, the consensus of the aforementioned authors is that the investor for the long run should invest each period so as to maximize the expected value of the logarithm of (1 + single period return). The early arguments for this "maximum-expected-log" (MEL) rule are most easily illustrated if we assume independent draws from the same probability distribution each period. Starting with a wealth of $W_0$, after $T$ periods the player's wealth is

$$W_T = W_0 \cdot \prod_{t=1}^{T} (1 + r_t) \tag{1}$$

where $r_t$ is the return on the portfolio in period $t$. Thus

$$\log(W_T / W_0) = \sum_{t=1}^{T} \log(1 + r_t) \tag{2}$$

If $\log(1 + r)$ has a finite mean and variance, the weak law of large numbers assues us that for any $\epsilon > 0$

$$\text{Prob}\left( \left| \frac{1}{T} \cdot \log(W_T / W_0) - E\log(1 + r) \right| > \epsilon \right) \to 0 \tag{3}$$

* IBM Thomas J. Watson Research Center, Yorktown Heights, New York.

and the strong law[1] assures us that

$$\lim_{T\to\infty} \frac{1}{T}\cdot\log(W_T/W_0)=E\log(1+r) \tag{4}$$

with probability 1.0. Thus if $E\log(1+r)$ for portfolio $A$ exceeds that for portfolio $B$, then the weak law assures us that, for sufficiently large $T$, portfolio $A$ has a probability as close to unity as you please of doing better than $B$ when time $=T$; and the strong law assures us that

$$W_T^A/W_T^B\to\infty \qquad \text{with probability one.} \tag{5}$$

Some authors have argued that the strategy which is optimal for the player for the long run is also a good rule for some or all real investors. My own interest in the subject stems from a different source. In tracing out the set of mean, variance $(E, V)$ efficient portfolios one passes through a portfolio which gives approximately maximum $E\log(1+r)$.[2] I argued that this "Kelly-Latané" point should be considered the upper limit for conservative choice among $E$, $V$ efficient portfolios, since portfolios with higher (arithmetic) mean give greater short-run variability with less return in the long run. A real investor might, however, perfer a smaller mean and variance, giving up return in the long run for stability in the short run.

Samuelson [14] and [15] objected to MEL as the solution to the problem posed in [1], [2], [7], [8], [9], [10]. Samuelson's objection may be illustrated as follows: suppose again that the same probability distributions of returns are available in each of $T$ periods, $t=1,2,\ldots,T$. (Samuelson has also treated the case in which $t$ is continuous; but his objections are asserted as well for the original discussion of discrete time. The latter, discrete time, analysis is the subject of the present paper.) Assume that the utility associated with a play of a game is

$$U=W_T^{\alpha}/\alpha \qquad \alpha\neq 0 \tag{6}$$

where $W_T$ is final wealth. Samuelson shows that, in order to maximize expected utility for the game as a whole, the same portfolio should be chosen each period. This always chosen portfolio is the one which maximizes single period

$$EU=E(1+r)^{\alpha}/\alpha. \tag{7}$$

Furthermore, if $EU_T^0$ is the expected return provided by this strategy for a $T$

1. In most cases the early literature on investment for the long run used the weak law of large numbers. The results in Breiman [1], however, specialize to a strong law of large numbers in the particular case of unchanging probability distributions. See also the Doob [4] reference cited by Breiman.

2. Markowitz [10] Chapters 6 and 13 conjectures, and Young and Trent [16] confirm that

$$E\log(1+r)\approx\log(1+E)-\tfrac{1}{2}\cdot(V/(1+E)^2)$$

for a wide class of actual ex post distributions of annual portfolio returns.

period game, and $EU_T^L$ is that provided by MEL, usually we will have

$$EU_T^0/EU_T^L \to \infty \qquad \text{as } T\to\infty \tag{8}$$

Thus, despite (3), (4) and (5), MEL does not appear to be asymptotically optimal for this apparently reasonable class of games.

Von Newman and Morgenstern [17] have directly and indirectly persuaded many, including Samuelson and myself, that, subject to certain caveats, the expected utility maxim is the correct criterion for rational choice among risky alternatives. Thus if it were true that the laws of large numbers implied the general superiority of MEL, but utility analysis contradicted this conclusion, I am among those who would accept the conclusions of utility analysis as the final authority. But not every model involving "expected utility" is a valid formalization of the subject purported to be analyzed. In particular I will argue that, on closer examination, utility analysis supports rather than contradicts MEL as a quite general solution to the problem of investment for the long run.

## II. The Sequence of Games

It is important to note that (8) is a result concerning a *sequence of games*. For fixed $T$, say $T=100, EU=EW_{100}^{\alpha}/\alpha$ is the expected utility (associated with a particular strategy) of a game involving precisely 100 periods. For $T=101, EW_{101}^{\alpha}/\alpha$ is the expected utility of a game lasting precisely 101 periods; and so on for $T=102,103,\ldots$.

That (8) is a statement about a sequence of games may be seen either from the statement of the problem or from the method of solution. In Samuelson's formulation $W_T$ is *final* wealth—wealth at the *end* of the game. If we let $T$ vary (as in "$T\to\infty$") we are talking about games of varying length.

Viewed differently, imagine computing the solution by dynamic programming starting from the last period and working in the direction of earlier periods. (Here we may ignore the fact that essentially the same solution reemerges in each step of the present dynamic program. Our problem here is not how to compute a solution economically, but what problem is being solved). If we allow our dynamic programming computer to run backwards in time for 100 periods, we arrive at the optimum first move, and the expected utility for the game as a whole given any initial $W_0$, for a game that is to last 100 moves. If we allow the computer to continue for an additional 100 periods we arrive at the optimum first move, and the expected utility for the game as a whole given any initial $W_0$, for a game that is to last for 200 moves; and so on for $T=201,202,\ldots$.

In particular, equation (8) is not a proposition about a single game that lasts forever. This particular point will be seen most clearly later in the paper when we formalize the utility analysis of unending games.

To explore the asymptotic optimality of MEL, we will need some notation concerning sequences of games in general. Let $T_1<T_2<T_3\cdots$ be a sequence of strictly increasing positive integers. In this paper[3] we will denote by $G_1, G_2, G_3\ldots$ a

3. A somewhat different, but equivalent, notation was used in [11].

sequence of games, where the $i$th game lasts $T_i$ moves. (In case the reader feels uncomfortable with the notion of a sequence of games, as did at least one of our colleges who read [11], perhaps the following remarks may help. The notion of a sequence of games is similar to the notion of a sequence of numbers, or a sequence of functions, or a sequence of probability distributions. In each case there is a first object (i.e., a first number or function or distribution or game) which we may denote as $G_1$; a second object (number, function, distribution, game) which we may denote by $G_2$; etc.).

In general we will not necessarily assume that the same opportunities are available in each of the $T_i$ periods of the game $G_i$. We will always assume that—as part of the rules that govern $G_i$—the game $G_i$ is to last exactly $T_i$ periods, and that the investor is to reinvest his entire wealth (without commissions, etc.) in each of the $T_i$ periods. Beyond this, specific assumptions are made in specific analyses.

In addition to a sequence of games, we shall speak of a sequence of strategies $s_1, s_2, s_3, \ldots$ where $s_i$ is a strategy (i.e., a complete rule of action) which is valid for (may be followed in) the game $G_i$. By convention, we treat the utility function as part of the specification of the rules of the game. The rules of $G_i$ and the strategy $s_i$ together imply an expected utility to playing that game in that manner.

## III. Alternate Sequence-of-Games Formalizations

Let $g$ equal the rate of return achieved during a play of the game $G_i$; i.e., writing $T$ for $T_i$:

$$W_T = W_0 \cdot (1+g)^T \tag{9}$$

or

$$g = (W_T / W_0)^{1/T} - 1. \tag{10}$$

In the Samuelson sequence of games, here denoted by $G_1, G_2, G_3, \ldots$, the utility function of each game $G_i$ was assumed to be

$$U = f(W_T) = W_T^{\alpha} / \alpha. \tag{11}$$

We can imagine another sequence of games—call them $H_1, H_2, H_3, \ldots$—which have the same number of moves and the same opportunities per move as $G_1, G_2, G_3, \ldots$, respectively, but have a different utility function. Specifically imagine that the utility associated with a play of each game $H_i$ is

$$U = V(g). \tag{11a}$$

for some increasing function of $g$. For a fixed game of length $T = T_i$, we can always find a function $V(g)$ which gives the same rankings of strategies as does some specific $f(W_T)$. For example, for fixed $T$ (11) associates the same $U$ to each possible play as does

$$U = V(g) = W_0^{\alpha} \cdot (1+g)^{\alpha T} / \alpha. \tag{11b}$$

Thus for a given $T$ it is of no consequence whether we assume that utility is a function of final wealth $W_T$ or of rate of return $g$.

On the other hand, the assumption that some utility function $V(g)$ remains constant in a sequence of games, as in $H_1, H_2, H_3, \ldots$ has quite different consequences than the assumption that some utility function $f(W_T)$ remains constant as in $G_1, G_2, \ldots$. Markowitz [11] shows that if $V(g)$ is continuous then

$$EV_T^L/EV_T^0 \to 1 \qquad \text{as } T\to\infty \tag{12a}$$

where $EV_T^L$ is the expected utility provided by the MEL strategy for the game $H_i$, and $EV_T^0$ is the expected utility provided by the optimum strategy (if such an optimum exists[4]); and if $V(g)$ is discontinuous then

$$EV_T^L/EV_T^0 \geqslant 1-\epsilon-\delta \tag{12b}$$

where $\delta$ is the largest jump in $V$ at a point of discontinuity, and $\epsilon\to 0$ as $T\to\infty$. (12a) and (12b) do not require the assumption that the same probability distributions are available each period. It is sufficient to assume that the return $r$ is bounded by two extremes $\underline{r}$ and $\bar{r}$:

$$-1<\underline{r}\leqslant r\leqslant \bar{r}<\infty \tag{13}$$

e.g., the investor is assumed to not lose more than, say, 99.99% nor make more than a million percent on any one move in any play of any game of the sequence. Note also that $V(g)$ is not required to be concave, nor strictly increasing nor differentiable; but of course it is allowed to be such.

Thus under quite general assumptions, if $V(g)$ is continuous MEL is asymptotically optimal in the sense of 12a. If $V(g)$ has small discontinuities, then MEL may possibly fail to be asymptotically optimal by small amounts as in 12b. These results are in contrast to (8), derived on the assumption of constant $U=f(W_T)$.

In [11] I argued that the assumption of constant $V(g)$ in a sequence of games is a more reasonable formalization of "investment for the long run" than is the assumption of constant $U(W_T)$. Given the basic assumptions of utility analysis, the choice between constant $V(g)$ and constant $U(W_T)$ is equivalent to deciding which of two types of questions would be more reasonable to ask (or determine from revealed preferences) of a rational player who invests for the long run in the sense under discussion.

Example of question of type I: what probability would make you indifferent between (a) a strategy which yields 6% with certainty in the long run: and (b) a strategy with a probability $\alpha$ of yielding 9% in the long run versus a probability of $1-\alpha$ of yielding 3% in the long run.

Example of question of type II: if your initial wealth is \$10,000.00, what

4. The assumptions of [11] do not necessarily imply that an optimum strategy exists. In any case (12a) and (12b) apply to any "other" strategy such that

$$EV_T^0 > EV_T^L \qquad \text{for all } T.$$

probability $\beta$ would make you indifferent between (a) a strategy which yields \$20,000 with certainty in the long run, versus (b) a strategy which yields \$25,000 with probability $\beta$ and \$15,000 with probability $(1-\beta)$ in the long run.

Question I has meaning if constant $V(g)$ is assumed; question II if constant $U(W_T)$ is assumed. It seemed to me (and still does) that preferences among probability distributions involving, e.g., 3%, 6% or 9% return in the indefinitely long run are more reasonable to assume than preferences among probability distributions involving a final wealth of, e.g., \$10,000, \$15,000 or \$20,000 in the long run. I will not try to further argue the case for constant $V(g)$ as opposed to constant $U(W_T)$ at this point, other than to encourage the reader to ask himself questions of type I and type II to judge.

In [11] I also argued that even if we were to assume constant $U(W_T)$ rather than constant $V(g)$, we would have to assume that $U$ was bounded (from above and below) in order to avoid paradoxes like those of Bernoulli [3] and Menger [13]. I then show that MEL is asymptotically optional for bounded $U(W_T)$. Merton and Samuelson [12] and Goldman [6] object to my definition of asymptotic optimality, although it is essentially the same as the criteria by which we judge, e.g. a statistic to be asymptotically efficient. Merton and Samuelson proposed, and Goldman adopted, an alternative criterion in terms of the "bribe" required to make a given strategy as good as the optimum strategy. But this bribe criteria seems to me unacceptable, since it violates a basic tenant of game theory—that the normalized form of a game (as described in [17]) is all that is needed for the comparison of strategies. It is not possible to infer the Samuelson-Merton-Goldman bribe from the normalized form of a game. Strategies Ia and Ib in game I may have the same expected utilities, respectively, as strategies IIa and IIb in game II; but a different bribe may be required to make Ia indifferent to Ib than is required to make IIa indifferent to IIb. Strategies IIIa and IIIb in a third game (not necessarily an investment game) may have the same pair of expected utilities as Ia and Ib in game I, or IIa and IIb in game II, but the notion of a bribe may have no meaning whatsoever in game III.[5] Thus unless we are prepared to reject the equivalence between the normalized and extensive form of a game in evaluating strategies, we must reject the Merton-Samuelson-Goldman bribe as part of a precise, formal definition of asymptotic optimality.

5. For example, suppose that strategy (a) has $EU_a=0$ and strategy (b) has $EU_b=\frac{1}{2}$. What bribe will make (a) as good as (b)? Consider the answer, e.g., for one period games I and II in which (a) accepts $W=\frac{1}{2}$ with certainty and (b) elects a 50-50 chance of $W=0$ versus $W=1$. In (I) suppose

$$U=\begin{cases} 0 & \text{for} \quad W<\frac{1}{2} \\ 10^*(W-\frac{1}{2}) & \text{for} \quad 0.5<W<0.6 \\ 1 & \quad\quad W>0.6 \end{cases}$$

while in II suppose

$$U=\begin{cases} 0 & \text{for} \quad W<0.9 \\ 10^*(W-0.9) & \quad\quad 0.9<W<1.0 \\ 1 & \quad\quad W>1.0 \end{cases}$$

In game I, (a) requires a bribe of 0.05; in game II (a) requires 0.45.

## IV. Unending Games

Even if we agree that a player playing a fixed finite game should maximize expected utility, we cannot determine whether MEL is asymptotically optimal for a given sequence of games $\{G_i\}$ unless we can agree on criteria for asymptotic optimality. What is needed is either "metacriteria" regarding how to choose criteria of asymptotic optimality, or else an alternate method of analyzing the desirability of strategies for the long run. This section presents such an alternate method, namely the utility analysis of unending games.

Consider a game $G_\infty$ which is like one of the games $G_i$ described above with this one exception: the game $G_\infty$ never terminates. Instead of having a first move, a second move, and so on through a $T_i$th move, we have an unending sequence of moves. As with a game $G_i$, a strategy for a $G_\infty$ is a rule specifying the choice of portfolio at each time $t$ as a function of the information available at that time. The only difference is that now the rule is defined for each positive integer $t=1,2,3,\ldots$ rather than only for $1 \leqslant t \leqslant T$.

Given a particular game $G_\infty$ and a strategy $(s)$, a play of the game involves an infinite sequence of "spins of the wheel" and results in an infinite sequence of wealths at each time:

$$(W_0, W_1, W_2, \ldots, W_t, \ldots) \tag{14}$$

where $W_0$ is initial wealth, and

$$W_t = W^*_{t-1}(1 + \text{return at time } t) \tag{15}$$

as in $G_i$.

The reader should find it no more unthinkable to imagine an infinite sequence of spins than to imagine drawing a uniformly distributed random variable. For example, if the same wheel is to be spun each time in an unending game, and if this wheel has ten equally probable stopping points, which we may label 0 through 9, then the infinite decimal expansion of a uniform [0, 1] random variable may be taken as the infinite sequence of random stopping points of the wheel.[6] If the wheel has sixteen stopping points, then the hexadecimal expansion of the random number may be used. In either case the infinite sequence of wealths $(W_0, W_1, W_2, \ldots)$ is implied by the rules of the game, the player's strategy, and the uniform random number drawn.

In general, a given $G_\infty$ and a given strategy imply a probability distribution of wealth-sequences $(W_0, W_1, W_2, \ldots)$.

Since $G_\infty$ has no "last period", we cannot speak of "final wealth". We can, however, assume that the player has preferences among alternate wealth-sequences: e.g., he may prefer the sequence of passbook entries provided by a savings account which compounds his money at 6%, starting with $W_0$, to one that compounds it at 3%. Given any two sequences:

$$W^a = (W_0, W^a_1, W^a_2, \ldots)$$

6. The fact that some numbers have two decimal expansions, like 0.4999... versus 0.5000..., may be resolved in any manner without effect on the analysis; since such numbers occur with zero probability.

and

$$W^b = (W_0, W_1^b, W_2^b, \ldots)$$

we may assume that the player either prefers $W^a$ to $W^b$, or $W^b$ to $W^a$ or is indifferent. Further, we may assume that given a choice between any two probability distributions among sequences of wealth

$$\Pr_A(W_0, W_1, W_2, \ldots)$$

versus

$$\Pr_B(W_0, W_1, W_2, \ldots)$$

he either prefers probability distribution $A$ to $B$, or $B$ to $A$, or is indifferent between the two probability distributions.

We shall not only assume that the player has such preferences, but also that he maximizes expected utility. In other words, we assume that he attaches a (finite) number

$$U(W_0, W_1, W_2, \ldots)$$

to each sequence of wealths, and chooses among alternate strategies so as to maximize $EU$.

The only additional assumption we make about the utility function $U(\ldots)$, is this:

If the sequence $W^a = (W_0, W_1^a, W_2^a, \ldots)$ eventually pulls even with, and then stays even with or ahead of the sequence

$$W^b = (W_0, W_1^b, W_2^b, \ldots)$$

then $W^a$ is at least as good as $W^b$; i.e., if there exists a $T$ such that

$$W_t^a \geqslant W_t^b \qquad \text{for } t \geqslant T \tag{16}$$

then $U(W^a) \geqslant U(W^b)$. This assumption expresses the basic notion that, in the sense that we have used the terms throughout this controversy, if player $A$ eventually gets and stays ahead of player $B$ (or at least stays even with him) then player $A$ has done at least as well as player $B$ "in the long run".

At first it may seem appropriate to make a stronger assumption that if $W_t^a$ eventually pulls ahead of $W_t^b$, and stays ahead, then the sequence $W^a$ is preferable to the sequence $W^b$. In other words, if there is a $T$ such that

$$W_t^a > W_t^b \qquad \text{for } t \geqslant T \tag{16a}$$

then

$$U(W^a) > U(W^b).$$

As shown in the footnote[7], this stronger assumption is too strong in that no utility function $U(W_0, W_1, W_2, \ldots)$ can have this property. Utility functions can however have the weaker requirement in (16).

The analysis of unending games is particularly easy if we assume that the same opportunities are available at each move, and that we only consider strategies which select the same probability distribution of returns each period. We shall make these assumptions at this point. Later we will summarize more general results derived in the appendix to this paper.

Without further loss of generality we will confine our discussion to just two strategies, namely, MEL and any other strategy, and will consider when the expected utility supplied by MEL is at least as great as that supplied by the other strategy. Letting $W_t^L$ and $W_t^0$ be the wealth at $t$ for a particular play of the game using MEL or the other strategy, respectively,

$$U(W_0, W_1^L, W_2^L, \ldots) \geqslant U(W_0, W_1^0, W_2^0, \ldots) \tag{17}$$

is implied if there is a $T$ such that

$$W_t^L \geqslant W_t^0 \qquad \text{for } t \geqslant T. \tag{18}$$

7. If $U$ orders all sequences $W=(W_0, W_1, W_2, \ldots)$ in a manner consistent with (16a), then in particular it orders sequences of the form

$$\begin{cases} W_0 - \text{given} \\ W_1 - \text{any positive number} \\ W_t = (1+\alpha)\cdot W_{t-1} \qquad \text{for } t > 2;\ \alpha > -1. \end{cases} \tag{N.1}$$

Since this family of sequences depends only on $W_1$ and $\alpha$, we may here write

$$V(W_1, \alpha) = U(W_0, W_1, W_2, \ldots). \tag{N.2}$$

Then (16a) requires

$$V(W_1^A, \alpha^A) > V(W_1^B, \alpha^B) \qquad \text{if either } \alpha^A > \alpha^B \text{ or}$$

$$\alpha^A = \alpha^B \qquad \text{and} \qquad W_1^A > W_1^B. \tag{N.3}$$

For any $\alpha$ let

$$U_{\text{low}}(\alpha) = GLB\ V(W_1, \alpha) \tag{N.4}$$

$$U^{\text{hi}}(\alpha) = LUB\ V(W_1, \alpha).$$

Then (N.3) implies

$$U_{\text{low}}(\alpha) < U^{\text{hi}}(\alpha) \qquad \text{for every } \alpha \tag{N.5a}$$

as well as

$$U_{\text{low}}(\alpha^A) > U^{\text{hi}}(\alpha^B) \qquad \text{if } \alpha^A > \alpha^B. \tag{N.5b}$$

But (N.5b) implies that we can have $U_{\text{low}}(\alpha) < U^{\text{hi}}(\alpha)$ for at most a countable number of values of $\alpha$, since at most a countable number of values of $\alpha$ can have $U^{\text{hi}}(\alpha) - U_{\text{low}}(\alpha) > 1/N$ for $N = 1,2,3,\ldots$. But this contradicts (N.5a).

Equation (2) implies that we have $W_t^L \geqslant W_t^0$ if and only if

$$\frac{1}{t}\sum_{i=1}^{t}\log(1+r_i^L) \geqslant \frac{1}{t}\sum_{i=1}^{t}\log(1+r_i^0). \tag{19}$$

Thus (18) will hold in any play of the game in which there exists a $T$ such that

$$\frac{1}{t}\sum_{i=1}^{t}\log(1+r_i^L) \geqslant \frac{1}{t}\sum_{i=1}^{t}\log(1+r_i^0) \qquad \text{for all } t \geqslant T. \tag{20}$$

Or, if we let

$$y_i = \log(1+r_i) \tag{21}$$

(20) may be written as

$$\frac{1}{t}\sum_{i=1}^{t} y_i^L \geqslant \frac{1}{t}\sum_{i=1}^{t} y_i^0 \qquad \text{for } t \geqslant T. \tag{22}$$

Under the present simplified assumptions

$$Ey^L > Ey^0 \tag{23}$$

by definition of MEL. But for random variables $y_1, y_2, \ldots$ with identical distributions and with (finite) expected value $\mu$, we have

$$\lim_{t\to\infty}\frac{1}{t}\sum_{i=1}^{t}(y_i-\mu)=0 \tag{24}$$

except for a set of probability measure zero. In other words

$$\lim_{t\to\infty}\frac{1}{t}\sum_{i=1}^{t} y_i=\mu \tag{25}$$

except for a set of sequences which have (in total) zero probability of occurrence (c.f. the strong law of large numbers in [4] or [5]). But (23) and (25) imply (as a simple corollary of the definition of the limit of sequence) that there exists $T$ such that

$$\frac{1}{t}\sum_{i=1}^{t} y_i^L > \frac{1}{t}\sum_{i=1}^{t} y_i^0 \qquad \text{for } t \geqslant T \tag{26}$$

except on a set of probability zero; hence (17) holds except on a set of measure zero. Since

$$EU = \int U(W_0, W_1, \ldots)\, dP(W_0, W_1, \ldots) \tag{27}$$

is not affected by arbitrarily changing the value of $U$ on a set of measure zero, we have

$$EU(W_0, W_1^L, W_2^L, \ldots) \geqslant EU(W_0, W_1^0, W_2^0, \ldots). \tag{28}$$

Thus, given our simplifying assumption of an unchanging probability distribution of returns for a given strategy, the superiority of MEL follows quite generally.

The case in which opportunities change from period to period and, whether or not opportunities change, strategies may select different distributions at different times, is treated in the appendix. It is shown there that if a certain continuity condition holds, then MEL is optimal quite generally. If this continuity condition does not hold, however, then there can exist games for which MEL is not optimal.

In this respect the results for the unending game are similar to those for the sequence of games with constant $V(g)$. In the latter case we found that MEL was asymptotically optimal for the sequence of games if $V(g)$ was continuous, but could fail to be so if $V(g)$ was discontinuous. In the case of the unending game, the theorem is not concerned with asymptotic optimality in a sequence of games, but optimality for a single game. Given a particular continuity condition, MEL is the optimum strategy.

## V. Conclusions

The analysis of investment for the long run in terms of the weak law of large numbers, Breiman's strong law analysis, and the utility analysis of unending games presented here each imply the superiority of MEL under broad assumptions for the hypothetical investor of [1], [2], [7], [8], [9], [10]. The acceptance or rejection of a similar conclusion for the sequence-of-games formalization depends on the definition of asymptotic optimality. For example, if constant $V(g)$ rather than constant $U(W_T)$ is assumed, as this writer believed plausible on *a priori* grounds, then the conclusion of the asymptotic analysis is approximately the same (even in terms of where MEL fails) as those of the unending game.

I conclude, therefore, that a portfolio analyst should not be faulted for warning an investor against choosing $E, V$ efficient portfolios with higher $E$ and $V$ but smaller $E\log(1+R)$, perhaps not even presenting that part of the $E, V$ curve which lies above the point with approximate maximum $E\log(1+R)$, on the grounds that such higher $E, V$ combinations have greater variability in the short run and less "return in the long run".

## Appendix

Using the notation of footnote 7, we will show that if $U_{\text{low}}(\alpha) = U^{\text{hi}}(\alpha)$ for all $\alpha$ then MEL is an optimum strategy quite generally; whereas, if $U^{\text{hi}}(\alpha) > U_{\text{low}}(\alpha)$ for some $\alpha_0$, then a game can be constructed in which MEL is not optimum. $U_{\text{low}}(\alpha) = U^{\text{hi}}(\alpha)$ for all $\alpha$ is the "continuity condition" referred to in the text.

For $v = L$ or 0, indicating the MEL strategy or some other strategy, respectively, we define

$$y_t^v = L_t^v + u_t^v \tag{29}$$

where

$$L_t^v = E\{y_t^v \mid L_1^v, L_2^v, \ldots, L_{t-1}^v, u_1^v, u_2^v, \ldots, u_{t-1}^v\} \tag{30}$$

is the expected value of $y_t^v$ given the events prior to time $t$. From this follows

$$E\{u_t^v \mid L_1^v, \ldots, u_{t:1}^v\} = 0. \tag{31}$$

The $u_t^v$ (for a given $v$) are thus what Doob [4] refers to as "orthogonal" random variables, and Feller [5] calls "completely fair" random variables. Therefore, writing var for variance,

$$\sum_{n=1}^{\infty} \frac{\operatorname{var}(u_n^v)}{n^2} < \infty \tag{32}$$

implies

$$\frac{1}{n}\sum_{t=1}^{n} u_t^v \qquad \text{converges to 0 almost always.} \tag{33}$$

(In particular, (32) holds if the var($u_n^v$) are bounded.) In addition to now assuming condition (32) we will also assume that the game is such that

$$\lim_{n\to\infty} \frac{1}{n}\sum_{i=1}^{n} L_i^L \qquad \text{exists almost always.} \tag{34}$$

This is the case, for example, when the same distributions are available each time, whether or not "the other" strategy uses a constant distribution. Since $L_i^L \geqslant L_i^0$ always, we have

$$\lim_n \frac{1}{n}\sum_{i=1}^{n} L_i^L = \limsup_n \frac{1}{n}\sum_{i=1}^{n} L_i^L \geqslant \limsup_n \frac{1}{n}\sum_{i=1}^{n} L_i^0 \qquad \text{always.} \tag{35}$$

Thus when (32) holds we have

$$\lim \frac{1}{n}\sum_{i=1}^{n} y_i^L \geqslant \limsup \frac{1}{n}\sum_{i=1}^{n} y_i^0 \qquad \text{almost always.} \tag{36}$$

In general,

$$\alpha = \limsup \frac{1}{n}\sum y_i^v \qquad \text{implies} \qquad U^{\text{hi}}(\alpha) \geqslant U(W_0, W_1^v, W_2^v, \ldots); \tag{37}$$

(since there always exists another series $y_1^*, y_2^*, \ldots$ such that

$$\alpha = \lim \frac{1}{n}\sum_{i=1}^{n} y_i^*$$

and

$$\frac{1}{n}\sum_{i=1}^{n} y_i^* \geqslant \frac{1}{n}\sum_{i=1}^{n} y_i^v \qquad \text{for all } n;$$

hence

$$U^{hi}(\alpha) \geqslant U(W_0, W_1^*, W_2^*, \ldots) \geqslant U(U_0, W_1^0, W_2^0, \ldots)).$$

If we now add to the assumptions expressed in equations (32) and (34), the assumption that $U^{hi}(\alpha) = U_{low}(\alpha)$ for all $\alpha$, we get directly from (36) and (37) that

$$EU^L \geqslant EU^0.$$

Conversely, the following is an example in which $U^{hi}(\alpha_0) > U_{low}(\alpha_0)$ and which MEL is not optimum: let $W_0 = 1$ and suppose that for some fixed positive $\alpha$ we have $U(1, 0.5, 0.5(1+\alpha), 0.5(1+\alpha)^2, \ldots)$ equals

$$U\big(1, (1+\alpha), (1+\alpha)^2, (1+\alpha)^3, \ldots\big) < U\big(1, 1.5, 1.5(1+\alpha), 1.5(1+\alpha)^2, \ldots\big).$$

With such a $U$-function it would be better to take a 50-50 chance of $W_1 = 0.5$ or 1.5 followed by $W_t = (1+\alpha)W_{t-1}, t \geqslant 2$, rather than have $W_t = (1+\alpha) \cdot W_{t-1}$ with certainty for $t \geqslant 1, \ldots$.

While the above shows that MEL can fail to be optimal when $U^{hi}(\alpha) > U_{low}(\alpha)$ for some $\alpha$, recall that we can have $U^{hi}(\alpha) > U_{low}(\alpha)$ for at most a countable number of values of $\alpha$. Thus MEL is optimal in a game in which

$$\alpha = \lim_n \frac{1}{n} \sum_{i=1}^{n} L_i^L$$

has a continuous distribution, or in which $\alpha$ has a discrete or mixed distribution but in which none of the points of discontinuity of the cumulative probability distribution of $\alpha$ have $U^{hi}(\alpha) > U_{low}(\alpha)$.

## REFERENCES


1. Leo Breiman. "Investment Policies for Expanding Businesses Optimal in a Long Run Sense," *Naval Research Logistics Quarterly*, 7:4, 1960, pp. 647–651.
2. ———. "Optimal Gambling Systems for Favorable Games," *Fourth Berkeley Symposium on Probability and Statistics, I*, 1961, pp. 65–78.
3. Daniel Bernoulli. "Exposition of a New Theory on the Measurement of Risk," *Econometrica*, XXII, January 1954, pp. 23–63. Translated by Louise Sommer—original 1738.
4. J. L. Doob. *Stochastic Processes*, John Wiley and Sons, New York, 1953.
5. William Feller. *An Introduction to Probabality Theory and Its Applications*, Volume II, John Wiley and Sons, New York, 1966.
6. M. B. Goldman. "A Negative Report on the 'Near Optimality' of the Max-Expected-Log Policy As Applied to Bounded Utilities for Long Lived Programs." *Journal of Financial Economics*, Vol. 1, No. 1, May 1974.
7. J. L. Kelly, Jr. "A New Interpretation of Information Rate," *Bell System Technical Journal*, pp. 917–926, 1956.
8. H. A. Latané. "Rational Decision Making in Portfolio Management," Ph.D. dissertation, University of North Carolina, 1957.
9. ———. "Criteria for Choice Among Risky Ventures," *Journal of Political Economy*, April 1959.
10. H. M. Markowitz. *Portfolio Selection: Efficient Diversification of Investments*, John Wiley and Sons, New York, 1959; Yale University Press, 1972.
11. ———. "Investment for the Long Run," Rodney L. White Center for Financial Research Working Paper no. 20-72 (University of Pennsylvania) 1972.

12. R. C. Merton and P. A. Samuelson. "Fallacy of the Log-Normal Approximation to Optimal Portfolio Decision-Making Over Many Periods," *Journal of Financial Economics*, Volume 1 No. 1, May 1974.
13. Karl Menger. "Das Unsicherheitsmoment in der Wertlehre. Betrachtungen im Anschluss an das sogenannte Petersburger Spiel," *Zeitschrift für Nationalökonomie*, Vol. 5, 1934. Translated in *Essays in Mathematical Economics in Honor of Oskar Morgenstern*, M. Shubik ed., Princeton University Press, 1967.
14. P. A. Samuelson. "Risk and Uncertainty: A Fallacy of Large Numbers," *Scientia*, 6th Series, 57th year, April–May 1963.
15. ———. "Lifetime Portfolio Selection by Dynamic Stochastic Programming," *Review of Economics and Statistics*, August 1969.
16. W. E. Yong and R. M. Trent. "Geometric Mean Approximation of Individual Security and Portfolio Performance," *Journal of Financial and Quantitative Analysis*, June 1969.
17. John von Neuman and Oskar Morgenstern. *Theory of Games and Economic Behavior*, Princeton University Press, 1944. John Wiley and Sons, 1967.

# Chapter 6

# Baruch College (CUNY) and Daiwa Securities

## Comments

The articles in this chapter are the results of a period in which I began as a professor at Baruch College, then, continuing as a professor, I came to head a group at Daiwa Securities Trust Company. All the articles in this chapter refer to portfolio theory and its use in one manner or another. Another output of this period was a 1987 book on mean-variance analysis. This book went out of print and was later reprinted with Chapter 13 rewritten by Peter Todd. The original Chapter 13 presented a critical line algorithm program written in SIMSCRIPT. The later version by Markowitz and Todd (2000) had the program rewritten in Visual Basic for Applications (VBA) at the request of the new publisher, Frank Fabozzi.

## References


Gew-rae Kim and Markowitz, H. M. (1989). *Investment Rules, Margin, and Market Volatility.* The Journal of Portfolio Management, Vol. 16, No. 1, pp. 45–52.

Markowitz, H. M. (1990). *Risk Adjustment.* Journal of Accounting, Auditing and Finance, Vol. 5, Nos. 1–2, Winter/Spring, pp. 213–225.

Markowitz, H. M. (1990). *Normative Portfolio Analysis: Past, Present, and Future.* Journal of Economics and Business, Vol. 42, No. 2, pp. 99–103.

Markowitz, H. M. (1991). *Individual versus Institutional Investing.* Financial Service Review, Vol. 1, pp. 1–9.

Markowitz, H. M. (1991). *Foundations of Portfolio Theory.* The Journal of Finance, Vol. 46, No. 2, pp. 469–477.

Markowitz, H. M., Todd, P., Xu, G. and Yamane, Y. (1992). *Fast Computation of Mean-Variance Efficient Sets Using Historical Covariances.* Journal of Financial Engineering, 1(2), pp. 117–132.

Markowitz, H. M., Todd, P., Xu, G. and Yamane, Y. (1993). *Computation of Mean-Semivariance Efficient Sets by the Critical Line Algorithm.* Annals of Operations Research, Vol. 45, pp. 307–317.

Markowitz, H. M. and Xu, G. (1994). *Data Mining Corrections.* The Journal of Portfolio Management, Vol. 21, No. 1, pp. 60–69.

# Investment rules, margin, and market volatility

*Depending on the relative number of portfolio insurers, crashes like that of October 1987 may not be uncommon.*

*Gew-rae Kim and Harry M. Markowitz*

The Brady Commission report [1988] ascribes the stock market break in mid-October 1987 in large part to "mechanical, price-insensitive selling by a number of institutions employing portfolio insurance strategies." Some dispute this view. For example, Leland and O'Brien argue in an interview in *Barron's* (Norris [1988]) that portfolio insurance could hardly have been a major cause of the break, because portfolio insurance transactions on October 19 amounted to only two-tenths of 1% of the value of outstanding stock.

Our article reports results of a discrete event simulation of a stock market composed of two kinds of participants: rebalancers and portfolio insurers. All participants hold only two assets: "stock" and cash. Rebalancers seek to maintain some fixed proportion between the two assets. Portfolio insurers follow a Constant Proportion Portfolio Insurance (CPPI) rule for shifting between stock and cash.[1] Simulation runs show how the volatility of the stock market depends on parameters such as the number of rebalancers versus CPPI investors, the level of the floor that the CPPI investors seek to protect, and whether the CPPI investors are allowed (and avail themselves of) margin.

The results we report are intended to show qualitatively the effects of margin and investment practice, rather than quantitatively reproduce current market volatility. We find that if margin is not used, the market is not explosive, but that the standard deviation of daily return increases manyfold as the number of CPPI investors increases from zero to 100 out of 150 investors. When 33% margin is allowed, a market with 75 CPPI investors and 75 rebalancers is explosive.

The simulation is asynchronous. That is, the model does not assume that stock price, and participant purchases and sales, evolve continuously in time, as do differential equation models, such as the closely related model of Anderson and Tutuncu [1988]. Nor does it assume that all investors' actions occur, synchronously, at discrete points in time. Rather, the "process" that simulates the actions of any one investor can schedule events at arbitrary future times.

In the model investors review their portfolios periodically, and perhaps place orders determined by their rebalancing or CPPI calculations. (The setting of price limits, and the changing of these limits or cancelling of orders, is described in the Appendix.) Depending on the limit price and the quantity ordered, an incoming buy (sell) order is executed immediately (perhaps only in part) if the list of current sell (buy) orders includes one or more orders that meets the price specification. Otherwise the uncompleted portion of the buy (sell) order is placed on the buy (sell) list. Thus prices and quantities are determined endogenously, by supply and demand.

The model's investors may be thought of as pension funds or investment companies. Each investor (pension fund) has deposits and withdrawals generated at random times in random amounts. All other quantities, such as trade-to-trade or day-to-day

GEW-RAE KIM is a doctoral student in the Department of Finance and Economics at Baruch College of the City University of New York (NY 10010). HARRY M. MARKOWITZ is Marvin Speiser Distinguished Professor of Finance and Economics at Baruch College.

changes in prices, are generated by the working of the simulated market.

This simulated market is highly simplified in several ways. First, it includes only two types of investors. Second, there are a limited number of these, 150 in total. The frequency and amount of deposits and withdrawals, and the lengths of insurance periods, are chosen so as to result in a reasonable level of transactions in this small market. Third, cash pays no interest; stocks pay no dividends; and stocks are bought and sold without commission.

The number of investors of each type can be changed by input to the model, as can parameters that specify the exact investment and trading rules. The inclusion of new types of investors would require new programming. For the most part this would require writing a SIMSCRIPT II.5 routine to describe the investment calculation of the new type of investor.[2] (A copy of the current program may be obtained through the authors.)



### THE SIMULATED MARKET

Table 1 lists parameters common to six runs described in Tables 2 through 5. A brief description of these parameters will serve as an introduction to the nature of the simulated market in general, as well as the specific simulations in Tables 2 through 5.

For ease of model input, each simulation run contains one or more investor "prototypes." An investor is of one or another of these prototypes. In all the runs reported there are two rebalancer prototypes and one CPPI investor prototype. All investors of a given prototype have the same initial attributes, e.g., the same starting value of portfolio, or the same mean time between deposits or withdrawals. This does not imply that each investor has the same simulated history, because the time and amount of deposits and withdrawals is random for each investor, as is the time when the investor reviews its portfolio.

One input to each simulation run is the number of individual investors of each prototype, to total 150 investors. The six runs reported in Table 2 have the number of CPPI investors as indicated. The balance of the 150 investors is split as evenly as possible between rebalancers of Prototype 1 and Prototype 2.

The price of the stock at the beginning of Day 1 is $100. As line 1 of the specifications indicates, each investor starts with $100,000. As line 2 indicates, CPPI investors put this half in stock and half in cash. The rebalancing investors differ only in one respect. Those

TABLE 1

Specifications: Base Case

| | Rebal1 | Rebal2 | CPPI |
|---|---|---|---|
| 1. Value of Starting Portfolio | $100,000 | $100,000 | $100,000 |
| 2. Initial Stock/(Stock + Cash) | 0.7 | 0.3 | 0.5 |
| 3. How Often Portfolio Reviewed (Days) | 5 | 5 | 5 |
| 4. How Often Deposit/Withdraw | 10 | 10 | 10 |
| 5. Minimum Deposit | −$8,000 | −$8,000 | −$8,000 |
| 6. Maximum Deposit | $9,000 | $9,000 | $9,000 |
| 7. Target Fraction of Stock/Total Assets | 0.50 | 0.50 | |
| 8. Rebalance if Actual is Less Than | 0.46 | 0.46 | |
| 9. Rebalance if Actual is Greater Than | 0.55 | 0.55 | |
| 10. Length of Insurance Plan | | | 65.0 |
| 11. Target Ratio of Stock to Cushion | | | 2.0 |
| 12. Buy Stock if Actual Ratio is: | | | 1.7 |
| 13. Sell Stock if Actual Ratio is: | | | 2.3 |
| 14. Cushion as a Fraction of Portfolio Value: At Start | | | 0.25 |
| 15. Cushion: Start of New Insurance Plan | | | 0.25 |
| 16. Maximum Ratio of Stock to Assets | | | 1.0 |

TABLE 2

Period Standard Deviation
Simulation Runs with Various Numbers
of CPPI Investors of 150 Total
CPPI Target Ratio: 2.0 Max Stock/Assets: 1.0

| | Number of CPPI Investors | | | | | |
|---|---|---|---|---|---|---|
| Quarter | 0 | 5 | 25 | 50 | 75 | 100 |
| 1 | 0.03216 | 0.02335 | 0.03810 | 0.06659 | 0.14575 | 0.12685 |
| 2 | 0.02622 | 0.02502 | 0.02565 | 0.09379 | 0.09231 | 0.25491 |
| 3 | 0.02874 | 0.01925 | 0.03433 | 0.07222 | 0.07599 | 0.08668 |
| 4 | 0.02768 | 0.02866 | 0.03283 | 0.05945 | 0.05915 | 0.12818 |
| 5 | 0.01770 | 0.02221 | 0.02430 | 0.04111 | 0.08425 | 0.09209 |
| 10 | 0.01873 | 0.01404 | 0.02349 | 0.05819 | 0.04411 | 0.10333 |
| 20 | 0.01835 | 0.01395 | 0.01309 | 0.03276 | 0.06275 | 0.06948 |
| 40 | 0.01294 | 0.00687 | 0.00953 | 0.04262 | 0.04703 | 0.06667 |
| 60 | 0.00756 | 0.00495 | 0.00692 | 0.04210 | 0.05174 | 0.04093 |
| 80 | 0.00278 | 0.00621 | 0.01299 | 0.03823 | 0.04987 | 0.07991 |
| 100 | 0.00214 | 0.00642 | 0.00939 | 0.02875 | 0.05267 | 0.05435 |

TABLE 3

Maximum Daily Return During Period
Simulation Runs with Various Numbers
of CPPI Investors of 150 Total
CPPI Target Ratio: 2.0 Max Stock/Assets: 1.0

| | Number of CPPI Investors | | | | | |
|---|---|---|---|---|---|---|
| Quarter | 0 | 5 | 25 | 50 | 75 | 100 |
| 1 | 0.09368 | 0.06458 | 0.10720 | 0.17767 | 0.55402 | 0.31322 |
| 2 | 0.08296 | 0.06226 | 0.12102 | 0.37494 | 0.32869 | 0.61738 |
| 3 | 0.07208 | 0.07192 | 0.12778 | 0.24484 | 0.17446 | 0.17895 |
| 4 | 0.07208 | 0.11216 | 0.10440 | 0.13248 | 0.17204 | 0.46847 |
| 5 | 0.03261 | 0.06205 | 0.07208 | 0.13971 | 0.33728 | 0.37759 |
| 10 | 0.06131 | 0.04327 | 0.08744 | 0.14959 | 0.19603 | 0.46354 |
| 20 | 0.09369 | 0.03030 | 0.04060 | 0.15229 | 0.19385 | 0.32129 |
| 40 | 0.05090 | 0.03030 | 0.04060 | 0.18129 | 0.13057 | 0.22841 |
| 60 | 0.03030 | 0.03030 | 0.02010 | 0.19603 | 0.19602 | 0.08883 |
| 80 | 0.01000 | 0.03030 | 0.07208 | 0.07485 | 0.21217 | 0.33387 |
| 100 | 0.01000 | 0.05101 | 0.07214 | 0.08275 | 0.22007 | 0.14936 |

TABLE 4

Minimum Daily Return During Period
Simulation Runs with Various Numbers
of CPPI Investors of 150 Total
CPPI Target Ratio: 2.0 Max Stock/Assets: 1.0

| | Number of CPPI Investors | | | | | |
|---|---|---|---|---|---|---|
| Quarter | 0 | 5 | 25 | 50 | 75 | 100 |
| 1 | −0.06149 | −0.06793 | −0.14915 | −0.19137 | −0.43072 | −0.34455 |
| 2 | −0.04915 | −0.06789 | −0.04133 | −0.21464 | −0.26330 | −0.62035 |
| 3 | −0.04925 | −0.04416 | −0.09098 | −0.17413 | −0.25283 | −0.31934 |
| 4 | −0.07661 | −0.08644 | −0.08443 | −0.17989 | −0.13404 | −0.25006 |
| 5 | −0.04901 | −0.06803 | −0.05936 | −0.18209 | −0.25853 | −0.18668 |
| 10 | −0.03960 | −0.02980 | −0.05795 | −0.20787 | −0.08653 | −0.24788 |
| 20 | −0.04901 | −0.04925 | −0.03950 | −0.10462 | −0.23914 | −0.19031 |
| 40 | −0.03950 | −0.02010 | −0.03902 | −0.17301 | −0.29226 | −0.21830 |
| 60 | −0.02980 | −0.01010 | −0.04910 | −0.10922 | −0.20682 | −0.19853 |
| 80 | −0.01010 | −0.02005 | −0.04901 | −0.22028 | −0.15701 | −0.28231 |
| 100 | −0.01000 | −0.01010 | −0.02010 | −0.16150 | −0.18510 | −0.20646 |

of Prototype 1 start with 70% in stock, and those of Prototype 2 start with 30% in stock. As each desires 50% in stock (see line 7), the economy starts in a state of disequilibrium.

Investors review their portfolios periodically to determine whether their rebalancing or portfolio insurance objectives are being met. The time between portfolio reviews is exponential, with mean time five days for all investors in all reported runs. Cash is added or withdrawn from the investor's account at random times, with an exponential distribution whose mean has been set to ten days for all investors in all runs. The amount deposited (+) or withdrawn (−) is generated by a uniform distribution with specified minimum and maximum. The minimum and maximum is set at −8,000 to +9,000 for all investors and runs. If the percent of stock to total value is less than 46% at the time of portfolio review, lines 7, 8, and 9 of the specifications indicate that all rebalancing investors will buy a number of shares (rounded to an integer) to raise their holding to 50%. Conversely, if holdings exceed 55%, rebalancing investors sell shares to reduce the proportion to 50%.

A CPPI insurance plan works as follows. Each plan has a start date and a duration. In all runs reported the duration of the CPPI plan is sixty-five (trading) days, considered to be one quarter. As the simulation opens, the number of days until the end of the current plan is drawn uniformly between one and sixty-five. When the current plan expires, another plan of one-quarter duration is started. Another is started when that one terminates, and so on.

While in fact CPPI plans are frequently of longer duration, we chose a one-quarter duration for our CPPI investors to increase activity in this 150-investor economy. The expiration dates of the various plans were spread out, so that simulation runs with large numbers of CPPI investors would not be destabilized by plans starting on the same day.

At the beginning of a plan, a floor is defined as

$$\text{Floor} = \text{Assets}_0 - \text{Cushion}_0$$

where $\text{Cushion}_0$ is a fraction, $\alpha$, of $\text{Assets}_0$. After the start day of the plan, the cushion varies with assets according to

$$\text{Cushion}_t = \text{Assets}_t - \text{Floor}.$$

The target stock position at time t is a multiple of the cushion:

$$\text{Target Value of Stock}_t = k \times \text{Cushion}_t.$$

The floor may be written as $(1 - \alpha)\ \text{Assets}_0$. During the course of the plan, the floor is increased by $(1 - \alpha)$ times value of deposit, for any deposit. A withdrawal is a negative deposit and reduces the floor.

An example will clarify the need for a floor adjustment mechanism. Suppose an investor (investment company) has $\alpha = 0.1$, $k = 5$, and $\text{Assets}_0 = \$100{,}000$. Floor then is \$90,000, and stock held is \$50,000. Suppose that, before the stock price changes, \$10,000 is withdrawn. If the floor is not adjusted, the cushion becomes zero, and the investor is directed to sell all its shares. With the adjustment, the floor becomes \$81,000, the cushion \$9,000, and the investor is directed to sell \$5,000 of its shares. In general the adjustment of the floor moderates the CPPI investor's actions. In practice the floor is also adjusted by an interest rate factor. The latter adjustment disappears if the interest rate is zero, as assumed in this model. (We are indebted to Lawrence Fisher for calling our attention to the need to adjust the floor for deposits and withdrawals.)

In the runs in Table 2, the CPPI investors use

TABLE 5

Period Trading Volume
Simulation Runs with Various Numbers
of CPPI Investors of 150 Total
CPPI Target Ratio: 2.0 Max Stock/Assets: 1.0

| | Number of CPPI Investors | | | | | |
|---|---|---|---|---|---|---|
| Quarter | 0 | 5 | 25 | 50 | 75 | 100 |
| 1 | 19,501 | 18,419 | 17,717 | 37,331 | 85,188 | 97,163 |
| 2 | 4,634 | 4,450 | 4,950 | 53,363 | 52,330 | 89,749 |
| 3 | 3,958 | 3,653 | 4,729 | 32,993 | 39,122 | 57,059 |
| 4 | 4,218 | 3,381 | 4,552 | 28,789 | 18,767 | 52,227 |
| 5 | 3,369 | 3,360 | 3,465 | 10,863 | 17,662 | 29,918 |
| 10 | 2,367 | 2,298 | 2,473 | 9,100 | 9,320 | 22,692 |
| 20 | 1,285 | 1,441 | 1,189 | 3,158 | 11,750 | 8,532 |
| 40 | 626 | 590 | 429 | 3,180 | 3,797 | 7,303 |
| 60 | 280 | 326 | 196 | 3,777 | 4,437 | 3,884 |
| 80 | 154 | 125 | 634 | 2,641 | 3,782 | 3,430 |
| 100 | 126 | 73 | 133 | 2,271 | 2,696 | 2,440 |

k = 2 and an initial cushion of one-fourth of assets. As a consequence, at the start of each plan the target value of stock is half of assets. At the beginning of the simulation we start the CPPI investors with actual stock positions equal to target positions. Thus any instability we see as we increase the proportion of CPPI investors in the population is not attributable to their own start-of-simulation trading requirements.

With $\alpha = 0.25$, the plan seeks to guarantee that the portfolio can lose no more than 25% of assets during the life of the plan. Cases with $\alpha = 0.10$, reported below, illustrate the effect on the market of different $\alpha$. For $\alpha = 0.10$, we let k = 5 so that initial stock assets remain at 0.5.

The CPPI formula may indicate that stock should exceed assets. This can be done if margin is available, but may be against the investor's policy. In the runs reported in Table 2 a maximum stock/portfolio value ratio of 1 is assumed. In other words, no margin is used. Other runs, reported below, allow a greater maximum stock/portfolio value when the CPPI formula indicates.



### EFFECT OF VARYING THE NUMBER OF PORTFOLIO INSURERS

#### Base Case

Tables 2 through 5 summarize six simulated runs, namely, one run each with the number of CPPI investors equal to 0, 5, 25, 50, 75, and 100 out of 150 investors and with inputs as tabulated in Table 1. Kim and Markowitz [1988] show that each of ten replications of the 0 CPPI and 50 CPPI cases produces the same qualitative conclusions. Table 2 shows the standard deviation of daily return, $R_t = (P_t - P_{t-1})/P_{t-1}$, on the stock during various quarters (sixty-five-day periods) in simulation runs with various numbers of CPPI investors. Tables 3 and 4 show the greatest one-day gain and the greatest one-day loss during the same quarters. Table 5 shows the volume of transactions for the entire quarter.

Recall that the simulation is started with half the rebalancers with substantially too much stock and half with too little. Not surprisingly, therefore, in the case with rebalancers only (0 CPPI investors) there is much more price volatility and trading volume in the first quarter than in subsequent quarters. Daily standard deviation of return falls almost monotonically from 0.032 in the first quarter to 0.002 in the one hundredth quarter. (The fall is not perfectly monotonic because this is the result of a single run with random components.) The range of returns, from greatest loss to greatest gain, shrinks similarly from (−0.061, + 0.094) in the first quarter to (−0.01, +0.01) in the hundredth. Volume falls from 19,501 shares during the first quarter to 126 shares during the hundredth.

As the proportion of CPPI investors increases, return volatility and volume generally increase. For example, for the run with 50 CPPI investors (i.e., one-third of the investor population), first-quarter standard deviation of return is twice that of the all-rebalancer case (0.0666 versus 0.0322). The standard deviation remains relatively high for the 50 CPPI case, ending about ten times as great as in the all-rebalancer case (0.0288 versus 0.0021). For the case with 50 CPPI investors, and for all the quarters reported in Table 4, a daily loss of at least 10% (for the day!) occurred in every quarter. The eightieth quarter, for example, included a day when the market dropped 22%, while the hundredth quarter included a day that lost 16%.

Thus in this test tube world with one-third CPPI investors and two-thirds rebalancers, crashes of the magnitude of October 19, 1987, happen almost every quarter. Market volatility increases as the number of CPPI investors out of the 150 is increased to 75 or 100. The volatility is especially great in the first quarter or two when disequilibrium among the rebalancers gives the market a bit of a kick. In the run with 100 CPPI investors, for example, the second quarter includes a day when the market dropped 62%!

Figure 1 plots the closing stock price for 160 days, for the simulation runs with 0, 50, and 100 CPPI investors. The first of these is a rather dull market. The second is relatively wild, even by October 1987 standards. For example, days 58 through 64 experienced consecutive losses of 9.5%, 9.9%, 19.1%, 7.6%, 2.2%, 13.2%, and 6.3%; days 86 and 87 had gains of 13.2% and 37.5%; days 94 through 96 had losses of 14.4%, 21.5%, and 19.1%. The third line, which represents prices for the run with 100 CPPI investors, shows even more spectacular rises and falls — with

FIGURE 1

NO, 50, AND 100 CPPI INVESTORS (TARGET: 2, MAX 1.0)

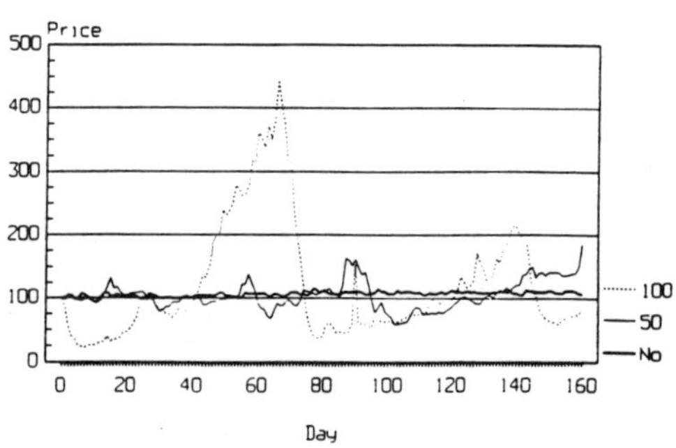

the stock price falling from 100 to 25, then rising above 400 and falling below 50 in the first eighty days.

#### Effects of Margin and Floor

Table 6 reports runs that are identical to those in Tables 2 through 5 except in one respect: CPPI investors are allowed a maximum ratio of stock to assets of 1.5. This is equivalent to 33.3% margin.

TABLE 6

Period Standard Deviation
Simulation Runs with Various Numbers
of CPPI Investors of 150 Total
CPPI Target Ratio: 2.0 Max Stock/Assets: 1.5

| | Number of CPPI Investors | | | | | |
|---|---|---|---|---|---|---|
| Quarter | 0 | 5 | 25 | 50 | 75 | 100 |
| 1 | 0.03216 | 0.02335 | 0.03810 | 0.06659 | 0.51475 | 0.31951 |
| 2 | 0.02622 | 0.02502 | 0.02565 | 0.09240 | ****** | ****** |
| 3 | 0.02874 | 0.01925 | 0.03433 | 0.08207 | ****** | ****** |
| 4 | 0.02768 | 0.02866 | 0.03283 | 0.06027 | ****** | ****** |
| 5 | 0.01770 | 0.02221 | 0.02430 | 0.04950 | ****** | ****** |
| 10 | 0.01873 | 0.01404 | 0.02349 | 0.03621 | ****** | ****** |
| 20 | 0.01835 | 0.01395 | 0.01309 | 0.05508 | ****** | ****** |
| 40 | 0.01294 | 0.00687 | 0.00953 | 0.02935 | ****** | ****** |
| 60 | 0.00756 | 0.00495 | 0.00692 | 0.03003 | ****** | ****** |
| 80 | 0.00278 | 0.00621 | 0.01299 | 0.04475 | ****** | ****** |
| 100 | 0.00214 | 0.00642 | 0.00939 | 0.04026 | ****** | ****** |

Note: Specifications are identical to those given in Table 1, except that the maximum ratio of stock to assets is 1.5.

This limited margin and a sufficient number of CPPI investors results in an explosive market. The simulation was halted when price reached $10^9$, which occurred during the second quarter in runs with 75 or 100 CPPI investors. Figure 2 shows closing prices for the economy with no CPPI investors, 75 CPPI investors with no margin, and 75 CPPI investors with 33% margin. In the latter case, specifications allow prices to increase sharply, expanding cushions, which leads to stock purchases that raise prices, in an explosive cycle.

FIGURE 2

NO AND 75 CPPI INVESTORS (TARGET: 2, MAX: 1.0 AND 1.5)

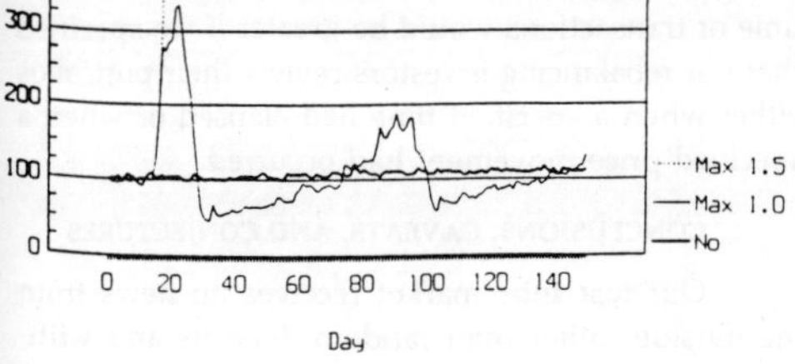


In runs in Table 6 with 50 or fewer CPPI investors, the market did not explode; price volatility increased little if at all as compared to cases without margin. Thus, at least in this test tube economy, the effect of permitting stock purchase on margin can differ widely, depending on the proportion of investors who buy when prices rise, as do CPPI investors, as opposed to those who sell as prices rise, as do rebalancers.

Table 7 shows the effect on the market if CPPI investors seek to protect a higher floor. In this case $\alpha = 0.1$, that is, CPPI investors seek to limit losses to at most 10% during any one plan period. The target ratio of stock to cushion is set at $k = 5$ so that the desired proportion of stock at the start of a plan is 50%. Note that price volatility is appreciably higher in this case than in the corresponding cases in Table 2 for small numbers of CPPI investors, especially 5 and 25 and especially in later quarters.

TABLE 7

Period Standard Deviation
Simulation Runs with Various Numbers
of CPPI Investors of 150 Total
CPPI Target Ratio: 5.0 Max Stock/Assets: 1.0

| | Number of CPPI Investors | | | | | |
|---|---|---|---|---|---|---|
| Quarter | 0 | 5 | 25 | 50 | 75 | 100 |
| 1 | 0.03216 | 0.04165 | 0.06616 | 0.08383 | 0.11016 | 0.13868 |
| 2 | 0.02622 | 0.03788 | 0.06187 | 0.08616 | 0.10812 | 0.12594 |
| 3 | 0.02874 | 0.04144 | 0.06127 | 0.07272 | 0.10191 | 0.12281 |
| 4 | 0.02768 | 0.03576 | 0.06629 | 0.06266 | 0.09154 | 0.11083 |
| 5 | 0.01770 | 0.03568 | 0.06224 | 0.07339 | 0.11980 | 0.10562 |
| 10 | 0.01873 | 0.03463 | 0.05204 | 0.06246 | 0.07571 | 0.08871 |
| 20 | 0.01835 | 0.02609 | 0.05417 | 0.05899 | 0.06237 | 0.08583 |
| 40 | 0.01294 | 0.03743 | 0.05428 | 0.06688 | 0.06389 | 0.05950 |
| 60 | 0.00756 | 0.03446 | 0.04532 | 0.03900 | 0.06478 | 0.04766 |
| 80 | 0.00278 | 0.03348 | 0.03986 | 0.05131 | 0.03709 | 0.05017 |
| 100 | 0.00214 | 0.03261 | 0.02714 | 0.04866 | 0.05126 | 0.04681 |

Note: Specifications differing from base case:

| | CPPI |
|---|---|
| 11. Target Ratio of Stock to Cushion | 5.0 |
| 12. Buy Stock if Actual Ratio is: | 4.7 |
| 13. Sell Stock if Actual Ratio is: | 5.3 |
| 14. Cushion as a Fraction of Portfolio Value: At Start | 0.1 |
| 15. Cushion: Start of New Insurance Plan | 0.1 |

Table 8 has $\alpha = 0.1$ and $k = 5$, as in Table 7, but with a maximum stock to asset ratio of 1.5. Runs with 50 or more CPPI investors are explosive. Runs with 5 and 25 CPPI investors are not explosive, but they are considerably more volatile, especially in the later quarters. In our test tube economy, the effect of introducing margin depends not only on the amount of margin permitted and the number of portfolio in-

TABLE 8

Period Standard Deviation
Simulation Runs with Various Numbers
of CPPI Investors of 150 Total
CPPI Target Ratio: 5.0 Max Stock/Assets: 1.5

| | Number of CPPI Investors | | | | | |
|---|---|---|---|---|---|---|
| Quarter | 0 | 5 | 25 | 50 | 75 | 100 |
| 1 | 0.03216 | 0.04165 | 0.06338 | 0.17581 | 0.30490 | 0.28043 |
| 2 | 0.02622 | 0.03788 | 0.07174 | ****** | ****** | ****** |
| 3 | 0.02874 | 0.04144 | 0.05250 | ****** | ****** | ****** |
| 4 | 0.02768 | 0.03576 | 0.05979 | ****** | ****** | ****** |
| 5 | 0.01770 | 0.03568 | 0.05931 | ****** | ****** | ****** |
| 10 | 0.01873 | 0.03463 | 0.05449 | ****** | ****** | ****** |
| 20 | 0.01835 | 0.02609 | 0.05255 | ****** | ****** | ****** |
| 40 | 0.01294 | 0.03080 | 0.04039 | ****** | ****** | ****** |
| 60 | 0.00756 | 0.03386 | 0.04440 | ****** | ****** | ****** |
| 80 | 0.00278 | 0.03212 | 0.04593 | ****** | ****** | ****** |
| 100 | 0.00214 | 0.04149 | 0.04644 | ****** | ****** | ****** |

Note: Specifications are identical to those given in Table 7, except that the maximum ratio of stock to assets is 1.5.

surers, but also on the level of the floor that the latter seek to protect.

**Volume of Transactions**

Leland and O'Brien in the 1988 Norris interview argue that portfolio insurance could hardly have caused the large drop in the market in October 1987, because the volume of transactions accounted for by portfolio insurance sales represented only two-tenths of 1% of the outstanding value of stock.[3] Figure 3 presents the closing price and volume of transactions for the simulation run with 50 CPPI investors with parameters as given in Table 1. The scale must be multiplied by 100 for volume. We will note two ways in which volume magnitude and pattern in this simulated run differ from that observed in fact, and suggest two model changes that may alleviate this discrepancy.

The number of shares in the simulated economy is constant at 75,000 throughout the run. Maximum volume of CPPI trades occurs on days 96, 97,

FIGURE 3

DAILY MARKET
(50 CPPI INVESTORS, TARGET: 2.0, MAX: 1.0)

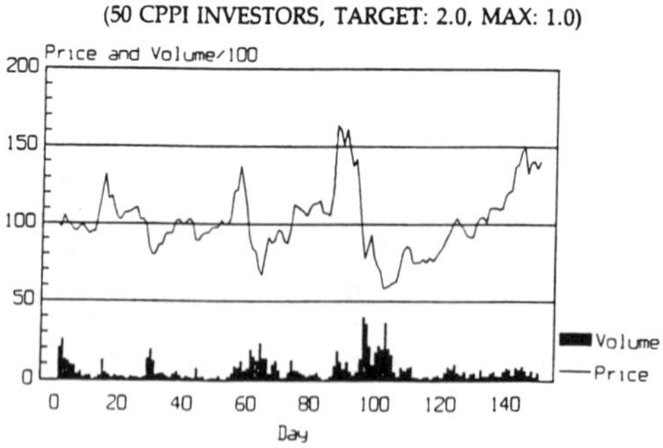


and 103 with 3,974, 3,548, and 3,619 shares trading, or about 5% of total volume. This is about twenty times as great as the portfolio insurance trading proportion observed on October 19, 1987.

This discrepancy can be cured by introducing a third type of investor: a buy-and hold investor. In the extreme, a buy-and-hold investor could start the simulation with a given stock position, and then make no transactions throughout the simulation period. No value reported in our tables and figures would be affected by the presence of these buy-and-hold investors, yet the volume of transactions as a proportion of outstanding stock would be reduced. If buy-and-hold investors held nineteen times as much stock as the rebalancer and CPPI investors, for example, then volume as a proportion of outstanding shares would be twentyfold smaller.

A second difference between the simulation and market behavior in October 1987 concerns the timing of high volume days. Of the three highest volume days, two are days on which price rose, following three or four days of large drops. This behavior differs from that experienced on October 19, 1987, when record volume was accompanied by record price decline.

It would seem as if the simulated rebalancers wisely wait for the portfolio insurers to finish dumping so they can pick up stock at bargain prices. Our rebalancers, though, have no such wisdom. When prices start falling from the peak, the cushions of the portfolio insurers shrink, so they seek to sell; prices are still high, so rebalancers who review their portfolios on these days are little inclined to buy. With a preponderance of sell orders to buy orders, price drops without record volume. As the price continues to decline, rebalancers find that buy orders are increasingly called for. Volume is greatest near the trough as rebalancers buy and insurers sell.

Maximum volume on days with maximum price falls (or rises) perhaps could be obtained if the model were modified as follows. In one model, rebalancers consider their portfolio periodically, at times that do not depend on the movements of the market. But on October 19, in fact, all or almost all large investors realized that large price movements had occurred, even before the day was over. The volume of transactions would be greater if we specified that our rebalancing investors review their portfolios either when a specified time had elapsed or when a specified price movement had occurred.

**CONCLUSIONS, CAVEATS, AND CONJECTURES**

Our test tube market receives no news from the outside, other than random deposits and withdrawals. The manyfold increase in price volatility accompanying an increase in the proportion of CPPI portfolio insurers is due solely to the working out of the rebalancer and CPPI formulas.

The other major brand of portfolio insurance (option replicating portfolio insurance) uses a different, more complicated, formula to determine purchases and sales. It is similar to CPPI in that it seeks to protect a floor by selling as prices drop and buying as prices rise (up to a point). The major exception to this pattern, as in the case of CPPI, is when one plan period ends and a new plan period begins. Further simulation would be required to determine whether option replicating portfolio insurance causes more or less market volatility than CPPI.

The man on the street (including Wall Street) has made much of the role of index futures and index arbitrage during the October 1987 break and thereafter. In fact, index futures have played an important role in facilitating portfolio insurance, but they would not play a similar role if we included them in our model. Portfolio insurance requires large changes in holdings of equities as the market fluctuates. In fact, commissions are cheaper, and execution problems usually reduced, if the equity position is varied by buying or selling the S&P 500 Futures Index rather than by buying or selling scores or hundreds of individual securities.

In the model, though, there is only one stock to be bought or sold, and there are no commissions. Introducing stock futures and their arbitrageurs would be like introducing two kinds of cash, one dollar bills and five dollar bills, which could be converted one to the other only every ten days, say. In the test tube world with zero interest, simulated arbitrageurs would keep the price of 100 five dollar bills from straying far from that of 500 one dollar bills — as they would keep a simulated stock futures price from straying far from the equivalent stock price.

(In the real world also, the enthusiasm for pressuring large firms to abandon the index arbitrage business seems misplaced. It is akin to rejecting the message by executing the messenger. The message on October 19, 1987, was that someone (especially portfolio insurers) was aggressively selling futures to lower their equity positions. This message was *transmitted* by index arbitrage in the manner spelled out in the Brady Commission report. When index arbitrage was interrupted — partly by policy and partly by deteriorating price information that hampered precise arbitrage — portfolio insurers switched from selling futures to selling stocks. This strongly suggests that, even if we could prevent everyone from arbitraging index futures and stocks — essentially the same commodity in two different markets — it would reduce stock market volatility little, as long as portfolio insurers are programmed to buy equities when prices rise and sell when prices fall.)

Portfolio insurers were not the only large sellers of stocks and futures on October 19. In addition to billions of dollars of futures and stock sales by portfolio insurers, "block sales by a few mutual funds accounted for about $900 million of stock sales."[4] Also, what the Brady Commission report refers to as "aggressive trading-oriented institutions" knew that portfolio insurers were programmed to sell, so sold first.

As the 1988 SEC report on the market break says, "we may never know what precise combination of investor psychology, economic developments and trading technologies caused the events of October." On the other hand, the working of the portfolio insurance formulas was surely a large component of the price movements of October 19. In particular, the Brady Commission report notes that the portfolio insurance formulas called for additional billions of dollars of stock or futures to be sold beyond the amount that was actually sold.

A frequent conclusion of research in any area is that more research is needed. This seems especially true in the present instance. Endless questions could be explored by varying parameters of our model. For example: What are the effects of varying the trading strategy parameters described in the Appendix? Of market size or of heterogeneity among investors? Other questions require modification to the model, particularly adding new types of investors. For your own research, send a floppy disk to us for a copy of the model.

## APPENDIX

### TRADING RULES AND PROCEDURES

This appendix presents the trading rules and certain other fine details of the market model. All numbers ("0.99," "1.01," . . .) are parameters that can be varied by input. The values given here were used in all simulation runs reported in this paper.

Before calculating a rebalancing or CPPI buy or sell, the investor obtains an estimate of the current stock price. If both bids and offers exist, price is estimated as midway between the best (highest) bid and the best (lowest) asked price. (The latter will exceed the former, or a trade would have occurred between the two.) In case there are no bids, price is estimated as 0.99 times the best offer; in case of no offers, it is estimated as 1.01 times the best bid. If there are neither bids nor offers, price is estimated as the last sale price. At the start of the simulation, last sale price is set at 100.

The parameters following may vary from one prototype to another, but we set them the same for all investors in all runs reported. As the text explains, the investor determines the number of shares of a buy or sell order by a rebalancing or CPPI calculation. In the case of a buy order, the investor initially bids at 1.01 times estimated stock price. This bid is reconsidered after 1.0 days if not filled before. Literally, the bid is withdrawn, the buy quantity recalculated at the current estimated price, and, if a buy order is still indicated, it is placed again at 1.01 times the new estimated price.

In the case of a sell order, the investor initially offers at 0.99 of estimated price, and reconsiders the offer (recomputes the sale quantity and resubmits the offer) after 1.0 days.

## REFERENCES

Anderson, R. W., and M. M. Tutuncu. "The Simple Price Dynamics of Portfolio Insurance and Program Trading." Columbia Futures Center Working Paper No. 173, 1988.

Black, F., and R. Jones. "Simplifying Portfolio Insurance." *Journal of Portfolio Management*, Fall 1988, pp. 48-51.

Kim, G., and H. M. Markowitz. "Investment Rules, Margin and Market Volatility." Baruch College Working Paper Series 88-21, 1988.

Kiviat, P. J., R. Villanueva, and H. M. Markowitz. "The SIMSCRIPT II.5 Programming Language." E. Russell, ed., CACI, 1983.

Leland, H. E. "Option Pricing and Replication with Transaction Costs." *Journal of Finance*, Vol. 40, No. 5, 1985, pp. 1283-1301.

——. "Who Should Buy Portfolio Insurance?" *Journal of Finance*, Vol. 35, No. 2, 1980, pp. 581-594.

Norris, R. "Maligned or Malign? Its Inventors Make the Case for Portfolio Insurance." *Barron's*, March 21, 1988.

"The October 1987 Market Break: A Report." Securities and Exchange Commission, Division of Market Regulation. Washington, DC: U.S. Government Printing Office, 1988.

Perold, A. F., and W. F. Sharpe. "Dynamic Strategies for Asset Allocation." *Financial Analysts Journal*, January-February 1988, pp. 16-27.

"Report of the Presidential Task Force on Market Mechanisms," Nicholas F. Brady, Chairman. (The Brady Commission Report.) Washington, DC: U.S. Government Printing Office, 1988.

Rubinstein, M. "Alternative Paths to Portfolio Insurance." *Financial Analysts Journal*, Vol. 41, No. 4, 1985, pp. 42-52.

Rubinstein, M., and H. E. Leland. "Replicating Options with Positions in Stock and Cash." *Financial Analysts Journal*, Vol. 37, No. 4, 1981, pp. 63-72.

[1] See Perold and Sharpe [1988] and Black and Jones [1987].

[2] SIMSCRIPT II.5 is the trademark of CACI, International. The language is described in Kiviat, Villanueva, and Markowitz [1983].

[3] Leland, O'Brien and Rubinstein sell "option replicating" portfolio insurance; see Leland [1980, 1985], Rubinstein [1985], and Rubinstein and Leland [1981]. The portfolio insurance we test in this paper is Constant Proportion Portfolio Insurance. The Brady Commission ascribes the break in mid-October in large part to the actions of portfolio insurers, without specifying technique. Further simulation would be required to determine if option replicating insurance is more benign than CPPI.

[4] The Brady Commission tells us that "this trading activity was concentrated in the hands of a surprisingly few institutions. On October 19, sell programs by three portfolio insurers accounted for just under $2 billion in the stock market; in the futures market three portfolio insurers accounted for the equivalent in value of $2.8 billion of stock."

# Risk Adjustment

HARRY M. MARKOWITZ*

As anyone knows who has shorted stocks in fact, the assumption that portfolios are chosen subject to the sole constraint $\Sigma X_i = 1$ is highly unrealistic. It would allow you to place \$1,000 with your broker, short a million dollars worth of stock A, and use the proceeds plus your \$1,000 to buy \$1,001,000 worth of stock B. If you do not know already, ask your broker how close this is to reality.

If we make the other usual CAPM assumptions but assume that the constraint set is a bounded polyhedron then the following conclusions typically fail to hold: (A) the market portfolio is an efficient portfolio; and (B) expected returns are linearly related to betas. Conclusions (A) and (B) fail to hold even if *some* investors are allowed $\Sigma X_i = 1$ as their sole constraint while others are subject to bounded constraint sets. Unfortunately, conclusion (B) is the basis of usual risk adjustment procedures.

These negative results have been reported before. The present paper considers how to risk adjust project returns in a CAPM world whose investors are constrained by any system of linear equality and/or inequality constraints in variables which may or may not be required to be nonnegative, with possibly heterogeneous investor constraint sets and beliefs. The remarkable thing about the results is that the risk adjustment formula for this general case is little more complex than the classical formula derived under highly restrictive assumptions.

The first of the following sections briefly reviews the failure of (A) and (B) for bounded constraints; the second derives a risk adjustment formula for the standard portfolio selection model with homogeneous beliefs and constraint sets. It also uses two "approximation assumptions." The next section drops one of the approximation assumptions, defends the other, and generalizes the analysis to heterogeneous beliefs and constraint sets. Then a final section comments on the results.

## Traditional CAPMs

The failure of (A) and (B) to hold if investors' constraint sets are bounded polyhedra is analyzed in Markowitz (1987). (Below, chapter numbers refer to chapters of the latter.) The failure of (A) and (B) may be illustrated in

*Department of Economics and Finance, Baruch College, City University of New York

Journal of Accounting, Auditing & Finance,
Conference on Productivity, Vol. 5, Nos. 1/2 (New Series),
Winter/Spring 1990.

terms of the standard 3-security mean-variance analysis with constraints $\sum_{i=1}^{3} X_i = 1$, $X_i \geq 0$, $i = 1,2,3$.

Markowitz (1952) analyzed the standard 3-security model graphically, plotting $X_1$ as the abscisa, $X_2$ as the ordinate and leaving $X_3 = 1 - X_1 - X_2$ implicit. As shown in Chapter 11, a frequently more revealing procedure is to transform this graph by

$$\begin{bmatrix} Z_1 \\ Z_2 \end{bmatrix} = \begin{bmatrix} \alpha_1 \\ \alpha_2 \end{bmatrix} + \begin{bmatrix} k_{11} & k_{12} \\ k_{21} & k_{22} \end{bmatrix} \begin{bmatrix} X_1 \\ X_2 \end{bmatrix} = \alpha + KX \tag{1}$$

where $\alpha$ and K are chosen so that V is minimized at $Z = 0$, and is of the form: $V_{MIN}$ plus a sum of squares. Ignoring the trivial case in which $V \equiv 0$, but permitting a singular covariance matrix, the transformation results in either

$$V = V_{MIN} + Z_1^2 + Z_2^2 \tag{2.1}$$

or, e.g.,

$$V = V_{MIN} + Z_1^2 \tag{2.2}$$

Here we will consider the case in which (2.1) results. In this case one can rotate the Z-coordinate system

$$Y = HZ \tag{3}$$

H orthogonal, so that portfolio expected return is

$$E = a_0 + a_1 Y_1 \tag{4.1}$$

$a_1 > 0$, as well as

$$V = V_{MIN} + Y_1^2 + Y_2^2 \tag{4.2}$$

This is always possible when (2.1) applies and the security expected returns are not all equal. A coordinate system in which (4.1) and (4.2) hold is referred to as a canonical form of the portfolio selection model.

We do not alter the set of EV efficient portfolios if we seek E*V rather than EV efficiency where

$$E^* = \frac{E - a_0}{a_1} \tag{5.1}$$

Then

$$E^* = Y_1 \tag{5.2}$$

Figure 1 presents standard 3-security portfolio analyses in canonical form, assuming that (2.1) applies. In the 1952 diagram the set of feasible portfolios is the area on and in the triangle

$$T = \{X \in R^2 : X_1 \geq 0, X_2 \geq 0, X_1 + X_2 \leq 1\} \tag{6}$$

Transformations (1) and (3) send T into a $\bar{T} = \overline{a\,b\,c}$. Any triangle $\bar{T}$ may be such a transformation.

The means, variances, and covariances which give rise to a particular $\bar{T}$, such as $\overline{a\,b\,c}$ in Figure 1a, may be determined as follows. Arbitrarily choose $a_0$, $a_1 > 0$, $V_{MIN} \geq 0$. From the coordinates $(Y_1^a, Y_2^a)$, etc., of the vertices a, b, c compute

$$\mu_a = a_0 + a_1 Y_1^a$$

$$V_a = V_{MIN} + (Y_1^a)^2 + (Y_2^a)^2$$

and similarly for b and c. Compute

$$\text{cov}(r_a, r_b) = V_{MIN} + Y_1^a Y_1^b + Y_2^a Y_2^b$$

and similarly for $\sigma_{bc}$ and $\sigma_{ac}$.

Since the V of a point is $V_{MIN}$ plus the square of the distance to the origin, among portfolios in a set S the one closest to the origin has minimum V. In particular, the Pythagorean theorem implies that, among feasible points with given E—that is, on a given vertical line—the one closest to the $Y_1$ – axis has minimum V.

In Figure 1a portfolio d, where the perpendicular from the origin meets ab, has minimum feasible variance. The set of efficient portfolios consists of line segments $\overline{d\,e}$, $\overline{e\,f}$, $\overline{f\,c}$. Suppose some investors buy portfolio g and the rest portfolio h. Then the market portfolio will be on the straight line connecting these; for example, at i. This point is not an efficient portfolio.

In Figure 1a the efficient set lies on segments of three critical lines. In practice, mean-variance efficient sets computed for large portfolios may contain segments from scores or hundreds of critical lines. This is not only true for the standard model but also for more complex models, for example, models which allow short positions but require collateral.

Suppose we make the usual CAPM assumptions that all investors have the same beliefs and are subject to the same constraints; but assume a standard constraint set, or other constraints involving inequalities or non-negative variables. Typically the efficient set will consist of many segments and, in such cases, typically the market will not be an efficient portfolio.

I say "typically" since "accidents" like that in Figure 1c are possible.

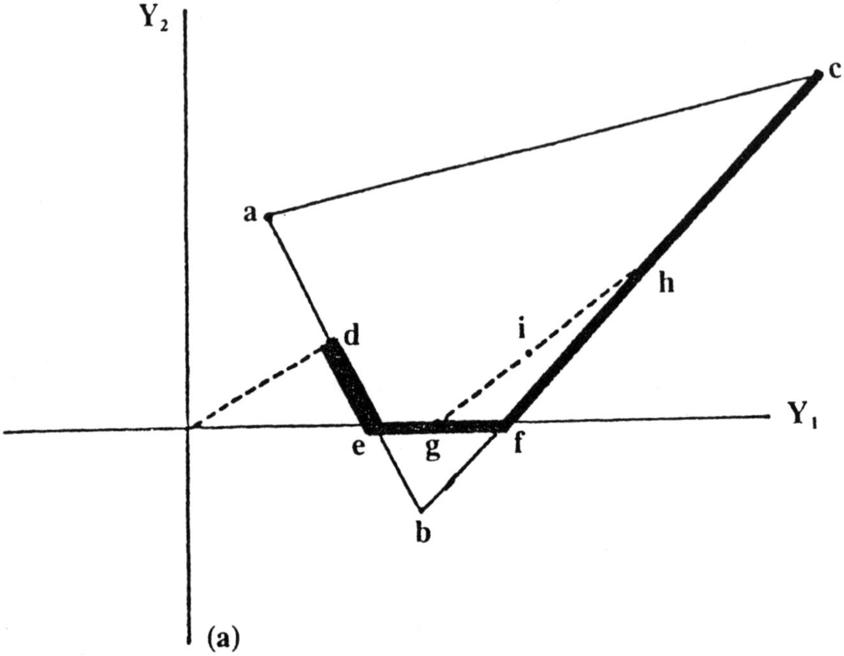


FIGURE 1a

As in Figure 1a, $\bar{T} = \overline{a\,b\,c}$ is the constraint set and d has smallest feasible V. In Figure 1c, the efficient segments are $\overline{d\,e}$, $\overline{e\,f}$, $\overline{f\,c}$. Suppose some investors choose portfolio g, the rest choose h. The market is on the line between g and h, possibly at i—which is efficient. Thus it is possible for the market to be an efficient portfolio even though investors select portfolios from more than one segment. But such is atypical; for example, a slight shift in investors between g and h will move the market off the efficient set. It is possible for the market to have almost *maximum* V for given E, such as portfolio g in 1b.

Chapter 12 shows that the $\mu_i$ and $\sigma_{ij}$ which give rise to figures 1a–c are consistent with market equilibrium. The following is an alternate, shorter demonstration. Suppose an economy of immortal investors has three securities, determines prices once a year, and for the last thousand years has arrived at the same prices for each security (say \$1 per share for each of the three issues). The same prices are sure to happen next year, therefore the only uncertainty is in dividend. We assume means, variances and covariances among dividends as shown in figure 1a (or 1b or 1c). We finally assume that the number of shares demanded with these parameters equal the fixed number of shares in existence. This provides a scenario by which any case in figure 1 is a long-run market equilibrium.

The reason why the Tobin (1968), Sharpe (1964), and Lintner (1965) model and the model with $\Sigma X_i = 1$ as its only constraint (analyzed by Roy (1952), Sharpe (1970), Merton (1972), and Black (1972)) imply that the market is an efficient portfolio is that both models imply that the set of efficient portfolios consists of at most one line segment. As long as the efficient set consists of one line segment then the market portfolio, which is a convex combination of points on this segment, must itself be on the segment, therefore efficient. But real world efficient sets consist of many line segments.

Another conclusion of the TSL and RSMB standard CAPMs is that

$$\mu_i = c_0 + c_1\beta_i \tag{7}$$

where $\mu_i$ is expected return of the ith security and $\beta_i$ is its regression against the return on the market portfolio. Chapter 12 shows that (7) holds if and only if the market portfolio lies on the $Y_1$ − axis.[1] As our figures show, this is atypical if the efficient set consists of more than one segment.

1. The "only if" part assumes that market portfolio has positive variance, else $\beta_i$ is undefined.

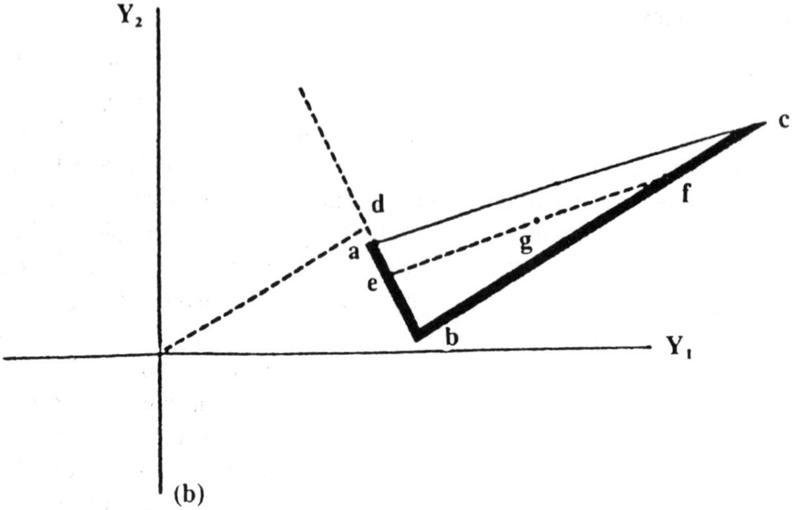


FIGURE 1b

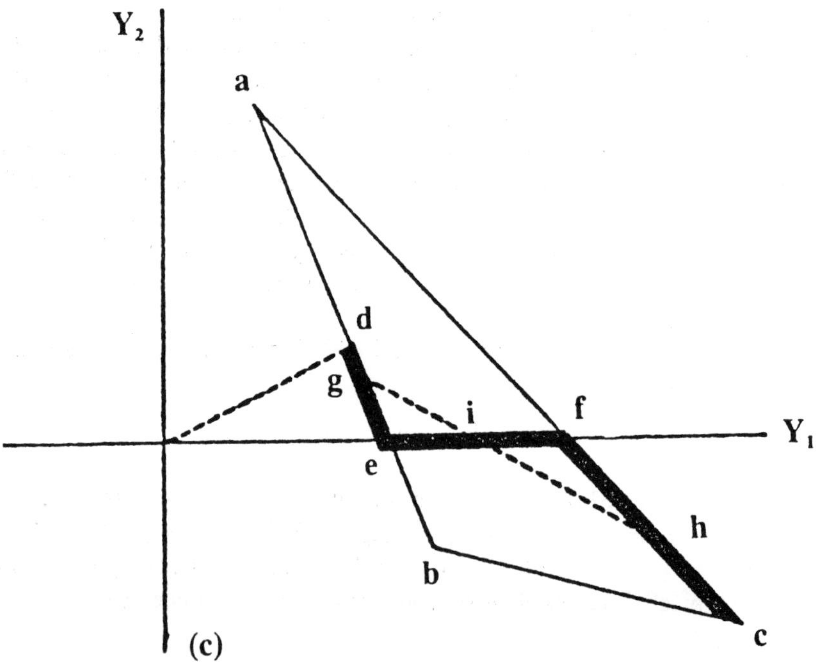


FIGURE 1c

## Risk Adjustment in the Standard, Homogeneous Model

Equation (7) is the basis for traditional risk adjustment procedures: a project is worthwhile if it adds more to a firm's $\mu_i$ than it does to $\beta_i$ times the market price of risk. If we reject traditional CAPM and therefore (7) how do we decide whether a project is worthwhile which adds to both the firm's $\mu_i$ and its systematic or unsystematic risk?

In the present section we answer this question assuming that all investors seek mean-variance efficiency, have the same beliefs and constraint sets and this common constraint set is the standard one:

$$\sum_{i=1}^{n} X_i = 1$$

$$X_i \geq 0 \quad i = 1, \ldots, n$$

In the next section we relax these assumptions.

We assume that a company has a fixed number of shares outstanding, and wishes to determine how a new project will effect the price of these shares. We also make two "approximation assumptions" which will be explained below.

From the investor's point of view we distinguish $\mu_i$, $\sigma_{ij}$—the expected return and covariance of return *per dollar* invested—from $\mu_i^s$, $\sigma_{ij}^s$, the same per share. Since return per dollar equals return per share divided by stock price P, we have

$$\mu_i = \mu_i^s / P_i \tag{8.1}$$

$$\sigma_{ij} = \sigma_{ij}^s / P_i P_j \tag{8.2}$$

The fraction of the I-th investor's wealth, $W^I$, in security i is

$$X_i^I = S_i^I P_i / W^I \tag{8.3}$$

where $S_i^I$ are shares held.

From the Kuhn-Tucker conditions for efficiency, if security i is "IN" the I-th portfolio then

$$\lambda_E^I \mu_i - \lambda_A^I - \sum_j X_j^I \sigma_{ij} = 0 \tag{9}$$

where $\lambda_E^I$, $\lambda_A^I$ are multipliers in the Lagrangian expression

$$L = \frac{1}{2}V - \lambda_E(\mu' X - E) + \lambda_A (\Sigma X_i - 1)$$

Substituting (8) into (9) and simplifying we get

$$W^I(\lambda_E^I \mu_i^s - \lambda_A^I P_i) - \Sigma S_j^I \sigma_{ij}^s = 0 \tag{10}$$

The last term

$$\begin{aligned}\psi_i^I &= \Sigma S_j^I \sigma_{ij}^s \\ &= \mathrm{cov}(r_i^s, \Sigma r_j^s S_j^I) \\ &= \mathrm{cov}(r_i^s, R^I)\end{aligned} \tag{11}$$

is the covariance between the return per share of the i-th security and the total dollar return $R^I$ of the I-th portfolio.

The left hand side of (9) is always non-positive, typically negative, if i is OUT of portfolio I. Define

$$A_i = \{I : X_i^I \epsilon \mathrm{IN}^I\}$$

as the clientele set of the i-th security. Sum (10) over investors in $A_i$ (abbreviated A below) to obtain

$$0 = (\sum_{I \epsilon A} W^I \lambda_E^I)\mu_i^s - (\sum_{I \epsilon A} W^I \lambda_A^I)P_i - \mathrm{cov}(r_i^s, \sum_{I \epsilon A} R^I) \tag{12.1}$$

Dividing by $\Sigma W^J$ and letting $w^I = W^I/\Sigma W^J$ we get

$$\begin{aligned}0 &= (\sum_{I \epsilon A} w^I \lambda_E^I)\mu_i^s - (\sum_{I \epsilon A} w^I \lambda_A^I)P_i - \mathrm{cov}(r_i^s, \sum_{I \epsilon A} R^I/\Sigma W^J) \\ &= a_i \mu_i^s - b_i P_i - \rho_i^T \sigma_i^s \sigma_i^T\end{aligned} \tag{12.2}$$

where $\mathrm{cov}(r_i^s, \sum_{I \epsilon A} R^I/\Sigma W^J)$ is rewritten as the standard deviation of the two, $\sigma_i^s$ and $\sigma_i^T$, times their correlation $\rho_i^T$. Assuming $b_i \neq 0$, from (12.2) we get

$$P_i = (a_i/b_i)\mu_i^s - \rho_i^T \sigma_i^s \sigma_i^T / b_i \tag{13}$$

where

$$a_i = \sum_{I \epsilon A} w^I \lambda_E^I \tag{14.1}$$

$$b_i = \sum_{I \epsilon A} w^I \lambda_A^I \tag{14.2}$$

are dollar weighted average $\lambda_E$ and $\lambda_A$ among $I \epsilon A_i$.

For now we make the following two "approximation assumptions." Later we will drop AA1 and defend AA2.

AA1: The adoption of the contemplated project by firm i does not change security i from $\mathrm{IN}^I$ to $\mathrm{OUT}^I$, or vice versa, for any investor I.

AA2: The adoption of the contemplated project by firm i may change $\mu_i^s$, $\sigma_i^s$, and $\rho_i^T$, but will leave $\sigma_i^T$, $a_i$ and $b_i$ [approximately] unchanged; since the change in $\mu_i^s$, $\sigma_i^s$, $\rho_i^T$ may induce (e.g.) CREF to buy more of stock i, but will have small effect on its efficient frontier or the location of the EV combination it chooses from it.

Under these assumptions the effect of changes in $\mu_i^s$, $\sigma_i^s$ and $\rho_i^T$ may be read from (13) in particular

$$\frac{\partial P_i}{\partial \mu_i^s} = a_i/b_i \tag{15.1}$$

$$\frac{\partial P_i}{\partial \sigma_i^s} = \rho_i^T \sigma_i^T/b_i \tag{15.2}$$

$$\frac{\partial P_i}{\partial \rho_i^T} = \sigma_i^s \sigma_i^T/b_i \tag{15.3}$$

## Observations and Extensions

1. Approximation Assumption 2 requires that $a_1$, $b_i$, and $\sigma_i^T$ be roughly unchanged as $\mu_i^s$, $\sigma_i^s$, and $\rho_i^T$ vary. For example, suppose in fact that $a_i$ varies but $b_i$ is constant. Then

$$\frac{\partial P_i}{\partial \mu_i^s} = (\frac{\partial a_i}{\partial \mu_i^s}\mu_i^s + a_i)/b_i$$

But if $\mu_i^S(\partial a_i/\partial \mu_i^S)$ is much smaller than $a_i$ our answer is almost the same as if we assumed $a_i$ constant. A similar comment applies to $\sigma_i^T$ and $b_i$, although the details are messier.

If we suppose that the changes in $\mu_i^S$, $\sigma_i^S$, and $\rho_i^T$—while they effect $P_i$ and perhaps the distribution of $S_i^I$ among investors—do not much change the shapes of investors' EV efficient sets or the location of the points chosen from them (since security i is a small part of large, diversified portfolios) then $\lambda_E^I$, $\lambda_A^I$ and $w^I$ will be little effected. It follows from (14) that $a_i$ and $b_i$ will change little. Similarly it seems plausible that the effects of adopting the project will have little effect on $\sigma_i^T$—the standard deviation on the return per dollar invested in all portfolios which contain i.

2. The essential difference between the results presented above and the classical CAPM results is the summation over the clientele set, $A_i$. In the TSL and RSMB models all i are IN for all I. If we weight (9) by $w^I$ and sum over *all* investors we obtain the classic result that the covariance between $r_i$ and the market portfolio is linearly related to $\mu_i$. If we then derive (13) as above we have the classic relationship in a new guise.

In the standard model (12) generally does not hold if we sum over all investors, since the left hand side of (9) is typically negative for I with i OUT. The net effect is that, rather than consider the relationship between the i – th security and *the* securities market, equation (13) bids us to consider the relationship between the i – th security and *its* market!

3. A review of the derivation of (12) and, from this, (13) and (15) will show that these equations do not depend on the assumption that the set A in (12.1) is the entire clientele set $A_i$. They depend only on the assumption that (9) and (10) hold for each I in A. Thus any non-null subset of $A_i$ will serve, provided that AA2 holds at least approximately. For example, A could consist of one large, well-diversified EV efficient portfolio. The result in (13) and (15) is independent of which such $I \in A_i$ we choose, since (9) holds for each of them.

Note also that, while the derivation of (12) assumes that the $I \in A_i$ are mean-variance efficient, it does not require that all investors seek mean-variance efficiency.

4. Approximation Assumption 1 requires that $X_i^I$ not change from IN to OUT or vice versa for any investor I. We have just observed that a subset of the clientele set $A_i$ will serve as A in (12). To assure AA1, use a subset A of investors who will remain in $A_i$ after the adoption of the project.

5. In the general portfolio selection model (Markowitz 1959, 1987) the investor seeks mean-variance efficiency subject to any number of linear equations in variables required to be nonnegative:

$$AX = d \tag{16.1}$$

$$X \geq 0 \tag{16.2}$$

where A is m × n, d m × 1. Standard linear programming procedures can be used to convert into form (16) any contraint set consisting of linear equalities and/or inequalities in variables which may or may not be required to be nonnegative. (See Chapter 2.)

First consider the case in which the ith component of X represents a long position in a security, and no short position is represented or allowed. If i is IN the I – th portfolio then, instead of (9), we have

$$\lambda_E^I \mu_i - \sum_{k=1}^{m} a_{ki} \lambda_k^I - \Sigma X_j^I \sigma_{ij} = 0 \tag{9'}$$

Following the same steps as before we again derive (12), except now

$$b_i = \sum_{I \in A} w^I \left( \sum_{k=1}^{m} a_{ki} \lambda_k^I \right) \tag{14.2'}$$

The various investors need not have the same right hand sides $d_k$ for the respective k – th equation. Differences in $d_k$ may be one reason, but not the sole reason, why the $\lambda_k^I$ vary from one investor to another. (The other reason is that, even with identical $d_k$, the $\lambda_k^I$ will vary with the point chosen from the efficient set.) If the I – th investor is not subject to equation k, this is equivalent to choosing $d_k^I$ so that $\lambda_k^I = 0$. As before, we assume that the new project will have little effect on $b_i$, now given by (14.2′).

Next suppose $X_i$ is allowed to be negative. Substitute $Y_i - Z_i$ for each appearance of $X_i$ and add the constraints $Y_i \geq 0$, $Z_i \geq 0$. $Y_i$ and $Z_i$ are the positive and negative parts of $X_i$. (See Chapter 2.) When this is done for all securities allowed to have short positions, a model results in which all variables are constrainted to be nonnegative.

$Y_i$ and $Z_i$ are now separate variables in the model and each have their clientele set, that is, those portfolios for which they are IN. The clientele set of $Z_i$ may be null, but that of $Y_i$ cannot, since long positions must exceed short positions by the amount of stock outstanding. It is sufficient to use the $Y_i$ clientele set, or some non-null subset of it, in (12) to derive (13) and (15) provided that AA2 applies.

6. In theory at least, heterogeneous beliefs are easily accommodated. Attach an I superscript on $\mu_i$ and $\sigma_{ij}$ in (9), proceed as before to arrive at (12.1) now written as

$$0 = (\sum_{I\epsilon A} W^I\lambda_E^I)\overline{\mu}_i^s - (\sum_{J\epsilon A} W^I(\sum_{k=1}^{m} a_{ki}\lambda_k^I))P_i - \sum_j S_j^T\overline{\sigma}_{ij}^s \qquad (12')$$

where

$$S_j^T = \sum_{I\epsilon A} S_j^I \qquad (17.1)$$

$$\overline{\sigma}_{ij}^s = \sum_{I\epsilon A} \sigma_{ij}^{sI}(S_j^I/S_j^T) \qquad (17.2)$$

$$\overline{\mu}_i^s = \sum_{I\epsilon A} \mu_i^{sI}(W^I\lambda_E^I/\sum_{J\epsilon A} W^J\lambda_E^J) \qquad (17.3)$$

Then proceed to (13) as before, slightly amended. Note that $\overline{\sigma}_{ij}^s$ is a share-weighted average whereas $\overline{\mu}^s$ is $(W\lambda_E)$ weighted.

The matrix $\overline{C} = (\overline{\sigma}_{ij}^s)$ will be a covariance matrix, that is, will be positive semi-definite, provided $S_j^I \geq 0$ (as in the $Y_i$ clientele set above) since, for $w^I \geq 0$,

$$\begin{aligned} X'\overline{C}X &= X'(\Sigma w^I C^I)X \\ &= \Sigma w^I X' C^I X \\ &\geq 0 \end{aligned}$$

7. Equations 12, 13, and 15, and their generalizations for heterogeneous beliefs, result from the fact that the Lagrangian equality (9) holds for all investors in the clientele set $A_i$. A similar condition holds in the models of Levy (1978) and Merton (1987). The three models differ in the source of

the clientele sets. In Levy's model "as a result of transaction costs, indivisibility of investment, or even the cost of keeping track of the new financial development of all securities, the κth investor decides to invest only in $n_\kappa$ securities." In the Merton model "each investor knows only about a subset of the available securities." In the model presented here investors differ in the sets of securities they hold because they choose portfolios from different segments of a piecewise linear set of efficient portfolios.

## Epilogue

The classic risk adjusting procedure is based on very special assumptions. Special assumptions can be justified on two grounds: (1) they are correct in fact, therefore greater generality is not needed, or (2) greater generality makes the problem intractible.

(1) The Tobin-Sharpe-Lintner assumption of unlimited borrowing opportunity at the riskless rate or, alternatively, the assumption that $\Sigma X_i = 1$ is the only constraint, are highly unrealistic. (They are assumptions that only a financial economist could pretend to believe.) It was a monumental achievement to first discover that relationships such as the classical risk adjustment formulas could be derived from any CAPM, albeit a simplified one to begin with. But it should surprise no one if such CAPMs fail some empirical test, given their surreal assumptions.

(2) In fact, the relationships become little more complicated if we allow heterogeneous beliefs, arbitrary systems of linear equalities or inequalities as constraints and differing constraint sets among investors. What is required is that certain sums be taken over a non-null subset of the $i-$th security's clientele set rather than over the entire market, that $\mu$'s and $\sigma_{ij}$'s be aggregated properly and that $b_i$ be interpreted suitably.

REFERENCES


1. Black, F. (1972). "Capital market equilibrium with restricted borrowing," *Journal of Business* (July).
2. Levy, H. (1978). "Equilibrium in an imperfect market: A constraint on the number of securities in the portfolio," *American Economic Review* 68(4) (September), 643–658.
3. Lintner, J. (1965). "The valuation of risk assets and the selection of risky investments in stock portfolios and capital budgets," *Review of Economics and Statistics* (February), 13–37.
4. Markowitz, H. M. (1952). "Portfolio selection," *The Journal of Finance* 7(1) (March), 77–91.
5. Markowitz, H. M. (1959). *Portfolio Selection: Efficient Diversification of Investments*. Wiley, Yale University Press, 1970.
6. Markowitz, H. M. (1987). *Mean-Variance Analysis in Portfolio Choice and Capital Markets*. Basil Blackwell, New York.
7. Merton, R. C. (1972). "An analytic derivation of the efficient portfolio frontier," *Journal of Financial and Quantitative Analysis* (September), 1851–1872.

8. Merton, R. C. (1987). "Presidential Address: A Simple Model of Capital Market Equilibrium with Incomplete Information," Forty-Fifth Annual Meeting, American Finance Association, New Orleans, Louisiana, December 28–30, 1986, *The Journal of Finance* 42(3) (July).
9. Roy, A. D. (1952). "Safety first and the holding of assets," *Econometrica* 20, 431–449.
10. Sharpe, W. F. (1964). "Capital asset prices: a theory of market equilibrium under conditions of risk," *The Journal of Finance* 19(3) (September).
11. Tobin, J. (1958). "Liquidity preference as behavior towards risk," *Review of Economic Studies* (February), 65–86.

J ECO BUSN
1990; 42:99-103

# Normative Portfolio Analysis: Past, Present, and Future

Harry M. Markowitz

The editors have invited me to contribute an essay on normative portfolio analysis to this issue of the *Journal of Economics and Business*. My comments are divided into three parts: 1) normative portfolio theory as of 1959, 2) a comparison of positive and normative theories, and 3) progress in normative analysis from 1959 to date, and beyond.

## I. Normative Portfolio Analysis as of 1959

Part IV of Markowitz (1959) presented my justification (then and, for the most part, now) for the use of mean-variance analysis in practice. This rationale differs substantially from that given in Markowitz (1952).

In Markowitz (1959) I presented a simpler version of the axioms of L. J. Savage (1954); argued that, if we delegate the management of our investments to a supercomputer with unlimited computing capability but limited information, we want the computer's choices to be consistent with these axioms; and concluded that the computer should act as if it sought to maximize expected utility for some game as a whole, using "personal probabilities" where objective probabilities are not known. Action in each period would therefore seek to maximize a single period utility function related to utility of the game as a whole, in the manner explained by Bellman (1957) (see also Mossin (1968), Samuelson (1969), and Ziemba and Vickson 1975).

Under certain assumptions, the single period utility function depends only on portfolio return. For this case I conjectured that, for real-world distributions and utility functions, if an investor (or the investor's computer) maximized a particular function of mean and variance of rate of return, namely,

$$f(E, V) = U(E) - U''(E)V/2, \tag{1}$$

then the investor would almost maximize expected utility. Sample calculations with a small number of distributions and utility functions supported this conjecture. Since then, a number of experiments with a variety of utility functions and historical or synthetic distributions of returns have—for the most part—confirmed the efficacy of the mean-variance approximation. See Markowitz (1959), Young and Trent (1969), Levy

Address requests for reprints to Harry M. Markowitz, Baruch College, Box 504, New York, New York 10010.



0148-6195/90/$03.50

and Markowitz (1979), Dexter, Yu, and Ziemba (1980), Pulley (1981), Pulley (1983), Kroll, Levy, and Markowitz (1984), Ederington (1986), Reid and Tew (1986), Simaan (1987), Grauer (1986), and Tew and Reid (1987). The exceptional cases concern highly risk-averse utility functions such as

$$U = -e^{-\alpha(1+R)},$$

for $\alpha \gtrsim 10$, where $R$ is rate of return. Levy and Markowitz (1979) argued that "few if any investors have preferences like" those of $U = -\exp(-\alpha(1+R))$ with $\alpha = 10$ or higher, since such an investor would prefer a 10% rate of return with certainty to a fifty-fifty chance of zero return (no gain, no loss) versus unlimited wealth. More precisely, the investor with $\alpha = 10$ would be indifferent between 1) a rate of return of 6.9% with certainty and 2) the fifty-fifty chance of no gain versus the unlimited "blank check." Markowitz, Reid, and Tew (1989) surveyed actual investors to determine the certainty-equivalent returns that investors consider as good as a 50–50 chance of a blank check. Their conclusions support the Levy–Markowitz view.

It may be wondered, if the theoretically correct course is to maximize $EU(R)$, why maximize $f(E, V)$ in equation (1), even if this is a fairly good approximation? The answer concerns cost of implementation. Normative portfolio analysis seeks desirable policies that can be followed in fact. Maximizing $EU$ is typically manyfold more difficult than performing a mean–variance analysis. While large institutional investors frequently employ quantative analysts, few have determined their actual utility function—by specifying their willingness to engage in various simple gambles, as described by von Neumann and Morgenstern (1944). Further, the entire joint distribution of returns must be estimated, which is generally a much larger requirement than estimating the means, variances, and covariances needed for a mean–variance analysis. Finally, the finding of the expected utility-maximizing portfolio typically requires many times more computational resources than does the critical line algorithm for tracing out a mean–variance efficient frontier.

While the above are reasons for using mean–variance analysis I do not wish to equate mean–variance analysis with normative portfolio analysis. In general, the latter seeks a "good" distribution of returns on the portfolio as a whole, ideally considering costs and perhaps state variables. In the future more quantative analysts associated with investing institutions may make explicit their institution's utility function, decide they can use historical returns, or else estimate parameters of joint distributions, and use increasingly more powerful computers and optimization techniques for actual portfolio selection. (Perhaps some of those who pursue this more rigorous analysis will find, in the end, that the mean–variance approximation is close enough.)

## II. Normative versus Positive Portfolio Analysis

The Capital Asset Pricing Models of Sharpe (1964), Lintner (1965), Mossin (1966), and Black (1972) are positive theories that seek to explain capital markets. They assume that investors seek mean–variance efficiency and have identical beliefs and identical constraints sets. They conclude that the market portfolio is an efficient portfolio, and that there is a linear relationship between the expected return of each security and its covariance with (or regression coefficient against) the market portfolio. An alternate model, the Arbitrage Pricing Theory (APT) of Ross (1976), dropped the assumption that investors seek mean–variance efficiency, and added an assumption concerning the joint distribution of security returns. It concluded that the market is an efficient portfolio (in the sense that some investor might wish to hold a portfolio with the same proportions as the market), and that there is a linear relationship between the expected return of each security and its covariances with various systematic sources of risk.

One major difference between the investors assumed in the positive (CAPM and APT) models and actual practice (e.g., by financial consultants using normative portfolio analysis) is the nature of the constraint used. The Sharpe–Lintner model permits unlimited borrowing. In fact, borrowing by individuals is limited, and is eschewed by large pension funds. The constraint set most frequently assumed in CAPMs is

$$\sum X_i = 1, \tag{2}$$

without nonnegativity requirements on the $X_i$. This constraint set is used by Roy (1952), Sharpe (1970, pp. 59–62), Merton (1972), Black, and the APT model. Negative $X_i$ are referred to as short positions; but equation (2) is not a realistic model of short position opportunities. According to equation (2) an investor could place $\$10^6$ at a broker, short $\$10^9$ of stock A, and use the proceeds plus account equity to buy $\$(10^6 + 10^9)$ of stock B. This is not the way the world works.

In normative portfolio analyses the constraint set presented to the optimizer can, and ideally does, represent the actual constraints of the investor, provided that these can be expressed as a system of linear equalities and/or weak inequalities in variables that may or may not be constrained to be nonnegative. A wide variety of real-world constraints can be modeled in this manner (see Markowitz (1987), chs. 1, 3).

Markowitz (1987) presents a CAPM that is identical to that of Sharpe–Lintner or Black except for the constraints on investors' choice. It assumes that the investor's constraint set is some *bounded* polyhedron. An example is the "standard" constraint set

$$\sum X_i = 1; \tag{3a}$$

$$X_i \geq 0, \quad i = 1, \ldots, n \tag{3b}$$

In the Sharpe–Lintner and Black CAPMs the market portfolio is a mean–variance efficient portfolio. This is not necessarily the case if equation (3) is the constraint set. In fact, it is quite easy to produce market equilibrium examples in which the market has almost maximum $V$ for given $E$, rather than minimum $V$. Also, it is typically not true that there is a linear relationship between the expected returns of securities and their betas. These negative results—that the market is not an efficient portfolio and there is no linear relationship between expected return and $\beta$—hold even if *some* investors have equation (3) and others have equation (2) for their constraint sets. The results hold whether or not the market contains a risk-fee asset.

Thus the implications of CAPMs depend on their highly simplified (one might say surrealistic) assumptions. In seeking plausible inputs for a normative analysis, the relationships implied by CAPMs should be tested like any other hypothesis. Their failure should cause no surprise.

## III. Progress and Opportunity in Normative Analysis

A great deal of progress has been made since 1959 in normative portfolio analysis. In the specific case of mean-variance analysis progress has been made, for example, in:

- understanding and estimating the structure of the covariance matrix (see, e.g., Sharpe (1963), King (1966), Blume (1971), and Rosenberg (1974));
- the computation of efficient sets for special covariance structures (see, e.g., Sharpe (1963), Elton, Gruber, and Padberg (1976, 1977, 1978), Markowitz and Perold (1981a, 1981b), and Perold (1984)); and
- the education of institutional investment management teams in concepts such as mean-variance efficiency.

The above list does not attempt to be exhaustive. Despite cited and uncited progress there is much left to accomplish. Some of this is in problem areas addressed by authors of papers that follow. At present I have seen only the abstracts of the papers—not quite all of them—so it would be risky of me to try to outline them in any specific way. But in general terms the authors offer to analyze, among other things, the efficacy of the mean-variance approximation, methods of going from historical data to ex ante estimates, conditions under which the distribution of portfolio returns may be taken as the objective without regard to its covariance with another state variable, criteria for recognizing good securities for adding to the universe to be presented to a mean-variance optimizer, the inclusion of the differential tax treatment between capital gains and dividends, the sensitivity of bank stocks to interest rate variation, and the selection of optimal international portfolios taking into account exchange risk.

I look forward to reading the papers. I share with the editors the hope that this volume will stimulate further work in this rewarding area.

## References

Bellman, R. E. 1957. *Dynamic Programming*. Princeton: Princeton University Press.

Black, F. July 1972. Capital market equilibrium with restricted borrowing. *Journal of Business* 45(3) 444-55.

Blume, M. March 1971. On the assessment of risk. *The Journal of Finance* 1-10.

Dexter, A. S., Yu, J. N. W., and Ziemba, W. T. 1980. Portfolio selection in a lognormal market when the investor has a power utility function: computational results. In *Stochastic Programming* (M. A. H. Dempster, ed.). New York: Academic Press, pp. 507-23.

Ederington, L. H. 1986. Mean-variance as an approximation to expected utility maximization. Working Paper 86-5. St. Louis: School of Business Administration, Washington University.

Elton, E. J., Gruber, M. J., and Padberg, M. W. December 1976. Simple criteria for optimal portfolio selection. *The Journal of Finance* 31(5):1341-57.

Elton, E. J., Gruber, M. J., and Padberg, M. W. November/December 1977. Simple criteria for optimal portfolio selection with upper bounds. *Operations Research* 25(6):952-67.

Elton, E. J., Gruber, M. J., and Padberg, M. W. March 1978. Simple criteria for optimal portfolio selection: tracing out the efficient frontier. *The Journal of Finance* 33(1):296-302.

Grauer, R. R. September 1986. Normality, solvency, and portfolio choice. *Journal of Financial and Quantitative Analysis* 21:265-78.

King. B. F. January 1966. Market and industry factors in stock price behavior. *Journal of Business* Supplement.

Kroll, Y., Levy, H., and Markowitz, H. M. March 1984. Mean-variance versus direct utility maximization. *The Journal of Finance* 39(1).

Levy, H., and Markowitz, H. M. June 1979. Approximating expected utility by a function of mean and variance. *American Economic Review* 69:308-17.

Lintner, J. February 1965. The valuation of risk assets and the selection of risky investments in stock portfolios and capital budgets. *Review of Economics and Statistics* 13-37.

Markowitz, H. M. March 1952. Portfolio selection. *The Journal of Finance* 7(1):77-91.
Markowitz, H. M. 1959. *Portfolio Selection: Efficient Diversification of Investments*. New Haven: Yale University Press, 1970.
Markowitz, H. M. 1987. *Mean-variance Analysis in Portfolio Choice and Capital Markets*. New York: Basil Blackwell.
Markowitz, H. M., and Perold, A. F. September 1981a. Portfolio analysis with factors and scenarios. *The Journal of Finance* 36(14).
Markowitz, H. M., and Perold, A. F. 1981b. Sparsity and piecewise linearity in large portfolio optimization problems. In *Sparse Matrices and Their Uses* (I. S. Duff, ed.). New York: Academic Press, 89-108.
Markowitz, H. M., Reid, D. W., and Tew, B. V. 1989. The value of a blank check. Baruch College Working Paper.
Merton, R. C. September 1972. An analytic derivation of the efficient portfolio frontier. *Journal of Financial and Quantitative Analysis* 1851-72.
Mossin, J. October 1966. Equilibrium in a capital asset market. *Econometrica* 34(4):768-83.
Mossin, J. April 1968. Optimal multiperiod portfolio policies. *Journal of Business*.
Perold, A. F. October 1984. Large-scale portfolio optimization. *Management Science* 30(10):1143-60.
Pulley, L. B. September 1981. A general mean-variance approximation to expected utility for short holding periods. *Journal of Financial and Quantitative Analysis* 16:361-73.
Pulley, L. B. July-August 1983. Mean-variance approximations to expected logarithmic utility. *Operations Research* 31:685-96.
Reid, D. W., and Tew, B. V. December 1986. Mean-variance versus direct utility maximization: a comment. *The Journal of Finance* 41:1177-79.
Rosenberg, B. March 1974. Extra-market components of covariance in security returns. *Journal of Financial and Quantitative Analysis*.
Ross, S. A. December 1976. The arbitrage theory of capital asset pricing. *Journal of Economic Theory* 13:341-360.
Roy, A. D. 1952. Safety first and the holding of assets. *Econometrica* 20: 431-49.
Samuelson, P. A. 1969. Lifetime portfolio selection by dynamic stochastic programming. *The Review of Economics and Statistics* 51:239-46.
Savage, L. J. 1954. *The Foundations of Statistics*. Wiley, New York: Dover, 1972.
Sharpe, W. F. January 1963. A simplified model for portfolio analysis. *Management Science* 9(2) 277-93.
Sharpe, W. F. September 1964. Capital asset prices: a theory of market equilibrium under conditions of risk. *The Journal of Finance* 19(3).
Sharpe, W. F. 1970. *Portfolio Theory and Capital Markets*. New York: McGraw-Hill.
Simaan, Y. 1987. Portfolio selection and capital asset pricing for a class of non-spherical distributions of asset returns. Dissertation. New York: Baruch College, The City University of New York.
Tew, B. V., and Reid, D. W. Fall 1987. More evidence on mean-variance versus direct utility maximization. *Journal of Financial Research* 10:249-57.
Von Neumann, J., and Morgenstern, O. 1944. *Theory of Games and Economic Behavior*. Princeton: Princeton University Press, 3rd ed., 1953.
Young, W. E., and Trent, R. H. June 1969. Geometric mean approximation of individual security and portfolio performance. *Journal of Financial and Quantitative Analysis* 4:179-99.
Ziemba, W. T., and Vickson, R. G., eds. 1975. *Stochastic Optimization Models in Finance*. New York: Academic Press.

FINANCIAL SERVICES REVIEW, 1(1):1-8
ISSN: 1057-0810


# Individual versus Institutional Investing

Harry M. Markowitz

*This paper first describes the analytic approach that Markowitz used in developing his portfolio theory. Developing a game-of-life simulation is a parallel approach for modelling individual financial management. To develop a realistic simulator will require deciding what goals are essential to the family planning process, formulating optimizable subproblems, using technology to interpret and record decisions, and developing decision rules which prove robust in the model and can be implemented in practice.*

Professor Mandell, editor of *Financial Services Review*, invited me to contribute an article related to financial research for the individual for the first issue of this journal. Since the subject is not my specialty, it was uncharacteristically risky of me to have accepted the invitation. But an evening of reflection convinced me that there were clear differences in the central features of investment for institutions and investment for individuals, that these differences suggest differences in desirable research methodology, and that a note on these differences may be of value.

As I thought about the subject further, on subsequent days, I found myself of two minds. On the one hand, surely financial decisions for the individual should be considered as part of the "game as a whole" which the individual plays out—"game" in the sense of von Neumann and Morgenstern (1944). Even reducing this game to its essentials, it has characteristics of situations for which simulation methods have proved to be the superior tool in practice. On the other hand, there are approximations to the individual's financial situation which seem good enough, and would allow us to solve analytically for optimal action.

Neither train of thought succeeded in defeating the other. Below, I present both views: the first in a section called "Thesis," and the second under "Antithesis," and attempt some reconciliation in a final section, "Synthesis."

**Harry M. Markowitz** • Marvin Speiser Distinguished Professor of Finance and Economics, Baruch College, CUNY, 17 Lexington Avenue, New York, NY 10010.

## Thesis

In Markowitz (1952) I conjecture that "Perhaps—for a great variety of investing institutions which consider yield to be a good thing; risk, a bad thing; gambling to be avoided—E, V efficiency is reasonable as a working hypothesis and a working maxim." The "investing institution" which I had most in mind when developing portfolio theory for my dissertation was the open-end investment company or "mutual fund." This familiarity was not from first hand experience (I was a student and son of a grocer) but from Wiesenberger & Company's *Investment Companies* (1944- ). It was plausible to assume for the mutual fund that its objective is to obtain a "good" probability distribution of year-to-year (or quarter-to-quarter) percent increase in its net asset value. In addition, I argued for mean and variance as criteria in judging "goodness." For the present discussion, the choice of mean-variance criteria is not the crux; rather it is the formulation of the problem as that of selecting a portfolio to achieve a good probability distribution of a single random variable: the return on the portfolio as a whole. This formulation turned out to be widely acceptable in practice as well as tractable analytically.

In the 1950s, I participated in attempts to develop advanced, but practical, methods for assisting manufacturing planning, particularly assisting equipment selection and production scheduling for job shops. We considered optimization techniques first, such as linear and dynamic programming, but found that too much reality had to be ignored to allow these techniques to be applied. Simulation techniques seemed promising, and were developed for real decision problems with real job shops. Experience confirmed the value of simulation analysis, but showed that programming the model was a bottleneck. This, and similar experiences in other application areas, stimulated development of the simulation programming languages of the early 1960s.

In the meantime, attempts continued to apply analytic techniques to shop scheduling problems. A recent survey (Lawler, Lenstra, Rinnooy Kan, and Shmoys, 1989) reports that some flow shop problems have been solved; others have been shown to be NP-hard (i.e., as hard to solve as the traveling salesman problem); but results for job shops are meager, leading the authors to end with a quote from Coffman, Hofri, and Weiss (1989), "there is a great need for new mathematical techniques useful for simplifying the derivation of results." In the meantime, simulation analysis is increasingly used in practice.

The difference between the investment company situation and that of a job shop is the number of state variables that need to be considered in a practical problem. For the investment company, it is plausible to assume that assets are liquid, therefore the state of the portfolio can be described by its total value. For the shop, its state description includes the contents of all its queues.

To judge whether the problem of financial planning for the individual is amenable to analytic solution, let us sketch what a "game-of-life" model might entail. We seek a model with sufficient realism as to be a guide to practice. For example, some economic theories find it convenient to assume that the individual is immortal, or that death is a Poisson process independent of the age of the individual. For actual financial planning, however, aging and mortality are salient facts that must be included in the model. On the other hand, many details of life which are important to the individual may be ignored for financial planning. For example, the model should include the probability of an accident or disease which will keep the individual from work for an extended period, the probability distribution of time to recover or die, costs of treatment and probability of relapse, since these possibilities are major factors in financial planning; but medical details are not required.

Since time and uncertainty are at the heart of the problem, I will sketch the model as if it were a simulation. This is for the purpose of model description, and does not itself preclude the possibility that the model could be solved analytically. The description will use the SIMSCRIPT worldview (Kiviat, Villanueva, and Markowitz, 1983). This says that, as of any instant in time, the model represents *entities* of various *entity types*; a given entity is characterized by the *values* of its *attributes*; also, it may own *sets* to which other entities belong, and belong to sets which other entities own. This status description changes at points in time called *events*.[1] One event may cause one or more subsequent events to occur after fixed or random time delays.

The essentials of the game-of-life is probably different for (a) the very wealthy, (b) the class of homeless that used to be called vagrants, and (c) most of my friends and relatives. I have the latter in mind as I sketch the model.

Among the types of entities of the model, we must distinguish between the *individual* (i.e., human person) and the (nuclear) *family*. Often, at some stage this family will "own" a set of individuals whose roles are husband, wife, children and perhaps residing elder. Frequently, in the course of events, the residing elder (if any) dies or is placed in a nursing facility; the children leave home to set up their own nuclear families; the original family (the subject of the model) then consists of husband and wife. When one dies the subject family consists of the survivor only. When the latter dies, the assets of the subject family are distributed to heirs, and the game-of-life is over for the subject family.

In the simplest case, assets may be thought of as belonging to the family rather than the individual, to be used by husband and wife and (at their discretion) by the children until, at the last, it is used to support the survivor and then distributed. It may be sufficient to characterize financial assets as the total value of the family's holdings in stocks, bonds, cash items and real estate [other than the family's home(s)]. Perhaps, upon further reflection, it may prove essential to disaggregate these items according to the maturity of the bonds, their tax exempt status, and the unrealized capital gains and losses of various assets. [Problem: must we distinguish many individual stocks in order to characterize available capital gains and losses for tax calculations?] Perhaps bonds and stocks may be treated as instantaneously marketable, perhaps with a small commission, but real estate requires greater (random) time and cost to sell.

Among other assets, the family may "own" [in the SIMSCRIPT sense, i.e., have associated with it] one or more residences. The residence may be owned (in the usual sense) or rented. If owned, the residence is characterized by original cost and a current market value; whether owned or rented, the residence has a value of owned furnishings. A home and its furnishings are clearly an illiquid asset, not only because of the time and cost to sell, but also that to move, and the mismatch between furniture needs of the old and new residences.

Attributes of individuals include those needed to characterize health, the employment or employability of husband and wife, and the educational objectives of each child. The assets of a residing elder can be characterized by associating with this entity his or her own nuclear family entity.

Events which change status include periodic events such as receiving a salary check, having a birthday, or the time when an income tax payment is due; and randomly occurring events such as becoming sick, becoming well, finding a job, losing a job, financing a house to buy, finding a buyer for a house to be sold. Changes in price levels, interest rates, and stock and real estate values could be computed periodically; e.g., increments in price levels and interest rates could be drawn from a joint distribution, then the change in real estate values could be computed as a function of the former increments and other random variables.

The simulated family must make decisions at various points in time, such as the level of (say) this week's nondurable consumption, transfers from cash to other liquid assets, the decision to search for and then buy a new house, and the decision of one of its members to retire. The simulated family makes these decisions according to decision rules. A major purpose of the model is to evaluate alternate decision rules.

The above is a partial sketch of a game-of-life model, rather than detailed specifications for one. The model should also include, as essential to evaluating family investment practice in fact, such things as IRAs, Keoghs, social security payments (or the individual's status with respect to future social security payments), status with respect to pension plans, various kinds of insurance, their costs and the kinds of events they insure against (e.g., house fire, car accident).

The model sketched above is, in certain ways, akin to the worksheets published as guides to families; see, e.g., *The Wall Street Journal* (1989). The model differs from the worksheet in that the model allows for many of life's random events—many more such events than one could take into account by filling out alternate, contingent worksheets. Since future status is random, the simulated family must follow adaptive decision rules rather than a single plan as expressed on a worksheet. As noted above, a major function of the model is to evaluate these decision rules.

This sketch of a game-of-life should suffice to convince one that the game is complex; most likely beyond analytic techniques. In contrast, using a good simulation language, it would not be difficult to program as a simulation model *once the specs of the model are decided.* It is unlikely that there will be general agreement as to what should be included in a game-of-life simulator, or how its output should be scored. Therefore it may be expected that there will be more than one game-of-life simulator; and it may be hoped that their respective assumptions will be clearly documented. The various simulators will allow us to see whether rules of behavior which work well in one model will prove robust when tried in alternate models. If so, this will encourage us to recommend them in practice.

In sum, I encourage readers with requisite skills to try building and using realistic game-of-life simulators; and editors to look kindly on the publication of their results.

## ANTITHESIS

The problem with simulation analysis is that it is not very good at finding near optimum decision rules. It takes many runs of the model to estimate the excellence of a given set of rules. Since the rules we seek may be adaptive—i.e., may recommend different allocations of resources under different circumstances, and "circumstances" admit to countless variations—it will not be feasible to search for optimum decision rules.

At various points in time in the game-of-life there are requirements for allocating resources among assets. If some of these can be formulated, at least approximately, as portfolio selection problems—where the problem is to get a good distribution of return on the allocation as a whole—then an optimum solution can be found for the approximate problem. If the approximation is a good one then, by definition of good approximation, the exact solution to the approximate problem will be part of a good solution for the more complex game.

For example, consider a family with a house, children a few years from college age, life insurance policies in place based on a separate calculation, which faces the question of whether to shift resources among asset classes such as equities, long term tax exempt bonds, short term tax exempt bonds, etc. Leaving aside, for the moment, the question of unrealized capital gains in the existing portfolio, and assuming that this family does not trade often enough to run up sizable brokerage commissions, then it is plausible to pretend that these assets are perfectly liquid, therefore the value of the portfolio at anytime is the sum of the market value of its constituents, and that the objective in choosing a portfolio of these assets is to get a good probability distribution of return (capital gain plus interest and dividends) for some period of analysis.

Whether the "goodness" of a probability distribution is to be measured by a utility function or by a mean-variance analysis, we must answer questions such as:

(1) How do we measure return? Clearly, the family wants return after taxes. First, if the family realizes capital gain by shifting out of an asset that has an unrealized gain at the beginning of the period, then the tax on this gain must be subtracted from holding period return. Second, if an asset produces income during the period, we must subtract the tax on this income from its return. Third, if an asset has a capital gain during the period then its value to the family is somewhere between its market value and the latter minus the tax if the gain is realized. For simplicity perhaps it is satisfactory to average these two values. Finally, it seems appropriate for the family to seek a good distribution of real rather than nominal return. This raises no problem for the optimizer. (In particular, see Markowitz 1987, chapter 1], concerning the treatment of real returns in a mean-variance analysis.)

(2) What constraints limit portfolio choice? Constraints should consist of those which are imposed by government agencies and brokerage houses on individuals, e.g., limited borrowing and short sales, plus perhaps self imposed constraints such as upper bounds on asset classes which are in fact less liquid than others.

(3) What period of analysis should be used? Do what we always do—pick one.

Admittedly, approximations (and guesses) must be made, but they can be made plausibly. Then the optimum solution can be found to the approximate model. If the approximation is satisfactory, this exact solution to the approximate problem should be part of a good solution for the real problem.

## Synthesis

The proposed optimization analysis is only an approximation. A realistic simulator could be used to test decision rules based on optimizing a simplified model as compared to rule-of-thumb decision rules. Also, it is not always clear how the approximation is to be made; e.g., what time period to use for the analysis, how the family should pick a portfolio from the mean-variance frontier, how to treat unrealized capital gains, whether it is sufficient to consider nominal returns or essential to consider real returns, and the like. The simulator could be used to evaluate such alternate methods of formulating the portfolio selection problem within the game-of-life model. Also, a number of investigators have evaluated the ability of a well chosen point from the mean-variance frontier to approximately maximize the expected value of a single period utility function.[2] Most have concluded that it does quite well for "reasonable" utility functions. This question could be re-examined within the framework of a game-of-life simulation analysis.

The exercise of building a realistic game-of-life simulator—deciding what is essential to the family planning process and incorporating it into a simulator without the severe constraint of producing an analytically tractable model—should be highly educational, especially to the model builders. So should the process of formulating optimizable subproblems and evaluating these within the simulator. Another challenge is to use modern computer technology to help understand and remember what has been done. I have in mind here the use of simulation/animation to display the workings of the simulated world (see CACI, 1988) and the use of some kind of database to allow one to browse the inputs and outputs of prior runs. Finally there is the process of deciding how the decision rules which prove robust in the simulated worlds can be explained and implemented in practice.

I admit that this all seems a lot harder than formulating a highly simplified model that can be solved analytically. But I believe it has more chance of producing credible decision rules for practice—just as simulation analysis continues to produce credible policy recommendations for manufacturing, while analytic methods are not yet available for most sufficiently realistic models in the latter area.

Obviously, results of realistic game-of-life simulators will not be ready for the next issue of this journal. In the short and the long run, we should expect that the *Financial Services Review* will publish research with various approaches to various aspects of its topic area. Such pluralism is desirable in research, as it is in politics and the marketplace.

## Notes

1. In programming, it is often convenient to bundle events together into *processes*; but for the present discussion it is more convenient to describe events.
2. Markowitz (1959); Young and Trent (1969); Levy and Markowitz (1979); Dexter, Yu, and Ziemba (1980); Pulley (1981); Pulley (1983); Levy and Markowitz (1984); Reid and Tew (1986); Simaam (1987), Grauer (1986); and Tew and Reid (1987).

## References

CACI. 1988. *Simgraphics: User's Guide and Case Book*. La Jolla, CA: CACI Products Co.

Coffman, Jr., E.G., M. Hofri, and G. Weiss. 1989. "Scheduling Stochastic Jobs with a Two Point Distribution on Two Parallel Machines," in *Probability Engineering and Information Science*, forthcoming.

Dexter, A.S., J.N.W. Yu, and W.T. Ziemba. 1980. "Portfolio Selection in a Lognormal Market When the Investor Has a Power Utility Function: Computational Results," pp. 507-523 in M.A.H. Dempster (ed.), *Stochastic Programming*. New York: Academic Press.

Ederington, L.H. 1986. "Mean-Variance as an Approximation to Expected Utility Maximization." Working Paper 86-5, School of Business Administration, Washington University, St. Louis, Missouri.

Grauer, R.R. 1986. "Normality, Solvency, and Portfolio Choice," *Journal of Financial and Quantitative Analysis*, 21: 265-278.

*Investment Companies*. 1944- . New York: Arthur Wiesenberger & Co.

Kiviat, P.J., R. Villanueva, and H.M. Markowitz. 1983. *The SIMSCRIPT II.5 Programming Language*, E. Russell (ed.). La Jolla, CA: CACI.

Kroll, Y., H. Levy, and H.M. Markowitz. 1984. "Mean-Variance versus Direct Utility Maximization," *Journal of Finance*, 39: 47-61.

Lawler, E.L., J.K. Lenstra, A.H.G. Rinnooy Kan, and D.B. Shmoys. 1989. "Sequencing and Scheduling Algorithms and Complexity," in *Handbooks in Operations Research and Management Science*. Vol. 4, *Logistics of Production and Inventory*, S.C. Graves, A.H.G. Rinnooy Kan, and P. Aipkin (eds.), forthcoming. New York: North Holland.

Levy, H., and H.M. Markowitz. 1979. "Approximating Expected Utility by a Function of Mean and Variance," *American Economic Review*, 69: 308-317.

Markowitz, H.M. 1952. "Portfolio Selection," *The Journal of Finance*, 7(1): 77-91.

Markowitz, H.M. 1959. *Portfolio Selection: Efficient Diversification of Investments*. New York: Wiley (Yale University Press, 1970, Basil Blackwell, 1991).

Markowitz, H.M. 1987. *Mean-Variance Analysis in Portfolio Choice and Capital Markets*. New York: Basil Blackwell.

Pulley, L.B. 1981. "A General Mean-Variance Approximation to Expected Utility for Short Holding Periods," *Journal of Financial and Quantitative Analysis*, 16: 361-373.

Pulley, L.B. 1983. "Mean-Variance Approximations to Expected Logarithmic Utility, *Operations Research*, 31: 685-696.

Reid, D.W., and B.V. Tew. 1986. "Mean-Variance versus Direct Utility Maximization: A Comment," *Journal of Finance*, 41: 1177-1179.

Simaan, Y. 1987. "Portfolio Selection and Capital Asset Pricing for a Class of Non-Spherical Distributions of Asset Returns." Dissertation, Baruch College, The City University of New York.

Tew, B.V., and D.W. Reid. 1987. "More Evidence on Mean-Variance versus Direct Utility Maximization," *Journal of Financial Research*, 10: 249-257.

Von Neumann, J., and O. Morgenstern. 1944. *Theory of Games and Economic Behavior*, 3rd edition, 1953. Princeton University Press.

*The Wall Street Journal*. 1989. "By the Numbers," in WSJ Reports: Early Retirement, December 8, R25-26.

Young, W.E., and R.H. Trent. 1969. "Geometric Mean Approximation of Individual Security and Portfolio Performance," *Journal of Financial and Quantitative Analysis*, 4: 179-189.

THE JOURNAL OF FINANCE • VOL. XLVI, NO. 2 • JUNE 1991

# Foundations of Portfolio Theory

HARRY M. MARKOWITZ*

WHEN I STUDIED MICROECONOMICS forty years ago, I was first taught how optimizing firms and consumers would behave, and then taught the nature of the economic equilibrium which would result from such behavior. Let me refer to this as part one and part two of my microeconomics course. My work on portfolio theory considers how an optimizing investor would behave, whereas the work by Sharpe and Lintner on the Capital Asset Pricing Model (CAPM for short) is concerned with economic equilibrium assuming all investors optimize in the particular manner I proposed. Thus my work on the one hand, and that of Sharpe and Lintner on the other, provide part one and part two of a microeconomics of capital markets.

Professor Sharpe will discuss CAPM, part two of the course. I will confine my remarks to part one, portfolio theory. There are three major ways in which portfolio theory differs from the theory of the firm and the theory of the consumer which I was taught. First, it is concerned with investors rather than manufacturing firms or consumers. Second, it is concerned with economic agents who act under uncertainty. Third, it is a theory which can be used to direct practice, at least by large (usually institutional) investors with sufficient computer and database resources. The fact that it deals with investors rather than producers or consumers needs no further comment. Let me expand on the second and third differences.

In my microeconomics course, the theory of the producer assumed that the competitive firm knows the price at which it will sell the goods it produces. In the real world there is a delay between the decision to produce, the time of production and the time of sale. The price of the product at the time of sale may differ from that which was expected when the production decision was made. This uncertainty of eventual sales price is important in actual production planning but, quite reasonably, was ignored in classical economic models. It was judged not essential to the problem at hand.

Uncertainty cannot be dismissed so easily in the analysis of optimizing investor behavior. An investor who knew future returns with certainty would invest in only one security, namely the one with the highest future return. If several securities had the same, highest, future return then the investor would be indifferent between any of these, or any combination of these. In no

*Marvin Speiser Distinguished Professor of Finance and Economics, Baruch College, CUNY and Director of Research, DAIWA Security Trust Company.

case would the investor actually prefer a diversified portfolio. But diversification is a common and reasonable investment practice. Why? To reduce uncertainty! Clearly, the existence of uncertainty is essential to the analysis of rational investment behavior.

In discussing uncertainty below, I will speak as if investors faced known probability distributions. Of course, none of us know probability distributions of security returns. But, I was convinced by Leonard J. Savage, one of my great teachers at the University of Chicago, that a rational agent acting under uncertainty would act according to "probability beliefs" where no objective probabilities are known; and these probability beliefs or "subjective probabilities" combine exactly as do objective probabilities. This assumed, it is not clear and not relevant whether the probabilities, expected values, etc., I speak of below are for subjective or objective distributions.

The basic principles of portfolio theory came to me one day while I was reading John Burr Williams, *The Theory of Investment Value*. Williams proposed that the value of a stock should equal the present value of its future dividend stream. But clearly dividends are uncertain, so I took William's recommendation to be to value a stock as the *expected value* of its discounted future dividend stream. But if the investor is concerned only with the expected values of securities, the investor must also be only interested in the expected value of the portfolio. To maximize the expected value of a portfolio, one need only invest in one security—the security with maximum expected return (or one such, if several tie for maximum). Thus action based on expected return only (like action based on certainty of the future) must be rejected as descriptive of actual or rational investment behavior.

It seemed obvious that investors are concerned with risk and return, and that these should be measured for the portfolio as a whole. Variance (or, equivalently, standard deviation), came to mind as a measure of risk of the portfolio. The fact that the variance of the portfolio, that is the variance of a weighted sum, involved all covariance terms added to the plausibility of the approach. Since there were two criteria—expected return and risk—the natural approach for an economics student was to imagine the investor selecting a point from the set of Pareto optimal expected return, variance of return combinations, now known as the efficient frontier. These were the basic elements of portfolio theory which appeared one day while reading Williams.

In subsequent months and years I filled in some details; and then others filled in many more. For example in 1956 I published the "critical line algorithm" for tracing out the efficient frontier given estimates of expected returns, variances and covariances, for any number of securities subject to various kinds of constraints. In my 1959 book I explored the relationship between my mean-variance analysis and the fundamental theories of action under risk and uncertainty of Von Neumann and Morgenstern and L. J. Savage.

Starting in the 1960s, Sharpe, Blume, King, and Rosenberg greatly clarified the problem of estimating covariances. This past September I attended the Berkeley Program in Finance at which several analysts reported success in using publicly available accounting figures, perhaps combined with security analysts' earnings estimates, to estimate expected returns. I do not mean that their estimates eliminate uncertainty—only that, on the average, securities with higher estimates outperform those with lower estimates.

So, equipped with database, computer algorithms and methods of estimation, the modern portfolio theorist is able to trace out mean-variance frontiers for large universes of securities. But is this the right thing to do for the investor? In particular, are mean and variance proper and sufficient criteria for portfolio choice?

To help answer this question, let us consider the theory of rational choice under uncertainty. In doing so, let us recall the third way in which portfolio theory is to differ from classical microeconomic theory of the firm or consumer. We seek a set of rules which investors can follow in fact—at least investors with sufficient computational resources. Thus we prefer an approximate method which is computationally feasible to a precise one which cannot be computed. I believe that this is the point at which Kenneth Arrow's work on the economics of uncertainty diverges from mine. He sought a precise and general solution. I sought as good an approximation as could be implemented. I believe that both lines of inquiry are valuable.

The discussion of principles of rational behavior under uncertainty in Part IV of my 1959 book starts with a variant of L. J. Savage's axioms. From such axioms it follows that one should choose a strategy which maximizes expected utility for a many-period game. This, in turn, implies that the investor should act each period so as to maximize the expected value of a single period utility function. This single period utility function may depend on portfolio return and perhaps other state variables. For now, assume that it depends only on portfolio return.

In this case, the crucial question is this: if an investor with a particular single period utility function acted only on the basis of expected return and variance, could the investor achieve almost maximum expected utility? Or, to put it another way, if you know the expected value and variance of a probability distribution of return on a portfolio can you guess fairly closely its expected utility?

A great deal of research has been done on this question, but more is needed. Let me briefly characterize some results, and some open questions. Table I is extracted from Levy and Markowitz. The rows of the table represent various utility functions. For example, the first row reports results for $U(R) = \log(1 + R)$ where $R$ is the rate of return on the portfolio; the second row reports results for $U(R) = (1 + R)^{0.1}$, etc., as indicated in the first column of the table. The second through fifth column of the table represent various sets of historical distributions of returns on portfolios. For example, the second column represents annual returns on 149 investment companies, 1958–67; the third column represents annual returns on 97 stocks.

The calculations associated with the second column in effect assume that an investor must choose one out of 149 portfolios, and his probability beliefs

**Table I**

**Correlation Between $EU$ and $f(E, V)$ for Four Historical Distributions**

| Utility Function | Annual Returns on 149 Mutual Funds[1] | Annual Returns on 97 Stocks[2] | Monthly returns on 97 Stocks[2] | Random Portfolios of 5 or 6 Stocks[3] |
|---|---|---|---|---|
| $\text{Log}(1 + R)$ | 0.997 | 0.880 | 0.995 | 0.998 |
| $(1 + R)^a$ | | | | |
| $a = 0.1$ | 0.998 | 0.895 | 0.996 | 0.998 |
| $a = 0.3$ | 0.999 | 0.932 | 0.998 | 0.999 |
| $a = 0.5$ | 0.999 | 0.968 | 0.999 | 0.999 |
| $a = 0.7$ | 0.999 | 0.991 | 0.999 | 0.999 |
| $a = 0.9$ | 0.999 | 0.999 | 0.999 | 0.999 |
| $-e^{b(1+R)}$ | | | | |
| $b = 0.1$ | 0.999 | 0.999 | 0.999 | 0.999 |
| $b = 0.5$ | 0.999 | 0.961 | 0.999 | 0.999 |
| $b = 1.0$ | 0.997 | 0.850 | 0.997 | 0.998 |
| $b = 3.0$ | 0.949 | 0.850 | 0.976 | 0.958 |
| $b = 5.0$ | 0.855 | 0.863 | 0.961 | 0.919 |
| $b = 10.$ | 0.449 | 0.659 | 0.899 | 0.768 |

[1]The annual rate of return of the 149 mutual funds are taken from the various annual issues of A. Wiesenberger and Company. All mutual funds whose rates of return are reported in Wiesenberger for the whole period 1958–67 are included in the analysis.

[2]This data base of 97 U.S. stocks, available at Hebrew University, had previously been obtained as follows: a sample of 100 stocks was randomly drawn from the CRSP (Center for Research in Security Prices, University of Chicago) tape, subject to the constraint that all had reported rates of return for the whole period 1948–68. Some mechanical problems reduced the usable sample size from 100 to 97. The inclusion only of stocks which had reported rates of return during the whole period may have introduced survival bias into the sample. This did not appear harmful for the purpose at hand.

[3]We randomly drew 5 stocks to constitute the first portfolio; 5 different stocks to constitute the second portfolio, etc. Since we have 97 stocks in our sample, the eighteenth and nineteenth portfolios include 6 stocks each. Repetition of this experiment with new random variables produced negligible variations in the numbers reported, except for the case of $U = -e^{-10(1+R)}$. A median figure is reported in the table for this case.

concerning returns on these portfolios are the same as historical returns. It is not that we recommend this as a way of forming beliefs; rather, we use this as an example of distributions of returns which occur in fact.

For each utility function, and for each of the 149 probability distributions of the second column, we computed its "expected" (that is, its mean) utility

$$EU = \textstyle\sum_{t=1}^{T} U(R_t)/T \qquad (1)$$

where $T$ is the number of periods in the sample, and $R_t$ the rate of return in period $t$. We also computed various approximations to $EU$ where the approximation depends only on the mean value $E$ and the variance $V$ of the distribution. Of the various approximations tried in Levy-Markowitz the one which did best, almost without exception, was essentially that suggested in Markowitz (1959), namely

$$f(E, V) = U(E) + 0.5U''(E)V \tag{2}$$

For example, if $U(R) = \log(1 + R)$,

$$f(E, V) = \log(1 + R) - 0.5V/(1 + E)^2. \tag{3}$$

Equation (2) may be thought of as a rule by which, if you know the $E$ and $V$ of a distribution, you can guess at its expected utility. The figures in Table I are for the Levy-Markowitz approximation which is essentially (2). The entry in the second column, first row reports that, over the 149 probability distributions, the correlation between $EU$ and $f(E, V)$ was 0.997 for $U = \log(1 + r)$. The remaining entries in the second column similarly show the correlation, over the 149 probability distributions, of $EU$ and $f(E, V)$ for the utility functions tested. In most cases the correlation was extremely high, usually exceeding 0.99. We will discuss an exceptional case shortly.

The third column shows the correlation between $EU$ and $f(E, V)$ for a sample of annual return on one-stock "portfolios". The correlations are clearly less than for the diversified investment company portfolios of the second column. The fourth column again considers undiversified, single stock portfolios, but this time for monthly holding period returns. The correlations are much higher than those of column three, usually as high or higher than those in column two. Thus, for the investor who revises his or her portfolio monthly, even for portfolios whose returns were as variable as those of individual stocks, $f(E, V)$ would be highly correlated with $EU$ for the utility functions considered.

The fifth column shows annual holding period returns, now for randomly selected portfolios with 5 or 6 securities each. The correlations are generally quite high again—comparable to those in the second column. Thus, at least, for these probability distributions and most of these utility functions, $f(E, V)$ approximates $EU$ quite well for diversified portfolios, even "slightly" diversified portfolios of size 5 and 6.

Not all expected maximizers are equally served by mean-variance approximations. For example, the investor with $U = -e^{-10(1+R)}$ will find mean-variance much less satisfactory than others presented in Table I. Levy and Markowitz have two observations concerning an expected utility maximizer with $U = -e^{-10(1+R)}$.

The first observation is that an investor who had $-e^{-10(1+R)}$ as his or her utility function would have some very strange preferences among probabilities of return. Reasonably enough, he or she would not insist on certainty of return. For example, the investor would prefer (a) a 50-50 chance of a 5 percent gain vs. a 25 percent gain rather than have (b) a 10 percent gain with certainty. On the other hand there is no $R$ which would induce the investor to take (a) a 50-50 chance of zero return (no gain, no loss) vs. a gain of $R$ rather than have (b) a 10 percent return with certainty. Thus a 50-50 chance of breaking even vs. a 100,000 percent return, would be considered less desirable than a 10 percent return with certainty. We believed that few if any investors had preferences anything like these.

A second observation was that even if some unusual investor did have the utility function in question, such an investor could determine in advance that f($E, V$) was not a good approximation for this $EU$. Table II shows the difference between $U(R)$ and the Taylor approximation upon which (2) is based, namely,

$$Q(R) = U(E) + U'(E)(R - E) + 0.5U''(E)(R - E)^2 \tag{4}$$

for $U = \log(1 + R)$ and $U = -1000e^{-10(1+R)}$, for $E = 0.10$. For the various $R$ listed in the first column, the second through fourth columns show $U(R)$, $Q(R)$ and $\Delta(R) = U(R) - Q(R)$ for $\log(1 + R)$; the following three columns show the same for $-1000e^{-10(1+R)}$. Since the choices implied by a utility function are unaffected by multiplying it by a positive constant, it is not the magnitude of the $\Delta(R)$s which are important. Rather it is the variation in $\Delta(R)$ as compared to that in $U(R)$. For example, Levy and Markowitz present a lower bound on the correlation between $U(R)$ and f($E, V$) as a function of the standard deviations of $U$ and $\Delta$. As we see in the table, as log(1 + $R$) goes from $-0.357$ at $R = -0.30$ to 0.470 at $R = 0.60$, $|\Delta|$ never exceeds 0.024. In contrast, as $-1000e^{-10(1+R)}$ goes from $-0.912$ to $-0.0001$, $|\Delta|$ often exceeds 0.03 and has a maximum of $-0.695$.[1] Thus, if an investor had $U = -e^{-10(1+R)}$ as a utility function, a comparison of $U(R)$, $Q(R)$, and $\Delta(R)$ would provide ample warning that mean-variance is not suitable.

Levy and Markowitz present other empirical results. They also explain the difference between assuming that an investor has a quadratic utility function

**Table II**

**Quadratic Approximation to Two Utility Functions**

**$E$ = 00.1**

| $R$ | log(1 + $R$) | $Q_L(R)$ | $\Delta_L$ | $-1000e^{10(1+R)}$ | $Q_E(R)$ | $\Delta_E$ |
|---|---|---|---|---|---|---|
| −.30 | −.35667 | −.33444 | −.02223 | −.91188 | −.21712 | −.69476 |
| −.20 | −.22314 | −.21461 | −.00854 | −.33546 | −.14196 | −.14950 |
| −.10 | −.10536 | −.10304 | −.00232 | −.12341 | −.08351 | −.03990 |
| .00 | .00000 | .00027 | −.00027 | −.04540 | −.04175 | −.00365 |
| .10 | .09531 | .09531 | .00000 | −.01670 | −.01670 | .00000 |
| .20 | .18232 | .18209 | .00023 | −.00614 | −.00835 | .00221 |
| .30 | .26236 | .26060 | .00176 | −.00226 | −.01670 | .01444 |
| .40 | .33647 | .33085 | .00563 | −.00083 | −.04175 | .04092 |
| .50 | .40546 | .39283 | .01263 | −.00031 | −.08351 | .08320 |
| .60 | .47000 | .44655 | .02345 | −.00011 | −.14196 | .14185 |

[1]Among the 149 mutual funds, those with $E$ near .10 all had annual returns between a 30% loss and a 60% gain. Specifically, 64 distributions had $0.08 \leq E \leq 0.12$, and all had returns within the range indicated.

versus using a quadratic approximation to a given utility function to develop an f($E$, $V$) approximation, such as that in (2). In particular, they show that f($E$, $V$) in (2) is not subject to the Arrow, Pratt objection to a quadratic utility function, that it has increasing risk aversion. Indeed, Levy and Markowitz show that a large class of f($E$, $V$) approximations, including (2), have the same risk aversion in the small as does the original *EU* maximizer.

I will not recount here these further Levy and Markowitz results, nor will I go into important results of many others. Chapter 3 of Markowitz (1987) includes a survey of the area up to that time. I will, however, briefly note results in two important unpublished papers.

Levy and Markowitz measure the efficacy of f($E$, $V$) by the correlation between it and *EU*. Y. Simaan defines the optimization premium to be the percent the investor would be just willing to pay out of the portfolio for the privilege of choosing the true *EU* maximizing portfolio rather than being confined to the mean-variance "second best". The reason for performing a mean-variance analysis in fact, rather than a theoretically correct expected utility analysis, is convenience, cost or feasibility. It is typically much more expensive to find a utility maximizing portfolio than to trace out an entire mean-variance frontier. The data requirements for an expected utility analysis can substantially exceed those of a mean-variance analysis, since estimates of first and second moments generally are not sufficient for the former. Finally, there is the problem of determining the investor's utility function. Simaan's criteria measures the worth, as a percent of the portfolio, paid out of the portfolio, of incurring the added expenses of finding an *EU* maximizing portfolio. He solves for this optimization premium analytically under certain assumptions.

L. Ederington evaluates *EU* approximations using thousands of synthetic time series generated by randomly selecting from actual time series. He evaluates approximations like (2), except that they use the first three of four moments, as well as (2) that uses the first two. It is all very well to point out theoretically that more moments are better than fewer. The practical question is: how much?

Ederington finds, as did Levy and Markowitz, that for some utility functions the mean-variance approximation is so good that there is virtually no room for improvement. Where the mean-variance approximation falters, Ederington finds that typically three moments provides little improvement to the approximation whereas four moments improves the approximation considerably.

Despite noteworthy results reported above, and many more that I have not described here, there is much to be done. Three examples will illustrate the need.

First, all the experimentation and analysis to date give us a rather spotty account of where mean-variance serves well and where it falters. Perhaps it is possible to develop a more systematic characterization of the utility functions and distributions for which the mean-variance approximation is good, bad and marginal.

Second, suppose that the investor has a utility function for which mean-variance provides a close approximation, but the investor does not know precisely which one. In this case the investor need not determine his or her utility function to obtain a near optimum portfolio. The investor need only pick carefully from the (one-dimensional) curve of efficient $EV$ combinations in the two dimensional $EV$ space. To pursue a similar approach when four moments are required, the investor must pick carefully from a three-dimensional surface in a four-dimensional space. This raises serious operational problems in itself, even if we overcome computational problems due to the nonconvexity of sets of portfolios with given third moment or better.

But perhaps there is an alternative. Perhaps some other measure of portfolio risk will serve in a two parameter analysis for some of the utility functions which are a problem to variance. For example, in Chapter 9 of Markowitz (1959) I proposed the "semi-variance" $S$ as a measure of risk where

$$S = E(\text{Min}\,(0, R - c)^2)$$

where $c = E(R)$ or $c$ is a constant independent of choice of portfolio. Semi-variance seems more plausible than variance as a measure of risk, since it is concerned only with adverse deviations. But, as far as I know, to date no one has determined whether there is a substantial class of utility functions for which mean-semi-variance succeeds while mean-variance fails to provide an adequate approximation to $EU$.

Third, in general the derived, single period utility functions can contain state-variables in addition to return (or end of period wealth). Expected utility, in this case, can be estimated from return and state-variable means, variances and covariances provided that utility is approximately quadratic in the relevant region. (Recall the Levy-Markowitz analysis of quadratic utility versus quadratic approximation in the relevant region.) To my knowledge, no one has investigated such quadratic approximation for cases in which state variables other than portfolio value are needed in practice.

In sum, it seems to me that the theory of rational behavior under uncertainty can continue to provide insights as to which practicable procedures provide near optimum results. In particular, it can further help evaluate the adequacy of mean and variance, or alternate practical measures, as criteria.

Finally, I would like to add a comment concerning portfolio theory as a part of the microeconomics of action under uncertainty. It has not always been considered so. For example, when I defended my dissertation as a student in the Economics Department of the University of Chicago, Professor Milton Friedman argued that portfolio theory was not Economics, and that they could not award me a Ph.D. degree in Economics for a dissertation which was not in Economics. I assume that he was only half serious, since they did award me the degree without long debate. As to the merits of his arguments, at this point I am quite willing to concede: at the time I defended my dissertation, portfolio theory was not part of Economics. But now it is.

**REFERENCES**

Arrow, K., 1965, *Aspects of the Theory of Risk Bearing*, (Helsinki).

Blume, M., 1971, On the assessment of risk, *Journal of Finance* 26, 1-10.

Ederington, L. H. 1986, Mean-variance as an approximation to expected utility maximization, Working paper 86-5, School of Business Administration, Washington University, St. Louis, Missouri.

King, B. F., 1966, Market and industry factors in stock price behavior, *Journal of Business* 39, 139-190.

Levy, H. and H. M. Markowitz, 1979, Approximating expected utility by a function of mean and variance, *American Economic Review* 69, 308-317.

Lintner, J., 1965, The valuation of risk assets and the selection of risky investments in stock portfolios and capital budgets, *Review of Economics and Statistics* February.

Markowitz, H. M., 1952, Portfolio selection, *The Journal of Finance* 7, 77-91.

——, 1956, The optimization of a quadratic function subject to linear constraints, *Naval Research Logistics Quarterly* 3.

——, 1959, *Portfolio Selection: Efficient Diversification of Investments*, (Wiley, Yale University Press, 1970, Basil Blackwell, 1991).

——, 1987, *Mean-Variance Analysis in Portfolio Choice and Capital Markets*, (Basil Blackwell, paperback edition, Basil Blackwell, 1990).

Pratt, J. W., 1964, Risk aversion in the small and in the large, *Econometrica* 32, 122-136.

Rosenberg, B., 1974, Extra-market components of covariance in security returns, *Journal of Financial and Quantitative Analysis* 29, 263-273.

Savage, L. J., 1954, *The Foundations of Statistics*, 2nd ed., (Wiley, Dover, 1972).

Sharpe, W. F., 1963, A simplified model for portfolio analysis, *Management Science* 9, 277-293.

——, 1964, Capital asset prices: A theory of market equilibrium under conditions of risk, *The Journal of Finance* 19, 425-442.

Simaan, Y., 1987, Portfolio selection and capital asset pricing for a class of non-spherical distributions of assets returns, dissertation, Baruch College, The City University of New York.

A. Wiesenberger and Company, *Investment Companies*, annual editions, New York.

Von Neumann, J., and O. Morgenstern, 1944, *Theory of Games and Economic Behavior*, 3rd ed., (Princeton University Press, 1953).

Williams, J. B., 1938, *The Theory of Investment Value*, (Harvard University Press, Cambridge).

*Journal of Financial Engineering—Volume 1—Number 2—Pages 117–132*

# Fast Computation of Mean–Variance Efficient Sets Using Historical Covariances

*Harry Markowitz*

*Peter Todd*

*Ganlin Xu*

*Yuji Yamane*

## ABSTRACT

The general mean–variance portfolio optimization problem seeks to determine the efficient frontier by solving a parametric quadratic programming problem that involves an arbitrary covariance matrix of security returns. In the case where historical covariances are used, and the number of securities in the optimization problem is much larger than the number of historical return observations, the problem can be reformulated so that a substantial saving in computer storage and execution time can be realized.

Harry Markowitz, Peter Todd, Ganlin Xu, and Yuji Yamane; Daiwa Securites Trust Company; Global Portfolio Research Department; 1 Evertrust Plaza, Jersey City, NJ 07302.

# I. INTRODUCTION

The general mean-variance portfolio optimization problem (Markowitz 1956, 1959, 1987) seeks to determine portfolios of securities that are efficient with respect to expected return (E) and variance (V), subject to any system of linear equalities in non-negative variables. Efficient portfolios have minimum V for a given E, and maximum E for a given V. The set of all efficient portfolios form the efficient frontier on the E-V graph. There is essentially only one algorithm for computing the efficient frontier, called the critical line algorithm (Markowitz 1956, 1959, 1987, Perold 1984). One input required for the portfolio optimization problem is the matrix of covariances of the returns of the securities (C). In practice, various methods are used to estimate C. One frequently used method is to compute C from historical returns data (see, e.g., Sharpe 1963, Haugen and Baker[1] 1990, 1991). Often T, the number of historical observations used to compute C, is less than n, the number of securities. In such cases C does not have full rank, although this does not cause any problem for the critical line algorithm. In this paper we show how the portfolio optimization problem can be reformulated to use historical covariance data without actually computing and storing C. When T is significantly less than n, a very substantial savings in computer storage and execution time can be realized.

Other methods besides computing covariances from historical returns are also used in practice to estimate the covariance matrix. In particular, C can be derived from factor models or scenario models (see, e.g., Sharpe 1963, Cohen and Pogue 1967, Resenberg 1974, Markowitz and Perold 1981). One advantage of the factor and scenario models has been that the mean-variance efficient frontier could be more quickly computed if these models were assumed and appropriate computational procedures used. Our objective here is not to argue for one estimation method over another, but to show how to compute more efficiently in the case where historical covariances are used.

## Statement of the Problem

The portfolio optimization problem can be stated as the parametric quadratic programming problem:

$$\text{minimize: } V = x'Cx$$

$$\begin{gathered} \text{subject to: } \mu' x = E \\ Ax = b \\ x \geq 0 \\ \text{for all } E \text{ in } [E_{min}, E_{max}] \end{gathered} \tag{1.1}$$

where x is the n-vector of security holdings representing the portfolio, $\mu$ is the n-vector of security expected returns, the A and b are an m,n-matrix and an m-vector specifying m linear constraints, $E_{max}$ is the maximum feasible E, and $E_{min}$ is the E of the efficient portfolio with minimum V.

## Review of the Critical Line Algorithm

The critical line algorithm computes the efficient frontier by solving an equivalent parametric quadratic programming problem:

$$\text{minimize: } \tfrac{1}{2}\, x'Cx - \lambda_E\, \mu' x$$

$$\begin{aligned} \text{subject to: } & Ax = b \\ & x \geq 0 \\ & \text{for all } \lambda_E \text{ in } [0,\infty) \end{aligned} \tag{1.2}$$

where $\lambda_E$ is the trade-off parameter between return and risk.[2] The algorithm iteratively traces out efficient portfolios along critical lines. For any given sets IN and OUT, the disjoint partitions of the universe of securities, a critical line $[X'(\lambda_E), \lambda'(\lambda_E), \eta''(\lambda_E)]' \in R^{2n+m}$ is defined by:

$$\begin{pmatrix} C_{IN} & A'_{IN} \\ A_{IN} & 0 \end{pmatrix} \begin{pmatrix} X_{IN} \\ \lambda \end{pmatrix} = \begin{pmatrix} 0 \\ b \end{pmatrix} + \lambda_E \begin{pmatrix} \mu_{IN} \\ 0 \end{pmatrix} \tag{1.3a}$$

$$x_i = 0, \qquad i \in OUT \tag{1.3b}$$

$$\eta = (C, A') \begin{pmatrix} x \\ \lambda \end{pmatrix} - \lambda_E\, \mu \tag{1.3c}$$

where $C_{IN}$ is the submatrix obtained by deleting OUT rows and OUT columns, $A_{IN}$ the matrix obtained by deleting OUT columns, and $\mu_{IN}$ is the vector obtained by deleting OUT rows. By the Kuhn-Tucker theorem, a point on a critical line is efficient if it satisfies:

$$\eta_i \geq 0 \qquad i \in OUT \tag{1.4a}$$

$$x_i \geq 0, \quad \eta_i = 0, \qquad i \in IN \tag{1.4b}$$

For a given IN we define:

$$M_{IN} \triangleq \begin{pmatrix} C_{IN} & A'_{IN} \\ A_{IN} & 0 \end{pmatrix} \qquad (1.5)$$

In practice we start with a particular IN set such that $M_{IN}$ is non-singular and Kuhn-Tuck conditions equations 1.3b and 1.4 are met for all $\lambda_E \in [\lambda_{low}{}^0, \infty)$; then trace out the frontier in the direction of decreasing $\lambda_E$. At each iteration the algorithm has an IN set such that $M_{IN}$ is non-singular and equations 1.3b and 1.4 hold for $\lambda_E \in [\lambda^j_{low}, \lambda^j_{high}]$, where $\lambda^j_{high} = \lambda^{j-1}_{low}$. Decreasing $\lambda_E$ below $\lambda^j_{low}$ would cause one of the inequalities in equation 1.4, denoted by index $\gamma$, to be violated. Then either $\{\gamma\}$ is deleted from OUT and added into IN, or $\{\gamma\}$ is deleted from IN and added into OUT. The new $M_{IN}$ matrix always remains non-singular, and the Kuhn-Tucker conditions continue to hold. The algorithm continues until $\lambda_E = 0$ is reached in finitely many number of steps. The portfolios reached at $\lambda^j_{low}$ are called **corner portfolios**.

To implement the critical line algorithm, we need to solve linear equation 1.3a efficiently. One way to do this is to store $M_{IN}{}^{-1}$ in the computer. When the IN set changes, $M_{IN}{}^{-1}$ is updated. Notice that the storage takes an order of $(\#IN + m)^2$. In practice, we do not shrink and expand the size of the stored $M_{IN}$ and $M_{IN}{}^{-1}$ matrices. Rather, we use flags and pointers to direct the computation to the IN part of M and $M_{IN}{}^{-1}$.

## II. REFORMULATION OF THE PROBLEM

Let $r_i$ denote the historical returns of security i. For each of the securities, we have T historical observations $r_{t,i}$, t = 1, . . ,T. From these observations, calculate historical mean return $\bar{r}_i$ of $i^{th}$ security by:

$$\bar{r}_i = \sum_{t=1}^{T} r_{t,i}/T, \qquad i = 1, \ldots, n \qquad (2.1)$$

and historical covariance $C_{ij}$ between $i^{th}$ security and $j^{th}$ security by:

$$C_{ij} = \sum_{t=1}^{T} \frac{(r_{t,i} - \bar{r}_i)(r_{t,j} - \bar{r}_j)}{T} \qquad (2.2)$$

We define a T X n matrix:

$$B = \frac{1}{\sqrt{T}}\begin{pmatrix} r_{11} & r_{12} & \cdot & \cdot & r_{1n} \\ \cdot & \cdot & \cdot & \cdot & \cdot \\ r_{T1} & r_{T2} & \cdot & \cdot & r_{Tn} \end{pmatrix} - \frac{1}{\sqrt{T}}\begin{pmatrix} \overline{r_1} & \cdot & \cdot & \overline{r_n} \\ \cdot & \cdot & \cdot & \cdot \\ \overline{r_1} & \cdot & \cdot & \overline{r_n} \end{pmatrix}$$

then write equation 2.2 in matrix form:

$$C = B' \cdot B \tag{2.3}$$

Instead of applying the critical line algorithm to model 1.2 directly with C computed by equation 2.2, we introduce T new variables:[3]

$$y = Bx \tag{2.4}$$

then;

$$V = y' \cdot y$$

Therefore, we can write the parametric quadratic programming problem 1.2 as:

$$\text{minimize: } \frac{1}{2}\begin{pmatrix} x' & y' \end{pmatrix}\begin{pmatrix} 0 & 0 \\ 0 & I \end{pmatrix}\begin{pmatrix} x \\ y \end{pmatrix} - \lambda_E(\mu',0)\begin{pmatrix} x \\ y \end{pmatrix} \tag{2.5}$$

$$\text{subject to: } \begin{pmatrix} B & -I \\ A & 0 \end{pmatrix}\begin{pmatrix} x \\ y \end{pmatrix} = \begin{pmatrix} 0 \\ b \end{pmatrix}$$

$$x \geq 0$$

With this new formulation, we can think of our universe of securities as consisting of n real securities and T fictitious securities with expected return $(\mu',0)' \in R^{n+T}$, and covariance matrix:

$$C \triangleq \begin{pmatrix} 0 & 0 \\ 0 & I \end{pmatrix}$$

# III. IMPLEMENTATION OF THE CRITICAL LINE ALGORITHM

Since there are no non-negativity constraints on the fictitious securities, they are always in the IN set of the reformulated model. It will be convenient below to still denote by IN and OUT the disjoint partitions of real securities. The corresponding critical line $[X'(\lambda_E), Y'(\lambda_E)', \lambda_a'(\lambda_E), \lambda_b'(\lambda_E), \eta'(\lambda_E)]' \in R^{2n+2T+m}$ and Kuhn–Tucker conditions are given by (see equations 1.3 and 1.4):

$$\begin{pmatrix} 0 & 0 & B_{IN}' & A_{IN}' \\ 0 & I & -I & 0 \\ B_{IN} & -I & 0 & 0 \\ A_{IN} & 0 & 0 & 0 \end{pmatrix} \begin{pmatrix} X_{IN} \\ Y \\ \lambda_a \\ \lambda_b \end{pmatrix} = \begin{pmatrix} 0 \\ 0 \\ 0 \\ b \end{pmatrix} + \lambda_E \begin{pmatrix} \mu_{IN} \\ 0 \\ 0 \\ 0 \end{pmatrix} \tag{3.1a}$$

$$\eta = \begin{pmatrix} 0 & 0 & B' & A' \\ 0 & I & -I & 0 \end{pmatrix} \begin{pmatrix} X \\ Y \\ \lambda_a \\ \lambda_b \end{pmatrix} - \lambda_E \begin{pmatrix} \mu \\ 0 \\ 0 \\ 0 \end{pmatrix} \tag{3.1b}$$

$$\begin{aligned} &\eta_i = 0, \quad && x_i \geq 0 \quad && i \in IN \\ &\eta_i \geq 0, && x_i = 0 && i \in OUT \\ &\eta_{n+t} = 0, && && t = 1, 2 \ldots T \end{aligned} \tag{3.1c}$$

The corresponding matrix $M_{IN}$ (see equation 1.5) is given by:

$$M_{IN} \triangleq \begin{pmatrix} 0 & 0 & B_{IN}' & A_{IN}' \\ 0 & I & -I & 0 \\ B_{IN} & -I & 0 & 0 \\ A_{IN} & 0 & 0 & 0 \end{pmatrix} \tag{3.2}$$

Therefore, to implement the critical line algorithm, we need to solve the system of equation 3.1a efficiently. For the rest of the paper, we will continue our discussion in terms of the following generic form of equation 3.1a:

$$M_{IN} \begin{pmatrix} X \\ Y \\ Z \\ W \end{pmatrix} = \begin{pmatrix} a_1 \\ a_2 \\ a_3 \\ a_4 \end{pmatrix} \tag{3.3}$$

Since equation 2.5 is a special case of the general portfolio selection model, we are guaranteed that $M_{IN}$ is non-singular. Writing out the system of equation 3.3, we have:

$$B_{IN}'Z + A_{IN}'W = a_1 \tag{3.4a}$$

$$Y - Z = a_2 \tag{3.4b}$$

$$B_{IN}X - Y = a_3 \tag{3.4c}$$

$$A_{IN}X = a_4 \tag{3.4d}$$

From equation 3.4b, we can solve Z in terms of Y:

$$Z = Y - a_2 \tag{3.5}$$

Substitute Z in to equation 3.4a, and we get:

$$B_{IN}'Y + A_{IN}'W = a_1 + B_{IN}'a_2$$

Combining this equation with equations 3.4c and 3.4d, we see that (X', Y', Z', W')' is a solution to equation 3.3 if and only if (Y', W', X')' is a solution to:

$$\begin{pmatrix} -I & 0 & B_{IN} \\ 0 & 0 & A_{IN} \\ B_{IN}' & A_{IN}' & 0 \end{pmatrix} \begin{pmatrix} Y \\ W \\ X \end{pmatrix} = \begin{pmatrix} a_3 \\ a_4 \\ a_1 + B_{IN}'a_2 \end{pmatrix} \tag{3.6}$$

and Z is given by equation 3.5. Now let us define:

$$M_{IN*} = \begin{pmatrix} -I & 0 & B_{IN} \\ 0 & 0 & A_{IN} \\ B'_{IN} & A'_{IN} & 0 \end{pmatrix} \tag{3.7}$$

Setting $a_1 = a_2 = a_3 = a_4 = 0$ in equation 3.6 and using equation 3.3, we see that $M_{IN*}(Y', W', X')' = 0$ implies that $M_{IN}\ (X', Y', Y', W')' = 0$. Since $M_{IN}$ is non-singular, we have $(Y', W', X')' = 0$. Thus, $M_{IN*}$ is non-singular too. Its inverse, $M_{IN*}^{-1}$, will be written as:

$$(M_{IN*})^{-1} = \begin{pmatrix} J & H' \\ H & G \end{pmatrix} = \begin{pmatrix} J & H'_1 & H'_2 \\ H_1 & G_{11} & G'_{21} \\ H_2 & G_{21} & G_{22} \end{pmatrix} \tag{3.8}$$

From:

$$M_{IN*} \bullet \begin{pmatrix} J & H' \\ H & G \end{pmatrix} = \begin{pmatrix} I & 0 & 0 \\ 0 & I & 0 \\ 0 & 0 & I \end{pmatrix}$$

we get:

$$\begin{aligned} H_1 &= G'_{21}\, B'_{IN} \\ H_2 &= G_{22}\, B'_{IN} \\ J &= B_{IN}\, G_{22}\, B'_{IN} - I \end{aligned} \tag{3.9}$$

**PROPOSITION 3.1.** The solution to the system of equations 3.3 can be expressed in terms of G by the following formulae:

$$X = G_{21}a_4 + G_{22}\,[a_1 + B_{IN}'\,(a_2 + a_3)]$$

$$Y = -a_3 + B_{IN}\,\{G_{21}a_4 + G_{22}\,[a_1 + B_{IN}'\,(a_2 + a_3)]\}$$

$$W = G_{11}a_4 + G_{21}' [a_1 + B_{IN}' (a_2 + a_3)] \tag{3.10}$$
$$Z = -a_3 - a_2 + B_{IN} \{G_{21}a_4 + G_{22} [a_1 + B_{IN}' (a_2 + a_3)]\}$$

Proof: From equation 3.6 and equation 3.8, we have:

$$\begin{pmatrix} Y \\ W \\ X \end{pmatrix} = \left(M_{IN*}\right)^{-1} \begin{pmatrix} a_3 \\ a_4 \\ a_1 + B'_{IN} a_2 \end{pmatrix} = \begin{pmatrix} J & H'_1 & H'_2 \\ H_1 & G_{11} & G'_{21} \\ H_2 & G_{21} & G_{22} \end{pmatrix} \begin{pmatrix} a_3 \\ a_4 \\ a_1 + B'_{IN} a_2 \end{pmatrix}$$

$$= \begin{pmatrix} Ja_3 + H'_1 a_4 + H'_2(a_1 + B'_{IN}a_2) \\ H_1 a_3 + G_{11}a_4 + G'_{21}(a_1 + B'_{IN}a_2) \\ H_2 a_3 + G_{21}a_4 + G'_{22}(a_1 + B'_{IN}a_2) \end{pmatrix}$$

Now equation 3.10 follows from this, equations 3.9 and 3.5.

Because of this Proposition, we need only store G, rather than $(M_{IN*})^{-1}$, during the computation process. Since $M_{IN*}$ is non-singular, therefore:

$$\text{The number of IN securities} = \text{Rank} \begin{pmatrix} B_{IN} \\ A_{IN} \\ 0 \end{pmatrix} = \text{Rank} \begin{pmatrix} B_{IN} \\ A_{IN} \end{pmatrix} \leq T + m.$$

This implies that the size of G is at most (2m + T). From this we see that if T is not large, there is not much requirement for storage, even if n is large. During the computation process, instead of computing $M_{IN*}^{-1}$ every time the IN set changes, we update G. We show how to do this in the appendix.

# IV. PERFORMANCE

We have developed a mean-variance optimizer program that can either use a covariance matrix (the "basic" algorithm) or historical returns (the "historical" algorithm described in this paper). We used this program to obtain a comparison of the performance of the two algorithms on a test problem involving 1008 securities and 60 historical monthly return observations. The only constraint was the budget constraint. The computed efficient frontier contained 62 corner portfolios.

Our test problem was run on an IBM RS/6000 Model 320 Computer. The program was compiled with the IBM AIX XL C compiler version 01.01.0001.0001. All floating-point computations use double precision arithmetic (8 bytes per number). Table 1 shows the execution time for the critical line algorithm, and the amount of storage allocated for arrays.

We recently rewrote the portions of the program that compute efficient frontiers using the historical algorithm to "optimize" for speed. Table 1 contains results for both the old un-optimized version of the program and the newer optimized version. It is likely that a similar increase in execution speed could be obtained by optimizing the code for the basic algorithm.

Our portfolio optimization program incorporates several additions to the basic optimization problem described by equation 1.1:

1. The ability to used lower bounds on x other than zero, and upper bounds on x, without increasing the number of constraint equations.
2. The ability to constrain portfolio turnover by introducing only one additional constraint equation, and no additional variables.
3. The ability to consider transaction costs for buying and selling securities in the optimization.

**Table 1**
**Performance Comparison**

| Algorithm | Exec. Time | Storage |
|---|---|---|
| Basic | 210.9 sec | 8,167 Kb |
| Historical, un-optimized | 25.8 sec | 781 Kb |
| Historical, optimized for speed | 11.0 sec | 1,257 Kb |

Although none of these features were used in the test problem presented here, their incorporation into the program causes a modest increase in execution times, even when they are not used.

# V. APPENDIX

**PROPOSITION 5.1.** Given two non-singular matrices M, N related by:

$$M = \begin{pmatrix} N & \alpha \\ \alpha' & \sigma \end{pmatrix} \tag{5.1}$$

where $\sigma$ is a constant, $\alpha$ is a vector. If:

$$M^{-1} = \begin{pmatrix} M^{*} & \xi \\ \xi' & \xi_K \end{pmatrix} \tag{5.2}$$

then $\xi_K \neq 0$, and:

$$N^{-1} = M^{*} - \xi \bullet \xi' / \xi_K \tag{5.3}$$

Furthermore, let:

$$\beta = N^{-1} \bullet \alpha \tag{5.4}$$

then $\beta_k^2 = \sigma - \alpha' \bullet \beta \neq 0$, and:

$$M^{-1} = \begin{pmatrix} N^{-1} + \dfrac{\beta \cdot \beta'}{\beta_K^2} & -\beta/\beta_K^2 \\ -\beta'/\beta_K^2 & 1/\beta_K^2 \end{pmatrix} \tag{5.5}$$

For proof, see Markowitz (1987, Chapter 13). Here we show how to apply the general formula to update G. First, let us consider the case of updating G by adding a security. Without loss of generality, we write:

$$M_{IN*}^{new} = \begin{pmatrix} -I & 0 & B_{IN} & b \\ 0 & 0 & A_{IN} & a \\ B'_{IN} & A'_{IN} & 0 & 0 \\ b' & a' & 0 & 0 \end{pmatrix} = \begin{pmatrix} M_{IN*} & & & b \\ & & & a \\ & & & 0 \\ b' & a' & 0 & 0 \end{pmatrix}$$

and:

$$(M_{IN*}^{new})^{-1} = \begin{pmatrix} J^{new} & (H^{new})' \\ H^{new} & G^{new} \end{pmatrix}$$

**PROPOSITION 5.2.** Let $\beta$ and $\beta_K$ be given by:

$$\beta = \begin{pmatrix} G_{11} & G'_{21} \\ G_{21} & G_{22} \end{pmatrix} \begin{pmatrix} a \\ B'_{IN} b \end{pmatrix} \tag{5.6}$$

$$\beta_K^2 = b'b - (a', b'B_{IN})\, G \begin{pmatrix} a \\ B'_{IN}\, b \end{pmatrix}$$

then we have $\beta_K{}^2 \neq 0$, and:

$$G^{new} = \begin{pmatrix} G + \dfrac{\beta \cdot \beta'}{\beta_K^2} & -\beta/\beta_K^2 \\ -\beta'/\beta_K^2 & 1/\beta_K^2 \end{pmatrix} \tag{5.8}$$

Proof: By equation 5.4, define:

$$\bar{X} = \begin{pmatrix} X \\ Y \\ Z \end{pmatrix} = (M_{IN*})^{-1} \begin{pmatrix} b \\ a \\ 0 \end{pmatrix}$$

and:

$$\zeta_K^2 = -(b' \ a' \ 0)\begin{pmatrix} X \\ Y \\ Z \end{pmatrix} = -b'x - a'y$$

Then Proposition 5.1 tells us that $\zeta_k^2 \neq 0$, and the general formula 5.5 gives us:

$$(M_{IN*}^{new})^{-1} = \begin{pmatrix} J^{new} & (H^{new})' \\ H^{new} & G^{new} \end{pmatrix} = \begin{pmatrix} (M_{IN})^{-1} + \overline{X} \bullet \overline{X}'/\zeta_K^2 & -\overline{X}/\zeta_K^2 \\ -\overline{X}'/\zeta_K^2 & 1/\zeta_K^2 \end{pmatrix}$$

From this and equation 3.8 we conclude that:

$$G^{new} = \begin{pmatrix} G + \begin{pmatrix} Y \\ Z \end{pmatrix}(Y' \ Z')/\zeta_K^2 & -\begin{pmatrix} Y \\ Z \end{pmatrix}/\zeta_K^2 \\ -(Y' \ Z')/\zeta_K^2 & 1/\zeta_K^2 \end{pmatrix}$$

Now it is just a matter of identifying $\beta$ with ( Y', Z')', and $\beta_K^2$ with $\zeta_k^2$. By equations 3.8 and 3.9, we have:

$$\begin{pmatrix} X \\ Y \\ Z \end{pmatrix} = (M_{IN*})^{-1}\begin{pmatrix} b \\ a \\ 0 \end{pmatrix} = \begin{pmatrix} Jb + H_1 a \\ H_1 b + G_{11} a \\ H_2 b + G_{21} a \end{pmatrix} = \begin{pmatrix} B_{IN} G_{22} B'_{IN} b & -b & +B_{IN} G_{21} a \\ G'_{21} B'_{IN} b & + G_{11} a & \\ G_{22} B'_{IN} b & + G_{21} a & \end{pmatrix}$$

Therefore:

$$\beta = \begin{pmatrix} Y \\ Z \end{pmatrix}, \qquad \beta_K^2 = \zeta_K^{-2}$$

Now let us consider the case of updating G by deleting a security. Without loss of generality, we write:

$$M_{IN*} = \begin{pmatrix} M_{IN*}^{new} & & & b \\ & & & a \\ & & & \underline{0} \\ b' & a' & \underline{0} & 0 \end{pmatrix}$$

and:

$$(M_{IN*}^{new})^{-1} = \begin{pmatrix} J^{new} & (H^{new})' \\ H^{new} & G^{new} \end{pmatrix} \tag{5.9}$$

**PROPOSITION 5.3.** Assume that:

$$G = \begin{pmatrix} G^{*} & g \\ g' & g_k \end{pmatrix}$$

then $g_k \neq 0$, and:

$$G^{new} = G^{*} - g \bullet g'/g_k \tag{5.10}$$

Proof: Let us write:

$$(M_{IN*})^{-1} = \begin{pmatrix} J & (H^{*})' & h \\ H^{*} & G^{*} & g \\ h' & g' & g_K \end{pmatrix}$$

By proposition 5.1, $g_K \neq 0$, and by equation 5.3:

$$(M_{IN*}^{new})^{-1} = \begin{pmatrix} J & (H*)' \\ H* & G* \end{pmatrix} - \begin{pmatrix} h \\ g \end{pmatrix} (h' \; g')/g_K$$

Now equation 5.10 follows clearly.

# NOTES

1. Haugen and Baker have at their disposal an adaptation of the Von Hohenbalken algorithm, developed by N. Baker and others at NISA (National Investment Services of America), which permits the rapid finding of one point on the efficient frontier when the historical covariance is used. Knowledge by us of the existence of this algorithm led to the development of the adaptation of the critical line algorithm reported here. Baker and Haugen each tell us that the basic insight in our algorithm differs from that in theirs. Since their algorithm is still a NISA secret, we do not know and cannot comment on the similarities or differences of the two algorithms.

2. There may be inefficient portfolios which satisfy equation 1.2 when $\lambda_E = 0$. However, the portfolio reached by the critical line algorithm at $\lambda_E = 0$ is efficient.

3. The idea of adding T new variables was developed independently by Konno and Suzuki (1991). However, the use made of this idea in the latter paper is quite different from that developed here. They recommend using piecewise linear approximations to the separable quadratic which results. The present paper shows how to trace out the entire efficient frontier, without resort to a piecewise linear or other approximation. The code based on the approach reported in this paper became operational in August 1990, but it has not been reported until now.

# REFERENCES

Cohen, K. J. and J. A. Pogue, "An Empirical Evaluation of Alternative Portfolio-Selection Models," *Journal of Business*, April 1967.

Haugen, R. A. and N. L. Baker, "Dedicated Stock Portfolios," *Journal of Portfolio Management*, 16:4, 1990, pp. 17–22.

Haugen, R. A. and N. L. Baker, "The Efficient Market Inefficiency of Capitalization-weighted Stock Portfolios," *Journal of Portfolio Management*, 17:3, 1991, pp. 35–40.

Konno, H. and K. Suzuki, "A Fast Algorithm for Solving Large Scale Mean-Variance Models by Compact Factorization of Covariance Matrices," Working Paper: IHSS 91-32, Institute of Human and Social Sciences, Tokyo Institute of Technology, February 1991.

Markowitz, H. M., "The Optimization of a Quadratic Function Subject to Linear Constraints," *Naval Research Logistics Quarterly*, 3, 1956, pp. 111-33.

Markowitz, H. M., *Portfolio Selection, Efficient Diversification of Investments,* Cowles Foundation Monograph 16, Yale University Press, 1959.

Markowitz, H. M., *Mean-Variance Analysis in Portfolio Choice and Capital Markets*, Cambridge: Basil Blackwell, 1987.

Markowitz, H. M. and A. F. Perold, "Portfolio Analysis with Factors and Scenarios," *Journal of Finance*, 32:4, 1981, pp. 871-77.

Perold, A. F., "Large-scale Portfolio Optimization," *Management Science*, October 1984, pp. 1143-60.

Rosenberg, B., "Extra-market Components of Covariance in Security Returns," *Journal of Financial and Quantitative Analysis*, March 1974.

Sharpe, W. F., "A Simplified Model for Portfolio Analysis," *Management Science*, January 1963, pp. 277-93.

Annals of Operations Research 45(1993)307–317



# Computation of mean-semivariance efficient sets by the Critical Line Algorithm

Harry Markowitz, Peter Todd, Ganlin Xu and Yuji Yamane
*Global Portfolio Research Department, Daiwa Securities Trust Company, One Evertrust Plaza, Jersey City, NJ 07302, USA*

The general mean-semivariance portfolio optimization problem seeks to determine the efficient frontier by solving a parametric non-quadratic programming problem. In this paper it is shown how to transform this problem into a general mean-variance optimization problem, hence the Critical Line Algorithm is applicable. This paper also discusses how to implement the critical line algorithm to save storage and reduce execution time.

**Keywords:** Mean-variance efficient frontier, mean-semivariance efficient frontier, historical returns, Critical Line Algorithm.

## 1. Introduction

One alternative to the general mean-variance portfolio optimization problem [2–4] is the mean-semivariance portfolio optimization problem proposed by Markowitz [3]. It seeks to determine portfolios of securities that are efficient with respect to expected return ($E$) and semivariance ($S$), subject to any system of linear equalities in non-negative variables. Efficient portfolios have minimum $S$ for a given $E$, and maximum $E$ for a given $S$. The set of all efficient portfolios form the efficient frontier on the $E$–$S$ graph.

Semivariance is like variance, except that it only counts downward deviation, not up and down deviations as does variance. Since an investor worries about underperformance rather than overperformance, semideviation is a more appropriate measure of investor's risk than variance[1]. Nevertheless semivariance

[1] (0.5 – semivariance/variance) could be taken as a measure of skewness. Positive skewness is generally thought to be a favorable characteristic of portfolio returns. This is sometimes listed as one of the justifications for using semivariance. One advantage semivariance $S$ has over the more traditional third moment $\mu_3$ is that it is much easier computationally to minimize $S$ than to maximize $\mu_3$. Markowitz [5, chapter 13] includes a comparison of mean-variance, mean-semivariance and other portfolio selection criteria in terms of implied approximating utility functions. Chapter 9 contains references to recent literature on semivariance.

is less used in practice. One reason is that it is more difficult to compute the $E-S$ frontier. In this paper, for the case where semivariance is estimated from historical data, we show how the mean-semivariance optimization problem can be reformulated as a mean-variance optimization problem by introducing additional variables. Therefore the Critical Line Algorithm is available for computing the $E-S$ frontier. This paper also discusses how to implement the critical line algorithm for the new formulation of the problem to get a quite fast algorithm. In Markowitz et al. [5] the same idea has been used to get a fast computation of the mean-variance efficient frontier when historical covariance is used.

Independently, King and Jensen [1] reported a similar approach to compute the mean-semivariance efficient frontier. They reformulated the problem as a classical mean-variance problem, then apply a fast parametric quadratic programming routine, developed at IBM, to trace out the efficient frontier.

For completeness and reference, we briefly review the mean-variance problem and the critical line algorithm.

## 2. Review of mean-variance model

### 2.1. THE MEAN-VARIANCE PROBLEM

The mean-variance portfolio optimization problem can be stated as the parametric quadratic programming problem:

$$\begin{aligned} \text{minimize:} \quad & V = X^{\mathrm{T}} C X \\ \text{subject to:} \quad & \mu^{\mathrm{T}} X = E, \\ & AX = b, \\ & X \geq 0, \\ \text{for all} \quad & E \text{ in } [E_{\min}, E_{\max}], \end{aligned} \tag{2.1}$$

where $X$ is the $n$-vector of security holdings representing the portfolio, $\mu$ is the $n$-vector of security expected returns, and the $A$ and $b$ are an $m, n$-matrix and an $m$-vector specifying $m$ linear constraints, $E_{\max}$ is the maximum feasible $E$ and $E_{\min}$ is the $E$ of the efficient portfolio with minimum $V$.

### 2.2. THE CRITICAL LINE ALGORITHM

The Critical Line Algorithm computes the efficient frontier by solving an equivalent parametric quadratic programming problem:

$$\begin{aligned} \text{minimize:} \quad & \tfrac{1}{2} X^{\mathrm{T}} C X - \lambda_E\, \mu^{\mathrm{T}} X \\ \text{subject to:} \quad & AX = b, \\ & X \geq 0, \\ \text{for all} \quad & \lambda_E \text{ in } [0, \infty), \end{aligned} \tag{2.2}$$

where $\lambda_E$ is the trade-off parameter between return and risk[2]. By the Kuhn–Tucker theorem, $X$ is an optimal solution of (2.2) if and only if there exist Lagrange multipliers $\lambda$ and $\eta$ such that

$$CX + A^{\mathrm{T}}\lambda - \lambda_E\, \mu - \eta = 0, \tag{2.3a}$$

$$AX = b, \tag{2.3b}$$

$$X_i \geq 0,\ \eta_i \geq 0,\ X_i \eta_i = 0. \tag{2.3c}$$

For any given sets $IN$ and $OUT$, disjoint partitions of the universe of securities, denote $C_{IN}$ to be the submatrix obtained by deleting $OUT$ rows and $OUT$ columns, $A_{IN}$ to be the matrix obtained by deleting $OUT$ columns, and $\mu_{IN}$, $\eta_{IN}$ and $X_{IN}$ ($X_{OUT}$) to be the vectors obtained by deleting $OUT$ rows ($IN$ rows, respectively). If there is a partition $IN$ and $OUT$ such that $X_{OUT} = \eta_{IN} = 0$ holds, then we can combine the $IN$ parts of eqs. (2.3a) and (2.3b) to obtain

$$\begin{pmatrix} C_{IN} & A_{IN}^{\mathrm{T}} \\ A_{IN} & 0 \end{pmatrix} \begin{pmatrix} X_{IN} \\ \lambda \end{pmatrix} = \begin{pmatrix} 0 \\ b \end{pmatrix} + \lambda_E \begin{pmatrix} \mu_{IN} \\ 0 \end{pmatrix} \tag{2.4}$$

Now, for any given partition $IN$ and $OUT$, a *critical line* $(X^{\mathrm{T}}(\lambda_E), \lambda^{\mathrm{T}}(\lambda_E), \eta^{\mathrm{T}}(\lambda_E))^{\mathrm{T}} \in \mathbb{R}^{2n+m}$ is defined by:

$$\begin{pmatrix} C_{IN} & A_{IN}^{\mathrm{T}} \\ A_{IN} & 0 \end{pmatrix} \begin{pmatrix} X_{IN} \\ \lambda \end{pmatrix} = \begin{pmatrix} 0 \\ b \end{pmatrix} + \lambda_E \begin{pmatrix} \mu_{IN} \\ 0 \end{pmatrix} \tag{2.5a}$$

$$X_i = 0, \quad i \in OUT, \tag{2.5b}$$

$$\eta = (C, A^{\mathrm{T}}) \begin{pmatrix} X \\ \lambda \end{pmatrix} - \lambda_E\, \mu. \tag{2.5c}$$

[2] There may be inefficient portfolios which satisfy (2.2) when $\lambda_E = 0$. However, the portfolio reached by the Critical Line Algorithm at $\lambda_E = 0$ is efficient.

It is clear from (2.3c) that a point on a critical line is efficient if it satisfies

$$\eta_i \geq 0, \quad i \in OUT, \tag{2.6a}$$

$$X_i \geq 0, \quad i \in IN. \tag{2.6b}$$

For a given $IN$ we define

$$M_{IN} \triangleq \begin{pmatrix} C_{IN} & A_{IN}^{\mathrm{T}} \\ A_{IN} & 0 \end{pmatrix}. \tag{2.7}$$

The Critical Line Algorithm iteratively traces out efficient portfolios along critical lines. In practice we start with a particular $IN$ set such that $M_{IN}$ is non-singular and Kuhn–Tucker conditions (2.5b) and (2.6) are met for all $\lambda_E \in [\lambda_{\text{low}}^0, \infty)$; then trace out the frontier in the direction of decreasing $\lambda_E$. At each iteration the algorithm has an $IN$ set such that $M_{IN}$ is non-singular and (2.5b) and (2.6) hold for $\lambda_E \in [\lambda_{\text{low}}^j, \lambda_{\text{high}}^j]$, where $\lambda_{\text{high}}^j = \lambda_{\text{low}}^{j-1}$. Decreasing $\lambda_E$ below $\lambda_{\text{low}}^j$ would cause one of the inequalities in (2.6), denoted by index $\gamma$, to be violated. Then either $\{\gamma\}$ is deleted from $OUT$ and added into $IN$, or $\{\gamma\}$ is deleted from $IN$ and added into $OUT$. The new $M_{IN}$ matrix always remains non-singular, and the Kuhn–Tucker conditions continue to hold. The algorithm continues until $\lambda_E = 0$ is reached in finitely many steps. The portfolios reached at $\lambda_{\text{low}}^j$ are called corner portfolios.

To implement the critical line algorithm, we need to solve linear eqs. (2.5a) efficiently. One way to do this is to store $M_{IN}^{-1}$ in the computer[3]. When the $IN$ set changes, $M_{IN}^{-1}$ is updated. Notice that the storage takes an order of $(\#IN + m)^2$. In practice, we do not shrink and expand the size of the stored $M_{IN}$ and $M_{IN}^{-1}$ matrices. Rather we use flags and pointers to direct the computation to the $IN$ part of $M$ and $M_{IN}^{-1}$.

## 3. Mean-semivariance model

### 3.1. THE MEAN-SEMIVARIANCE PROBLEM

Let $r_{ti}$ be the return of security $i$ for historical period $t$ for $i = 1$ to $n$ and $t = 1$ to $T$, and $bm_t$ be the return of a benchmark at period $t$ for $t = 1$ to $T$. Possibly $bm_t = c$, a constant for all $t$. For any portfolio vector $X$ and $E$, we define semi-

[3] The standard Critical Line Algorithm stores $M^{-1}$, but since each iteration alters only one row and column of the symmetric matrix $M$, the inverse is updated with approximately $(m+n)^2/2$ additions and multiplications.

variance versus mean as:

$$S_E(X) = \frac{1}{T}\sum_{t=1}^{T}\left\{\left(\sum_{i=1}^{n} r_{ti}X_i - E\right)^-\right\}^2, \tag{3.1a}$$

and semivariance versus the benchmark as:

$$S_{bm}(X) = \frac{1}{T}\sum_{t=1}^{T}\left\{\left(\sum_{i=1}^{n} r_{ti}X_i - bm_t\right)^-\right\}^2, \tag{3.1b}$$

where

$$z^- = \begin{cases} |z|, & \text{if } z < 0, \\ 0, & \text{if } z \geq 0. \end{cases}$$

There are two versions of the mean-semivariance problem. One is:

$$\begin{aligned} \text{minimize:} \quad & S_E(X) \\ \text{subject to:} \quad & \mu^{\mathrm{T}} X = E, \\ & AX = b, \\ & X \geq 0, \\ \text{for all} \quad & E \text{ in } [E_{\min}, E_{\max}], \end{aligned} \tag{3.2}$$

where all the parameters have the same meaning as in (2.1) except that $E_{\min}$ is the $E$ of the efficient portfolio with minimum $S$. The other version of the semivariance problem is:

$$\begin{aligned} \text{minimize:} \quad & S_{bm}(X) \\ \text{subject to:} \quad & \mu^{\mathrm{T}} X = E, \\ & AX = b, \\ & X \geq 0, \\ \text{for all} \quad & E \text{ in } [E_{\min}, E_{\max}], \end{aligned} \tag{3.3}$$

where all the parameters have the same meanings as above.

### 3.2. REFORMULATION OF THE PROBLEM

In this section we will discuss problem (3.2) first. The necessary changes for problem (3.3) will be stated at the end of this section. Since $S_E$ is not in quadratic form as $V$ is, we cannot apply the Critical Line Algorithm to problem (3.2) directly. However, we can reformulate the problem into the form of (2.1) by introducing $2T$ new variables.

First let

$$Y_t = \sum_{i=1}^{n} \frac{1}{\sqrt{T}}(r_{ti} - \mu_i)X_i, \tag{3.4}$$

for $t = 1$ to $T$, and substitute constraint $\mu^T X = E$ into the objective function $S_E(X)$, we see that problem (3.2) is equivalent to

$$\left.\begin{aligned} \text{minimize:} \quad & \sum_{t=1}^{T} (Y_t^-)^2 \\ \text{subject to:} \quad & \mu^{\mathrm{T}} X = E, \\ & AX = b, \\ & X \geq 0, \\ \text{for all} \quad & E \text{ in } [E_{\min}, E_{\max}], \end{aligned}\right\} \tag{3.5}$$

and with additional constraints (3.4). To get a quadratic objective function, let

$$Z_t = Y_t^- \tag{3.6}$$

for $t = 1$ to $T$. It is now clear that problem (3.2) is equivalent to

$$\left.\begin{aligned} \text{minimize:} \quad & \sum_{t=1}^{T} Z_t^2 \\ \text{subject to:} \quad & \mu^{\mathrm{T}} X = E, \\ & BX - Y + Z = 0, \\ & AX \qquad = b, \\ & X \geq 0,\ Y \geq 0,\ Z \geq 0, \\ \text{for all} \quad & E \text{ in } [E_{\min}, E_{\max}], \end{aligned}\right\} \tag{3.7}$$

where $B$ is a $T$ by $n$ matrix defined by

$$B = \frac{1}{\sqrt{T}} \begin{pmatrix} r_{11} & r_{12} & \cdots & r_{1n} \\ \cdots & \cdots & & \cdots \\ r_{T1} & r_{T2} & \cdots & r_{Tn} \end{pmatrix} - \frac{1}{\sqrt{T}} \begin{pmatrix} \mu_1 & \cdots & \mu_n \\ \cdots & & \cdots \\ \mu_1 & \cdots & \mu_n \end{pmatrix}. \tag{3.8}$$

With this new formulation, we can think of our universe of securities as consisting of $n$ real securities and $2T$ fictitious securities with expected return $(\mu^{\mathrm{T}},0,0)^{\mathrm{T}} \in \mathbb{R}^{n+2T}$, and convariance matrix

$$C \triangleq \begin{pmatrix} 0 & 0 & 0 \\ 0 & 0 & 0 \\ 0 & 0 & I \end{pmatrix}.$$

Now we can apply the Critical Line Algorithm to the new problem (3.7). For problem (3.3) the only change we need to make is to define $B$ as:

$$B = \frac{1}{\sqrt{T}} \begin{pmatrix} r_{11} & r_{12} & \cdots & r_{1n} \\ \cdots & \cdots & & \cdots \\ r_{T1} & r_{T2} & \cdots & r_{Tn} \end{pmatrix} - \frac{1}{\sqrt{T}} \begin{pmatrix} bm_1 & \cdots & bm_1 \\ \cdots & & \cdots \\ bm_T & \cdots & bm_T \end{pmatrix}. \tag{3.8}$$

## 4. Implementation of the Critical Line Algorithm

The direct application of the Critical Line Algorithm to the reformulated problem (3.7) will not be efficient computationally because of the introduction of the $2T$ new variables. To be efficient, one would have to take the advantage of the sparsity of the resulting $M_{IN}$ matrix. The readers are referred to Markowitz et al. [5] for details. Here we will show how the algorithm can be designed without introducing the $z$ variables.

As we see that the only reason for introducing the $z$ variables is to get a quadratic objective function, since the Critical Line Algorithm was developed for the mean-variance optimization problem. However, the principle behind the Critical Line Algorithm is applicable to much more general objective functions including functions like $S_E(X)$ and $S_{bm}(X)$ in the mean-semivariance optimization problem. Let us recall eq. (2.5), the definition of the critical line. Keep in mind that the main eq. (2.5a) has two parts. One part is the original linear constraints. The other part is the first order condition of the Lagrange function, holding $X_{OUT}$ constant zero. Suppose that $V$, in the mean-variance problem (2.1) or (2.2), is replaced by a general function $f(x)$, then the natural definition of "critical line" will be

$$\nabla_{IN} f(X) + A_{IN}^{\mathrm{T}} \lambda = \lambda_E \mu_{IN}, \tag{4.1a}$$

$$A_{IN} X_{IN} = b \tag{4.1b}$$

$$X_i = 0, \quad \text{for } i \in OUT, \tag{4.1c}$$

$$\eta = \nabla f(X) + A^{\mathrm{T}} \lambda - \lambda_E \mu, \tag{4.1d}$$

where $\nabla_{IN} f$ is the gradient of $f$ with respect to the $IN$ variables. By the Kuhn–Tucker condition, a point on the critical line is optimal if

$$\eta_i \geq 0, \quad i \in OUT, \tag{4.2a}$$

$$X_i \geq 0, \quad i \in IN. \tag{4.2b}$$

Now let us see how to apply the above definition to the reformulated mean semivariance problem (3.5) and (3.4) (not (3.7)!). Let $f(X, Y) = \frac{1}{2}\sum_{t=1}^{T}(Y_t^-)^2$. Since there are no constraints on $Y$, all the $Y$ variables will be $IN$. For convenience, let us still denote $IN$ and $OUT$ the partition of the real securities only. Let $\lambda_1$ ($\lambda_2$, respectively) be the Lagrange multiplier corresponding to constraints (3.4) ($AX = b$, respectively). Then the corresponding eqs. (4.1a) and (4.1b) are:

$$\nabla_{IN} f(X, Y) + B_{IN}^{\mathrm{T}} \lambda_1 + A_{IN}^{\mathrm{T}} \lambda_2 = \lambda_E \mu_{IN}, \tag{4.3a}$$

$$\nabla_y f(X, Y) - \lambda_1 = 0, \tag{4.3b}$$

$$B_{IN} X_{IN} - Y = 0, \tag{4.3c}$$

$$A_{IN} X_{IN} = b. \tag{4.3d}$$

Since $f_{\dot{x}} = 0$ and $f_y = -Y^-$, we can write out the above equations more explicitly:

$$B_{IN}^{\mathrm{T}} \lambda_1 + A_{IN}^{\mathrm{T}} \lambda_2 = \lambda_E \mu_{IN}, \tag{4.4a}$$

$$-Y^- - \lambda_1 = 0, \tag{4.4b}$$

$$B_{IN} X_{IN} - Y = 0, \tag{4.4c}$$

$$A_{IN} X_{IN} = b. \tag{4.4d}$$

Therefore, to get an efficient algorithm, we need to solve the above system of equations efficiently. Let $Y_u$ contain $Y_t$'s such that $Y_t \geq 0$ and let $Y_d$ contain $Y_t$'s such that $Y_t < 0$. Define

$$\lambda_1 = \begin{pmatrix} \lambda_{1u} \\ \lambda_{1d} \end{pmatrix}, \quad B_{IN} = \begin{pmatrix} B_{INu} \\ B_{INd} \end{pmatrix}.$$

From eq. (4.4b), we have

$$\lambda_{1u} = 0, \tag{4.5a}$$

and

$$\lambda_{1d} = Y_d. \tag{4.5b}$$

By substituting eqs. (4.5) into (4.4a) and (4.4c), and by splitting eq. (4.4c), we have:

$$(B_{INd})^{\mathrm{T}} Y_d + A_{IN}^{\mathrm{T}} \lambda_2 = \lambda_E \mu_{IN}, \tag{4.6a}$$

$$B_{INd} X_{IN} - Y_d = 0, \tag{4.6b}$$

$$A_{IN} X_{IN} = b, \tag{4.6c}$$

$$B_{INu} X_{IN} - Y_u = 0. \tag{4.6d}$$

Now we see that to solve eqs. (4.4), we can solve (4.6a) – (4.6c) first to get ($X_{IN}$, $Y_d$, $\lambda_2$), and the rest of the unknowns can be obtained easily from eqs. (4.6d) and (4.5). Finally, eqs. (4.6a) – (4.6c) can be written in matrix form

$$\begin{pmatrix} -I_d & 0 & B_{INd} \\ 0 & 0 & A_{IN} \\ B_{INd}^{\mathrm{T}} & A_{IN}^{\mathrm{T}} & 0 \end{pmatrix} \begin{pmatrix} Y_d \\ \lambda_2 \\ X_{IN} \end{pmatrix} = \begin{pmatrix} 0 \\ b \\ \lambda_E \mu_{IN} \end{pmatrix}.$$

An efficient algorithm to solve the above system of equations has been presented in Markowitz et al. [5], which is one of the key points to get a fast computation of the mean-variance efficient frontier.

## 5. Performance

We have developed an optimization program that can compute the mean-variance and mean-semivariance efficient frontiers using the historical returns based on the ideas presented in this paper and in Markowitz et al. [5]. We used this pro-

TABLE 1
Performance comparison.

| Frontier | Corner portfolios | Exec. time | Storage |
|---|---|---|---|
| Mean-variance | 62 | 9.44 sec | 1257 Kb |
| Mean-semivariance | 130 | 17.67 sec | 1257 Kb |

gram to obtain a comparison on a test problem involving 1008 securities and 60 historical monthly returns observations. The only constraint was the budget constraint.

Our test problem was run on an IBM RS/6000 Model 320 computer. The program was compiled with the IBM AIX XL C compiler version 01.01.0003.0013. All floating-point computations use double precision arithmetic (8 bytes per number). Table 1 shows the execution time for the critical line algorithm, and the amount of storage allocated for arrays.

The higher execution time for the mean-semivariance optimization in table 1 is primarily due to a greater number of corner portfolios being produced for our test problem. In our experience, a mean-semivariance optimization can produce either a larger or smaller number of corner portfolios than a mean-variance optimization, depending on the particular problem. The execution time per corner portfolio is normally quite close for the two optimization methods.

Our portfolio optimization program incorporates several additions to the basic optimization problem described by eqs. (2.1) and (3.2):

(1) The ability to use lower bounds on $X$ other than zero, and upper bounds on $X$, without increasing the number of constraint equations.
(2) The ability to constrain portfolio turnover by introducing only one additional constraint equation, and no additional variables.
(3) The ability to consider transaction costs for buying and selling securities in the optimization.

Although none of these features were used in the test problem presented here, their incorporation into the program causes a modest increase in execution times, even when they are not used.

## References

[1] A.J. King and D.L. Jensen, Linear–quadratic efficient frontiers for portfolio optimization, RC 16524, IBM Research Report (1991).

[2] H.M. Markowitz, The optimization of the quadratic function subject to linear constraints, Naval Res. Log. Quarterly 3 (1956) 111–133.

[3] H.M. Markowitz, *Portfolio Selection, Efficiency Diversification of Investments,* Cowles Foundation Monograph 16 (Yale University Press, 1959 2nd ed.: Basil Blackwell, Cambridge, 1991).

[4] H.M. Markowitz, *Mean-Variance Analysis in Portfolio Choice and Capital Markets* (Basil Blackwell, Cambridge, 1987).

[5] H.M. Markowitz, P. Todd, G.L. Xu and Y. Yamane, Fast computation of mean-variance efficient sets using historical covariance, J. Fin. Eng. (1991), to appear.

# Data Mining Corrections

*Simple and plausible.*

Harry M. Markowitz and Gan Lin Xu

It is common practice, in financial research and other fields, to compute how well alternate policies would have worked if they had been followed in the past. The policy that would have worked best in the past is then recommended for use in the future.

The past performance of this best policy is often used as an estimate of its future performance. We shall see that this is not a good estimate under the circumstances, and that, under certain assumptions, a better estimate can be provided.

Suppose, for simplicity, that a research team is interested in one criterion — let's say, the geometric mean (GM) of return for various portfolio selection methods. These portfolio selection methods may be composed of methods for estimating expected returns and variances and covariances of return, constraints on portfolio turnover and on holdings of individual securities, and a rule for selecting a portfolio from the mean-variance efficient frontier.

Suppose that historical simulation or "backtesting" of hundreds of portfolio selection methods, each a combination of the various measures, shows that the method that worked best during the past T periods had a geometric return of $GM_b$. Since it will be convenient to work with the logarithm of GM, we define

$$g_b = \log_e (1 + GM_b)$$

and ask, "how should we estimate the future perfor-

**HARRY M. MARKOWITZ** is president of Harry Markowitz Co. in San Diego (CA 92109), and a consultant at Daiwa Securities Trust Co. in Jersey City (NJ 07302).

**GAN LIN XU** is in the global portfolio research department at Daiwa Securities Trust Company in Jersey City.

mance of this historically best portfolio selection method?"

If the rates of return $R_t$, $t = 1, ..., T$, for the method were i.i.d. (independent, identically distributed) draws from some population, if we were willing to assume that future draws are from the same population, and if it were true that we had looked at only one method, then

$$g_b = \sum_{t=1}^{T} \log_e(1 + R_t)/T \tag{1}$$

would be the best (i.e., least square) unbiased estimate of the true, underlying population $g_b$. This is no longer so when more than one method is examined and $g_b$ is the historically best one.

The practice of examining many methods and recommending the one that historically did best is commonly referred to as "data mining." It seems to us that data mining is unavoidable, as it is unreasonable to expect a financial research team to set up a data base and program a simulator for the purpose of evaluating one method, without any variations or alternatives.

A Bayesian rational investor with limited knowledge of the world but unlimited computational ability would have no compunctions about data mining — that is, about considering the implications of data for many policies.[1] This hypothetical investor would use data to update probability beliefs about hypotheses in a space of hypotheses about how security returns are generated, and would evaluate alternate portfolio selection methods in terms of a posteriori beliefs.

This is fine for our hypothetical rational investor, but seems too difficult for humans to apply even with our latest computers and data bases. In this article we propose simplified models of this process, and show that these lead to simple, plausible "data mining corrections."[2]

We propose three models of different degrees of generality, then present data mining corrections for these models and tests of significance; that is, tests of the hypothesis that all methods are equally good and that observed differences are attributable to "noise."

## THE MODELS

Let $y_{it}$ for $i = 1, ..., n$ and $t = 1, ..., T$ be the logarithm of one plus return for an $i^{th}$ portfolio selection method in period t. Appropriate data mining corrections and tests of significance depend on assumptions made as to how the $y_{it}$ are generated. We present three models of increasing complexity, models I, II, and III.

### Model I

The return of $y_{it}$ for method i in period t is assumed to be of the form

$$y_{it} = \mu_i + z_t + \varepsilon_{it} \tag{2A}$$

where $\mu_i$ is a "model effect," $z_t$ a period effect, and $\varepsilon_{it}$ a random deviation. In model I we assume that $z_t$ is observable; e.g., $z_t$ could be assumed to be the return on some broad market index. We can therefore write (2A) as

$$r_{it} = \mu_i + \varepsilon_{it} \tag{2B}$$

where

$$r_{it} = y_{it} - z_{it} \tag{2C}$$

and where $\varepsilon_{it}$ has mean zero and is uncorrelated with $\mu_i$ and every other $\varepsilon_{js}$:

$$E\varepsilon_{it} = 0 \tag{2D}$$

$$\text{cov}(\mu_i, \varepsilon_{jt}) = 0 \text{ for all j and t} \tag{2E}$$

$$\text{cov}(\varepsilon_{it}, \varepsilon_{js}) = 0 \text{ for all } i \neq j \text{ or } s \neq t \tag{2F}$$

For the most part, we will treat the $\mu_i$ (as well as the $\varepsilon_{it}$) as random variables drawn from some population. As we will see later, we may view them as unknown parameters concerning which we hold Bayesian priors.

### Model II

As with model I, we assume that

$$y_{it} = \mu_i + z_t + \varepsilon_{it} \tag{3A}$$

In model II, however, we do not assume that $z_t$ is observable. We do assume that $\mu_i$, $z_t$, and $\varepsilon_{it}$ are uncorrelated, and $\varepsilon_{it}$ has zero mean:

$$\text{cov}(\mu_i, z_t) = \text{cov}(\mu_i, \varepsilon_{it})$$

$$= \text{cov}(z_t, \varepsilon_{it}) = 0 \quad \text{(3B)}$$

$$E\varepsilon_{it} = 0 \quad \text{for all i and t} \quad \text{(3C)}$$

One problem we consider is whether the observed $y_{it}$ is consistent with the hypothesis that all the $\mu_i$ are equal, i.e., that the population of $\mu_i$ has zero variance. We also provide estimates of the unknown $\mu_i$.

In both models I and II, we find that the appropriate estimate of $\mu_i$ is not the average return

$$\bar{r}_i = \sum_t y_{it} \Big/ T \quad \text{(4A)}$$

Rather, we "regress" the estimate of $\mu_i$ back toward the grand average

$$\bar{r} = \sum_i \bar{r}_i \Big/ n \quad \text{(4B)}$$

by a formula of the form

$$\hat{\mu} = \bar{r} + \beta(\bar{r}_i - \bar{r}) \quad \text{(4C)}$$

where $\beta$ is between 0 and 1. The formula for $\beta$ differs between models I and II. In both models the estimation procedure can be given a Bayesian interpretation.

**Model III**

Similar to Equation (2B) of model I, model III is

$$y_{it} = \mu_i + \varepsilon_{it} \quad \text{(5A)}$$

requiring

$$E\varepsilon_{it} = 0 \quad \text{(5B)}$$

In this model, however, we do not require that $\varepsilon_{it}$ and $\varepsilon_{jt}$ be uncorrelated. At a given time t, the $\varepsilon_{it}$ may have any pattern of correlations. They are assumed to be uncorrelated between time periods:

$$\text{cov}(\varepsilon_{it}, \varepsilon_{js}) = 0 \quad \text{for } s \neq t \quad \text{(5C)}$$

Models I and II are special cases of model III.

Since model III is the most general, always using it rather than one of the other two would seem desirable. But the procedures for estimating $\mu_i$, and for testing the hypothesis that all $\mu_i$ are the same, are considerably more complex for model III than for models I and II.

The three models are examples of analysis of variance models. See Kendall, Stuart, and Ord [1983] for further details and alternatives.

## ESTIMATION OF β FOR MODEL I

First, we consider $\mu_i$ in Equation (2B), as well as $\varepsilon_{it}$, to be random; that is, we imagine that portfolio selection method i with expected $r_{it}$ equal to $\mu_i$ has been drawn at random from some population of methods; or, that "nature" has randomly assigned a $\mu_i$ to a method. Our job is to estimate its value.

If we knew the joint distribution of $\mu$ and $\varepsilon$, then the best linear estimate of an unknown $\mu_i$, given its observed $r_{it}$, is[3]

$$\hat{\mu}_i = E\mu + \beta(\bar{r}_i - E\mu) \quad \text{(6A)}$$

where

$$\beta = \frac{\text{cov}(\bar{r}_i, \mu)}{\text{Var}(\bar{r}_i)} = \frac{\text{Var}(\mu)}{\text{Var}(\mu) + \text{Var}(\varepsilon)/T} \quad \text{(6B)}$$

Here $\beta$ is the regression coefficient of $\mu_i$ against $\bar{r}_i$ that, if $\text{Var}(\varepsilon) > 0$, satisfies $\beta < 1$ as compared to the regression coefficient of $\bar{r}_i$ against $\mu_i$, which is always 1.0.

We present two methods of estimating $\beta$ for model I from a sample of n portfolio selection methods and T periods. With each of these, $\mu_i$ is then estimated by

$$\mu_i^{est} = \bar{r} + \beta^{est}(\bar{r}_i - \bar{r}) \quad \text{(7)}$$

Compare this with Equation (6A).

The first method of estimating $\beta$ uses unbiased estimates of $V(\mu)$ and $V(\mu) + V(\varepsilon)/T$ in (6B). We refer to this as the RUE method (ratio of unbiased estimates). It makes no assumption concerning the joint distribution of $\mu$ and $\varepsilon$ other than those in Equations

(2D)-(2F). Our second method assumes $\mu$ and $\varepsilon$ to be joint normally distributed and solves for the ML estimate (maximum likelihood).

The two methods produce estimates that are the same except for a term in both numerator and denominator, which goes to zero as $n \to \infty$. A comparison of the two methods of estimation is of importance to us for two reasons. First, the ML method is theoretically superior, but relies on the normality assumption. Second, an explicit ML formula is not available for models II and III. It is reassuring that the two estimates rapidly approach each other as n increases.

We use the RUE calculation for model II, relying in part on the fact that RUE and ML are close for model I where both are available.

Theorem 1. For model I, let

$$N^u = \left(1 + \frac{1}{n-1} + \frac{1}{T-1}\right) \times$$

$$B - A/(T-1) \quad (8A)$$

$$D^u = \left(1 + \frac{1}{n-1}\right) B \quad (8B)$$

where

$$A = \frac{1}{nT}\sum_{i,t}\left(r_{it} - \bar{r}\right)^2, \; B = \frac{1}{n}\sum_i \left(\bar{r}_i - \bar{r}\right)^2 \quad (8C)$$

Then $N^u$ and $D^u$ are unbiased estimates of $V(\mu)$ and $V(\mu) + V(\varepsilon)/T$, respectively.

*Proof*: From (A-3) and (A-5) of the proof of Theorem A in the appendix, we have

$$E(A - B) = \left(1 - \frac{1}{T}\right)\sigma_\varepsilon^2$$

$$EB = \left(1 - \frac{1}{n}\right)\sigma_\mu^2 + \left(1 - \frac{1}{n}\right)\frac{1}{T}\sigma_\varepsilon^2$$

It is now easy to check that $E(N^u) = \text{var}(\mu)$ and $E(D^u) = \text{var}(\mu) + \text{var}(\varepsilon)/T$. Q.E.D.

In light of Theorem 1 we define

$$\beta^u = N^u/D^u \quad (8D)$$

as the RUE method of estimating $\beta$. The term B in (8) is the observed variance of the method averages $\bar{r}_i$. Term A is the observed total observed variability of the $r_{it}$. Later we test the hypothesis that all $\mu$ are equal, which uses this observed total variability versus the variability of method averages.

It is possible to have $N^u < 0$ in a particular sample, despite $EN^u = V(\mu) \geq 0$. If one encounters a sample with $N^u < 0$, one is tempted to estimate $V(\mu) = 0$, because $V(\mu) < 0$ is impossible. But when $N^u < 0$ has positive probability,

$$E(N) = E[\max(N^u, 0)] > EN^u = V(\mu)$$

So N is not unbiased. This is a standard problem with certain analysis of variance estimates.

Theorem 2. For model I, if $\mu$ and $\varepsilon$ are independent, joint normally distributed random variables, then the maximum likelihood estimate of $\beta$ in (7A) is

$$\beta^m = \begin{cases} N^m/D^m & \text{if } B \geq A/T \\ 0 & \text{otherwise} \end{cases} \quad (9)$$

where

$$N^m = \left(1 + \frac{1}{T-1}\right)B - A/(T-1)$$

$$D^m = B$$

and A and B are defined in Equation (8C).

*Proof*: Log-likelihood is

$$\phi(\mu, \rho, \sigma) = \frac{nT}{2}\log(2\pi) -$$

$$\frac{1}{2}\left[nT\log\sigma^2 + n\log\left(1 + T\rho^2\right)\right]$$

$$-\frac{1}{2}\sum_{i=1}^n \sigma^{-2}\left(r_i - \mu e\right)'\left(I - \frac{\rho^2}{1+\rho^2 T}E\right)\left(r_i - \mu e\right)$$

where $e' = (1, 1, ..., 1) \varepsilon R^T$, $r'_i = (r_{i1}, r_{i2}, ..., r_{iT})$, I is the $T \times T$ identity matrix, $E = ee'$, $\sigma^2 = \text{Var}(\varepsilon)$, and $\rho^2$ is defined as Var (μ)/Var (ε). Because of the non-negativity constraints on ρ and σ, the first-order conditions are

$$\mu \sum_{i=1}^{n} e'\left(I - \frac{\rho^2}{1 + \rho^2 T} E\right) e -$$

$$\sum_{i=1}^{n} r_i'\left(I - \frac{\rho^2}{1 + \rho^2 T} E\right) e = 0$$

$$nT\sigma^2 - \sum_{i=1}^{n} (r_i - \mu e)'\left(I - \frac{\rho^2}{1 + \rho^2 T} E\right) \times$$

$$(r_i - \mu e) = 0$$

$$nT\left(1 + T\rho^2\right) - \sigma^{-2} \sum_{i=1}^{n} (r_i - \mu e)' \times$$

$$E(r_i - \mu e)\begin{cases} \geq 0, \text{ if } \rho = 0 \\ = 0, \text{ if } \rho > 0 \end{cases}$$

The unique solutions to these equations are given by

$$\hat{\mu} = \bar{r};$$

$$\hat{\sigma}^2 = \begin{cases} T(A - B)/(T - 1), \text{ if } TB \geq A \\ A, \text{ otherwise} \end{cases}$$

$$\hat{\rho}^2 = \begin{cases} \dfrac{TB - A}{T(A - B)}, \text{ if } TB > A \\ 0, \text{ otherwise} \end{cases}$$

Q.E.D.

In (9), when $B < A/T$, the observed variability in $\bar{r}_i$ is so low as compared to the total observed variability that $\beta = 0$ is the ML estimate. We then estimate that all $\mu_i$ are the same and use the grand mean $\bar{r}$ as the estimate of each $\mu_i^{est}$. In the case of $B \geq A/T$, comparison of Equations (8) and (9) shows that $\beta^u$ and $\beta^m$ differ by a term of $B/(n - 1)$ in the numerator and denominator. This is relatively small for moderate to large n, and goes to zero as n increases.

It may be shown that $A \geq B$; hence

$$1 \geq \beta^m \geq 0$$

The case of $\beta^m = 1$ occurs only if there is no observed variability in the $r_{it}$. Otherwise the estimate of $\mu_i$ is lowered for $\bar{r}_i > \bar{r}$ and raised for $\bar{r}_i < \bar{r}$.

## ESTIMATION OF β FOR MODEL II

If we knew the joint distribution of μ, z, and ε in model II, then the best linear estimate $\hat{\mu}_i$ of $\mu_i$, given $y_{kt}$, would be the complicated formula:

$$\hat{\mu}_i = E(\mu + z) + \frac{\text{Var}(\mu)}{\text{Var}(\mu) + \text{Var}(\varepsilon)/T}(\bar{y}_i - \bar{y}) + \frac{T\text{Var}(\mu)}{\text{Var}(\varepsilon) + T\text{Var}(\mu) + n\text{Var}(z)} \times [\bar{y} - E(\mu + z)] \quad (10A)$$

where

$$\bar{y}_i = \frac{1}{T}\sum_{t=1}^{T} y_{it}, \quad \bar{y} = \frac{1}{n}\sum_{i=1}^{n} \bar{y}_i \quad (10B)$$

In practice we propose to estimate E (μ + z), Var (μ), Var (z), and Var (ε) from a sample. Since the best estimate of E (μ + z) is $\bar{y}$, the last term in (10A) is zero, and the estimate of $\mu_i$ now has the simple form

$$\mu_i^{est} = \bar{y} + \beta^{est}(\bar{y}_i - \bar{y}) \quad (10C)$$

where $\beta^{est}$ is again given by Equation (6B), with estimated Var (μ) and Var (μ) + Var (ε)/T. It is remarkable that the estimates of $\mu_i$ in both models follow the same formula; the formulas for estimating Var (μ) and Var (ε) are different for models I and II.

We give only the RUE estimate of $\beta^{est}$ in model II because the ML estimation method does not have an explicit solution. We saw that for model I the RUE and

ML estimates of $\mu_i$ are close, providing some support for the use of either for model I and the use of RUE for model II.

Theorem 3. For model II let

$$N^* = \frac{n}{(T-1)(n-1)}\left(TB^* + C^* - A^*\right) \quad (11A)$$

$$D^* = \left(1 + \frac{1}{n-1}\right)B^* \quad (11B)$$

where

$$A^* = \frac{1}{nT}\sum_{i,t}\left(y_{it} - \bar{y}\right)^2, \quad B^* = \frac{1}{n}\sum_i\left(\bar{y}_i - \bar{y}\right)^2$$

$$C^* = \frac{1}{T}\sum_t\left(\bar{y}^t - \bar{y}\right)^2, \quad \bar{y}^t = \frac{1}{n}\sum_i y_{it} \quad (11C)$$

Then $N^*$ and $D^*$ are unbiased estimates of V $(\mu)$ and V $(\mu)$ + V $(\varepsilon)/T$. Therefore $N^*/D^*$ is an RUE estimate of $\beta$.

*Proof*: This is a special case of theorem A in the appendix. Q.E.D.

As with model I, the estimate of $\beta$ depends on total observed variability, $A^*$, and variability among methods, $B^*$. In model II it also depends on $C^*$, the variability among time periods. As with model I, for a particular sample we could have a negative numerator, $N^*$. Since V $(\mu) < 0$ is impossible, we suggest

$$\beta^* = \begin{cases} N^*/D^*, \text{ if } N^* > 0 \\ 0, \text{ otherwise} \end{cases} \quad (12)$$

for model II.

## TESTS OF SIGNIFICANCE

Even when $\beta^* > 0$ in model II or $\beta^u > 0$ in model I, we may wish to know if this positive estimate could reasonably be expected by chance when in fact V $(\mu) = 0$; i.e., is it statistically significant? Theorem 4 provides a test of this. The test takes into account that many methods have been tried. It is to be contrasted with the usual procedure of trying many methods, picking out the best one, then pretending that the only possibilities are the chosen one and a null hypothesis. With a sufficiently large and varied set of methods, the latter procedure will declare the observed difference to be significant even if all $\mu_i$ are equal.

Theorem 4. Suppose that Var $(\mu) = 0$, i.e., $\mu_i = \mu$ are constants, and z and $\varepsilon$ are normal, independently distributed in models I and II. Let $F^*$ and F be

$$F^* = \frac{(T-1)B^*}{A^* - B^* - C^*} \quad (13A)$$

$$F = \frac{(T-1)B}{A-B}\frac{n}{n-1} \quad (13B)$$

Then $F^*$ and F are distributed as F with $[n-1, (n-1)(T-1)]$ and $[n-1, n(T-1)]$ degrees of freedom, respectively.

*Proof*: Complete proof appears in Kendall, Stuart, and Ord [1983] (35.8, 35.25, and 36.16). Here we note only that, under the assumption of the theorem, each term of $B^*$ is independent of

$$\left(y_{it} - \bar{y}_i - \bar{y}^t + \bar{y}\right)^2$$

and these sum to nT $(A^* - B^* - C^*)$. Q.E.D.

Notice that $\beta^u = N^u/D^u = 1 - 1/F$, and $\beta^* = N^*/D^* = 1 - 1/F^*$. This implies that $N^u > 0$ $(N^* > 0)$ if and only if $F > 1$ $(F^* > 1)$. Since the critical values of F and $F^*$ (above which the Var $(\mu) = 0$ hypothesis is rejected, e.g., at the 10% level or lower) exceed one, the critical levels of $\beta^*$ and $\beta^u$ are positive.

## ESTIMATION FOR MODEL III

Estimation procedures for model III are more complex. We describe them briefly for the reader who knows matrix notation.

If we knew the joint distribution of $\mu$ and $\varepsilon$ in model III, the best linear estimate of the $\mu$ vector is

$$\hat{\mu} = E(\mu)e + \text{Var}(\mu)\left[\frac{1}{T}C + \text{Var}(\mu)I\right]^{-1} \times$$

$$\left(\tilde{y} - E(\mu) e\right) \tag{14}$$

where $e' = (1, 1, \ldots, 1)$, $\tilde{y} = (\bar{y}_1, \bar{y}_2, \ldots, \bar{y}_n)$, I is the $n \times n$ identity matrix, and $C = [\mathrm{cov}\ (\varepsilon_i, \varepsilon_j)]$ is the covariance matrix.

In practice we use unbiased estimates of Var $(\mu)$ and C/T + Var $(\mu)$ I. This procedure is a generalization of the RUE method in models I and II. Since the estimate of $\mu_i$ in (14) depends on the entire $\tilde{y} - E(\mu) e$ vector, it is possible that the method with the highest observed $\bar{y}_i$ will not be the method with the highest estimated $\mu_i$.

To apply Equation (14), we replace E $(\mu)$ by grand mean $\bar{y}$, and replace Var $(\mu)$ by $N^*$ in (11A); see Theorem A in the appendix. Theorem 5 provides an estimate of the covariance matrix C.

Theorem 5. Let

$$\hat{\sigma}_{ij} = \frac{\sum_{t=1}^{T} \left(y_{it} - \bar{y}_i\right)\left(y_{jt} - \bar{y}_j\right)}{T - 1}$$

Then it is an unbiased estimate of $\sigma_{ij}$.

*Proof*. The theorem follows from the equality

$$E \sum_{t=1}^{T} \left(y_{it} - \bar{y}_i\right)\left(y_{jt} - \bar{y}_j\right) =$$

$$E \sum_{t=1}^{T} \left(\varepsilon_{it} - \bar{\varepsilon}_i\right)\left(\varepsilon_{jt} - \bar{\varepsilon}_j\right) =$$

$$(T - 1)\sigma_{ij}$$

Q.E.D.

A test of the hypothesis that all $\mu_i$ are equal in model III is given in Rao [1973].

## A BAYESIAN VIEW OF THE METHODS

Our estimation equations are related to a Bayesian approach. Specifically, in the case of model III, under certain assumptions concerning prior beliefs, the posterior beliefs (after the sample) have the same expected values for the $r_i$ as the estimates of $\mu_i$ given in Equation (14). Assumptions include normality and that the covariance matrix C is known in advance. Equations (6) and (10) for models I and II are special cases with particular C.

Theorem 6. For model III, assume that $\varepsilon_t$ is normally distributed with known non-singular covariance matrix C, and that the prior distribution of $\mu$ is normal with mean vector E $(\mu)$ e and covariance matrix $\sigma_\mu^2$ I. Then the posterior distribution of $\mu$ after observations y is normally distributed with mean $\hat{\mu}$ as given in (14) and covariance matrix

$$\left(TC^{-1} + \sigma_\mu^{-2} I\right)^{-1}$$

*Proof*. Since density function

$$\rho(\tilde{\mu}) = \frac{1}{\left(2\pi\sigma_\mu^2\right)^{n/2}} \exp\left\{-\frac{1}{2}\sum_{i=1}^{n} \frac{\left(\mu_i - E(\mu)\right)^2}{\sigma_\mu^2}\right\}$$

$$\rho(y \mid \tilde{\mu}) = \frac{1}{\left(2\pi \det C\right)^{T/2}} \times$$

$$\exp\left\{-\frac{1}{2}\sum_{t=1}^{T} \left(\tilde{y}_t - \tilde{\mu}\right)' C^{-1} \left(\tilde{y}_t - \tilde{\mu}\right)\right\}$$

Manipulation of formulas gives us

$$\rho(\tilde{\mu} \mid y) \propto \rho(\tilde{\mu})\rho(y \mid \tilde{\mu})$$

$$\propto \exp\left\{-\frac{1}{2}\left(\tilde{\mu} - \hat{\mu}\right)'\left(TC^{-1} + \sigma_\mu^{-2} I\right)\left(\tilde{\mu} - \hat{\mu}\right)\right\}$$

Q.E.D.

## EXPERIENCE WITH MODELS I, II, AND III

The data mining correction procedures we have described were developed to address an immediate need of the Global Portfolio Research Department (GPRD) of Daiwa Securities Trust Company, a subsidiary of Daiwa Securities, Japan. GPRD was founded in February 1990. Most of 1990 was spent writing optimization, simulation, and data management programs; acquiring and checking data and incorporating it into the GPRD data base; and backtesting methods of managing equity portfolios.

In the first instance, we focused on managing

portfolios of Japanese stocks. In particular, John Guerard proposed a number of variations on the "composite model" methodology for estimating expected returns that he had developed for U.S. securities; see Guerard [1987] and Guerard and Stone [1992]. These were backtested for the Japanese market.

The first fund managed using GPRD methodology was Fund Academy, a Japanese investment company started in January 1991. In the months prior to the start of Fund Academy, the GPRD backtests revealed methods that quite substantially outperformed both the capitalization-weighted Tokyo Price Index (TOPIX) and an equal-weighted benchmark. These backtests were controlled, as best could be, for transaction costs and look-ahead bias. Outperformance was less, but still substantial, when liquidity was considered.

It occurred to us that perhaps the outperformance could be explained by the fact that we had tried a great variety of methods and picked the one that had performed best in the past. While the problem of data mining is widely recognized, we knew of no standard procedure for correcting for it. Models I, II, and III are the result of our attempt to develop data mining corrections and tests of significance, based on as realistic assumptions as we could manage in practice.

Our first application was to backtests of a large number of methods for the period January 1975 through most (later, all) of 1990. Using model I with the TOPIX as $z_t$, we found no statistically significant observed differences in methods for the sixteen-year (192-month) simulation. With model II, the differences were statistically significant, with $\beta = 0.59$. At that point we abandoned model I. (We still use it for expository purposes and to compare RUE with ML.)

Tests with small numbers of methods indicated that models II and III gave similar estimates. Since model III requires much more computer time for large numbers of methods, and its results are harder to interpret, we used model II. We present model III here because some results apply to model III generally (as well as model II in particular), and model III may be needed in some future applications.

In addition to the 192-month backtests, we performed some 68-month backtests to see if there were an advantage to using forecast earnings rather than historical earnings. (The backtest period was shortened because of forecast earnings data availability.) The backtests showed differences, but the model II calculation judged them not statistically significant. More recently, with two or three additional years of data and a somewhat different way of forecasting earnings, the differences were significant.

Further information about the GPRD methods and backtests is presented in Bloch, Guerard, Markowitz, Todd, and Xu [1993]. A backtest is presented there using seventeen methods: four using univariate expected return estimates (earnings-to-price, book-to-price, etc.), twelve using variations of the composite methodology, and one benchmark. A 192-month data mining analysis of these seventeen methods was statistically significant; a 60-month analysis was not.

The greatest surprise for us from our data mining calculations is the inability for five or six years' worth of data to reject hypotheses that differences in methods are due to chance. We do not say that five years is never enough and sixteen years is always sufficient. Rather, we say one should give the data some opportunity to accept or reject the hypothesis that observed differences are due to chance.

## WHY NO HOLDOUT PERIOD?

The tests we describe use the entire period permitted by data availability. A common alternate is to save recent data as a "holdout period." We use the former rather than the latter approach.

One problem with a holdout period is that it is routinely "data mined"; that is, if a method that did well in the base period does poorly in the holdout period, the researcher does not abandon the project. Rather he or she tries another method that did well in the base period, until one is found that also does well in the holdout period. Such a procedure will eventually produce a "successful" method, even if all methods are equally good.

Second, holdout periods tend to be short — usually too short for any statistically significant test. Also, often all that is looked at about the holdout period is whether the proposed method beat some benchmark. Even a randomly chosen method has roughly a fifty-fifty chance of doing so.

Finally, the holdout period reduces the number of observations used in the base period. There may appear to be enough observations in the base period if we perform tests of significance incorrectly — testing the best method against a null method, as if it were the only one ever tested. But, as we have seen, if we take into account that many methods were tested, then the entire analysis may be judged insignificant even for lengths of analysis that are now frequently thought to be adequate. Reducing the length of the base period, to provide for a holdout period, may rob the results of statistical significance when the test is performed properly.

## SUMMARY

The policy that worked best in the past may not work as well in the future; it may have been best in the past in part because it was "lucky" rather than because it is best in fact. The problem is to adjust past performance in estimating future performance to reflect such data mining.

If we assume that the policies tested are drawn at random from a number of policies, plausible data mining corrections emerge. The exact correction depends on the assumed model of how expected returns, and then realized returns, are drawn.

## APPENDIX

Theorem A. Given the assumptions of model III, let $N^*$ and $D^*$ be defined as in (11); then

$$E\left(N^*\right) = \sigma_\mu^2 \tag{A-1}$$

$$E\left(D^*\right) = \sigma_\mu^2 + \frac{1}{T(n-1)}\left(\sum_i \sigma_{ii} - \sum_{i,j} \sigma_{ij}/n\right) \tag{A-2}$$

*Proof*: Define

$$\bar{\varepsilon}_i = \frac{1}{T}\sum_{t=1}^{T} \varepsilon_{it}, \qquad \bar{\varepsilon} = \frac{1}{nT}\sum_{i=1}^{n}\sum_{t=1}^{T} \varepsilon_{it}$$

$$\bar{\varepsilon}^t = \frac{1}{n}\sum_{i=1}^{n} \varepsilon_{it}, \qquad \bar{\mu} = \frac{1}{n}\sum_{i=1}^{n} \mu_i$$

First compute $E(B^*)$ and $E(C^*)$.

$$\begin{aligned} E(nB^*) &= E\sum_{i=1}^{n} \left(\bar{y}_i - \bar{y}\right)^2 \\ &= E\sum_{i=1}^{n} \left(\mu_i - \bar{\mu} + \bar{\varepsilon}_i - \bar{\varepsilon}\right)^2 \\ &= E\sum_{i=1}^{n} \left(\mu_i - \bar{\mu}\right)^2 + E\left(\sum_{i=1}^{n} \bar{\varepsilon}_i^2 - n\bar{\varepsilon}^2\right) \qquad \text{(A-3)} \\ &= (n-1)\sigma_\mu^2 + \frac{1}{T}\left(\sum_{i=1}^{n} \sigma_{ii} - \sum_{i,j} \sigma_{ij}/n\right) \end{aligned}$$

$$\begin{aligned} E(TC^*) &= E\sum_{t=1}^{T} \left(\bar{y}^t - \bar{y}\right)^2 = E\sum_{t=1}^{T} \left(\bar{\varepsilon}^t - \bar{\varepsilon}\right)^2 \\ &= E\left(\sum_{t=1}^{T} \bar{\varepsilon}_i^2 - T\bar{\varepsilon}^2\right) = \frac{T-1}{n}\sum_{i,j} \sigma_{ij}/n \qquad \text{(A-4)} \end{aligned}$$

To finish the proof, we need to compute $E(A^* - B^*)$. Since identity

$$\begin{aligned} nT\left(A^* - B^*\right) &= \sum_{i,t} \left(\bar{y}_{it} - \bar{y}\right)^2 - T\sum_{i=1}^{n} \left(\bar{y}_i - \bar{y}\right)^2 \\ &= \sum_{i,t} \left(y_{it} - \bar{y}_i\right)^2 = \sum_{i,t} \varepsilon_{it}^2 - T\sum_{i=1}^{n} \bar{\varepsilon}_i^2 \end{aligned}$$

holds, it follows that

$$\begin{aligned} EnT\left(A^* - B^*\right) &= E\left(\sum_{i,t} \varepsilon_{it}^2 - T\sum_{i=1}^{n} \bar{\varepsilon}_i^2\right) \\ &= (T-1)\sum_{i=1}^{n} \sigma_{ii} \qquad \text{(A-5)} \end{aligned}$$

Q.E.D.

## ENDNOTES

[1]For discussions of the Bayesian decision-maker, see Bawa, Brown, and Klein [1979], Markowitz [1991], Savage [1972], and Zellner [1987].

[2]The data mining correction formula presented here is similar to Vasicek's [1973] Bayesian estimation of security betas. Other closely related data mining problems have been studied by Leamer [1978], Lakonishok and Smidt [1988], Merton [1987], and Lo and MacKinlay [1990]. The simplified process used in this article is sometimes referred to as the empirical Bayes approach.

[3]The best linear estimate $\hat{\mu}_i$ is the estimate that minimizes $E\left(\mu_i - \hat{\mu}_i\right)^2$ among those $\hat{\mu}_i$ of form

$$x_0 + \sum_{i,t} x_{it} r_{it}\,.$$

## REFERENCES


Bawa, V.S., S.J. Brown, and R.W. Klein. "Estimation Risk and Optimal Portfolio Choice." In *Studies in Bayesian Econometrics*, New York: North Holland, 1979.

Bloch, M., J. Guerard, H. Markowitz, P. Todd, and G. Xu. "A Comparison of Some Aspects of the U.S. and Japanese Equity Markets." *Japan and the World Economy*, Vol. 5 (1993), pp. 3-27.

Guerard, J.B., Jr. "Linear Constraints, Robust-Weighting, and Efficient Composite Modeling." *Journal of Forecasting*, 6 (1987), pp. 193-199.

Guerard, J.B., Jr., and B.K. Stone. "Composite Forecasting of Annual Corporate Earnings." In A. Chen, ed., *Research in Finance 10*. Greenwich, CT: JAI Press, 1992, pp. 205-230.

Kendall, M., A. Stuart, and J.K. Ord. *The Advanced Theory of Statistics*. London: Charles Griffin & Company Limited, 1983, Vol. 3, 4th edition.

Lakonishok, J., and S. Smidt. "Are Seasonal Anomalies Real? A Ninety-Year Perspective." *Journal of Financial Studies*, Vol. 1 (1988), pp. 403-426.

Leamer, E. *Specification Searches*. New York: John Wiley & Sons, 1978.

Lo, A.W., and A.C. MacKinlay. "Data-Snooping Biases in Tests of Financial Asset Pricing Models." *Review of Financial Studies*, Vol. 3, No. 3 (1990), pp. 431-467.

Markowitz, H.M. *Portfolio Selection: Efficient Diversification of Investments*. Cambridge, MA: Basil Blackwell, 1991, 2nd edition.

Merton, R. "On the Current State of the Stock Market Rationality Hypothesis." In R. Dornbusch, S. Fisher, and J. Bossons, eds., *Macroeconomics and Finance: Essays in Honor of Franco Modigliani*. Cambridge, MA: MIT Press, 1987.

Rao, C.R. *Linear Statistical Inference and Its Applications*. New York: John Wiley & Sons, 1973, 2nd edition.

Savage, L.J. *The Foundations of Statistics*. New York: Dover, 1972, 2nd edition.

Vasicek, O. "A Note on Using Cross-Sectional Information on Bayesian Estimation of Security Betas." *Journal of Finance*, Vol. 28 (1973), pp. 1233-1239.

Zellner, A. *An Introduction to Bayesian Inference in Econometrics*. New York: John Wiley & Sons, 1987.

# Chapter 7

# Harry Markowitz Company

## Comments

The article on *Single-period Mean-variance Analysis in a Changing World* concerns how to solve dynamic programming problems which one encounters in portfolio selection over time, but applies to any dynamic programming problem. The great problem with dynamic programming is what is called the "curse of dimensionality," that is, it becomes increasingly difficult or impossible in fact to solve such problems when they take more than a few variables to describe the state of the system at any one time. The article by Van Dijk and me presents a heuristic which, in the examples we present, does almost as well as the optimum solution as well as better than alternate heuristics. It does not prove that the proposed heuristic, "the quadratic surrogate heuristic," will in fact do well with larger portfolio selection problems. Further experimentation along this line is reported in an article by Kritzman, Myrgren and Page (2007) which says the following:

> "Our tests reveal that the quadratic heuristic provides solutions that are remarkably close to the dynamic programming solution for those cases in which dynamic programming is feasible and far superior to solutions based on standard industry heuristics. In the case of five assets, in fact, it performs better than dynamic programming due to approximations required to implement the dynamic programming algorithm. Moreover, unlike the dynamic programming solution, the quadric heuristic is scalable to as many as several hundred assets."

The next two articles are the result of collaboration with Bruce Jacobs and Ken Levy. One concerns an asynchronous simulation of financial markets. Further information concerning this matter can be found on the Jacobs and Levy website www.jacobslevy.com. The second concerns a matter indicated in the title and the article.

The next to the last article is on portfolio theory. The last article is a retrospective that includes portfolio theory, sparse matrices and SIMSCRIPT. It forms a gung-ho for the book.

## References

Markowitz, H. M. and Usmen, N. (1996a). *The Likelihood of Various Stock Market Return Distributions, Part 1: Principles of Inference.* Journal of Risk and Uncertainty, Vol. 13, pp. 207–219.

Markowitz, H. M. and Usmen, N. (1996b). *The Likelihood of Various Stock Market Return Distributions, Part 2: Empirical Results.* Journal of Risk and Uncertainty, Vol. 13, pp. 221–247.

Markowitz, H. M. and Usmen, N. (2003). *Resampled Frontiers Versus Diffuse Bayes: An Experiment.* Journal of Investment Management, 1(4), pp. 9–25.

Markowitz, H. M. (1997). *On Socks, Ties and Extended Outcomes.* In "Economic and Environmental Risk and Uncertainty: New Models and Methods", pp. 219–226. Kluwer Academic Publishers.

Markowitz, H. M. and van Dijk, E. (2003). *Single-Period Mean-Variance Analysis in a Changing World.* Financial Analysts Journal, March/April 2003, Vol. 59, No. 2, pp. 30–43.

Jacobs, B., Levy, K. and Markowitz, H. M. (2004). *Financial Market Simulation.* The Journal of Portfolio Management, 30th Anniversary Issue, pp. 1–10.

Jacobs, B., Levy, K. and Markowitz, H. M. (2005). *Portfolio Optimization with Factors, Scenarios, and Realistic Short Positions.* Operations Research, Vol. 53, No. 4, July–August, pp. 586–599.

Markowitz, H. M. (2005). *Market Efficiency: A Theoretical Distinction and So What?* Financial Analysts Journal, September/October, Vol. 61, No. 5, pp. 17–30.

Markowitz, H. M. (2002). *Efficient Portfolios, Sparse Matrices, and Entities: A Retrospective.* Operations Research, January–February, Vol. 50, No. 1, pp. 154–160.

Markowitz, H. M. (2006). *De Finetti Scoops Markowitz.* Journal of Investment Management, Vol. 4, No. 3, pp. 1–27.

Markowitz, H. M. (2008). *CAPM Investors Do Not Get Paid for Bearing Risks.* The Journal of Portfolio Management, Vol. 34, No. 2, pp. 91–94.

Journal of Risk and Uncertainty, 13:207–219 (1996)


# The Likelihood of Various Stock Market Return Distributions, Part 1: Principles of Inference

HARRY M. MARKOWITZ
*Daiwa Securities Trust Company, 1010 Turquoise Street, Suite 245, San Diego, CA 92109*

NILUFER USMEN
*Rutgers University*

## Abstract

This is the first of two articles which apply certain principles of inference to a practical, financial question. The present article argues and cites arguments which contend that decision making should be Bayesian, that classical (R. A. Fisher, Neyman–Pearson) inference can be highly misleading for Bayesians as can the use of diffuse priors, and that Bayesian statisticians should show remote clients with a variety of priors how a sample implies shifts in their beliefs. We also consider practical implications of the fact that human decision makers and their statisticians cannot fully emulate Savage's rational decision maker.

**Key words:** bayesian inference, remote clients, Lindley's paradox

**JEL Classification:** D80

A companion article (Markowitz and Usmen, 1996) which follows in this issue describes how remote Bayesian clients should shift beliefs among various hypotheses concerning the probability distribution of daily changes in the Standard and Poor (S&P) 500 Index of stock prices, given a particular sample. The original motivation for the study was methodological. We wanted to see if useful data analysis could be performed on practical financial problems within constraints imposed by certain philosophical principles, namely, that

- financial research is primarily to improve financial decisions;
- rational decision making is Bayesian;
- the commonly used, commonly published classical (R.A. Fisher, Neyman–Pearson) inference methods are highly unreliable guides to Bayesians;
- the conjugate or diffuse priors frequently assumed in Bayesian studies are either restrictive (few hypotheses admit nontrivial conjugates) or are possibly highly misleading; and
- the Bayesian statistician should show remote clients, with a wide variety of priors, the extent to which they should *shift* beliefs, from prior to posterior, given the sample at hand (as opposed to assuming a single set of priors and showing the posterior beliefs implied by the sample and these priors).

The present article discusses the above principles. The thesis of the two articles is that not only are the principles of inference presented here correct, as argued in the present article, but useful, as illustrated in the next.

## 1. Financial research supports financial decision making: An example

We assume that research into financial phenomena, such as the probability distribution of daily changes in the S&P 500 Index of stock prices, has as its ultimate consumer financial decision makers who use it to guide action. As concrete examples in later discussions, in this and the companion article, we will briefly sketch some actions which could make use of a model of daily changes in the S&P 500 Index. To begin, let us consider a hypothetical investment management team which acts in the following simplified manner. The team periodically chooses a portfolio whose securities are any or all of the following: cash, a particular S&P 500 Index fund, and purchases of puts or calls on the S&P 500 Index with various striking (settlement) prices but all having the same expiration date. The management team analyzes the situation as if a portfolio of these securities must be bought now and held until the common expiration date; and, in addition, a mean-variance approximation is deemed sufficient.[1]

Suppose further, for a moment, that the team is willing to assume that

$$y_t = \log (Y_t/Y_{t-1}) \tag{1}$$

where $Y_t$ is the level of S&P 500 Index on the $t$-th trading day, is generated by one specific model, e.g., by independent draws from a distribution with density $f(y; \theta_{\Delta\tau(t)})$, where $\Delta\tau(t)$ is the number of calendar days between the close of trading day $t - 1$ and $t$, and $\theta$ is a vector of parameters depending on $\Delta\tau$. Then, the probability distribution of the change in log $Y_t$ over $T$ trading days, log $(Y_T/Y_0)$, can be computed by concatenating (numerically, if necessary) the right number of draws from $f(y; \theta_1)$, $f(y, \theta_2)$,.... Since profit or loss (after expenses) on any of the assets of the portfolio depends only on $Y_T/Y_0$, the probability distribution of log $(Y_T/Y_0)$ can be used to calculate the means, standard deviations, and covariances required to generate a mean-variance efficient frontier from which a portfolio is selected.

Next, suppose that the team finds several models of the return generating process plausible—either with the same functional form and different parameters, or with alternate functional forms—and is willing to state subjective probabilities for each. Then, a probability distribution of $\log(Y_T/Y_0)$ can be determined for each model, and a final probability distribution computed as a convex combination of the individual distributions with the probabilities of models as weights. From this, the required means, variances, and covariances of assets are computed. One purpose of the research described in the companion article is to help the above management team decide which models of S&P 500 Index changes to use and what probabilities to assign them.

The analysis becomes more complex if the potential assets of the portfolio include the writing (selling) of puts and calls, and long or short futures positions. With such securities, it is possible that the fund will have to use the portfolio's cash, or borrow, or liquidate part of the portfolio when cash and borrowing power are gone, as these positions are "marked to market" each day. (For the rules of the game, see Securities Training Corporation, 1995). To analyze this situation formally, the team would need to model prices of the puts, calls, and futures as well as the change in the Index, since the daily "mark to market" process looks at the gains and losses in various positions, including price changes in futures and written options. It would also be necessary to model the price process for puts and calls if the portfolio were restricted to the securities first listed, but the analysis considered the possibility of portfolio revision between now and expiration, or considered options with different expiration dates.

If we lived in the world analyzed by Black and Scholes (1973), the price of each option would be uniquely determined by the value of the Index, the time to expiration, and the parameters of the Index change process. But there are assumptions of the Black–Scholes world which may not be appropriate here. For one, the Black–Scholes model assumes that $y_t$ is normally distributed. The management team may assign a probability less than one to that hypothesis. For another, a plot of successive pairs of S&P 500 Index, value-of-option combinations suggests that there is not a unique value of the option for a given time to expiration and value of Index. Rather, there is noise in the relationship. In fact, one reason for shifting the composition of the portfolio is to take advantage of times when one option is "rich" or "lean," relative to others.

A list of "combined" models, each model providing a random process for generating daily changes in the Index and in the prices of the futures and options plus explicit or implicit probabilities for each model, could be used to evaluate dynamic strategies in various more or less formal ways. A possible formal analysis could involve utilities of outcomes as well as probabilities of models which might be beyond analytic solution. Therefore Monte Carlo analysis would be needed to evaluate expected utility for a given strategy and model. The results for different models would be weighted by their probabilities to estimate the overall expected utility for each strategy tried. A less formal analysis might involve inspecting printouts or summaries of simulation runs for different strategies, then picking a best strategy subjectively.

The probabilities made explicit in the formal analysis and implicit in the less formal one are the subjective probabilities assigned to joint hypotheses about Index changes *and* changes in options and futures prices. The companion article addresses only part of this problem, namely, the (marginal) probabilities assigned to the Index change generation component of the joint hypotheses.

## 2. Rational (coherent) decision making is Bayesian

The axiomatic approach in Savage (1954) convinced many that the proposals of Ramsey (1926, 1928) and DeFinnetti (1937) were correct, namely, that rational (coherent) choice under uncertainty requires a decision maker to act as if she or he attaches probability beliefs to states or events for which objective probabilities are not known (if objective probabilities exist at all), and to update these probability beliefs according to Bayes's rule as evidence accumulates. Fishburn (1981) surveys many subsequent axiom systems which imply action according to probability beliefs. These axiom systems also imply that rational (coherent) decision making requires maximizing expected utility. Machina and Schmeidler (1992) present an axiom system which implies probability beliefs updated by Bayes's rule, but which does not necessarily require expected utility maximization. The reason why one of the present authors is a Bayesian has changed little since Markowitz (1959, chapters 10–13). Others (e.g., Shafer, 1982) dispute the rationality of probability beliefs updated by Bayes's rule. We will not try to review the arguments for and against the axiom systems. Rather, we state as a premise of this and the companion article that we assume the reader to be convinced (e.g., by Savage, 1954) that coherent decision making requires probability beliefs updated by Bayes's rule.

Savage (1954, p. 7) asserts that his axioms describe "a highly idealized theory of the behavior of a 'rational' person with respect to decisions." It will be relevant below to briefly review two characteristics of this idealized rational decision maker (RDM) which no real human or real computer can share. First, the RDM makes no mistakes in logic or arithmetic. "In particular, such a person cannot be uncertain about decidable mathematical propositions" (Savage, above-cited page). For example, two RDMs would not bet on the value of some not yet computed (by humans) decimal place of $\pi = 3.14.....$, since each can instantly compute the answer.

Second, an RDM does not make up hypotheses as she or he goes along. This is perhaps more easily discussed using the setup of Markowitz (1959, chapter 12) rather than Savage (1954). In Markowitz (1959, p. 258), the basic concepts used in describing the RDM consist of "outcome," "decision," and "nature of the world." "We shall assume that there is a finite (though perhaps extremely large) number of hypotheses about the world, one and only one of which is true.... Each such hypothesis will be referred to as a possible *Nature of the World*.... [W]e admit the possibility that objective random variables and probability distributions may be used in the definition of a Nature of the World." In the Savage setup, the state of the world corresponds to one of the above hypotheses plus draws of random variables given the hypothesis. In particular, in the Savage setup an act—i.e., choice of strategy—determines outcomes as a single valued function of states; in the Markowitz setup a strategy (decision) plus a hypothesis (nature of the world) determines a probability distribution of outcomes.

A specific hypothesis about the generation of changes in the S&P 500 Index typically consists of a model plus specification of parameters in the model. For example, we noted that the Black–Scholes model assumes that $y_t$ is normally distributed, and, therefore, has two parameters. Other hypothesized models for $y_t$ include Student's $t$ distribution with three parameters (Blattberg and Gonedes, 1974) and the stable Paretian family (Mandelbrot, 1963; Fama, 1963) with four parameters. These models assume that the successive draws are independent with identical distributions (i.i.d.). In the companion article, we consider models in which draws are independent but in which distribution depends on $\Delta\tau$. Bollerslev, Chou, and Kroner (1992) survey models with conditional heteroskedasticity including ARCH, GARCH, EGARCH, IGARCH and ARCH-M. These models come in varying versions, e.g., GARCH (1,1), GARCH (2,1), GARCH (1,2) ..., with the numbers of parameters depending on the version. These models could be adapted to treat weekdays, weekends, one-day holidays, and three-day weekends differently.

The RDM knows of all these models a priori, as well as of others which humans have proposed in the past, will propose in the future, and will never propose. The RDM acts as if she or he assigns probability beliefs to each of these. For convenience here, assume that nonzero beliefs are assigned to only a (large but) finite number of hypotheses, including, e.g., GARCH $(i, j)$ versions, and, for each, a (large but) finite mesh of parameter values. As information accumulates, the RDM updates these beliefs according to Bayes's rule. When faced with a decision, such as a choice of an options and futures strategy, the RDM chooses a policy to maximize a function of the probabilities of final outcomes. For the expected utility maximizing RDM, this is a linear function, with utilities of outcomes as coefficients, with probabilities of final outcomes equal to weighted averages of the probabilities of outcomes given the various hypotheses, and with the current (posterior) probabilities of hypotheses as weights.

## 3. Classical statistics is an unreliable indicator of how Bayesians should shift beliefs

Commonly used "objective" statistical procedures of the R.A. Fisher or Neyman–Pearson schools can grossly mislead as to how a Bayesian should shift beliefs given evidence. As a simple example, suppose that there are two hypotheses $H_1$, $H_2$, and three possible observations $O_1$, $O_2$, $O_3$, with $P(O|H)$ given in the following table.

| | $H_1$ | $H_2$ |
|---|---|---|
| $O_1$ | .01 | .98 |
| $O_2$ | .04 | .01 |
| $O_3$ | .95 | .01 |

The set $\{O_1, O_2\}$ is a Neyman–Pearson test region for $H_1$ with size = 5%. Thus, if $O_2$ is observed, the test rejects $H_1$ at the 5% level. To some this might suggest that a Bayesian should shift belief by a factor of 19:1 against $H_1$ in favor of $H_2$. But $P(O_2|H_1)/P(O_2|H_2) = 4$ implies that beliefs should be shifted by a factor of four against $H_2$ *in favor* of $H_1$! A principal difference between the Bayesian calculation and classical inference is that the former uses only $P(O|H)$ for the observation which actually occurred, i.e., $P(O_2|H_1)$ and $P(O_2|H_2)$ in the example. This is the "likelihood principle" which Birnbaum (1962, 1968, 1969) argues follows from simpler and more immediate premises than those used to derive personal probability. Neyman–Pearson and R. A. Fisher procedures, on the other hand, depend in part on the probabilities of outcomes that did not occur, e.g., $P(\{O_1, O_2\}|H_1) = P(O_1|H_1) + P(O_2|H_1)$.

The above example has the virtue of being simple, and the drawback of not being "realistic," that is, not drawn from actual practice. Edwards, Lindman, and Savage (1963, p. 230) examine tests used in practice, find that the Bayesian shift in odds ratio is typically much less than the classical $p$-value or size of test, and illustrate "how classically significant values of $t$ can, in realistic cases, be based on data that actually favor the null hypothesis" (also, see Berger and Sellke, 1987). In their rejoinder to a comment by Pratt, the latter say that they "are in complete agreement that Edwards, Lindman, and Savage (1963) (EL&S) contained the essence of our article. Indeed, had EL&S not been so mysteriously ignored for so long, our contribution would have been mainly...."

Lindley (1957, p. 190) shows that there is no limit to how misleading a $p$-value or test size can be. He considers a sample of size $n$ drawn from a normal population with known variance $\sigma^2$ and unknown mean $\theta$, and a test of the hypothesis that $\theta = \theta_0$. The Bayesian is assumed to attach a probability $c$ to $\theta = \theta_0$ and to have a prior density $p(\theta)$ on a finite interval I that includes $\theta_0$. Bounded $p(\theta)$ is sufficient but not necessary for the Lindley argument. Suppose that the observation is just significant, i.e., just rejects the hypothesis $\theta = \theta_0$ at the $\alpha$ % level. Depending on $n$, the posterior belief $\bar{c}$ of the Bayesian that $\theta = \theta_0$ may be arbitrarily close to 1. Lindley provides a numerical example, with $\sigma = 1$, $c = 1/2$ and the observation just significant at the 5% level of a two-sided test. "[F]or small samples ($n \leq 10$) the probability of $\theta_0$ has decreased appreciably from its initial value of 1/2 ... For medium samples ($10 < n < 100$) the probability has only decreased a little ... By the time $n$ has reached a value of 300 $\bar{c}$ is equal to $c$." For larger $n$, the classical statistician has rejected $\theta = \theta_0$ at the 5% level, whereas the Bayesian has increased the probability attached to that hypothesis.[2]

## 4. Remote Bayesian clients

The companion article which follows is concerned with advising readers as to which models of stock market return distributions to use, and how to weight their results in applications such as the choice of strategy for buying and selling options on the S&P 500 Index. These readers are "remote clients" in the sense of Hildreth (1963). We do not know their priors, which presumably vary from one to another. Since they are only pale imitations of RDMs, they typically have only vague ideas of their own priors.

We will not tell the remote clients what their posterior beliefs should be given our data. Rather, we will advise on shifts in beliefs. According to Bayes's rule, if $H_1$ and $H_2$ are two hypotheses, given a sample $S$, their posterior odds ratio $P(H_1|S)/P(H_2|S)$ is related to their a priori odds $P(H_1)/P(H_2)$ by

$$\frac{P(H_1|S)}{P(H_2|S)} = \frac{P(S|H_1)}{P(S|H_2)} \bullet \frac{P(H_1)}{P(H_2)}.$$

$P(S|H_1)/P(S|H_2)$ is known as the Bayes factor. We shall also refer to it as the shift in belief $\psi(H_1, H_2)$ from $H_2$ to $H_1$.

When $H_1$ and $H_2$ are simple hypotheses, $\psi$ does not depend on priors. For example, confining ourselves for the present to $t$ with $\Delta\tau = 1$ (weekdays), let $H_1 = \{y_t$ is i.i.d. normal with mean $\mu = 5.45\ (10^{-4})$ and standard deviation $\sigma = 7.68\ (10^{-3})\}$.

$H_2 = \{y_t$ is i.i.d. Student's $t$ with mean and variance as in $H_1$ and $M_4 = 12\}$, where $M_4$ is the normalized fourth moment $E(y - \mu_1)^4/\sigma^4$, a.k.a. $\beta_2$ or $\gamma_2 + 3$. Then, for a sample to be described in the companion article, $\psi(H_1, H_2) = 10^{-74}$. If an RDM had considered $H_1$ $10^9$ more probable prior to the sample, she or he would think $H_2$ $10^{65}$ more probable after.

If $H_3 = \{$i.i.d. Pearson Type IV with mean, variance and $M_4$ as in $H_2$ and $M_3 = 0.038\}$, where $M_3$, the normalized third moment is $E(y - \mu_1)^3/\sigma^3$ (a.k.a. $\gamma_1 = \sqrt{\beta_2}$), then $\psi(H_3, H_2) = 1.06$. An RDM would shift the odds ratio by a factor of 1.06 in favor of the slightly skewed Type IV distribution against the symmetric Student's $t$.

Throughout we shall speak of shifts of belief rather than accepting or rejecting a hypothesis. In general, even if the odds ratio between any two hypotheses $H$ and $K$ should be shifted by a factor of $\psi(H, K) = .95/.05 = 19$ (as the classical rejection of $K$ at the 5% level *sounds* as if it implies), $K$ should not necessarily be treated as if it had a posterior probability of zero. Suppose $H$ was a normal distribution and $K$ a long-tailed distribution such as $H_2$ or $H_3$ above. Suppose that the prior odds ratio was 1:1, and, therefore, the posterior odds 19:1. For some strategies with options and futures, a many-$\sigma$ move in the Index can be disastrous; while, for other strategies, a many-$\sigma$ move is beneficial or benign. Consequently, expected utilities of different strategies may rank quite differently for posterior probabilities $(p_H, p_K) = (1, 0)$ than for (.95, .05).

The problem is not alleviated by choosing a different significance level. It might seem that if a shift of 19:1 is not sufficient to reject, i.e., henceforth to ignore, the unfavored hypothesis, then perhaps a larger shift, say 99:1, should be used. But the statement that $K$ has been rejected at the 1% level—now defined as $\psi(H, K) \geq 99$—does not tell us whether $\psi = 101$ or $10^{70}$; and non-rejection doesn't tell us whether $\psi = 98$ or 1.06 or $10^{-70}$. Assuming a prior of 1:1 to be specific, a posterior odds ratio of $10^{70}$:1 may have different practical consequences than 101:1; and an odds ratio of $10^{-70}$:1 different practical consequences than 98:1. Finally, it will typically make little practical difference whether the posterior odds ratio is 101:1 or 98:1, though the former would reject and the latter accept at the 1% level. Thus, the information that a hypothesis has been rejected or not at some preassigned level is usually much less informative than the value of $\psi$, or even the order of magnitude of $\psi$.

For compound hypotheseses $H$ and $K$, $\psi$ depends on priors within $H \cup K$ but not outside it. Specifically,

$$\psi(H, K) = \frac{\int_H P(S|\theta)dP(\theta)}{\int_K P(S|\theta)dP(\theta)}, \tag{2}$$

where $H$ and $K$ may be subsets of the same or different parameter spaces.

Suppose, for example, that

$$H = \{y_t \text{ is i.i.d. normal}\}$$

$$K = \{y_t \text{ is i.i.d. Pearson with } M_4 = 12\}$$

The shift $\psi(H, K)$ depends on the prior distribution $\tilde{p}(M_3) = p(\mu_0, \sigma_0, M_3, 12)$ for any given $\mu_0$, $\sigma_0$. A frequent practice is to assume, with Jeffreys (1939), that priors are diffuse,[3] e.g.,

$$\tilde{p}(M_3) \equiv 1. \tag{3}$$

This is an improper prior in that $\int \tilde{p}(M_3)dM_3 = \infty$. Improper priors can lead to strange results (see, for example, Buehler, 1959, and Lindley, 1972). To avoid such strange results, it is sometimes argued that the implications for action of an improper prior should be considered as the limit of the implications for action of increasingly diffuse but proper priors such as

$$\tilde{p}(M_3) = 1/(m_{\mathrm{HI}} - m_{\mathrm{LOW}}) \text{ on } [m_{\mathrm{LOW}}, m_{\mathrm{HI}}] \tag{4}$$

as $m_{\mathrm{HI}} \to \infty$, $m_{\mathrm{LOW}} \to -\infty$.

Even though for a wide range of plausible $M_3$ likelihood $P(S|\mu_0, \sigma_0, M_3, 12) >> P(S|\mu_0, \sigma_0, 0, 3)$, prior (3) implies $\psi(H, K) = \infty$. The posterior odds is 1:0 in favor of the normal. For priors of the form (4), $\psi \to \infty$ as either $m_{\mathrm{LOW}} \to -\infty$ or $m_{\mathrm{HI}} \to \infty$ or both. On the other hand, we conjecture that most readers have priors which assign a non-negligible, positive probability to some moderately sized interval $M_3 \in [\underline{m}, \overline{m}]$. Specifically, we propose in the companion article that the reader has a prior probability of *at least* .01 for the interval [−2.667, 2.667]. To visualize this interval, note that −2.667 and 2.667 are the $M_3$ values for binomial distributions with $p = .1$ and .9, respectively. Based on this assumption, and other assumptions which favor the normal distribution, i.e., will not overstate the shift against the normal, we compute that the shift for $K$ against the normal hypothesis $H$ is at least $10^{69}$. Thus, for readers who find the proposed inequality acceptable, (3) and (4) imply the wrong conclusion concerning $\psi(H, K)$.

An approach frequently recommended for Bayesian inference is to use a conjugate prior (see DeGroot, 1970; Zellner, 1971; Berger, 1985). But no nontrivial conjugate prior is available for the entire Pearson family. Another approach is to compute worst (or best) case shifts; i.e., to determine the prior which maximizes the shift for $H$ against $K$, or for $K$ against $H$ (Berger and Sellke, 1987). This is highly commendable if the best case and worst case $\psi$ are close; but in our example the best case for $H$ is the diffuse prior with $\psi(H, K) = \infty$; whereas the best case for $K$ concentrates $M_3$ near .04 and has $\psi(H, K)$ about $10^{-74}$.

Still another approach, "stable estimation" recommended by EL&S, will play an important role in the companion article with respect to both simple and compound hypotheses. In particular, we will note that if the principle of stable estimation applies to $H$ in (2) and, separately, to $K$ as well, then $\psi$ in (2) takes on a relatively simple form even when $H$ and $K$ are of high dimension. This is of general methodolical interest, but it is convenient to postpone its discussion until the companion article's section on shifts of belief between compound hypotheses.

## 5. Human approximation to an RDM

The following are some practical consequences of the fact that neither Bayesian statisticians nor their remote clients are RDMs.

(1) Given a sample $S$, the RDM updates beliefs simultaneously for a space of "all" hypotheses. The human statistician is forced to evaluate a relatively small subset of hypotheses. Later, the same or different statisticians can evaluate additional hypotheses relative to each other and relative to the best of the first batch. Still later, a third batch can be evaluated, and so on. It is not necessary to estimate the shift in beliefs between every member of the first batch and every member of the second, since (for a given sample $S$) the shift $\psi$ between two hypotheses in the first batch is not affected by the existence of the second. Further, if $H_1$ and $H_2$ are hypotheses (simple or compound) in the first batch and $H_3$ a hypothesis in the second batch, then

$$\psi(H_1, H_3) = \psi(H_1, H_2) \cdot \psi(H_2, H_3).$$

Thus, if $H_3$ "beats" the winners of the first group, it a fortiori beats the losers.

In particular, in the companion article, we consider various hypotheses that assume $y_t$ to be i.i.d. for a given $\Delta\tau$. It is plausible that among the many proposed non-i.i.d. hypotheses some will explain the data better than i.i.d. hypotheses; that is, $S$ will shift belief from the latter to the former. But at the least, the results establish a contender to be bested, namely, the i.i.d. Pearson Type IV distributions with (as we shall see) any of a wide range of $M_4$ and with $\mu$, $\sigma$, and $M_3$ depending on $\Delta\tau$.

(2) Suppose an RDM considers only 1,000 different models possible and of equal prior probability. Suppose further that, for a given sample $S$, one of these models (say $H_1$) is very successful; specifically, its posterior probability is 100-fold greater than its prior probability; ten hypotheses ($H_2$, ....., $H_{11}$) are moderately successful and increase their probability ten-fold; 100 ($H_{12}$, ....., $H_{111}$) hold their own, neither gaining nor losing probability; and the remaining 889 lose probability. Then, after $S$, the RDM's probability beliefs assign one chance in ten that $H_1$ is true, one in ten that the truth lies in $H_2$, ....., $H_{11}$; one in ten that it lies in $H_{12}$, ... $H_{111}$; and seven in ten that it is one of the poorer

performing hypotheses $H_{112}$, ..... $H_{1000}$. In general, then, it is not necessarily a good approximation to RDM behavior to assume that the truth lies in one (or a few) likelihood-maximizing hypotheses.

(3) Even if the statistician evaluates a thousand hypotheses, these are not all that have been proposed, will be proposed, or could be proposed. One stimulus for a search for a new hypothesis is dissatisfaction with the fit of the best of the existing ones. For example, suppose that the only hypotheses considered in some analysis are Gaussian and, for the best fitting of them, $H$, four and five standard deviation observations are not extremely rare. It then seems plausible that there exists some long-tailed distribution $K$ such that the shift $\psi$ towards $K$ away from $H$ is substantial. More generally, Box (1980) proposes that if the sample $S$ falls in a region $R$ of least likely samples given $H$, then one should search for an alternate hypothesis. As discussed above, $P(R|H) = .01$ and $S \in R$ does *not* imply a shift in belief of 99 : 1 against $H$ in favor of "all other"; but $S \in R$ is a plausible heuristic for starting a search for an alternative $K$. Box also considers how $R$ should be formed depending on general characteristics of likely competing hypotheses. Good (1956) presents an alternate heuristic for initiating a search for some $K$.

(4) What should the HDM do while the statistician searches for a better fitting hypothesis, especially considering that even a well fitting one is not to be trusted completely. One approach is to (a) evaluate alternate investment strategies in terms of the models considered and—in addition—(b) consider scenarios $y^* = (y_1, y_2, ....., y_T)$, which would be particularly disastrous for otherwise attractive strategies. The scenario $y^*$ might involve a single extremely large in magnitude positive or negative $y_t$, or an unusual number of positive or negative $y_t$, etc., depending on the nature of the otherwise favored strategies. A scenario such as $y^*$ might be unlikely given the models considered, but still seem to have non-negligible probability for models not yet explored. The HDM then intuits a subjective probability for a scenario as disastrous as $y^*$. Rather than trying only to intuit $P(H_i)$ for models considered, the HDM also produces a subjective estimate for $P(R) = \int P(y \in R|H)dP(H)$ where $R$ is a region of disastrous scenarios represented by $y^*$. The choice of strategy then proceeds, formally or informally, including the utilities[4] and probabilities of the disastrous scenarios as well as the utilities and probabilities of scenarios from the favored hypothesis.

## 6. Summary

This article has argued, or referred to arguments contending, that rational decision making is Bayesian, classical statistics and improper priors are poor guides for Bayesians, and Bayesian statisticians should seek to serve remote Bayesian clients with a wide variety of prior beliefs. These principles are illustrated in an application of practical significance in a companion article, Markowitz and Usmen (1996), which follows.

## Acknowledgments

The authors appreciate very much the valuable suggestions of Mark Machina and a referee.

## Notes

1. A common misimpression is that mean-variance analysis is applicable only if probability distributions are normal or utility is quadratic. This is not true even if one requires a precisely optimum solution. See Chamberlain (1983). More important, various authors (Markowitz, 1959; Young and Trent, 1969; Levy and Markowitz, 1979; Dexter, Yu, and Ziemba, 1980; Pulley, 1981, 1983; Kroll, Levy, and Markowitz, 1984; Simaan, 1993) report that properly chosen mean-variance efficient portfolios give almost maximum expected utility for a wide variety of utility functions and return distributions like those of diversified investment portfolios. This includes portfolios of puts and calls, though not for portfolios consisting of a single put or call (Hlawitschka, 1994). The mean-variance approximation breaks down if the decision maker is pathologically risk-averse, i.e., would prefer a few percent return with certainty to a fifty-fifty chance of breakeven versus a blank check (Markowitz, Reid, and Tew, 1994).
2. Alternate views of Lindley's paradox are presented in Shafer (1982). Regarding some points made therein, in Lindley's original article, note (a) the acknowledgment of Jeffreys (1948) and comparison of Lindley's treatment with that of the latter; and (b) Lindley's treatment of a fairly general prior on I as opposed to requiring a flat one.
3. On the one hand, Jeffreys (1939) recommends the use of improper priors in estimation. "Two rules appear to cover the entire ground. If the parameter may have any value in a finite range, or from $-\infty$ to $+\infty$, its prior probability should be taken as uniformly distributed. If it arises in such a way that it may conceivably have any value from 0 to $\infty$, the prior probability of its logarithm should be taken as uniformly distributed. I have not found any case of estimation so far where any other rule appears to be needed ..." (p. 96, section 3.1). On the other hand, in chapter V, "Significance Test: One New Parameter," he recognizes that an unbounded uniform distribution for $\alpha$, as an alternate to the null hypothesis $\alpha = a$, will always lead to accepting the latter. Jeffreys asserts that "In practice there is always a limit to the possible range of these values, within which the prior probability of $\alpha$ may be taken as uniformly distributed." In our problem, however, there is not a specific value of $M_3$ at which prior plausibility goes from something to nothing.
4. An RDM satisfying the Savage axioms will maximize expected utility for life as a whole. An HDM must focus on some aspect of life restricted with respect to time or scope. In Savage's words, the HDM must consider a "small world." But, for an RDM, while the expected utility maxim applies to the game of life as a whole, it may not apply to a small world (Markowitz, 1959, chapter 11; Machina, 1984).

## References

Berger, James. (1985). *Statistical Decision Theory and Bayesian Analysis*, 2nd ed., New York: Springer-Verlag.

Berger, James, and Thomas Sellke. (1987). "Testing a Point Null Hypothesis: The Irreconcilability of P Values and Evidence," *Journal of the American Statistical Association* 82, 112–139.

Birnbaum, Allan. (1962). "On the Foundations of Statistical Inference," *Journal of the American Statistical Association* 57, 269–326.

Birnbaum, Allan. (1968). "Likelihood," *International Encyclopedia of the Social Sciences*, 299–301.

Birnbaum Allan. (1969). "Concepts of Statistical Evidence." In *Essays in Honor of Ernest Nagel: Philosophy, Science and Method*. New York: St. Martin's Press, pp. 112–143.

Black, Fischer, and Myron Scholes. (1973). "The Pricing of Options and Corporate Liabilities," *Journal of Political Economy* (May/June), 637–654.

Blattberg, Robert, and Nicholas Gonedes. (1974). "A Comparison of the Stable and Student Distributions as Statistical Models for Stock Prices," *Journal of Business* 47, 244–280.

Bollerslev, Tim, Ray Chou, and Kenneth Kroner. (1992). "ARCH Modeling in Finance: A Review of the Theory and Empirical Evidence," *Journal of Econometrics* 52, 5–59.

Box, George. (1980). "Sampling and Bayes' Inference in Scientific Modeling and Robustness," (with discussion), *Journal of the Royal Statistical Society* 143, Part 4, 383–430.

Buehler, Robert. (1959). "Some Validity Criteria for Statistical Inferences," *Annals of Mathematical Statistics* 30, 845–863.

Chamberlain, Gary. (1983). "A Characterization of the Distributions that Imply Mean-Variance Utility Functions," *Journal of Economic Theory* 29, 185–201.

deFinnetti, Bruno. (1937). "La Prevision: Ses Lois Logiques, Ses Sources Subjectives," *Annals de l'Institute Henri Poincaré* 7. English translation in H. E. Kyburg, Jr. and H. G. Smokler (eds.), *Studies in Subjective Probability*. New York: John Wiley & Sons. (1964).

DeGroot, Morris. (1970). *Optimal Statistical Decisions*. New York: McGraw-Hill.

Dexter, Albert, J. Yu, and William Ziemba. (1980). "Portfolio Selection in a Lognormal Market When the Investor Has a Power Utility Function: Computational Results." In M. A. H. Dempster (ed.), *Stochastic Programming*. New York: Academic Press, pp. 507–523.

Edwards, Ward, Harold Lindman, and Leonard Savage. (1963). "Bayesian Statistical Inference for Psychological Research," *Psychological Review* 70, 193–242.

Fama, Eugene. (1963). "Mandelbrot and the Stable Paretian Hypothesis," *Journal of Business* 36, 420–429.

Fishburn, Peter. (1981). "Subjective Expected Utility: A Review of Normative Theories," *Theory and Decision* 13, 139–199.

Good, I. J. (1956). "The Surprise Index for the Multivariate Normal Distribution," *The Annals of Mathematical Statistics* 27, 1130–1135; 28 (1957), 1055.

Hildreth, Clifford. (1963). "Bayesian Statisticians and Remote Clients," *Econometrica* 31, 422–438.

Hlawitschka, Walter. (1994). "The Empirical Nature of Taylor-Series Approximations to Expected Utility," *American Economic Review* 84, 713–719.

Jeffreys, H. (1939). *Theory of Probability*. Oxford University Press; (1948), 2nd ed.

Kroll, Yoram, Haim Levy, and Harry Markowitz. (1984). "Mean Variance versus Direct Utility Maximization," *Journal of Finance* 39, 47–61.

Levy, Haim, and Harry Markowitz. (1979). "Approximating Expected Utility by a Function of Mean and Variance," *American Economic Review* 69, 308–317.

Lindley, D. V. (1957). "A Statistical Paradox," *Biometrika* 44, 187–192.

Lindley, D. V. (1972). *Bayesian Statistics, A Review*. Philadelphia: Society for Industrial and Applied Mathematics.

Machina, Mark. (1984). "Temporal Risk and the Nature of Induced Preferences," *Journal of Economic Theory* 33, 199–231.

Machina, Mark, and David Schmeidler. (1992). "A More Robust Definition of Subjective Probability," *Econometrica* 60, 745–780.

Mandelbrot, Benoit. (1963). "The Variation of Certain Speculative Prices," *Journal of Business* 36, 394–419.

Markowitz, Harry. (1959). *Portfolio Selection: Efficient Diversification of Investments*. New York: John Wiley & Sons; (1991), 2nd. ed. Basil Blackwell Cambridge, MA.

Markowitz, Harry, Donald Reid, and Bernard Tew. (1994). "The Value of a Blank Check," *Journal of Portfolio Management* 20, 82–91.

Markowitz, Harry, and Nilufer Usmen. (1996). "The Likelihood of Various Stock Market Return Distributions, Part 2: Empirical Results," *Journal of Risk and Uncertainty* 13, 221–247.

Pulley, Lawrence. (1981). "A General Mean-Variance Approximation to Expected Utility for Short Holding Periods," *Journal of Financial and Quantitative Analysis* 16, 361–373.

Pulley, Lawrence. (1983). "Mean-Variance Approximations to Expected Logarithmic Utility," *Operations Research* 31, 685–696.

Ramsey, Frank. (1931). "Truth and Probability" (1926); "Further Considerations" (1928). In *The Foundations of Mathematics and Other Logical Essays*. London: Kegan Paul, and New York: Harcourt, Brace and Co.

Savage, Leonard. (1954). *The Foundations of Statistics*. New York: John Wiley Sons; (1972), 2nd ed., Dover.
Securities Training Corporation. (1995). *National Commodities Futures*, Series 3.
Shafer, Glenn. (1982). "Lindley's Paradox." *Journal of the American Statistical Association* 77, 325–351.
Simaan, Yusif. (1993). "What is the Opportunity Cost of Mean-variance Investment Strategies?" *Management Science* 39, 578–587.
Young, William, and Robert Trent. (1969). "Geometric Mean Approximation of Individual Security and Portfolio Performance." *Journal of Financial Quantitative Analysis* 4, 179–199.
Zellner, Arnold. (1971). *An Introduction to Bayesian Inference in Econometrics*. New York: John Wiley & Sons.

Journal of Risk and Uncertainty, 13:221–247 (1996)


# The Likelihood of Various Stock Market Return Distributions, Part 2: Empirical Results

HARRY M. MARKOWITZ
*Daiwa Securities Trust Company, 1010 Turquoise Street, Suite 245, San Diego, CA 92109*

NILUFER USMEN
*Rutgers University*

*Abstract*

The present article shows how Bayesians should shift beliefs among a family of models concerning the prob ability distribution of daily changes in the Standard & Poor 500 Index, given a particular sample. The preceding article in this issue showed that classical (R.A. Fisher, Neyman–Pearson) inference can be highly misleading for Bayesians, as can the assumption of a diffuse prior. The present article discusses how to bound Bayesian shifts in belief for compound hypotheses generally, as well as the specific shifts in beliefs among simple and compound hypotheses implied by the particular sample.

**Key words:** stock market returns, Bayesian inference, remote clients

**JEL Classification:** D80

The preceding article in this issue, Markowitz and Usmen (1996), presents basic principles of inference. The present article applies these principles to hypotheses concerning the probability distribution of daily changes in the Standard and Poor (S&P) 500 Index of stock prices, given a particular sample.

The preceding article accepts the Savage (1954) thesis that rational action is Bayesian; argues, and cites arguments which contend, that classical (R.A. Fisher and Neyman–Pearson) statistical procedures can be highly misleading for the Bayesian, as may the assumption of a diffuse prior; and argues that Bayesian statisticians should show remote clients (in the sense of Hildreth, 1963), how their beliefs should shift given a particular sample. The problem of how to justify bounds on these shifts in the case of compound hypotheses was postponed until section 2 of the present article. The preceding article also noted practical implications of the fact that a human decision maker (HDM) cannot fully emulate Savage's rational decision maker (RDM). In particular, an RDM updates beliefs for a space of "all" hypotheses simultaneously, whereas a human statistician must necessarily analyze shifts among a limited class of hypotheses and later analyze shifts among additional hypotheses, etc.

Section 1 of the present article presents implied shifts in beliefs among a family of simple hypotheses; section 2 presents shifts among compound hypotheses; section 3 discusses possible next steps beyond the family of hypotheses discussed in sections 1 and 2; and section 4 summarizes.

## 1. Simple hypotheses

In this section, we present the family (F) of hypotheses whose likelihoods will be considered, the sample from which likelihoods will be derived, and the shifts in beliefs which these likelihoods imply for various simple hypotheses.

### *1.1. Hypotheses*

The hypotheses $h \in F$ assume that each $y_t$ is generated independently of the others; and that all $y_t$ with the same

$$\Delta\tau(t) = \tau(t) - \tau(t-1) \quad (1)$$

have the same distribution, where $t$ is the number of trading days and $\tau$ the number of calendar days since some given date. For example, $\Delta\tau(t) = 1$ for a non-holiday Tuesday through Friday; $\Delta\tau(t) = 3$ for a non-holiday Monday, etc. We further assume that $y_t$ is generated by a Pearson family distribution (see Stuart and Ord, 1994, volume I, chapter 6; Elderton and Johnson, 1969). The Pearson family consists of all distributions which satisfy a certain differential equation. Included as special cases are the normal, gamma (therefore chi-square), beta distributions of the first and second kind, Student's $t$ (including Cauchy), uniform, Pareto, and exponential. One without a popular name, known simply as Type IV, turns out to have a distinguished role in the analysis. Student's $t$ (Type VII) is the symmetric special case of Type IV. Our approach provides a level playing field among these various forms of probability distribution. For Pearson distributions whose first four moments are finite, there is a one-to-one corespondence between distributions and combinations of these moments. In particular, the type of the distribution can be determined from the four moments.

### *1.2. The sample*

Our sample $S$ contains daily returns from 7/2/62 to 12/31/83 of the S&P 500 Composite Index (variable DSPR on the CRSP tapes distributed by the Center for Research in Security Prices at the University of Chicago). This data was current at the time that we began our research. Calculations and analysis took a while: our description of principles and results was much delayed in publication (not due to the present journal).[1] We have decided not to further delay publication by recalculating results with current data, since, in addition to being anxious to have our views known, the sample in hand will serve to illustrate the principles stated in the preceding article, and use of this sample provides an ample "holdout period." In particular, October 19, 1987 was subsequent to the data set.

$S$ is partitioned into four subsamples according to $\Delta\tau(t)$,

$$S = S_1 \cup S_2 \cup S_3 \cup S_4.$$

One four-day market close, $\Delta\tau = 5$, was deleted from $S$. Table 1 presents various sample statistics for the subsamples. Note, for example, that the mean value of $y_t$ is positive for weekdays ($S_1$) and negative for weekends ($S_3$).

Some methods of data analysis discard, or weight lightly, extreme values of observations (Tukey, 1960, 1962; Hoaglin, Mosteller, and Tukey, 1983, 1985). One reason for underweighting outliers is that in some applications extreme deviations are not only rare but also of little importance. This was so in the bomber machine-gun-fire examples which Tukey (1960) describes as the origin of research into "contaminated" distributions: "But a very aberrant deviation will never be assigned a large probability or large effectiveness, so that details of the tails of the expression to be averaged will never greatly matter."

But large deviations in finance can have substantial effects. A personal story about the very aberrant deviation in the S&P 500 Index that occurred on October 19, 1987, "Black Monday," illustrates two points relevant to the present and preceding articles. In the early 1970s, one of the authors was portfolio manager of, and subsequently consultant to, Arbitrage Management Company (AMC), which invested in portfolios of related securities such as puts and calls on the same underlying stock. AMC continued in business, and did quite well, until October 19, 1987 at which time it went bankrupt. The problem was that on the preceding Friday the owner/portfolio manager took over a put position at a "very attractive price" with the intention of hedging the position as soon as possible on Monday. In the turmoil of Monday, the hedge was never established, and AMC lost millions on a position that was intended to make tens of thousands.

The first moral of the story is that outliers can have large effects. The second is that there are problems of execution which we have not discussed, of which those of Black Monday are an extreme but not sole example.

*Table 1.* Sample statistics for subsamples $S_1$–$S_4$

| | Weekdays $S_1$ | One-day holidays $S_2$ | Two-day weekends $S_3$ | Three-day weekends $S_4$ |
|---|---|---|---|---|
| mean | 0.000545 | 0.002241 | −0.001285 | −0.001072 |
| stddev | 0.007683 | 0.007504 | 0.008928 | 0.008505 |
| $M_3$ | 0.369890 | 0.369194 | −0.126860 | 0.581082 |
| $M_4$ | 5.638431 | 5.686108 | 5.044298 | 6.645554 |
| min | −0.030024 | −0.020403 | −0.040786 | −0.022851 |
| max | 0.049003 | 0.030113 | 0.041014 | 0.039022 |
| count | 4186 | 91 | 1002 | 118 |

Another reason for ignoring or underweighting extreme observations is that, in many contexts, they often do not represent the phenomenon to be explained, but are the results of a transcription error, or the like, by some anonymous clerk at some unheralded moment. But large changes in the S&P 500 Index are highly public events. We have therefore not removed "outliers" from our sample.

Figure 1 shows the relative frequency (on the vertical axis) of various values of $y_t$ (on the horizontal axis) for the four subsamples. The curves for weekdays ($S_1$) and two-day weekends ($S_3$) are smoother, because they are larger subsamples. However, it is difficult to see outliers in figures 1a and 1b for $S_1$ and $S_3$. These are tabulated in table 2, which shows the frequency in $S_1$, "Weekdays," and $S_3$, "Two-Day Weekends," of various values of $|y_t| \geq .02$. It also notes the value of $z = (y - \text{Mean})/\text{StdDev}$. For example, the $y = -0.04$ observation in $S_3$ is 4.34 subsample standard deviations below the mean of that subsample; there are seven more observations in $S_3$ which are at least three subsample standard deviations below the subsample mean, and five observations in $S_3$ which are three or more standard deviations above the mean, out of a total of 1,002 observations. $S_1$ has 12 observations, $3\sigma$ below and 14 $4\sigma$ above the subsample mean, including seven $5\sigma$ above. Clearly, normal hypotheses are very poor explanations of these observations. We will see if the Pearson family offers better ones.

### 1.3. Computation of likelihood

The likelihood calculations are performed with observations grouped into intervals of width $\Delta y = .001$.[2] Writing $k$ for $\Delta\tau(t)$, the likelihood $LH_k$ of $S_k$ for given, finite $\mu^k$, $\sigma^k$, $M_3^k$, $M_4^k$ is approximately

$$LH(S_k) \cong \prod_{t \in S_k} f(y_t; \mu^k, \sigma^k, M_3^k, M_4^k)\Delta y$$

$$= \prod_{t \in S_k} f(y_t; \theta^k)\Delta y \qquad (2a)$$

where $y_t$ is the midpoint of the interval of the $t$-th observation, $f$ is the appropriate Pearson density, and $\mu^k$, $\sigma^k$, $M_3^k$, $M_4^k$ are the mean, standard deviation, normalized third, and normalized fourth moments, as defined in the preceding article. The likelihood of $S$ is

$$LH(S) = LH(S_1) \cdot LH(S_2) \cdot LH(S_3) \cdot L(S_4). \qquad (2b)$$

It facilitates interpretation to work with the common logarithm of the likelihood function, $LLH_k = \log_{10} LH_k$. From (2a),

$$LLH_k \cong \sum_{t=1}^{n_k} \log_{10} f(y_t; \theta^k) + n_k \log_{10}\Delta y \qquad (3)$$

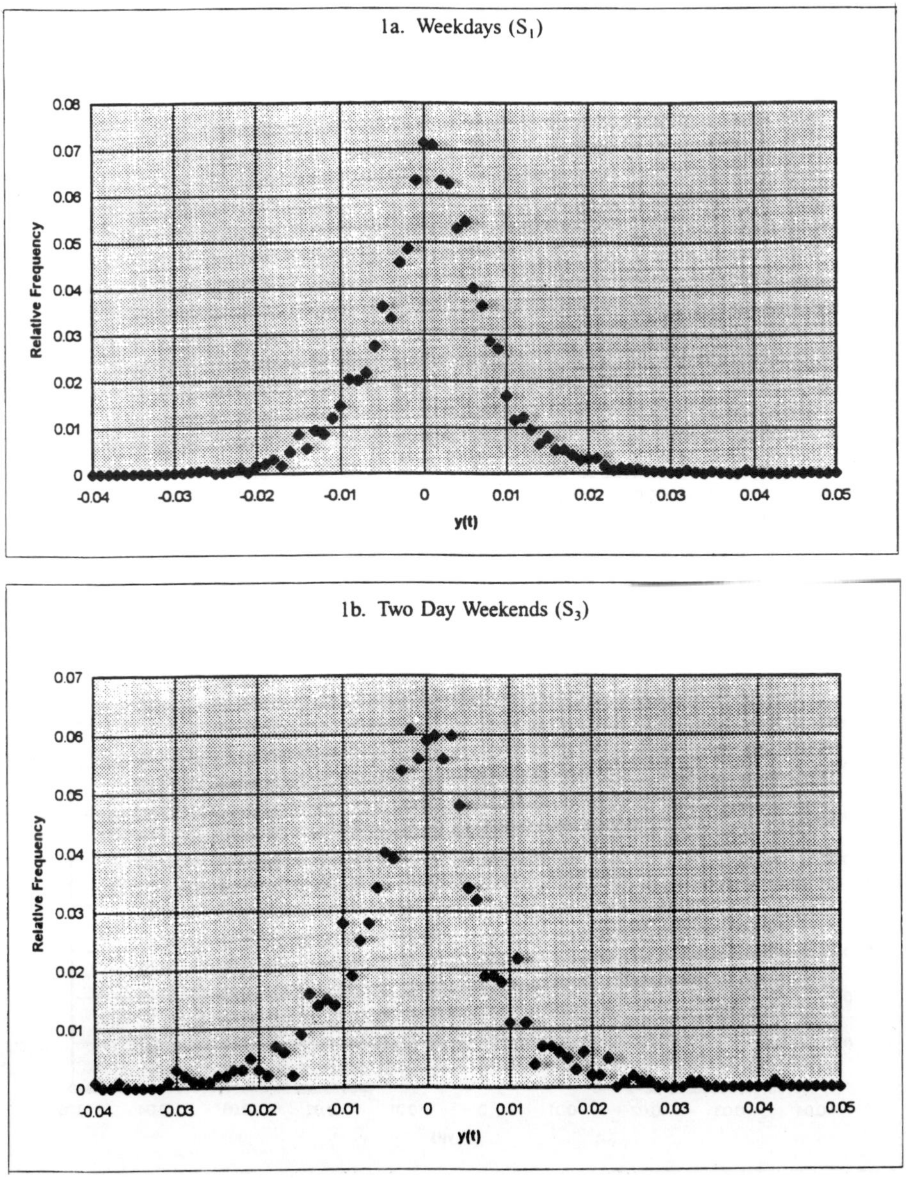


*Figure 1.* Relative frequency of $y(t)$ for subsamples $S_1$–$S_4$.

where $n_k$ is the number of observations in $S(k)$. The log likelihood for the entire sample is $LLH = \sum_k LLH_k$. Since we will be concerned only with ratios of $LH$ or differences of $LLH$, we may add an arbitrary constant $c_k$ to $LLH_k$ to make it of a convenient magnitude.

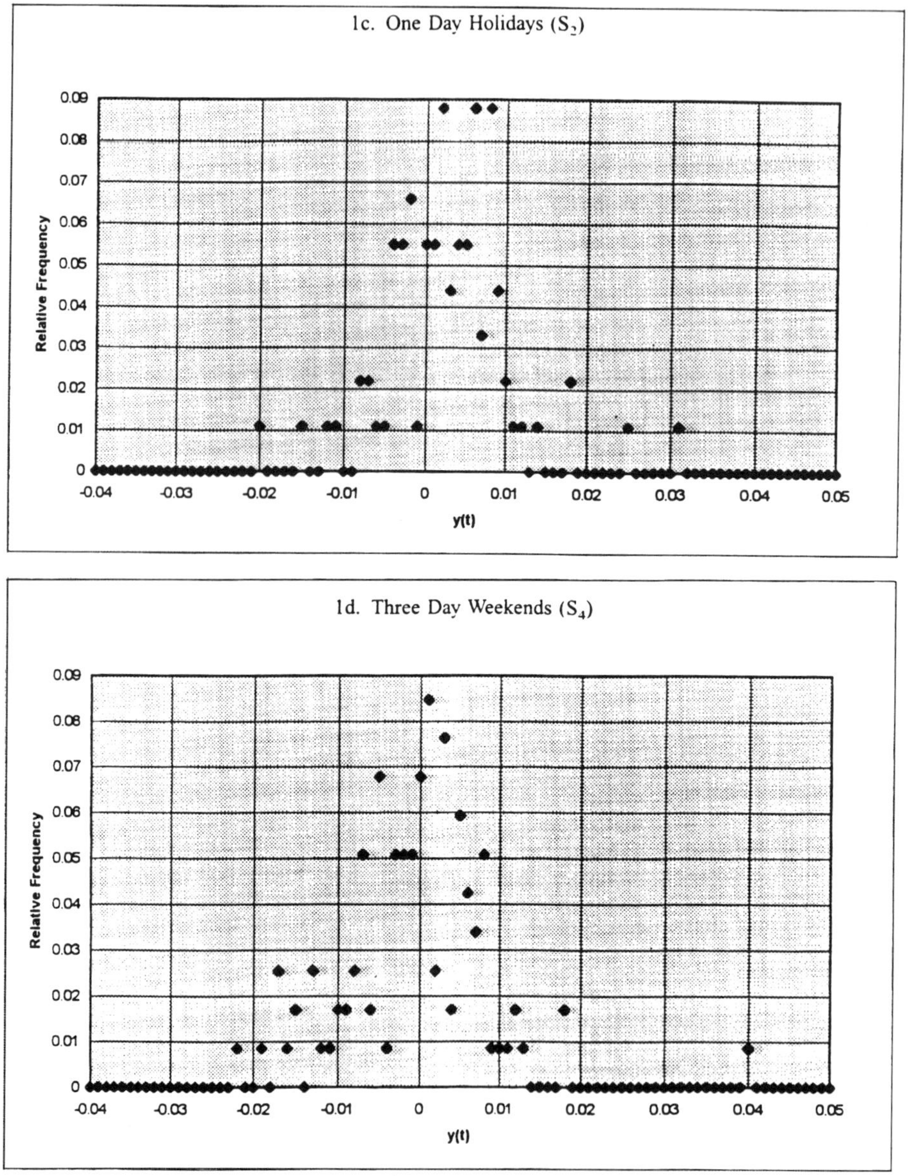


*Figure 1.* (continued)

## *1.4. Shifts in beliefs*

We assume that the reader is an HDM (human decision maker) who wishes to emulate an *RDM* (rational decision maker), who attaches a nonzero prior to the set $F^* \subset F$ of Pearson family i.i.d. distributions with *finite* first four moments for each subsample,

*Table 2.* Large deviations in subsamples $S_1$ and $S_3$

| Negative deviations | | | | | Positive deviations | | | | |
|---|---|---|---|---|---|---|---|---|---|
| | Weekdays ($S_1$) | | Two-day weekends ($S_3$) | | | Weekdays ($S_1$) | | Two-day weekends ($S_3$) | |
| $y(t)$ | Sigma | Count | Sigma | Count | $y(t)$ | Sigma | Count | Sigma | Count |
| −0.040 | −5.278 | 0 | −4.339 | 1 | 0.020 | 2.532 | 13 | 2.385 | 2 |
| −0.039 | −5.148 | 0 | −4.227 | 0 | 0.021 | 2.663 | 14 | 2.497 | 2 |
| −0.038 | −5.017 | 0 | −4.115 | 0 | 0.022 | 2.793 | 7 | 2.61 | 5 |
| −0.037 | −4.887 | 0 | −4.003 | 1 | 0.023 | 2.923 | 3 | 2.722 | 0 |
| −0.036 | −4.757 | 0 | −3.890 | 0 | 0.024 | 3.053 | 5 | 2.834 | 1 |
| −0.035 | −4.627 | 0 | −3.778 | 0 | 0.025 | 3.183 | 5 | 2.946 | 2 |
| −0.034 | −4.497 | 0 | −3.666 | 0 | 0.026 | 3.313 | 4 | 3.058 | 1 |
| −0.033 | −4.367 | 0 | −3.554 | 0 | 0.027 | 3.444 | 2 | 3.17 | 1 |
| −0.032 | −4.236 | 0 | −3.442 | 0 | 0.028 | 3.574 | 2 | 3.282 | 0 |
| −0.031 | −4.106 | 0 | −3.330 | 1 | 0.029 | 3.704 | 2 | 3.394 | 0 |
| −0.030 | −3.976 | 1 | −3.218 | 3 | 0.030 | 3.834 | 1 | 3.506 | 0 |
| −0.029 | −3.846 | 1 | −3.106 | 2 | 0.031 | 3.964 | 1 | 3.618 | 0 |
| −0.028 | −3.716 | 2 | −2.994 | 1 | 0.032 | 4.094 | 3 | 3.73 | 1 |
| −0.027 | −3.586 | 2 | −2.882 | 1 | 0.033 | 4.225 | 1 | 3.842 | 1 |
| −0.026 | −3.455 | 3 | −2.770 | 1 | 0.034 | 4.355 | 0 | 3.954 | 0 |
| −0.025 | −3.325 | 0 | −2.658 | 2 | 0.035 | 4.485 | 2 | 4.066 | 0 |
| −0.024 | −3.195 | 1 | −2.546 | 2 | 0.036 | 4.615 | 1 | 4.178 | 0 |
| −0.023 | −3.065 | 2 | −2.434 | 3 | 0.037 | 4.745 | 0 | 4.291 | 0 |
| −0.022 | −2.935 | 5 | −2.321 | 3 | 0.038 | 4.875 | 0 | 4.403 | 0 |
| −0.021 | −2.804 | 1 | −2.209 | 5 | 0.039 | 5.006 | 3 | 4.515 | 0 |
| −0.020 | −2.674 | 7 | −2.097 | 3 | 0.040 | 5.136 | 1 | 4.627 | 0 |
| | | | | | 0.041 | 5.266 | 0 | 4.739 | 0 |
| | | | | | 0.042 | 5.396 | 0 | 4.851 | 1 |
| | | | | | 0.043 | 5.526 | 0 | 4.963 | 0 |
| | | | | | 0.044 | 5.656 | 0 | 5.075 | 0 |
| | | | | | 0.045 | 5.787 | 1 | 5.187 | 0 |
| | | | | | 0.046 | 5.917 | 0 | 5.299 | 0 |
| | | | | | 0.047 | 6.047 | 1 | 5.411 | 0 |
| | | | | | 0.048 | 6.177 | 0 | 5.523 | 0 |
| | | | | | 0.049 | 6.307 | 0 | 5.635 | 0 |
| | | | | | 0.050 | 6.437 | 1 | 5.747 | 0 |

$(\mu^1, \sigma^1, \ldots, M_3^4, M_4^4) \in R^{16}$. In the first instance, we consider shifts in beliefs among distributions in $F^*$. Unless otherwise stated, we assume that, within $F^*$, prior beliefs can be described by a continuous density function $p(\theta)$ of $\theta = (\mu^1, \ldots, M_4^4)$. In particular, zero prior is attached to any specific $(\mu^1, \ldots, M_4^4) = (\mu^{1*}, \ldots, M_4^{4*})$. By a "simple hypothesis" we shall mean a small region enclosing $(\mu^{1*}, \ldots, M_4^{4*})$. Similarly, in section 2, we speak of "approximately normal," "$M_4 \cong 12$" and the like to refer to appropriate regions of nonzero Lesbegue and $P$-measure for compound hypotheses.

An RDM would shift beliefs simultaneously for all 16 parameters $(\mu^1, \ldots, M_4^4)$ on the basis of $S$. We find that for a human understanding of the shifts implied by $S$, it is preferable to first consider shifts in $(\mu^1, \sigma^1, M_3^1, M_4^1)$ as if the large subsample $S_1$ were the only information available, then consider shifts in beliefs about ($\mu^1, \sigma^1$, $M_3^1$, $M_4^1$, $\mu^3$, $\sigma^3$, $M_3^3$, $M_4^3$, implied by $S_1 \cup S_3$, and then add the relatively small subsamples $S_2$ and $S_4$.

The row labeled $M_4 = 3$ in table 3a shows the values of mean and standard deviation which maximize likelihood for the subsample $S_1$ among normal distributions. Rows labeled sample (i.e., subsample $M_4^1 = 5.64$), 12, 18, 1,000, 100,000, show likelihood maximizing values of mean, standard deviation, and $M_3$ for the specified values of $M_4$. $LLH + c_1$, Maximum $LLH$ (among those shown) minus the given $LLH$, and the Bayes factor, namely, Max $LH/LH$, are also shown. For example, an *RDM* would shift belief by a factor of $10^{74}$ against the normal distribution in favor of the Pearson-family distribution with $M_4 = 12$ and the parameters shown; would shift towards the latter by a factor of 1500 against the likelihood maximizing combination with the subsample $M_4$; and by less than a factor of four against $M_4 = 100{,}000$.

Recall that the first four moments determine the type of a Pearson-family distribution. Except for the normal distribution in the first line of results, all moment combinations in table 3a are of Pearson Type IV with frequency function

$$\log(f) = \log k + (\beta/2)\log((y+\gamma)^2 + \delta^2) - (\beta\gamma/\delta)\arctan((y+\gamma)/\delta), \tag{4}$$

where $\beta$, $\gamma$, $\delta$ can be computed from the moments, and $k$ determined so that $\int f(y)dy = 1$. More generally, all non-normal likelihood maximizing combinations for all subsamples in tables 3a–3d and all entries in tables 4–6 discussed below are of Type IV. Figure 2 shows Pearson types as a function of $M_3$ and $M_4$. The asterisk represents the likelihood-maximizing combination for $S_1$. The numbers represent the numbers of orders of magnitudes which an *RDM* would shift against the combination, in favor of the likelihood-maximizing combination, with mean and standard deviation set at their respective *LH*-maximizing values. Clearly, Type IV is the "winner" among Pearson-family distributions.

Table 4a illustrates how $LH(S_1)$ varies with $\mu^1$ with the values of $\sigma^1$, $M_3^1$ and $M_4^1$ held at their likelihood-maximizing values of table 3a. In contrast to table 3a, in which likelihood is rather insensitive to $M_4$ for $M_4 \geq 12$, table 4a shows that likelihood declines at an increasing rate as $\mu^1$ moves from its likelihood-maximizing value. Similarly, tables 5a and 6a illustrate how $LH(S_1)$ varies with $\sigma^1$ and $M_3^1$.[3]

$LH(S_1)$ as a function of ($\mu^1, \sigma^1, M_3^1, M_4^1$) is roughly a five-dimensional version of the Blue Ridge Mountains of the southeastern United States. One drives many miles along the highway at the top of the ridge in roughly a north-south direction with little change in altitude; whereas looking from the ridge in an east-west direction, one sees the mountain fall away rapidly. If one imagines driving the ridge in ($\mu^1, \sigma^1, M_3^1, M_4^1$) $\times$ $LH(S_1)$ space in the direction of increasing $M_4$, the ridge rises rapidly at $M_4^1 = 3$ and is still rising at the sample $M_4^1$. By $M_4 = 12$, it has leveled off, sloping gently upward and then downward. Looking in the direction of $\mu^1$, $\sigma^1$ or $M_3^1$ one sees $LH(S_1)$ falling at an increasing rate as far as the eye can see (i.e., as far as our computations extend).

*Table 3.* Likelihood-maximizing parameter combinations for various values of $M_4$

| | 3a. Weekdays ($S_1$) | | | | | | 3b. Two-day weekends ($S_3$) | | | | | |
|---|---|---|---|---|---|---|---|---|---|---|---|---|
| $M_4$ | Mean × 1,000 | Std dev × 1,000 | $M_3$ | $LLH$ | Max$LLH$ − $LLH$ | Max$LH$/$LH$ | Mean × 1,000 | Std dev × 1,000 | $M_3$ | $LLH$ | Max$LLH$ − $LLH$ | Max$LH$/$LH$ |
| 3 | 0.545 | 7.68 | *0.000 | −29.839 | 74.254 | $>10^{74}$ | −1.29 | 8.93 | *0.000 | 35.713 | 18.111 | $>10^{18}$ |
| Sample | 0.515 | 7.51 | 0.068 | 41.227 | 3.188 | 1,542.00 | −1.29 | 8.93 | −0.127 | 51.450 | 2.374 | 236.61 |
| 12 | 0.500 | 7.79 | 0.038 | 44.415 | 0.000 | 1.00 | −1.29 | 8.93 | −0.254 | 53.649 | 0.175 | 1.50 |
| 18 | 0.500 | 7.91 | 0.034 | 44.413 | 0.002 | 1.01 | −1.29 | 8.93 | −0.381 | 53.701 | 0.123 | 1.33 |
| 1,000 | 0.510 | 8.06 | 0.022 | 43.859 | 0.556 | 3.60 | −1.29 | 9.47 | −0.507 | 53.822 | 0.002 | 1.01 |
| 100,000 | 0.500 | 8.06 | 0.022 | 43.844 | 0.571 | 3.72 | −1.29 | 9.47 | −0.507 | 53.824 | 0.000 | 1.00 |
| | 3c. One-day holidays ($S_2$) | | | | | | 3d. Three-day weekends ($S_4$) | | | | | |
| 3 | 2.24 | 7.50 | *0.000 | 37.250 | 2.068 | 116.95 | −1.07 | 8.51 | *0.000 | 71.700 | 2.230 | 169.82 |
| Sample | 2.24 | 7.50 | 0.123 | 39.117 | 0.201 | 1.59 | −1.07 | 8.51 | −0.291 | 73.913 | 0.017 | 1.04 |
| 12 | 2.24 | 7.50 | 0.049 | 39.306 | 0.012 | 1.03 | −1.07 | 8.51 | −0.581 | 73.930 | 0.000 | 1.00 |
| 18 | 2.24 | 7.50 | 0.049 | 39.318 | 0.000 | 1.00 | −1.07 | 8.51 | −0.581 | 73.901 | 0.029 | 1.07 |
| 1000 | 2.24 | 7.50 | −0.025 | 39.301 | 0.017 | 1.04 | −1.07 | 8.51 | −0.872 | 73.781 | 0.149 | 1.41 |
| 10,000 | 2.24 | 7.50 | −0.025 | 39.300 | 0.018 | 1.04 | −1.07 | 8.51 | −0.872 | 73.778 | 0.152 | 1.42 |

*This line represents the normal (Gaussian) case. Thus, $M_3 = 0$ was the only value tested for this line.

*Table 4*. Likelihood as a function of mean with values of standard deviation, $M_3$, and $M_4$ set at their likelihood-maximizing value in table 3 for each subsample

| 4a. Weekdays ($S_1$) | | | | 4b. Two-day weekends ($S_3$) | | | |
|---|---|---|---|---|---|---|---|
| Mean × 1,000 | *LLH* | Max*LLH* − *LLH* | Max*LH*/ *LH* | Mean × 1,000 | *LLH* | Max*LLH* − *LLH* | Max*LH*/ *LH* |
| −0.182 | 35.371 | 9.044 | $1.1 \times 10^9$ | −3.01 | 43.793 | 10.031 | $1.1 \times 10^{10}$ |
| 0.000 | 39.542 | 4.873 | 74,645.00 | −2.58 | 48.262 | 5.562 | $3.6 \times 10^5$ |
| 0.182 | 42.435 | 1.983 | 96.16 | −2.15 | 51.427 | 2.397 | 249.46 |
| 0.363 | 44.045 | 0.370 | 2.34 | −1.72 | 53.283 | 0.541 | 3.48 |
| @0.500 | 44.415 | 0.000 | 1.00 | *@−1.29 | 53.824 | 0.000 | 1.00 |
| *0.545 | 44.375 | 0.040 | 1.10 | −0.86 | 53.057 | 0.767 | 5.85 |
| 0.727 | 43.420 | 0.995 | 10.00 | −0.43 | 50.983 | 2.841 | 693.43 |
| 0.908 | 41.177 | 3.238 | 1,730.00 | 0.00 | 47.616 | 6.208 | $1.6 \times 10^6$ |
| 1.090 | 37.656 | 6.759 | $5.7 \times 10^6$ | 0.43 | 42.965 | 10.859 | $7.2 \times 10^{10}$ |
| 1.272 | 32.850 | 11.565 | $3.7 \times 10^{11}$ | | | | |
| **4c. One-day holidays ($S_2$)** | | | | **4d. Three-day weekends ($S_4$)** | | | |
| −2.24 | 30.522 | 8.796 | $6.3 \times 10^8$ | −6.42 | 61.691 | 12.239 | $1.7 \times 10^{12}$ |
| −1.12 | 34.341 | 4.977 | 94,841.85 | −5.35 | 66.136 | 7.794 | $6.2 \times 10^7$ |
| 0.00 | 37.121 | 2.197 | 157.40 | −4.28 | 69.619 | 4.311 | 20,464.45 |
| 1.12 | 38.794 | 0.524 | 3.34 | −3.21 | 72.095 | 1.835 | 68.39 |
| *@2.24 | 39.318 | 0.000 | 1.00 | −2.14 | 73.537 | 0.393 | 2.47 |
| 3.36 | 38.671 | 0.647 | 4.44 | *@−1.07 | 73.930 | 0.000 | 1.00 |
| 4.48 | 36.860 | 2.458 | 287.08 | 0.00 | 73.280 | 0.650 | 4.47 |
| 5.60 | 33.916 | 5.402 | $2.5 \times 10^5$ | 1.07 | 71.605 | 2.325 | 211.35 |
| 6.72 | 29.899 | 9.419 | $2.6 \times 10^9$ | 2.14 | 68.940 | 4.990 | 97,723.72 |
| | | | | 3.21 | 65.336 | 8.594 | $3.9 \times 10^8$ |
| | | | | 4.28 | 60.853 | 13.077 | $1.2 \times 10^{13}$ |

@Maximum likelihood.
*Sample moment.

In table 3a we saw that, for various $M_4^1 \in (3,100000]$, $LH(S_1)$ is maximized by some $M_3 > 0$; but, in table 6a (and in other cases not shown here), we see that the $LH(S_1)$-maximizing combinations with $M_3^1 = 0$ gives almost, though not quite, maximum *LH*. The column labeled *S*1 in table 7 shows $LH(S_1)$ for various $M_4^1$, for $M_3^1 = 0$ (i.e., Student's *t*), and for the *LH*-maximizing values of μ and σ given these $M_3^1$ and $M_4^1$.[4] On the basis of tables 3a and 6a, we assume that the *LH* in table 7 are close to the maximum $LH(S_1)$ for the various $M_4$. For the Student's *t* distribution, the relationship between $M_4$ and degrees of freedom (d.f.) is

$$\text{df} = 4 + 6/(M_4 - 3) \tag{5}$$

provided d.f. > 4.0. Table 7 fills in table 3 as to the shape of maximum *LH* as a function of $M_4$, and extends the results to distributions with infinite $M_4$. In particular, we see that Max *LH* exceeds *LH* by less than two orders of magnitude from somewhere in the interval

*Table 5.* Likelihood as a function of standard deviation with values of mean, $M_3$, and $M_4$ set at their likelihood-maximizing value in table 3 for each subsample

| 5a. Weekdays ($S_1$) | | | | 5b. Two-day weekends ($S_3$) | | | |
|---|---|---|---|---|---|---|---|
| Std dev × 1.000 | *LLH* | Max *LLH* − *LLH* | Max *LH*/ *LH* | Std dev × 1,000 | *LLH* | Max *LLH* − *LLH* | Max *LH*/ *LH* |
| 6.96 | 30.308 | 14.107 | $1.3 \times 10^{14}$ | 7.73 | 44.692 | 9.132 | $1.4 \times 10^{9}$ |
| 7.21 | 37.757 | 6.658 | $4.6 \times 10^{6}$ | 8.35 | 50.703 | 3.121 | 1,321.00 |
| 7.45 | 42.194 | 2.221 | 166.00 | *8.93 | 53.363 | 0.461 | 2.89 |
| *7.68 | 44.189 | 0.226 | 1.68 | @9.47 | 53.824 | 0.000 | 1.00 |
| @7.79 | 44.415 | 0.000 | 1.00 | 9.98 | 52.794 | 1.030 | 10.72 |
| 7.91 | 44.185 | 0.230 | 1.70 | 10.47 | 50.730 | 3.094 | 1,242.00 |
| 8.14 | 42.530 | 1.885 | 76.70 | 10.94 | 47.939 | 5.885 | $7.7 \times 10^{5}$ |
| 8.33 | 39.514 | 4.901 | 79,616.00 | 11.39 | 44.632 | 9.192 | $1.6 \times 10^{9}$ |
| 8.54 | 35.362 | 9.053 | $1.1 \times 10^{9}$ | | | | |
| **5c. One-day holidays ($S_2$)** | | | | **5d. Three-day weekends ($S_4$)** | | | |
| 4.04 | 28.618 | 10.700 | $5.0 \times 10^{10}$ | 4.73 | 60.606 | 13.324 | $2.1 \times 10^{13}$ |
| 5.13 | 35.416 | 3.902 | 7,979.95 | 5.69 | 67.793 | 6.137 | $1.4 \times 10^{6}$ |
| 6.02 | 38.030 | 1.288 | 19.41 | 6.51 | 71.220 | 2.710 | 512.86 |
| *@7.50 | 39.318 | 0.000 | 1.00 | *@8.51 | 73.930 | 0.000 | 1.00 |
| 8.73 | 38.852 | 0.466 | 2.92 | 10.12 | 73.131 | 0.799 | 6.30 |
| 9.81 | 37.837 | 1.481 | 30.27 | 11.51 | 71.438 | 2.492 | 310.46 |
| 10.78 | 36.640 | 2.678 | 476.43 | 12.74 | 69.509 | 4.421 | 26,363.31 |
| 11.67 | 35.396 | 3.922 | 8,356.03 | 13.87 | 67.558 | 6.372 | $2.3 \times 10^{6}$ |
| 12.50 | 34.163 | 5.155 | $1.4 \times 10^{5}$ | 14.91 | 65.663 | 8.267 | $1.9 \times 10^{8}$ |
| 13.27 | 32.965 | 6.353 | $2.3 \times 10^{6}$ | | | | |
| 14.01 | 31.812 | 7.506 | $3.2 \times 10^{7}$ | | | | |

@ Maximum likelihood
*Sample moment

$5.5 < \text{d.f.} < 6.0$, i.e., $6 < M_4 < 7$, to somewhere in $3.50 < \text{d.f.} < 4.00$, in which interval $M_4 = \infty$.

Table 3b shows likelihood-maximizing parameter settings and the associated likelihoods among normal distributions, Pearson-family distributions with sample $M_4^3 = 5.04$ and among distributions with $M_4^3 = 12$, 18, 1,000 and 100,000 for subsample $S_3$. Perhaps, in part because of the smaller sample size, the shift against the maximizing normal distribution is only 18 orders of magnitude, as compared to 74 orders in $S_1$. Still, a factor of 1,000,000,000,000,000,000 is a substantial shift in belief against the normal distribution. In the case of $S_3$, the Bayes factor calls for a slight shift away from the likelihood-maximizing distribution with $M_4^3 = 12$, in favor of that with $M_4^3 = 100{,}000$, as compared to the shift from 100,000 towards 12 in $S_1$. Table 7 shows that, on the path with $M_3 = 0$, $LH(S_3)$ is largest at $M_4^3 = 1{,}000$ (d.f. $= 4.006$) among the $M_4$ and d.f. values examined.

Independence of the random draws $y_1, y_2 \ldots$ does not imply independence of prior or posterior beliefs about $(\mu^1, \sigma^1, M_3^1, M_4^1)$ versus $(\mu^3, \sigma^3, M_3^3, M_4^3)$. In particular, different RDMs may attach different prior beliefs to statements such as $y_t \in S_3$ is distributed like the sum of one, two, or three draws from $y_t \in S_1$. Tables 1 and 3 show that both the sample

*Table 6.* Likelihood as a function of $M_3$ with values of mean, standard deviation, and $M_4$ set at their likelihood-maximizing value in table 3 for each subsample

| 6a. Weekdays ($S_1$) | | | | 6b. Two-day weekends ($S_3$) | | | |
|---|---|---|---|---|---|---|---|
| $M_3$ | *LLH* | Max*LLH* − *LLH* | Max*LH*/*LH* | $M_3$ | *LLH* | Max*LLH* − *LLH* | Max*LH*/*LH* |
| −0.740 | 32.902 | 11.513 | $3.3 \times 10^{11}$ | −2.410 | 45.159 | 8.665 | $4.6 \times 10^{8}$ |
| −0.555 | 38.227 | 6.188 | $1.5 \times 10^{6}$ | −2.029 | 48.621 | 5.203 | $1.6 \times 10^{5}$ |
| −0.370 | 41.639 | 2.776 | 597.00 | −1.522 | 51.651 | 2.173 | 149.00 |
| −0.185 | 43.612 | 0.803 | 6.35 | −1.015 | 53.298 | 0.526 | 3.36 |
| 0.000 | 44.391 | 0.024 | 1.06 | @−0.507 | 53.824 | 0.000 | 1.00 |
| @0.038 | 44.415 | 0.000 | 1.00 | *−0.127 | 53.564 | 0.260 | 1.82 |
| 0.185 | 44.066 | 0.349 | 2.23 | 0.000 | 53.358 | 0.466 | 2.92 |
| *0.370 | 42.605 | 1.810 | 65.00 | 0.508 | 51.926 | 1.898 | 79.00 |
| 0.555 | 39.841 | 4.574 | 37,497.00 | 1.015 | 49.47 | 4.354 | 22,594.00 |
| 0.740 | 35.405 | 9.010 | $1.0 \times 10^{9}$ | 1.523 | 45.819 | 8.005 | $1.0 \times 10^{8}$ |
| | | | | 1.776 | 43.455 | 10.369 | $2.3 \times 10^{10}$ |
| **6c. One-day holidays ($S_2$)** | | | | **6d. Three-day weekends ($S_4$)** | | | |
| −2.215 | 30.812 | 8.506 | $3.2 \times 10^{8}$ | −1.918 | 58.822 | 15.108 | $1.3 \times 10^{15}$ |
| −1.846 | 36.865 | 2.453 | 283.79 | −1.743 | 69.835 | 4.095 | 12,445.15 |
| −1.477 | 38.286 | 1.032 | 10.76 | −1.453 | 72.795 | 1.135 | 13.64 |
| −1.107 | 38.865 | 0.453 | 2.84 | −1.162 | 73.579 | 0.351 | 2.24 |
| −0.738 | 39.144 | 0.174 | 1.49 | −0.872 | 73.853 | 0.077 | 1.19 |
| −0.369 | 39.276 | 0.042 | 1.10 | @−0.581 | 73.930 | 0.000 | 1.00 |
| @0.000 | 39.318 | 0.000 | 1.00 | −0.291 | 73.906 | 0.024 | 1.06 |
| @0.049 | 39.318 | 0.000 | 1.00 | 0.000 | 73.811 | 0.119 | 1.32 |
| *0.369 | 39.285 | 0.033 | 1.08 | 0.291 | 73.653 | 0.277 | 1.89 |
| 0.738 | 39.170 | 0.148 | 1.41 | *0.581 | 73.413 | 0.517 | 3.29 |
| 1.107 | 38.927 | 0.391 | 2.46 | 0.872 | 73.048 | 0.882 | 7.62 |
| 1.477 | 38.432 | 0.886 | 7.69 | 1.162 | 72.443 | 1.487 | 30.69 |
| 1.846 | 37.240 | 2.078 | 119.67 | 1.453 | 71.258 | 2.672 | 469.89 |
| 2.215 | 32.253 | 7.065 | $1.1 \times 10^{7}$ | 1.743 | 67.834 | 6.096 | $1.2 \times 10^{6}$ |

@Maximum likelihood.
*Sample moment.

and *LH*-maximizing values of mean μ were positive for subsample $S_1$ and negative for $S_3$. Table 4 and similar computations with other values of σ, $M_3$, and $M_4$ imply large shifts against hypotheses with $\mu^1 = \mu^3$. In particular, $LLH\{(\mu^1, \sigma^1, M_3^1, M_4^1) = (5.0, 7.79, 0.038, 12)$ *and* $(\mu^3, \sigma^3, M_3^3, M_4^3) = (-1.29, 9.47, -.507, 100000)\}$

$$= 44.415 + C_1 + 53.824 + C_2$$

$$= 98.239 + (C_1 + C_2);$$

whereas $LLH\{(\mu^1, \sigma^1, M_3^1, M_4^1) = (0.0, 7.79, 0.038, 12)$ *and* $(\mu^3, \sigma^3, M_3^3, M_4^3) = (0.0, 9.47, -.507, 100000)\}$

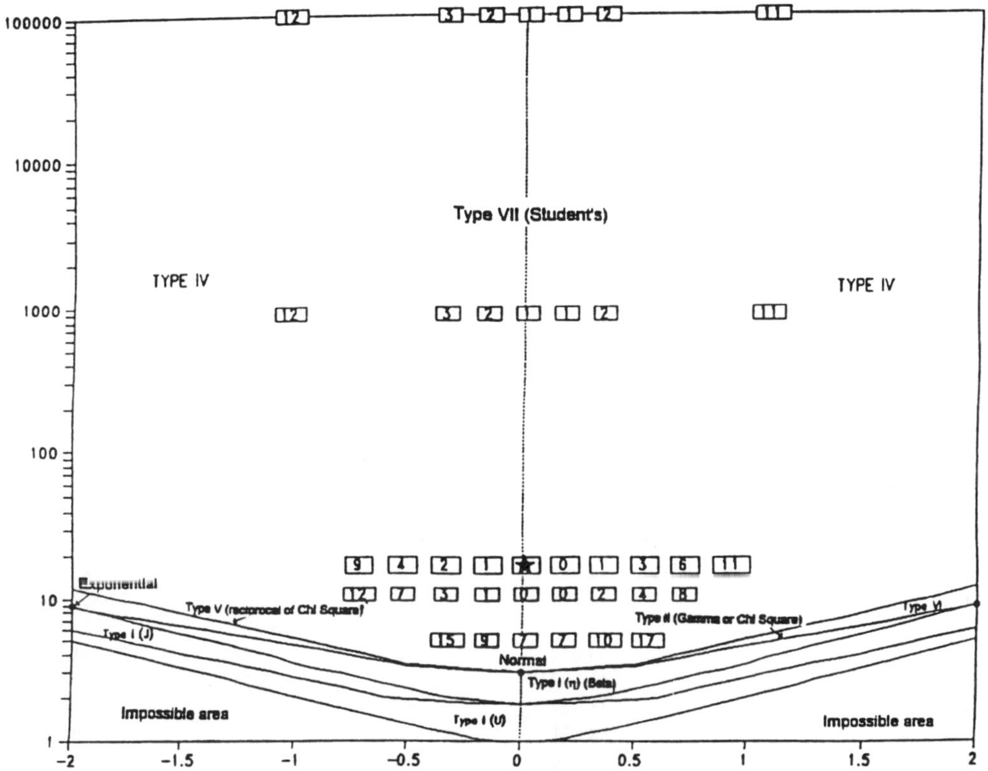


*Figure 2.* Areas and bounding curves associated with Pearson type distributions; numbers indicate Max *LLH*-*LLH* for Sample $S_1$ with $\mu$ and $\sigma$ set at their likelihood-maximizing values. ★Distribution with maximum LH (S.h) among the cases computed. Adopted from Johnson, Nixon, and Amos (1963).

$$= 39.542 + C_1 + 47.616 + C_2$$
$$= 87.158 + (C_1 + C_2),$$

implying an 11 order-of-magnitude shift towards the former joint hypothesis against the latter. Table 6 shows that $S_1 \cup S_3$ shifts belief somewhat towards hypotheses with $M_3^1 > 0$ and $M_3^3 < 0$, and table 3 shows some shift towards $M_4^3 > M_4^1$. However, these are not as massive as those towards $\mu^1 > 0 > \mu^3$.

Parts c and d of tables 3–7 present likelihood information for $S_2$ and $S_4$. Because of the smaller sample sizes, *LH* as a function of $(\mu, \sigma, M_3, M_4)$ tends to be flatter than for $S_1$ and $S_3$. For example, the shift in belief indicated in table 3c from the normal to the Type IV distribution with $M_4 = 18$ is a mere 117-fold. On the other hand, if a remote client has a strong prior to the effect that if $y_t \in S_1$ is not Gaussian, then $y_t \in S_2$ is probably not Gaussian, then $S_1 \cup S_2$ would shift posterior beliefs strongly toward hypotheses with $y_t \in S_2$ not Gaussian.

*Table 7.* LLH of Student's—*t* (Type VII) distributions with various degrees of freedom and with means and standard deviations set at their *LLH*-maximizing values

| d.f. | $M_4$ | Weekdays $S_1$ | Two-day weekends $S_3$ | One-day holidays $S_2$ | Three-day weekends $S_4$ |
|---|---|---|---|---|---|
| 1.00 | Infinite | −142.515 | 12.499 | 34.546 | 67.404 |
| 1.50 | Infinite | −38.205 | 36.255 | 37.089 | 70.812 |
| 2.00 | Infinite | 5.382 | 45.867 | 38.378 | 72.050 |
| 2.50 | Infinite | 26.169 | 50.186 | 38.925 | 72.973 |
| 3.00 | Infinite | 36.518 | 52.205 | 39.174 | 73.375 |
| 3.50 | Infinite | 41.584 | 53:118 | 39.196 | 73.607 |
| 4.00 | Infinite | 43.669 | 53.405 | 39.300 | 73.629 |
| 4.00006 | 100,000 | 43.811 | 53.405 | 39.300 | 73.629 |
| 4.006 | 1,000 | 43.822 | 53.649 | 39.301 | 73.632 |
| 4.01 | 603 | 43.833 | 53.406 | 39.301 | 73.634 |
| 4.10 | 63 | 44.015 | 53.401 | 39.310 | 73.673 |
| 4.20 | 33 | 44.200 | 53.366 | 39.316 | 73.709 |
| 4.30 | 23 | 44.309 | 53.308 | 39.319 | 73.739 |
| 4.40 | 18 | 44.379 | 53.255 | 39.318 | 73.764 |
| 4.50 | 15 | 44.402 | 53.266 | 39.315 | 73.785 |
| 4.60 | 13 | 44.398 | 53.257 | 39.309 | 73.802 |
| 4.75 | 11 | 44.314 | 53.213 | 39.298 | 73.821 |
| 5.00 | 9 | 44.027 | 53.076 | 39.274 | 73.841 |
| 5.50 | 7 | 43.108 | 52.660 | 39.213 | 73.850 |
| 6.00 | 6 | 41.827 | 52.156 | 39.144 | 73.834 |
| 7.00 | 5 | 38.831 | 51.290 | 39.007 | 73.770 |
| 10.00 | 4 | 29.576 | 48.833 | 38.672 | 73.527 |
| Infinite | 3 | −29.839 | 35.713 | 37.250 | 71.700 |

The *LH*-maximizing values of ($\mu$, $\sigma$, $M_3$) for given $M_4$ reported in table 3 for $S_2$ and $S_4$ seem reasonable, roughly, as compared to the values for $S_1$ and $S_3$. For example, since $y_t \in S_3$ (Two-Day Weekends) have *LH*-maximizing $\mu^3 < 0$ and $M_3^3 < 0$, it is not surprising that $y_t \in S_4$ (Three-Day Weekends) do so as well. If it seems implausible that the magnitude of $\mu$ is smaller for $S_4$ than $S_3$, note that table 4d shows that the meager evidence of $S_4$ does not shift much probability away from hypotheses in which $|\mu^4|$ equals or slightly exceeds $|\mu^3|$.

Edwards, Lindman, and Savage (1963) define the "principle of stable estimation" in terms of conditions under which the posterior distribution is approximately proportional to the likelihood function: that is, to a reasonable approximation, the prior distribution may be assumed uniform over some finite range beyond which *LH* is so small that $\int LH(\theta)\, p(\theta)\, d\theta$ may be assumed negligible. Briefly, "To ignore the departures from uniformity, it suffices that your actual prior density change gently in the region favored by the data and not itself too strongly favor some other region." Edwards, Lindman, and Savage (1963) then spell this out more formally. In the present case, we recommend the principle of stable estimation for approximating $p(\theta|S)$ as a function of $\mu$, $\sigma$, $M_3$ *for any fixed* $M_4$, at least for subsamples $S_1$ and $S_3$.

## 2. Compound hypotheses

The preceding section considered simple hypotheses—sufficiently small neighborhoods that negligible error results if one assumes $p(\theta)$ and $LH(\theta)$ to be constant therein. The present section considers compound hypotheses—sets in which $p(\theta)$ and $LH(\theta)$ may vary considerably. As noted in the preceding article, the difficulty with dealing with compound hypotheses is that the shift $\psi(H, K)$ of probability belief from $K$ to $H$ depends on priors within $H \cup K$. The subsections of this section consider whether $S_1$ is normal; consider whether it is Student's $t$; comment on Kon's evidence that it is distributed like a contaminated normal; and note generally that when the principle of stable estimation can be applied within $H$ and $K$ separately, the problem of estimating $\psi(H, K)$ is simplified considerably, no matter what the dimensionality of $H$ or $K$ may be.

### 2.1. *Is $y_t$ Gaussian?*

At the time we did the analysis, we felt that the most urgent question involving compound hypotheses was what the shift in belief $\psi(H_N, H_K)$ should be between

$$H_N = \{y_t \in S_1 \text{ is approximately Gaussian; i.e., } M_3 \cong 0, M_4 \cong 3\}$$

and

$$H_K = \{y_t \in S_1 \text{ has high kurtosis, specifically } M_4 \geq 12\}.$$

*First, texts* on Bayesian inference (such as DeGroot, 1970; Zellner, 1971; Berger, 1985) devote much space to procedures using normal variates, since these are relatively easy to work with. Second, the dominant theory at the time (and, to a lesser extent, even now) concerning the fair value of an option is based on the Black–Scholes model, which assumes $y_t$ to be normal. This assumption came to be regarded more critically after October 19, 1987, Black Monday, not only by academicians but also by the marketplace. If the Black–Scholes model were correct, then the "representative investor's" estimate of the volatility of the S&P Index could be inferred from the price, striking price, expiration date, and type of an option on it. If we hold expiration date and option type constant and consider options with different striking prices, then the market prices of the various options should imply the same volatility for the Index. As reviewed in Rubinstein (1994), this was approximately true before Black Monday and clearly not so since. Subsequent to Black Monday, the market has priced options as if the underlying distribution has higher kurtosis than the normal. As a matter of style, we present our assumptions, analysis, and conclusions in the present tense, but those of this section are the same as in the original article submitted for publication prior to Black Monday.

In this section, we state *inequalities* on priors which either (a) we believe most or all readers will find apply to their own priors, or (b) will tend to favor $H_N$. We show that these inequalities plus the observed $LH$ imply a massive shift against $H_N$ in favor of $H_K$.

*Assumptions concerning beliefs about $M_3$*

Let

$$I = \{-2.667 \leq M_3 \leq 2.667\}.$$

For comparison, 2.667 and −2.667 are the $M_3$ values for single draws from a binomial distribution with $p = .1$ and .9, respectively. We pick these values, expecting most readers to agree that a prior of $P(M_3 \in I) \geq .5$ is reasonable. However, we compute a bound on the shift $\psi$ assuming that, for any given $\mu$, $\sigma$, $M_4$,

$$P(M_3 \in I) \geq .01, \tag{A1}$$

i.e., there is *at least* 1% prior probability that $-2.667 \leq M_3 \leq 2.667$.

Also, we calculate $\psi$ assuming that

prior probability is uniformly distributed on $I$. (A2)

This assumption is "favorable to $H_N$"—will tend to understate the shift towards $H_K$ away from $H_N$, as compared to unimodal distributions with mode near $M_3 = 0$—since $LH$ is greatest near $M_3 = 0$.

*Assumption on priors for $\mu$ and $\sigma$*

We calculate $\psi$ assuming that, for a given $M_3$ and $M_4$, the "shape" of the prior density as a function of $\mu$ and $\sigma$ is the same as that for the normal, i.e.,

$$\text{p}(\mu, \sigma, M_3, M_4) = k(M_3, M_4)p(\mu, \sigma, 0, 3), \tag{B1}$$

and that

the prior probability $P(\mu, \sigma, 0, 3)$ is concentrated at the values of $\mu$, $\sigma$ (B2)

which maximize $LH$ for the normal distribution, i.e.,

$$\mu = \mu^* = 5.45\ (10^{-4}) \text{ and } \sigma = \sigma^* = 7.68\ (10^{-3}).$$

These assumptions are not made because they are plausible, but because they will tend to understate the shift of belief from $H_N$ to $H_K$.

*Other assumptions*

We also assume that *given* $M_4 \geq 12$, at least some small conditional probability (1% is used in the calculation) is attached to $M_4 \in [12, 100000]$ i.e.,

$$P\{M_4 \in [12, 100000]\} \geq .01\ P\{M_4 \geq 12\}. \tag{C1}$$

Finally, there is

an interpolation assumption concerning $LH$ (C2)

that certain properties of $LH$ observed at specific values of parameters also hold for intermediate values.

*Calculation of shift between $H_N$ and $H_K$*

We first consider how to compute a bound on the shift in probability from $H_N$ towards $\{M_4 \cong M_4^*\}$ for any specific $M_4^*$. We illustrate the computation for $M_4^* = 18$. By (B1) and (B2),

$$R^* = P(M_4 \cong 18|S)/P(\cong \text{normal}|S)$$
$$\geq \frac{\int LH(\mu^*, \sigma^*, M_3, 18)p(M_3|M_4 = 18)dM_3 P(M4 \cong 18)}{LH(\mu^*, \sigma^*, 0, 3)P(\cong \text{Normal})} \tag{6}$$

Therefore, using (A1) and (A2),

$$\psi(\{M_4 \cong 18\}, H_N) \geq \frac{.01}{5.334} \int_{-2.667}^{2.667} \frac{LH(\mu^*, \sigma^*, M_3, 18)}{LH(\mu^*, \sigma^*, 0, 3)} dM_3 \tag{7a}$$

$$\geq \frac{.01}{5.334} \int_{-.185}^{.370} \frac{LH(\mu^*, \sigma^*, M_3, 18)}{LH(\mu^*, \sigma^*, 0, 3)} dM_3. \tag{7b}$$

As reported in table 8, in the interval [−.185, .370]

*Table 8.* Comparison of $LLH$ and $LH$ for the normal versus $M_4 = 18$ and various $M_3$

| $M_3$ | $LLH(\mu^*, \sigma^*, M_3, 18)$ | $LLH(\mu^*, \sigma^*, 0, 3)$ | Difference $\Delta$ | $10^{\Delta}$ |
|---|---|---|---|---|
| −0.740 | 34.822 | −29.839 | 64.661 | $>10^{64}$ |
| −0.555 | 38.736 | −29.839 | 68.575 | $>10^{68}$ |
| −0.370 | 41.374 | −29.839 | 71.213 | $>10^{71}$ |
| −0.185 | 42.956 | −29.839 | 72.795 | $>10^{72}$ |
| 0.000 | 43.605 | −29.839 | 73.444 | $>10^{73}$ |
| 0.185 | 43.376 | −29.839 | 73.215 | $>10^{73}$ |
| 0.370 | 42.242 | −29.839 | 72.081 | $>10^{72}$ |
| 0.555 | 40.111 | −29.839 | 69.950 | $>10^{69}$ |
| 0.740 | 36.802 | −29.839 | 66.641 | $>10^{66}$ |

$$\frac{LH(\mu^*, \sigma^*, M_3, 18)}{LH(\mu^*, \sigma^*, 0, 3)} > 10^{72}.$$

Thus, using (C2)

$$\psi > (.01/5.334)(.555)10^{72}$$
$$> 10^{69}. \tag{7c}$$

Similarly, the $LH(\mu^*, \sigma^*, M_3, M_4)$ in table 9 imply

$$\psi(\{M_4 \cong 12\}, H_N) > 10^{69}. \tag{8a}$$

$$\psi(\{M_4 \cong 1000\}, H_N) > 10^{66} \tag{8b}$$

$$\psi(\{M_4 \cong 100000\}, H_N) > 10^{66}. \tag{8c}$$

Hence, on the basis of tables 8 and 9, and (C2), we conclude

$$\psi(\{M_4 \in [12. 100000]\}, H_N) > 10^{66} \tag{9}$$

whatever the prior distribution within $M_4 \in [12, 100000]$. This and (C1) imply

$$\psi(\{M_4 \geq 12\}, H_N) > 10^{64}. \tag{10}$$

*Comment*

Our assumptions do not restrict the form of the prior distribution. For example, the marginal prior distribution of $M_3$ is not itself required to have a third moment, or even a first. It is only required to assign *at least* a modicum of probability to a reasonably wide interval of $M_3$.

The qualitative conclusion is quite insensitive to the numbers assumed in (A), (B), and (C1). For example. if we had assumed $P(M_3 \in I) \geq .001$ in (A1), then still $\psi \geq 10^{68}$ at (7c), and similarly for the effect of varying .01 in (C1) on the conclusion (10).

*Table 9.* LLH for $\mu^* = 5.45$. $\sigma^* = 7.68$. and various $M_3$ and $M_4$

| $M_4$ / $M_3$ | 12 | 18 | 1000 | 100000 |
|---|---|---|---|---|
| −0.185 | 43.212 | 42.956 | 40.806 | 40.750 |
| 0.000 | 44.105 | 43.605 | 41.156 | 41.092 |
| 0.185 | 43.880 | 43.376 | 40.913 | 40.851 |
| 0.370 | 42.497 | 42.242 | 40.069 | 40.012 |

There is a problem as to whose priors we are talking about. Some potential end-user HDMs were not born at the beginning of the sample period. Therefore the prior $P$ in (A), (B), and (C1) cannot literally be their beliefs prior to the time of the sample, nor can it refer to the beliefs which they held before reading this analysis, subsequent to the sample period (since those beliefs may have been influenced by stock market graphs and lore which already reflect information in the sample).

The intended ultimate consumer of this study is an HDM who seeks to emulate an RDM. If an HDM judges that the RDM to be emulated would have had priors satisfying (A), (B), and (C1) prior to the sample, then the HDM is assured that the RDM's shift in beliefs satisfies (10). If not, table 6a gives some idea as to how dispersed the RDM's beliefs about $M_3$ must be for $\psi$ to shift from $H_K$ towards $H_N$.

An analysis of $S_3$, like the above for $S_1$, would imply an extremely large shift of belief from $H_N$ to $H_K$, but of much less magnitude than for $S_1$. $S_2$ and $S_4$ are too small to produce equally decisive results.

*2.2. Is $y_t$ generated by a student's t distribution?*

Assumptions (A), (B), and (C) imply a massive shift in belief from normal distributions to Type IV distributions with larger $M_4$. If an RDM were certain that $M_3 = 0$, then the shift would be to Type VII, Student's $t$, the symmetric case of Type IV. Our own beliefs concerning the beliefs of RDMs, which most HDMs would want to emulate, are that they are less certain about $M_3$. Among tractable assumptions, we recommend the stable estimation assumption of Edwards, Lindman, and Savage (1963) as the most plausible for the distribution of ($\mu$, $\sigma$, $M_3$) for given $M_4$.

In this case, posterior belief for given $M_4 = M_4$* in the relevant neighborhood of the $LH$ = maximizing ($\mu$, $\sigma$, $M_3$) is approximately the likelihood function $LH(\mu, \sigma, M_3, M_4{*})$. Since $LH$ is continuous here, there is zero probability that $M_3 = 0$ exactly. However, $LH$ as a function of $M_3$ as in table 6—now considered to be approximately the posterior distribution for given $M_4$—may have $M_3$ "practically" equal to zero. That is, if we evaluate investment questions involving the S & P 500 Index, the answer may be the same whether we assume $M_3 = 0$ or use other values of $M_3$ with nonnegligible $LH$. Whether this is so may depend on the problem analyzed. For example, perhaps, for the option strategies discussed in the preceding article, it makes little difference whether simulations draw the appropriate numbers of $y_t$ in $S_1$ and $S_3$ from distributions with the $LH$ = maximizing values of $M_3$ or with $M_3^1 = M_3^3 = 0$; whereas, if the question is whether a brokerage house should be especially cautious not to start a weekend with a large imbalance in the aggregate of their S&P 500 Index options and futures, $M_3^3 = 0.38$ versus $M_3^3 = 0.0$ might be relevant. We say "might": the simulations would tell the practical importance of this much skewness.

### 2.3. Is $y_t$ contaminated normal?

Kon (1984) concludes that a mixture of two or more normal distributions is a more likely explanation for daily returns than is the Student's $t$ distribution. We review Kon's methods, since they relate to the present article in two crucial ways: (1) whether the Student's $t$ distribution, which does so well in our analysis, has already been bested by another form of distribution, and (2) how one answers such questions generally for compound hypotheses. In the latter connection, we also review the proposals and results of Akaike (1974, 1977, 1979) and of Schwarz (1978).

Kon's Table V, partly reproduced in table 10 here, shows

$$\Lambda_{is} = LH_i(\hat{\theta}_i)/LH_S(\hat{\theta}_S)$$

for 30 individual securities (ID = 1, ..., 30) and three indexes (ID = 31, 32, 33) for July 2, 1962 through December 31, 1980. In particular, our $y_t$ is ID = 31. Hypothesis $H_S$ is that daily returns are drawn from i.i.d. Student's $t$. Hypothesis $H_i$ is a mixture of $i$ normal distributions with $LH_i$ computed *as if*[5] $y_t = \log(r_t/r_{t-1})$ is generated i.i.d. as follows:

> each day a number $j$ between 1 and $i$ is drawn at random with probabilities $\lambda_1, \lambda_2 \ldots \lambda_i$,
> then a random variable $y_t = \log(r_t/r_{t-1})$ is drawn normally from $N(\mu_j, \sigma_j)$.

Kon assumes in effect, that the shift between hypotheses $H_a$ and $H_b$ is[6]

$$\psi_{a,b} = LH_a(\hat{\theta}_a)/LH_b(\hat{\theta}_b), \tag{11}$$

*Table 10.* Extract from Kon (1984) table V: "Log-likelihood ratios for comparing the student and discrete mixture of normal distribution models"

| ID | $\text{Log}\Lambda_{1S}$ | $\text{Log}\Lambda_{2S}$ | $\text{Log}\Lambda_{3S}$ | $\text{Log}\Lambda_{4S}$ | $\text{Log}\Lambda_{5S}$ |
|---|---|---|---|---|---|
| 1 | −280.981 | −2.959 | N.A. | N.A. | N.A. |
| 2 | −198.195 | −13.396 | 12.717 | 19.665 | N.A. |
| 3 | −300.484 | −6.342 | 11.370 | 13.759 | N.A. |
| . | | | | | |
| . | | | | | |
| . | | | | | |
| 31 | −222.506 | −4.701 | 20.457 | 21.485 | 23.219 |
| 32 | −250.715 | 0.543 | 27.575 | 29.601 | N.A. |
| 33 | −448.294 | −3.482 | 35.848 | 38.297 | N.A. |
| Nr. > 0 | 0 | 8 | 18 | 13 | 1 |
| Out of | 33 | 33 | 18 | 13 | 1 |

where $\hat{\theta}$ is the combination of parameters which maximizes $LH(\theta)$. He concludes that Student's $t$ is a better explanation than the mixture of two normals but poorer than the mixture of three.

One problem with Kon's analysis is that it does not take into account the different dimensionalities of the models. Since $\Sigma\lambda_j = 1$, $H_i$ has $3i - 1$ free parameters. $H_s$ with three parameters nevertheless scores better than $H_2$ with five, though worse than $H_3$ with eight. $\text{Log}\psi(H_j, H_k)$ equals $\log\Lambda_{jS} - \log\Lambda_{kS}$; i.e., the difference between the entry in the $j$th and $k$th column. Since $H_{i-1}$ is a special case of $H_i$, $LH_i(\theta_i)$ cannot decrease with $i$. In the table, it is strictly increasing. If shift in belief equaled (11), then belief among the $H_i$ would always shift toward the one with maximum $i$—which does not seem correct.

Akaike (1974, 1977, 1979) and Schwarz (1978) propose criteria, depending on the number of parameters to be fitted, for choosing among alternate models. Let us consider whether these criteria allow us to estimate $\psi$ in accordance with the principles presented in the preceding article.

For a model (compound hypothesis) with $k$ parameters which may be independently varied, the Akaike information criterion is

$$\text{AIC} = -2\log(LH(\hat{\theta})) + 2k,$$

where $\theta \in R^k$, and AIC is a measure of the "badness of fit" of the model. Given several models (typically two or more of a series with differing $k$, like number of terms in a polynomial regression), the one with the minimum AIC estimate (MAICE) is favored. "MAICE may be viewed as the mode of the posterior distribution corresponding to the prior distribution defined by the product of the improper uniform distribution of the parameters within the model and the probability of the model defined to be proportionate to the exponential of minus the number of parameters." (Akaike, 1977). Given our observations in the preceding article on the use of the improper uniform distribution, clearly AIC does not fit here as the estimate of $\psi$.

Schwarz (1978) shows that, for a certain class of models[7] such that the parameter vector $\theta_j$ for model $j$ is confined to a $k_j$ dimensional subspace of a common $K$-dimensional space, the posterior probability of model $j$ given sample $S$ may be written as

$$\log P(\text{model } j|S) = \log(LH_j(\hat{\theta}_j)) - 1/2k_j \log n + R, \tag{12a}$$

where the remainder $R$ is bounded as $n \longrightarrow \infty$. Since the prior $P$ of model $j$ does not depend on $n$, both the log of the posterior odds ratio and the log of the shift $\psi$ can be written as

$$\log\psi = \log(LH(\hat{\theta}_j)) - \log(LH(\hat{\theta}_i)) - 1/2(k_j - k_i)\log n + R'. \tag{12b}$$

The Schwarz result is extremely valuable as an example of the asymptotic behavior of $\psi$. However, even when the assumptions of the analysis are met, $R'$ can be substantial, even for samples as large as our $S_1$. This suggests that we should seek some procedure which makes greater use of the actual $LH$ at hand. The next section presents our proposal along these lines.

### 2.4. On estimating $\psi$

As in section 2.1, in general we seek plausible assumptions about prior probabilities $P$ which, when combined with knowledge of $LH$, imply useful facts about $\psi$. "Useful facts" include either (a) for essentially any plausible prior, there is a very large shift in favor of one of the hypotheses against the other, or, conversely, (b) for some plausible priors, the shift is in one direction; for others it is in the other direction. In case (a), the HDM can ignore one of the hypotheses. In case (b), ideally the HDM should evaluate policies under both.

We propose the following: begin by exploring $LH(\theta)$ for $\theta \in H_1$ and for $\theta \in H_2$ for some sample $S$ in the manner illustrated in section 1 of this article. (E.g., for one of Kon's $H_i$, explore $LH_i(\theta)$ for $S_I$ for comparison with the Pearson family.) In particular, determine $\hat{\theta}_1$ and $\hat{\theta}_2$; see if either model has a ridge or a higher dimensional "plateau." If there is no ridge, confirm that stable estimation seems plausible for each model.

In the case in which the Edwards, Lindman, and Savage (1963) principle of stable estimation applies, separately, to neighborhoods $N_1$ of $\hat{\theta}_1$ and $N_2$ of $\hat{\theta}_2$, we assume that $p(\theta) \cong p_i$ in $N_i$ and $\int LH(\theta)p(\theta)d\theta$ may be neglected beyond $N_i$. Therefore, the shift is

$$\psi(H_1, H_2) = \frac{p_1}{P(H_1)} \cdot \int_{\theta \in N_1} LH_1(\theta)d\theta \Big/ \frac{p_2}{P(H_2)} \cdot \int_{\theta \in N_2} LH_2(\theta)d\theta$$

$$= \frac{p_1/P(H_1)}{p_2/P(H_2)} \cdot \frac{\int_{\theta \in N_1} LH_1(\theta)d\theta}{\int_{\theta \in N_2} LH_2(\theta)d\theta}. \tag{13}$$

If $V_i$ is the volume of $N_i$, then (13) can be written as

$$\psi(H_1, H_2) = \frac{P(N_1|H_1)}{P(N_2|H_2)} \cdot \frac{A_1}{A_2}$$

$$= \Gamma_{12} A_1/A_2 \tag{14a}$$

where

$$A_i = \int_{\theta \in N_i} LH_i(\theta)d\theta / V_i \tag{14b}$$

and

$$P(N_i|H_i) = V_i p_i / P(H_i). \tag{14c}$$

$A_i$, the average value of $LH$ in $N_i$, may be approximated by quadrature or Monte Carlo methods. Thus, in the present case, the problem of producing useful bounds on $\psi$ reduces to one of finding plausible and useful bounds on the scalar $\Gamma_{12}$: the prior probability (of an RDM to be emulated) of $\theta \in N_1$ if $H_1$ is true, divided by that of $H_2$.

Comparing (13) with (11), we see that the correct calculation when stable estimation applies involves an area under a surface $\int LH(\theta)d\theta$ rather than the height of the highest point $LH(\hat{\theta})$. Equation (12b) provides a higher order estimate of $\psi$ than (11). But it is derived for a particular system of compound hypotheses, and even then the remainder term $R$ can be quite large for samples as large as $S_1$. At the least, the factor $A_1/A_2$ can (and we believe should) be computed from $LH$.

The relationship between dimensionality and $\Gamma_{12}$ is relatively simple when $N_i$ is a cube and $P(\theta)$ is uniformly distributed on a larger cube $M_i \supset N_i$. Then each dimension will have shrunk by a factor $\lambda_i$ between $M_i$ and $N_i$. If $\lambda_1 = \lambda_2$, then $\Gamma_{12} = \lambda^{m-n}$ where $m = \dim(N_1)$ and $n = \dim(N_2)$. For example, if $H_1$ is a six-parameter model, $H_2$ a three-parameter model, and dimension has shrunk by $\lambda = 1/10$ between the $M_i$ and $N_i$, then $\Gamma_{12} = 1/1000$. $A_1/A_2 > 1{,}000$ is required for the shift to favor $H_1$. $N_1$ was, a priori, one of a million equally likely cubes, whereas $N_2$ was one of a thousand.

Justification for bounds on $\Gamma_{12}$ depends on the specific hypotheses $H_1$ and $H_2$. We recommend that one first calculate $A_1/A_2$. This shows the ranges of values of $\Gamma_{12}$ which would produce conclusive results of one or another kind. You are on your own as to which of these ranges is defensible and how to defend it. In case of a ridge or plateau, the analysis must be repeated starting from various points on the ridge or plateau.

The estimation of a bound on $\psi(H_N, H_K)$ in section 2.1 can be used to illustrate this approach. The analysis presented in section 2.1 was the one used when the data was first analyzed, before Black Monday. A similar result could have been obtained using the method of the present section as follows: for a given $M_4$, choose rectangles $R_1 = [\mu_L, \mu_H] \times [\sigma_L, \sigma_H]$ and $R_2 = R_1 \times [M_{3L}, M_{3H}]$ such that $LH(\theta)/LH(\hat{\theta}) < 10^{-9}$ for $\theta \notin R_i$. Compute $A_1/A_2$. Argue that, for any given $M_3$* and the given $M_4$*, presumably

$$P((\mu, \sigma) \in R_1|M_3 = 0, M_4 = 3) \cong P((\mu, \sigma) \in R_1|M_3^*, M_4^*),$$

but assume only (something like)

$$P((\mu, \sigma) \in R_1|M_3^*, M_4^*) > (.01)\, P((\mu, \sigma) \in R_1|M_3 = 0, M_4 = 3).$$

This, plus an assumption such as $P(M_3 \in [M_{3L}, M_{3H}]) \geq .01$, implies $\Gamma_{NK} < 10{,}000$. Since $A_K$ is dozens of orders of magnitudes greater than $A_N$, $\Gamma_{NK} A_N/A_K$ implies a massive shift against $H_N$. Vary $M_4$* and argue that a similar shift against $H_N$ is implied for all $M_4 \in [M_{12}, M_{100000}]$. Assume, e.g., $P(M_4 \in [M_{12}, M_{100000}]) > .01\, P(M_4 \geq 12)$ and reach the same conclusion as in section 2.1.

## 3. What next?

Whenever we present or reexamine our results, further research suggests itself or is suggested to us, some of which has been incorporated into the present article. Further analyses remain in queue; but must await other projects and articles—at least some, we hope, by other authors. The present section briefly reviews some analyses in queue.

One type of analysis in queue is the comparison of Pearson-family i.i.d. models with other i.i.d. families, such as Kon's mixtures of normals or the stable Paretian family (Mandelbrot, 1963; Fama, 1963). What is basically needed is either density functions for any member of the family, or an algorithm to approximate this, such as the series approximation to the stable Paretian density in Feller (1971). Various computations are performed with these densities as the basis, including summing logs to compute *LLH*, searching for max *LLH* or for ridges, searching a "large enough" area around the max or from the ridge, and tabulating and plotting results, etc. Much or all of this can now be done, with little programming, using a full-featured spreadsheet which accepts the density computation as a user-defined function, such as Visual Basic functions in EXCEL. It would be convenient if density functions programmed by one research team were readily available to others. This suggests a website on the Internet. A more commercial, less do-it-yourself solution would be a statistical package or service which provides a large variety of density functions, analysis, and display routines and, ideally, data access. Such a product line is not especially complicated as compared to existing statistical libraries. The market will probably supply it once a demand manifests itself.

The situation for non-i.i.d. families of models is the same as for i.i.d. families, except that *LLH* is no longer simply the sum of the *LLH* for individual observations. The log density of the sample can, once again, be programmed as a user-defined function for a spreadsheet or in a commercial statistical package. Based on the results in section 1 of this article, existing non-i.i.d. models should be enhanced to reflect whether the $y_t$ to be predicted will be from $S_1$ or $S_3$, and whether the observed, past $y_t$ was from $S_1$ or $S_3$.

Another type of extension is to random vectors $y_t = (y_{1t},\ldots, y_{nt})$ such as the joint distribution of a few asset classes or thousands of individual securities. The latter poses especially challenging but, we believe, not insurmountable computational and modeling problems.[8]

## 4. Summary and conclusions

The research reported here was an empirical investigation into questions at two levels. At a methodological level, the question was whether interesting empirical research could be conducted in accordance with certain principles including: (1) that the Bayesian statistician should show remote clients with a wide variety of priors the extent to which they should shift beliefs given a sample: and (2) that classical inference methods or Bayesian methods using diffuse priors can be highly unreliable guides for (1). At a substantive level, we considered how a Bayesian should shift beliefs among probability distribution of $y_t$ which are Pearson-family i.i.d. for a given number of calendar days between trading days. Conclusions include:

- The Bayesian should shift massively against normal distributions in favor of Pearson Type IV including Student's *t* (Type VII) as the symmetric special case. (This is based on a sample which ended before Black Monday. Prior to Black Monday, the market for S&P 500 futures acted as if $y_t$ were normally distributed.)
- More generally, belief should shift massively away from all the many other forms of distribution encompassed by the Pearson system towards Pearson Type IV. E.g., belief should shift away from gamma distributions towards Type IV, away from beta of the first and second kind, etc.
- Belief should shift away from hypotheses that a draw of $y_t$ from subsample $S_3$, Two-Day Weekends is distributed like the sum of one, two, or three draws from $S_1$, Weekdays. Specifically, belief is drawn heavily towards hypotheses in which mean is positive for $y_t \in S_1$ and negative for $y_t \in S_3$.

We find the substantive conclusions interesting, and relevant for action, and therefore draw an affirmative conclusion to the methodological question.

## Acknowledgments

The authors very much appreciate the valuable suggestions of Mark Machina and a referee.

## Notes

1. We submitted our first version as a single article to a prominent finance journal, received a fairly favorable referee review, but the article was adamantly rejected by the editor mostly for being too dogmatically Bayesian. The article sat in a drawer for a few years, then was revised, partly on the basis of comments by referee and editor, to seem more appropriate for a finance journal. In particular, the first version of what later became the first of two articles was compressed into a long footnote. The revised article was politely rejected by the editor of a different prominent finance journal after a process that included a favorable referee report; revision of the article in accordance with the referee recommendations; and acceptance by the referee but rejection by a new, second referee. This "tie" was eventually broken, unfavorably, by a third and fourth referee.
2. $\Delta y$ was chosen small enough so that the first four moments are about the same whether computed from grouped or ungrouped data. In particular, Sheppard's corrections are negligible.
3. Some step sizes of the "independent variables" in tables 4–6 may seem puzzling. This is due to the nature of the search of *LLH* as a function of ($\mu$, $\sigma$, $M_3$, $M_4$). For each subsample and for each $M_4$ in table 3, we explored a grid of ($\mu$, $\sigma$, $M_3$) combinations intended to span *LH* within roughly nine orders of magnitude of maximum *LH* for that $M_4$, with about eight to ten grid steps in each dimension. In one direction, steps of equal variance were used, but the results are reported here in terms of standard deviation (as one referee wisely suggested). Each dimension of the grid included the sample moment ($\mu$, $\sigma$ or $M_3$), but frequently

skipped over the (unknown) $LH$-maximizing value. More refined computation approximated the $LH$-maximizing values, and these were added to tables 4–6. In some cases, a grid size used to represent $LH$ as a function of some moment worked well for most of the table, but led to a large jump in max $LH/LH$ at one end of the table. In some such cases, one or more intermediate values were added, and sometimes an original entry with Max $LH/LH \gg 10^9$ was dropped.

4. The computation of maximum $LH(S_1)$ for various $M_4$ with $M_3 = 0$ first appeared in Kim, Markowitz, and Usmen (1988).
5. Kon speaks as if $y_t = \log(r_t/r_{t-1})$ is generated by $N(m_1, \sigma_1)$ for a while, then by $N(m_2, \sigma_2)$ for a while, then by $N(m_1, \sigma_1)$, etc. However, his actual computation of $LH$ is as described in the text. Specifically, Kon's computation would produce the same value if the observations $y_1, y_2 \ldots$ were arbitrarily permuted.
6. Blattberg and Gonedes (1974) refer to (11) for large samples as the asymptotic posterior odds of $H_1$ relative to $H_2$.
7. "[O]bservations come from a Koopman–Darmois family, i.e., relative to some fixed measure on the sample space, they possess a density of the form

$$f(\chi, \theta) = \exp(\theta \cdot y(\chi) - b(\theta)),$$

where $\theta$ ranges over the natural parameter space $\Theta$, a convex subset of the $K$-dimensional Euclidean space, and $y$ is the sufficient $K$-dimensional statistic. The competing models are given by sets of the form $m_j \cap \Theta$, where each $m_j$ is a $k_j$-dimensional linear submanifold of $K$-dimensional space" Schwarz (1978).
8. For example, suppose that one family of models to be analyzed is the one-factor model for generating returns on thousands of securities with, say, Type IV distributions for the idiosyncratic term. Rather than consider a specific hypothesis as having a many-thousand dimensional parameter $B$, including the $\alpha$s, $\beta$s, and residual distribution parameters of each security, it is more manageable intuitively to model $B$ as generated by some distribution $f(B; \theta)$, where $\theta$ is of much smaller dimension. The likelihood of the sample $y_t$ for a given parameter vector $\theta$ is

$$I = \int_B LH(y_t|B)p(B; \theta)dB.$$

One can imagine estimating $I$ by randomly drawing $B$ according to $p(B; \theta)$ and calculating $LH(y_t|B)$. This is very inefficient, since it will rarely generate a $B$ in the part of the $B$-space which makes non-negligible contributions to $I$, since, for example, most generated $B$s will assign low $\beta_i$ to many securities whose observations "must" have come from a high $\beta$, and high $\beta_i$ to many securities which "must" be low $\beta$. A much more efficient procedure is to determine a relatively small region $R$ of $B$-space, which includes essentially all nonnegligible $LH(y_t|B)$, then sample from $R$ and compute $LH(y_t|B)p(B|\theta)$ for each sampled $B$.

## References


Akaike, Hirotugu. (1974). "A New Look at the Statistical Model Identification," *IEEE Transactions on Automatic Control* AO-19, 716–723.

Akaike, Hirotugu. (1977). "Entropy Maximization Principle," *Applications of Statistics*. Amsterdam: North-Holland Publishing Co.

Akaike, Hirotugu. (1979). "A Bayesian Extension of the Minimum AIC Procedure of Autoregressive Model Fitting," *Biometrika* 66, 237–242.

Berger, James. (1985). *Statistical Decision Theory and Bayesian Analysis*, 2nd ed., New York: Springer-Verlag.

DeGroot, Morris. (1970). *Optimal Statistical Decisions*. New York: McGraw-Hill.

Edwards, Ward, Harold Lindman, and Leonard Savage. (1963). "Bayesian Statistical Inference for Psychological Research," *Psychological Review* 70.

Elderton, William, and Norman Johnson. (1969). *Systems of Frequency Curves*. Cambridge, UK: Cambridge University Press.

Fama, Eugene. (1963). "Mandelbrot and the Stable Paretian Hypothesis," *Journal of Business* 36, 420–429.

Feller, W. (1971). *An Introduction to Probability Theory and Its Applications*, Vol. 2, 2nd ed. New York: John Wiley & Sons.
Hildreth, Clifford. (1963). "Bayesian Statisticians and Remote Clients," *Econometrica* 31, 422–438.
Hoaglin, David, Frederick Mosteller, and John Tukey, eds. (1983). *Understanding Robust and Exploratory Data Analysis*. New York: John Wiley & Sons.
Hoaglin, David, Frederick Mosteller, and John Tukey. (1985). *Exploring Data Tables Trends and Shapes*. New York: John Wiley & Sons.
Johnson, Norman, E. Nixon, and D. Amos (1963). "Table of Percentage Points of Pearson Curves, For Given $\sqrt{\beta_1}$ and $\beta_2$ Expressed in Standard Measure," *Biometrika* 50, 459–498.
Kim, Gew-Rae, Harry Markowitz, and Nilufer Usmen. (1988). "The Likelihood of Various Stock Market Hypotheses: Student Distributions with Low Degrees of Freedom." Baruch College Working Paper Series No. 88-11, Economics and Finance.
Kon, Stanley. (1984). "Models of Stock Returns—A Comparison," *Journal of Finance* 39, 147–165.
Mandelbrot, Benoit. (1963). "The Variation of Certain Speculative Prices," *Journal of Business* 36, 394–419.
Markowitz, Harry, and Nilufer Usmen. (1996). "The Likelihood of Various Stock Market Return Distributions, Part 1: Principles of Inference," *Journal of Risk and Uncertainty* 13, 207–219.
Rubinstein, Mark. (1994). "Implied Binomial Trees," *The Journal of Finance* 69, 771–818.
Savage, Leonard. (1954). *The Foundations of Statistics*, 2nd ed. New York: John Wiley & Sons; (1972), Dover.
Schwarz, Gideon. (1978). "Estimating the Dimension of a Model," *The Annals of Statistics* 6, 461–464.
Stuart, A., and K. Ord. (1994). *Kendall's Advanced Theory of Statistics, Distribution Theory*, Vol. 1. 6th ed. London: Edward Arnold.
Tukey, John. (1960). "A Survey of Sampling from Contaminated Distributions," *Contributions to Probability and Statistics: Essays in Honor of Harold Hotelling*. Stanford University Press.
Tukey, John. (1962). "The Future of Data Analysis," *Annals of Mathematical Statistics*.
Zellner Arnold. (1971). *An Introduction to Bayesian Inference in Econometrics*. New York: John Wiley & Sons.

JOURNAL OF INVESTMENT MANAGEMENT, Vol. 1, No. 4, (2003), pp. 9–25


www.joim.com

# RESAMPLED FRONTIERS VERSUS DIFFUSE BAYES: AN EXPERIMENT

*Harry M. Markowitz*[a,*] *and Nilufer Usmen*[b]

## Introduction

This paper reports on an experiment which compares two methods of handling the fact that empirically observed means, variances, and covariances, for a mean–variance analysis, are themselves noisy. One method is Bayes inference using diffuse priors which the present authors, among many others, have recommended. (Markowitz and Usmen, 1996a,b). The other is the method of Resampled Efficient Frontiers™ recommended by Richard O. Michaud (Michaud, 1998).[1]

The experiment is a computer simulation "game" with two players and a referee. In the game the referee generates 10 "truths" about eight asset classes. For each truth the referee draws 100 different possible "histories" of 216 monthly observations. (We chose eight asset classes and 216 months to keep the experiment as close as possible to that of Michaud.)

[a]Harry Markowitz Company, San Diego, California, USA.

[b]School of Business, Montclair State University, Upper Montclair, New Jersey, USA.

*Corresponding author. Harry Markowitz Company, 1010 Turquoise Street Suite 245, San Diego, CA 92109, USA. Tel.: (858) 488-7212.

Each history is presented to each player. The players know that the truth is a joint normal distribution with unchanging means, variances, and covariances but do not know the parameter values. The Michaud player uses the observed history to generate a resampled frontier. That is, for a given history the player randomly generates many mean–variance efficient frontiers and averages these. The Bayes player uses the observed history to update beliefs, from prior to posterior, then uses these beliefs to compute one efficient frontier. Because of the high dimensionality of the "hypothesis-space," Monte Carlo sampling must be used to approximate the Bayes player's *ex post* means, variances, and covariances. Given their respective frontiers each player picks three portfolios, namely, the portfolios which each player believes maximizes

$$EU = E - \lambda V \qquad (1)$$

for $\lambda = 0.5, 1.0, 2.0$, where $E$ and $V$ are the portfolio mean and variance. The referee notes the player's actual expected utility using the true means, variances, and covariances—known only to the referee. The referee also notes each player's estimate of its expected utility. This is repeated for the 100 randomly drawn histories for a given truth and the 10 truths of the game.

The assumption of normality and unchanging distributions may be unrealistic, but both players are apprised of the rules of the game. It is not obvious that the assumptions favor one methodology over the other. The authors expected the Bayesian approach with diffuse priors to do better than the resampled frontier approach. In fact, the opposite turned out to be the case. Section 1 of this paper describes how the referee generates truths and, from these, the histories "observed" by the players; Section 2 describes the actions of the Michaud player; Section 3 describes the actions of the diffuse Bayesian player; Section 4 presents the results of the experiment; Section 5 points out some questions raised by these results; Section 6 summarizes.

## 1 The referee and the game

The experiment ("game") is outlined in Exhibit A. The referee generates 100 histories from 10 "truths," each history consisting of returns on eight asset classes during 216 consecutive months. Each truth is itself randomly generated by the referee by computing the means, variances and covariances of 216 draws of eight returns each from a "seed" distribution. This seed distribution is normally distributed with means, variances, and covariances equal to the historic excess return over the US 30-day T-bill rate of the eight asset classes listed in Table 1 for the 216 months from January 1978 through December 1995, as in Michaud (1998).

**Exhibit A** The Experiment.

Referee chooses First/Next "Truth"

"Truth" is a joint normal return distribution with fixed mean vector $\mu$ and covariance matrix $C$ not known to the players.

Referee draws First/Next historical sample randomly from Truth

For Player = {Bayesian, Resampler}

Referee gives historical sample to Player.

Player applies its procedure to sample. (See write-ups of respective procedures.)

For the given sample and for each utility function (specifically, for $EU = E - \lambda V$ for $\lambda = \frac{1}{2}$, 1, and 2) the Player returns:

Selected Portfolio
Estimate of its Expected Utility

For each (Player, Utility function):

Referee computes True expected utility.

Repeat for Next Historical Sample

After all historical samples have been generated and processed, and with Truth still fixed:

For each utility function, see which player had higher $EU$ on average.
Compare $EU$ achieved versus $EU$ anticipated on average.

Repeat for Next Truth

Did one of the players do better for most Truths or on average?

**Table 1** Asset classes used in experiment.[a]

| Asset class | Data source |
|---|---|
| Canadian Equities | Morgan Stanley Capital International[b] |
| French Equities | Morgan Stanley Capital International[b] |
| German Equities | Morgan Stanley Capital International[b] |
| Japanese Equities | Morgan Stanley Capital International[b] |
| United Kingdom Equities | Morgan Stanley Capital International[b] |
| United States Equities | S & P 500 Index total return |
| United States Bonds | Lehman Brothers[c] |
| Euro Bonds | Lehman Brothers[d] |

[a]Source: Michaud (1998) p. 13, footnote 16.
[b]Dollar return indexes net of withholding taxes.
[c]Government/Corporate US bond index.
[d]Eurobond global index.

Having thus established a truth, the referee generates a 216 month "history" from this truth by sampling joint normally from the truth's mean vector and covariance matrix. Each history is presented to each of the two players. Each player tells the referee, for each history, the portfolio which the player believes maximizes $EU$ in (1) for $\lambda = 0.5, 1.0, 2.0$, respectively. The player also provides the referee with the player's own estimate of $EU$. The referee computes the actual value of $EU$ from the truth, known only to the referee. The referee tabulates the actual value and the players' estimates of this value for the two players. This is repeated for 100 histories per truth and 10 truths for the experiment.

## 2 The Michaud Player

Michaud proposes the following procedure to handle the fact that observed means, variances, and covariances are not the true parameters but contain noise. In private conversations with the present authors, Michaud points out that more sophisticated procedures could be incorporated into the resampling philosophy. We grant this, but note that it would be difficult to formulate an experiment that encompasses all the possible nuances of both the resampling and Bayesian approaches. The experiment we report here, admittedly, compares "vanilla" resampled frontiers with diffuse Bayes implemented by a particular Monte Carlo analysis.

Following Michaud (1998), the "Michaud player" in our experiment proceeds as follows: given a specific history $O$ ("$O$" for "Observation") generated by the referee with its means, variances, and covariances, the Michaud player draws 500 new samples of returns on the eight asset classes for 216 months, drawing these from a joint normally distributed i.i.d. random process with the same means, variances, and covariances as $O$. For each of these 500 samples the Michaud player generates an efficient frontier and then averages these 500 efficient frontiers. Specifically, it notes the first, second, third.... 101st points on the frontier spaced by equal increments of standard deviation. The first point is the one with the highest expected return on the frontier; the 101st point is the one with the lowest standard deviation. The "resampled frontier" has as its first portfolio the average holdings of the first portfolios of the 500 particular frontiers, its second portfolio is the average holdings of 500 second portfolios, etc.

The portfolio mean and variance ascribed to each of the 101 portfolios of the resampled frontier are computed using the original means, variances, and covariances of the observation $O$. (The present authors thank Richard and Robert Michaud for clarification on this point.) The task that each of the players is assigned is to provide portfolios which maximize the expected value of (1). Therefore, for a given history the Michaud player picks from his resampled frontier the points which maximize the expected value of its estimated $EU$ for $\lambda = 0.5, 1.0$, and 2.0. This process is repeated for each of the 100 randomly drawn histories for each of the 10 truths presented to the player by the referee.

## 3 The diffuse Bayes player

### 3.1 Basics

At any moment in time (say $t = 0$) the Bayesian rational decision maker (RDM) acts as if it ascribes a probability distribution $P_0(h)$ to hypotheses $h$ in some space $H$ of possible hypotheses. In the present discussion, a hypothesis is a vector of eight means and 36 distinct variances and covariances:

$$h' = (\mu_1^h, \ldots, \mu_8^h, \sigma_{11}^h, \sigma_{12}^h, \ldots, \sigma_{88}^h) \quad (2)$$

plus the assertion that the variables

$$r' = (r_1, \ldots, r_8) \quad (3)$$

are joint normally distributed with these parameters. The hypothesis space $H$ may be taken as all possible values of $h$:

$$H = R^{44} \quad (4)$$

It is inconsequential whether we restrict $H$ to the set $H^*$ of 44-tuples that can possibly be parameters of a joint normal distribution, or define it as in (4) and understand that

$$P_0(R^{44} - H^*) = 0 \quad (5)$$

The probability distribution $P_t(H)$ changes over time, as we review below. We assume that, as of any time $t$, the RDM chooses an action $\alpha$ so as to maximize a single-period utility function

$$EU = \underset{h}{E}[E(U(r;\alpha)|h)] \quad (6)$$

In other words, the action $\alpha$ is chosen so as to maximize $EU$ where $U$ depends on returns $r$ and action $\alpha$, and the expected return in (6) is computed as if Nature randomly drew a hypothesis $h$ using probability distribution $P_t$, then drew $r$ given $h$. In the present experiment the action $\alpha$ is the choice of a portfolio.[2]

To pick a portfolio which maximizes $EU$ in (6) using the utility function in (1), the RDM uses only its estimated portfolio mean ($E$) and portfolio variance ($V$) which depend only on its estimated means $\mu_i$ of securities and the covariances $\sigma_{ij}$ (including variances $V_i = \sigma_{ii}$) between pairs of securities. These are given by

$$\begin{aligned} \mu_i &= E(r_i) = \underset{h}{E}(E(r_i)|h) \\ &= \underset{h}{E}\,\mu_i^h \\ &= \text{Avg}\,\mu_i^h \end{aligned} \quad (7)$$

$$\begin{aligned} \sigma_{ij} &= E(r_i - \mu_i)(r_j - \mu_j) \\ &= \text{E}(r_i - \mu_i^h + \mu_i^h - \mu_i) \\ &\quad \times (r_j - \mu_j^h + \mu_j^h - \mu_j) \\ &= \text{E}(\sigma_{ij}^h) - E(\mu_i^h - \mu_i)(\mu_j^h - \mu_j) \\ &= \text{Avg}\,\sigma_{ij}^h - \text{cov}(\mu_i^h, \mu_j^h) \end{aligned} \quad (8)$$

since, e.g.

$$\underset{h}{E}[(r_i - \mu_i^h)(\mu_j^h - \mu_j)|h] = 0$$

In particular, for $i = j$ (Eq. 8) says

$$V_i = \text{Avg}\, V_i^h - \text{Var}\,(\mu_i^h) \quad (9)$$

The last line of (7) and (8) are mnemonics for the immediately preceding lines. These formulas tell us that, for the Bayesian RDM, the expected value of $r_i$ at time $t$ is the average, using $P_t(h)$ over $h \in H$, of $\mu_i^h$; whereas covariance between $r_i$ and $r_j$ is the average $\sigma_{ij}^h$ plus the covariance between $\mu_i^h$ and $\mu_j^h$. In particular, the variance of $r_i$ is the average $V_i^h$ *plus* the variance of $\mu_i^h$.

As evidence accumulates, $P_t(h)$ changes over time, according to the Bayes rule. If $P_t(H)$ has a probability density function $p_t(h)$, and $O$ is an observation taken between $t$ and $t+1$ (e.g. $O$ is the set of monthly returns $r_{it}$ for $i = 1, \ldots, 8, t = 1$ to 216 as described before), with $L(O|h)$ the probability density of $O$ given hypothesis $h$, then,

$$p_{t+1}(h) = \frac{p_t(h)\, L(O|h)}{\int_H p_t(h)\, L(O|h)\, dh} \tag{10}$$

The human decision maker (HDM) who wishes to emulate an RDM sometimes avoids the burden of specifying $p_t(h)$ by assuming that

$$p_t(h) = 1/\text{vol}(\Omega^*) \quad \text{for all } h \in \Omega^* \tag{11}$$

where "vol" stands for volume and $\Omega^* \subset H$ is assumed to be sufficiently large that

$$\int_{H-\Omega^*} p_t(h)\, L(O|h)\, dh \tag{12}$$

is negligible. With (11) assumed, the updated beliefs of (10) become

$$p_{t+1}(h) = L(O|h)/D \tag{13}$$

where

$$D = \int_{\Omega^*} L(O|h)\, dh \tag{14}$$

and the expected value [with respect to $P_{t+1}(h)$] of any integrable function $v(h)$ is

$$\underset{H}{E}\, v(h) = N/D \tag{15}$$

where

$$N = \int_{\Omega^*} v(h)\, L(O|h)\, dh$$

In principle, (15) can be used to compute $\underset{H}{E}\, \mu_i^h, \underset{H}{E}\, \sigma_{ij}^h$, and $\underset{H}{E}\, \mu_i^h \mu_j^h$, which are necessary to compute Avg $\mu_i^h$, Avg $\sigma_{ij}^h$ and cov $(\mu_i^h, \mu_j^h)$ in (7) and (8). The practical problem is that $N$ and $D$ are integrals over 44-dimensional spaces. As is often done, we will use Monte Carlo analysis to approximate a high-dimensional integral. The specifics of how we do this are described in a following subsection. First, we discuss the fact that a hypothesis space can often be parameterized in different ways, and present a parameterization of the present situation that will be very convenient for the Monte Carlo analysis that follows.

### 3.2 *Diffuse priors*

Suppose, for the moment, that there was only one unknown parameter, an expected value $\mu$ of one random variable $r$. Then, the standard diffuse prior spreads probability belief concerning $\mu$ uniformly over some large interval:

$$p(\mu) = \begin{cases} \dfrac{1}{2\Delta} & \text{for } \mu \in [-\Delta, \Delta] \\ 0 & \text{otherwise} \end{cases} \tag{16}$$

The choice of $\Delta$ is not important as long as $\Delta$ is sufficiently large, since the contribution to $E(r)$ becomes negligible beyond a sufficiently large $\Delta$. Admittedly, this is often not a very plausible prior. For example, if $r$ is the return on an asset class it is not plausible for the asset class to have a large constant-through-time negative expected return. Such an asset class would disappear. However, the use of (16) is justified as convenient because it saves making a decision as to the exact form to be used for prior beliefs. In effect, it assumes that posterior beliefs are proportional to the likelihood function $L(O|h)$. One justification for assuming posterior beliefs are proportional to $L(O|h)$ is the Edwards *et al.* (1963) principle of stable estimation. "To ignore the departures from uniformity, it suffices that your actual prior density change gently in the region favored by the data and not itself too strongly favor some other region" (p. 202). In particular, it suffices if the likelihood function is much more concentrated than the prior beliefs are, and prior beliefs do not strongly favor any region.

Next, suppose that there are two parameters to be estimated, namely an expected return $\mu$ and a standard deviation $\sigma$. Now, there are competing choices for a diffuse prior such as

$$p(\mu,\sigma) = \begin{cases} \dfrac{1}{2\Delta_1\Delta_2} & \text{for } \mu \in [-\Delta_1, \Delta_1] \\ & \quad \sigma \in [0, \Delta_2] \\ 0 & \text{otherwise} \end{cases} \tag{17}$$

$$p(\mu, V) = \begin{cases} \dfrac{1}{2\Delta_1\Delta_2} & \text{for } \mu \in [-\Delta_1, \Delta_1] \\ & \quad V \in [0, \Delta_2] \\ 0 & \text{otherwise} \end{cases} \tag{18}$$

$$p(\mu, \log\sigma) = \begin{cases} \dfrac{1}{4\Delta_1\Delta_2} & \text{for } \mu \in [-\Delta_1, \Delta_1] \\ & \quad \log\sigma \in [-\Delta_2, \Delta_2] \\ 0 & \text{otherwise} \end{cases} \tag{19}$$

Since $\log\sigma = \frac{1}{2}\log V$, a similar expression for $p(\mu, \log V)$ would not be a new alternative. Since the use of (16) is justified by convenience and the principle of stable estimation, even when not plausible, one should be permitted the choice between (17), (18), and (19) on the basis of convenience, since the principle of stable estimation would seem to apply about equally to any of them.

With two normally distributed random variables, $r = (r_1, r_2)$, the hypothesis space would most naturally include the choice of

$$h' = (\mu_1, \mu_2, \sigma_1, \sigma_2, \sigma_{12}, \text{or } \rho_{12}).$$

One way of forming diffuse priors for the above is to assume that $\mu_1, \sigma_1$ and $\mu_2, \sigma_2$ each have as priors (17), (18), or (19) and that $\rho_{12}$ is independently drawn with a prior density of

$$p(\rho) = \begin{cases} \frac{1}{2} & \text{for } -1 \leq \rho \leq 1 \\ 0 & \text{otherwise} \end{cases} \tag{20}$$

It might seem that one could repeat the process for $n = 8$ with $\mu_i, \sigma_i, i = 1, \ldots, 8$ having priors (17), (18), or (19) and with each $\rho_{ij}$ independently having (20) as a prior for $i = 1, \ldots, 7, j = i+1, \ldots, 8$. One problem with this is that it assigns positive probabilities to correlation matrixes which are logically impossible. For example, it is impossible to have $\rho_{ij} < -\frac{1}{7}$ for every $i \neq j$ for eight returns.

We use a different "diffuse approach" which avoids the above difficulty and is computationally quite convenient for the Monte Carlo analysis described below. This approach uses priors equivalent to nature drawing $r_{it}$ according to

$$p(r_{it}) = \begin{cases} \dfrac{1}{2\Delta} & \text{for } r_{it} \in [-\Delta, \Delta] \\ 0 & \text{otherwise} \end{cases} \tag{21}$$

independently for $i = 1, \ldots, 8$, $t = 1, \ldots, 216$, then computing $\mu_i$, $\sigma_i$, and $\rho_{ij}$ as the means, standard deviations, and the correlations of the randomly drawn $r_i$. The distribution of $(\mu_1, \mu_2, \ldots, \sigma_{88})$ is implicit. In other words, we will find it most convenient to assume that prior probability distribution of $(\mu_1, \ldots, \sigma_{88})$ is the same as that of the sample statistics of random variables $r_1, \ldots, r_8$ drawn uniformly and independently, for sample size $T = 216$. For example, for a very large $\Delta$ in (21) the distribution of $\mu_1$ is approximately normally distributed with a large standard deviation.

### 3.3 *Importance sampling*

Let

$$K = R^8 \times R^{216} \tag{22}$$

be the space of $8 \times 216$ real matrices. Examples of members of $K$ include $O$, the historical observation handed to each player, and $k_1, \ldots, k_{500}$, the 500 histories which the Michaud player generates. Recall, $H$ is defined in (4) as $R^{44}$. Members of $H$ include $h$ in (2), the parameters of a joint normal distribution of $(r_1, \ldots, r_8)$.

Let $f_{KH}$ be a function $f_{KH} : K \to H$ which associates with each point $k \in K$ the $(\mu_1, \ldots, \sigma_{88})$ vector $h \in H$ obtained by computing these parameters from the returns matrix $k$. For two points $k_1$, and $k_2$ in $K$ we define

$$L(k_1|k_2) = L[k_1|f_{KH}(k_2)] = \prod_{t=1}^{216} N[r^t; f_{KH}(k_2)] \quad (23)$$

where $N(r; h)$ is the normal density of the random vector $r$ given the parameters $h$. In other words, (23) defines the likelihood of $k_1$ given $k_2$ to mean the likelihood of getting the sample $k_1$ from a normal distribution with parameters $f_{KH}(k_2)$.

Let

$$K^* = \{r \in K | \quad |r_{it}| \leq \Delta \forall i, t\} \quad (24)$$

for some large $\Delta$. We assume that the prior density is uniformly distributed over this set, $K^*$. To evaluate an expected value as in (15) by integration would require integration over a large rectangle in an $8 \times 216$-dimensional space. This is not feasible. On the other hand, an estimate of $E(v)$ by Monte Carlo, for randomly drawn $v$, depends on sample size and the moments of $v$ rather than on the dimensionality of $K$.

Given any function $v(k)$ of the sample point $k$, in principle, one could estimate the Bayes player's $E(v)$ given $O$, by sampling $k$ from $K^*$ with probability

$$p(k) = L(O|k)/D \quad (25a)$$

where

$$D = \int_{K^*} L(O|k) \mathrm{d}k \quad (25b)$$

Instead, we will have the Bayes player use the same 500 samples from $K$ which the Michaud player uses to compute its resampled frontier. We must keep in mind that these 500 samples were drawn with probability density

$$q(k) = L(k|O) \quad (26)$$

That is, the $8 \times 216$ matrices $(r_{it})$ in Michaud's samples are drawn joint normally assuming the parameters of the "historical" observation $O$. Observe that $L(k|O)$ in (26) is not the same as $L(O|k)/D$ in (25a).

There is a standard correction applicable when we wish to estimate an expected value

$$E(v) = \int_{K^*} p(k)\, v(k)\, \mathrm{d}k \quad (27)$$

and we draw a sample, e.g. $v_1, \ldots, v_{500}$, with probability $q(k)$ rather than $p(k)$. The sample average

$$v^* = \frac{1}{500} \sum_{i=1}^{500} v_i \quad (28)$$

has expected value

$$E(v^*) = \int_{K^*} q(k)\, v(k)\, \mathrm{d}k \quad (29)$$

which may differ from $E(v)$ in (27). Instead, we may use a weighted average

$$\bar{v} = \frac{1}{500} \sum [p(k)/q(k)]\, v(k_i) \quad (30)$$

This has expected value

$$E(\bar{v}) = \int q(k)\, [p(k)/q(k)]\, v(k)\, \mathrm{d}k = E(v) \quad (31)$$

as desired.

The weights in (30) correct for sample probabilities $q(k)$ provided $q(k) > 0$ when $p(k) > 0$. This does not mean that all sampling distributions $q(k)$ are equally good. Since all are adjusted to have the correct $E(v)$, the variance of $v$ depends only on

$$E(v^2) = \int q(k)\left(\frac{p(k)\cdot v(k)}{q(k)}\right)^2 dk$$
$$= \int \frac{[p(k)v(k)]^2}{q(k)} dk \tag{32}$$

The minimum of this subject to

$$\int q(k)dk = 1.0 \tag{33}$$

is

$$q(k) = p(k)|v(k)| \tag{34}$$

Since our sample will serve to estimate the expected value of many different $v(k)$, perhaps all we can conclude from (34) is that it is best to avoid large $q(k)$ where $p(k)$ is relatively small. $q(k) = p(k)$ seems at least good and perhaps ideal.

To compute $p(k)$ we need $D$ from (25b). We now discuss how we approximate this. Let "Vol" be the volume of $K^*$ in (24). Assuming $L(O|k)$may be ignored in $K - K^*$, Eq. (25b) may be written as

$$D = \text{Vol} \int_{K^*} \frac{L(O|k)dk}{\text{Vol}} \tag{35}$$

This is the volume of $K^*$ times the expected value of $L(O|k)$ in $K^*$ when $k$ is drawn uniformly from it; i.e. with probability density

$$p(k) = 1/\text{Vol} \tag{36}$$

We can approximate this expected value using the sample of 500 $k_i$ drawn with probability density $L(k_i|O)$ by weighing the observed $L(O|k_i)$ by the ratio of desired density (1/Vol) to the density used $L(k_i|O)$. That is, we can approximate the integral (expected value) in (35) by

$$\overline{L} = \frac{1}{500}\sum \frac{L(O|k_i)}{L(k_i|O)\cdot \text{Vol}} \tag{37}$$

But $D$ in (35) is "Vol" times the integral, so the approximation to $D$ is

$$\overline{D} = \frac{1}{500}\sum \frac{L(O|k_i)}{L(k_i|O)} \tag{38}$$

i.e. the observed average ratio of $L(O|k_i)$ to $L(k_i|O)$. It may be objected that if $\overline{D}$ is substituted for $D$ in (25a), $p(k)$ is a ratio of unbiased estimators, which is not necessarily unbiased. On the other hand, with $\overline{D}$ thus used for $D$ in (25a), the weights in (30) sum to one. In this case $\bar{v}$ is a weighted average of $v(k)$, which seems attractive.

Our procedure then is as follows. To approximate the RDMs posterior mean vector $\mu$ and covariance matrix $C$ we approximate the expected values of variables $v$, such as $Er_i$ and $Er_ir_j$ for all $i, j$, then use the relationships in (7) and (8). To estimate $E(v)$, we evaluate $v$ for each of the Michaud samples from $O$, namely $k_1, \ldots, k_{500}$, then form the weighted average $\bar{v}$ of the $v(k_i)$ where the weights are shown in (30) with $q(k)$ defined in (26) and $p(k)$ defined in (25a) and (38).

For the one case we checked, most weights $q(k)/p(k)$ are close to unity. Table 2 shows the deciles of the 500 weights computed for Truth 1 History 1. All weights were greater than 0.80 and not greater than 1.025. Ninety percent of the weights were between 0.96 and 1.025. This says that the Michaud sample is a good one for the present

**Table 2** Distribution of weights $p(k)/q(k)$ for Truth 1 History 1.

| Deciles | From | To |
|---|---|---|
| 1st | 0.809 | 0.961 |
| 2nd | 0.961 | 0.984 |
| 3rd | 0.984 | 0.999 |
| 4th | 0.999 | 1.005 |
| 5th | 1.005 | 1.012 |
| 6th | 1.012 | 1.015 |
| 7th | 1.015 | 1.019 |
| 8th | 1.019 | 1.021 |
| 9th | 1.021 | 1.023 |
| 10th | 1.023 | 1.025 |

purpose, according to (34) and the discussion that follows it.

## 4 Results

The results of the experiment are presented in Tables 3 and 4. The first panel of Table 3 shows averages of estimated and actual expected utility achieved by the two players. Specifically, for $\lambda = 0.5, 1.0$, and 2.0, as indicated by the row labeled "Lambda", and for each player, as indicated by the row labeled "Player", the table presents two columns of information. The first column is the average (over the 100 histories generated for a truth) of the players' estimate of expected utility. The second column is the average of actual utility as evaluated by the referee. For example, on the line labeled Truth 1 we see that, on average over the 100 histories generated for Truth 1, the Bayesian player believed it had achieved an expected utility of 0.01181 whereas the average of its actual *EU* was 0.00712. The comparable numbers for the Michaud player are 0.01032 and 0.00753. Thus, both players overestimated how well they did, but the Michaud player overestimated less and achieved more. On the next nine lines similar numbers are reported for Truth 2 through Truth 10. The final three lines of the panel summarize results for the 10 truths. In particular, the average over the 10 truths of the Bayesian player's estimate was 0.01383 but it actually achieved 0.00861. In the average over all 10 truths, again, the Michaud player overestimated less and achieved more. In fact, comparing the average *EU* each player achieved in each of the 10 truths, the average over the 100 histories was greater for the Michaud player than the Bayes player in the case of each of the 10 truths, as noted in the last line of Panel A.

A similar story holds for $\lambda = 1.0$ and 2.0. Looking at the last row of Panel A for the actual *EU* achieved by the two players for these cases we see that the Michaud player achieved a higher average (over the 100 histories for a given Truth) in 10 out of 10 truths for $\lambda = 1.0$ and 2.0.

For some individual histories of the 100 histories of a given truth, the Bayes player had a higher *EU* than the Michaud player. In fact, in Panel B of Table 1 the entry for $\lambda = 0.5$, Bayes player, Truth 1 reports that the Bayes player achieved a higher *EU* than the Michaud player in 54 out of the 100 histories, despite having a lower average over the 100. Sticking with Truth 1, the Bayes player also "won" 54 out of 100 times for $\lambda = 1.0$, and 50 out of 100 for $\lambda = 2.0$. The Bayes player's "win count" was even more favorable in the case of Truth 6. In this case, the Bayes player "beat" the Michaud player 62 times out of 100 for $\lambda = 0.5$, 60 for $\lambda = 1.0$ and 66 for $\lambda = 2.0$. Nevertheless, the average *EU* achieved, averaged over the 100 histories, was higher for the Michaud player in each of these Truths.

Panel C of Table 1 shows that, for a given Truth, the standard deviation of the achieved *EU* was higher for the Bayesian than the Michaud player. For example, for Truth 1, $\lambda = 0.5$, the standard deviation of *EU* for the Bayesian player was 0.00210 as compared to 0.00132 for the Michaud player. In fact, for all three values of $\lambda$ and all 10 truths, the variance of the actual *EU* was lower for the Michaud player than the Bayes player.

Most significant for our purpose is the fact that the Michaud strategy delivered higher average *EU* in 10 out of 10 truths for three out of three values of $\lambda$. Thus, the Michaud player did a better job of achieving the objective, namely high *EU*.

Table 4 displays the results of a slightly different game. In this second game, for each history and each truth each player computes an efficient frontier as in the first game. But instead of picking a point from the frontier for each $\lambda$, the player passes its entire frontier to the referee. For each $\lambda$ the referee picks the point on the player's frontier that has the

**Table 3** Player's choice of portfolio.

| λ: | 0.5 | 0.5 | 0.5 | 0.5 | 1.0 | 1.0 | 1.0 | 1.0 | 2.0 | 2.0 | 2.0 | 2.0 |
|---|---|---|---|---|---|---|---|---|---|---|---|---|
| Player: | Bayes | Bayes | Michaud | Michaud | Bayes | Bayes | Michaud | Michaud | Bayes | Bayes | Michaud | Michaud |
| Eval. by: | Player | Referee | Player | Referee | Player | Referee | Player | Referee | Player | Referee | Player | Referee |
| *Panel A:* EU *averaged over 100 histories, for each of 10 truths* | | | | | | | | | | | | |
| Truth 1 | 0.01181 | 0.00712 | 0.01032 | 0.00753 | 0.01004 | 0.00564 | 0.00886 | 0.00594 | 0.00754 | 0.00394 | 0.00678 | 0.00426 |
| Truth 2 | 0.01528 | 0.00885 | 0.01194 | 0.00901 | 0.01389 | 0.00783 | 0.01085 | 0.00801 | 0.01160 | 0.00616 | 0.00902 | 0.00664 |
| Truth 3 | 0.01011 | 0.00614 | 0.01009 | 0.00737 | 0.00887 | 0.00534 | 0.00904 | 0.00636 | 0.00692 | 0.00410 | 0.00721 | 0.00481 |
| Truth 4 | 0.01457 | 0.00850 | 0.01147 | 0.00862 | 0.01324 | 0.00746 | 0.01041 | 0.00763 | 0.01094 | 0.00573 | 0.00849 | 0.00600 |
| Truth 5 | 0.01170 | 0.00641 | 0.00984 | 0.00694 | 0.00988 | 0.00480 | 0.00846 | 0.00549 | 0.00706 | 0.00282 | 0.00612 | 0.00322 |
| Truth 6 | 0.01646 | 0.01056 | 0.01304 | 0.01078 | 0.01462 | 0.00890 | 0.01149 | 0.00914 | 0.01173 | 0.00670 | 0.00911 | 0.00700 |
| Truth 7 | 0.01590 | 0.01147 | 0.01408 | 0.01152 | 0.01412 | 0.00989 | 0.01271 | 0.01015 | 0.01124 | 0.00758 | 0.01036 | 0.00793 |
| Truth 8 | 0.01502 | 0.00956 | 0.01261 | 0.01005 | 0.01329 | 0.00811 | 0.01119 | 0.00861 | 0.01053 | 0.00578 | 0.00866 | 0.00610 |
| Truth 9 | 0.01402 | 0.00906 | 0.01241 | 0.00961 | 0.01204 | 0.00719 | 0.01087 | 0.00798 | 0.00892 | 0.00462 | 0.00812 | 0.00521 |
| Truth 10 | 0.01343 | 0.00846 | 0.01130 | 0.00900 | 0.01176 | 0.00676 | 0.00975 | 0.00735 | 0.00909 | 0.00402 | 0.00712 | 0.00453 |
| Grand mean | 0.01383 | 0.00861 | 0.01171 | 0.00904 | 0.01217 | 0.00719 | 0.01036 | 0.00767 | 0.00956 | 0.00514 | 0.00810 | 0.00557 |
| Std Dev | 0.00205 | 0.00171 | 0.00138 | 0.00150 | 0.00200 | 0.00161 | 0.00133 | 0.00145 | 0.00190 | 0.00148 | 0.00129 | 0.00142 |
| No. times better | | 0 | | 10 | | 0 | | 10 | | 0 | | 10 |
| *Panel B: Number of "wins" out of 100 histories, for each of 10 truths* | | | | | | | | | | | | |
| Truth 1 | | 54 | | 46 | | 54 | | 46 | | 50 | | 50 |
| Truth 2 | | 52 | | 48 | | 54 | | 46 | | 65 | | 35 |
| Truth 3 | | 46 | | 54 | | 43 | | 57 | | 41 | | 59 |
| Truth 4 | | 57 | | 43 | | 61 | | 39 | | 64 | | 36 |
| Truth 5 | | 43 | | 57 | | 27 | | 73 | | 30 | | 70 |
| Truth 6 | | 62 | | 38 | | 60 | | 40 | | 66 | | 34 |
| Truth 7 | | 57 | | 43 | | 53 | | 47 | | 42 | | 58 |
| Truth 8 | | 54 | | 46 | | 48 | | 52 | | 41 | | 59 |
| Truth 9 | | 32 | | 68 | | 28 | | 72 | | 27 | | 73 |
| Truth 10 | | 61 | | 39 | | 49 | | 51 | | 52 | | 48 |
| Avg No. wins | | 51.80 | | 48.20 | | 47.70 | | 52.30 | | 47.80 | | 52.20 |
| No. times better | | 7 | | 3 | | 5 | | 5 | | 5 | | 5 |

*Panel C: Standard deviation of* EU *over 100 histories, for each of 10 truths*

| | | | | | | | | | | | | |
|---|---|---|---|---|---|---|---|---|---|---|---|---|
| Truth 1 | 0.00516 | 0.00210 | 0.00401 | 0.00132 | 0.00470 | 0.00160 | 0.00359 | 0.00102 | 0.00356 | 0.00116 | 0.00265 | 0.00077 |
| Truth 2 | 0.00445 | 0.00149 | 0.00395 | 0.00084 | 0.00416 | 0.00185 | 0.00372 | 0.00113 | 0.00364 | 0.00258 | 0.00331 | 0.00121 |
| Truth 3 | 0.00354 | 0.00244 | 0.00339 | 0.00109 | 0.00331 | 0.00231 | 0.00321 | 0.00092 | 0.00286 | 0.00178 | 0.00284 | 0.00080 |
| Truth 4 | 0.00413 | 0.00203 | 0.00369 | 0.00101 | 0.00398 | 0.00221 | 0.00356 | 0.00118 | 0.00377 | 0.00235 | 0.00331 | 0.00117 |
| Truth 5 | 0.00514 | 0.00161 | 0.00347 | 0.00077 | 0.00475 | 0.00126 | 0.00329 | 0.00064 | 0.00396 | 0.00077 | 0.00275 | 0.00058 |
| Truth 6 | 0.00469 | 0.00223 | 0.00476 | 0.00114 | 0.00427 | 0.00227 | 0.00438 | 0.00101 | 0.00379 | 0.00242 | 0.00373 | 0.00117 |
| Truth 7 | 0.00544 | 0.00178 | 0.00347 | 0.00088 | 0.00515 | 0.00155 | 0.00331 | 0.00086 | 0.00447 | 0.00111 | 0.00296 | 0.00073 |
| Truth 8 | 0.00399 | 0.00217 | 0.00413 | 0.00119 | 0.00391 | 0.00169 | 0.00398 | 0.00112 | 0.00376 | 0.00117 | 0.00360 | 0.00073 |
| Truth 9 | 0.00584 | 0.00144 | 0.00371 | 0.00056 | 0.00551 | 0.00130 | 0.00359 | 0.00070 | 0.00467 | 0.00089 | 0.00319 | 0.00065 |
| Truth 10 | 0.00399 | 0.00283 | 0.00445 | 0.00155 | 0.00391 | 0.00212 | 0.00422 | 0.00130 | 0.00360 | 0.00166 | 0.00353 | 0.00077 |
| Avg Std Dev | | 0.00201 | | 0.00104 | | 0.00182 | | 0.00099 | | 0.00159 | | 0.00086 |
| No. times better | | 0 | | 10 | | 0 | | 10 | | 0 | | 10 |

**Table 4** Referee's choice of portfolio.

| λ: | 0.5 | 0.5 | 1 | 1 | 2 | 2 |
|---|---|---|---|---|---|---|
| Player: | Bayes | Michaud | Bayes | Michaud | Bayes | Michaud |
| Eval. by: | Referee | Referee | Referee | Referee | Referee | Referee |
| *Panel A:* EU *averaged over 100 histories, for 10 truths* | | | | | | |
| Truth 1 | 0.007811 | 0.007709 | 0.006303 | 0.006253 | 0.004852 | 0.004899 |
| Truth 2 | 0.009625 | 0.009407 | 0.008641 | 0.008594 | 0.006967 | 0.007104 |
| Truth 3 | 0.007111 | 0.007552 | 0.006198 | 0.006647 | 0.004741 | 0.005139 |
| Truth 4 | 0.009157 | 0.008915 | 0.008109 | 0.008049 | 0.006220 | 0.006395 |
| Truth 5 | 0.006721 | 0.007008 | 0.005200 | 0.005662 | 0.003504 | 0.003661 |
| Truth 6 | 0.011378 | 0.011183 | 0.009781 | 0.009608 | 0.007486 | 0.007481 |
| Truth 7 | 0.011935 | 0.011571 | 0.010425 | 0.010303 | 0.008178 | 0.008260 |
| Truth 8 | 0.009674 | 0.010071 | 0.008225 | 0.008714 | 0.005799 | 0.006309 |
| Truth 9 | 0.009423 | 0.009641 | 0.007712 | 0.008169 | 0.005112 | 0.005576 |
| Truth 10 | 0.008193 | 0.008854 | 0.006665 | 0.007339 | 0.004151 | 0.004718 |
| Grand mean | 0.009103 | 0.009191 | 0.007726 | 0.007934 | 0.005701 | 0.005954 |
| Std Dev | 0.001701 | 0.001509 | 0.001654 | 0.001473 | 0.001506 | 0.001414 |
| No. times better | 5 | 5 | 5 | 5 | 1 | 9 |
| *Panel B: Number of wins out of 100 histories, for 10 truths* | | | | | | |
| Truth 1 | 65 | | 60 | | 53 | |
| Truth 2 | 66 | | 64 | | 58 | |
| Truth 3 | 56 | | 52 | | 47 | |
| Truth 4 | 69 | | 67 | | 59 | |
| Truth 5 | 52 | | 32 | | 40 | |
| Truth 6 | 67 | | 60 | | 60 | |
| Truth 7 | 74 | | 64 | | 54 | |
| Truth 8 | 44 | | 45 | | 37 | |
| Truth 9 | 55 | | 43 | | 27 | |
| Truth 10 | 58 | | 58 | | 41 | |
| Avg wins | 60.6 | | 54.5 | | 47.6 | |
| No. times greater | 9 | | 7 | | 5 | |
| *Panel C: Std Dev of* EU *over 100 histories, for 10 truths* | | | | | | |
| Truth 1 | 0.00169 | 0.00112 | 0.00113 | 0.00082 | 0.00063 | 0.00043 |
| Truth 2 | 0.00110 | 0.00058 | 0.00128 | 0.00060 | 0.00127 | 0.00061 |
| Truth 3 | 0.00186 | 0.00083 | 0.00163 | 0.00071 | 0.00117 | 0.00060 |
| Truth 4 | 0.00143 | 0.00062 | 0.00153 | 0.00060 | 0.00139 | 0.00056 |
| Truth 5 | 0.00145 | 0.00053 | 0.00106 | 0.00042 | 0.00054 | 0.00027 |
| Truth 6 | 0.00106 | 0.00086 | 0.00074 | 0.00058 | 0.00099 | 0.00059 |
| Truth 7 | 0.00131 | 0.00084 | 0.00102 | 0.00070 | 0.00090 | 0.00054 |

Table 4 *(continued)*

| $\lambda$: | 0.5 | 0.5 | 1 | 1 | 2 | 2 |
|---|---|---|---|---|---|---|
| Player: | Bayes | Michaud | Bayes | Michaud | Bayes | Michaud |
| Eval. by: | Referee | Referee | Referee | Referee | Referee | Referee |
| Truth 8 | 0.00137 | 0.00089 | 0.00121 | 0.00076 | 0.00093 | 0.00051 |
| Truth 9 | 0.00137 | 0.00053 | 0.00125 | 0.00049 | 0.00082 | 0.00039 |
| Truth 10 | 0.00274 | 0.00145 | 0.00223 | 0.00114 | 0.00120 | 0.00067 |
| Avg Std Dev | 0.00154 | 0.00082 | 0.00131 | 0.00068 | 0.00098 | 0.00052 |
| No. times lower | 0 | 10 | 0 | 10 | 0 | 10 |

highest true *EU*. Game 2 thus addresses the question of whether the superiority of the Michaud player over the diffuse Bayesian player in the first game is due to a better frontier or to a better pick from an equally good frontier.

The Bayes player does much better in Game 2 than it did in Game 1. In particular, for $\lambda = 0.5$ and 1.0 Panel A of Table 4 shows that with five out of 10 truths the Bayesian player achieves higher average *EU* than the Michaud player as compared to 0 out of 10 in Game 1. Also, Panel B shows that for $\lambda = 0.5$ and 1.0 the Bayesian player has a higher *EU* in many more histories for a given Truth than the Michaud player. On the other hand, the Michaud player comes out ahead overall. In particular, for every $\lambda$ the "Grand Mean" of achieved *EU* averaged over all truths is greater for the Michaud player than the Bayesian player. However, the outperformance of the Michaud player over the Bayes player is smaller in the second game than in the first. In particular, for $\lambda = 0.5$ the difference in performance between the two players is only about 20% as great in the second game as it is in the first ($0.000088 = 0.009191 - 0.009103$ versus $0.00043 = 0.00904 - 0.00861$), about 44% as great when $\lambda = 1.0$ and 59% as great when $\lambda = 2.0$.

As explained in the next section, for $\lambda = 0.5$, *EU* in (1) is approximately[3] $E(\ln(1+r))$. This, in turn, is $\ln(1+g)$ where $g$ is the geometric mean or growth rate. We can, therefore, give the results in Tables 3 and 4 a more concrete interpretation for the case of $\lambda = \frac{1}{2}$. Annualizing, the Bayes player believes it can achieve an "average"[4] annual growth rate of 18.05% ($0.180548 = \exp(12 \cdot (0.01383)) - 1$), whereas the portfolios it chose had an average actual growth rate of 10.89% and the best from its frontier averaged a growth rate of 11.54%. The Michaud player thought it could achieve an average annual growth of 15.09%; the portfolios it chose had an average growth rate of 11.46%; the actual average highest growth portfolio on its frontier was 11.66%. Thus, in game 1, the Michaud methodology adds 0.57 to the average growth rate. In game 2 it adds 0.12.

The relatively better performance of the Bayesian player in Game 2 (as compared to its performance in Game 1) suggests that the Game 1 superiority of the Michaud player is more due to a wise pick from its frontier than due to a superior frontier, though the latter reason is also applicable.

## 5 Questions

The preceding results raise questions for portfolio theory and practice. In particular, the results represent something of a crisis for the theoretical foundations of portfolio theory as presented in Part IV of Markowitz (1959), Chapters 10–13. Chapters 10 through 12 present introductory accounts of utility analysis as justified by Von Neumann and Morgenstern (1944), personal probability as justified by Savage (1954), and dynamic programming as presented by Bellman (1957). Chapter 13 applies these principles to the problem of selecting a portfolio. Specifically, mean–variance analysis is justified as an approximation to the single-period "derived" utility function always associated with many-period utility maximization. It is argued that the mean–variance approximation should be good as long as the probability distribution of return is not spread out too much. Calculations—by Markowitz (1959), Young and Trent (1969), Levy and Markowitz (1979), Dexter *et al.* (1980), Pulley (1981, 1983), Kroll *et al.* (1984), Simaan (1987) and Hlawitschka (1994)—show that, for most utility functions proposed for practice, the mean–variance approximation to expected utility is quite robust. As Levy and Markowitz conclude

> If Mr. X can carefully pick the E,V efficient portfolio which is best for him then Mr. X, who still does not know his current utility function, has nevertheless selected a portfolio with maximum or almost maximum expected utility.

In addition, Markowitz and van Dijk (2003) illustrate the ability of a suitably constructed "single-period" mean–variance analysis to give near-optimum results in the case of transaction costs and changing probability distributions. One caveat however: as Grauer (1986) illustrates, the return distributions from highly levered portfolios are too spread out for mean–variance approximations to do well. However, for unlevered return distributions as considered in the present paper, computations have generally shown mean–variance to be quite good.

Thus, until now, calculations seem to support the theoretical foundations for mean–variance analysis presented in Part IV of Markowitz (1959). An integral part of these foundations is that a RDM will use probability beliefs where objective probabilities are not known, and will update these beliefs according to the Bayes rule as evidence accumulates. Usually, when Bayesian inference is tried in practice it is assumed that, prior to the sample in hand, beliefs are "diffuse"—i.e. "neutral" in some sense with respect to which hypothesis is true—as recommended by Jefferies (1948) or Edwards *et al.* (1963).

Given this background, the results presented in this paper are badly in need of an explanation. Such explanation could be in terms of why Bayesian updating did not do better, or why the Michaud estimation did so well.

Concerning why Bayesian updating did not do better: it may have to do with the difference between the computation which we performed and which a RDM would perform. The latter is an integration over a high-dimensional space, well beyond foreseeable human computational abilities. We approximated this integral by Monte Carlo sampling. (Note the distinction between the sample which the referee handed both players, and the sample we used to approximately compute the integral which the RDM computes exactly.) If this—exact versus approximate calculation of updated beliefs—is the source of difficulty with the Bayesian approach taken here, then, maybe the conclusion will be that Bayesian inference is ideal for the RDM but not for the human, at least at the level of computational effort spent by the Bayesian and Michaud players in the reported experiment.

Alternatively, perhaps the problem with the approach taken here is the priors used. Perhaps "diffuse prior" should be defined differently. Or, perhaps, an informed prior should be used like those of Black and Litterman (1990)—but updating the priors using history rather than user estimates as in Black and Litterman.[5]

Expected utility and Bayesian inference were originally proposed, by Daniel Bernoulli and Thomas Bayes in the Eighteenth Century, as plausible rules for action when the future is unknown (see Bernoulli, 1954; Bayes, 1958). Von Neumann and Morgenstern (1944) and Savage (1954) derive these rules from more basic principles of rational behavior. The resampled frontier as presented by Michaud (1998) is a plausible procedure which, we find, works quite well. But how does it relate to the theory of rational behavior? Does it contradict one or more of Savage's axioms? If so, is this a black mark against the method or against the axioms? Or does Michaud's procedure somehow satisfy the Savage axioms? We would very much like to know the answers to some or all of these questions.

Practical questions, raised by the success of the Michaud method in the experiments reported here, include those of costs and benefits. In particular, how much expected return do these procedures add for a given level of risk—in practice. This may involve transaction costs, changing probability distributions, non-normal distributions—all assumed away in the current experiments. Historical backtests might shed some light on these matters.

Concerning costs, computation costs may or may not be a problem. It does not take long or cost much these days to generate a set of 500 frontiers and average these. But it might still be computationally burdensome to compute many such resampled frontiers in a backtest with many monthly re-optimizations, with the backtest frequently repeated to see the effects of alternate parameter settings. However, a Bayesian update of beliefs would also be computationally burdensome in such a case.

Finally, the cost of using a resampled efficient frontier depends on what the patent holder charges for the use of this patented procedure (see note 1).

## 6 Conclusions

This paper reports the results of an experiment comparing two procedures for dealing with sampling error in the inputs to a mean–variance analysis. One procedure is the Bayesian updating of diffuse priors. The other is Michaud's resampled efficient frontier. In the experiment a referee generates 10 "truths" at random from a "seed" distribution. From each "truth" the referee randomly generates 100 histories. Each history is presented to a Bayesian player and a Michaud player. Each player follows its prescribed procedure to determine which portfolio would provide highest $E - \lambda V$ for $\lambda = 0.5, 1.0$, and 2.0. Sometimes one player, sometimes the other picks a portfolio with higher $E - \lambda V$. But in the case of each truth and each value of $\lambda$, the average of the 100 values of $E - \lambda V$ is higher for the Michaud player than the Bayes player. However, the Bayes player does almost as well as the Michaud player when each player presents its entire efficient frontier to the referee, and the referee picks the player's best portfolio from the frontier. This suggests that the chief problem with the Bayesian player's choice of portfolio is that the latter is more over-optimistic than is the Michaud player in estimating achievable portfolio mean and variance.

This result has practical implications for the estimation of inputs to a mean–variance analysis, even for methods other than the two considered explicitly here. For example, in practice, mean–variance analysis is often performed at an asset class level with estimates of means based partly on judgment, but using historical variances and covariances. The results of this paper imply that these variance estimates are too low. First, if you accept the theory of rational behavior under uncertainty developed by

Savage (1954), as explained by Markowitz (1959) Chapter 12, then you should not use historical variance, nor even an average variance—averaged over possible explanations of history. Rather, you should use the latter *plus* a term reflecting your uncertainty in your estimate of the mean. Furthermore, the results of the present paper imply that, for reasons unknown to us, when this theoretical correction is made, the investor is still too optimistic for his or her own best interest.

## Acknowledgments

The authors thank Anthony Tessitore for extremely valuable suggestions.

## Notes

1 Resampled efficiency, as described in Michaud (1998, Chapters 6 and 7), was co-invented by Richard Michaud and Robert Michaud and is a US patented procedure, #6,003,018, December 1999, patent pending worldwide. New Frontier Advisors, LLC, has exclusive licensing rights worldwide.

2 The assumption of a single-period utility function is not less general than the assumption of many-period or continuous-time utility maximization, since many-period or continuous-time utility maximization may be reduced to a series of one-period or instantaneous utility maximizations using a "derived" utility function, as described by Bellman (1957). In general, the time-varying derived utility function $U_t$ may be a complicated function that includes state variables as well as returns, and depends on what has gone before. Our specific assumption, that $U_t$ is given by (1), is a vast simplification which we justify on the grounds that our objective is not to solve the dynamic programming problem for some many-period or continuous-time investment model, but to take a reading on the ability of two alternate methods to handle uncertainty.

3 For other values of $\lambda$, $EU$ in (1) approximates the expected value of other utility functions. The choices made by a Bernoulli/Von Neumann and Morgenstern utility function are not affected by adding a constant or multiplying by a positive constant. That is, the same decisions maximize $E[a + bU(r)]$, $b > 0$, as those that maximize $EU(r)$. It is, therefore, essential to the validity of the comparisons made in the text—e.g. that the difference in performance is only 20% as great in game 2 as game 1 when $\lambda = 0.5$—that this comparison is in fact unaffected by the arbitrary choice of $a$ and $b > 0$.

4 The "average" referred to here is the antilog of an average logarithm, therefore, a geometric mean.

5 Harvey *et al.* (2003) reports the results of an experiment in which Bayes outperforms Michaud when conjugate priors are used.

## References

Bayes, T. (1958). "Essay Toward Solving a Problem in the Doctrine of Chances: with a biographical note by G. A. Barnard." *Biometrika* 45, 293–315. (Also published separately by the Biometrika Office, University College, London.) *Philosophical Transactions of the Royal Society* 370–418, 1763.

Bellman, R. E. (1957). *Dynamic Programming.* Princeton, New Jersey: Princeton University Press.

Bernoulli, D. (1954). "Specimen theoriae novae de mensura sortis. Exposition of a New Theory on the Measurement of Risk" (English translation by Louise Sommer). *Econometrica* 22, 23–26. (Originally published in 1738. *Comm. Acad. Sci. Imp. Petropolitanae* 5, 175–192.)

Black, F. and Litterman, R. (1990). "Asset Allocation: Combining Investor Views with Market Equilibrium." *Journal of Fixed Income* Goldman Sachs, September.

Dexter, A. S., Yu, J. N. W., and Ziemba, W. T. (1980). "Portfolio Selection in a Lognormal Market When the Investor Has a Power Utility Function: Computational Results." In: M. A. H. Dempster (ed.), *Stochastic Programming.* New York: Academic Press, 507–523.

Edwards, W., Lindman, H., and Savage, L. J. (1963). "Bayesian Statistical Inference for Psychological Research." *Psychological Review* 70(3), 193–242.

Grauer, R. R. (1986). "Normality, Solvency, and Portfolio Choice." *Journal of Financial and Quantitative Analysis* 21(3), 265–278.

Harvey, C. R., Liechty, J. C., Leichty, M. W. and Muller, P. (2003). Portfolio Selection with Higher Moments, working paper, Duke University, Durham, NC.

Hlawitschka, W. (1994). "The Empirical Nature of Taylor-Series Approximations to Expected Utility." *The American Economic Review* 84(3), 713–719.

Jeffreys, H. (1948). *Theory of Probability.* Oxford: Clarendon Press.

Kroll, Y., Levy, H., and Markowitz, H. M. (1984). "Mean Variance Versus Direct Utility Maximization." *Journal of Finance* 39(1) 47–61.

Levy, H. and Markowitz, H. M. (1979). "Approximating Expected Utility by a Function of Mean and Variance." *American Economic Review* 69(3), 308–317.

Markowitz, H. M. (1959). *Portfolio Selection: Efficient Diversification of Investments.* John Wiley & Sons New York: 2nd edn., Basil Blackwell, Cambridge, MA.

Markowitz, H. M. and Usmen, N. (1996a). "The Likelihood of Various Stock Market Return Distributions, Part 1: Principles of Inference." *Journal of Risk and Uncertainty* 13, 207–219.

Markowitz, H. M. and Usmen, N. (1996b). "The Likelihood of Various Stock Market Return Distributions, Part 2: Empirical Results." *Journal of Risk and Uncertainty* 13, 221–247.

Markowitz, H. M. and van Dijk, E. L. (2003). "Single-Period Mean–Variance Analysis in a Changing World." *Financial Analysts Journal* 59(2), 30–44.

Michaud, R. O. (1998). *Efficient Asset Management: A Practical Guide to Stock Portfolio Optimization and Asset Allocation.* Boston, MA: Harvard Business School Press.

Pulley, L. M. (1981). "A General Mean–variance Approximation to Expected Utility for Short Holding Periods." *Journal of Financial and Quantitative Analysis* 16, 361–373.

Pulley, L. M. (1983). "Mean-Variance Approximations to Expected Logarithmic Utility." *Operations Research* 31(4), 685–696.

Savage, L. J. (1954). *The Foundations of Statistic*, 2nd edn. Dover Publications, Inc., New York: John Wiley & Sons.

Simaan, Y. (1987). "Portfolio Selection and Capital Asset Pricing for a Class of Non-Spherical Distributions of Assets Returns." PhD Thesis, Baruch College, The City University of New York.

Von Neumann, J. and Morgenstern, O. (1944). *Theory of Games and Economic Behavior, Princeton University Press*, 3rd ed., Wiley, 1967.

Young, W. E. and Trent, R. H. (1969). "Geometric Mean Approximation of Individual Security and Portfolio Performance." *Journal of Financial and Quantitative Analysis* 4, 179–199.

*Keywords*: Resampled frontier; Bayesian analysis; diffuse Bayes; mean–variance analysis; sampling errors; Michaud.

**On Socks, Ties and Extended Outcomes**

Harry M. Markowitz
*Harry Markowitz Company*
*1010 Turquoise Street, Suite 245*
*San Diego, California 92109*

As a result of conversations with Robin Pope during FUR VII, I now realize that the socks versus tie example of Markowitz 1959, Chapter 10, cannot be resolved in the way suggested there. I report below the additional problem, which I had not previously realized, and its resolution.

Chapter 10 describes examples which seem to be contradictions of the expected utility rule. It is argued there that they are not contradictions at all, but cases in which the set of outcomes has been misspecified. One example says that I would prefer a tie to socks for my birthday, but would not like to know which in advance. In other words, I would prefer a fifty-fifty chance of socks versus tie to knowing in advance which I will get.

This would seem to be a contradiction of the expected utility maxim since, with expected utility, if outcome A is preferred to outcome B, it is preferred to a fifty-fifty chance of A or B. The solution to this apparent contradiction is that "outcomes" here -- i.e., that which the decision maker seeks -- consists not only of socks or tie, but states of anticipation, surprise and the fun of the birthday ceremony.

To resolve the paradox it is sufficient to distinguish four outcomes depending on the decision maker's states in two time intervals:

$$0_1 = (KS, GS)$$
$$0_2 = (KT, GT)$$
$$0_3 = (FF, GS)$$
$$0_4 = (FF, GT)$$

"On Socks, Ties and Extended Outcomes", Economic and Environmental Risk and Uncertainty: New Models and Methods, Kluwer Academic Publishers, 1997, pp. 219-226.

| where: | is short for: |
|---|---|
| KS | "know socks" |
| GS | "get socks" |
| KT | "know tie" |
| GT | "get tie" |
| FF | have a fifty-fifty chance of socks or tie. |

GS represents both the receipt of socks and their subsequent use. GT, KS, KT and FF similarly may represent a sequence of states during one of the two time intervals distinguished here. The various states during the time interval need not be separately denoted for this analysis.

Since there are four outcomes -- $0_1$, $0_2$, $0_3$, $0_4$ -- the Von Neumann and Morgenstern (VN-M) axioms[1] assume, among other things, that the decision maker has a preference ordering among probability distributions of these outcomes: $(p_1, p_2, p_3, p_4)$ where $p_i$ is the probability of the i-th outcome. The VN-M axioms imply that there exist numbers $u_1, u_2, u_3, u_4$ such that $P = (p_1, p_2, p_3, p_4)$ is preferred to $Q = (q_1, q_2, q_3, q_4)$ if and only if

$$\Sigma u_i p_i > \Sigma u_i q. \tag{1}$$

With outcomes defined as above, the decision maker's preferences do not contradict expected utility. They only require that

$$1/2\, u_3 + 1/2\, u_4 > u_2 > u_1 \tag{2}$$

It is usually assumed that the decision maker can rank any two probability distributions P and Q. There is a problem with this not recognized in Markowitz (1959). Consider, for example, the probability distribution (0, 0, .3, .7). This is a thirty percent chance that the probability is fifty-fifty, and socks result; and a seventy percent chance that the probability is fifty-fifty, and a tie results. Ranking this probability distribution makes no more sense than ranking a fifty-fifty chance that two equals one.

It follows that the only probability distributions which can be contemplated, much less ranked, are those of the form $(p_1, p_2, .5(1-p_1-p_2), .5(1-p_1-p_2))$[2].

The probability distributions P which are of this form constitute a convex set S. One can postulate the axioms of expected utility for probability distributions P, Q, R... from the set S. In this case there exist $u_1$, $u_2$, $u_3$, $u_4$ which order P, Q, R... according to preferences, but these $u_i$ are not unique. As usual, the origin and scale of utility is arbitrary; e.g. since $0_2$ is preferred to $0_1$, we may arbitrarily assign $u_1 = 0$, $u_2 = 1$. Usually, once a zero and unit are thus decided, all other $u_i$ are uniquely determined. In the present case, if ($u_1$, $u_2$, $u_3$, $u_4$) describe preferences then so do $u_1$, $u_2$, $u_3 + c$, $u_4 - c$ for any real number c; since

$$u_1p_1 + u_2p_2 + .5(1\text{-}p_1\text{-}p_2)u_3 + .5(1\text{-}p_1\text{-}p_2)u_4$$
$$= u_1p_1 + u_2p_2 + .5(1\text{-}p_1\text{-}p_2)(u_3{+}c) + .5(1\text{-}p_1\text{-}p_2)(u_4\text{-}c) \quad (3)$$

One way to resolve the ambiguity in choice of c is to assume that the utility of the game is the sum of the contribution from the having of extra socks, having a new tie, and playing the birthday game. Thus,

$$1/2U(FF,GS) + 1/2U(FF,GT) = U(G) + 1/2U(KS,GS) + 1/2U(KT,GT) \quad (4a)$$

$$U(FF,GS) = U(G) + U(KS,GS) \quad (4b)$$

$$U(FF,GT) + U(G) + U(KT,GT) \quad (4c)$$

Here we write U(KS,GS) for $u_1$, etc.; and U(G) is the utility of playing the birthday game. For example, if we let $u_1 = 0$, $u_2 = 1$ by convention, and if $1/2u_3 + 1/2u_4 = 2$ -- because the decision maker is indifferent between $0_2$ with certainty (i.e., (0,1,0,0)) and a 50-50 chance of $0_1$ = (1,0,0,0) versus (0,0,1/2,1/2) -- then

$$2 = U(G) + 1/2; \text{ i.e., } U(G) = 1.5 \quad (5a)$$

$$u_3 = 1.5 + 0 = 1.5 \quad (5b)$$

$$u_4 = 1.5 + 1 = 2.5 \quad (5c)$$

Therefore $1/2u_3 + 1/2u_4 = 2.0$ as required.

## Extensions and Reflections

1. As John Pratt explained during the Roundtable, the above remarks readily generalize to two or more chance situations. These chance situations may be (a) alternates, or (b) offered at different times. An example of "(a) alternates" would be either

(i) I play the slot machines, or
(ii) I play roulette.

To simplify the example, I don't have time for both. An example of "(b) chance situations offered at different times" would be one in which one strategy I could follow is to

(1) play roulette on Saturday (and follow a particular betting rule), then
(2) go to the races on Sunday (and follow a particular betting rule there).

An outcome for the latter situation would, as before, be a sequence of states during different time intervals. For example, one outcome might be:

(spend Saturday playing roulette winning/losing x dollars,
spend Sunday at the tracks winning/losing y dollars,
therefore wake up Monday with z dollars in pocket.)

Such (a) alternate gambles and/or (b) successive gambles can be nicely represented by the tree diagrams which John Pratt presented at the Roundtable. For ease of presentation, John assumed that one sure outcome had higher utility and one had lower utility than all gambles. This assumption is not necessary, as illustrated in the example at the end of the previous section.

2. The above argument suggests that outcomes can always be elaborated to fit gambling preferences. Robin Pope is not convinced. I don't want to argue that matter right here. Rather, I want to report a private discussion (on the bus from hotel to conference) between Mark Machina and me.

Mark argued that even assuming that outcomes can always be elaborated to represent gambling (i.e., risk situation) preferences, it may not always be most convenient to do so. The analysis may be "neater" in a space of outcomes where expected utility does not apply than in a more complex one where it does. I agreed with Mark, and still do.[3]

3. Let's return to an example of a single chance situation, like "socks versus tie". I would prefer to see the home team win than lose; but I would prefer not to know in advance, before I go out to the ballpark. A game in which the home team was so strong and the visitors so weak that the former was sure to win, would be little fun to watch.

The resolution of this "paradox" is as in the "socks and tie" example. As in that example, we can solve for the utility of the game itself U(G), which here I write as $U(G_{.5})$ for a game with a .5 chance. But it is not necessary that we will get the same U(G) from formula (4) from a 90-10 game as we do from a 50-50 game (using the appropriate probabilities in (4)). Perhaps a 90-10 game is duller than a 50-50 game. Thus $U(G_p)$ which solves

$$pU(G_p,W) + (1-p)U(G_p,L) = U(G_p) + pU(W) + (1-p)U(L)$$

may be a function of p (where U(W) is utility of a win, $U(G_p,W)$ is the utility of watching a game with probability p of winning and seeing the home team win, etc.).

Various views may be illustrated by this example:

(a) $U(G_p,W)$ and $U(G_p,L)$ are, I believe, what R. Pope refers to as probability dependent utilities.

(b) One may amplify the definition of outcome so that EU still holds.

Then outcomes include, for example:

$(G_{.3}, W)$

$(G_{.3}, L)$

$(G_{.4}, W)$

etc.

If, for simplicity, we only allow a finite number of values of p, then there are a finite number of outcomes which can be labeled $0_1, \ldots, 0_n$ and analyzed as above. As before, only a convex subset of probability distributions $(p_1, \ldots, p_n)$ are thinkable. The situation can be illustrated by J. Pratt's trees.

If we allow any p in the interval $0 \le p \le 1.0$ then we have to use a different notation, sprinkle the word "measurable" here and there like holy water, but nothing essential changes.

(c) Finally, perhaps there exists a simpler space than the set of all

$$\left( G_p, \begin{Bmatrix} W \\ L \end{Bmatrix} \right)$$

such that preferences can be easily described but are not EU in the simpler space.

## Appendix
## Axioms for Multiperiod Analysis

Markowitz (1959), Chapter 11 on "Utility Analysis Over Time" specifies that: "An *outcome*, in the present discussion, is a time pattern of consumption $(C_1, C_2, \ldots, C_T)$. If the $C_t$ represent amounts of money devoted to consumption, an outcome is a 'history' of consumption expenditures. If the $C_t$ are vectors of goods and services enjoyed, then consumption is a time series of such enjoyments". (The inclusion of a bequest value at the end of the outcome vector makes no essential change to the analysis.) Time series of enjoyments postulated as outcomes in the body of the present paper are special cases of Markowitz (1959) Chapter 11 outcomes.

For convenience, Chapter 11 assumes that there are only a finite number of possible outcomes thus defined; e.g., a finite number of possible consumption numbers or vectors $C_t$ as of any time t and a finite (T) number of periods, hence a finite number of outcomes, i.e., time series

$$O_j = (C_{1j}, C_{2j}, \ldots, C_{tj}) \quad j = 1, \ldots, n. \tag{6}$$

In other words "we can therefore label possible outcomes with numbers 1, 2, . . ., j, . . ., n." Then a probability distribution of possible outcomes -- i.e. a probability distribution over all possible histories of consumption -- is a vector

$$(p_1, \ldots, p_n)$$

(where, of course, $p_j \geq 0$ and $\Sigma p_j = 1$).

The remainder of Chapter 11 notes that the formal axioms which were applied in Chapter 10 to probability distributions of outcomes at one point in time could also be asserted for probability distributions of Chapter 11 types of outcomes; considers "whether formal relationships expressed in the axioms appear as plausible when applied to probability distributions of outcomes spread over time as they do when applied to outcomes resulting from single period choice situations"; and points out that (obviously) if preferences among the probability vectors $P = (p_1, \ldots, p_n)$ in Chapter 11 satisfy the same axioms as the probability distributions $P = (p_1, \ldots, p_n)$ in Chapter 10, then they too can be ordered by attaching a number $u_i$ to outcome

$$O_i = (C_{1i}, \ldots, C_{Ti})$$

and choosing P to $Q = (q_1, \ldots, q_n)$

if and only if

$$\Sigma p_i u_i > \Sigma q_i u_i.$$

Chapter 11 uses Bellman's dynamic programming arguments to reduce the many-period choice of strategy to a sequence of single period expected utility maximizations. Chapter 11 then discusses other matters related to the $u_i$ and dynamic expected utility analysis; e.g. it points out that $u_i$ is <u>not necessarily</u> of the form $u_i = \Sigma U_t(C_{ti})$, etc.

It was my memory, or assumption, as of August 1, 1994, that the treatment of an "outcome" as a time series as in (6) was standard by 1959. I looked in Von Neumann and Morgenstern (1967[4]) so that I could cite them as considering their axioms applicable to the four outcomes in the socks versus tie example. However, in their discussion of expected utility, Von Neumann and Morgenstern state in Section 3.3.3, that "it would be an unnecessary complication, as far as our present objectives are concerned, to get entangled with the problems of the preferences between events in different periods in the future. It seems, however, that such difficulties can be obviated by locating all 'events' in which we are interested at one and the same, standardized moment, preferably in the immediate future". Thus, in their discussion of expected utility, Von Neumann and Morgenstern deal only with the single period case. (Most of their book is concerned with games which may have many moves, but have a single payoff to each player at the end of the game.)

I am somewhat at a loss as to who to cite on this matter. The examples by Bellman that I looked at treat the dynamics of games, like those of VN-M, with a single scalar payoff. Perhaps the Savage (1954) discussion of Small Worlds (Section 5.5) is the proper reference. Markowitz (1959) Chapter 11 may be considered a spelling out of the finite case of the matter as Savage discusses it there.

## References

Bellman, R. E. (1957) *Dynamic Programming*, Princeton University Press.

Markowitz, H. M. (1959) *Portfolio Selection: Efficient Diversification of Investments*, Wiley, Yale University Press, 1970, Basil Blackwell, 1991.

Savage, L. J. (1954) *The Foundations of Statistics*, Wiley, 2nd edition, Dover, 1972.

Von Neumann, J. and Morgenstern, O. (1944) *Theory of Games and Economic Behavior*, Princeton University Press, 3rd edition, Wiley, 1967.

## Endnotes

[1]At least the version of the VN-M axioms used in Chapter 11 of Markowitz (1959) allows "outcomes" to be sequences and assumes that the decision maker orders probability distributions of outcomes thus defined. See the Appendix on "Axioms for Multiperiod Analysis".

[2] This assumes that the decision maker can choose randomly between $0_1$, $0_2$ and a 50-50 chance of $0_3$, $0_4$, with negligible emotional effect of this randomization. For example, perhaps this emotionally neutral randomization consists of typing values for $p_1$ and $p_2$

into a computer, pressing enter, and the random choice appears immediately on the screen.

[3]On rereading Markowitz (1959) I find that my view then (p. 225) was that "the expected utility maxim can be extended to include such considerations; but a large number of such 'extensions' transform the maxim from a convenient rule to a useless formality."

[4]The preface to the second edition dated September 1946, says that "we have added an Appendix containing an axiomatic derivation of numerical utility. This subject was discussed in considerable detail [in the first edition], but in the main qualitatively, in Section 3."

"On Socks, Ties and Extended Outcomes", Economic and Environmental Risk and Uncertainty: New Models and Methods, Kluwer Academic Publishers, 1997, pp. 219-226.

# Single-Period Mean–Variance Analysis in a Changing World

Harry M. Markowitz and Erik L. van Dijk

*Ideally, financial analysts would like to be able to optimize a consumption–investment game with many securities, many time periods, transaction costs, and changing probability distributions. We cannot. For a small optimizable version of such a game, we consider in this article how much would be lost by following one or another heuristic that could be easily scaled to handle large games. For the games considered, a particular mean–variance heuristic does almost as well as the optimum strategy.*

Consider a model of financial decision making along the following lines: An investor acts at discrete points in time that are a fixed interval apart (e.g., a year, quarter, month, day, or millisecond). At each point in time, the investor chooses a portfolio from a universe of many securities (e.g., a dozen asset classes or 1,000 individual securities). Changes in the portfolio from one point in time to the next incur costs. The probability distribution of holding-period security returns may change over time. Perhaps the investor (or investment management team) is uncertain as to which hypothesis about security returns is correct—the investor's beliefs shifting in a Bayesian fashion as evidence accumulates. The investment portfolio is subject to deposits and withdrawals and occasionally distributes dividends. The investor seeks to maximize the expected value of some "total utility function" of the stream of present and future dividends (e.g., the sum of the present values of the single-period utility of each future dividend).

A "strategy" for this model, in the sense of Von Neumann and Morgenstern (1953), is a rule that specifies, for all circumstances that can arise during the course of this investment "game," the action to be taken should that circumstance arise. An optimum strategy maximizes the expected value of total utility.

Calculating the optimum strategy for such a game is well beyond foreseeable computing capabilities. The game could be formulated as a dynamic programming problem (Bellman 1957), but it has too many state variables to solve in relevant time, even with the fastest parallel computers. If the model were restricted to a finite number of periods, such as the number of months in a human lifetime, it could be formulated as a stochastic programming problem (Dantzig 1955; Dantzig and Infanger 1993). But the set of possible paths would fan out so rapidly that the problem would be beyond foreseeable storage, as well as computational, capabilities. The most one could use, in general, to analyze such a "realistic" investment model would be a detailed Monte Carlo simulation (see Levy, Levy, and Solomon 2000). Such a simulation model could be used to evaluate alternative heuristic strategies.

We seek in this article to form some idea of how well one particular type of heuristic strategy might perform in comparison with an optimum strategy and with other heuristics. The heuristic of interest is the "mean–variance surrogate for the 'derived' utility function," or for short, "MV surrogate heuristic."

We know from dynamic programming that a multiperiod (or unending) investment game may be solved by solving a sequence of one-period games. For example, number the points in time when decisions can be made as $t = 0, 1, \ldots$, and the intervals between them as $1, 2, \ldots$. Thus, time interval $t$ lies between time points $t-1$ and $t$. Suppose that total utility is the sum of the present values of one-period utilities:

$$U = \sum_{t=1}^{\infty} d^{t-1} u(D_t), \tag{1}$$

where

$d$ = discount factor ($d < 1$)

$u(D_t)$ = interval $t$'s contribution to expected utility

$D_t$ = "dividend" paid during time interval $t$

*Harry M. Markowitz is president of Harry Markowitz Company, San Diego, California. Erik L. van Dijk is visiting lecturer at the Free University of Amsterdam and Nyenrode Business School, Breukelen, Netherlands.*

In this case, the optimum action at point-in-time $t-1$, given that the system is in state $S_{t-1}$, is the one that maximizes the expected value of

$$u(D_t) + dW(S_t), \tag{2}$$

where $W(S_t)$, the derived utility function, is the expected value of

$$\sum_{t=1}^{\infty} d^{i-1} u(D_{t+i}), \tag{3}$$

given that the investor is in state $S_t$ at time point $t$ and plays the game optimally then and thereafter.

Bellman showed how to compute $W$ and, hence, solve such games when the dimensionality of $S$ is not too great. An MV surrogate heuristic replaces $W$ by a function of the mean and variance of the investor's portfolio. We will restrict ourselves to a simple linear surrogate function,

$$W^S = E_{wt} E_P + V_{wt} V_P, \tag{4}$$

where $E_P$ and $V_P$ are, respectively, the mean and variance of the portfolio and $E_{wt}$ and $V_{wt}$ are their respective weights in the linear surrogate function. The definition in Equation 4 omits a constant term as irrelevant because its inclusion would not affect the optimum choice of action if $W^S$ were substituted for $W$ in Equation 2.[1]

We assume that the investor (presumably a large institutional investor) has a detailed simulation model with which to explore such questions as: What $E_P$ and $V_P$ should be used? For example, if the portfolio is reviewed monthly, should $E_P$ be the expected return on the chosen portfolio for its first month, quarter, year—assuming the portfolio is held at least that long without further changes? Other questions the investing institution might explore with its simulation model are: What $E_{wt}$ and $V_{wt}$ should be used in Equation 4? How well does the MV surrogate heuristic compare with other heuristics? The important question the simulator cannot solve is: Given a good choice of $E_P$ and $V_P$ and good $E_{wt}$ and $V_{wt}$ weights, how well does the MV surrogate do when compared with an optimum strategy?

Our general approach will be to define a simple dynamic investment model—one for which an optimum (or nearly optimum) strategy can be computed—and evaluate how well MV surrogate heuristics do in this model relative to other heuristics and the optimum strategy. The MV surrogate and other heuristics may be scaled to larger problems for which optimum solutions cannot be computed. The object of the experiment is to provide a reading on how well the MV surrogate might do for such realistic problems when compared with the optimum solution and other heuristics.

Specifically, we consider an investment game with two assets—stock and cash. The investor's portfolio can be in one of 11 states—0 percent stock, 10 percent, 20 percent, . . ., 100 percent stock. Transaction costs are incurred when the investor changes the portfolio state. The investor has a model for forecasting stock returns that can be in one of five prediction states: (1) very optimistic, (2) optimistic, (3) neutral, (4) pessimistic, (5) very pessimistic. Thus, the system as a whole can be in any of 55 states. The game is unending, with a utility function as in Equation 1. The assumption of an unending game simplifies the analysis because time then remaining is not a state variable. We made specific assumptions concerning the ability of the forecasting model to predict and concerning the probability of transition from one prediction state to another. In particular, we assumed the forecasting model had enough predictive ability that—in the absence of transaction costs—the optimum portfolio to hold would differ considerably from one forecast state to another.

The optimum strategy for this game may be written as an $11 \times 5$ *action matrix*, $\tilde{\mathbf{A}}$, that specifies choice of next portfolio as a function of the current portfolio and prediction state. For example, if the portfolio is in Portfolio State 6 (50 percent stock) and the predictive model is in Prediction State 1 (very optimistic), the (6, 1) entry of the action matrix specifies which portfolio should be selected. Associated with action matrix $\tilde{\mathbf{A}}$ is an $11 \times 5$ expected discounted utility matrix, $\tilde{\mathbf{W}}$. The value $\tilde{\mathbf{W}}(i,j)$ is the expected value of discounted utility (Equation 1) if the game starts in (portfolio, prediction) states $(i,j)$ and follows action matrix $\tilde{\mathbf{A}}$ henceforth.

We seek a (nearly) optimum action matrix for two levels of transaction costs—50 bps and 200 bps. For a given model (e.g., a model with cost $c = 0.005$), the action matrix $\mathbf{A}^*$ we obtain using dynamic programming will not necessarily be the optimum $\tilde{\mathbf{A}}$ but its associated expected discounted utility matrix, $\mathbf{W}^*$, will be within $\varepsilon$ of $\tilde{\mathbf{W}}$. That is,

$$\mathbf{W}^*(i,j) \geq \tilde{\mathbf{W}}(i,j) - \varepsilon \text{ for all } i, j, \tag{5}$$

where $\varepsilon$ will be chosen so that the difference between $\tilde{\mathbf{W}}$ and $\mathbf{W}^*$ is negligible.

If we are given an arbitrary action matrix $\mathbf{A}$, optimal or other, we can solve for its $\mathbf{W_A}$ matrix—namely, the expected utility of the game starting from each of the 55 portfolio and prediction states and following the given action matrix then and thereafter—by solving a certain system of 55 simultaneous equations. This system of 55 equations—with action matrix $\mathbf{A}$ as input and expected total utilities $\mathbf{W}$ as output—is used to evaluate various heuristics, including MV surrogate heuristics for various choices of $E_{wt}$ and $V_{wt}$ and for various definitions of $E$ and $V$, as well as other heuristics. This process allows us to say—at least for this simplified world—how well a good MV surrogate heuristic would do in competition with other heuristics and with an optimum strategy. (In more complex situations, we would need to use simulations to estimate $\mathbf{W_A}$ and would not have an optimum $\mathbf{W^*}$ with which to compare it.)

Multiperiod and continuous-time portfolio selection with transaction costs has been studied by Zabel (1973), Kamin (1975), Magill and Constantinides (1976), Constantinides (1979), and Davis and Norman (1990). The model analyzed by Constantinides in 1979 is closest to the model presented here.[2] The general conclusions of these papers for the two-asset case are as follows: First, there is an interval in which no action is taken; outside this interval, the investor buys or sells to bring the portfolio position to the closest boundary of the no-action interval. Second, computing the location of the no-action interval is usually not easy. Third, the many-security portfolio selection problem with transaction costs also has a no-action region that is practically impossible to compute.

We approach the problem from the other direction. We start with heuristics that are easy to compute and ask what is lost by using one or another such heuristic rather than doing the estimation and optimization required to seek an optimum.

Balduzzi and Lynch (1999) and Lynch and Balduzzi (2000) computed optimum consumption and investment for a two-asset dynamic model with transaction costs that incorporated "the empirically documented predictability of asset returns" (Balduzzi and Lynch, p. 47). Our objective is different. We do not claim that the changing forecasts in our model reflect the true predictably of asset returns. Rather, they are part of an experiment to study the performance of portfolio choice heuristics that can be easily scaled to larger models with optimization procedures that cannot be scaled. The forecasting model within the experiment is such that if there were no transaction costs, there would be considerable shifting of portfolio assets. This model feature provides a substantial challenge for the heuristics tested.

The mean–variance surrogate heuristics discussed here should be distinguished from mean–variance *approximations* discussed in Markowitz (1959, Ch. 6 and Ch. 13), Young and Trent (1969), Levy and Markowitz (1979), Dexter, Yu, and Ziemba (1980), Pulley (1981, 1983), Kroll, Levy, and Markowitz (1984), Grauer (1986), Ederington (1986), Simaan (1987), and Markowitz, Reid, and Tew (1994). In this literature, a single-period $U(R)$ function is given and a function $f(E,V)$ of mean and variance is sought to approximate expected $U(R)$. In the type of multiperiod situation we address here, we know that a single-period "derived" utility function, albeit a complex function of many state variables, exists. We did not seek to approximate it, because in complex cases, we would not know what it is. Rather, we sought a mean–variance surrogate to stand in its place. In general, our proposal for real-world, dynamic portfolio selection problems with illiquidities was to use simulation analysis to seek a mean–variance surrogate that does as well as one can find in terms of the game as a whole. The purpose of the experiments was to get an initial reading on how good this surrogate might be.

## The Model

We have summarized the model in **Exhibit 1**. In the first two items of Exhibit 1, we assume a one-month interval between portfolio reviews and a constant risk-free rate of 0.4 of 1 percent (i.e., 40 bps a month). The table in Item 3 shows, for example, that if the predictive model is in State 1—most optimistic—the expected return for stock for the forthcoming month is 64 bps with a standard deviation of $\sqrt{0.000592} \approx 0.024$. The next four columns provide the mean and variance of return for the forthcoming month given that the predictive model is in one of States 2 through 5. We discuss later in this section the suitability of these numbers for the present experiment and the meaning of other numbers in the table in Item 3.

The entries in the table in Item 4 show assumed probabilities $P(j,h)$ that the prediction state $h$ listed at the top of the table will occur at time $t + 1$ given that the prediction state $j$ listed in the left column of the table is true at time $t$. Given this Markov matrix, one can solve for the long-run, "ergodic," steady-state probabilities of the various states. These long-run probabilities are listed in the last column of this table in Item 4. (Returning to Item 3, note that the last column shows the mean and variance of a random number drawn by first drawing a prediction state according to the steady-state distribution and then drawing a stock return given that prediction state.)

Item 5 states that the investor seeks to maximize the expected value of discounted future utility functions, with discount factor $d = 0.99$ (or a discount rate of 0.01 a month). Thus, for example, utility 10 years from now counts only 30 percent as much as utility next month—but it still counts.

Let $p_{t-1}$ be the fraction of the investor's portfolio held in stocks at time $t - 1$ and $p_t$ be the fraction the investor decides at time $t - 1$ to accumulate up to (or sell down to or continue to hold) for time $t$.

**Exhibit 1. Summary of Investment Model**

1. Interval between portfolio reviews: 1 month
2. Risk-free rate, $r_f$, per month (assumed constant): 0.004
3. Monthly stock return means, $E^1, \ldots, E^5$, and variances, $V^1, \ldots, V^5$, for various prediction states:

| Variable | State 1 | State 2 | State 3 | State 4 | State 5 | Steady State |
|---|---|---|---|---|---|---|
| $E$ | 0.0064 | 0.0050 | 0.0042 | 0.0038 | 0.0027 | 0.0044 |
| $V$ | 0.000592 | 0.000556 | 0.000538 | 0.000539 | 0.000567 | 0.000559 |
| Opt$X$ | 4.05 | 1.80 | 0.37 | –0.37 | –2.29 | 0.72 |

*Note*: Opt$X$ is the investment that would be optimal in each state if there were no transaction costs.

4. From–to transition probabilities, $P$, between predictive states:

| | New State | | | | | |
|---|---|---|---|---|---|---|
| Old State | 1 | 2 | 3 | 4 | 5 | Steady State |
| 1 | 0.702 | 0.298 | 0 | 0 | 0 | 0.1608 |
| 2 | 0.173 | 0.643 | 0.133 | 0.051 | 0 | 0.2771 |
| 3 | 0 | 0.260 | 0.370 | 0.348 | 0.022 | 0.1363 |
| 4 | 0 | 0.065 | 0.179 | 0.615 | 0.141 | 0.2393 |
| 5 | 0 | 0 | 0.033 | 0.164 | 0.803 | 0.1865 |

5. Total utility is

$$U = \sum_{t=1}^{\infty} d^{t-1} u(D_t).$$

The investor's objective is to maximize expected total utility, $EU$.
For the cases reported, we used a monthly discount factor of $d = 0.99$.

6. We usually refer to "portfolio state"

$i = 1, \ldots, 11$

representing fraction invested

$p = 0.0, 0.1, \ldots, 1.0.$

Note that

$$p = \frac{i-1}{10}.$$

Thus, for example, $p = 0.2$ in the present discussion corresponds to $i = 3$ elsewhere.
In computing utility for a period, we defined the "effective fraction" invested in stocks during time interval $t$, $p_t^e$, as

$$p_t^e = \theta_p p_{t-1} + (1 - \theta_p) p_t,$$

where $p_{t-1}$ and $p_t$ are the fraction invested in stock at, respectively, the beginning and end of the time interval. In particular, the conditional expected return and variance of return (at time point $t-1$) on the portfolio during time interval $t$ given state $j$, current stock fraction $p_{t-1}$, and chosen stock fraction $p_t$ are

$$E_t^P = p_t^e E_{t-1}^j + (1 - p_t^e) r_f$$

and

$$V_t^P = (p_t^e)^2 V_{t-1}^j,$$

where $E_{t-1}^j$ and $V_{t-1}^j$ are the mean and variance of stock return for time interval $t$ given the prediction state at $t-1$.

Runs reported use $\theta_p = 1/2$.

7. Transaction cost incurred during time interval $t$ is $c\,|p_t - p_{t-1}|$.

We report cases with $c = 0.005$ and $c = 0.020$.

8. We assumed that

$$Eu(D_t) = E(D_t) - kV(D_t),$$

where $u(D_t)$ is as defined for Equation 1 and $k$ reflects risk aversion. For the cases reported, $k = 0.5$.

9. We assumed that the entire return on the portfolio for the month net of costs was "distributed." Thus, the conditional expected utility for time interval $t$ at time point $t-1$ given prediction state $j$, current portfolio state $i$, and selected portfolio state $g$ is

$$E(u\,|\,i,j,g) = E_t^P - c\,|p_t - p_{t-1}| - kV_t^P,$$

where $p_{t-1} = 0.1(i-1)$, $p_t = 0.1(g-1)$, and $E_t^P$ and $V_t^P$ are defined in Item 6. Note that $E_t^P$ and $V_t^P$ for time interval $t$ depend on the prediction state at $t-1$.

10. We assumed that the return on equities is bounded by some (very large) number $M$.

What fraction should one assume the investor holds in the interval between these two points in time, during which the random return on equities is determined? Should we assume $p_{t-1}$, $p_t$, or something in between? Item 6 defines an "effective fraction" assumed to be held during the interval as determined by parameter $\theta_p$, set to 0.5 in these experiments. Item 6 notes the consequent portfolio expected return and variance of return for the month at time $t-1$ given that the prediction state is $j$, the current stock fraction is $p_{t-1}$, and the next fraction is $p_t$. The fraction invested is related to the integer "portfolio state" $i$ by $p = (i-1)/10$.

As noted in Item 7, we will report the results of two experiments—one with transaction costs, including market impact and bid–ask spread, equal to $c = 0.005$ and the other with $c = 0.02$. Thus, we consider Exhibit 1 to be defining two specific models. If we abstract from the particular parameters given, Exhibit 1 defines a "general" model.

As noted in Item 8, we assume that $u(D_t)$ can be adequately defined by a mean–variance approximation. In the runs reported, we let mean–variance trade-off $k$ equal 0.5, which made $Eu(D_t)$ approximately $E[\log(1 + \text{Return})]$. As noted in the introduction, this approximation has proved quite good in the case of historical portfolio returns.[3] Larger values of $k$ approximate more risk-averse utility functions. The assumption that the monthly contribution to total expected utility, $Eu(D_t)$, may be approximated well by a function of mean and variance does not obviously imply that a mean–variance surrogate will serve well in place of derived utility function $W$. Such a conclusion remains to be seen from the experiment. The assumption in Item 8 does conveniently imply that, given each prediction state $j$, we need only specify the mean and variance of the stock return distributions.

Returning to the table in Item 3, note that the last row shows the investment that is optimal in each state in the absence of transaction costs, with $\theta_p = 0$, $k = 0.5$ (as assumed), and stock shorting and leverage being permitted. Specifically,

$$\hat{X} = \frac{E^j - r_f}{2kV^j} = \frac{E^j - r_f}{V^j}. \tag{6}$$

We see that if the predictive model is very optimistic, for example, and there are no transaction costs, the optimal strategy is to borrow roughly \$3 for every dollar of equity and invest the \$4 in stock. Conversely, if the model is very pessimistic, going short 229 percent is optimal. These differences in optimal positions would be less extreme if $k$ were larger, the differences between $E^j$ and $r_f$ were smaller, or $V^j$ were larger.

The present work is the result of a collaboration of two authors—one especially interested in predictive models and the other interested primarily in optimization. We do not claim here that we can offer the reader a model that can actually predict expected returns of 24 bps more than the risk-free rate in some circumstances and 13 bps less in others with the errors of estimation given in the table in Item 3. Rather, we are considering the question of how one should proceed if one had such a predictive model and higher or lower transaction costs. We chose the parameters in Item 3 based on a judgment that a predictive model that implied much greater leverage when it is optimistic, or a much greater short position when it is pessimistic, would not be plausible and a predictive model that implied much smaller moves in the absence of transaction costs would not much challenge the alternative heuristics.

Similarly, we chose the transition probabilities in Item 4 on the assumption that good months would tend to be followed by good months, bad months by bad months, and the steady-state distribution would imply roughly that the familiar 60/40 or 70/30 stock/cash mix is optimal. As it is, as shown in the last column of Item 3, the chosen parameters imply that a 72/28 mix is optimal in the case of the steady-state distribution. Thus, one might say that, given the chosen parameters, the "strategic" optimal portfolio is 72 percent in stock. We will explore whether this strategic solution is related to the dynamic optimum in some way.

Item 9 confesses that we assumed that all returns, net of cost, are "distributed" each month. In particular, we assumed that losses are collected from the investors as well as gains being distributed to them. This assumption is expressed in two ways—in the calculation of the contribution to expected utility for the period (as shown in Item 9) and, implicitly, in the assumption that if at time $t-1$ the investor selects fraction $p_t$, then, in fact, $p_t$ will be the fraction invested in stocks when time $t$ occurs. The assumption that returns are distributed helps keep the state space small and, specifically, greatly simplifies the computation of the optimal policy. Although unrealistic, this assumption seems innocuous for present purposes because this method of scoring does not particularly favor the MV heuristic over other heuristics or the optimal solution.

Item 10 assumes that security returns are bounded by some huge number.

As noted, we selected the model in Exhibit 1, including parameter settings, as a simple, easily computed, but challenging setting within which to test heuristics. Once we selected the model and began testing, we made no further changes in the model.

## Computation of Expected Discounted Utility

In this section, we consider action matrixes that represent heuristics, such as an MV surrogate or a heuristic that immediately moves to a 60/40 mix of stocks and cash and stays at this mix forever. We will report the expected value of $U$ in Equation 1, $EU$, if the action matrix is followed starting in portfolio state $i$ and prediction state $j$. We describe the evaluation of $E\left[\sum_{t=1}^{\infty} d^{t-1}u(D_t)\right]$, where $D_t$ is current return on the investor's portfolio minus transaction costs.

For a given action matrix $\mathbf{A}$, let

$$\mathbf{W}_{\mathbf{A}}^{T}(i,j) = E\left[\sum_{t=1}^{T} d^{t-1}u(D_t)\right] \tag{7}$$

be the expected utility of a $T$-period game that starts in portfolio state $i$ and prediction state $j$ and in which action matrix $\mathbf{A}$ is followed at every move. Then,

$$\mathbf{W}_{\mathbf{A}}^{T+1}(i,j) = \mathbf{u}_{ij}^{\mathbf{A}} + d\sum_{g,h}\tilde{\mathbf{P}}^{\mathbf{A}}(i,j,g,h)\mathbf{W}_{\mathbf{A}}^{T}(g,h), \tag{8}$$

where

$$\mathbf{u}_{ij}^{\mathbf{A}} = E[u \mid i,j,\mathbf{A}(i,j)] \tag{9}$$

(see Item 9 in Exhibit 1) and $\tilde{\mathbf{P}}^{\mathbf{A}}(i,j,g,h)$ is the probability that state $(g,h)$ will occur at $t$ if state $(i,j)$ holds at time $t-1$. Specifically, $\tilde{\mathbf{P}}^{\mathbf{A}}(i,j,g,h)$ is zero unless $g = \mathbf{A}(i,j)$. With this choice of state $g$, $\tilde{\mathbf{P}}^{\mathbf{A}}(i,j,g,h)$ equals the probability $P(j,h)$ of transition from prediction state $j$ to prediction state $h$ given in Item 4.

It will be convenient to write $\mathbf{W}_{\mathbf{A}}^{T}$ and $\mathbf{u}^{\mathbf{A}}$ as vectors with 55 components rather than 11 × 5 matrixes and write $\tilde{\mathbf{P}}^{\mathbf{A}}$ as a 55 × 55 matrix rather than a four-dimensional object. Toward this end, we write

$$\mathbf{W}_{T}^{\mathbf{A}}(m) = \mathbf{W}_{T}^{\mathbf{A}}(i,j), \tag{10a}$$

$$\mathbf{u}^{\mathbf{A}}(m) = \mathbf{u}^{\mathbf{A}}(i,j), \tag{10b}$$

and

$$\tilde{\mathbf{P}}^{\mathbf{A}}(m,n) = \tilde{\mathbf{P}}^{\mathbf{A}}(i,j,g,h), \tag{10c}$$

where

$$m = 11 \times (j-1) + i \tag{11a}$$

and

$$n = 11 \times (h-1) + g. \tag{11b}$$

Then, Equation 8 becomes

$$\mathbf{W}_{\mathbf{A}}^{T+1} = \mathbf{u}^{\mathbf{A}} + d\tilde{\mathbf{P}}^{\mathbf{A}}\mathbf{W}_{\mathbf{A}}^{T}, \tag{12}$$

where $\mathbf{W}_{\mathbf{A}}^{T}$ equals $\mathbf{W}_{\mathbf{A}}^{T}(m)$, and so on.

Starting with $\mathbf{W}_{\mathbf{A}}^{1} = \mathbf{u}^{\mathbf{A}}$, Equation 12 presents an iterative scheme for computing $\mathbf{W}_{\mathbf{A}}^{T}$ expressed as a 55-component vector. Because $d\tilde{\mathbf{P}}^{\mathbf{A}}$ is a contraction mapping, the sequence in Equation 12 is convergent:

$$\mathbf{W}_{\mathbf{A}}^{T} \to \mathbf{W}_{\mathbf{A}} \text{ as } T \to \infty, \tag{13a}$$

where

$$\mathbf{W}_{\mathbf{A}} = \lim_{T\to\infty} E\left[\sum_{t=1}^{T} d^{t-1}u(D_t)\right] \tag{13b}$$

when action matrix $\mathbf{A}$ is followed each move. It is also true that

$$\mathbf{W}_{\mathbf{A}} = E\left[\lim_{T\to\infty}\sum_{t=1}^{T} d^{t-1}u(D_t)\right] \tag{14}$$

because Item 10 of Exhibit 1 implies that for every path generated and for $S > T$,

$$\left|\sum_{t=1}^{S} d^{t-1}\mathbf{u}^{\mathbf{A}}(D_t) - \sum_{t=1}^{T} d^{t-1}\mathbf{u}^{\mathbf{A}}(D_t)\right| < M\frac{d^T}{1-d}. \tag{15}$$

Thus, every sequence converges and is bounded by the integrable function $M/(1-d)$. Therefore, Lesbegue's bounded convergence theorem applies and allows interchange of the operators.

Equation 12 implies that the limiting vector $\mathbf{W}_{\mathbf{A}}$ is the fixed point satisfying

$$\mathbf{W}_{\mathbf{A}} = \mathbf{u}^{\mathbf{A}} + d\tilde{\mathbf{P}}^{\mathbf{A}}\mathbf{W}_{\mathbf{A}}. \tag{16}$$

This system of 55 equations in 55 unknowns is sparse; it contains, at most, 5 nonzero coefficients per row. The system can be readily solved by MatLab's sparse matrix solver, which is how the various $\mathbf{W}_{\mathbf{A}}$ we will report were obtained.

## Computation of (Nearly) Optimal Strategy

In the manner explained by Bellman, we approximated the optimal solution to the unending game by a sequence of $T$-period games for $T = 1, 2, \ldots, S$ for some large $S$. In this section, we present notation for the $T$-period game, review Bellman's dynamic programming procedure as it applies to our model, and establish an upper bound on

$$\|\tilde{\mathbf{W}} - \mathbf{W}^*\| = \max_{i,j}|\tilde{\mathbf{W}}(i,j) - \mathbf{W}^*(i,j)|, \tag{17}$$

where $\tilde{\mathbf{W}}(i,j)$ is the expected discounted utility of the unending game starting in state $(i,j)$ if an optimal strategy is followed and $\mathbf{W}^* = \tilde{\mathbf{W}}^T$ is the expected discounted utility matrix for the dynamic programming solution after $T$ iterations for some large $T$.

The $T$-period game involves $T + 1$ time points and $T$ time intervals (or periods) aligned as follows:

| Time point | 0 | | 1 | | 2 | | . . . $T-1$ | | $T$ |
|---|---|---|---|---|---|---|---|---|---|
| Time interval | | 1 | | 2 | | | | $T$ | |

A strategy for the $T$-period game is a rule that specifies for each decision point $t = 0, \ldots, T-1$ and for each state $(i,j)$ the next portfolio selected. Such a strategy can be represented as an action matrix subscripted by time point $t$:

$$g = A_t^T(i,j) \text{ for } t = 0, \ldots, T-1. \tag{18}$$

The optimum strategy, $\tilde{A}_t^T$, maximizes the expected value of

$$U = \sum_{t=1}^{T} d^{t-1} u(D[t]). \tag{19}$$

This optimum expected discounted utility for the $T$-period game as a function of starting state $(i, j)$ is denoted $\tilde{W}^T(i,j)$.

The optimum first and only decision in a one-period game ($T = 1$), namely, $g = \tilde{A}_0^1(i,j)$, is found by computing

$$\tilde{W}^1(i,j) = \max_g E(u|i,j,g). \tag{20}$$

Given that optimum strategy $\tilde{A}_t^T(i,j)$ and the expected utility it provides, $\tilde{W}^T(i,j)$, have been determined up to some $T$, the optimum first move for the $(T+1)$-period game, $g = \tilde{A}_0^{T+1}(i,j)$, is determined by finding

$$\tilde{W}^{T+1}(i,j) = \max_g \left\{ E(u|i,j,g) + d \underset{h}{E}[\tilde{W}^T(g,h)|j] \right\}, \tag{21a}$$

where

$$\underset{h}{E}[\tilde{W}^T(g,h)|j] = \sum_{h=1}^{5} P(j,h)\tilde{W}^T(g,h). \tag{21b}$$

For $t > 0$,

$$\tilde{A}_t^{T+1}(i,j) = \tilde{A}_{t-1}^T(i,j) \tag{22}$$

(i.e., the two strategies are the same when there are the same number of periods to go).

For a large value of $T$ (actually, $T$ = 1,200—that is, a hundred years worth of months), we used Equations 20 and 21 to compute $\tilde{W}^T$ and $\tilde{A}_0^T$ for each of our two specific models—$c = 0.005$ and $c = 0.02$.

For large $T$, $\tilde{W}^T(i,j)$ cannot be much less than $\tilde{W}(i,j)$, as can be seen as follows. Let

$$u_{\max} = \max_{i,j,g} E(u|i,j,g). \tag{23}$$

Write the expected utility of the unending game as

$$\begin{aligned} EU &= E\left[\sum_{t=1}^{T} d^{t-1} u(D_t)\right] + E\left[\sum_{t=T+1}^{\infty} d^{t-1} u(D_t)\right] \\ &\le \tilde{W}^T + \sum_{t=T+1}^{\infty} d^{t-1} u_{\max} \\ &= \tilde{W}^T + \frac{d^T u_{\max}}{1-d} \end{aligned} \tag{24}$$

because $\tilde{W}^T$ is the maximum expected value of

$$\sum_{t=1}^{T} d^{t-1} u(D_t) \text{ and } Eu(D_t) \le u_{\max}. \tag{25}$$

For both specific models,

$$u_{\max} = E(u|11, 1, 11) = 0.0061.$$

This is $Eu(D_t)$ given that $i = 11$ (fully invested), $j = 1$ (very optimistic), and $k = 11$ (no change in the portfolio). Thus, for both specific models,

$$\tilde{W}^{1200} \ge \tilde{W} - \varepsilon, \tag{26a}$$

where

$$\varepsilon = d^{1200} \frac{u_{\max}}{1-d} = 3.53 \times 10^{-6}. \tag{26b}$$

A strategy for the unending game may base action on $t$ as well as $(i,j)$, as expressed by a $t$-subscripted action matrix, $A_t(i,j)$. But the optimum strategy uses an unchanging action matrix, $\tilde{A}_t = \tilde{A}$ (if $\tilde{A}$ is unique; otherwise, there exists an optimum). This fact may be seen as follows. As of time $T$, write $EU$ as

$$\sum_{t=1}^{T} d^{t-1} u(D_t) + d^T E \sum_{t=1}^{\infty} d^{t-1} u(D_{T+t}).$$

The first sum has happened and cannot be changed. Given the state $(i,j)$ at time $T$, the maximization of the second expected discounted sum is isomorphic to the original problem and has, therefore, the same optimum initial action matrix $\tilde{A}_T(i,j) = \tilde{A}_0(i,j)$. Because $T$ is arbitrary, $\tilde{A}_T$ does not depend on $T$:

$$\tilde{A}_T = \tilde{A}_0 = \tilde{A} \text{ for all } T. \tag{27}$$

For the two specific models of Exhibit 1, we used $A^* = A_0^{1200}$ as an approximation to $\tilde{A}$. For reasons illustrated in Appendix A, we cannot guarantee that $A^* = \tilde{A}$. We can, however, guarantee a lower bound on how close the associated $W^*$ is to $\tilde{W}$ for each specific model. For $c = 0.005$, for example, given the $A^*$ action matrix, we computed $W^*$ from the 55 equations described in the previous section. It turns out that

$$\mathbf{W}^*(i,j) > \tilde{\mathbf{W}}^{1200}(i,j) \text{ for all } (i,j), \tag{28}$$

and we conclude from Equation 26a that

$$\mathbf{W}^*(i,j) > \tilde{\mathbf{W}}(i,j) - \varepsilon \tag{29}$$

for the $\varepsilon$ in Equation 26b. Equation 29 thus provides the bound on the norm in Equation 17. The same holds when $c = 0.02$. Inequality 28 is consistent with the hypotheses that $\mathbf{W}^* = \tilde{\mathbf{W}}$ in fact and the remainder of the game $t > 1{,}200$ makes a positive contribution to total expected utility.

## (Nearly) Optimum Action Matrixes

Panel A in **Table 1** presents the $\mathbf{A}^* = \tilde{\mathbf{A}}^{1200}$ action matrix for $c = 0.005$. This action matrix became optimum at iteration 17 of the dynamic programming calculation and stayed optimum thereafter. In other words, for any finite game of length $T = 17$, 18, . . ., 1,200 months, the $\mathbf{A}^*$ in Panel A is the optimum first-action matrix. In particular, for the finite game with $T = 1{,}200$, $\mathbf{A}^*$ is the optimum action matrix for time $t = 1$ through 1,183.

**Table 1. Optimum Action Matrixes: Fraction of Stock at *t***
(1,200 iterations)

| Stock Fraction at $t-1$ | Prediction State 1 | 2 | 3 | 4 | 5 |
|---|---|---|---|---|---|
| *A. Cost = 0.005;* $\mathbf{A}^*$ *optimum from iteration 17* | | | | | |
| 0.0 | 1.0 | 0.3 | 0.0 | 0.0 | 0.0 |
| 0.1 | 1.0 | 0.3 | 0.1 | 0.1 | 0.0 |
| 0.2 | 1.0 | 0.3 | 0.2 | 0.2 | 0.0 |
| 0.3 | 1.0 | 0.3 | 0.3 | 0.3 | 0.0 |
| 0.4 | 1.0 | 0.4 | 0.4 | 0.4 | 0.0 |
| 0.5 | 1.0 | 0.5 | 0.5 | 0.5 | 0.0 |
| 0.6 | 1.0 | 0.6 | 0.6 | 0.6 | 0.0 |
| 0.7 | 1.0 | 0.7 | 0.7 | 0.7 | 0.0 |
| 0.8 | 1.0 | 0.8 | 0.8 | 0.8 | 0.0 |
| 0.9 | 1.0 | 0.9 | 0.9 | 0.9 | 0.0 |
| 1.0 | 1.0 | 1.0 | 1.0 | 1.0 | 0.0 |
| *B. Cost = 0.02;* $\mathbf{A}^*$ *optimum from iteration 211* | | | | | |
| 0.0 | 0.6 | 0.0 | 0.0 | 0.0 | 0.0 |
| 0.1 | 0.6 | 0.1 | 0.1 | 0.1 | 0.1 |
| 0.2 | 0.6 | 0.2 | 0.2 | 0.2 | 0.2 |
| 0.3 | 0.6 | 0.3 | 0.3 | 0.3 | 0.3 |
| 0.4 | 0.6 | 0.4 | 0.4 | 0.4 | 0.4 |
| 0.5 | 0.6 | 0.5 | 0.5 | 0.5 | 0.5 |
| 0.6 | 0.6 | 0.6 | 0.6 | 0.6 | 0.6 |
| 0.7 | 0.7 | 0.7 | 0.7 | 0.7 | 0.7 |
| 0.8 | 0.8 | 0.8 | 0.8 | 0.8 | 0.8 |
| 0.9 | 0.9 | 0.9 | 0.9 | 0.9 | 0.9 |
| 1.0 | 1.0 | 1.0 | 1.0 | 1.0 | 0.9 |

As we illustrate in Appendix A, we cannot conclude from the preceding that $\mathbf{A}^*$ in Table 1, Panel A, is the optimum action matrix for the unending game with $c = 0.005$. All we can guarantee is that for each starting combination of portfolio state and prediction state, the expected utility of the unending game provided by $\mathbf{A}^*$ is within $(3.53)(10^{-6})$ of the utility provided by optimum action matrix $\tilde{\mathbf{A}}$. We would be surprised if $\tilde{\mathbf{A}}^{1200}$ were not optimal in fact, but all we can guarantee is that it is "nearly optimal" in the sense described.

Panel B of Table 1 presents the $\mathbf{A}^* = \tilde{\mathbf{A}}^{1200}$ action matrix for $c = 0.02$. This action matrix became optimum at iteration 211 and remained so. That is, a finite game had to have $T \geq 211$ months to go (rather than $T \geq 17$ as in the case of $c = 0.005$) for the $\mathbf{A}^*$ in Panel B to be the optimum first-action matrix. It is plausible that more dynamic programming iterations would be needed to reach an optimum solution when $c = 0.02$ than are needed when $c = 0.005$ (assuming both $\mathbf{A}^*$ matrixes are optimum). With the higher cost, the finite game has to be longer to justify a move to that which is optimum for the unending game.

As would be expected, $\mathbf{A}^*$ for $c = 0.005$ (i.e., $\mathbf{A}^*_{0.005}$) shows greater activity than does $\mathbf{A}^*$ for $c = 0.02$. Specifically, with Prediction State 1 (very optimistic), $\mathbf{A}^*_{0.005}$ specifies shifting to 100 percent stock at $t$ no matter what the portfolio state is at $t-1$. In the same prediction state, $\mathbf{A}^*_{0.02}$ specifies moving to a stock position of 60 percent if the portfolio has less than that percentage at $t-1$; otherwise, it recommends not changing the portfolio. In case of Prediction State 5 (very pessimistic), $\mathbf{A}^*_{0.005}$ converts to 100 percent cash no matter what the start-of-period portfolio whereas $\mathbf{A}^*_{0.02}$ recommends no change unless the starting portfolio is 100 percent stock, and then it only moves stock down to 90 percent. Action matrix $\mathbf{A}^*_{0.005}$ is also somewhat more active than $\mathbf{A}^*_{0.02}$ in the case of prediction State 2. Neither $\mathbf{A}^*_{0.005}$ nor $\mathbf{A}^*_{0.02}$ takes action in Prediction States 3 and 4.

The composition of a portfolio for an investor following action matrix $\mathbf{A}^*_{0.005}$ will vary between 0 and 100 percent invested over time. If an investor following action matrix $\mathbf{A}^*_{0.02}$ starts with less than 60 percent in stock, it will bring stock holdings to this level the first time Prediction State 1 occurs, then hold that level forever. If it starts with stock holdings of 100 percent, it will move stocks to 90 percent in Prediction State 5 and hold that level forever. If it starts with between 60 percent and 90 percent in stocks, it never changes. The steady-state optimum of $p = 0.72$ in Item 3 of Exhibit 1 is within this never-change zone of $\mathbf{A}^*_{0.02}$.

## MV Heuristic

For the model with $c$ = 0.005, for example, if we knew $\tilde{\mathbf{A}}_{0.005}$, we could compute $\tilde{\mathbf{W}}(i,j)$ for $c = 0.005$ by using Equation 16 with $\mathbf{A} = \tilde{\mathbf{A}}_{0.005}$. Conversely, if we knew $\tilde{\mathbf{W}}(i,j)$ for $c = 0.005$, we could determine $\tilde{\mathbf{A}}_{0.005}$ because $g = \tilde{\mathbf{A}}_{0.005}(i,j)$ is the value of $g$ that maximizes

$$E(u|i,j,g) + d\underset{h}{E}[\tilde{\mathbf{W}}(g,h)|j]. \tag{30}$$

The MV heuristic presented here replaces $\tilde{\mathbf{W}}$ in Expression 30 by a linear function of portfolio mean $E_p$ and variance $V_p$ as in Equation 4. Specifically, $\mathbf{A}^{MV}_{0.005}(i,j)$ is the value of $g$ that maximizes

$$E(u|i,j,g) + d\left[\underset{h}{E}(E_{wt}E_p + V_{wt}V_p|j)\right] \tag{31}$$

for a "good" choice of $E_{wt}$, $V_{wt}$, and the measures $E_p$ and $V_p$.

We experimented with the definitions of $E_p$ and $V_p$ and found that a quite simple definition works surprisingly well. Specifically, in the results reported here, $E_p(i,j)$ and $V_p(i,j)$ are the mean and variance of portfolio return for a single month if $j$ is the prediction state at the beginning of the month ($j = j_{t-1}$) and $i$ is the beginning and ending portfolio state ($i = i_{t-1} = i_t$). With $E_p$ and $V_p$ thus defined, $\mathbf{A}^{MV}(i,j)$ is the value of $g$ that maximizes

$$\begin{aligned}\psi(i,j) = {} & E(u|i,j,g) + d\underset{h}{E}[E_{wt}E_p(g,h) \\ & + V_{wt}V_p(g,h)|j].\end{aligned} \tag{32}$$

The procedure for finding a "good" set of $E_{wt}$ and $V_{wt}$ weights was as follows: For a given set of weights $E_{wt}$ and $V_{wt}$, the calculation we have presented so far determined an action matrix $\mathbf{A}^{MV}$. We solved Equation 16 to determine the expected present value of utility, $\mathbf{W}^{MV}$, of the unending game if action table $\mathbf{A}^{MV}$ were followed forever starting in state $(i,j)$. A figure of merit, *FOM*, was assigned the weights $E_{wt}$ and $V_{wt}$ by summing the entries in the $\mathbf{W}^{MV}$ matrix:

$$FOM(E_{wt}, V_{wt}) = \sum_i \sum_j \mathbf{W}^{MV}(i,j). \tag{33}$$

(The $\mathbf{W}^{MV}$ matrix should be labeled with the $E_{wt}$ and $V_{wt}$ used to generate it; these labels have been suppressed here.) We varied weights $E_{wt}$ and $V_{wt}$ ($E_{wt}$ by steps of 0.1, $V_{wt}$ by steps of 0.01) to maximize the *FOM*.

For $c$ = 0.005, one set of *FOM*-maximizing weights, is

$$\varphi_{0.005} = 4.4E_p - 0.44V_p. \tag{34a}$$

Actually, $E_{wt}$ = 4.4 and any $V_{wt}$ from –0.35 through –0.52 provides maximum *FOM*. Varying $E_{wt}$ by 0.1 in either direction reduces the *FOM*. The $\mathbf{A}^{MV}_{0.005}$ action table obtained by using the weights in Equation 34a is presented in Panel A of **Table 2**. It is identical to the (nearly) optimum action table $\mathbf{A}^*_{0.005}$ except for 0.2 rather than 0.3 at $\mathbf{A}(2,2)$ and $\mathbf{A}(3,2)$. We examine the effect on expected discounted utility of these differences in the action matrixes in the next section.

**Table 2. Action Matrix for MV Heuristic: Fraction of Stock at $t$**

| Fraction of Stock at $t-1$ | Prediction State 1 | 2 | 3 | 4 | 5 |
|---|---|---|---|---|---|
| *A. Cost = 0.005; $E_{wt}$ = 4.40; and $V_{wt}$ = –0.44*[a] | | | | | |
| 0.0 | 1.0 | 0.3 | 0.0 | 0.0 | 0.0 |
| 0.1 | 1.0 | 0.2 | 0.1 | 0.1 | 0.0 |
| 0.2 | 1.0 | 0.2 | 0.2 | 0.2 | 0.0 |
| 0.3 | 1.0 | 0.3 | 0.3 | 0.3 | 0.0 |
| 0.4 | 1.0 | 0.4 | 0.4 | 0.4 | 0.0 |
| 0.5 | 1.0 | 0.5 | 0.5 | 0.5 | 0.0 |
| 0.6 | 1.0 | 0.6 | 0.6 | 0.6 | 0.0 |
| 0.7 | 1.0 | 0.7 | 0.7 | 0.7 | 0.0 |
| 0.8 | 1.0 | 0.8 | 0.8 | 0.8 | 0.0 |
| 0.9 | 1.0 | 0.9 | 0.9 | 0.9 | 0.0 |
| 1.0 | 1.0 | 1.0 | 1.0 | 1.0 | 0.0 |
| *B. Cost = 0.02; $E_{wt}$ = 10.0; and $V_{wt}$ = –1.0*[b] | | | | | |
| 0.0 | 0.6 | 0.0 | 0.0 | 0.0 | 0.0 |
| 0.1 | 0.6 | 0.1 | 0.1 | 0.1 | 0.1 |
| 0.2 | 0.6 | 0.2 | 0.2 | 0.2 | 0.2 |
| 0.3 | 0.6 | 0.3 | 0.3 | 0.3 | 0.3 |
| 0.4 | 0.6 | 0.4 | 0.4 | 0.4 | 0.4 |
| 0.5 | 0.6 | 0.5 | 0.5 | 0.5 | 0.5 |
| 0.6 | 0.6 | 0.6 | 0.6 | 0.6 | 0.6 |
| 0.7 | 0.7 | 0.7 | 0.7 | 0.7 | 0.7 |
| 0.8 | 0.8 | 0.8 | 0.8 | 0.8 | 0.8 |
| 0.9 | 0.9 | 0.9 | 0.9 | 0.9 | 0.9 |
| 1.0 | 1.0 | 1.0 | 1.0 | 1.0 | 1.0 |

[a] $E_{wt}$ = 4.40 and any $V_{wt} \in (-0.52, -0.35)$ produces the same result.
[b] $E_{wt}$ = 10.0 and any $V_{wt} \in (-1.06, -0.99)$ produces the same result.

The same exercise performed for $c$ = 0.02 produced best $E_{wt}$ and $V_{wt}$ of

$$\varphi_{0.02} = 10.0E_p - 1.0V_p. \tag{34b}$$

Again, the maximum *FOM* was reached for $E_{wt}$ = 10.0 and a range of $V_{wt}$, namely, $V_{wt} \in (-1.06, -0.99)$. The $\mathbf{A}^{MV}_{0.02}$ is presented in Panel B of Table 2. It is the same as $\mathbf{A}^*_{0.02}$ except for 1.0 rather than 0.9 in $\mathbf{A}(11,5)$. We discuss the expected discounted utility for $\mathbf{A}^{MV}$, $\mathbf{A}^*$, and certain other heuristics in the next section.

Note that for both $c$ = 0.005 and $c$ = 0.02, the rate of substitution between $E$ and $V$ in $\mathbf{W}^{MV}$ is 10:–1 whereas the rate of substitution in the underlying single-period utility is 2:–1. This rate contrasts with the model without transaction costs analyzed by Mossin (1968) and Samuelson (1969), in which the investor seeks to maximize the expected value of a function $U$ of terminal wealth $w_T$. They concluded that if utility of final wealth is the logarithm or a power function, then the derived utility functions, $J_t(w_t)$, will be the same function (ignoring inconsequential location and scale constants). This conclusion implies that the mean–variance approximation to the derived utility functions, $J_t$, will have the same trade-off between mean and variance as does the given final utility function, $U$. Because the results of the model we have presented here are quite different and the efficacy of the mean–variance heuristic as reported in the next section is quite remarkable, we tried to minimize the probability that the differences are the result of some bug somewhere in our programs by computing each major result by at least two distinct methods.[4]

When we substitute the definitions for $E(u \mid i,j,g)$ and $E_h(\ldots \mid j)$ (from Item 9 in Exhibit 1 and Equation 21b) and use $p_{t-1}$ and $p_t$ in place of the corresponding $j$ and $g$, then $\psi(i,j)$ in Equation 32 can be spelled out as

$$\psi(i,j) = \max_{p_t}\left[E^j_{t-1}p^e + r_f(1-p^e)\right] - k(p^e)^2 V^j_{t-1} - c\left|p_t - p_{t-1}\right| + (0.99)\left\langle E_{wt}\left\{p_t\left[\sum_{h=1}^{5} P(j,h)E^h\right] + (1-p_t)r_f\right\} + V_{wt}p_t^2\sum_{h=1}^{5}P(j,h)V^h\right\rangle, \quad (35a)$$

where

$$p^e \equiv \theta_p p_{t-1} + (1-\theta_p)p_t. \quad (35b)$$

For computation, we substituted Expression 35b for $p^e$ in Equation 35a to express $\psi(i,j)$ as a function of one variable, $p_t$, and constants $p_{t-1}$, $E_{wt}$, $V_{wt}$, $E^h$, and so on. Equation 35a can be rewritten so that the expression to be maximized is a weighted sum of $E'$ and $V'$, where $E'$ is a weighted average of current-period and potential next-period expected values minus a transaction cost term and $V'$ is a weighted average of present-period and potential next-period variances.[5]

## Comparison of Expected Utilities

**Figure 1** shows the expected discounted utility for the unending game with cost $c$ of 0.005 for the nearly optimum $\mathbf{A}^*_{0.005}$ action matrix and various heuristics. The horizontal axis represents the 55 possible initial conditions for the game. From left to right, the first 11 points represent starting conditions with prediction state $j = 1$ and portfolio state $i = 1, 2, \ldots, 11$ (i.e., respectively, stock fraction $p = 0.0, 0.1, \ldots, 1.0$). The next 11 points represent initial conditions with prediction state $j = 2$ and portfolio state $i = 1, \ldots, 11$, and so on, through the final 11 points for prediction state $j = 5$ and the 11 portfolio states.

The curves labeled "All Cash," "All Stock," and "60/40 Mix" show expected discounted utility for action matrixes that move immediately to the indicated mix and stay there. For example, the "60/40 Mix" action matrix has $\mathbf{A}(i,j) = 7 (p = 0.6)$ for all $(i,j)$. An unending game played using this matrix would move to a 60/40 mix as its first action and never move again.

The curve labeled "No Action" never changes its allocation. Note, for example, that the expected discounted utility of "No Action" is the same as that of "All Cash" (with portfolio state $p = 0.0$) at points 1, 12, 23, 34, and 45 and is the same as "All Stock" in cases where initial portfolio state $p$ is 1.0. The "Very Active" curve shows the expected discounted utility of a heuristic that, given the current portfolio and prediction states $(i,j)$, chooses the next portfolio state, $g$, that maximizes the expected utility in time period $t$—assuming zero transaction costs (and with $\theta_p$ of zero).

*The expected discounted utility curves for the optimum and mean–variance heuristic strategies are virtually indistinguishable and dominate the curves of the other heuristics!*

**Figure 2** shows the same information for $c = 0.02$. In this case, the very active heuristic performs far below the others. The inactive strategy is actually optimum if the initial portfolio state is $p \in (0.6, 0.9)$, as noted at the end of the "(Nearly) Optimum Action Matrixes" section, and virtually so for $p \in (0.5, 1.0)$. As in the case with $c = 0.005$, the expected discounted utility provided by the MV heuristic is almost indistinguishable from the optimum expected total utility for all $(i,j)$.

## Where Next?

As it stands, the model presented here is as useless for practice as the airplane the Wright brothers flew at Kitty Hawk. The importance of this model, if it is to have any, will come from what happens next. We sketch some possible future uses of the method of mean–variance surrogates.

The general version of the model provided in Exhibit 1 could be used with other parameter settings to represent investment situations with (1) large numbers of securities and (2) other forecast models, such as ARCH-type models.[6]

**Figure 1. Expected Discounted Utility for Various Strategies: Unending Game, Cost 0.005**

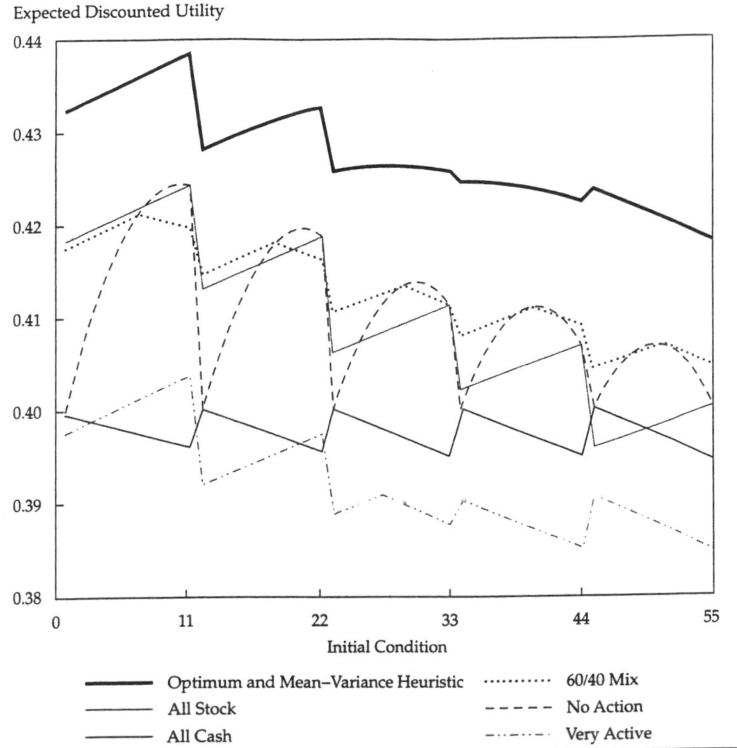


ARCH- and GARCH-type models would imply greater differences in variances among prediction states and little or none in returns in Exhibit 1, Item 3.[7] This result would imply smaller differences in optimal investment. We conjecture that this application would, therefore, make it easier for appropriate heuristics to approximate optimum performance. The methodology of the preceding sections could be used to test the conjecture for a toy model with two securities. For larger numbers of securities or asset classes, Monte Carlo runs could be used to select (or fine-tune) $E_{wt}$ and $V_{wt}$ and evaluate the discounted expected utility of the mean–variance surrogate versus other heuristics.

Generally, an optimum solution cannot be computed for higher-dimensional problems. The simulation runs can only compare alternative heuristics. As long as optimization is out of the question, a more realistic consumption assumption than Item 9 in Exhibit 1 should not be a problem for the heuristics, including mean–variance surrogate heuristics.

Beyond the model in Exhibit 1, an important area where illiquidities are paramount is portfolio management for taxable entities. Unrealized capital gains are a much greater source of illiquidity than the transaction costs considered here. Some aspects of the problem of choosing the correct action in the presence of unrealized capital gains are well within current computing capabilities. It does not take long to trace out a mean–variance-efficient frontier for a problem with thousands of decision variables, where the variables may or may not have lower and/or upper bounds and may be subject perhaps to hundreds of other linear equality or inequality constraints (see Perold 1984 or Markowitz and Todd 2000). In particular, it would not strain current computing capacity to include sales from individual lots as decision variables and have the budget equation recognize that the proceeds of such sales depend on the amount of unrealized capital gains or losses in the lot.

The hard part has to do with the objective of such analysis. For example, we could seek efficiency in terms of the mean and variance of end-of-period net asset value. One approach would assume that a dollar's worth of unrealized capital

**Figure 2. Expected Discounted Utility for Various Strategies: Unending Game, Cost 0.02**

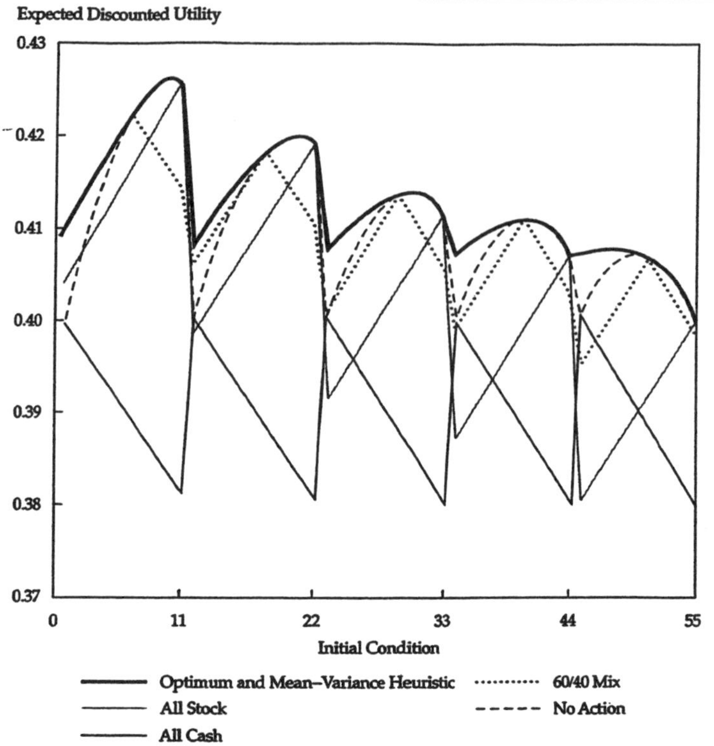


*Note*: The Very Active line is below 0.25 in all cases and is not shown.

gain is worth a dollar's worth of cash. An alternative approach would assume that a dollar's worth of unrealized capital gain is worth some fraction times as much as a dollar's worth of cash. Thus, the objective of the analysis could be efficiency in terms of the mean and variance of a weighted sum of holdings in which unrealized gains and losses would be weighted differently from after-tax funds. A weighted average of the mean and variance of a portfolio computed in this manner could be the mean–variance surrogate and would stand in place of the complex, unknown derived utility function of this problem.

## Conclusion

Consider again the model described in the first paragraph of this article—with transaction costs and many securities whose joint distribution of returns changes over time. We have examined two examples of a simple version of such a model—examples in which the optimum solutions can be calculated—and we found that the discounted expected utility supplied by a mean–variance surrogate heuristic in these particular cases is practically indistinguishable from the optimum value. This finding suggests that the same may be true in more-complex models. At least the finding justifies an effort to test such rules in realistic investment simulations.

Most, if not all, real-world portfolio management situations involve some illiquidities, changing beliefs, and more securities or asset classes than permit optimization. Thus, the portfolio manager or managers must use some heuristic procedure for changing their portfolios. Their procedure may be something they make up as they go along, or they may follow some formal rule. They may adopt a formal rule because it seems plausible or because it performs better than other heuristics when tested in a Monte Carlo simulation or backtest. We argue that the mean–variance surrogate heuristics should be among those rules tried.[8]

## Appendix A. Why A* May Not Equal $\tilde{\mathbf{A}}$

In this appendix, we use a simpler example of a model of the Exhibit 1 structure to illustrate why we cannot conclude that **A*** equals $\tilde{\mathbf{A}}$ even when $\mathbf{A}^* = \mathbf{A}_t^T$ for $t = 0, \ldots, S$ for large $S \le T$.

Let the number of prediction states equal the number of portfolio states and be 2; let the risk-free rate, $r_f$, be 0; and let the prediction-state transition matrix be

$$\mathbf{P} = \begin{pmatrix} 1/2 & 1/2 \\ 1/2 & 1/2 \end{pmatrix}$$

Label the two portfolio states "0" and "1." The 0 state is all cash, and the 1 state is all stock. The two prediction states will also be labeled "0" and "1"; Prediction State 0 is pessimistic, and Prediction State 1 is optimistic. The model assumes that $\theta_p = 1$ (see Item 6 of Exhibit 1); thus, $Eu(D_t)$ depends only on $p_{t-1}$ (and the prediction state), not on $p_t$. Assume that if the portfolio state is 1 at $t - 1$, then

$$Eu(D_t) = \left\{ \begin{matrix} 0 & \text{if prediction state} = 0 \\ 0.0001 & \text{if prediction state} = 1 \end{matrix} \right\}.$$

Because stocks do as well as cash in this model in Prediction State 0 and stocks do better than cash in Prediction State 1, switching from all stocks to all cash is never optimum. Whether it is optimum to switch from cash to stocks depends on transaction cost $c$ but, because of the assumed $p$ and $t_p$, does not depend on the current prediction state. Thus, the two possible optimum action matrixes are

$$\mathbf{A}_N = \begin{pmatrix} 0 & 0 \\ 1 & 1 \end{pmatrix} \quad \text{(A1)}$$

and

$$\mathbf{A}_Y = \begin{pmatrix} 1 & 1 \\ 1 & 1 \end{pmatrix}. \quad \text{(A2)}$$

For both $\mathbf{A}_N$ and $\mathbf{A}_Y$, if the game starts in Portfolio State 1, the expected value of the game is

$$\frac{0.0001}{1-d} = 0.01.$$

If the game starts in Portfolio State 0, the values of the game for $\mathbf{A}_N$ and $\mathbf{A}_Y$ are, respectively,

$$E_N = 0$$

and

$$E_Y = 0.0001\frac{d}{1-d} - c.$$

Thus, whether the initial prediction state is 0 or 1, $\mathbf{A}_Y$ is better than $\mathbf{A}_N$ if and only if

$$c < 0.0001\frac{d}{1-d} = 0.0099.$$

For a finite game of length $T$, as in the infinite game, if the initial portfolio state is 0, it is optimum to switch immediately or never switch. The expected utilities of the two alternatives are

$$E_N^T = 0$$

and

$$E_Y^T = 0.0001\frac{d - d^{T+1}}{1-d} - c.$$

Thus, for a game of length $T$, $\mathbf{A}_Y$ is better than $\mathbf{A}_N$ if and only if

$$c < 0.0001\frac{d - d^{T+1}}{1+d}$$

$$< 0.0001\frac{d}{1-d}.$$

To construct an example in which

$$\mathbf{A}_t^T = \mathbf{A}_N \text{ for } t = 1, \ldots, S = 10^7$$

but

$$\mathbf{A}_N \neq \tilde{\mathbf{A}},$$

let $u = 10^8$ and

$$c = \frac{0.0001(d - d^{u+1})}{1-d}.$$

In this case, $\mathbf{A}_N$ will be the optimum choice at every move for a game of, for example, length $T = 1{,}000{,}000$ but will not be the long-run optimum **A**.

## Notes

1. We will see that, because of transaction costs, we cannot further simplify Equation 4 at this stage by dividing through by $E_{wt}$, as is often justified.
2. As explained in Note 5, although Constantinides presented a remarkably general discrete-time analysis, the model we present here is not a special case of his model. In particular, an "inexpensive" (in that it did not increase the dimensionality of the problem) bit of realism in our model makes it no longer true that if current equity position $X$ is, for example, less than some level $\underline{X}$ that depends on the state of the system, then the optimum action is to buy $\underline{X} - X$ no matter the value of $X < \underline{X}$.

3. See especially Young and Trent regarding mean–variance approximations to expected log or, equivalently, geometric mean return.
4. In particular, we computed the action matrixes for the MV heuristic for various $E_{wt}$ and $V_{wt}$ by using MatLab and recomputed the action matrixes for the winning weights shown in Table 2 by using Excel. The Excel and MatLab answers were the same except for the $(i,j) = (2,2)$ entry for $c = 0.005$. The Excel computation gave almost identical scores to $g = \mathbf{A}(2,2) = 2$ and $g = \mathbf{A}(2,2) = 3$ but slightly favored $g = 3$. For $c = 0.005$ and 0.02, we used dynamic programming to compute $\tilde{\mathbf{W}}^{1200}$ and used 55 simultaneous equations to compute $\mathbf{W}^*$. From Equation 26a, $\tilde{\mathbf{W}}^{1200} < \mathbf{W}^* < \tilde{\mathbf{W}}^{1200} + \varepsilon$ for $\varepsilon$ in Equation 26b, lending confidence to both programs. The similarity between $\mathbf{W}^*$ and $\mathbf{W}^{MV}$ is to be expected because of the similarity of their action matrixes. Finally, we will note later that $\mathbf{W_A}$ for certain heuristic action matrixes, or certain relationships among the $\mathbf{W_A}$, can be determined with little or no calculation. In every instance, these results confirmed the simultaneous equation calculations.
5. In Table 2, Prediction State 2 in Panel A violates the condition that there is an $\underline{X}$ (which varies here with prediction state) such that if $X_{t-1} < \underline{X}$ then $X_t = \underline{X}$. The reason this can happen in the present model is that, after substituting its definition for $p^e$, Equation 35a includes the term $-k\theta_p(1 - \theta_p)p_{t-1}p_t$. Thus, the marginal cost of, for example, increasing $p_t$ from a contemplated $p_t = 2$ to a contemplated $p_t = 3$ depends on $p_{t-1}$—in the exact optimization calculation as well as in the MV approximation. This term appears to be of "second order," because the violation of the no-action principle does not occur in $\tilde{\mathbf{A}}$ and is inconsequential to $\mathbf{A}^{MV}$ (see Note 4). The term disappears if either $\theta_p = 0$ or $\theta_p = 1$—that is, if the new position is attained instantly either at the beginning or the end of the month.
6. ARCH = autoregressive conditional heteroscedasticity.
7. GARCH = generalized autoregressive conditional heteroscedasticity.
8. The idea of using single-period mean–variance analysis in place of the derived utility function of dynamic programming appeared in Chapter 13 of *Portfolio Selection*. Markowitz asserted, "The value of the dynamic programming analysis [for us] does not lie in its immediate application to the selection of portfolios. It lies, rather, in the insight it supplies concerning optimum strategies and how, in principle, they could be computed. It emphasizes the use of the single period utility function" (p. 279). See also p. 299 on "Changing Distributions," which is in contrast to the usual textbook characterization of mean–variance analysis as static. The 1959 book did not attempt to test the efficacy of this approach, however, as we have done in this article.

## References


Balduzzi, P., and A.W. Lynch. 1999. "Transaction Costs and Predictability: Some Utility Cost Calculations." *Journal of Financial Economics*, vol. 52, no. 1 (April):47–78.

Bellman, R.E. 1957. *Dynamic Programming*. Princeton, NJ: Princeton University Press.

Constantinides, George M. 1979. "Multiperiod Consumption and Investment Behavior with Convex Transactions Costs." *Management Science*, vol. 25, no. 11 (November):1127–37.

Dantzig, G.B. 1955. "Linear Programming under Uncertainty." *Management Science*, vol. 1, nos. 3–4 (April–July):197–206.

Dantzig, G.B., and G. Infanger. 1993. "Multi-Stage Stochastic Linear Programs for Portfolio Optimization." *Annals of Operations Research*, vol. 45:59–76.

Davis, M.H.A., and A.R. Norman. 1990. "Portfolio Selection with Transaction Costs." *Mathematics of Operations Research*, vol. 15, no. 4 (November):676–713.

Dexter, A.S., J.N.W. Yu, and W.T. Ziemba. 1980. "Portfolio Selection in a Lognormal Market When the Investor Has a Power Utility Function: Computational Results." In *Stochastic Programming*. Edited by M.A.H. Dempster. New York: Academic Press.

Ederington, L.H. 1986. "Mean–Variance as an Approximation to Expected Utility Maximization." Working Paper 86-5, School of Business Administration, Washington University.

Grauer, R.R. 1986. "Normality, Solvency, and Portfolio Choice." *Journal of Financial and Quantitative Analysis*, vol. 21, no. 3 (September):265–278.

Kamin, J.H. 1975. "Optimal Portfolio Revision with a Proportional Transaction Cost." *Management Science*, vol. 21, no. 11:1263–71.

Kroll, Y., H. Levy, and H.M. Markowitz. 1984. "Mean Variance versus Direct Utility Maximization." *Journal of Finance*, vol. 39, no. 1 (March):47–61.

Levy, Haim, and Harry Markowitz. 1979. "Approximating Expected Utility by a Function of Mean and Variance." *American Economic Review*, vol. 69, no. 3 (June):308–317.

Levy, Moshe, Haim Levy, and Sorin Solomon. 2000. *Microscopic Simulation of Financial Markets*. New York: Academic Press.

Lynch, A.W., and P. Balduzzi. 2000. "Predictability and Transaction Costs: The Impact on Rebalancing Rules and Behavior," vol. 55, no. 5 (October):2285–94.

Magill, M.J.P., and G.M. Constantinides. 1976. "Portfolio Selection with Transactions Costs." *Journal of Economic Theory*, vol. 13, no. 2 (October):245–263.

Markowitz, H.M. 1959. *Portfolio Selection: Efficient Diversification of Investments*. New York: John Wiley & Sons. 1991. 2nd ed. Cambridge, MA: Basil Blackwell.

Markowitz, H.M. and P. Todd. 2000. *Mean–Variance Analysis in Portfolio Choice and Capital Markets*. New Hope, PA: Frank Fabozzi and Associates.

Markowitz, H.M., D.W. Reid, and B.V. Tew. 1994. "The Value of a Blank Check." *Journal of Portfolio Management*, vol. 20, no. 4 (Summer):82–91.

Mossin, Jan. 1968. "Optimal Multiperiod Portfolio Policies." *Journal of Business*, vol. 41, no. 2 (April):215–229.

Perold, A.F. 1984. "Large-Scale Portfolio Optimization." *Management Science*, vol. 30, no. 10 (October):1143–60.

Pulley, L.M. 1981. "A General Mean–Variance Approximation to Expected Utility for Short Holding Periods." *Journal of Financial and Quantitative Analysis*, vol. 16, no. 3 (September):361–373.

———. 1983. "Mean–Variance Approximations to Expected Logarithmic Utility." *Operations Research*, vol. 31, no. 4 (July–August):685–696.

Samuelson, P.A. 1969. "Lifetime Portfolio Selection." *Review of Economics and Statistics*, vol. 51, no. 3 (August):239–246.

# Financial Market Simulation

BRUCE I. JACOBS, KENNETH N. LEVY,
AND HARRY M. MARKOWITZ

When they want to see how complex systems work, scientists often turn to asynchronous-time simulation models, which allow processes to change sporadically over time, typically at irregular intervals. While rarely used in finance today, such models may turn out to be valuable tools for understanding how markets respond to changes in the participation rates of different types of investors, for example, or to changes in regulatory or investment policies. The asynchronous, discrete-event, stock market simulator described here allows users to create a model of the market, using their own inputs. Users can vary the numbers of investors, traders, portfolio analysts, and securities, as well as their own investing and trading decision rules. Such a simulation may be able to provide a more realistic picture of complex markets.

# Financial Market Simulation

*In the 21st century.*

Bruce I. Jacobs, Kenneth N. Levy, and Harry M. Markowitz

Large and detailed asynchronous (event-based) simulation models are widely used in the planning and analysis of systems such as manufacturing, logistics, and warfare. They have been used relatively little in financial analysis, especially as compared with continuous-time models. We believe the situation will change, starting early in the 21st century.

Here we describe asynchronous simulation generally, and illustrate its capabilities in terms of a particular asynchronous and discrete-event stock market simulator, the JLM Market Simulator (JLM Sim).

The JLM Sim is not a model of a market per se. Rather, it is a tool investors can use to create a model of the market using their own inputs. JLM Sim users can vary the number of securities, statisticians, portfolio analysts, investors, and traders in the simulated market and the decision rules they use. Prices are set endogenously, as new orders encounter already placed limit orders. At present, all JLM Sim investors are mean-variance investors; more advanced versions of JLM Sim later may accommodate additional investor types.

Dynamic models may be described as synchronous, asynchronous, or continuous. Asynchronous simulation has some advantages over synchronous or continuous-time financial models. We summarize the salient features of the current JLM Sim and some desirable features that may be added to more advanced versions.

**Bruce Jacobs** is principal at Jacobs Levy Equity Management in Florham Park, NJ.
**bruce.jacobs@jacobslevy.com**

**Kenneth Levy** is principal at Jacobs Levy Equity Management in Florham Park, NJ.

**Harry Markowitz** is principal at Harry Markowitz Company in San Diego, CA.

## TYPES OF DYNAMIC MODELS

Dynamic models can be divided between discrete and continuous. In discrete models, time advances in discrete increments, while in continuous models the system changes continuously over time. Discrete-time models can be broken down further into synchronous and asynchronous models. In synchronous-time models, the clock advances by fixed increments such as a day or a year. Typically, the status of all system constituents is updated at each increment of time.

In asynchronous models, time advances, typically by uneven increments, to the next scheduled event. The next event might be Investor A's placement of initial orders after reoptimization, followed perhaps a split second later by the review of an order placed previously by Trader B on behalf of Investor C, followed perhaps seconds later by the end-of-day event when accounts with short or leveraged positions are marked to market.[1]

In an asynchronous model, events often involve only one or a few entities—one investor and a trader, or an investor selling and one or more buyers with orders on the books. An event may also involve many entities, as the end-of-day event does.

It is possible to solve some asynchronous-time models analytically; certain queuing models are a case in point (see Haverkort [1998] for examples on both sides). Most large and detailed asynchronous models that attempt to model some complex system fairly literally usually require computer simulation to derive their implications.[2]

## JLM SIMULATOR

In a JLM Sim simulation run, as of any instant of simulated time, the simulated market has a state or *status*. This status changes at points in time called *events*.

### Status

The status of the JLM Sim simulated market is described in terms of how many *entities* of different *entity types* there are; the values of their *attributes*; and the members of the *sets* they own. Entity types represented in JLM Sim include securities, investors, traders, order slips, portfolio analysts, and statisticians. A given simulation may include many individual entities of each type. Attributes include the price of a security and the volume traded so far today, the current wealth of an investor, and the buy or sell attributes of an order slip. As of any instant, an attribute of an (individual) entity has one and only one value.

Sets include the set of buy orders and the set of sell orders for a given security. The security's buy orders are sorted from high to low according to the limit price attribute of the order slip or, among order slips with the same limit price, according to an arrival time (actually, a take-a-number) attribute of the order slip. The sell order set is similarly sorted, but from low to high. We say that each security *owns* a buy-order set and a sell-order set. Zero, one, or more order slips *belong to*, or are *members of*, each set. (In a computer program, the sets would be named buy_orders and sell_orders sets, since in a computer program a hyphen (as in buy-orders) would be treated as a minus sign.)

Exhibit 1 summarizes some of the JLM Sim EAS (entity, attribute, and set) structure. We want to convey the general idea of the JLM Sim EAS structure, rather than provide a user's manual.

The entity types are listed first. The names of the attributes are in the second column, and the names of sets owned by the entities of the entity types are in the third. The last column, headed member or data type, shows such information as whether an attribute is an integer or real (decimal) number and what type of entity belongs to the named set.

The first entity type listed in Exhibit 1 is the system as a whole. Attributes of the system include the current lending and borrowing rates of interest. Sets owned by the system include all the securities, statisticians, and portfolio analysts in the simulation, as well as the kept trading days, kept months, investor templates, and trader templates.

Each simulated investor is created from some *investor template* and identified by the template it came from and a sequence number. For example, an investor could be the 456th investor from template 3. One attribute of an investor template is the number of investors that are to be generated from this template. One attribute of the investor is the template from which it comes.

All investors from a given template share certain attributes, such as reoptimization frequency and risk aversion parameter K. These are stored as attributes of the investor template. Investors from a given template differ with respect to attributes such as starting wealth and time at which they reoptimize. The starting wealth of an investor is drawn randomly at the beginning of the simulation from a lognormal distribution whose parameters are attributes of the investor template. Individual investors from a given template have differing experiences during a simulation run, depending in part on circumstances such as when they reoptimize and when their traders try to execute the resulting buy and sell orders.

When an investor wants to reoptimize, it chooses an "ideal" portfolio from a mean-variance efficient frontier that is based on estimates by one or another statistician. The choice of portfolio depends on the investor's risk aversion. The investor may seek to move only partway from current to ideal portfolio, depending on turnover constraints. To accomplish this, the investor places buy and sell orders with its trader.

Every trader is generated from a *trader template*. The number of traders generated from a given trader template is

## EXHIBIT 1
**Entity, Attribute, and Set Structure of JLM Sim (extract)**

| ENTITY TYPES | ATTRIBUTES | SETS OWNED | Member or Data Type |
|---|---|---|---|
| TheSystem | | | |
| | SimTime | | Real |
| | RFLendRatePerDay | | Real |
| | BrokerRatePerDay | | Real |
| | Liquidation_trader_nr | | Integer |
| | | Securities | Security |
| | | KeptTradingDays | Day |
| | | KeptMonths | Month |
| | | Statisticians | Statistician |
| | | PortfolioAnalysts | PortfolioAnalyst |
| | | InvestorsTemplates | Investor_template |
| Security | | | |
| | LastTradePrice | | Real |
| | Price | | Real |
| | StartOfDayPrice | | Real |
| | StartOfMonthPrice | | Real |
| | VolumeSoFarToday | | Integer |
| | | Buy_orders | Order |
| | | Sell_orders | Order |
| Security_X_Day | | | |
| | DailyReturn | | Real |
| | DailyVolume | | Integer |
| | DailyClosePrice | | Real |
| Security_X_Month | | | |
| | MonthlyReturn | | Real |
| | MonthlyVolume | | Integer |
| | MonthlyClosePrice | | Real |
| Statistician | | | |
| | EstMethodForMeans | | Enumeration |
| | EstMethodForCovs | | Enumeration |
| Statistician_X_Security | | | |
| | AnnualizedMean | | Real |
| Statistician_X_Security_X_Security | | | |
| | AnnualizedCov | | Real |
| PortfolioAnalyst | | | |
| | StatisticianNr | | Integer |
| | | EfficientSet | EfficientSegment |

not fixed. Rather, each investor template specifies, as an attribute, which trader template its investors will use. Traders from this trader template are created as needed to service investors who need to trade.

Attributes of a trader template include the alpha and the beta of the buy candidate (buy_alpha, buy_beta). To execute a buy order for a security, the trader initially places an order at a limit price, determined as follows:

$$\text{Limit Price} = \text{Buy-Alpha} + \text{Buy-Beta} \times \text{Price} \qquad (1)$$

For example, if buy-alpha equals –0.02 and buy-beta equals 1.0, the trader bids two cents less than current price. If buy-alpha equals 0.0 and buy-beta equals 1.01, the trader bids 1 percent more than current price (unless an offer is available at a lower price). Here "price" may equal the average of bid and asked prices, the bid or the asked price, or the last transaction price, depending on the availability of these prices and certain relationships among them.

If the security's sell-order set includes a sell order at the buyer's limit price or less, a transaction takes place for the lesser of the buyer's desired trade size (amount_to_do) and the seller's desired trade size (amount_to_do), where the "amounts to do" are attributes of the order slip entity. If this trade does not complete the buyer's order, the amount to do on the buyer's order slip is reduced by the amount of the security pur-

**EXHIBIT 1** (continued)
**Entity, Attribute, and Set Structure of JLM Sim (extract)**

| ENTITY TYPES | ATTRIBUTES | SETS OWNED | Member or Data Type |
|---|---|---|---|
| EfficientSegment | | | |
| | HighE | | Real |
| | HighV | | Real |
| | LowE | | Real |
| | LowV | | Real |
| | HighPortfolio | | CornerPortfolio |
| | LowPortfolio | | CornerPortfolio |
| CornerPortfolio | | | |
| | Cp_nr | | Integer |
| | E | | Real |
| | V | | Real |
| Corner_Portfolio_X_Security | | | |
| | X | | Real |
| InvestorTemplate | | | |
| | Nr_investors | | Integer |
| | Portfolio_analyst_nr | | Integer |
| | Trader_template_nr | | Integer |
| | Mean_log10_init_wealth | | Real |
| | Sigma_log10_init_wealth | | Real |
| | K | | Real |
| | Reoptimization_frequency | | Enumeration |
| | | Investors | Investor |
| InvestorTemplate_X_Security | | | |
| | Total_bought_today | | Integer |
| | Nr_of_buyers | | Integer |
| | Seq_nr_of_largest_buyer | | Integer |
| | Purchase_of_largest_buyer | | Integer |
| | Total_sold_today | | Integer |
| | Nr_of_sellers | | Integer |
| | Seq_nr_of_largest_seller | | Integer |
| | Sale_of_largest_seller | | Integer |
| Investor | | | |
| | Seq_nr | | Integer |
| | Investor_template_nr | | Integer |
| | StartingWealth | | Real |
| | Deposits_received | | Real |
| | Withdrawals_paid | | Real |
| | Withdrawals_owed | | Real |
| | Collateral_for_short_positons | | Real |
| | CurrentWealth | | Real |

chased, and the process is repeated. If the buyer's remaining order cannot be filled by orders from the sell-orders set, it is entered into the buy-orders set.

The buyer waits a specified time (buy_first_time_wait) before attempting to complete the order by raising its bid. It adds specified increments to alpha and beta (buy_alpha_inc, buy_beta_inc), and recomputes a new limit price using Equation (1) but substituting the old limit price for price. If this does not result in a purchase, the trader waits a further specified time (buy_following_time_wait) before again recomputing the limit price.

This process is repeated a specified number of times (buy_max_nr_price_changes). The buyer then waits a specified time (buy_last_time_wait) before canceling any uncompleted order. A similar procedure applies for sell orders using the attributes of the trader template.

Entity types listed in Exhibit 1 include "security_X_day," "security_X_month," "statistician_X_security," and so on. These are called *compound entity types*. "Investor_X_security," for example, is an investor-security combination. A "statistician_X_security_X_security" is a statistician-security-security triplet. Compound entities can have attributes and own sets. For example, "buy_or_sell_ amount" is an attribute of "trader_X_security"; that is, each trader-

**EXHIBIT 1** (continued)
**Entity, Attribute, and Set Structure of JLM Sim (extract)**

| ENTITY TYPES | ATTRIBUTES | SETS OWNED | Member or Data Type |
|---|---|---|---|
| Investor_X_Security | | | |
| | X_units | | Real |
| Trader_template | | | |
| | Buy_Alpha | | Real |
| | Buy_Beta | | Real |
| | Buy_Alpha_inc | | Real |
| | Buy_Beta_inc | | Real |
| | Buy_First_time_wait | | Real |
| | Buy_Following_time_wait | | Real |
| | Buy_Last_time_wait | | Real |
| | Buy_Max_nr_price_changes | | Integer |
| | Sell_Alpha | | Real |
| | Sell_Beta | | Real |
| | Sell_Alpha_inc | | Real |
| | Sell_Beta_inc | | Real |
| | Sell_First_time_wait | | Real |
| | Sell_Following_time_wait | | Real |
| | Sell_Last_time_wait | | Real |
| | Sell_Max_nr_price_changes | | Integer |
| Trader | | | |
| | Trader_template_nr | | Integer |
| | Investor_being_served | | Investor_ID |
| Trader_X_Security | | | |
| | Buy_or_sell_amount | | Integer |
| | Amount_on_order | | Integer |
| | | Orders_against_amount | Order_slip |
| Order_slip | | | |
| | Buy_or_sell | | Enumeration |
| | Trader_placing_order | | Trader_ID |
| | Security_ordered | | Integer |
| | Limit_price | | Real |
| | Amount_to_do | | Integer |
| | Order_status | | Enumeration |

security pair has a negative, zero, or positive amount associated with it, which indicates the number of units (shares or bonds) that are to be sold (if negative) or bought (if positive). "Amount_on_order" is another attribute of trader-security. The amount on order may be less than the buy or sell amount, because buy orders are sometimes delayed until cash is raised from sell orders.

"X_units" is the number of units (shares of stock or face value of bonds) attribute of investor-security; "estimated_annualized_mean" is an attribute of statistician-security; "estimated_annualized_cov(ariance)" is an attribute of statistician-security-security.

"Day" is an entity type in JLM Sim that has no attributes on its own, but appears in compound entity types. Specifically, "daily return," "daily volume," and "daily close price" are attributes of security-day. The system owns sets of recent "kept days" and "kept months" that provide data needed by statisticians or for other purposes.

In general, then, as of any instance in simulated time in a JLM Sim run, the status of the system is described by the values of the attributes and members of sets owned by individuals of entity types (including compound entity types) such as those listed in Exhibit 1.

### Events

Events change the status of the simulation and cause future event occurrences. Exhibit 2 lists the principal event types of JLM Sim and the actions they bring about. The initialization routines are not events per se, but are included in the exhibit because they change the status of the simulation (from nothing to something) and cause the first event occurrences.[3]

***Initialization.*** The initialization routines create statisticians, portfolio analysts, investor templates, and trader templates with the attributes specified by the user. In the current version of JLM Sim, statisticians use historical returns to estimate covariances, and, depending on the expected return estimation procedure specified, may use historical returns to estimate expected returns. The returns the statisticians find in *kept days* and *kept months* at the start of the simulation are randomly generated by a factor model specified by the user.

The number of investors to be generated from a given investor template is an attribute specified by the JLM Sim user. When each investor is created during initialization, its starting wealth is drawn randomly from a lognormal distribution whose user-specified parameters are attributes of the investor template.

Another attribute governs whether investors from a given investor template reoptimize daily, monthly, quarterly, or annually. If investors from an investor template reoptimize monthly, for example, an initialization routine determines when during the first month of the simulation a particular investor will first reoptimize.[4]

***Reoptimization.*** The reoptimization event cancels any buy or sell orders the investor has outstanding. The investor's portfolio analyst selects an ideal portfolio that, if there were no transaction costs, would maximize:

$$\text{(Portfolio Expected Return)} - K\text{(Portfolio Variance)} \quad (2)$$

where K, the investor's risk aversion, is an attribute of the investor template. The investor computes the buy and sell orders needed to move the portfolio toward the ideal portfolio. The amount of movement from current toward ideal can be regulated by constraints intended to reduce costly turnover.

Desired purchases and sales are handed to the trader to execute as we have described. The trader first attempts to execute each order to sell long positions or to cover short positions. Before placing an order to buy or to sell short, the trader considers the possibility that the purchase or short sale will be executed before the long sale or the short cover, putting the investor in violation of some regulation or self-imposed investor constraint. In that case, some or all purchases or short sales may be postponed until sufficient long sales or short covering transactions are completed.

When an order is placed, JLM Sim determines whether there is an order on the books at the limit price or better. If there is, a transaction takes place. If the transaction does not complete the new order, further transactions are sought from the book; any uncompleted portion is finally "placed on the books"—entered into the set of buy orders or sell orders—and an order review is scheduled. If, before this time, the order is filled by a transaction emanating from some other investor's reoptimization or order review, the order review is cancelled.

***Order Review.*** In the order review event, the order is either repriced or cancelled. If the order is repriced, JLM Sim considers whether there are one or more matching orders. If the order is not filled, it is placed on the books again, and a new order review is scheduled. If a transaction takes place, and the matching investor has raised cash by selling or has covered a short position, then the matching investor may take further action.

***End of Day.*** The end-of-day event updates daily statistics and perhaps monthly statistics. It checks to see if accounts with leverage or short sales have violated maintenance margin requirements. Any account that is found in violation of these requirements is turned over to a liquidation trader, who trades to get the account back into compliance.

One trader template is designated by the JLM Sim user as the liquidation trader template (an attribute of the system). Liquidation traders are created as needed, and use the liquidation trader template's parameters (such as sell-alpha and sell-beta). The parameters of the liquidation trader template, specified by the JLM Sim user, are presumably more aggressive than those of other traders, in that they are slanted toward quick execution rather than favorable prices.

## EXHIBIT 2
**JLM Sim Events**

| Event | Actions |
|---|---|
| Initialize | Creates and initializes status in accord with the JLM Sim user's instructions. |
| Reoptimize | May randomly generate deposits and withdrawals for the particular investor. Has investor's portfolio analyst compute the investor's ideal portfolio, using estimates supplied by its statistician. Investor computes how far it should move from its current portfolio toward the ideal portfolio, considering turnover constraints. Places orders with trader. Trader executes orders if there are matching orders on other side, and places balance of order on books. If trades are executed, trader for the other party may take further actions. |
| Review Order | Changes the limit price or cancels the order. If limit price is changed, actions may occur similar to those that occur when an order is placed during reoptimization. |
| End of Day | Updates daily and perhaps monthly statistics. Marks to market accounts with leverage or short positions. Accounts that violate maintenance margin requirements are turned over to a liquidation trader. |

## OBJECTIVES AND EXTENSIONS

The JLM Market Simulator is not a model of a market per se; rather, it is a tool that allows its user to model a market by supplying certain components. When JLM Sim users specify the numbers of entities of different types and their attributes, they have in effect created a model of a market. This model is run for the length of time specified by the user—and can be run repeatedly with different initial random number seeds—in order to estimate the probable outcomes.

At first, the user will need to experiment with the model to get it to reflect aspects of a real-world market. Not only should the components of the model imitate, to some extent, their real-world counterparts, but the resulting price behavior should be reasonably comparable to its real-world counterpart. When the model has achieved some plausibility in terms of inputs and outputs, it can be used to test investment and trading policies, or regulatory policies such as the efficacy of the uptick rule and the relationship between changes in interest rates and market response.

We expect to enhance JLM Sim, in part based on the experience of users of the current version. Below are two areas in which enhancements seem attractive.

### Alternative Investor and Trader Behaviors

In the current JLM Sim, all investors are mean-variance investors. Other investors may use other criteria and procedures, such as downside risk, mean-absolute deviation, or resampled frontiers. Behavioral finance specialists have described nonrational market behavior. We would like to make any and all such investor behaviors available to the JLM Sim modeler.[5]

One way to do so is for us to program such alternative investor behaviors into JLM Sim and let the user specify which behavior is to be used by investors of a given investor template. The advantage of this approach is that it puts the given behavior at the user's disposal without requiring that the user program it. The disadvantage is that the user is limited to the behaviors the originators programmed.

Another approach would be to allow users to program their own proposals for investor behavior. The most natural way to do this would be for the user or user's programmer to program in C++, the computer language in which JLM Sim is programmed. This would make use of C++'s ability to have one version of an object (entity) extend or override another version of the object, making use of public information about other entities (such as the bid and asked prices for a stock) without being able to directly modify such information.

The advantage of this approach is that it is open-ended; the disadvantage is that it requires the user to be or have access to a C++ programmer.

### Model Size

JLM Sim runs fairly fast, at a few thousand events per second, on a 2.4 GHz PC, primarily for two reasons. One reason is that JLM Sim stores the simulated status (entities, attributes, and sets) and the calendar of coming event notices in the computer's random access memory rather than the longer-to-access disk memory. The second reason is that JLM Sim uses a particular software to file, remove, and find members of very large, ordered sets, such as the set of coming events, that currently handles only sets stored in RAM.[6]

For these two reasons, a JLM Sim simulation run on a personal computer must fit into the PC's virtual RAM, and will run especially fast if it fits into the PC's real RAM for the most part. As a rough rule of thumb, this sets an upper limit for JLM Sim at roughly a few tens of thousands of investors if there are only a few securities (say, ten or fewer), or at fewer investors if there are many securities. These limits should be sufficient for experimenting with markets that have features of real markets, but are smaller.

For example, in our own tests of JLM Sim features, we think of a run's thousands of investors as thousands of investment companies with random deposits and withdrawals rather than as a market of individuals.

If we tried to build a life-sized market on a PC by using disk memory as if it were a large RAM, we would slow the simulator by orders of magnitude because of the orders-of-magnitude difference between the access times of RAM and disk. We believe this size bottleneck can be overcome, and that life-sized markets can be simulated economically with the hardware currently available, but programming beyond that of JLM Sim will be required.

Part of the solution to the size bottleneck is to keep on disk only, not in RAM, data that probably won't be needed for some time, such as information about investors who optimize monthly or quarterly when it is not their time to reoptimize. Thus RAM would be used as a large cache.

Another part of the solution is to use multiple PCs, perhaps linked by the Internet. Such a distributed simulation is not uncommon now (see Fujimoto [1998]). Market simulation should be well suited to this mode of computing, because of the intrinsically decentralized nature of much market activity.

## ADVANTAGES OF ASYNCHRONOUS FINANCE MODELS

The most common dynamic financial models assume that security prices follow a continuous-time process, which is either a Brownian motion or a function of a Brownian motion. Physicists, too, are sometimes content to describe the movements of particles suspended in liquid as a Brownian motion or similar continuous random path. For other purposes, however, they look behind the scenes at the molecules that jostle the particles and cause their erratic motion, or the atoms that bind together to form molecules, or the electrons, protons, and neutrons that form atoms, or the quarks that constitute the neutrons and protons.

In finance as well, it may be sufficient for some purposes to assume that prices follow a given random process. For other purposes, it is essential to look behind the scenes at what causes price movements.

A major advantage of continuous-time models that represent price movements by given random processes is that some of them can be solved explicitly. This allows the analytic evaluation of investment strategies, or of the values of prices derived from the given price series, such as the price of an option, given the price process of the underlying security. But the assumption of a given and fixed price process is sometimes questionable.

For example, investment actions may change the price process, or a change in the composition of the behind-the-scenes agents may change the price process. Continuous-time models may also be insufficient when the question to be analyzed is whether micro theories about the behavior of investors add up to the observed macro phenomena of the market.

Consider liquidity, which has proved to be a problem for large investors. One extreme case is Black Monday, October 19, 1987, which many believe was exacerbated by an option-based strategy known as portfolio insurance. Other extreme cases include the collapse of hedge funds that have found the liquidity of their positions drying up just when they needed to liquidate them.[7]

In these and in many less extreme cases, investment policies should be evaluated taking into account the fact that the large investor is not a price-taker; rather its own actions affect the price process. Our view is that an asynchronous market simulator such as JLM Sim is best equipped to handle this by representing the agents and market mechanisms behind the observed prices.

Kim and Markowitz [1989] present an asynchronous simulation whose investors are either rebalancers or (constant-proportion) portfolio insurers. They show that the behavior of the market changes radically as the proportion of one kind of investor varies in relation to another. Similarly, debugging runs during the development of JLM Sim show that the behavior of the market varies with the composition of the mean-variance investors in the market, depending on their estimation procedures and risk aversion.

Prior to the 1987 market break, there was an increasing use of portfolio insurance. Perhaps a model that incorporated the actual procedures of the parties in the market, portfolio insurers and others, could have anticipated the consequences of this shifting composition of market participants. More generally, it seems to us that, in order to predict the consequences of future trends in the composition of the market, one needs a model that reflects the differing procedures of different kinds of participants. Asynchronous simulation is well suited to this purpose.

Various hypotheses have been put forth as to the behavior of individual or institutional investors. Some of these hypotheses postulate optimizing behavior of one sort or another. Other hypotheses postulate not-necessarily-rational investor reactions to events. Both the optimizing and the behavioral hypotheses admit different versions, and the market surely includes a mixture of investors with differing investment patterns. One function of a detailed asynchronous simulation is to determine the consequences of a proposed population of investors characterized by one or more behavior patterns, to see whether the postulated behavior patterns add up to observed market processes. A model that starts by *assuming* some random price process cannot *deduce* this process from investor characteristics.

An alternative method of dynamic modeling is synchronous discrete, as opposed to continuous or asynchronous discrete modeling. For example, the microscopic simulation model of Levy, Levy, and Solomon [2000] is a discrete synchronous model. In each period, a market equilibrium price is computed from the demand and supply curves of all investors. These demand curves are based on optimizing or behavioral considerations similar to those that can be incorporated into an asynchronous model.

But a world where prices are set by an equilibrium calculation based on all investors' demand and supply curves is a world different from one where investors can enter and leave the market at any time and may or may not find other investors waiting for them. When we look at actual markets, we see the latter. We contend, therefore, that a model such as the JLM Sim is capable of a more literal representation of the world than the LLS microscopic simulation model.

More generally, asynchronous simulation, allowing action at any time by anyone, is capable of a more literal representation of actual markets than discrete synchronous models, which typically have to make some special assumption about the effect of everyone acting at the same time at assumed discrete times.

## CAVEAT

Before we can use a model for prediction and policy evaluation, we must verify that its implications approximate reality under circumstances observed in the past. We should not expect it to be an easy task to build a complex asynchronous simulation with reasonably realistic properties. A lesson we learned from debugging runs of JLM Sim illustrates this.

In early runs, with two securities and 4,000 investors, we found that the price of stocks doubled and redoubled or fell by comparable amounts in the course of a day, even in a market with no news. Since these stocks were substantially priced, not penny stocks, such frequent explosive price movements were obviously unrealistic.

One problem, we found, was that portfolio analysis that uses historical averages for expected returns is not sensitive to large short-term changes in price. A second problem was that the simulated traders (following our bidding rules) had no sense of recent price levels.

Of our 4,000 investors in these runs, 1,000 each of two different investor templates reoptimized daily. Suppose that, on a particular day, investors of one of these daily reoptimizing templates are inclined to buy security A. At various times during the day, individual investors of this template would instruct their traders to purchase the desired numbers of shares. In some cases, the traders' decision rule was to bid slightly higher than the current price (which may be the average of the bid and ask, or possibly just the bid if there is no ask). Once the supply offered on the books was used up, one trader raised the current bid a bit, and the next a bit more, the next a bit more, unmindful of the fact the stock that they had just bid $200 for sold for $100 earlier in the day.

Part of the cure for the explosive consequences of the initial models was to allow the JLM Sim user to specify "anchoring rules" for traders. For example, anchoring rules provided in JLM Sim instruct the trader to bid as we have described, but to not bid more than the average or the maximum of the recent closing price plus some percentage or plus some number of standard deviations, and, similarly, not to offer less than the average or minimum of the recent closing price minus some percentage or some number of standard deviations.

The JLM Sim modeler can specify different anchoring rules or different parameter values, such as the number of days to be used in computing recent minimums, maximums, or averages, for different trader templates. These anchoring rules help eliminate the problem of frequent explosions and implosions.

The idea that traders have some sense of how much to pay for a security, and that this price is somehow related to the prices at which the security has recently traded, is perhaps too obvious to mention. But if you do not tell this to the model, the model does not know.

What else of importance have we forgotten to tell the model? Inevitably, many things.

Whether we have left out something essential will be determined, in large part, by comparing the implications of the model with the behavior of actual markets. This is an iterative process. When we have a model that imitates the coarse features of the world, then we seek one that imitates its finer features. At some point in this process, we are willing to give some weight to the model's answers to questions such as, within the world as modeled, what are better and worse policies, and what are the effects of changing conditions.

## CONCLUSION

Asynchronous-time, discrete-event simulation is widely used to model complex systems, but has seldom been used to model financial markets. We have sketched out the JLM Sim to illustrate the nature and capabilities of asynchronous market simulation.

The status of a JLM Sim simulation is given by the attribute values and set memberships of entities such as investors, traders, portfolio analysts, statisticians, and securities. Events such as reoptimization, order review, and end of day change the status of the simulation and cause subsequent event occurrences. In an asynchronous simulation, time advances to the next most imminent event, typically in varying increments.

Certain continuous-time financial models, which assume particular price processes, have the advantage of being analytically solvable, but such models are inappropriate when the policies to be evaluated change the price process, perhaps in some non-obvious way; the changing composition of market participants changes the price process; or the issue at hand is whether the postulated micro behavior of financial agents, plus market mechanisms, implies observed market macro behavior. Asynchronous simulation is well suited to such analyses.

## ENDNOTES

[1]Synchronous versus asynchronous simulation is sometimes referred to as time-based versus event-based simulation (see Haverkort [1998, pp. 412-417]), or as fixed-increment time advance versus next-event time advance (see Law and Kelton [2000]). For examples of applications of simulation methodology, see Banks [1998], particularly the articles on manufacturing simulation (Rohrer [1998] and Ulgen and Gunal [1998]); logistics and transportation systems simulation (Manivannan [1998]); computer and communications systems simulation (Hartmann and Schwetman [1998]); and military simulation (Kang and Roland [1998]).

Discrete-event simulations of financial markets are surveyed by Levy, Levy, and Solomon [2000]. These are mostly synchronous simulation models. Continuous-time models in finance are surveyed in Merton [1990].

The JLM Market Simulator will soon be available, gratis, on the web. See www.jacobslevy.com for availability announcements.

[2]See Jacobs and Levy [1989] for a discussion of a taxonomy used in the sciences to classify systems into three types—ordered, complex, and random—and its application to the stock market.

[3]Simulation programming languages such as SIMULA (see Dahl and Nygaard [1966]) and SIMSCRIPT II.5 [1987] describe status change in terms of processes instead of or in addition to event occurrences. That is, instead of an order review event, we could speak of a trading process that includes a wait statement between the order placing action and the order review action. This process view is often helpful in modeling. At the implementation level, however, the simulation is in fact an asynchronous event simula-

tion; end-of-wait coming-event notices are placed on the calendar like events in an event-oriented simulation. In the case of JLM Sim, which is implemented in C++ using the EAS-E.org package for handling large ranked sets, this is easiest to program as an event-oriented simulator.

[4]What actually happens within the computer is that a reoptimization coming-event notice is filed into the calendar, which is a set of coming events ordered by time of occurrence. After initialization is completed, the timing routine is called upon. The timing routine removes the first, most imminent, event from the calendar, calls upon the appropriate event routine, and hands the coming-event notice to the event routine. When the reoptimization coming-event notice for a particular investor is removed from the front of the calendar set, the reoptimization event is called, and is given the event-notice that remembers the investor to be reoptimized.

[5]Mean-absolute deviation is discussed in Konno and Yamazaki [1991]. The resampled frontier is discussed in Michaud [1998] and Markowitz and Usmen [2003]. Downside risk, also know as semi-deviation, is the subject of an issue of *The Journal of Investing* [1994]. Behavioral finance is surveyed in Shefrin [2001].

[6]The EAS-E software is available free at www.eas-e.org.

[7]See Jacobs [1999, 2004], Jacobs and Levy [2005], Kim and Markowitz [1989], and *Report of the Presidential Task Force* [1988].

## REFERENCES

Banks, Jerry, ed. *Handbook of Simulation: Principles, Methodology, Advances, Applications, and Practice.* New York: John Wiley & Sons, 1998.

Dahl, O., and K. Nygaard. "SIMULA—an ALGOL-Based Simulation Language." *Comm. ACM*, 9, No. 9 (September 1966), pp. 671-678.

Fujimoto, Richard M. "Parallel and Distributed Simulation." Chapter 12 in Banks [1998], pp. 429-464.

Hartmann, Alfred, and Herb Schwetman. "Discrete-Event Simulation of Computer and Communication Systems." Chapter 20 in Banks [1998], pp. 659-676.

Haverkort, Boudewijn R. *Performance of Computer Communication Systems: A Model-Based Approach.* New York: John Wiley & Sons, 1998.

Jacobs, Bruce I. *Capital Ideas and Market Realities: Option Replication, Investor Behavior, and Stock Market Crashes.* Oxford: Blackwell, 1999.

——. "Risk Avoidance and Market Fragility." *Financial Analysts Journal*, January/February 2004.

Jacobs, Bruce I., and Kenneth N. Levy. "The Complexity of the Stock Market." *The Journal of Portfolio Management*, 16, No. 1 (Fall 1989), pp. 19-27.

——. "A Tale of Two Hedge Funds." In Bruce I. Jacobs and Kenneth N. Levy, eds., *Market Neutral Strategies.* New York: John Wiley & Sons, 2005.

Kang, Keebon, and Ronald J. Roland. "Military Simulation Systems." Chapter 19 in Banks [1998], pp. 645-658.

Kim, G., and Harry M. Markowitz. "Investment Rules, Margin and Market Volatility." *The Journal of Portfolio Management*, 16, No. 1 (Fall 1989), pp. 45-52.

Konno, H., and H. Yamazaki. "Mean-Absolute Deviation Portfolio Optimization Model and its Applications to Tokyo Stock Market." *Management Science*, 37, No. 5 (May 1991).

Law, Averill M., and W. David Kelton. *Simulation Modeling and Analysis*, 3rd ed. New York: McGraw-Hill, 2000.

Levy, Moshe, Haim Levy, and Sorin Solomon. *Microscopic Simulation of Financial Markets: From Investor Behavior to Market Phenomena.* Berkeley, CA: Academic Press, 2000.

Manivannan, Mani S. "Simulation of Logistics and Transportation Systems." Chapter 16 in Banks [1998], pp. 571-604.

Markowitz, Harry M., and Nilufer Usmen. "Resampled Frontiers versus Diffuse Bayes: An Experiment." *Journal of Investment Management*, 4, No. 1 (2003), pp. 9-25.

Merton, Robert C. *Continuous-Time Finance.* Cambridge: Blackwell, 1990.

Michaud, Richard O. *Efficient Asset Management: A Practical Guide to Stock Portfolio Optimization and Asset Allocation.* Boston: Harvard Business School Press, 1998.

*Report of the Presidential Task Force on Market Mechanisms* (The Brady Commission). Government Printing Office, January 1988.

Rohrer, Matthew W. "Simulation of Manufacturing and Material Handling Systems." Chapter 14 in Banks [1998], pp. 519-545.

Shefrin, Hersh, ed. *Behavioral Finance*, Vol. 1. Northampton, MA: Edward Elgar, 2001.

SIMSCRIPT II.5 Programming Language. CACI Products Company, 1987.

Ulgen, Onur, and Ali Gunal. "Simulation in the Automobile Industry." Chapter 15 in Banks [1998], pp. 547-570.

OPERATIONS RESEARCH
Vol. 53, No. 4, July–August 2005, pp. 586–599
ISSN 0030-364X | EISSN 1526-5463 | 05 | 5304 | 0586

DOI 10.1287/opre.1050.0212


# Portfolio Optimization with Factors, Scenarios, and Realistic Short Positions[1]

**Bruce I. Jacobs, Kenneth N. Levy**
Jacobs Levy Equity Management, 100 Campus Drive, P.O. Box 650, Florham Park, New Jersey 07932-0650
{bruce.jacobs@jlem.com, ken.levy@jlem.com}

**Harry M. Markowitz**
Harry Markowitz Company, 1010 Turquoise Street, Suite 245, San Diego, California 92109, harryhmm@aol.com

This paper presents fast algorithms for calculating mean-variance efficient frontiers when the investor can sell securities short as well as buy long, and when a factor and/or scenario model of covariance is assumed. Currently, fast algorithms for factor, scenario, or mixed (factor and scenario) models exist, but (except for a special case of the results reported here) apply only to portfolios of long positions. Factor and scenario models are used widely in applied portfolio analysis, and short sales have been used increasingly as part of large institutional portfolios. Generally, the critical line algorithm (CLA) traces out mean-variance efficient sets when the investor's choice is subject to any system of linear equality or inequality constraints. Versions of CLA that take advantage of factor and/or scenario models of covariance gain speed by greatly simplifying the equations for segments of the efficient set. These same algorithms can be used, unchanged, for the long-short portfolio selection problem provided a certain condition on the constraint set holds. This condition usually holds in practice.

*Subject classifications*: finance, portfolio: optimization with short sales.
*Area of review*: Financial Engineering.
*History*: Received April 2002; revisions received June 2003, May 2004; accepted July 2004.

## 1. Introduction

This paper presents fast methods for computing the set of "mean-variance efficient" portfolios for an investor who can sell securities short as well as buy them long, provided that certain conditions are satisfied. One might think that ever-faster computers obviate the need for such fast algorithms. However, analyses with large numbers of securities, users waiting for answers in real time, Monte Carlo simulation runs that require many reoptimizations, and simulation experiments requiring many simulation runs, make speedy computation of efficient frontiers still prized. (Parkinson's Law continues to outpace Moore's law.)

A *feasible* portfolio is one that meets specified constraints. A mean-variance *efficient* portfolio is one that provides minimum variance among feasible portfolios with a given (or greater) expected return, and maximum expected return for given (or less) variance. The expected return and variance provided by an efficient portfolio is called an efficient mean-variance (EV) combination. The set of all efficient EV combinations is called the *efficient frontier*.

The critical line algorithm (CLA) traces out a piecewise linear set of efficient portfolios that provide the efficient frontier, subject to any system of linear equality or weak inequality constraints. In general, the inputs to the CLA are constraint parameters, the means and variances of securities, and the covariances between pairs of securities.

The CLA is especially fast if the covariances between securities are described by a "factor model." A factor model assumes that the return on a security depends linearly on the movement of one or more factors common to many securities (e.g., a general market factor, industry factors, a flight-to-quality factor) plus the security's independent "idiosyncratic" term. The use of a factor model not only accelerates computation, it also reduces input requirements. Furthermore, factor model inputs (including regression coefficients of security returns against factors, and variances of underlying factors) are more easily understood, and more easily adjusted to reflect changing conditions, than are the coefficients of a full covariance matrix.

In fast efficient-set algorithms using factor models, "fictitious securities" are introduced into the model, one for each common factor (see Sharpe 1963, Cohen and Pogue 1967). The "amount invested" in each fictitious security is constrained to be a linear combination of the investments in the real securities. With the model thus augmented, the covariance matrix becomes diagonal, or nearly so, and the equations for the pieces of the efficient set become much easier to solve.

Scenario models provide an alternative to factor models for describing the relationships among security returns. A scenario model enumerates different scenarios that can occur in the future and estimates the mean and variance of each security's return under each scenario. Fast efficient-set algorithms using scenario models are similar to those using factor models. Fast algorithms (albeit not quite as fast) also exist that combine factor and scenario models of covariance.

Fast computational methods are also available for covariances computed from historical returns with many more securities than observations. Applicable cases encountered in practice include ones with thousands of securities, but only dozens of months or hundreds of days worth of observations.

This paper presents fast algorithms for tracing out efficient sets when factor, scenario, or certain historical models are assumed, and when the investor is allowed to short securities.

Some capital asset pricing models (CAPMs) assume, in effect, that one can sell a security short without limit and use the proceeds to buy securities long. This is a mathematically convenient assumption for hypothetical models of the economy, but it is unrealistic. Actual constraints on long-short portfolios change over time and, at a given instant, vary from broker to broker and from client to client. Thus, the portfolio analyst charged with generating an efficient frontier for a particular investor must model the specific constraints to which that investor's choice is subject, including constraints the investor itself imposes as a matter of policy. To our knowledge all such constraints—whether imposed by regulators, brokers, or self-imposed—are expressible as linear equalities or weak inequalities and therefore can be incorporated into the general portfolio selection model. Later, we will give examples of current real-world constraints, but our results are not restricted to some particular constraint set.

A portfolio optimization with $n$ securities, which can be bought long or sold short, may be set up as a model with $n$ variables representing long positions and another $n$ variables representing short positions. The types of constraints noted in the preceding paragraph are easily expressed in terms of the $2n$ variables. However, even if a factor or scenario model holds for the $n$ securities held long, it does not hold for the $2n$-variable model representing short and long positions. Specifically, the $2n$-variable long-short model violates the assumption that the idiosyncratic terms are uncorrelated. Nevertheless, under certain assumptions, if the requisite information (e.g., regression coefficients and idiosyncratic variances for the factor model for the $2n$ variables) is fed into the appropriate factor or scenario program, a correct efficient frontier results.

The principal result of this paper is a sufficient condition that assures that an existing (originally long-only) factor or scenario code will compute the correct answer to the long-short problem. We refer to this condition as "Property P." Property P does not hold in general for an arbitrary long-short portfolio selection model, but it appears to be widely satisfied in practice. When a factor or scenario model of covariance is assumed and Property P is satisfied, a fast algorithm for the long-short model is readily at hand. No new programming is needed. The long-only program produces the correct answer to the $2n$-variable long-short problem, despite the "error" in assumption. Also, the fast algorithm for historical covariance matrices (when the number of securities greatly exceeds the number of observations) produces correct answers to the $2n$-variable long-short problem, whether or not Property P holds.

The results reported in this paper generalize a result due to Alexander (1993) and Kwan (1995). Their results apply to the Elton et al. (1976) algorithm. The Elton et al. algorithm assumes only one constraint equation—namely, a budget constraint—and makes special assumptions about the factor structure of a factor model.

Section 2 defines the "general" mean-variance problem. Section 3 summarizes its solution by CLA. Section 4 describes how the covariance matrix can be (almost) diagonalized if a factor, scenario, or historical model of covariance is used. Section 5 outlines short sales in the real world. Section 6 presents notation for portfolio optimization with short sales and a diagonizable model of covariance. Section 7 derives fast methods for solving the latter problem. Section 8 illustrates the results. Section 9 summarizes.

## 2. The General Mean-Variance Problem

Suppose that the return $R_P$ on the portfolio over some forthcoming period is a weighted sum of the $n$ security returns $R=[r_1, r_2, \ldots, r_n]'$,

$$R_P = R'X, \tag{1}$$

where the weights $X = [X_1, \ldots, X_n]'$ are chosen by the investor. Assuming that the $r_i$ are random variables with finite means and variances,

$$E_P = \sum_{i=1}^{n} \mu_i X_i = \mu' X, \tag{2}$$

$$V_P = \sum_{i=1}^{n}\sum_{j=1}^{n} \sigma_{ij} X_i X_j = X'CX, \tag{3}$$

where $E_P$ and $V_P$ are the expected return and variance of the portfolio, $\mu = [\mu_1, \ldots, \mu_n]'$ are the expected returns on the $n$ securities, $\sigma_{ij}$ is the covariance between $r_i$ and $r_j$, and $C$ is the covariance matrix $(\sigma_{ij})$. Markowitz (1959) assumes that $X$ is chosen subject to the following constraints:

$$AX = b, \tag{4}$$

$$X \geqslant 0, \tag{5}$$

where $A$ is an $m \times n$ constraint matrix and $b$ an $m$ component "right-hand side" vector.

As in linear programming, constraints (4) and (5) can represent weak linear inequalities ($\geqslant$ or $\leqslant$) by use of slack variables. For example,

$$\sum_j a_{ij} X_j \leqslant b \tag{6}$$

is written as

$$\sum_j a_{ij}X_j + X_s = b, \quad X_s \geqslant 0. \tag{7}$$

Also, a variable $X_i$ not required to be nonnegative is handled in (4) and (5) by substituting for it

$$X_i = X_{ip} - X_{in}, \quad X_{ip} \geqslant 0,\ X_{in} \geqslant 0, \tag{8}$$

where $X_{ip}$ and $X_{in}$ are the positive and negative parts of $X_i$.

It is not required that the covariance matrix $C$ in (3) be nonsingular. This is essential, because $X$ may include risk-free securities, slack variables, and pairs of securities representing short and long positions. Also, sometimes $C$ is estimated from historical returns with less periods than there are securities. Any of these circumstances will result in $\det(C) = 0$. In addition, it is desirable for the computational procedure not to fail if $A$ in (4) is not of full rank.

A portfolio $X$ is said to be *feasible* if it satisfies constraints (4) and (5). A pair of real numbers $(E_P, V_P)$ is said to be a feasible EV combination if $E_P$ and $V_P$ satisfy (2) and (3) for some feasible portfolio $X$. A feasible $(E_P, V_P)$ pair is *inefficient* if some other feasible pair $(E_P^*, V_P^*)$ dominates it; that is, has higher expected return, $E_P^* > E_P$, but no higher variance, $V_P^* \leqslant V_P$; or, has lower variance, $V_P^* < V_P$, but no lower expected return, $E_P^* \geqslant E_P$. If $(E_P, V_P)$ is not thus dominated, it is called an *efficient* EV combination. A feasible portfolio $X$ is efficient or inefficient according to whether its $(E_P, V_P)$ is efficient or inefficient.

The general (single-period) mean-variance portfolio selection problem is to find all efficient EV combinations, and feasible portfolios that yield these, for all possible $A, b, \mu$, and $C$ in (2), (3), (4), and (5). Problems with weak linear inequalities and variables not required to be nonnegative can be converted into this form.

## 3. Solution to the General Problem

It is possible that, for a given $A$ and $b$, the model is infeasible, that is, no portfolio $X$ satisfies (4) and (5). It is also possible for a model to be feasible and yet have no mean-variance efficient portfolios. In this case, if $\hat{X}$ is feasible with minimum $V_P$ and with expected return $E$, there is another feasible portfolio $X^*$ with the same $V$ and with $E^* > E$. This can occur if $C$ is singular and the constraint set unbounded. Below, we assume that the model is feasible and has efficient portfolios.

Next, we summarize (without proof) certain properties and formulas of efficient sets.[2] The set of efficient EV combinations is piecewise parabolic. In general, there may be more than one efficient portfolio $X$ for a given efficient EV combination. When the set of efficient portfolios is unique—with only one feasible portfolio $X$ for any given efficient EV combination—the set of efficient portfolios is piecewise linear. The formula for an efficient segment (of the piecewise linear efficient set) is given below. When the set of efficient portfolios is not unique there is nevertheless a "complete, nonredundant" set of efficient portfolios that satisfy the equations below. By "complete, nonredundant" we mean a set of efficient portfolios with one and only one $X$ for each efficient EV combination. The CLA provides such a complete, nonredundant set of efficient portfolios whether or not the set of efficient portfolios is unique.

The Lagrangian expression for the general model is

$$L = V/2 + \sum_{k=1}^{m} \lambda_k \left( \sum_{j=1}^{n} a_{kj} X_j \right) - \lambda_E \sum_{i=1}^{n} \mu_i X_i. \tag{9}$$

Let

$$\eta = \frac{\partial L}{\partial X} = [C \quad A' \quad \mu] \begin{bmatrix} X \\ \lambda \\ -\lambda_E \end{bmatrix}, \tag{10}$$

where $\lambda = [\lambda_1, \ldots, \lambda_m]'$. For the moment, to develop a definition, arbitrarily select a nonempty subset of $\{1, 2, \ldots, n\}$ and designate this subset as the IN variables, and its complement as the OUT variables. Let

$$M = \begin{bmatrix} C & A' \\ A & O \end{bmatrix} \tag{11}$$

and let $M_{\text{IN}}$ be the $M$ matrix with the rows and columns that correspond to OUT variables deleted. Similarly, let $\mu_{\text{IN}}$ and $X_{\text{IN}}$ be the $\mu$ and $X$ vectors with OUT components deleted, and $0_{\text{IN}}$ be a zero vector of the same size as $\mu_{\text{IN}}$. If $M_{\text{IN}}$ is nonsingular, we say that the arbitrarily chosen IN set has an associated *critical line* satisfying

$$X_i = 0 \quad \text{for } i \in \text{OUT} \tag{12}$$

and

$$M_{\text{IN}} \begin{bmatrix} X_{\text{IN}} \\ \lambda \end{bmatrix} = \begin{bmatrix} 0_{\text{IN}} \\ b \end{bmatrix} + \begin{bmatrix} \mu_{\text{IN}} \\ 0 \end{bmatrix} \lambda_E. \tag{13}$$

Multiplying through by $M_{\text{IN}}^{-1}$ solves (13) for $X_{\text{IN}}$ and $\lambda$ as linear functions of $\lambda_E$:

$$\begin{bmatrix} X_{\text{IN}} \\ \lambda \end{bmatrix} = \alpha_{\text{IN}} + \beta_{\text{IN}} \lambda_E. \tag{14}$$

If we substitute (14) into (10), we find that the $\eta$ vector is also a linear function of $\lambda_E$:

$$\eta = \gamma_{\text{IN}} + \delta_{\text{IN}} \lambda_E. \tag{15}$$

Conditions (13) imply

$$\eta_i = 0 \quad \text{for } i \in \text{IN}. \tag{16}$$

In light of (12) and (16), if a point on the critical line also satisfies

$$X_i \geqslant 0 \quad \text{for } i \in \text{IN}, \tag{17}$$

$$\eta_i \geqslant 0 \quad \text{for } i \in \text{OUT}, \tag{18}$$

$$\lambda_E > 0, \tag{19}$$

then the point is efficient, by the Kuhn-Tucker theorem. If any point on the critical line is efficient, then there will be an interval of that line (possibly open-ended) all of whose points are efficient. We refer to such an interval as an *efficient segment*.

Because there are $2^n - 1$ nonnull subsets of $\{1, \ldots, n\}$, it is impractical to enumerate them all, determine which have nonsingular $M_{\mathrm{IN}}$, among these determine which contain efficient segments, then piece these together to form a complete, nonredundant set of efficient portfolios. The CLA produces such a set without searching among irrelevant IN sets.

The CLA proceeds as follows.[3] It traces out the efficient set from high to low $\lambda_E$. At a typical step, we have in hand a critical line with an efficient segment, and with IN-set $\mathrm{IN}_t$; we also have in hand the corresponding $M_{\mathrm{IN}}$ and $M_{\mathrm{IN}}^{-1}$. We can then solve for $\alpha_{\mathrm{IN}}$, $\beta_{\mathrm{IN}}$, $\gamma_{\mathrm{IN}}$, and $\delta_{\mathrm{IN}}$, from which it is easy to determine which occurs first as $\lambda_E$ is reduced:

$$\text{an } X_i \downarrow 0 \quad \text{for } i \text{ IN}, \tag{20}$$

$$\text{an } \eta_i \downarrow 0 \quad \text{for } i \text{ OUT}, \tag{21}$$

$$\text{or } \lambda_E \downarrow 0. \tag{22}$$

In case $\lambda_E \downarrow 0$ first, we have reached the efficient portfolio with minimum feasible $V$, and the algorithm stops.[4]

If $X_i \downarrow 0$ first, then $i$ moves from IN to OUT on the next ("adjacent") efficient segment. It is shown that $\eta_i$ will increase on this next segment. On the other hand, if $\eta_i \downarrow 0$ first, then $i$ moves from OUT to IN in the new IN set, $\mathrm{IN}_{t+1}$, and $X_i$ will increase on the new segment.[5] If the algorithm has not stopped, because $\lambda_E \downarrow 0$ has not been reached, the new $M$ matrix, $M_{\mathrm{IN}(t+1)}$, is nonsingular. It is obtained from the old by adding or deleting one column and the corresponding row. This allows us to update $M^{-1}$ relatively inexpensively, and use it to solve for $\alpha$, $\beta$, $\gamma$, $\delta$, etc., as before. The algorithm ends, with $\lambda_E \downarrow 0$, in a finite number of iterations.[6]

## 4. Diagonizable Models of Covariance

**Factor Models.** In the introduction, we referred to "fast algorithms" based on certain models of covariance. In this section, we summarize such algorithms for the factor and scenario models and for models with historical covariance matrices when there are more securities than observations. In problems with a large number of securities, computation time may differ by orders of magnitude between using a dense covariance matrix and using the diagonal or nearly diagonal covariance matrices permitted by the aforementioned models of covariance.

For the present, we are concerned with portfolios of long positions, which we denote as

$$X^{\mathrm{L}} = [X_1, \ldots, X_\nu, \ldots, X_n]'.$$

We write its constraints as

$$A^{\mathrm{L}} X^{\mathrm{L}} = b^{\mathrm{L}}, \tag{23}$$

$$X^{\mathrm{L}} \geqslant 0. \tag{24}$$

The portfolio may include zero-variance "securities" such as cash or dummy variables. We assume that

$$\begin{aligned} V_i > 0 \quad & \text{for } i \in [1, \nu], \\ V_i = 0 \quad & \text{for } i \in [\nu+1, n]. \end{aligned} \tag{25}$$

If $\nu = n$, then $[\nu+1, n]$ is empty.

A factor model of covariance assumes that security returns are related to each other because they are related to common underlying factors. Specifically, it assumes that

$$r_i = \alpha_i + \sum_{k=1}^{K} \beta_{ik} f_k + u_i, \quad i = 1, \ldots, n, \tag{26}$$

where $K$ is the number of common factors, $f_k$ is the $k$th common factor, and $u_i$ is an idiosyncratic term assumed uncorrelated with $f_k$, $k = 1, \ldots, K$, and all $u_j$ for $i \neq j$. In matrix notation,

$$R = \alpha + BF + U, \tag{27}$$

where $\alpha = [\alpha_1, \ldots, \alpha_n]'$, $B = [\beta_{ik}]$ is $n \times K$, $F = [f_i, \ldots, f_K]'$, and $U = [u_1, \ldots, u_n]'$. From (27) and (1), we see that

$$R_P = \alpha' X^{\mathrm{L}} + F'B'X^{\mathrm{L}} + U'X^{\mathrm{L}}. \tag{28}$$

Because $F$ and $U$ are uncorrelated, the above implies

$$V_P = (X^{\mathrm{L}})' B Q_f B' X^{\mathrm{L}} + (X^{\mathrm{L}})' Q_u X^{\mathrm{L}}, \tag{29}$$

where $Q_f$ and $Q_u$ are the covariance matrices of $F$ and $U$, respectively. By assumption, $Q_u$ is diagonal with $i$th diagonal term $V(u_i)$. $Q_f$ is not necessarily diagonal.

Define $K$ "fictitious" investments in terms of "real" investments,

$$B'X^{\mathrm{L}} - \begin{bmatrix} X_{n+1} \\ \vdots \\ X_{n+K} \end{bmatrix} = 0. \tag{30}$$

We let $X^{\mathrm{LA}} = [X_1, \ldots, X_{n+K}]'$ and

$$A^{\mathrm{LA}} X^{\mathrm{LA}} = b^{\mathrm{LA}} \tag{31}$$

be constraints (23) with (30) appended. We may write (29) as

$$V_P = (X^{\mathrm{LA}})' C^{\mathrm{LA}} X^{\mathrm{LA}}, \tag{32}$$

where

$$C^{LA} = \begin{bmatrix} Q_u & 0 \\ 0 & Q_f \end{bmatrix}.$$

The original problem may be restated as finding EV-efficient $X^{LA}$ subject to (31) and (24) with portfolio variance defined as in (32). The $M$-matrix now is

$$M^{LA} = \begin{bmatrix} C^{LA} & (A^{LA})' \\ A^{LA} & 0 \end{bmatrix}. \quad (33)$$

Because $X_i \geqslant 0$ is not required for the fictitious securities, $i \in [n+1, n+K]$, it is convenient to permit $X_i < 0$ for these variables (rather than separate them into positive and negative parts as in (8)). Then, for $i > n$, $X_i$ is IN on all critical lines. We refer to the portfolio selection model with constraints (31) and covariance matrix (32) as the "diagonalized version" of the factor model. (Strictly speaking, we mean "almost diagonalized" because $Q_f$ is not necessarily diagonal.) It is assumed that all risky securities, $i \in [1, \nu]$, have positive idiosyncratic risk:

$$V(u_i) > 0, \quad i \in [1, \nu]. \quad (34)$$

Typically, $\nu \gg n + K - \nu$; therefore $M^{LA}$ is quite sparse and well structured. This is the basis for fast CLAs for factor models.[7] If $Q_f$ is known to be diagonal, the algorithm can be further streamlined.

**Scenario Models.** A scenario model analyzed by Markowitz and Perold (1981a, b) assumes that one of $S$ mutually exclusive scenarios will occur with probability $P_s$, $s = 1, \ldots, S$. If scenario $s$ occurs, then the return on the $i$th security is

$$r_i = \mu_{is} + u_{is}, \quad (35)$$

where $\mathrm{E}(u_{is}) = \mathrm{cov}(u_{is}, u_{js}) = 0$ for $i \neq j$. Let $V_{is} = \mathrm{E}(u_{is}^2 \mid s)$. The expected return $E$ of the portfolio is still given by (2), provided that the $\mu$ in (2) are computed as follows:

$$\mu_i = \sum_{s=1}^{S} P_s \mu_{is}. \quad (36)$$

These $\mu_i$ can be computed in advance of the optimization calculation. Let

$$X_{n+s} = \sum_{i=1}^{n} X_i(\mu_{is} - \mu_i) \quad \forall s \in [1, S]. \quad (37)$$

This equals the expected value $E_s$ of the portfolio, given that scenario $s$ occurs, less portfolio grand mean $E$. The variance of the portfolio is

$$V_P = \mathrm{E}(R_P - E_P)^2 = \sum_{s=1}^{S} P_s \mathrm{E}(R_P - E_s + E_s - E)^2$$
$$= \sum_{i=1}^{n+S} X_i^2 \widetilde{V}_i, \quad (38)$$

where

$$\widetilde{V}_i = \sum_{s=1}^{S} P_s V_{is}, \quad i = 1, \ldots, n,$$

$$\widetilde{V}_{n+s} = P_s, \quad s = 1, \ldots, S.$$

Thus, $V_P$ can be expressed as a positively weighted sum of squares in the $n$ original variables and $S$ new, fictitious variables that are linearly related to the original variables by (37).

Apart from notation (e.g., using $S$ for $K$ and (37) for (30)), the scenario model is formally the same as the factor model with $Q_f$ diagonal. That is, the meanings of the coefficients are different but, with change of notation, the portfolio selection problem with a scenario model of covariance has an $M^{LA}$ matrix as in (33), with diagonal $Q_f$. We refer to the portfolio selection problem with constraints (37) appended to the given constraints, and variance expressed as in (38), as the diagonalized version of the scenario model (35).[8]

**Historical Covariance Matrices.** Consider the case in which $T$ historical periods (e.g., months or days) are used to estimate covariances among $n$ securities. Let

$$X_{n+t} = \sum_{i=1}^{n} X_i(r_{it} - m_i), \quad t = 1, \ldots, T, \quad (39)$$

where $r_{it}$ is the return on the $i$th security during period $t$, and $m_i$ is the $i$th security's historical average return:

$$m_i = \frac{1}{T} \sum_{t=1}^{T} r_{it}.$$

The $m_i$ do not necessarily equal the estimated expected return $\mu_i$ in (2). Then, $X_{n+t}$ is the difference between portfolio return in the $t$th period and the portfolio's average return. Therefore, the historical variance of a portfolio is a constant times

$$V_P = \sum_{t=1}^{T} X_{n+t}^2. \quad (40)$$

This is a sum of squares in new, fictitious securities that are linearly related to the old. Once again, the problem can be expressed as a portfolio selection problem with $M^{LA}$ matrix as in (33). In the present case, we have

$$Q_u = 0. \quad (41)$$

$M^{LA}$ is again sparse and well structured but, because of (41), requires different handling than in the fast algorithms for the factor and scenario models.[9]

We refer to the portfolio selection model with constraints (39) appended and variance expressed as in (40) as the diagonalized version of the historical covariance model. We refer to the three models described in this section as "diagonalizable" models. For large problems, the above models afford a reduction in computation requirements roughly proportional to the reduction in the number of nonzero entries between $M^L$ and $M^{LA}$.

Jacobs et al.: *Portfolio Optimization with Factors, Scenarios, and Realistic Short Positions*
Operations Research 53(4), pp. 586–599, © 2005 INFORMS 

## 5. Short Sales in Practice

Capital asset pricing models (CAPMs) frequently assume that the investor chooses a portfolio subject only to the constraint

$$\sum_{i=1}^{n} X_i = 1, \tag{42}$$

without constraint on the sign of $X_i$. Negative $X_i$ are interpreted as short positions. In particular, (42) permits $(x, 1-x, 0, \ldots, 0)$ as feasible for all real $x$. For example, (42) would permit an investor to deposit \$1,000 with her broker, short \$1,000,000 of Stock A, and use the proceeds plus the original deposit to purchase \$1,001,000 of Stock B. This is not how short positions work in fact.

No single constraint set applies to all long-short investors. The portfolio analyst must model the specific constraint set for the particular client. To illustrate what this may involve, we outline a few real-world short-sale constraints (see also Jacobs and Levy 2000).

To sell short for any customer, a broker must borrow the stock to be sold, and actually sell it. The brokerage firm may borrow the stock from itself, typically from customer stock held "in street name" in margin accounts. Alternatively, the broker may borrow the stock from another investor, typically a large institutional investor. Some intermediary may facilitate the process of bringing together demand and supply of stock-to-lend. Sometimes a lender cannot be found for the desired stock. In this case, the stock cannot be sold short. Furthermore, the lender retains the right to call back the stock; if he does, and another lender is not promptly available, the investor must cover the short position (i.e., buy back the stock) and deliver it to the lender.

The proceeds of a short sale are used as collateral for the lender of the stock. In fact, if the stock is borrowed from another investor, the broker must put up more than 100% of the proceeds of the sale as collateral, usually about 105%. (Note that this is required of the broker to protect the stock lender, as opposed to the requirement on the short seller discussed in the next paragraph.) The proceeds of the stock sale are invested in "cash instruments" such as short-term Treasury bills. The broker and the stock lender retain a portion of the interest earned on the proceeds. A large institutional investor that shorts stock typically receives a portion of the interest (referred to as a "short rebate"). A small retail customer who sells short typically receives no part of the interest.

The short seller is subject to Regulation (Reg) T. Reg T covers common stock, convertible bonds, and equity mutual funds; securities such as U.S. Treasury bonds or bond funds and municipal bonds or bond funds are exempt from Reg T. Reg T requires that the sum of the long positions plus the sum of the (absolute value of) short positions must not exceed twice the equity in the account. If we normalize so that "1" represents the equity in the account, then Reg T requires

$$\sum_{i=1}^{2n} X_i \leqslant H, \tag{43}$$

where $X_i$ represents a long position for $i \in [1, n]$, a short position for $i \in [n+1, 2n]$, and currently Reg T specifies $H = 2$. This inequality, of course, can be converted to an equality by introduction of a slack variable.

As a matter of policy, the broker or investor may set $H$ at a lower level. There may be additional constraints on the choice of

$$X^{LS} = [X_1, \ldots, X_{2n}]'. \tag{44}$$

For example, some securities are hard to borrow. The broker may therefore limit the amount of the short position or not permit short positions in the particular security.

Constraint (43) with $H = 2$ is referred to as a "50% margin requirement" on both short and long positions. In practice, the nature of this margin requirement is different for short and long positions. In the case of long positions, the customer may borrow as much as 50% of the value of the position from the broker. In the case of a short position, the customer does not borrow money from the broker; the margin requirement is a collateral requirement. Furthermore, the Reg T requirements are for "initial margin"—the equity required in the account to establish initial positions. It does not constrain the value of the positions maintained after they are established. However, there are "maintenance margin" requirements imposed by securities exchanges and by brokers. Consequently, one motive of the investor in setting her or his own $H$ in (43) is to reduce the probability of needing additional cash for maintenance margin.[10]

Reg T can be circumvented in several ways. For example, hedge funds often set up offshore accounts, which are not subject to Reg T. Alternatively, a large hedge fund can set up as a broker-dealer, with a "real" broker-dealer acting as the "back office." In this case, the hedge fund, as broker-dealer, is subject to broker-dealer capital requirements rather than Reg T requirements. This permits much more leverage than Reg T. In the extreme, the only constraint is what the broker imposes on the hedge fund's portfolio to assure that, in the case of unfavorable market movements, the broker is secure. A hedge fund could also circumvent Reg T by having a broker set up a proprietary trading account of its own that is managed by the fund. Gains and losses in the proprietary trading account are transferred to the hedge fund via prearranged swap contracts. The only constraint imposed by this arrangement is the broker's own capital requirements, plus whatever constraints the broker imposes.[11]

Also lying outside Reg T are certain arrangements that allow the investor to use noncash collateral, including existing long positions, to collateralize the shares borrowed to sell short, freeing up the proceeds from short sales to be used for further purchases and short sales. In all these cases, the broker-dealer imposes its own requirements for its own security. The portfolio analyst must model the situation as she or he finds it.[12]

## 6. Modeling Short Sales

We assume that the choice of $X^{LS}$ is subject to some system of linear constraints in nonnegative variables

$$A^{LS}X^{LS}=b^{LS}, \tag{45}$$

$$X^{LS}\geqslant 0. \tag{46}$$

Portfolio return is

$$R_P=\sum_{i=1}^{n} r_iX_i+\sum_{i=n+1}^{2n}(-r_{i-n})X_i+r_c\sum_{i=n+1}^{2n}h_{i-n}X_i. \tag{47}$$

The first term on the right of Equation (47) represents the return contribution of the securities held long. The second term represents the contribution of the securities sold short. The third term represents the short rebate, where

$$h_i\leqslant 1,\quad i=1,\ldots,n. \tag{48}$$

Usually, $h_i\geqslant 0$, but this condition is sometimes violated for hard to borrow stocks, and is not required for our results.[13] $r_c$ is the return on "cash" or "collateral." Cash is also a risk-free security that can be held long, i.e., $c\in[\nu+1,n]$. In particular, we assume that $\nu<n$.[14] Let

$$R^{LS}=\begin{bmatrix}\tilde r_1\\ \vdots\\ \tilde r_{2n}\end{bmatrix}=\begin{bmatrix}R^L\\ -R^L+hr_c\end{bmatrix}, \tag{49}$$

$$\mu^{LS}=\mathrm{E}(R^{LS})=\begin{bmatrix}\mu^L\\ -\mu^L+hr_c\end{bmatrix}. \tag{50}$$

The expected return and variance of the long-short portfolio are

$$E=(\mu^{LS})'X^{LS}, \tag{51}$$

$$V_P=(X^{LS})'C^{LS}X^{LS}, \tag{52}$$

where

$$C^{LS}=\begin{bmatrix}C^L & -C^L\\ -C^L & C^L\end{bmatrix}, \tag{53}$$

and where $C^L$ is the long-only covariance matrix.

If we assume a multifactor model with returns $r_i$ given by (26) and (27), then (49) implies

$$R^{LS}=\begin{bmatrix}\alpha\\ -\alpha\end{bmatrix}+\begin{bmatrix}B\\ -B\end{bmatrix}F+\begin{bmatrix}U\\ -U\end{bmatrix}+\begin{bmatrix}0_n\\ h\end{bmatrix}r_c, \tag{54}$$

where $0_n$ is an $n$-vector of zeros. Hence, the covariance matrix of $R^{LS}$ is

$$C^{LSA}=\begin{bmatrix}B\\ -B\end{bmatrix}Q_f\begin{bmatrix}B\\ -B\end{bmatrix}'+\begin{bmatrix}Q_u & -Q_u\\ -Q_u & Q_u\end{bmatrix}. \tag{55}$$

Thus, if we define

$$(X^{LSA})'=[(X^L)',(X^S)',(X^A)'], \tag{56}$$

where

$$X^A=[X_{2n+1},\ldots,X_{2n+K}]'$$

are portfolio betas, then $X^{LSA}$ is chosen subject to constraints (46) and

$$A^{LSA}X^{LSA}=b^{LSA}. \tag{57}$$

The latter are constraints (45) with

$$X^A=[B,-B]X^{LS}$$

appended.

Define the vector

$$\delta=\begin{bmatrix}B\\ -B\end{bmatrix}'\begin{bmatrix}X^L\\ X^S\end{bmatrix}=B'X^L-B'X^S. \tag{58}$$

The $k$th entries of the vectors $B'X^L$ and $B'X^S$ are the contributions of the long and short portions of the portfolio, respectively, to the $k$th "fictitious security." Thus, $\delta$ is a vector of the differences between the contributions of the long and short portions of the portfolio. Using Equations (55), (58), and definitions, we obtain portfolio variance as

$$V_P=\delta'Q_f\delta+\begin{bmatrix}X^L\\ X^S\end{bmatrix}'\begin{bmatrix}Q_u & -Q_u\\ -Q_u & Q_u\end{bmatrix}\begin{bmatrix}X^L\\ X^S\end{bmatrix}. \tag{59}$$

Recalling that $Q_u$ is diagonal, this can be written as

$$V_P=\delta'Q_f\delta+\sum_{i=1}^{2n}X_i^2V_i-2\sum_{i=1}^{n}X_iX_{n+i}V_i. \tag{60}$$

## 7. Solution to Long-Short Model

Equation (60) is the same as the diagonalized form (32) of the diagonalizable models of §4 *except* for the inclusion of the last sum of cross-product terms. Fortunately, for certain models the sum of cross products can be ignored. For these models, a portfolio optimizer that assumes that variance is

$$V_P'(X)=\delta^TQ_f\delta+\sum_{i=1}^{2n}X_i^2V_i \tag{61}$$

instead of that given in Equation (60), will still produce a correct mean-variance efficient frontier.

Note that if

$$X_i X_{n+i} = 0, \quad i = 1, \ldots, n \tag{62}$$

holds (i.e., the investor is not long and short the same security), then $V_P$ in (60) equals $V_P'$ in (61). We shall refer to a portfolio that satisfies (62) as *trim*; otherwise it is *untrim*. We will refer to the portfolio selection model with $E$, $V_P$, and constraints given by (51), (60), (57), and (46) as the *original* model; and that with (60) replaced by (61) as the *modified* model. In this section, we consider conditions under which the efficient set for the modified model is an efficient set for the original model.

Clearly, $V_P'(X) \equiv V_P(X)$ if $V_i = 0$ for all $i \in [1, 2n]$, as is the case for the diagonalized historical model. We will denote this simple but useful result as a theorem.

THEOREM 1. *An efficient set for the modified historical model provides an efficient set for the original historical model.*

PROOF. See the preceding paragraph. □

Theorem 1 makes no assumption concerning the constraint set or expected returns other than the background assumptions that the model is feasible and has efficient portfolios. The other diagonalizable models of §4 require a further assumption to reach a similar conclusion. The following assumption is sufficient.

PROPERTY P. If in the original model $X$ is a feasible portfolio with $X_i X_{n+i} > 0$ for some specific $i$, then there is a feasible portfolio $Y$ with

$$\begin{aligned} &Y_i = X_i - \theta_i, \\ &Y_{n+i} = X_{n+i} - \theta_i, \\ &Y_j = X_j, \quad j \neq i, n+i, \; j \in [1, \nu] \cup [n+1, n+\nu] \end{aligned} \tag{63}$$

for $\theta_i = \min\{X_i, X_{n+i}\}$. Also, $Y$ has the same or greater mean as $X$.

In other words, if $X$ has a positive long and a positive short position in the same security, it is feasible to subtract the above $\theta_i$ from both positions, keeping all other risky securities unchanged, adjusting only zero-variance $X_i$, without reducing portfolio expected return. Note that $Y_i Y_{n+i} = 0$.

While Property P is not necessarily true, it does hold for a wide variety of constraint sets met in practice. Suppose, for example, that choice of a long-short portfolio is subject to any or all of the following constraints: (A) a Reg T type of constraint as in (43), perhaps with $H > 2$ for an investor not subject to Reg T; (B) upper bounds on individual long or short positions; (C) the requirement that the value of long positions be close to the value of short positions—specifically,

$$\tau \geqslant \sum_{i=1}^{\nu} X_i - \sum_{i=1}^{\nu} X_{n+i} \geqslant -\tau \tag{64}$$

for some given tolerance level $\tau$;[15] as well as the nonnegativity requirement (46), and a budget constraint

$$\sum_{i=1}^{\nu} X_i + X_c - X_b \leqslant 1, \tag{65}$$

where $X_c$ is a cash balance and $X_b$ is an amount borrowed. (Note that the sum in (65) is through $\nu$, i.e., it includes risky *long* positions only. Recall that, unlike investors in CAPMs with (42) as their only constraint, Reg T-constrained investors do not get to spend the proceeds from selling short, although they may share the interest collected on these proceeds.) If $X$ is any feasible portfolio (i.e., meets each of the above constraints) with $X_i X_{n+i} > 0$, then $Y$ with

$$\begin{aligned} &Y_i = X_i - \theta_i, \\ &Y_{n+i} = X_i - \theta_i, \\ &Y_c = X_c + \theta_i, \\ &Y_j = X_j \quad \text{for } j \in \{[1, \nu] \cup [n+1, n+\nu]\} \backslash \{i, n+i\}, \\ &\theta_i = \min\{X_i, X_{n+i}\} \end{aligned} \tag{66}$$

meets the constraints. When the constraints are written as equalities, as in (7), then zero-variance slack variables are adjusted to maintain the equalities. Also, from (47) and (48),

$$E_Y = E_X + \theta_i(1 - h_i) r_c \geqslant E_X. \tag{67}$$

Thus, a constraint set consisting of (46), (65), and some or all of (A), (B), and (C) does satisfy Property P. Note that Property P only requires $Y$ to be feasible, not necessarily efficient; thus we need not be concerned, in checking Property P that, say, $Y$ might be improved by reducing $X_b$ rather than increasing $X_c$ in case $X_b > 0$.

On the other hand, if there is an upper bound on the holding of cash,

$$X_c \leqslant u_c, \tag{68}$$

then Property P may not be satisfied. If, for example, there are no upper bounds on the other $X_i$, then

$$\begin{aligned} &X_1 = 1 - u_c, \\ &X_{n+1} = 1 - u_c, \\ &X_c = u_c, \\ &X_i = 0 \quad \text{otherwise} \end{aligned} \tag{69}$$

is feasible, but $X_1$ and $X_{n+1}$ cannot be reduced by adjusting zero-variance variables, including $X_c$ in (65), in the manner required by Property P without violating (68).

THEOREM 2. *If Property* P *holds in the original model, then for each efficient* $(E, V_P)$ *combination there is one and only one trim portfolio* $Y$ *with the same* $(E, V_P)$.

(There may also be untrim efficient portfolios with this $(E, V_P)$.)

Proof. Because $(E, V_P)$ is feasible, there is a portfolio $X$ that provides it. If $X$ is untrim, successive transformations (63) for each $i$ in turn with $X_iX_{n+i} > 0$ yields a trim, feasible $Y$ with the same or greater $E$ than $X$ and, from Equations (58) and (59), the same $V_P$. If $Y$ has greater $E$, then $X$ could not be efficient, whereas if $Y$ has the same $E$ as $X$, then $Y$ too is efficient. Thus, for any efficient $(E, V_P)$ combination, there exists a trim feasible $Y$ that provides it.

To show that $Y$ is unique, let us suppose that another trim feasible (therefore efficient) portfolio $Z$ supplies $(E, V_P)$. Let

$$W = \xi Y + (1-\xi)Z. \tag{70}$$

If we show that $V_P$ as a function of $\xi$ is strictly convex, then $(1/2)Y + (1/2)Z$ is feasible (because the constraint set is convex), has the same $E$ (because $E$ is linear), and less $V_P$ than $Y$ or $Z$, contradicting the assumption that $Y$ and $Z$ are efficient. To see that $V_P$ is a strictly convex function of $\xi$, first confirm that the first term on the right-hand side of (60) is constant or a convex function of $\xi$ (because $Q_f$ is positive semidefinite and $\delta$ is linear in $\xi$). Next, note that the last two terms of Equation (60) may be obtained by substituting $\widetilde{X}_i = X_i - X_{n+i}$ into $\sum_{i=1}^{n} V_i\widetilde{X}_i^2$. As a function of $\xi$, this is a sum of terms that are either constant (i.e., for those $i$ with $\widetilde{Y}_i = \widetilde{Z}_i$, which implies $Y_i = Z_i$ because $Y$ and $Z$ are trim) or strictly convex. It follows that $V_P$ is a strictly convex function of $\xi$ provided $Y \neq Z$. □

From (60) and (61), we see that

$$V_P' - V_P = 2\sum_{i=1}^{n} X_iX_{n+i}V_i. \tag{71}$$

Thus, $V_P' = V_P$ for trim portfolios, and $V_P' > V_P$ for untrim ones.

Theorem 3. *If Property* P *holds in the original model, then the modified model has the same set of efficient $(E, V_P)$ combinations as does the original model. Also, it has a unique set of efficient portfolios (one for each efficient $(E, V_P)$ combination) that is the same as the unique set of trim efficient portfolios in the original model.*

Proof. First, we show that all efficient $(E, V_P)$ combinations in the original model, and all trim efficient portfolios in the original model, are efficient $(E, V_P')$ combinations and portfolios for the modified model. Then, we show that no additional portfolios or $(E, V_P')$ combinations are efficient for the latter model. Because $V_P = V_P'$ for trim portfolios, and each efficient $(E, V_P)$ combination in the original model can be supplied by a trim portfolio $X$, each efficient $(E, V_P)$ combination of the original model is feasible in the modified model. It will also be efficient in the modified model unless some other feasible portfolio $Y$ dominates it in that model (i.e., has greater $E$ for the same or less $V_P'$, or less $V_P'$ for the same or greater $E$). Because the models have the same feasible sets and expected returns, and because $V_P(Y) \leqslant V_P'(Y)$, $V_P'(X) = V_P(X)$, if $Y$ dominated $X$ in the modified model (e.g., with $\mathrm{E}(Y) \geqslant \mathrm{E}(X)$ and $V_P'(Y) < V_P'(X)$), then it would also dominate it in the original model, contradicting the hypothesis that $X$ is efficient in the original model. Thus, all trim portfolios that are efficient in the original model are efficient in the modified model.

We now show that no other portfolios are efficient in the modified model. If the constraint set is bounded, therefore compact as well as closed, then the efficient $(E, V_P)$ combinations of the original model span a closed interval $[\underline{E}, \bar{E}]$ of expected returns, where $\bar{E}$ is the maximum feasible expected return and $\underline{E}$ is the expected return of the efficient portfolio with minimum $V_P$. According to Theorem 2, if $X$ is a trim, efficient portfolio and $Y$ is another efficient portfolio with the same $(E, V_P)$, then $Y$ is untrim. Therefore, (71) implies $V_P'(Y) > V_P(Y) = V_P'(X)$. Thus, $Y$ is not efficient in the modified model. This, plus the fact that $V_P' = V_P$ for trim portfolios, and the uniqueness statement in Theorem 2, implies that the efficient set for the modified model is unique for $E \in [\underline{E}, \bar{E}]$. Nor can the modified model have an efficient portfolio with $E$ outside $[\underline{E}, \bar{E}]$, for then the modified model will either have a feasible portfolio with greater $E$ than $\bar{E}$, which is impossible because the two models have the same feasible sets and expected returns, or have smaller $V$ than the minimum feasible $V$ in the original model, which is impossible because $V_P' \geqslant V_P$.

If $E$ is not bounded above, then the preceding argument applies except that it is unnecessary to check for an efficient portfolio in the modified model with $E > \bar{E}$. □

Theorem 3 assures us that we can naively use a factor or scenario portfolio optimizer, ignoring the negative correlation between $u_i$ and $u_{n+i}$, and get a correct answer to the long-short portfolio selection problem when Property P holds. This is not necessarily the case if Property P does not hold. For example, consider any diagonalized model with a Reg T constraint (with $H = 2$), a budget constraint (65), and an upper bound ($u_c < 1.0$) on cash. Assume that $V_i > 0$ for all $i \in [1, \nu]$. In the original model, consider the portfolio with

$$X_1 = X_{n+1} = 1, \qquad X_i = 0 \quad \text{otherwise.}$$

This portfolio is feasible and has zero variance. Thus, zero variance is feasible; therefore, some portfolio (not necessarily the above portfolio) has zero variance and is efficient. However, the modified version of this model has no feasible zero-variance portfolios: The upper bound on cash implies that $X_i > 0$ for some $i \in [1, \nu]$, which implies $V_P' > 0$, because $\mathrm{cov}(r_i, r_j) = 0$ for risky securities in the modified model. Thus, absent some assumption such as Property P, it is possible that an efficient set for the modified model may not be an efficient set for the original model.

## 8. Example

Tables 1 through 8 illustrate the content and purpose of the theorems of the preceding section. We consider a three-security, one-factor model subject only to Reg T, the budget constraint, and nonnegativity constraints. In this case, (26) may be written as

$$r_i = \alpha_i + \beta_i f + u_i, \quad i = 1, 2, 3. \tag{72}$$

Table 1 presents inputs to such a model for three hypothetical securities. In all the tables, long positions in the three securities are labeled 1L, 2L, and 3L. Table 1 shows for each of these long positions, the expected return $\mu_i$, beta $\beta_i$, idiosyncratic variance $V_i = \text{var}(u_i)$, and rebate fraction $h_i$. The latter is needed to compute the expected return of the corresponding short position. Table 1 also shows the lending rate, the borrowing rate, and the variance of the underlying factor $f$.

The betas of the securities, their idiosyncratic variances, and the variance of the underlying factor could be used to compute the covariances among the long positions according to the formulas

$$\text{cov}(r_i, r_j) = \beta_i \beta_j V(f), \quad i \neq j, \tag{73a}$$

$$V(r_i) = \beta_i^2 V(f) + V(u_i), \quad i = 1, 2, 3. \tag{73b}$$

The result of this calculation for the present example is shown in Table 2.

As Sharpe (1963) explains for a long-only portfolio analysis, the covariance matrix for a one-factor model can be transformed into a sum of squares by introducing a new variable constrained to be the portfolio beta, as in (30). Table 3 contains the covariance matrix for this four-security version of the three-security single-factor model. The algorithm presented in Sharpe (1963) takes advantage of the fact that the covariance matrix is diagonal, with nonzero entries on the diagonal, rather than a dense arbitrary covariance matrix (i.e., an arbitrary positive, semidefinite matrix) as the general critical line algorithm permits. Sharpe's diagonalized version of the $n$-security one-factor model is frequently referred to as *the* diagonal model.

The advantage of thus diagonalizing the covariance matrix increases with the number of securities in the portfolio analysis. Column 2 of Table 4 presents the number of input coefficients required by the diagonal model of covariance: namely $n$ betas, $n$ idiosyncratic variances, and one factor $f$ variance. The third column of Table 4 presents the number of unique covariances needed by a computation expecting an arbitrary covariance matrix, namely $n(n+1)/2$. Specifically, with three securities there are actually more coefficients in the diagonal model than in the nondiagonalized version. With 5,000 securities, the diagonal model works with about 10,000 coefficients, whereas the 5,000-by-5,000 covariance matrix of the general model has over 12 million unique covariances (counting $\sigma_{ij} = \sigma_{ji}$ as *one* covariance). Both versions of the model also need $n$ expected returns.

Both versions of the model will go through the same number of iterations and come out with the same efficient frontier. The work per iteration depends on how many securities are IN as well as the total number of securities. For moderate to large-size analyses, much less work is required by the diagonal model per iteration.[16]

Table 5 presents the expected returns, betas, and idiosyncratic variances for both the long and short securities corresponding to the long securities in Table 1. Short positions are labeled 1S, 2S, 3S. The expected returns for the short positions are computed according to Equation (50). The betas of the short position are the negative of those for the long position, whereas the idiosyncratic variances are the same for the short position as for the corresponding long position.

**Table 1.** Illustrative three-security one-factor model.

| Security $i$ | Expected return $\mu(i)$ | Beta $\beta(i)$ | Idiosyncratic variance $V(i)$ | Rebate fraction $h(i)$ |
|---|---|---|---|---|
| 1L | 0.10 | 0.80 | 0.0768 | 0.5 |
| 2L | 0.12 | 1.00 | 0.1200 | 0.5 |
| 3L | 0.16 | 1.25 | 0.1875 | 0.5 |
| Lend | 0.03 | 0.00 | 0.0000 | NA |
| Borrow | 0.05 | 0.00 | 0.0000 | NA |
| Variance of factor | | 0.0400 | | |

*Notes.* This table shows inputs to a three-security, one-factor long-short model. These consist of the expected return, beta against the factor, and idiosyncratic variance of each long position. Also needed are the rebate fraction of each security (for computing expected returns of short positions), the rates at which the investor can borrow and lend, and the variance of the underlying factor.

**Table 2.** Covariances among long positions.

| Security | 1L | 2L | 3L |
|---|---|---|---|
| 1L | 0.1024 | 0.0320 | 0.0400 |
| 2L | 0.0320 | 0.1600 | 0.0500 |
| 3L | 0.0400 | 0.0500 | 0.2500 |

*Notes.* This table shows covariances among long positions, computed from their betas, idiosyncratic variances, and the variance of the underlying factor.

**Table 3.** Covariances when dummy security is included.

| Security | 1L | 2L | 3L | PB |
|---|---|---|---|---|
| 1L | 0.0768 | 0 | 0 | 0 |
| 2L | 0 | 0.1200 | 0 | 0 |
| 3L | 0 | 0 | 0.1875 | 0 |
| PB | 0 | 0 | 0 | 0.0400 |

*Notes.* In a model with long positions only, the introduction of "portfolio beta" as a fourth (dummy) "security" diagonalizes the covariance matrix. An added equation is needed to constrain PB to equal portfolio beta.

**Table 4.** Number of unique coefficients required by model of covariance.

| Number of securities | With dummy security | Without dummy security |
|---|---|---|
| 3 | 7 | 6 |
| 20 | 41 | 210 |
| 100 | 201 | 5,050 |
| 500 | 1,001 | 125,250 |
| 1,000 | 2,001 | 500,500 |
| 3,000 | 6,001 | 4,501,500 |
| 5,000 | 10,001 | 12,502,500 |

*Notes.* This table shows the number of coefficients needed to characterize the covariance structure when the dummy variable of Table 3 is or is not added to the model. Since $\text{cov}(i, j) = \text{cov}(j, i)$ these are counted only once.

The covariances among long and short positions, presented in Table 6, are derived from Table 2 using Equation (53). We could compute an efficient frontier for the short-long model using the expected returns in Table 5, and the covariance matrix in Table 6, using a general portfolio analysis program that permits an arbitrary covariance matrix. If we perform the Sharpe (1963) trick of expressing return as a linear function of amount invested in the factor, plus amounts invested in the idiosyncratic terms as in our Equations (54) and (57), then the covariance matrix for the long-short model is as presented in Table 7. Note that the covariance matrix is no longer diagonalized because, for example, the 1L idiosyncratic term has a −1.0 correlation with 1S.

If we present the data in Table 5 to the Sharpe (1963) algorithm, it will assume that the covariance matrix is in fact diagonal, such as that in Table 8. Theorem 3 assures us

**Table 5.** Illustrative three-security one-factor model with long (L) and short (S) positions.

| Security $i$ | Expected return $\mu(i)$ | Beta $\beta(i)$ | Idiosyncratic variance $V(i)$ |
|---|---|---|---|
| 1L | 0.100 | 0.80 | 0.0768 |
| 2L | 0.120 | 1.00 | 0.1200 |
| 3L | 0.160 | 1.25 | 0.1875 |
| 1S | −0.085 | −0.80 | 0.0768 |
| 2S | −0.105 | −1.00 | 0.1200 |
| 3S | −0.145 | −1.25 | 0.1875 |
| Lend | 0.030 | 0.00 | 0.0000 |
| Borrow | 0.050 | 0.00 | 0.0000 |
| Variance of factor | | 0.0400 | |

*Notes.* The table shows properties of a three-security one-factor long-short model, derived from Table 1. Here the expected returns of the short positions are the negative of those of the corresponding long positions, plus short rebate interest on the proceeds. The betas of the short positions are the negative of the long positions; the idiosyncratic variances are the same as those of the long positions. Not noted in the table is the fact that the covariances between the idiosyncratic terms of $n$S and $n$L are not zero.

**Table 6.** Covariances among long and short positions.

| Security | 1L | 2L | 3L | 1S | 2S | 3S |
|---|---|---|---|---|---|---|
| 1L | 0.1024 | 0.0320 | 0.0400 | −0.1024 | −0.0320 | −0.0400 |
| 2L | 0.0320 | 0.1600 | 0.0500 | −0.0320 | −0.1600 | −0.0500 |
| 3L | 0.0400 | 0.0500 | 0.2500 | −0.0400 | −0.0500 | −0.2500 |
| 1S | −0.1024 | −0.0320 | −0.0400 | 0.1024 | 0.0320 | 0.0400 |
| 2S | −0.0320 | −0.1600 | −0.0500 | 0.0320 | 0.1600 | 0.0500 |
| 3S | −0.0400 | −0.0500 | −0.2500 | 0.0400 | 0.0500 | 0.2500 |

*Notes.* The table shows covariances among short and long positions. These entries are of the same magnitude as the long-only covariances in Table 2, with the same sign in case of long-long or short-short covariances, and opposite sign in case of long-short or short-long covariances.

that the efficient frontier computed assuming the diagonal covariance matrix in Table 8 is the same as the efficient frontier computed using the correct covariance matrix in Table 7. It also assures us that, for any number of securities, we get the correct result if we ignore the correlations among the idiosyncratic terms for a many-factor model, scenario model, or a mixed factor and scenario model. It further assures us that the efficient frontier is correctly computed if additional constraints are imposed on the choice of portfolio, provided that the constraint set satisfies Property P. In particular, we may present the requisite parameters to the Markowitz-Perold (1981a, b) algorithm for the scenario model or mixed-scenario models, ignoring the correlation between the short and long idiosyncratic terms, for any system of constraints that satisfies Property P.

In the case of the $n$-security one-factor model, the advantage of using the diagonal model (as permitted by Theorem 3) rather than a general model is again given by Table 4 and Endnote 13, except that now an $n$-security long-short model has $2n$ "securities." For example, if there are 500 securities in the universe, then the diagonal model will be told that there are 1,001 securities whose covariance structure is described by 2,001 coefficients, whereas the general model will require 500,500 unique (arbitrary, as far as it knows) covariances.

## 9. Summary

CAPMs frequently assume, in effect, that an investor can sell a security short without limit and invest the proceeds of the short sale in some other stock. In fact, this is not the case. This paper describes some actual short-sale arrangements. However, short-sale requirements vary from time to time, broker to broker, and investor to investor. Thus, the portfolio analyst must model the sale requirements of the specific client as she or he finds them.

The CLA traces out a piecewise linear set of efficient portfolios subject to any finite system of linear equality or inequality constraints, for any covariance matrix and expected return vector. Because the covariance matrix is arbitrary, the CLA can trace out efficient sets for long-short

**Table 7.** Covariances when dummy security is included.

| Security | 1L | 2L | 3L | 1S | 2S | 3S | PB |
|---|---|---|---|---|---|---|---|
| 1L | 0.0768 | 0 | 0 | −0.0768 | 0 | 0 | 0 |
| 2L | 0 | 0.1200 | 0 | 0 | −0.1200 | 0 | 0 |
| 3L | 0 | 0 | 0.1875 | 0 | 0 | −0.1875 | 0 |
| 1S | −0.0768 | 0 | 0 | 0.0768 | 0 | 0 | 0 |
| 2S | 0 | −0.1200 | 0 | 0 | 0.1200 | 0 | 0 |
| 3S | 0 | 0 | −0.1875 | 0 | 0 | 0.1875 | 0 |
| PB | 0 | 0 | 0 | 0 | 0 | 0 | 0.0400 |

*Notes.* Above are the covariances among long, short, and the dummy security, PB, when portfolio beta is introduced as a seventh dummy security. An equation is added to constrain PB to be portfolio beta. Unlike the long-only case in Table 3, the covariance matrix is no longer diagonal.

portfolio selection problems provided that the constraints on choice of portfolio are linear equalities or weak inequalities. Examples of such constraints include a budget constraint, the Reg T "margin requirement constraint," upper bounds on long or short positions in individual or groups of assets, or the requirement that the sum (or a weighted sum) of long positions not differ "too much" from the sum (or the weighted sum) of short positions.

While the CLA may be applied to an arbitrary covariance matrix, it is especially fast for models in which covariances are implied by a factor or scenario model. In this case, an equivalent model can be written, including new "fictitious" securities whose magnitudes are linearly related to the magnitudes of the "real" securities, so that the covariance matrix becomes diagonal or almost so. Special programs exist to exploit the resultant sparse, well-structured efficient-set equations.

A portfolio selection problem in which securities can be held short or long can be modeled as a $2n$-security problem, in which a first $n$ represents long positions, and another $n$ short positions, and all $2n$ are required to have nonnegative values. Even if long positions in $n$ securities satisfy the assumptions of the factor or scenario model, the $2n$-variable long-short model does not satisfy these same assumptions, because idiosyncratic terms are not uncorrelated. Nevertheless, if the information for the $2n$ variables is fed into a factor or scenario program, a correct answer is computed—provided that a certain condition ("Property P") holds.

Property P essentially requires that if a portfolio with short and long positions in the same stock is feasible, then it is also feasible to reduce both positions, keeping the holdings of all other risky stocks the same; and this reduction in both the short and long positions in the same stock does not decrease the expected return of the portfolio. When this condition is met, then the $2n$-variable version of the long-short problem can be run on the appropriate factor or scenario model program. The correct answer is produced despite the violation of the assumption that the idiosyncratic terms are uncorrelated.

A fast CLA also exists for the situation in which historical covariances are used, but there are many more securities than time periods. This algorithm produces the correct answer when applied to the $2n$-variable version of the long-short problem, whether or not Property P holds.

The speed-up in computation that results from the use of "diagonalized" versions of factor, scenario, or historical models is approximately equal to the ratio of nonzero coefficients in the equations of the two models. For large problems, this timesaving can be considerable.

**Table 8.** Covariances based on Theorem 3.

| Security | 1L | 2L | 3L | 1S | 2S | 3S | PB |
|---|---|---|---|---|---|---|---|
| 1L | 0.0768 | 0 | 0 | 0 | 0 | 0 | 0 |
| 2L | 0 | 0.1200 | 0 | 0 | 0 | 0 | 0 |
| 3L | 0 | 0 | 0.1875 | 0 | 0 | 0 | 0 |
| 1S | 0 | 0 | 0 | 0.0768 | 0 | 0 | 0 |
| 2S | 0 | 0 | 0 | 0 | 0.1200 | 0 | 0 |
| 3S | 0 | 0 | 0 | 0 | 0 | 0.1875 | 0 |
| PB | 0 | 0 | 0 | 0 | 0 | 0 | 0.0400 |

*Notes.* If the data in Table 5 are presented to a standard factor model portfolio optimizer, the program will assume that the model has the covariance structure in this table, with a diagonal covariance matrix, rather than the correct one, that in Table 7. Theorem 3 assures us that the optimizer will nevertheless compute the efficient frontier correctly. Theorem 3 further assures us that this is so for a many-factor model, a scenario model, or a mixed factor-scenario model of covariance; and remains true for any system of linear equality or (weak) inequality constraints that satisfy Property P.

## Endnotes

1. The results reported in this paper were first circulated in a Jacobs Levy Equity Management working paper (Jacobs et al. 2001).
2. For proofs and further details, see Markowitz (1959), Appendix A, Perold (1984), Markowitz (1987), or Markowitz and Todd (2000).
3. See Markowitz and Todd (2000), Chapter 8, for how to get a first critical line.
4. If $C$ is singular, there may be more than one portfolio with minimum feasible $V$. Because the $V$-minimizing portfolios may have different $E$s, they may not all be efficient, but it is shown that the portfolio reached by the CLA when $\lambda_E \downarrow 0$ is efficient as well as $V$-minimizing.
5. See Markowitz and Todd (2000), Chapter 9, for what to do in case of ties.
6. Note that the CLA as presented in Markowitz (1956) is an example of a linear complementary algorithm as defined in Wolfe (1959).
7. See Sharpe (1963) in particular, and Markowitz and Perold (1981b) in general, for details.
8. For models that combine both scenarios and factors, see Markowitz and Perold (1981a, b).
9. For details, see Markowitz et al. (1992).
10. See Fortune (2000) for details on initial and maintenance margin requirements for long and short positions on exempt and nonexempt securities. Also see www.federalreserve.gov/regulations, 12 CFR 220, Credit by Brokers and Dealers (Regulation T). See Jacobs and Levy (1993) on margin requirements and cash needed for liquidity.

Equation (43) can also be written as

$$\sum_{i=1}^{2n} 0.5X_i \leqslant 1, \tag{10.1}$$

reflecting a 50% margin on short and long positions. Actually, the Reg T initial short margin requirement is stated as 150%—of which 100% out of the 150% is supplied by the proceeds of the sale of the borrowed stock. Constraint (10.1) is a special case of

$$\sum_{i=1}^{2n} m_i X_i \leqslant 1, \tag{10.2}$$

where $m_i$ here represents the net (after proceeds, where applicable) margin requirement of the $i$th position. Constraint (10.2) is more general than (43) or (10.1) in that it permits, in particular, (a) a net short margin requirement that differs from the long margin requirement, and (b) securities that are exempt from Reg T requirements. We use (43) in examples, but Theorems 1, 2, and 3 apply to *any* system of constraints (4) and (5) with properties specified in theorems.

11. Rule 15c3-1 of the Securities Exchange Act of 1934 governs capital requirements for broker-dealers, including the provision that indebtedness cannot exceed 1,500% of net capital (800% for 12 months after commencing business as a broker or dealer).
12. Noncash collateral typically consists of letters of credit or securities. It is usually 100% to 105% of the amount borrowed. The gains and losses on the collateral belong to the borrower, and the lender is generally paid a fee. The collateral is marked to market and augmented by the borrower if necessary.
13. Usually $h_i < 1$. However, the case of $h_i = 1$ is conceivable, and is covered by our theorems. Large institutional investors often perform mean-variance analysis at an asset class level and then implement the asset class allocations using either index funds or using internal or external fund managers. If, say, an internal market neutral fund borrows shares from, say, an internal large-cap or small-cap fund, the allocation of interest on the proceeds between borrowing fund and lending fund is arbitrary. The institution's policy might allocate all the interest to the borrowing fund, because the institution's policy might prohibit external stock lending, so that the particular interest income would not exist except for the internal market neutral fund's activities.

If no zero-variance variable is ever held short, we may write (47) as

$$R_P = \sum_{i=1}^{n} r_i X_i + \sum_{i=n+1}^{n+\nu} (-r_{i-n}) X_i + r_c \sum_{i=n+1}^{n+\nu} h_{i-n} X_i. \tag{13.1}$$

Alternatively, we can leave it *as is* in (47) and assume that (45) contains equations of the form

$$X_{n+i} = 0 \tag{13.2}$$

for $i \in [\nu + 1, n]$. Generally, if a security cannot be sold short (e.g., because it cannot be borrowed), then this can be represented either by including a constraint of the form (13.2) or by omitting $n + i$ from the analysis. The latter approach is advisable in practice; the former is notationally convenient here.

14. Equation (47) does not include tax considerations, and therefore would be applicable to tax-exempt organizations such as university endowments and corporate pension plans.
15. Jacobs et al. (1998, 1999) address the conditions under which optimal portfolios that are constrained to hold roughly equal amounts in long and short positions are equivalent to optimal portfolios without this constraint. In practice, long-short portfolios are often managed in this "market-neutral" fashion.
16. If $n_I$ securities are IN, then the Sharpe (1963) algorithm requires a few more than $3n + 7n_I$ multiplications and divisions plus $3n + 5n_I$ additions, whereas the general algorithm requires $2n_I n + 5n + 2n_I^2 - n_I$ multiplications and divisions, and $2n_I n + 3n + 2n_I^2 - 2n_I$ additions. Thus, if $n = 1{,}000$ and $n_I = 10$, as at the high end of the frontier,

or $n_I = 100$ as might occur at the low end of the frontier, then the diagonal model requires 3,070 or 3,700 multiplications and divisions for the iteration, whereas the general algorithm requires 25,190 or 269,900.

## References


Alexander, Gordon J. 1993. Short selling and efficient sets. *J. Finance* **48** 1497–1506.

Cohen, K. J., J. A. Pogue. 1967. An empirical evaluation of alternative portfolio selection models. *J. Bus.* **40** 166–193.

Elton, Edwin J., Martin J. Gruber, Manfred W. Padberg. 1976. Simple criteria for optimal portfolio selection. *J. Finance* **31** 1341–1357.

Fortune, Peter. 2000. Margin requirements, margin loans, and margin rates: Practice and principles. *New England Econom. Rev.* **2000** (September/October) 19–44.

Jacobs, Bruce I., Kenneth N. Levy. 1993. The generality of long-short equitized strategies: A correction. *Financial Analysts J.* **49**(March/April) 22.

Jacobs, Bruce I., Kenneth N. Levy. 2000. *Equity Management: Quantitative Analysis for Stock Selection.* McGraw-Hill, New York.

Jacobs, Bruce I., Kenneth N. Levy, David Starer. 1998. On the optimality of long-short strategies. *Financial Analysts J.* **54**(March/April) 40–51.

Jacobs, Bruce I., Kenneth N. Levy, David Starer. 1999. Long-short portfolio management: An integrated approach. *J. Portfolio Management* **25**(Winter) 23–32.

Jacobs, Bruce I., Kenneth N. Levy, Harry M. Markowitz, David Starer. 2001. Optimization and neutrality of long-short portfolios. Jacobs Levy Equity Management, Florham Park, NJ.

Kwan, Clarence C. Y. 1995. Optimal portfolio selection under institutional procedures for short selling. *J. Banking Finance* **19** 871–889.

Markowitz, Harry M. 1956. The optimization of a quadratic function subject to linear constraints. *Naval Res. Logistics Quart.* **3** 111–133.

Markowitz, Harry M. 1959. *Portfolio Selection: Efficient Diversification of Investments.* John Wiley and Sons, New York, and 1991 2nd ed., Basil Blackwell, Cambridge, MA.

Markowitz, Harry M. 1987. *Mean-Variance Analysis in Portfolio Choice and Capital Markets.* Basil Blackwell, Cambridge, MA.

Markowitz, Harry M., André F. Perold. 1981a. Portfolio analysis with factors and scenarios. *J. Finance* **36** 871–877.

Markowitz, Harry M., André F. Perold. 1981b. Sparsity and piecewise linearity in large portfolio optimization problems. I. S. Duff, ed. *Sparse Matrices and Their Uses.* Academic Press, London, UK, 89–108.

Markowitz, Harry M., Peter Todd. 2000. *Mean-Variance Analysis in Portfolio Choice and Capital Markets.* Revised reissue of Markowitz (1987) with chapter by Peter Todd, Frank J. Fabozzi Associates, New Hope, PA.

Markowitz, Harry M., Peter Todd, Gan Lin Xu, Yuji Yamane. 1992. Fast computation of mean-variance efficient sets using historical covariances. *J. Financial Engrg.* **2** 117–132.

Perold, André F. 1984. Large-scale portfolio optimization. *Management Sci.* **10** 1143–1160.

Sharpe, William F. 1963. A simplified model for portfolio analysis. *Management Sci.* **9** 277–293.

Wolfe, Philip. 1959. The simplex method for quadratic programming. *Econometrica* **3** 382–398.

FINANCIAL ANALYSTS JOURNAL®
Reflections

# Market Efficiency: A Theoretical Distinction and So What?

Harry M. Markowitz

The capital asset pricing model (CAPM) is an elegant theory. With the aid of some simplifying assumptions, it comes to dramatic conclusions about practical matters, such as how to choose an investment portfolio, how to forecast the expected return of a security or asset class, how to price a new security, or how to price risky assets in a merger or acquisition.

The CAPM starts with some assumptions about investors and markets and deduces its dramatic conclusions from these assumptions. First, it assumes that investors seek mean–variance efficient portfolios; in other words, it assumes that investors seek low volatility and high return on average. Different investors may have different trade-offs between these two, depending on their aversion to risk. Second, the CAPM assumes that taxes, transaction costs, and other illiquidities can be ignored for the purposes of this analysis. In effect, it assumes that such illiquidities may impede the market's approach to the CAPM solution but do not change the general tendency of the market. A third CAPM assumption is that all investors have the same predictions for the expected returns, volatilities, and correlations of securities. This assumption is usually not critical.[1] Finally, the CAPM makes assumptions about what portfolios the investor can select. The original Sharpe (1964)–Lintner (1965) CAPM considered long positions only and assumed that the investor could borrow without limit at the risk-free rate. From this assumption, and the three in the preceding paragraph, one can deduce conclusions of the sort outlined in the first paragraph.

The assumption that the investor can borrow without limit is crucial to the Sharpe–Lintner model's conclusions. As illustrated later in this article, if we accept the other three CAPM assumptions but assume limited (or no) borrowing, the Sharpe–Lintner conclusions no longer follow. For example, if the four premises of the Sharpe–Lintner original CAPM were true, then the "market portfolio"—a portfolio whose amounts invested are proportional to each security's market capitalization—would be an efficient portfolio. We could not find a portfolio with greater return (on average) without greater volatility. In fact, if the four premises of the Sharpe–Lintner original CAPM were true, the market portfolio, plus perhaps borrowing and lending, would be the *only* efficient portfolio. If, however, we assume the first three premises of the Sharpe–Lintner CAPM but take into account the fact that investors have limited borrowing capacity, then it no longer follows that the market portfolio is efficient. As this article will illustrate, this inefficiency of the market portfolio could be substantial and it would not be arbitraged away even if some investors could borrow without limit.

> When one clearly unrealistic assumption of the capital asset pricing model is replaced by a real-world version, some of the dramatic CAPM conclusions no longer follow.

---

*Harry M. Markowitz is president of Harry Markowitz Company, San Diego, California.*

 

FINANCIAL ANALYSTS JOURNAL
*Reflections*

Before the CAPM, conventional wisdom was that some investments were suitable for widows and orphans whereas others were suitable only for those prepared to take on "a businessman's risk." The CAPM convinced many that this conventional wisdom was wrong; the market portfolio is the proper mix among risky securities for everyone. The portfolios of the widow and businessman should differ only in the amount of cash or leverage used. As we will see, however, an analysis that takes into account limited borrowing capacity implies that the pre-CAPM conventional wisdom is probably correct.

An alternate version of the CAPM speaks of investors holding short as well as long positions. But the portfolios this alternate CAPM permits are as unrealistic as those of the Sharpe–Lintner CAPM with unlimited borrowing. The alternate CAPM assumes that the proceeds of a short sale can be used, without limit, to buy securities long. For example, the alternate CAPM assumes that an investor could deposit $1,000 with a broker, short $1,000,000 worth of Stock A, then use the proceeds and the original deposit to buy $1,001,000 of Stock B. The world does not work this way.

Like the original CAPM, the alternate CAPM implies that the market portfolio is an efficient portfolio, although not the only one (as in the original CAPM). If one takes into account real-world constraints on the holding of short and long positions, however, the efficiency of the market portfolio no longer follows, as will be illustrated.

Both the original CAPM, with unlimited borrowing, and the alternate CAPM, with unrealistic short rules, imply that the expected return of a stock depends in a simple (linear) way on its beta, and only on its beta. This conclusion has been used for estimating expected returns, but it has lost favor for this use because of poor predictive results. It is still used routinely in "risk adjustment," however, for valuing assets and analyzing investment strategies on a "risk-adjusted basis." I will show here that the conclusion that expected returns are linear functions of beta does not hold when real-world limits on permitted portfolio holdings are introduced into the CAPM. This discussion will call into question the frequent use of beta in risk adjustment.

I will discuss the assumptions and conclusions of the CAPM formally and then illustrate the effect on CAPM conclusions of varying the CAPM assumptions concerning the investor's constraint set. Afterward, I will sketch how the points illustrated in the simple examples generalize to more complex cases. Finally, I will discuss the implications of the analysis for financial theory, practice, and pedagogy.

## A Distinction

We should distinguish between the statement that "the market is efficient," in the sense that market participants have accurate information and use it correctly to their benefit and the statement that "the market portfolio is an efficient portfolio." Under some conditions, the former implies the latter. In particular, if one makes the following assumptions,

A1. transaction costs and other illiquidities can be ignored (as I will do throughout this article),

A2. all investors hold mean–variance efficient portfolios,

A3. all investors hold the same (correct) beliefs about means, variances, and covariances of securities, and—*in addition*—

A4. every investor can lend all she or he has or can borrow all she or he wants at the risk-free rate,

*then* Conclusion 1 follows:

C1. The market portfolio is a mean–variance efficient portfolio.

C1 also follows if A4 is replaced by A4′:

A4′. Investors can sell short without limit and use the proceeds of the sale to buy long positions.

In particular, A4′ says that any investor can deposit $1,000 with a broker, short $1,000,000 worth of one security, and buy long $1,001,000 worth of another security.

Neither A4 nor A4′ is realistic. Regarding A4, when an investor borrows, not only does the investor pay more than when the U.S. government borrows, but (a point of equal or greater importance here) the amount of credit extended is limited to what the lender believes the borrower has a reasonable probability of repaying. Regarding A4′, if the investor deposits $1,000 with a broker, Federal Reserve Regulation T permits the investor to buy a $2,000 long position *or* take on a $2,000 short position *or* take on a $1,000 long and a $1,000 short position, but it does not allow an unlimited amount short plus the same unlimited amount long, as assumed in A4′.

If one replaces A4 or A4′ with a more realistic description of the investor's investment constraints, then C1 usually no longer follows; even though all

 

FINANCIAL ANALYSTS JOURNAL®
Reflections

investors share the same beliefs and each holds a mean–variance efficient portfolio, the market portfolio need not be an efficient portfolio. This departure from efficiency can be quite substantial. In fact, the market portfolio can have almost *maximum* variance among feasible portfolios with the same expected value rather than *minimum* such variance; that is, the market portfolio can be about as *inefficient* as a feasible portfolio can get (see Chapter 11 of Markowitz 1987 or Markowitz and Todd 2000).

In addition to C1, A1 through A4 (or A1 through A4′) imply

C2. In equilibrium, the expected return for each security depends only on its beta (the regression of its returns against the return on the market). This relationship between the security's expected return and its beta is a simple, linear relationship.

C2 is the basis for the CAPM's prescriptions for risk adjustment and asset valuation. Like the first conclusion, C2 does not follow from assumptions A1 through A3 if A4 (or A4′) is replaced by a more realistic description of the investor's investment constraints.

Often, financial research attempts to determine "market efficiency" by testing whether C2 holds. But the failure of C2 to hold empirically does not prove that the market is not efficient in the general sense of possessing correct information and using it advantageously. Nor does the failure of C2 to hold empirically prove that the market is not efficient in the narrower sense of A2 and A3—namely, that participants hold mean–variance efficient portfolios in terms of commonly held correct beliefs. I will not argue here that A2 and A3 are true—or argue that they are false. I argue only that, in the absence of either A4 or A4′, the empirical refutation of C2 is not an empirical refutation of A1 through A3.[2]

**Example.** In this first example, I assume that investors cannot sell short or borrow (but I note subsequently that the same results hold if investors *can* borrow limited amounts or *can* sell short but are subject to Reg T or some similar constraint). The example assumes A1 through A3; that is, it ignores taxes, transaction costs, and other illiquidities; it assumes that all investors have the same beliefs about the means, variances, and covariances of security returns; and it assumes that each investor holds a portfolio that is mean–variance efficient in terms of these beliefs.

This example consists of long positions in three risky securities with the expected returns and standard deviations shown in **Table 1**. To keep things simple, we will assume that returns are uncorrelated. However, the results also hold for correlated returns.

**Table 1. Expected Returns and Standard Deviations of Three Risky Securities**

| Security | Expected Return | Standard Deviation |
|---|---|---|
| 1 | 0.15% | 0.18% |
| 2 | 0.10 | 0.12 |
| 3 | 0.20 | 0.30 |

Let $X_1$, $X_2$, and $X_3$ represent the fraction of her wealth that some investor invests in, respectively, Securities 1, 2, and 3. Assume that the investor can choose any portfolio that meets the following constraints:

$$X_1 + X_2 + X_3 = 1.0 \tag{1}$$

and

$$X_1 \geq 0, X_2 \geq 0, X_3 \geq 0. \tag{2}$$

The first of these is a budget equation; the second is a requirement that none of the investments be negative. We will contrast the efficient set and market portfolio we get with Budget Equation 1 and Nonnegativity Requirement 2 as constraints with the set and portfolio we get if we assume A4′ (that is, if we assume that Budget Equation 1 is the only constraint). In **Figure 1**, $X_1$—the fraction invested in Security 1—is plotted on the horizontal axis; $X_2$—the fraction in Security 2—is plotted on the vertical axis; and $X_3$—the fraction invested in the third security—is given implicitly by the relationship

$$X_3 = 1 - X_1 - X_2. \tag{3}$$

Figure 1 should be thought of as extended without limits in all directions. Every point (portfolio) on this extended page is feasible according to assumption A4′. For example, the point with $X_1 = 93$ and $X_2 = -106$ (therefore, $X_3 = 14$ according to Equation 3) is feasible according to assumption A4′ because it satisfies Equation 1. It is not feasible when Budget Equation 1 *and* Nonnegativity Requirement 2 are required because it does not satisfy $X_2 \geq 0$.

The only points (portfolios) in Figure 1 that satisfy Budget Equation 1 and Nonnegativity Requirement 2 are on and in the triangle whose vertices are the points (1,0), (0,1), and (0,0). The first

FINANCIAL ANALYSTS JOURNAL®
Reflections
60 YEARS 1945–2005

**Figure 1. Efficient Sets with and without Nonnegativity Constraints**

of these points represents an undiversified portfolio with 100 percent invested in Security 1 ($X_1 = 1.0$); the second, a 100 percent investment in Security 2 ($X_2 = 1.0$); the third, a 100 percent investment in Security 3 ($X_3 = 1.0$). The diagonal side of the triangle connecting points (1,0) and (0,1) includes investments in Securities 1 and 2 but not Security 3; the horizontal side connecting (0,0) and (1,0) has investments in Securities 1 and 3 but not in Security 2; the side connecting (0,0) and (0,1) has $X_1 = 0$. Points within the triangle represent portfolios with positive investments in all three securities. The vertices, sides, and interior of the triangle all meet Budget Equation 1 and Nonnegativity Requirement 2.

If assumption A4′ holds (therefore Budget Equation 1 is the only constraint), then all the portfolios with the least standard deviation for various levels of expected return lie on the straight line labeled $\ell\ell'$ in Figure 1. Because two points determine a line, we know the whole line if we know two points on it. One point on the line is the portfolio that minimizes standard deviation among all portfolios on the extended page (i.e., among all portfolios that satisfy Equation 1). When returns are uncorrelated, this risk-minimizing portfolio satisfies:

$$X_1 = \frac{K_c}{V_1}, \tag{4a}$$

$$X_2 = \frac{K_c}{V_2}, \tag{4b}$$

and

$$X_3 = \frac{K_c}{V_3}, \tag{4c}$$

where $V_1$, $V_2$, $V_3$ are the variances (standard deviations squared) of the three securities and $K_c$ is chosen so that Equation 1 is satisfied; that is,

$$K_c = \frac{1}{(1/V_1)+(1/V_2)+(1/V_3)}. \tag{5}$$

Thus, when returns are uncorrelated, the variance-minimizing portfolio is always within the triangle. For the current example,

$$X_1 = 0.28,$$

$$X_2 = 0.62,$$

and

$$X_3 = 0.10.$$

This point is the point labeled "*c*" in Figure 1.

When returns are uncorrelated, another point on the line that minimizes portfolio variance for various levels of portfolio expected return is

$$X_1 = \frac{K_a E_1}{V_1}, \tag{6a}$$

$$X_2 = \frac{K_a E_2}{V_2}, \tag{6b}$$

and

$$X_3 = \frac{K_a E_3}{V_3}, \tag{6c}$$

where $E_1$, $E_2$, and $E_3$ are the expected returns of the three securities and $K_a$ is chosen to satisfy Equation 1. In our example, this is the portfolio

$$X_1 = 0.34,$$

$$X_2 = 0.50,$$

and

$$X_3 = 0.16.$$

It is the point labeled "*a*" in Figure 1.

If we continue to assume Budget Equation 1 as the only constraint, all points on the straight line through *a* and *c* minimize portfolio variance for various levels of portfolio expected return. However, not all these points are *efficient* portfolios. Efficient portfolios are those encountered if we start at *c* and move continuously in the direction of *a*,

FINANCIAL ANALYSTS JOURNAL®
Reflections

and beyond, without stop. As we move away from $c$ in this direction, portfolio expected return, $E_P$, and portfolio variance, $V_P$, increase. All portfolios encountered provide minimum $V_P$ for the given $E_P$—*or greater* $E_P$—among all portfolios that satisfy Budget Equation 1. In contrast, if we start at $c$ and move in the other direction, we do not encounter efficient portfolios (other than $c$) because $V_P$ increases but $E_P$ decreases. The same $V_P$ but greater $E_P$ can be found elsewhere on $\ell\ell'$.

Thus, in this example, if Budget Equation 1 is the only constraint, the set of efficient portfolios is the "ray" that starts at $c$ and moves in a straight line through $a$ and beyond.

As one moves on the line $\ell\ell'$ in the direction of $a$ and beyond, at some point the line $\ell\ell'$ leaves the triangle. In the present example, this is the point labeled "$b$" in Figure 1, with

$$X_1 = 0.58,$$

$$X_2 = 0.00,$$

and

$$X_3 = 0.42.$$

Portfolio $b$ still satisfies the constraints (Budget Equation 1 and Nonnegativity Requirement 2), but points beyond $b$ on the line $\ell\ell'$ no longer satisfy Nonnegativity Requirement 2 because they violate the requirement that $X_2 \geq 0$. Beyond point $b$, therefore, the efficient set when Budget Equation 1 *and* Nonnegativity Requirement 2 are required departs from the efficient set when Budget Equation 1 only is required.

At point $b$, investment in Security 2 is zero ($X_2 = 0$). For efficient portfolios with higher expected return, the efficient set moves along the horizontal edge of the triangle, from $b$ to (0,0), where an undiversified portfolio is invested only in Security 3 ($X_3 = 1$), the security with the highest expected return in the example.

We will see that, quite generally, a set of mean–variance efficient portfolios is "piecewise linear"; that is, it is made up of one or more straight-line segments that meet at points called "corner portfolios." When Equation 1 is the only constraint, the efficient set contains only one corner portfolio—namely, point $c$ in Figure 1—and only one line "segment"—namely, the segment that starts at $c$ and moves without end in the direction of increasing $E_p$. When nonnegativity constraints are imposed, the set of efficient portfolios typically has more than one segment and more than one corner portfolio. In Figure 1, this set of efficient portfolios consists of two line segments connecting three corner portfolios—$c$, $b$, and (0,0).

**The Two-Fund Separation Theorem.** The fact that two points determine a line is known in financial theory as the "two-fund separation theorem." In particular, all the portfolios on $\ell\ell'$ in Figure 1 can be obtained by (positive or negative) investments in portfolios $a$ and $c$ subject only to the constraint

$$X_a + X_c = 1, \tag{7}$$

where $X_a$ and $X_c$ are the "fractions" of the portfolio allocated to, respectively, subportfolios $a$ and $c$. Note that Equation 7 permits the investor to short one portfolio and use the proceeds to invest more than 100 percent in the other portfolio. If both $X_a$ and $X_c$ are positive, then the resulting portfolio lies within the interval connecting $a$ and $c$ in Figure 1. If $X_c$ is negative, then $X_a > 1$ and the resulting portfolio lies outside the interval, beyond $a$. Similarly, if $X_a < 0$ and $X_c > 1$, the portfolio lies outside the interval beyond $c$.

What is true in particular on $\ell\ell'$ is true in general for any two distinct points on any line in portfolio space. All points on the line can be represented by investments $X_a$ and $X_c$ in two distinct subportfolios on the line, where $X_a$ and $X_c$ satisfy Equation 7. I will use this relationship between points and lines several times.

**The Market Portfolio.** Consider a market in which investors must satisfy Budget Equation 1 and Nonnegativity Requirement 2. I show in the next section and in Appendix A that—in this case—beliefs about means, variances, and covariances that imply the efficient set in Figure 1 are consistent with market equilibrium.

Assume there are two types of investors in this market: cautious investors who select the portfolio at $d$ = (0.40, 0.37) in Figure 1 and aggressive investors who select the portfolio at $e$ = (0.20, 0.00). Similar conclusions would be reached if we specified two other portfolios as long as one of the portfolios were on one of the segments and the other portfolio were on the other segment of the efficient set. Similar conclusions would also be reached if there were more than two types of investors as long as some were on one segment and some on the other.

FINANCIAL ANALYSTS JOURNAL®
*Reflections*

According to the two-fund separation theorem, the market portfolio lies on the straight line connecting $d$ and $e$ [for example, at $M = (0.30, 0.19)$]. The market is efficient, in that each participant holds an efficient portfolio, but note that the *market portfolio, M,* is not an efficient portfolio. It is not on either segment of the efficient set when Budget Equation 1 and Nonnegativity Requirement 2 are the constraints (nor is it, incidentally, on the ray that is the efficient set when Budget Equation 1 only is the constraint).

**A Simple Market.** The preceding shows that if investors selected portfolios subject to the constraints of Budget Equation 1 and Nonnegativity Requirement 2, all held the beliefs in Table 1, and some preferred portfolios on one segment of the efficient set and others preferred a portfolio on the other, then the market portfolio would not be a mean–variance efficient portfolio. This section shows that means, variances, and covariances that imply Figure 1 are consistent with economic equilibrium when shorting and borrowing are unavailable.

Imagine an economy in which the inhabitants live on coconuts and the produce of their own gardens. The economy has three enterprises, namely, three coconut farms. Once a year, a market convenes to trade the shares of the three coconut farms. Each year, the resulting prices of shares turn out to be the same as those of preceding years because the number of people with given endowments and risk aversion is the same each year (perhaps because of overlapping generations rather than immortal participants). Thus, the only source of uncertainty of return is the dividend each stock pays during the year—which is the stock's pro rata share of the farm's production.

It is shown in Appendix A that means, variances and covariances of coconut production exist that imply the efficient set in Figure 1—or any other three-security efficient set that we cite. If we insist that coconut production be nonnegative, it may be necessary to add a constant to all expected returns (the same constant to each). Doing so will increase the expected returns of each portfolio but not change the set of efficient portfolios. It is then possible to find a probability distribution of coconut production, with production always nonnegative, for the given (slightly modified) means, variances, and covariances and, therefore, for the given set of efficient portfolios.

With such a probability distribution of returns, the market is rational, in the sense that each participant knows the true probability distribution of returns and each seeks and achieves mean–variance efficiency. Nevertheless, in contrast to the usual CAPM conclusion, the market portfolio is not an efficient portfolio. It follows that there is no representative investor, because no investor wants to hold the market portfolio. Also, as we will see in a subsequent section, expected returns are not linearly related to betas.

**Arbitrage.** Suppose that most investors are subject to Nonnegativity Requirement 2 but that one investor can short, in the CAPM sense—that is, is subject only to Budget Equation 1. (Perhaps the CAPM investor has surreptitious access to a vault containing stock certificates that he or she can "borrow" temporarily without posting collateral.) Would this CAPM investor arbitrage away the inefficiency in the market portfolio?

If there were a Portfolio $P$ on $\ell\ell'$ that beat Market Portfolio $M$ with certainty, then the CAPM investor could short any amount of $M$, use the proceeds to buy P and make an arbitrarily large gain with certainty. But $P$ does not beat $M$ with certainty; it simply offers a better probability distribution. In fact, the investor with Equation 1 as the only constraint is better off picking a point on $\ell\ell'$ and ignoring $M$. **Figure 2** illustrates this idea. If $P$ is any point *on* the line $\ell\ell'$ and $M$ is any point *off* the line $\ell\ell'$, then according to the two-fund separation theorem, the portfolio produced by shorting $M$ and using the proceeds (plus the original "\$1") to buy $P$ lies on the straight line connecting $M$ and $P$. Specifically, it lies on the far side from $M$, beyond $P$, such as $Q$ in Figure 2. But Portfolio $Q$ is not efficient for the investor with Equation 1 as the only constraint. Some Portfolio $R$ on $\ell\ell'$ (not shown in Figure 2) supplies a higher mean and lower variance.

Now that we have seen that an investor subject only to Equation 1 will choose a portfolio from $\ell\ell'$ without regard to the market portfolio, let us consider market equilibrium when some investors are subject to Equation 1 only and some to Equation 1 and Nonnegativity Requirement 2. Suppose that, as in Figure 1, the average holdings (weighted by investor wealth) of investors subject to Budget Equation 1 and Nonnegativity Requirement 2 is the point $M$. It would be the market portfolio if these were the only investors. Suppose further that the wealth-weighted average of the one or more investors subject only to Equation 1 is

 

FINANCIAL ANALYSTS JOURNAL®
Reflections

**Figure 2. Effect of Trying to Arbitrage an Inefficient Market Portfolio**

point $P$ in **Figure 3**. As in Figure 2, $P$ must lie on $\ell\ell'$, whereas $M$ typically lies off $\ell\ell'$. When both types of investors are present, the market portfolio lies on the straight line between the two averages, $M$ and $P$, such as point $M^a$ in Figure 3. The position of $M^a$ depends on the relative wealth of the two types of investors, but in any case, it is off $\ell\ell'$; therefore, it is not efficient for investors subject to Equation 1 only.

Whether it is efficient for investors subject to both Budget Equation 1 and Nonnegativity Requirement 2 is a more complicated story. The portfolios $M^a$, $M^b$, and $M^c$ lie on the straight line connecting $P$ and $M$. Portfolio $M^c$ cannot be a market equilibrium because it implies a negative total demand for Security 2. If $M^b$ is the market portfolio, then the market portfolio is efficient for investors with Budget Equation 1 and Nonnegativity Requirement 2 as constraints but there is zero net demand for shares of Security 2. For an equilibrium with positive net demand for all three securities, the market must be within the constraint triangle, as $M^a$ is in Figure 3. But such a combination of $M$ and $P$ is inefficient for investors subject to Budget Equation 1 and Nonnegativity Requirement 2, as well as for those subject only to Budget Equation 1.

**Expected Returns and Betas.** If Assumptions 1–4 or 1–4′ are true, then Conclusion 2 follows: Expected returns are linearly related to the betas of each security. That is, for some choice of numbers $a$ and $b$, the following three equations hold:

$$E_1 = a + b\beta_1, \tag{8a}$$

$$E_2 = a + b\beta_2, \tag{8b}$$

and

$$E_3 = a + b\beta_3, \tag{8c}$$

where $\beta_i$ is the coefficient of regression of the return on the $i$th security against the return on the market. But these equations do not necessarily hold if A1–A3 are true but neither A4 nor A4 is true.

In particular, C2 will typically not be true if investors satisfy A1–A3 but are subject to Equation 1 and Nonnegativity Requirement 2 as constraints. I will illustrate this statement in terms of the three-security example in Figure 3.

The first column of **Table 2** shows the fraction $P_i$ of security $i$ in Portfolio $P$ on the $\ell\ell'$ line in Figure 3. The second column states the covariance between each security and $P$. Given our current assumption that the returns on the three securities are uncorrelated, the covariance between security $i$ and Portfolio $P$ depends only on how much of the security is in the portfolio, and it is given by the formula

$$\operatorname{cov}(R_i,P) = P_iV_i, \text{ for } i = 1, 2, 3. \tag{9}$$

**Figure 3. Market Portfolios with and without Nonnegativity Constraints**

FINANCIAL ANALYSTS JOURNAL®
Reflections

**Table 2. Three Risky Securities in Portfolio *P* of Figure 3**

| Security | Percent in $P$ | $\text{cov}_{i,P} = P_i V_i$ | $\beta_{i,P}$ |
|---|---|---|---|
| 1 | 0.70% | 0.0227 | 0.52 |
| 2 | –0.25 | –0.0036 | –0.08 |
| 3 | 0.55 | 0.0495 | 1.12 |

*Note*: var($P$) = 0.0440; $\beta_{i,P} = \text{cov}_{i,P}/\text{var}(P)$.

**Table 3. Three Risky Securities in Market Portfolio *M* of Figure 3**

| Security | Percent in $M$ | $\text{cov}_{i,M} = M_i V_i$ | $\beta_{i,M}$ |
|---|---|---|---|
| 1 | 0.30% | 0.0097 | 0.36 |
| 2 | 0.19 | 0.0027 | 0.10 |
| 3 | 0.51 | 0.0459 | 1.71 |

*Note*: var($M$) = 0.0268; $\beta_{i,M} = \text{cov}_{i,M}/\text{var}(M)$.

The beta of any security return regressed against any Portfolio $P$ is defined to be

$$\beta_{i,P} = \frac{\text{cov}_{i,P}}{\text{var}(P)}. \tag{10}$$

These betas are listed in the last column of Table 2.

Similarly, **Table 3** shows the fraction held in Market Portfolio $M$, the covariance between each security and Portfolio $M$, and the beta of each security return regressed against the return on $M$, where $M$ in Figure 3 is also the market portfolio $M$ in Figure 1.

In **Figure 4**, the points labeled 1 vs. $P$, 2 vs. $P$, and 3 vs. $P$ show expected return on the vertical axis against $\beta_{i,P}$ plotted on the horizontal axis. The points labeled 1 vs. $M$, 2 vs. $M$, and 3 vs. $M$ show the same expected returns plotted against $\beta_{i,M}$. The three observations for each case are connected by lines. We see that the three points that represent expected returns and betas-versus-$P$ lie on a single straight line whereas the three points representing expected returns and betas-versus-$M$ do not lie on a straight line. The implication is that there is a linear relationship between expected returns and betas-versus-$P$ but no such relationship between expected returns and betas-versus-$M$. In other words, for some choice of $a$ and $b$, Equation 8 holds if the betas in Equation 8 are from regressions against $P$ but no such $a$ and $b$ exists when the betas are from regressions against $M$. More generally, if Market Portfolio $M$ is any point on $\ell\ell'$, then a linear relationship exists between expected return and beta. In contrast, if $M$ is any point off $\ell\ell'$, there is no such relationship (see Roll 1977; Markowitz 1987; Markowitz and Todd).

**Figure 4. Relationship between Expected Returns and Betas versus an Efficient and an Inefficient Market Portfolio**

 

FINANCIAL ANALYSTS JOURNAL®
Reflections

**Limited Borrowing.** In this section, I introduce a risk-free asset into the discussion. The Sharpe–Lintner CAPM assumes A1–A4 including unlimited borrowing at the risk-free rate. These assumptions imply that the market portfolio is a mean–variance efficient portfolio and that expected returns are linearly related to betas against the market portfolio. In this section, I illustrate that this conclusion no longer follows if borrowing is either not permitted or permitted but limited.

To illustrate this idea, the example in Table 1 is modified so that Security 3 now has 0 variance and a (risk-free) return of $r_0$ = 3 percent, as shown in **Table 4**. We continue to assume the budget constraint (Equation 1) and

$$X_1 \geq 0 \text{ and } X_2 \geq 0. \tag{11a}$$

$X_3 > 0$ represents lending at the risk-free rate; $X_3 < 0$ represents borrowing at the same rate. Prohibited borrowing would be represented by the constraint

$$X_3 \geq 0. \tag{11b}$$

Borrowing limited to, for example, the equity in the account would be represented by

$$X_3 \geq -1.0. \tag{11c}$$

Unlimited borrowing would be represented by the constraints of Budget Equation 1 and Nonnegativity Requirement 11a, with no constraint on $X_3$.

**Table 4. Expected Returns and Standard Deviations of Three Securities Including Cash**

| Security | Expected Return | Standard Deviation |
|---|---|---|
| 1 | 0.15% | 0.18% |
| 2 | 0.10 | 0.12 |
| 3 | 0.03 | 0.00 |

In **Figure 5**, as in Figure 1, the horizontal axis represents $X_1$, the fraction of the portfolio invested in Security 1; the vertical axis represents $X_2$, the fraction invested in Security 2. As before, $X_3$ is given implicitly by Equation 3. If borrowing is forbidden, then the set of feasible portfolios is, as before, on and in the triangle with vertices (0,0), (1,0) and (0,1). If no more than 100 percent borrowing is permitted, the set of feasible portfolios is the points on and in the triangle whose vertices are (0,0), (2,0) and (0,2). If unlimited borrowing is permitted, the set of feasible portfolios is the entire positive quadrant.

**Figure 5. Market Portfolio when Borrowing Permitted but Limited**

In our example assuming uncorrelated returns, when borrowing is unconstrained, the set of efficient portfolios is the set of portfolios that satisfies

$$X_1 = \frac{h(E_1 - r_0)}{V_1} \tag{12a}$$

and

$$X_2 = \frac{h(E_2 - r_0)}{V_2} \tag{12b}$$

for any zero or positive choice of $h$. This line is the ray that starts at the origin (0,0)—the all-cash portfolio—and proceeds into the positive quadrant along line $\ell\ell'$ in Figure 5 passing through the point (0.43, 0.57) for the example in Table 4. When borrowing is not limited, the efficient set proceeds along $\ell\ell'$ without bounds. If investors cannot borrow more than 100 percent of equity, then the efficient set cannot go beyond the line connecting (2,0) and (0,2); that is, it cannot go beyond point $b$ = (0.86, 1.14). From that point, under Nonnegativity Requirement 11c, the efficient set moves along the line connecting (0,2) and (2, 0) until it reaches the point (2,0), representing the portfolio that is 200 percent invested in the highest yielding security, namely, Security 1 in the present example.

Suppose some investors choose portfolio $d$ = (0.39, 0.51) on one segment in Figure 5 and all others choose portfolio $e$ = (1.40, 0.60) on the other segment.

FINANCIAL ANALYSTS JOURNAL®
Reflections

Then, "the market"—including cash or borrowing—is a point between them, such as $M'$. Portfolio $M$ is $M'$ "normalized" so that the "market portfolio" adds up to 100 percent. Neither $M$ nor $M'$ is an efficient portfolio. Nor is there a linear relationship between expected returns and betas regressed against either $M$ or $M'$. Such a relationship exists only if $M$ is on $\ell\ell'$. The fact that the market is inefficient implies that there is no representative investor. No rational investor holds either $M$ or $M$.

## Generalizations

Mean–variance efficient sets are computed in practice for models ranging in size from toy problems with two, three, or four assets to small problems with a dozen or so asset classes to large problems containing thousands of securities. To calculate an efficient frontier, the "critical line algorithm" (CLA) accepts as inputs any vector of expected return estimates, any matrix of covariance estimates (even a singular covariance matrix), and any linear equality or inequality constraints on the choice of portfolio (such as upper bounds on individual security holdings, sums of security holdings, or weighted sums of security holdings). From these inputs, the CLA produces a piecewise linear set of efficient portfolios. This set of portfolios "looks like" the ones in Figures 1–5, except that now the sets are difficult to draw because a portfolio in an analysis with 1,000 securities requires approximately a 1,000-dimensional space to be plotted. (When portfolio choice is subject to a budget constraint, a 999-dimensional space is sufficient.) Although we cannot plot points on a 999-dimensional blackboard, the basic mathematical properties of points, lines, and efficient sets in 999-dimensional space are the same as those in 2-dimensional space. The diagrams in Figures 1–5, which illustrate these properties, can help our intuition as to the nature of the properties in higher dimensional spaces.

One property of points and lines that is the same in 999-dimensional space as it is in 2-dimensional space is that two points determine a line. In particular, all portfolios that lie on a straight line in an any-dimensional space may be obtained by investing amounts $X_a$ and $X_c$ in Portfolios $P_a$ and $P_c$ on the line. $P_a$ and $P_c$ may be any two fixed, *different* portfolios on the line. As in the three-security case, $X_a$ and $X_c$ are subject to constraining Equation 7 and either may be negative. If $X_a$ is negative, then $X_c > 1.0$ and Point (Portfolio) $P$ obtained by allocating $X_a$ to $P_a$ and $X_c$ to $P_c$ lies outside the interval connecting $P_a$ and $P_c$, beyond $P_c$. The other cases—with $X_c < 0$ or with $X_a \geq 0$ and $X_c \geq 0$—are as described in the discussion of the two-fund separation theorem for the three-security case.

Suppose that there are $n$ securities (for $n = 3$ or 30 or 3,000), that not all expected returns are the same, and that the $n$ securities have a nonsingular covariance matrix. If the only constraint on the choice of portfolio is

$$\sum_{i=1}^{n} X_i = 1, \qquad (13)$$

then the portfolios that minimize portfolio variance $V_P$ for various values of portfolio expected return $E_P$ lie on a single straight line $\ell\ell'$ in $(n-1)$-dimensional portfolio space. Expected return increases as one moves in one direction on this line; decreases, in the other direction. The set of efficient portfolios in this case is the ray that starts at the $V_P$-minimizing portfolio and moves on $\ell\ell'$ in the direction of increasing $E_P$. Repeated use of the two-fund separation theorem shows that if all investors hold portfolios somewhere on this ray, the market portfolio will also be on this ray and, therefore, will also be efficient. Thus, the efficiency of the market portfolio when $n = 3$ and Equation 1 is the only constraint generalizes to any $n$ with Equation 13 as the only constraint.

Next, consider an investor subject to a no-shorting constraint:

$$X_i \geq 0, i = 1, \ldots, n \qquad (14)$$

as well as a budget constraint (Equation 13). For simplicity, assume that one security has the greatest expected return (albeit, perhaps, just slightly more than the second-greatest expected return). When Budget Equation 13 *and* Nonnegativity Requirement 14 are the constraints, and the only constraints, on portfolio choice, the critical line algorithm begins with the portfolio with highest expected return, namely, the portfolio that is 100 percent invested in the security with highest expected return. The CLA traces out the set of efficient portfolios, from top to bottom (i.e., from this portfolio with maximum expected return down to the portfolio with minimum variance). The computation proceeds in a series of iterations. Each iteration computes one piece (one linear segment) of the piecewise linear efficient set. Each successive segment has either one more or one less security than the preceding segment. If the analysis includes a risk-free asset (or, equivalently, risk-free lending), the last segment to be computed (the one with the lowest portfolio mean and variance) is the one and only segment that contains the risk-free asset (Tobin 1958).

This characterization of efficient sets remains true if limited borrowing is allowed, as illustrated in Figure 5. It also remains true when short selling is permitted but is subject to a Reg T or similar constraint (see Jacobs, Levy, and Markowitz 2005). In this case, if no other constraints are included (such as upper bounds on holdings), then short sales subject to Reg T can be modeled by an analysis with $2n + 3$ variables. The first $n$ variables represent long positions; the second $n$ variables represent short positions; and the final three variables represent, respectively, lending, borrowing, and slack in the Reg T constraint. These variables are subject to the following constraints:

$$\sum_{i=1}^{n} X_i + X_{2n+1} - X_{2n+2} = 1, \tag{15a}$$

$$\sum_{i=1}^{2n} X_i + X_{2n+3} = 2, \tag{15b}$$

and

$$X_i \geq 0, \text{ with } i = 1, \ldots, 2n + 3. \tag{15c}$$

The portfolio with maximum expected return typically contains two variables at positive levels (perhaps a short or long position plus borrowing). As in the case without short positions, CLA traces out the efficient frontier in a series of iterations—each iteration producing one piece of the piecewise linear efficient set, each piece having one more or (occasionally) one less nonzero variable than did the preceding piece.

A great variety of mean–variance efficient sets are computed in practice. For example, some are computed for asset classes; some of these results are then implemented by index funds. Other efficient set analyses are performed at the individual-security level. Among the latter, analyses differ as to which securities constitute "the universe" of securities from which the portfolio optimizer is to select for its portfolios. Some permit short positions; some do not.

For comparability with the classic CAPM, let us assume here that all investors perform their mean–variance analyses in terms of individual securities rather than asset classes, all use the same universe of "all marketable securities," and either all include short sales (subject to a Reg T–like constraint) or all exclude short sales.

Even so, there properly should be a variety of portfolio analyses generating a variety of frontiers. Because different institutions have different liability structures, they properly have different efficient sets of marketable securities. For example, an insurance company or pension fund, with liabilities determined outside the portfolio analysis, should choose portfolios that are efficient in terms of the mean and variance of assets minus liabilities. When different investors properly have different efficient sets, the question of whether the market portfolio is a mean–variance efficient portfolio raises the question: efficient for whom?

For comparability with the CAPM, let us assume that all investors may properly ignore their particular liability structure in computing the efficient frontier; each uses the same mean, variance, and covariance estimates for the same universe of marketable securities; and each is subject to the same constraints. In other words, we assume that they all generate and select portfolios from the same mean–variance efficient frontier.

In tracing out this frontier, CLA starts at the high end with an undiversified portfolio. It proceeds in a sequence of iterations that generate "lower" segments of the piecewise linear efficient frontier (i.e., segments with lower portfolio mean and lower portfolio variance). Each successive segment adds or deletes one security (or possibly a short position) to the list of active securities. Thus, if the universe consists of, say, 10,000 securities, then if all securities are to be demanded by someone, this universal efficient frontier must contain at least 10,000 segments. If investors have sufficiently diverse risk tolerances, they will choose portfolios on different segments. Some will prefer portfolios on one or another of the typically less diversified high-risk/high-return segments. Others will select portfolios on one or another of the typically more diversified lower-risk segments. The market is an average, weighted by investor wealth, of portfolios selected from these diverse segments. Although it is mathematically possible for this average to accidentally fall on the efficient frontier, such an outcome is extremely unlikely.

Thus, in this world that is like the CAPM but has realistic constraints, the market portfolio is typically not an efficient portfolio. Therefore, there is no representative investor and expected return is not a linear function of regressions of security returns against the market.

FINANCIAL ANALYSTS JOURNAL®
*Reflections*

## So What?

This section presents some implications of the preceding analysis.

**So What #1.** A frequent explanation of why observed expected returns do not appear to be linearly related to betas is that the measures of market return used in the tests do not measure the true, universal market portfolio that appears in the CAPM. The conclusion is that to test the CAPM, we need to measure returns on a cap-weighted world portfolio. The preceding discussion implies, however, that before spending vast resources on ever finer approximations to returns on this cap-weighted universal portfolio, we should note that CAPM Conclusion 2 (that expected returns are linearly related to betas) is not likely to be true if real-world constraints are substituted for Assumption 4 or Assumption 4′.

**So What #2.** Traditionally, some investments were thought of as businessmen's risks while others were thought appropriate for widows and orphans. The CAPM, assuming A1–A4, concludes that one and only one portfolio is efficient for all investors. The only difference should be the amount of cash or borrowing with which the portfolio is combined. In contrast, when borrowing is limited and short sales are prohibited or subject to real-world constraints, the composition of the portfolio of risky securities changes radically from one end to the other of the efficient frontier. At the high end, it contains few securities, usually with a predominance of those with high expected return. At the low end, it tends to be more diversified, with a more-than-proportional presence of the less volatile securities. In other words, the high end of the frontier will indeed tend to be dominated by businessman-risk securities; whereas the low end, although perhaps spiced up and diversified with some more volatile securities, will typically have more than its proportionate share of widow-and-orphan securities.

**So What #3.** The linear relationship between expected returns and betas (against the market portfolio return) that is implied by the CAPM is the basis for a standard "risk-adjustment" calculation. This calculation is used, for example, to determine which of two projects that a company might pursue would best enhance its stock market value or which of two securities, groups of securities, or investment strategies has performed best. Because the existence of a linear relationship between expected returns and betas is questionable, the reliability of its use in risk adjustment must be questioned.

It might seem at first that the use of the CAPM risk-adjustment formula is indispensable for decisions like those I just described because there is no alternative. This is not the case. In particular, concerning the desirability of an asset class with a particular return pattern, a frequent practice now is to run an efficient frontier with and without the asset class. (This practice is subject to the essential caveat that future returns are not necessarily like those of the past, but the CAPM adjustment is subject to this same caveat.) The comparison of frontiers with and without the asset class avoids Assumptions 4 and 4′ and Conclusion 2.

Concerning the choice between two projects, I previously considered their effect on a company's stock price under the assumption that the stock appears in some but not all segments of investors' efficient frontiers (Markowitz 1990). The resulting computation is similar to that of the CAPM but involves only investors who own the company's stock. In other words, the calculation takes into account the company's clientele. For estimating the effects of investment policy in a dynamic world with mean–variance investors holding different beliefs and real-world constraints, Jacobs et al. (2004) proposed detailed, asynchronous simulation. Potentially, the simulated market could also include investors other than those with mean–variance objectives.[3] In sum, the position that "there is no alternative" to the CAPM for risk-adjustment calculations was never completely true and is certainly not true now.

**So What #4.** The implications of the CAPM are taught to MBA students and CFA charterholders. The lack of realism in A4 and A4′ is rarely pointed out, and the consequences of replacing these assumptions with more realistic assumptions are rarely (if ever) discussed. Worse, often the distinction between the CAPM and mean–variance analysis is confused. Not only do some say or suggest that if investors use mean–variance analysis, C1 and C2 will follow; some say or suggest that if an investor uses mean–variance analysis, C2 should be assumed to hold among inputs.

Despite its drawbacks as illustrated here, the CAPM should be taught. It is like studying the motion of objects on Earth under the assumption that the Earth has no air. The calculations and results are much simpler if this assumption is made. But at some point, the obvious fact that, on Earth, cannon balls and feathers do not fall at the same rate should be noted and explained to some extent. Similarly, at some point, the finance student should be shown the effect of replacing A4 or A4′ with more realistic constraints and the "so what" should be explained.

About 30 years ago, Fama (1976), in Chapter 8, explained the main points of the present article: that A4 or A4′ are not realistic and that, if more realistic assumptions are substituted, C1 and C2 no longer follow. The two principal differences between Fama's presentation then and the current presentation are (1) my use of certain ("portfolio space") diagrams to illustrate the source and possible extent of the market portfolio inefficiency and (2) our respective conclusions concerning "so what." Fama's conclusion at the time was that what one could say about models with more realistic versions of A4 or A4′ was that they "fall substantially short of interesting and testable propositions about the nature capital market equilibrium. For such propositions, we have to rely on" the CAPM (p. 305). My own conclusion is that it is time to move on.

## Conclusion

The CAPM is a thing of beauty. Thanks to one or another counterfactual assumption, it achieves clean and simple conclusions. Sharpe did not claim that investors can, in fact, borrow all they want at the risk-free rate. Rather, he argued:

> In order to derive conditions for equilibrium in the capital market we invoke two assumptions. First, we assume a common pure rate of interest, with all investors able to borrow [without limit] or lend funds on equal terms. Second, we assume homogeneity of investor expectations. Needless to say, these are highly restrictive and undoubtedly unrealistic assumptions. However, since the proper test of a theory is not the realism of its assumptions but the acceptability of its implications, and since these assumptions imply equilibrium conditions which form a major part of classical financial doctrine, it is far from clear that this formulation should be rejected—especially in view of the dearth of alternative models leading to similar results. (pp. 433–434)

Now, 40 years later, in the face of the empirical problems with the implications of the model, we should be cognizant of the consequences of varying its convenient but unrealistic assumptions. In particular, we should be cognizant of what more realistic assumptions concerning investment constraints imply about how we should invest, value assets, and adjust for risk.

## Appendix A. Finding a Probability Distribution for a Given Efficient Set

To construct a probability distribution of coconut production whose means, variances, and covariances imply a specific three-security efficient set, you may proceed as follows. The simplest distribution to construct with the requisite efficient set is a finite population with $S$ equally likely sample points, $s = 1, \ldots, S$, with $r_i^s$ as the return on security $i$ if sample point $s$ occurs. The procedure is as follows:

First, use the procedure described in Chapter 11 of Markowitz (1987) or Markowitz and Todd to produce an expected return vector, $\boldsymbol{\mu}$, and a covariance matrix, $\mathbf{C}$, that gives rise to the specified efficient set. (There are always many $\boldsymbol{\mu}$'s and $\mathbf{C}$'s that will serve. Start with any.) This step is not necessary if a $\boldsymbol{\mu}$ and a $\mathbf{C}$ are already given, as in the example in the text.

Second, by using a program that finds the eigenvalues and eigenroots of $\mathbf{C}$, you can find a matrix $\mathbf{B}$ such that $\mathbf{C} = \mathbf{B'B}$.[4]

Third, let $\mathbf{R}^a$ be a matrix containing a finite sample space for three random variables with 0 mean and covariance matrix $\mathbf{I}$. For example,

$$\mathbf{R}^a = \frac{\sqrt{2}}{4}\begin{pmatrix} 1 & 1 & 1 & 1 & -1 & -1 & -1 & -1 \\ 1 & 1 & -1 & -1 & 1 & 1 & -1 & -1 \\ 1 & -1 & 1 & -1 & 1 & -1 & 1 & -1 \end{pmatrix}$$
$$= r_{i,j}^a,$$

where $r_{i,j}^a$ is the value of the $i$th random variable in state (sample point) $j$.

Then, $\mathbf{R}^b = \mathbf{BR}^a$ is the matrix of a sample space of three random variables with 0 mean and covariance matrix $\mathbf{C}$. And $\mathbf{R}^c = \left(r_{i,j}^b + \mu_i\right)$ has covariance $\mathbf{C}$ and expected return $\mu$. If $\mathbf{R}^c$ has any negative entries and if $k$ is the magnitude of the largest in magnitude negative $r_{i,j}^c$, then $\mathbf{R}^d = (r_{i,j}^c + k)$ is the matrix of a sample space of hypothetical coconut production with nonnegative output and with the specified efficient set.

FINANCIAL ANALYSTS JOURNAL®
Reflections
60 YEARS 1945-2005

## Notes

1. Some conclusions remain unchanged if we assume heterogeneous rather than homogeneous beliefs; other conclusions apply to average predictions rather than unique predictions.
2. A3 asserts that the market is strong-form efficient in the Fama (1970) taxonomy. Thus, what I will show is that, even if the market is strong-form efficient, the market portfolio is not necessarily a mean–variance efficient portfolio.
3. Simulation analysis as presented by Jacobs et al. (2004) would hardly have been feasible in 1964 when Sharpe presented the CAPM. Computer and software development since that time makes such simulation quite manageable.
4. For example, see Franklin (2000). The formula is a corollary of Section 4.7, Theorem 3. Also, see Section 7.3, Equation 21 for an alternative factorization of **C**.

## References

Fama, Eugene F. 1970. "Efficient Capital Markets: A Review of Theory and Empirical Work." *Journal of Finance*, vol. 25, no. 2 (May):383–417.

———. 1976. *Foundations of Finance: Portfolio Decisions and Securities Prices.*" New York: Basic Books.

Franklin, Joel N. 2000. *Matrix Theory*. Mineola, NY: Dover Publications.

Jacobs, Bruce I., Kenneth N. Levy, and Harry M. Markowitz. 2004. "Financial Market Simulation." *Journal of Portfolio Management* (30th Anniversary):142–152.

———. Forthcoming 2005. "Portfolio Optimization with Factors, Scenarios, and Realistic Short Positions." *Operations Research.*

Lintner, John. 1965. "The Valuation of Risk Assets and the Selection of Risky Investments in Stock Portfolios and Capital Budgets." *Review of Economics and Statistics*, vol. 47, no. 1 (February):13–37.

Markowitz, Harry M. 1987. *Mean–Variance Analysis in Portfolio Choice and Capital Markets.* Oxford, U.K.: Basil Blackwell.

———. 1990. "Risk Adjustment." *Journal of Accounting, Auditing and Finance*, vol. 5 (Winter/Spring):213–225.

Markowitz, Harry M., and Peter Todd. 2000. *Mean–Variance Analysis in Portfolio Choice and Capital Markets* (revised reissue with chapter by Peter Todd). New Hope, PA: Frank J. Fabozzi Associates.

Roll, Richard. 1977. "A Critique of the Asset Pricing Theory's Tests, Part I: On Past and Potential Testability of the Theory." *Journal of Financial Economics*, vol. 4, no. 2 (March):129–176.

Sharpe, William F. 1964. "Capital Asset Prices: A Theory of Market Equilibrium under Conditions of Risk." *Journal of Finance*, vol. 14, no. 3 (September):425–441.

Tobin, J. 1958. "Liquidity Preference as Behavior towards Risk." *Review of Economic Studies*, vol. 25, no. 2 (February):65–86.

 

Date Created/Modified:
8/3/05

**CFA**
Nov/Dec FAJ
Markowitz

**Figure 1**

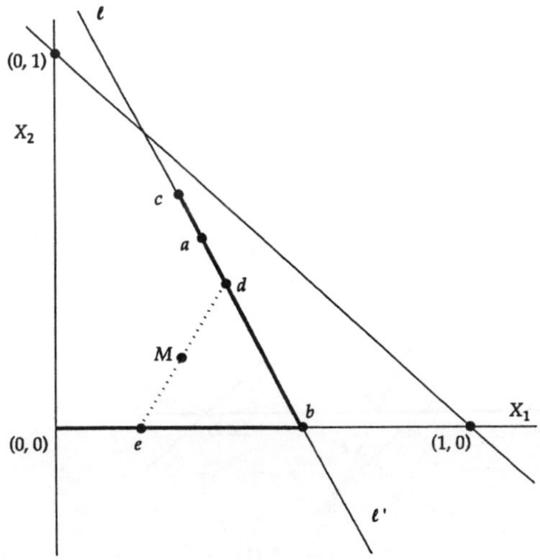

Date Created/Modified:
8/3/05

**CFA**
Nov/Dec FAJ
Markowitz

**Figure 2**

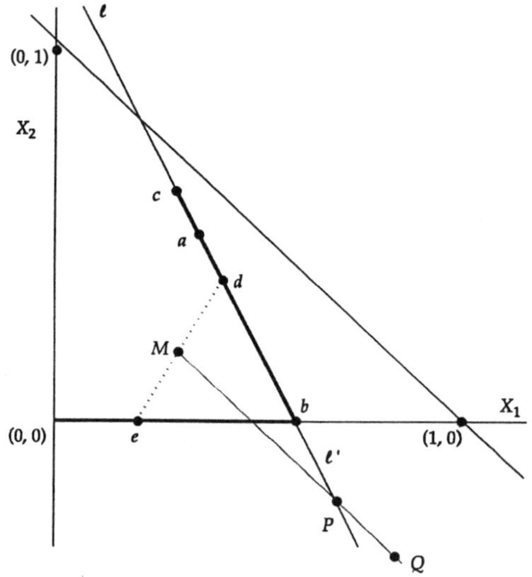

Date Created/Modified:
8/3/05

**CFA**
Nov/Dec FAJ
Markowitz

**Figure 3**

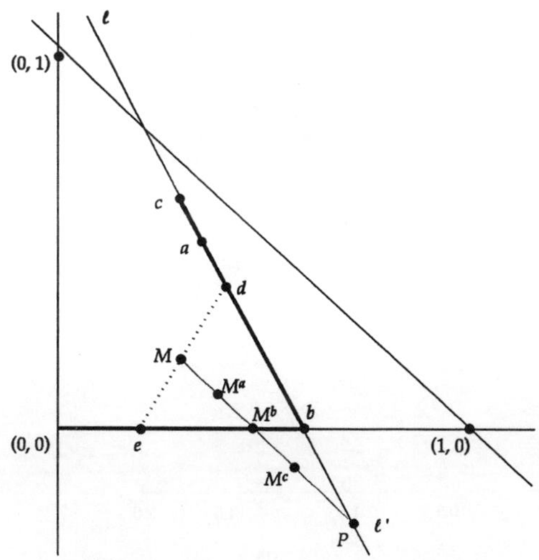

Date Created / Modified:
8/3/05

**CFA**
Nov / Dec FAJ
Markowitz

**Figure 4**

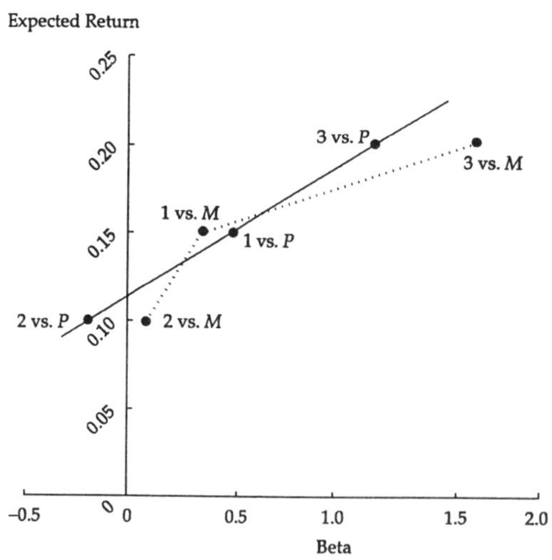

Date Created / Modified:
8/3/05

**CFA**
Nov / Dec FAJ
Markowitz

**Figure 5**

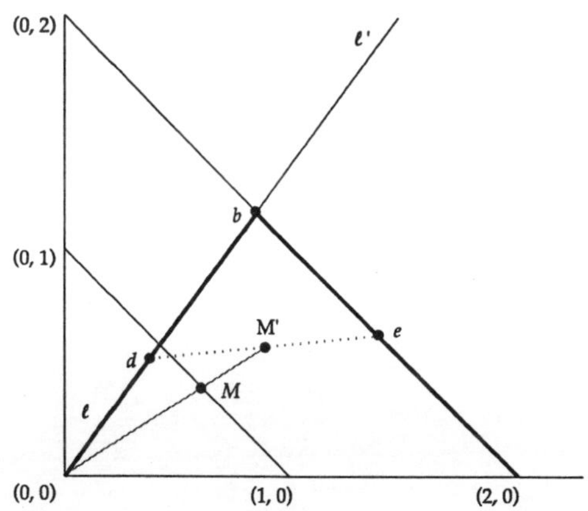

# EFFICIENT PORTFOLIOS, SPARSE MATRICES, AND ENTITIES: A RETROSPECTIVE

HARRY M. MARKOWITZ

*1010 Turquoise Street, Suite 245, San Diego, California 92109*

In 1989 I was pleased and honored to be awarded the ORSA/TIMS (now INFORMS) John von Neumann Theory Prize for my work in portfolio theory, sparse matrices, and SIMSCRIPT. The following is a retrospective on my work in these fields.

## PORTFOLIO THEORY

**The Epiphany.** One day in 1950, in the library of the Business School of the University of Chicago, I was checking out the possibility of writing my Ph.D. dissertation for the Economics Department in some area which applied mathematical or statistical techniques to "the stock market." My adviser, Professor Jacob Marschak, had sent me to Professor Marshall Ketchum in the Business School for a reading list. I had taken no courses in finance but, following Professor Ketchum's reading list, I had worked my way through Graham and Dodd's (1951) *Security Analysis*, had examined Wiesenberger's (1941) survey of *Investment Companies*, and now began reading John Burr Williams (1938) *The Theory of Investment Value*.

Williams asserted that the value of a stock is the expected present value of its future dividends. It struck me that if the investor, or investing institution, was concerned only with the expected value of a stock, it must also be only concerned with the expected value of its portfolio-as-a-whole. But maximizing the expected value of a portfolio requires investment in only one stock. Diversification is a common practice, as I saw in Wiesenberger's survey of investment companies, and makes sense since, as everyone knew even then, one should "not put all one's eggs in one basket." Clearly, investors diversify to avoid risk. What was missing from Williams' analysis was the notion of the risk of the portfolio-as-a-whole. An obvious measure which came to mind was variance or, equivalently, standard deviation. A check with a probability book on the library shelf showed that the variance of a weighted sum of random variables (e.g., the return on a portfolio of securities) involves the covariances or correlations among the random variables, as well as the weights and individual variances. This made good sense. It said that the riskiness of the portfolio had to do not only with the riskiness of the individual securities therein, but also to what extent they moved up and down together.

Since there were two quantities involved, I drew a graph with risk (standard deviation) on one axis and return (i.e., expected return) on the other. I was currently taking T. J. Koopmans' course on "activity analysis" in which he distinguished between efficient and inefficient combinations of production activities (Koopmans 1951). I accordingly labeled as efficient those combinations of risk and return which were not dominated by some other such combination.

The above thought process occurred while I was still in the middle of reading the Williams' book. Later in the book Williams deals with the question of risk essentially by advising the investor to invest in a large number of securities, presumably all with roughly maximum expected return. I felt that the theory which I had now sketched was more satisfactory than that which Williams presented, since he made no mention of avoiding covariance, combining securities efficiently into portfolios, or the risk-return trade-off of the portfolio-as-a-whole. I proposed to the Economics Department that I write a dissertation on "portfolio selection," and this was accepted.

The inputs to a portfolio analysis consist of the means and variances of individual securities and correlations between these securities. The outputs consist of efficient risk-return combinations and the portfolios which give rise to them. I assumed that it was the job of the security analyst to provide the inputs to the portfolio analysis. I took it as my task to figure out how to derive efficient sets from these inputs. It turns out that the set of efficient portfolios is piecewise linear, consisting of a finite number of pieces. Thus the entire efficient frontier can be written down by characterizing these finite number of pieces. The critical line algorithm for tracing out the efficient frontier is presented in Markowitz (1956). The basic tool used is the Kuhn-Tucker (1951) theorem. Markowitz (1959) Appendix A, shows that the algorithm works even if the covariance matrix is singular. Markowitz (1952) presented a geometric description of the set of efficient portfolios. Perold (1984), Markowitz (1987), and Markowitz and Todd (2000) discuss the shapes, properties, and computation of mean-variance efficient sets.

*Subject classifications:* Finance, portfolio: origins of portfolio theory. Professional: comments on. Programming, linear: sparse matrices. Simulation, languages: SIMSCRIPT.
*Area of review:* ANNIVERSARY ISSUE (SPECIAL).

**Operations Research** © 2002 INFORMS
Vol. 50, No. 1, January–February 2002, pp. 154–160

0030-364X/02/5001-0154 $05.00
1526-5463 electronic ISSN

**Dantzig's Influence.** I left Chicago in the Fall of 1951 for the RAND Corporation in Santa Monica with my course work finished but my dissertation still to be written. It was my good fortune to have George Dantzig join RAND about a year after I got there. George was the focal point of those doing work on linear programming. In the next section I will discuss my one noteworthy contribution to this area. In this subsection I note ways in which Dantzig's work contributed to the solution of the portfolio selection problem.

Markowitz (1956) defines the portfolio selection problem as that of finding mean-variance efficient portfolios subject to linear equality and inequality constraints. This is the same constraint set as that of linear programming, but with mean-variance efficiency rather than the optimization of a linear function as the objective. The portfolio with maximum expected return, when it exists, is the natural starting point in tracing out the set of efficient portfolios. Since expected return is a linear function of portfolio investments, finding the portfolio with maximum expected return is a linear-programming problem. Dantzig's simplex algorithm not only provides the solution to this problem, but also provides the critical line algorithm with various other services. In particular, it determines whether or not the constraint set specified by the analyst is feasible, whether or not feasible portfolio expected return is bounded and, if the model is rank deficient, it provides an equivalent model which is not rank deficient.

Along any piece of the set of efficient portfolios, security holdings $X_i$, $i = 1, \ldots, n$, vary linearly. Also the $\eta_i$ (the partial of a Lagrangian with respect to the $X_i$) vary linearly. The algorithm knows when a corner portfolio is reached, where one efficient segment of the efficient set meets the next segment of the efficient set, by determining which event happens first: an $X_i \downarrow 0$ for an IN variable or an $\eta_i \downarrow 0$ for an OUT variable. (IN variables may vary on a critical line; OUT variables may not.) But what should be done if two or more of these variables go to zero simultaneously? Usually it is sufficient to break ties arbitrarily, but conceivably this could get one into trouble. The solution to this problem presented in Markowitz (1956) is an adaptation of the Dantzig solution to a comparable problem for the simplex algorithm.

**Wolfe's Generalization.** My work at the RAND Corporation did not include "portfolio analysis." But no one objected to my taking the time to write my 1952 and 1956 articles. I submitted the latter to the *Naval Research Logistics Quarterly* edited by Alan J. Hoffman. Elsewhere, Phil Wolfe had been working on the quadratic-programming problem, to minimize a quadratic function ($Q - \lambda L$, Q positive semidefinite, L linear) subject to linear constraints. Wolfe also submitted his work to *NRLQ*. Hoffman sent Wolfe's paper to me and my paper to Wolfe for refereeing. We both recommended that the other paper be published, and both were. As a by-product of tracing out the efficient frontier, the critical line algorithm minimizes $Q - \lambda L$ (for variance Q and expected return L) for all $\lambda \geqslant 0$. Thus the critical line algorithm is , incidentally, a quadratic-programming algorithm. It struck Phil Wolfe that the critical line algorithm solves the quadratic-programming problem in a sequence of steps which are precisely the same as the steps by which the simplex algorithm solves the linear-programming problem, with one exception. The variables of the quadratic program come in pairs $X_i$, $\eta_i$. When one of these pairs is IN the linear programming "basis," the other is OUT. Wolfe thus defined quadratic programming as an example of linear complimentarity programming. At first it seemed that the practical use of this observation was to easily convert a linear-programming code into a quadratic-programming (or portfolio selection) code. Subsequently, it was found that other problems satisfied the linear complimentarity format, e.g., non-zero-sum games (Lemke 1965).

**Tobin's Invitation.** While at the University of Chicago, I was a student member of the Cowles Commission for Research in Economics under the guidance of its head, Tjalling Koopmans, and its former head Jacob Marschak. Former members of this relatively small organization have included Nobel Laureates Kenneth Arrow, Gerard Debreu, Lawrence Klein, Tjalling Koopmans, Harry Markowitz, Franco Modigliani, James Tobin, and other leaders in their fields. See Arrow et al. (1991).

At RAND in 1954 or 1955 I received a call from Professor James Tobin of Yale who explained that Tjalling Koopmans had decided that it was time for him to give up the administrative responsibilities of being head of the Cowles Commission. The search for a suitable new head of Cowles had the following result: The Cowles *Commission* was to move from Chicago to Yale, change its name from "Cowles Commission for Research in Economics at *Chicago*" to "The Cowles *Foundation* for Research in Economics at *Yale*." Professor Tobin would become its new head. Tobin invited me to spend the 1955/56 academic year at Yale writing a Cowles Foundation monograph on portfolio theory. A draft was finished during this period, subsequently reviewed by Tobin and Debreu with helpful suggestions from each, was revised and finally appeared as Markowitz (1959).

The 1955/56 academic year provided me the opportunity to focus on writing about portfolio theory and filling in some gaps in the theory as I saw them. In particular, Markowitz (1959) includes observations on how one or another model of covariance can be used in lieu of estimating individual correlation coefficients, as well as the definition of and computational procedures for using what I called "semivariance" (now sometimes referred to as "downside risk") in lieu of variance in risk-return analysis.

A principal preoccupation during this period was to reconcile portfolio theory with the theory of rational behavior under uncertainty as developed by von Neumann and Morgenstern (1944), Leonard J. Savage (1954), and others. Specifically, the problem was to reconcile the use of single-period mean-variance analysis by (or on behalf of) an investor who should maximize a many-period utility function. My answer lay in the observation that for many utility functions and for probability distributions of portfolio returns "like" those observed in fact, one can closely approximate expected value of the (Bellman 1957 "derived") utility function knowing only the mean and variance of the distribution. For details see Markowitz (1959) Part IV, Levy and Markowitz (1979), and Hlawitschka (1994); also Young and Trent (1969), Dexter et al. (1980), Pulley (1981, 1983), Kroll et al. (1984), Markowitz et al. (1994). Distinguish this view from the view that a Gaussian return distribution or a quadratic utility function is required for the use of mean-variance analysis.

**After 1959.** With the publication of Markowitz (1959) I had said what I had to say about portfolio theory, and published nothing further in the field for quite some time. My last substantial contribution to the early development of portfolio theory was to advise a young colleague at the RAND Corporation who was considering writing his dissertation (for UCLA) on portfolio theory. This led to his first publication: Sharpe (1963).

## SPARSE MATRICES

**Linear Programming at RAND.** Shortly after I arrived at RAND I was approached by a small team of economists who wanted to apply linear programming to some RAND problem. They asked me to read Dantzig (1951) and supervise the programming and running of RAND's first simplex code. The programmer assigned to the project was Clifford Shaw, who later participated in the pioneering work on artificial intelligence with Herbert Simon and Alan Newell. The linear-programming problem posed by my economist colleagues was small by modern standards, with perhaps about 30 or 40 equations. If I recall correctly, the computer was called a "card-programmed calculator." I do recall Cliff Shaw often telling me that we did another iteration or two yesterday. He expressed optimism that some day we could do four iterations per day. I said that I would believe it when I saw it. We finally obtained an optimum solution.

George Dantzig arrived at RAND in 1952, about a year after I did. Linear programming and related technology made rapid advances during the 1950s, partly due to increased computer speeds and size and, in equal part at least, due to improved algorithms. See Dantzig (1963) for details.

**Process Analysis and Sparse Matrices.** Dantzig, Ford, Fulkerson, and Johnson were in RAND's math department; Alan Manne and I were in the economics department elsewhere in the RAND building in Santa Monica. Alan and I together with others, including Thomas Marschak at RAND and Tibor Fabian and Alan Rowe at UCLA, became interested in building industry-wide and multi-industry "process analysis" models of economic capability. Our objectives were similar to that of the Leontief (1951) "input-output" model, but our assumptions and methods were different. See Manne and Markowitz (1963) or Markowitz (1954).

Our models were constrained by the then-current size limitations on general LP models (about 200 equations max). It struck me that the constraint matrices for the models we had built, or were likely to build, consisted of mostly zero entries; that one could solve a modest-size system of equations by hand, without great difficulty, using Gaussian elimination if the matrix contained mostly zeros, and one carefully picked pivot elements so as to fill in as few as possible non-zeros when eliminating the row and column of the pivot; and perhaps this could be used to solve large linear programs with relatively few non-zero coefficients. I dubbed these "sparse matrices."

My solution to how to use sparsity in linear programming was based, in two ways, on the "product form of inverse" which Dantzig (1963) ascribes to a suggestion of Alex Orden. First, the product form illustrated that if one wanted to solve systems of equations with the same matrix and different right-hand sides, it was not necessary to actually have "the" inverse matrix. It was sufficient, and sometimes more convenient, to store the coefficients of a sequence of linear transformations which would transform any RHS to the solution of the equations. I viewed my procedure as developing a sparse such sequence of transformations which I called the "elimination form of the inverse." Second, in the simplex algorithm you do not solve several systems of equations with the same ("basis") matrix. Rather, you solve a system of equations using a given basis matrix $A_1$, then a system using its transpose $A_1'$, then a system of equations with matrix $A_2$ that is identical to $A_1$ except for one altered column, etc. The purpose of the product form was to allow one to economically compute $(A_2^{-1}b)$ if one had $A_1^{-1}$. In particular, if one has a sparse constraint matrix one need not reinvert the basis matrix every iteration (or only solve specific equations without storing the inverse transformations), but can reinvert the basis now and then when the accumulated product form transformations grow too large.

I presented my sparse matrix methods in Markowitz (1957), including the results of an implementation programmed by William Orchard-Hayes. After that I lost track of what happened to sparse matrices until I visited IBM Research in Yorktown Heights, NY sometime in the early 1970s. Alan Hoffman told me that there had recently been a *second* conference there on sparse matrices, and that there was considerable work in the area (Willoughby 1969, Rose and Willoughby 1972; later, Duff 1981). I learned recently that the "Markowitz rule," for choosing pivots so as to keep the remaining matrix sparse after the Gaussian elimination step, is still used in at least one large production code for solving sparse matrix systems.

## ENTITIES, ATTRIBUTES, SETS, AND EVENTS

**SIMSCRIPT [I].** What we now refer to as SIMSCRIPT I (then referred to simply as SIMSCRIPT) was developed at the beginning of the 1960s at the RAND Corporation and made available without charge through the SHARE organization, see Markowitz et al. (1963). SIMSCRIPT was designed to facilitate the programming of "discrete event" simulation models, especially "asynchronous" discrete event simulators, as compared to continuous time or difference equation models.

The objective of SIMSCRIPT was to allow the simulation programmer to describe the world to be simulated, and relieve said programmer from implementation details insofar as we could. The SIMSCRIPT world view is as follows: As of an instant in time the system to be simulated has a *status* that changes at points in time called *events*. Status is described in terms of how many of various types of *entities* exist, what are the values of their *attributes*, and what entities belong to the *sets* which other entities own. Early 21st Century programming languages are likely to refer to Entities, Attributes, and Sets as Objects, Properties, and Collections (or Child-Parent relationships). Programming languages at the beginning of the 1960s spoke instead of variables and arrays.

The SIMSCRIPT [I] programmer described the entities, attributes, and sets of the system to be simulated on a Definition Form. In those days, the computer input was typically the punched card. The data written on the Definition Form, to be keypunched and placed in the SIMSCRIPT source program deck, included names of entity types; names of attributes, their data types, and precision information; the names of sets plus information as to what type of entity owns the set, what type belongs to it, and how the set is organized, e.g., FIFO, LIFO, or RANKED by one or more attributes of the members.

Changes in status were described in event routines written in the SIMSCRIPT programming language. The language included commands to CREATE and DESTROY entities, FILE entities into or REMOVE them from sets, FIND set members meeting specified tests, DO some action(s) FOR EACH member of sets, CAUSE or CANCEL subsequent event occurrences, etc., as well as perform arithmetic operations on attributes. We sought to make the commands English-like, "self-documenting." For example, to take specified actions on a set, like the jobs in the queue of some machine group, MG, one wrote

```
FOR EACH JOB IN QUEUE(MG)
        ⋮
REPEAT
```

where QUEUE(MG) is read "Queue of MG" just as $f(x)$ is read "$f$ of $x$." In addition to the Definition Form and event routine language, SIMSCRIPT [I] had an Initialization Form for describing the initial status of the system and a Report Generator Form which provided a WYSIWYG output specification.

**Prelude to SIMSCRIPT.** SIMSCRIPT'S Entity, Attribute, Set, and Event view evolved gradually as the result of a great deal of simulation programming.

When I joined RAND in 1951, computer simulation was being used to evaluate bomber attrition given various offense-defense configurations, and other "war-game" situations. During the early to mid-1950s I sat in on discussions at UCLA of a group, including Alan J. Rowe and headed by Melvin Salveson, that sought to apply advanced analysis techniques to manufacturing planning. The consensus was that shop-wide and larger-scale manufacturing analysis was not amenable to analytic solution or optimization algorithms. Simulation analysis was required. Later in the 1950s, RAND created a Logistics Laboratory. Its first major project, LP1, was a man-machine simulation in which actual air force logistics officers played the role of air force logistics officers. The computer flew simulated missions, generated part failures and other maintenance requirements, and kept track of parts supplies and aircraft status. My job in LP1 was to coordinate the programming of the computer models.

Some time after LP1 was finished I got an offer I couldn't refuse from the General Electric Company, eventually moving to its Manufacturing Services at GE Headquarters in New York City. Alan Rowe was already there and had just supervised the programming of a large, detailed job shop simulator programmed in assembly language. Alan designed this simulator with a particular GE shop as an initial application, but with many input parameters whose specification were to tailor the model to other shops. But when Alan went to apply the model to a next GE shop it turned out not to be as general purpose as hoped.

Upon reviewing Alan's experience my own theory at the time was to reduce programming time and increase flexibility by building the next big shop simulator in FORTRAN II, constructing it as a system of reusable subroutines. I soon had the opportunity to test this theory. Together with a team of GE Transformer Department manufacturing engineers, and Mort Allen who programmed the model, I participated in the development of GEMS, the General Electric Manufacturing Simulator.

GEMS was well received within General Electric. Manufacturing Services conducted internal GE seminars on MSS (Manufacturing Systems Simulation) with GEMS as a principal topic, and we soon got a request to simulate another GE shop. In the process of building a simulator for this next GE shop it became apparent that GEMS was not as flexible as I had hoped. As to GEMS' reusable subroutines, the routines which proved most reusable performed basic actions like linking jobs into queues, or events into the calendar of coming events.

My next hypothesis was that such facilities could be placed at the disposal of the simulation programmer more conveniently as part of a simulation language rather than as subroutines. I didn't want to write this simulation language at General Electric, since it seemed (given my situation within GE at the time) that it might be deemed proprietary, for internal GE use only, and I wanted to see the ideas disseminated. I went into the job market seeking a new home for me and my nascent simulation language and ended up returning to RAND.

I search my memories but cannot recall when and under what circumstances—between the time of reviewing GEMS' experience and the time of designing SIMSCRIPT forms and commands—that the SIMSCRIPT mantra emerged, that "the world consists of entities, attributes and sets and changes with events."

**SIMSCRIPT I.5.** The SIMSCRIPT [I] language was implemented as a preprocessor into FORTRAN. Bernie Hausner programmed the preprocessor; Herb Karr wrote the programming manual (Markowitz et al. 1963). The three of us jointly designed the fine details of the language.

In 1963 Herb persuaded me to join him in forming California Analysis Centers, Inc., later Consolidated Analysis Centers, Inc., always CACI. Initially, we gave SIMSCRIPT [I] courses, and looked for contract work related to SIMSCRIPT applications. Separately, (that is, not through CACI), I consulted for RAND on SIMSCRIPT II development and other matters; Herb consulted for Douglas Aircraft.

CACI got a contract from IBM to make a version of SIMSCRIPT [I] for a new operating system. We used this as an opportunity to implement SIMSCRIPT as a compiler into assembly language rather than a preprocessor into FORTRAN, using an entity-attribute-set view of the compiling process which was concurrently being used at RAND in building the SIMSCRIPT II compiler. The new CACI version of SIMSCRIPT [I] also removed certain language restrictions that had been imposed on the original RAND SIMSCRIPT due to it being a preprocessor into FORTRAN, or due to our inexperience. We called the resulting product SIMSCRIPT I.5, since it was an advance over the original SIMSCRIPT but definitely not what was being developed as SIMSCRIPT II. One large CACI line-of-business for the next few years was the building of SIMSCRIPT I.5 compilers for various computers and operating systems.

**SIMSCRIPT II.** In 1968 or 1969 RAND released SIMSCRIPT II to SHARE and published its programming manual (Kiviat et al. 1968).[1] SIMSCRIPT II was not presented primarily as a simulation language, as was SIMSCRIPT [I], but as a general purpose language with a simulation programming capability. One conspicuous difference between SIMSCRIPT I and II is that the latter dispensed with the forms used by the former. For example, instead of the Definition Form, entities, attributes, and sets in SIMSCRIPT II are defined by statements such as EVERY MACHINE-GROUP HAS A NR_FREE_MACHINES AND OWNS A QUEUE. This was to avoid the logistics problems of supplying forms to users. If a new SIMSCRIPT were designed today, the Definition Form might be back—as part of a (now common) GUI (Graphical User Interface).

The original plan for SIMSCRIPT II was that it be documented and, to a certain extent, implemented in "seven levels." Kiviat and Villanueva summarize in their Preface to Kiviat et al. (1969) the functions of the first five levels as follows.

> Level 1: a simple teaching language ... Level 2: A language roughly comparable in power with FORTRAN ... Level 3: A language roughly comparable in power to ALGOL or PL/I ... Level 4: That part of SIMSCRIPT II that contains the entity-attribute-set features of SIMSCRIPT ··· Level 5: The simulation-oriented part of SIMSCRIPT II ...

Level 6 was to contain the SIMSCRIPT II database management facilities. The premise is that not only *simulated* worlds can be characterized by entities-attributes-sets and events, but so too the "real world" as represented by databases. Level 7 was intended to make the SIMSCRIPT II "language writing language" available to the sophisticated user to make special purpose extensions of the language. (See Markowitz 1979.)

The "SIMSCRIPT II" which RAND released to SHARE and documented in Kiviat et al. (1969) implemented Levels 1 through 5. Levels 1, 2, 3, 4, and 6 (5 omitted) were implemented within IBM under the name EAS-E. EAS-E is documented in Markowitz et al. (1984), Malhotra et al. (1983), and Pazel et al. (1983). We considered EAS-E to be very successful in terms of efficiency of execution and ease of programming as demonstrated by one real-world internal IBM application. However, I failed to convince IBM that it should release EAS-E as a product. IBM had recently converted from the hierarchical IMS database system to the relational System R with its SQL front end, and wasn't interested in launching a programming system based on a different data model at the same time.

**Prelude to SIMSCRIPT II.** SIMSCRIPT II improvements to SIMSCRIPT's simulation capabilities were primarily due to intensive use within RAND of SIMSCRIPT [I] for logistics system simulation. For example, SIMSCRIPT [I] had an ACCUMULATE statement which could be used instead of an assignment statement. Not only would ACCUMULATE assign the new value of the variable, but would also accumulate statistics, such as the min, max, and time-weighted mean and standard deviation of the updated variable. Inspection of the first real simulation programs written in SIMSCRIPT [I] showed a sizable fraction of the coding devoted to ACCUMULATE statements. It also made it clear that coding could be greatly reduced if the statistics to be accumulated were specified once, at Definition Time, rather than with each assignment.

SIMSCRIPT II was already on the drawing board before SIMSCRIPT [I]'s manual and preprocessor were completed. We knew we wanted SIMSCRIPT II to be rid of the SIMSCRIPT [I] forms, and to compile into assembly language rather than preprocess into FORTRAN. While we were at it, we wanted to remove some restrictions imposed on SIMSCRIPT [I] by implementation considerations, and restyle the language a bit to make it still more "self-documenting." SIMSCRIPT II is not an easy language for which to write a compiler. Consider, for example, how one can concatenate control phrases in SIMSCRIPT II, as in

```
FOR EACH MACHINE_GROUP IN SHOP
WITH NR_FREE_MACHINES
  (MACHINE_GROUP)> 0,
FOR EACH JOB IN QUEUE (MACHINE_GROUP)
WITH DUE_DATE (JOB) < TODAY
```

It is not required that FOR and WITH phrases be on separate lines as above. Line breaks and spacing on the page are inessential (except on the "form lines" following "PRINT *n* LINES THUS ...," which replaced SIMSCRIPT [I]'s Report Generator). FOR, WITH, WHILE, UNTIL, and UNLESS phrases can be combined in any meaningful manner, used to control single statements or blocks of statements demarked by DO ... LOOP statements; or they can be incorporated into FIND or COMPUTE statements. The latter computes MIN, MAX, SUM, MEAN, or STD_DEV at an instant of time, as distinguished from the ACCUMULATE statement which accumulates statistics across time.

I had heard that the compiler for some programming language was written in that language itself (perhaps JOVIAL in JOVIAL, but I am not sure I recall correctly). The idea of a SIMSCRIPT II compiler programmed in SIMSCRIPT II intrigued me. Of course, one would have to "bootstrap" a first version from SIMSCRIPT [I], but after that one could program more advanced versions in SIMSCRIPT II itself. While Bernie Hausner and Herb Karr finished the implementation and documentation of SIMSCRIPT [I], I made a first draft of the statements of SIMSCRIPT II and basic design of what the "SIMSCRIPT II-in-SIMSCRIPT II" compiler would be like. The latter includes an entity, attribute, and set description of the status of the compiler, and a language-writing-language wherein more complex commands, like FIND and COMPUTE, could be defined in terms of more basic commands. See Markowitz (1979) for details.

The first programmer that RAND assigned to the SIMSCRIPT II project struggled. After about a year Bernie Hausner returned to RAND from an extended leave and world travels. Bernie started the compiler over from scratch, programmed for a year or so until SIMSCRIPT II was capable of compiling SIMSCRIPT II. He then turned SIMSCRIPT II compiler development over to Richard Villanueva, assuring me that Richard was capable of completing the SIMSCRIPT II compiler, and he was. Bernie left RAND, joined the United Nations to supervise the building of a database system, and eventually became a U.N. diplomat. Meanwhile, back at RAND, Phil Kiviat agreed to write the SIMCRIPT II programming manual. At first the three of us—Kiviat, Villanueva and me—served as design team to specify remaining language details. Eventually, as my duties at CACI increasingly distracted me from the RAND SIMSCRIPT II development, the final language specifications as well as compiler development and the writing of Kiviat et al. were completed by Kiviat and Villanueva.

Eventually CACI established its own version, SIMSCRIPT II.5. Under the able guidance of Ed Russell this became the workhorse of the SIMSCRIPT simulation community for many years. All this was after CACI and I parted company.

**March 15, 1968, and After.** By the beginning of 1968 CACI had grown from Herb and me to a small but growing company planning to "go public." CACI's initial public offering did in fact take place during the second half of 1968. That was the good news. The bad news was that Herb and I had a major disagreement over the pricing of a new product, then a disagreement over how to settle disagreements. This was finally settled on March 15—the Ides of March—of 1968 when Herb Karr, with about 47% of CACI stock and Jim Berkson, vice president of finance, with about 5% of the stock, fired me with about 47% of the stock.

Currently, 33 years later, CACI continues to support SIMSCRIPT II.5. The February 2001 issue of *ORMS Today* (Swain 2001) includes a "Software Product Listing" for "Power Tools for Visualization and Decision Making." A line of the table for SIMSCRIPT II.5 includes the following entries. "Typical Applications of the Software: Building large, complex, high-fidelity, discrete event simulation models, with Interactive 2D graphics and built-in animation." The graphics and animation are new, i.e., were not part of SIMSCRIPT II, circa 1968. "Primary Markets for which the software is applied: Military theater-level simulations, telecommunications, factory simulations, hospital processes." "Price: PC Windows: $25,000. Unix: $35,000." "Vendor's Comments: Simscript II.5 Integrated development tools are based on famous language Simscript used worldwide for building portable, robust commercial quality simulation packages."

If it were up to me I would add database (i.e., Level 6) and distributed (e.g., internet) entities, and cut the price.

## ENDNOTE

[1] Kiviat et al. has a 1968 copyright date and a Preface dated April 1969.

## REFERENCES

Arrow, K. J., G. Debreu, E. Malinvaud, R. M. Solow. 1991. *Cowles Fiftieth Anniversary: Four Essays and an Index of Publications*. Yale University Printing Service, New Haven, CT.

Bellman, R. E. 1957. *Dynamic Programming*. Princeton University Press, Princeton, NJ.

Dantzig, G. B. 1951. Maximization of a linear function of variables subject to linear inequalities. T. C. Koopmans, ed. *Activity Analysis of Production and Allocation*. John Wiley & Sons, Inc., New York, 339–347.

——. 1963. *Linear Programming and Extensions*. Princeton University Press, Princeton, NJ.

Dexter, A. S., J. N. W. Yu, W. T. Ziemba. 1980. Portfolio selection in a lognormal market when the investor has a power utility function: Computational results. M. A. H. Dempster, ed. *Stochastic Programming*. Academic Press, New York, 507–523.

Duff, I. S. ed. 1981. *Sparse Matrices and Their Uses*. Academic Press, New York.

Graham, B., D. L. Dodd. 1951. *Security Analysis*, 3rd ed. McGraw-Hill Book Company, New York.

Hlawitschka, Walter. 1994. The empirical nature of Taylor-series approximations to expected utility. *Amer. Econom. Rev.* **84**(3) 713–719.

Kiviat, P. J., R. Villanueva, H. M. Markowitz. 1969. *The SIMSCRIPT II Programming Language*. Prentice Hall, Englewood Cliffs, NJ.

Koopmans, T. C. 1951. Analysis of production as an efficient combination of activities. T. C. Koopmans, ed. 1971. *Activity Analysis of Production and Allocation*, 7th ed. Yale University Press, New Haven, CT.

Kroll, Y., H. Levy, H. M. Markowitz. 1984. Mean variance versus direct utility maximization. *J. Finance* **39**(1) 47–61.

Kuhn, H. W., A. W. Tucker. 1951. Nonlinear programming. J. Neyman, ed. *Proc. Second Berkeley Sympos. on Math. Statist. Probab.* University of California Press, Berkeley, CA, 481–492.

Lemke, C. E. 1965. Bimatrix equilibrium points and mathematical programming. *Management Sci.* **11**(7) 681–689.

Leontief, W. 1951. *The Structure of American Economy, 1919–1931*. Oxford University Press, New York.

Levy, H., H. M. Markowitz. 1979. Approximating expected utility by a function of mean and variance. *Amer. Econ. Rev.* **69**(3) 308–317.

Malhotra, A., ——, D. P. Pazel. 1983. EAS-E; an integrated approach to application development. *ACM Trans. Database Systems* **8**(4) 515–542.

Manne, A. S., ——. 1963. *Studies in Process Analysis: Economy-wide Production Capabilities*. John Wiley and Sons, New York.

Markowitz, H. M. 1952. Portfolio selection. *J. Finance* **7**(1) 77–91.

——. 1954. Industry-wide, multi-industry and economy-wide process analysis. T. Barna, ed. *The Structural Interdependence of the Economy*. John Wiley and Sons, New York.

——. 1956. The optimization of a quadratic function subject to linear constraints. *Naval Res. Logist. Quart.* **3** 111–133.

——. 1957. The elimination form of the inverse and its application to linear programming. *Management Sci.* **3** 255–269.

——. 1959. *Portfolio Selection: Efficient Diversification of Investments*, 2nd ed. Basil Blackwell, Cambridge, MA.

——. 1979. SIMSCRIPT. J. Belzer, A. G. Holzman, A. Kent, eds. *Encyclopedia of Computer Science and Technology*, Vol. 13. Marcel Dekker, Inc., New York.

——. 1987. *Mean-Variance Analysis in Portfolio Choice and Capital Markets*. Basil Blackwell Ltd., Oxford, U.K.

——, Peter Todd. 2000. *Mean-Variance Analysis in Portfolio Choice and Capital Markets*. With Chapter 13 by G. Peter Todd. First published in 1987 by Basil Blackwell. Revised reissue by Frank Fabozzi and Associates, New Hope, PA.

——, B. Hausner, H. Karr. 1963. *SIMSCRIPT: A Simulation Programming Language*. Prentice Hall, Englewood Cliffs, NJ.

——, A. Malhotra, A. Pazel. 1984. The EAS-E application development system: Principles and language summary. *Comm. Assoc. Comput. Mach.* **27**(8) 785–799.

——, D. Reid, B. Tew. 1994. The value of a blank check. *J. Portfolio Management* **20**(4) 82–91.

Pazel, D. P., A. Malhotra, H. M. Markowitz. 1983. The system architecture of EAS-E: An integrated programming and data base language. *IBM Systems J.* **22**(3) 188–198.

Perold, A. 1984. Large-scale portfolio optimization. *Management Sci.* **30**(10) 1143–1160.

Pulley, L. M. 1981. A general mean-variance approximation to expected utility for short holding periods. *J. Financial Quant. Anal.* **16** 361–373.

——. 1983. Mean-Variance approximations to expected logarithmic utility. *Oper. Res.* **31**(4) 685–696.

Rose, D. J., R. A. Willoughby, eds. 1972. *Sparse Matrices and Their Applications. Proc. Sympos. IBM Res. Center*, September 9–10, 1971. Plenum Press, New York.

Savage, L. J. 1954. *The Foundations of Statistics*. Wiley, New York. 2nd ed., Dover, New York.

Sharpe, W. F. 1963. A simplified model for portfolio analysis. *Management Sci.* **IX**(2) 277–293.

Swain, J. J. 2001. Power tools for visualization and decision-making. *OR/MS Today* **28**(1) 52–63.

Von Neumann, J., O. Morgenstern. 1944. *Theory of Games and Economic Behavior*, 3rd ed. (1953), Princeton University Press, Princeton, NJ.

Wiesenberger, A., and Company. 1941. *Investment Companies*, Annual editions since 1941. New York.

Williams, J. B. 1938. *The Theory of Investment Value*. Harvard University Press, Cambridge, MA.

Willoughby, R. A., ed. 1969. *Proc. Sympos. Sparse Matrices Their Appl.* IBM Report RAI (No. 11707). Yorktown Heights, NY.

Young, W. E., R. H. Trent. 1969. Geometric mean approximation of individual security and portfolio performance. *J. Financial Quant. Anal.* **4**(June) 179–199.

June 21, 2005

# De Finetti Scoops Markowitz

Harry M. Markowitz

## The News

It has recently come to my attention that, in the context of choosing optimum reinsurance levels, de Finetti (1940) essentially proposed mean-variance portfolio analysis using *correlated* risks. About the same time John Burr Williams (1938), the leading financial theorist of the time, claimed that sufficient diversification can make portfolio risk negligible, implicitly assuming that security returns are uncorrelated. Williams did not realize that when security returns are positively correlated, even unlimited diversification leaves a substantial level of portfolio risk. It was not until Roy (1952) and Markowitz (1952) that the notions of portfolio optimization with correlated risks entered the financial literature.

De Finetti's analysis included what we now call "the efficient frontier" which provides minimum risk for given expected return on the reinsured portfolio as a whole. De Finetti did not *solve* the problem of computing mean-variance efficient reinsurance frontiers with correlated risks. He does solve it for the case with uncorrelated risks. For this case he finds that the set of efficient reinsurance portfolios is a series of connected straight line segments, and provides formulas for these successive segments. He explains why the reinsurance problem with correlated risks is more complicated, and solves special cases of it. One of these special cases is that of small correlations, namely, correlations small enough that the same sequence of efficient line segments occur – with their exact positions shifted a bit – as occur in the uncorrelated case. The other special case he solves makes restrictive assumptions on both the covariance matrix and expected returns.

The de Finetti, Roy and Markowitz (1952) models are special cases of a class of models for which a computing procedure, called the "critical line algorithm" (CLA), is provided in Markowitz (1956). While de Finetti does not solve the reinsurance problem with correlated risks, he outlines some of its properties. He tells where the efficient set starts; how it traces out a sequence of connected straight line segments; and then describes how it ends, i.e., the general location of the last segment of the path. I will refer to de Finetti's statement on the latter matter as "de Finetti's last segment conjecture." It is not correct. The de Finetti model does, however, admit of an interesting "correct last segment proposition."

Many properties of the solution to the de Finetti model are already present when there are only two policies to be reinsured. The next section of this paper defines the de Finetti model, presents graphs of all possible solutions to the two-policy case, and sketches how the properties exhibited in these graphs generalize to cases with any number of policies. It also shows which possible two-policy solutions contradict de Finetti's last segment conjecture, and states the correct last segment proposition for the de Finetti model.

The second section below solves a specific two-policy de Finetti problem, and sketches how the principles illustrated in the example generalize to the solution of problems for any number of policies. The "computation" of the two-policy efficient set in the example involves very little number crunching. It mostly illustrates how certain properties of the solution guide CLA in tracing out the efficient frontier.

The text is followed by a Technical Supplement whose three sections describe: how to solve the de Finetti model for any number of policies; a proof of the correct last segment proposition for the de Finetti model; and some observations on developments in optimization techniques during the intervening years between de Finetti (1940) and Markowitz (1956).

A translation of Chapter 1 of de Finetti by Luca Barone is presented later in this volume. Also in this volume is a historical note by Mark Rubinstein on Bruno de Finetti himself and on de Finetti (1940).

**The De Finetti Model**

De Finetti, Markowitz (1952) and Roy seek mean-variance efficient portfolios subject to different constraints on the choice of portfolio. Roy requires that the "fractions" invested in various securities sum to one:

$$X_1 + X_2 + \ldots + X_n = 1 \tag{1}$$

Roy permits these "fractions" to be positive or negative. Markowitz (1952) requires the fractions invested to be nonnegative; i.e., requires

$$X_i \geq 0 \quad i = 1, \ldots, n \tag{2}$$

as well as (1). De Finetti's constraints on the choice of reinsurance portfolio do *not* include the budget constraint (1), but do include upper bounds

$$X_i \leq 1 \quad i = 1, \ldots, n \tag{3}$$

as well as lower bounds (2), where $X_i$ here represents the fraction of an insurance policy (or block of business) retained, i.e., not reinsured. In addition to this formulation in terms of *fractions* of policies retained, de Finetti also presents a version in terms of *values* of policies retained. We will stick with the formulation in terms of fractions retained.

The notation used here differs from that in de Finetti. In particular, my $X_i$ is de Finetti's $a_i$. This is because the writer, and maybe the reader, feels more comfortable in a tradition in which X is a variable and "a" represents a constant.

In applications of the Markowitz (1952) or Roy models, it is usually assumed that the return, $r_i$, on the i-th security is measured as a rate of return per dollar invested. This is treated as a random variable in the two models. In the de Finetti model $r_i$ represents the number of "dollars" (or other monetary units) earned on the i-th policy if retained. Strictly speaking, since it may be of value to the insurance company to sell an insurance policy and then completely reinsure it, $r_i$ should be defined as the amount earned if fully retained over and above the amount that would be earned if the policy were completely reinsured. In this model, as in the previous two, $r_i$ is assumed to be random. It is also assumed that if the fraction $X_i$ is retained then the amount $X_i r_i$ will be earned (over and above how much would be earned if the policy were completely reinsured). These *interpretations* of the numbers involved do not affect our main concern in this paper, namely, the correctness of computing procedures by which we find efficient frontiers given the requisite numbers (that is, given the means, variances and correlations of the $r_i$).

Unless otherwise specified, when I refer below to the "de Finetti reinsurance model" I refer to a mean-variance portfolio selection model which satisfies the following conditions:

C1. Inequalities (2) and (3) are the only constraints on the choice of portfolio. In particular, the model does not include a budget constraint (1).

C2. All expected returns are positive.

C3. The "zero portfolio", with $X_i = 0$ for all *i*, is the only portfolio with zero variance.

The Technical Supplement also notes properties and procedures when condition C1 is kept, but conditions C2 and/or C3 are removed.

When there are only two policies a portfolio in the de Finetti model can be represented by a point P in Figure 1. The horizontal coordinate of P represents $X_1$, the fraction retained of policy 1; the vertical coordinate represents $X_2$, the fraction retained of policy 2. A reinsurance program is feasible (can be done) if it is represented by a point on or in the square with corners at (0,0) and (1,1). All points (portfolios) on and in this square satisfy requirements (2) and (3), that the fractions invested be between zero and one. Portfolios outside the square violate this requirement.

If there are three securities then we need a three-dimensional drawing to represent a portfolio in the de Finetti model. The set of feasible portfolios would be the cube with corners at (0, 0, 0) and (1, 1, 1). Similarly, for any *n*, a portfolio in the de Finetti model is represented by a point in n-dimensional space. The set of feasible portfolios is represented by an n-dimensional cube, or "hypercube", with corners (0, 0, …, 0, 0) and (1, 1, …, 1, 1).

The portfolio with maximum expected return is at the corner of the square, cube or hypercube with all $X_i = 1$. In particular in the two-policy case it is at (1, 1). This portfolio supplies highest expected return, but also highest variance among mean-variance efficient portfolios. This is the start of a path which traces out the set of efficient portfolios from that with highest mean to that with lowest variance.

In the general n-dimensional case the set of efficient portfolios, starting at the corner with all $X_i = 1$, moves along one edge of the square, or (hyper)cube, with one $X_i$ less than one and the remainder remaining at one. This is illustrated in Figures 2a and 2b for the two-policy case. In Figure 2a the efficient set starts at (1,1) and moves down the right side of the feasible square. As it does $X_1$ remains equal to one and $X_2$ falls. As $X_2$ falls efficient portfolios with lower *E and* lower *V* are traversed. In Figure 2b, the efficient set moves along the top of efficient square, where $X_2$ remains equal to 1.0 and $X_1$ declines.

De Finetti asserts that this first segment of the efficient set is the start of a "continuous broken line which links the starting point… to the origin (0, 0, …, 0)." In Figure 2a, for example, the piecewise linear efficient set consists of the segment from (1,1) to the point a, followed by the segment from a to (0,0). The latter is the unique minimum-variance portfolio.

De Finetti asserts that the last segment of this piecewise linear path -- the segment that ends at the origin -- is "inside the hypercube." In other words, along this last segment all $X_i$ are strictly between 0 and 1. In the two-policy case this asserts that the last segment of the efficient set approaches (0,0) from inside the square. As we shall see, this is not correct. In general, for $n \geq 2$, it is possible for one or more $X_i$ to already be zero throughout the last segment. For $n = 2$, the difference may be seen by contrasting Figure 2a with 3a, or 2b with 3b. In 2a the last segment approaches (0, 0) from within the square. But, as we will show with a numerical example in the next section, the path 3a is also a possible set of efficient portfolios. In this case the first segment starts at (1, 1) and ends at (1, 0). The last segment approaches (0, 0) along the horizontal axis on which $X_2 = 0$. Figure 3b shows a similar case except with the efficient set moving first along the top of the feasible square, with $X_2$ fixed at one, reaching the point (0,1), then approaching (0,0) along the left side of the square.

Thus, in the de Finetti model with correlated risks, unlike the case with uncorrelated risks, it is possible for the last segment to approach the zero portfolio along the edge of the square or along the face or edge of the cube or hypercube, rather than through its interior.

There is, however, a correct "last segment proposition" for the de Finetti model. Some background is needed, however, before we state it. On any segment of the piecewise linear set of efficient portfolios, three kinds of variables may be present, namely, those with

(1) $X_i$ fixed at 1.0 throughout the segment;

(2) $X_i$ fixed at $0.0$; and

(3) $0 < X_i < 1.0$ *within* the segment (i.e., except possibly at one or the other end points).

We will refer to these as the "UP", the "Down" and the "IN" variables, respectively. In figure 4, for example, we have labeled the point at (1,1) as (UP, UP) since both variables equal one there. The right side of the square between (1,1) and (1,0) is labeled (UP, IN) since $X_1 = 1$ and $X_2$ is between 0 and 1. The interior of the square is labeled (IN, IN) since both variables are between zero and one at each point there; etc. The various faces, edges and interior of a (hyper)cube can be similarly labeled.

A correct last segment proposition for the de Finetti model is the following: as we trace out the set of efficient segments, from that with the highest mean to that with lowest variance, *the first efficient segment we encounter with no UP variables is the last segment.* For $n = 2$, as Figures 2 and 3 illustrate, once we encounter an efficient segment with no variable UP, the segment heads directly towards the risk-free final portfolio.

The placement of $a$ or $a'$ in Figures 2a or 2b depends on the specific values of the input parameters. The efficient set in Figure 5 is also possible. In this case the efficient set consists of one segment proceeding directly from the maximum expected return portfolio (1, 1) to the variance minimizing portfolio (0, 0). In this case both $X_1$ and $X_2$ change from being UP at (1, 1) to being IN on the efficient segment that traverses the interior of the square. The critical line algorithm considers this a "degenerate case", and treats it as case 2a with an imperceptible distance between (1, 1) and the point $a$ or, equivalently, case 2b with an imperceptible distance between (1, 1) and $a'$. Thus, as CLA traces out the set of efficient portfolios, only one variable at a time changes its state (e.g., from UP to IN or from IN to Down) on successive segments of the efficient set, until the zero-portfolio with minimum variance is reached and the algorithm stops. The five patterns in Figures 2, 3 and 5 are the only possible patterns of efficient portfolios in the two-policy de Finetti model assuming C1, C2, C3.

**Numerical Example**

Suppose that two policies have the means (m), standard deviations (SD) and correlation (corr) in the following table.

**Table 1**

| Policy | Expected Return (m) | Standard Deviation (SD) | Correlation (corr) |
|---|---|---|---|
| 1 | 1 | 1 | 0.6 |
| 2 | 1 | 2 | |

The general formulas for the mean (a.k.a. expected) value of the portfolio, $E_P$, and the variance (a.k.a standard-deviation squared) of the portfolio, $V_P$, is given below, followed by specific numbers from the Table.

$$\begin{aligned} E_P &= m_1 X_1 + m_2 X_2 \\ &= X_1 + X_2 \end{aligned} \tag{4}$$

$$\begin{aligned} V_P &= (SD_1)^2 X_1^2 + (SD_2)^2 X_2^2 \\ &\qquad + 2(corr)(SD_1)(SD_2) X_1 X_2 \\ &= X_1^2 + 4X_2^2 + 2.4 X_1 X_2 \end{aligned} \tag{5}$$

When we trace out the set of efficient portfolios, rather than thinking of these portfolios as minimizing $V_P$ for various $E_P$, or as maximizing $E_P$ for various $V_P$, we will think of them as minimizing

$$L = \frac{1}{2} V_P - w\, E_P \tag{6}$$

for various $w$. For any value of $w$, we are to minimize L in (6) subject to the upper and lower bound constraints (2) and (3). "$w$" is the weight we place, for the moment, on making $E_p$ large versus $V_p$ small.

In tracing out the set of efficient portfolios in a two-policy example we need to keep track of four numbers (as well as $w$). The first two numbers are $X_1$ and $X_2$, the fractions retained of the first and second policies. The third and fourth numbers are $e_1$ and $e_2$, where $e_1$ is the change in L per small change in $X_1$ (i.e., $\partial L / \partial X_1$ or the extra L we get per extra unit of $X_1$); and similarly for $e_2$. The general formulas for $e_1$ and $e_2$ in the two-policy case, and their numerical values in the present example, are

$$\begin{aligned} e_1 &= X_1(SD_1)^2 + X_2(corr)(SD_1)(SD_2) - wm_1 \\ &= X_1 \qquad\quad +1.2X_2 \qquad\qquad\qquad -w \end{aligned} \tag{7a}$$

$$\begin{aligned} e_2 &= X_1(corr)(SD_1)(SD_2) + X_2(SD_2)^2 - wm_2 \\ &= 1.2X_1 \qquad\qquad\quad +4X_2 \qquad -w \end{aligned} \tag{7b}$$

Consider a portfolio which minimizes L for some fixed $w$ subject to constraints (2) and (3). If $X_1 = 0$ in this portfolio then $e_1$ cannot be negative. Because (since $e_1$ is the change in L per small increase in $X_1$) if $e_1$ were negative we could reduce L by *increasing* $X_1$ a bit, contradicting the assumption that $X_1 = 0$ is part of a portfolio that minimizes L. (We cannot *decrease* $X_1$, in this case, since it is already at its lower bound.) Similarly, if instead we had $X_1 = 1.0$, then $e_1$ cannot be positive, for otherwise a permitted decrease in $X_1$ would reduce L. Finally, if $X_1$ is strictly between 0 and 1 then we must have $e_1 = 0$; for otherwise either a permitted small increase or a permitted small decrease could reduce L. Similar statements connect $X_2$ and $e_2$.

In sum, a portfolio which minimizes L for some $w \geq 0$ must have, for $i = 1$ and 2,

$$e_i \geq 0 \text{ if } X_i = 0 \qquad \text{(Policy } i \text{ Down)} \tag{8a}$$

$$e_i = 0 \text{ if } 0 < X_i < 1 \qquad \text{(Policy i IN)} \tag{8b}$$

$$e_i \leq 0 \text{ if } X_i = 1 \qquad \text{(Policy } i \text{ UP)} \tag{8c}$$

CLA uses this connection between $X_i$ and $e_i$ to tell when a particular policy must move from UP to IN along the efficient frontier.

We start with the $E_P$ maximizing portfolio with $X_1 = 1$, $X_2 = 1$. Substituting these values of $X_1$ and $X_2$ into (7a) and (7b) we find that, at the $E_P$ maximizing portfolio we have

$$e_1 = 2.2 - w \tag{9a}$$

$$e_2 = 5.2 - w \tag{9b}$$

Note that if $w$ is large enough then $e_1$ and $e_2$ are negative, confirming that the $E_P$ maximizing portfolio does minimize L when $E_P$ is given a sufficiently large weight in (6). This is not surprising and should, in fact, be obvious from examining (6). What is less obvious is the answer to this question: below what value of $w$ is (1, 1) no longer the L minimizing portfolio? Equations (9a) and (9b) show us that $e_1$ goes from negative to positive when $w$ goes below 2.2; and $e_2$ does so when $w$ falls below 5.2. The $E_P$ maximizing portfolio still satisfies (8c) at $w = 5.2$, but will fail that condition for $i = 2$ for $w$ below 5.2. The $E_p$ maximizing portfolio (1,1) no longer minimizes L if the weight, $w$, placed on $E_p$ in (6) is less than 5.2.

At $w = 5.2$, policy 2 must go from UP to IN, while $X_1$ stays UP, in order to continue to satisfy optimality conditions (8). The efficient set moves down the right side of the square, as in 2a or 3a. Since policy 1 is UP and 2 is IN, conditions (8) tell us that, as we move down the right side of the feasible square, we must have $X_1 = 1.0$ and $e_2 = 0$; therefore (from 7b)

$$\begin{aligned} 1.2 + 4X_2 - w = 0 \\ w = 1.2 + 4X_2 \end{aligned} \tag{10}$$

Finally, if we substitute $X_1 = 1$, and (10) for $w$ into (7a), and simplify, we get

$$e_1 = -0.2 - 2.8X_2 \tag{11}$$

Whether or not this is an example of Figure 2a or Figure 3a depends on which goes to zero first--$e_1$ or $X_2$--as we reduce $w$. If $e_1$ goes to zero first then $X_1$ moves from UP to IN. If $X_2$ goes to zero first then $X_2$ moves from IN to Down. In the present example we see from (11) that when $X_2$ reaches zero $e_1$ is still negative, therefore $X_2$ goes to zero before $e_1$. On the next "segment" $X_2$ is Down as well as $X_1$ UP.

The single point (1, 0) is treated like a "segment" since it minimizes L for a range of $w$. Substituting $X_1 = 1$ and $X_2 = 0$ into (7a) and (7b) we see that at the point (1, 0), with policy 1 UP and policy 2 Down we have

$$e_1 = 1 - w \tag{12a}$$

$$e_2 = 1.2 - w \tag{12b}$$

From (8a), (8c), (12a) and (12b) it follows that the point (1, 0) minimizes L for all $w$ in the interval between 1.0 and 1.2. At $w = 1.0$, $e_1$ reaches zero therefore $X_1$ goes IN.

The efficient set next moves along the horizontal axis as in Figure 3a. We could continue the CLA calculation, now with $X_2 = 0$ and $e_1 = 0$; but at this point we note that no variable is UP, and invoke the "correct last segment proposition" which assures us that the set of efficient portfolios now heads directly toward the (0, 0) portfolio.

*Generalizations* In order for a two-plan example to have Figure 3a as its diagram, the point (1,0) must be efficient. In order for (1, 0) to be efficient we must have

$$e_1 = (SD_1)^2 - m_1 w \le 0$$
$$e_2 = (corr)(SD_1)(SD_2) - m_2 w \ge 0$$

In other words, there must be a range of $w$ values so that

$$\frac{(SD_1)^2}{m_1} \le w \le \frac{(corr)(SD_1)(SD_2)}{m_2}$$

This will occur if and only if

$$\frac{SD_1}{m_1} \le \frac{(corr)SD_2}{m_2}$$

We see, in particular, that it never occurs if returns on the two policies are uncorrelated.

For more than two policies, the computation proceeds very much as illustrated in the two-policy case. We trace out the efficient set by minimizing L for various values of $w$ in (6). We keep track of retention levels $X_1, \ldots, X_n$ and the extra L per change in $X_i$, namely, $e_1, \ldots, e_n$. The $E_P$-maximizing portfolio is (1, 1, …, 1, 1) with all $X_i = 1$. If $w$ is sufficiently large then all the $e_i$ are negative. We calculate which $e_i$ reaches zero first as w is reduced. This determines which policy $i$ moves from UP to IN on the efficient segment that starts at (1, 1, …, 1, 1). The computation proceeds to generate one efficient segment after another, always checking for which happens first, an $e_i$ goes to zero or an

$X_i$ reaches a boundary (usually 0, or possibly 1 when n > 2). This determines which $i$ will move from UP to IN, or IN to Down (or possibly IN to UP or Down to IN when n > 2). In tracing out the set of efficient segments, a given combination of UPs, Downs and INs never appears twice. The computation stops when the $V_p$-minimizing portfolio (0, 0, …, 0, 0) is reached.

If, on a particular segment, $k > 1$ policies are IN then it is necessary to solve $k$ equations in $k$ unknowns to determine how the IN variables change with $w$ along the segment. The actual computation is reduced by taking advantage of the fact that only one policy is added or one is deleted between successive efficient segments. Formulas for the $n$-policy case are provided in the Technical Supplement to this paper.

**Summary**

De Finetti (1940) essentially proposed the use of mean-variance analysis in deciding how much of various blocks of insurance to retain or reinsure. He emphasized the importance of considering correlated risks in the analysis, over a decade before Roy (1952) and Markowitz (1952) proposed the same for choosing portfolios of financial assets. For the case of uncorrelated risks, de Finetti solved the problem of computing the set of mean-variance efficient portfolios. He does not solve this in general for correlated risks. The present paper notes that the de Finetti problem is a special case of a class of problems solved by the "critical line algorithm" (CLA) presented in Markowitz (1956). The paper illustrates the use of CLA to solve the de Finetti problem.

## Technical Supplement

**Solution to the de Finetti problem.**

In this section we present without proof, the application of CLA to the de Finetti problem. For proofs see Perold (1984), Markowitz (1987), or Markowitz and Todd (2000). We will refer to the latter two as "M or MT."

The notation in the body of this paper was chosen in deference to the reader who considers any formula with a Greek letter to be complicated. The notation in this Technical Supplement is that of M or MT, with two exceptions. First, upper bounds are modeled explicitly as in M or MT Exercise 7.3, rather than implicitly as just another linear inequality constraint. Secondly, Chapter 2 of M or MT says that it will be convenient to assume that the constraint set includes at least one linear equality constraint "to avoid having to distinguish … between cases in which the A matrix exists and those in which it does not" (page 27). This convention is not convenient in the present case, since inequalities (2) and (3) are the only constraints on choice of portfolio in the deFinetti model. As in M or MT, we let $\mu = (\mu_1, \ldots, \mu_n)'$, $C = (\sigma_{ij})$, be an expected return vector and covariance matrix. Unless specified, we do *not* require $\mu > 0$ or $|C| \neq 0$. (Markowitz (1959) Appendix A shows that CLA works when C is singular.) The expected return E and variance V of the portfolio $X = (X_1, \ldots, X_n)'$ is $E = \mu' X$ and $V = X'CX$ respectively. For any scalar $\lambda_E \in (0, \infty)$, a feasible portfolio which minimizes

$$L = \tfrac{1}{2}V - \lambda_E E \qquad \text{(A.1)}$$

must be efficient. We return below to the case of $\lambda_E = 0$. Let

$$\begin{aligned} \eta &= (\eta_1, \ldots, \eta_n)' \\ &= \left( \frac{\partial L}{\partial X_1}, \ldots, \frac{\partial L}{\partial X_n} \right)' \\ &= CX - \lambda_E \mu \end{aligned} \tag{A.2}$$

A necessary and sufficient condition for $X$ to minimize L for $\lambda_E \in (0, \infty)$ is that

$$\eta_i \geq 0 \text{ for } X_i = 0 \tag{A.3a}$$

$$\eta_i = 0 \text{ for } 0 < X_i < 1 \tag{A.3b}$$

$$\eta_i \leq 0 \text{ for } X_i = 1 \tag{A.3c}$$

For the de Finetti model CLA produces a piecewise linear path in $(X, \eta, \lambda_E)$-space along which (A.3) is satisfied at all points. The path is defined for all $\lambda_E \in [0, \infty)$. When $C$ is singular (A.3) does not guarantee portfolio efficiency when $\lambda_E = 0$; since a portfolio may have minimum $V$, but not maximize $E$ among feasible portfolios with this $V$. *However*, the portfolio produced by CLA for $\lambda_E = 0$ *is* efficient.

On any linear segment of the piecewise linear efficient set, the numbers $\{1, \ldots, n\}$ are partitioned into three sets: UP, Down and IN. Throughout the segment $X_i = 1$ for $i \in UP$, and $X_i = 0$ for $i \in Down$; whereas the $X_i$ $i \in IN$ vary so as to satisfy (A.3b). Specifically, let $X_{IN}$, $\mu_{IN}$, $C_{IN}$ represent, respectively, a vector of portfolio holdings, a vector of expected returns and a covariance matrix for only those variables which are "IN" on a particular segment. If no variable is IN, then think of $X_{IN}$, $\mu_{IN}$ and $C_{IN}$ as null vectors and matrix. If $C_{IN}$ is not null it will be nonsingular, even if $C$ is singular.

Let $b$ be an $n$-component vector whose *i-th* component is

$$b_i = \sum_{j \in UP} \sigma_{ij} \tag{A.4}$$

and $b_{IN}$ be the $n_{IN}$ component vector with $b_i$ as in (A.4) for $i \in IN$.

Condition (A.3b) specifies that

$$C_{IN} X_{IN} = -b_{IN} + \lambda_E \mu_{IN} \tag{A.5}$$

Thus, along a linear segment of the efficient set $X_{IN}$ varies with $\lambda_E$ according to the formula

$$\begin{aligned} X_{IN} &= -C_{IN}^{-1} b_{IN} + \lambda_E C_{IN}^{-1} \mu_{IN} \\ &= \alpha_{IN} + \beta_{IN} \lambda_E \end{aligned} \tag{A.6}$$

where $\alpha_{IN}$ and $\beta_{IN}$ equal $-C_{IN}^{-1} b_{IN}$ and $\mathrm{C}_{\mathrm{IN}}^{-1} \mu_{IN}$. Thus, along the efficient segment $X$ varies linearly

$$X = \alpha^{IN} + \beta^{IN} \lambda_E \tag{A.7}$$

where $\beta_i = 0$ for $i \in \mathrm{UP}$ or Down, $\alpha_i = 0$ for $i \in$ Down, $\alpha_i = 1$ for $i \in$ UP; and $\alpha_i, \beta_i$ are given by (A6) for $i \in \mathrm{IN}$. Substituting (A.7) into (A.2) we find that $\eta$ also varies linearly with $\lambda_E$ :

$$\eta = \gamma^{IN} + \delta^{IN} \lambda_E \tag{A.8}$$

For the moment we postpone discussing how the algorithm starts. We pick up the story at the point in iterative cycle $t$ at which the new IN set, $IN_t$, has been derived from $IN_{t-1}$ by one of the following actions

(E1) an $i$ moves from UP to IN;

(E2) an $i$ moves from IN to Down;

(E3) an $i$ moves from IN to UP;

(E4) an $i$ moves from Down to IN.

(E3) and (E4) may seem like unusual events in the de Finetti model; but we have no theorem that says they cannot happen, so the algorithm should be prepared to handle them. In particular, below we exhibit a two-policy example with $\mu_i < 0$, in which (E4) occurs.

The end of the cycle for the efficient segment with $IN = IN_{t-1}$ supplies two additional pieces of information for the start of iteration $t$, namely,

- $\lambda_{t-1}^{LOW} = \lambda_t^{HI}$
- $C_{IN}^{-1}$ for $IN = IN_{t-1}$

The first of these, the lowest value of $\lambda_E$ at which the segment with $IN = IN_{t-1}$ is efficient, equals the highest value of $\lambda_E$ at which the segment with $IN = IN_t$ is efficient. This new segment will be efficient for a closed interval $\left[\lambda_t^{LOW}, \lambda_t^{HI}\right]$ with $\lambda_t^{LOW}$ to be determined.

Write $C_t$ for "$C_{IN}$ with $IN = IN_t$". $C_t$ is the same as $C_{t-1}$ except for the insertion or the deletion of *one* row and the corresponding column. See M or MT, Chapter 13 for formulas for deriving $C_t^{-1}$ from $C_{t-1}^{-1}$ efficiently. With $C_t^{-1}$ in hand, (A.6) and (A.2) are used to determine the $\alpha_i, \beta_i, \gamma_i$ and $\delta_i$ in (A.7) and (A.8). $\lambda_t^{LOW}$ is the largest of the following

- $-\alpha_i/\beta_i$ for $i \in$ IN and $\beta_i > 0$
- $(1-\alpha_i)/\beta_i$ for $i \in$ IN and $\beta_i < 0$ (A.9)
- $-\gamma_i/\delta_i$ for $i \in$ UP and $\delta_i < 0$ or
  for $i \in$ Down and $\delta_i > 0$
- 0.0

Accordingly, $i$ switches from IN to Down, IN to UP, UP to IN or Down to IN, or the V minimizing efficient portfolio has been reached. Except in the last case, we have now determined $IN_{t+1}$ and are ready for the next iteration.

It is possible for the $IN_t$ set to be empty. This is always true at the starting portfolio $\bar{X}$ with maximum expected return, but can also happen at an intermediate step such as the point (1, 0) in the example in the text. The $X_i$ do not change with $\lambda_E$ on such a segment, but the "segment" is indeed a segment if plotted in $(X, \lambda_E, \eta)$ space, and the computation proceeds as described above.

If we assume, with de Finetti, that $\mu_i > 0 \; \forall i$ then the unique E maximizing portfolio is $\bar{X} = (1,1,...,1)$. If some $\mu_i < 0$, then the unique E maximizing portfolio has

$$\bar{X}_i = 1 \text{ for } \mu_i > 0$$
$$\bar{X}_i = 0 \text{ for } \mu_i < 0$$

If any $\mu_i = 0$ the case is "degenerate" in the sense that more than one portfolio maximizes E. We will not treat that case here. See M or MT Chapter 9 concerning degenerate cases generally. Even though $\mu_i < 0$, we can have $X_i > 0$ in some efficient portfolios. For an example, solve the two-policy problem with

$$\mu_1 = 1,\ \mu_2 = -1,\ \sigma_1 = 1,\ \sigma_2 = 2,\ \rho = -0.6.$$

Assuming $\mu_i \neq 0\ \forall i$, the critical line algorithm starts with the unique E maximizing portfolio $\bar{X}$. $\bar{X}$ is a "segment" in which all $i$ are UP or Down. $\lambda_{HI}$ for this first "segment" is $\infty$. $\lambda_{LOW}$ is the largest value of $\lambda_E$ at which an $\eta_i \uparrow 0$ for $i$ UP or $\eta_i \downarrow 0$ for $i$ Down as $\lambda_E \downarrow \lambda_{Low}$. This determines the first $i$ to go IN, and the algorithm proceeds as described above. CLA stops when $\lambda_E = 0$. The same partition into UP, Down, IN is never encountered twice; therefore CLA stops in a finite number of steps.

**A Correct Final Segment Theorem**

The following holds for the de Finetti reinsurance model, whether or not $\mu > 0$ or $|C| \neq 0$.

*Theorem. If an efficient segment has no UP variables, then the low end of the segment has $\lambda_E = 0.0$ and portfolio $\underline{X} = (0, 0 \ldots, 0)$.*

Proof. Since no $i$ is UP, equation (A.4) implies that $b = 0$ in (A.5) therefore

$$\begin{aligned} X_{IN} &= \lambda_E C_{IN}^{-1} \mu_{IN} \\ &= \beta_{IN} \lambda_E \end{aligned} \tag{A.10}$$

Therefore

$$X = \beta \lambda_E \tag{A.11}$$

with $\beta_i = 0$ for $i$ not IN. Substituting this into (A.2) we find that $\eta$ is of the form

$$\eta = \delta \lambda_E \tag{A.12}$$

(A.11) and (A.12) imply that no nonzero $X_i$ or $\eta_i$ can become 0, and no $0 \leq X_i < 1$ can become one, in the interval

$$0 < \lambda_E < \lambda_{IN}^t.$$

The theorem follows.

The theorem tells us that *if* a segment has no *i* UP variables *then* it is the last segment, ending in the zero vector. It does not tell us that this is, in fact, how the minimum variance efficient portfolio is reached. If *C* is nonsingular, then the zero vector is indeed the unique minimum variance efficient portfolio. But when *C* is singular the zero vector may or may not be efficient. For example, consider again the means and standard deviations in Table 1 of the text. Now first suppose $\rho = +1$; then suppose $\rho = -1$. In the first case, a portfolio has zero variance if and only if $X_1 = -2X_2$. But among such $(X_1, X_2)$ combinations, only (0, 0) is feasible. Therefore $X = 0$ is efficient. If $\rho = -1$, any portfolio with $X_1 = 2X_2$ has zero variance. Among these, (1, ½) has the same variance and higher mean than (0,0); therefore the latter is not efficient[1].

**Advances in Mathematical Programming (1940-1956)**

A revolution in mathematical programming occurred between the time of de Finetti and Markowitz (1956). Methods and results of this period were used by Markowitz in the development of CLA for the "general" portfolio selection problem. This section reviews these methods and results, and concludes that it was not their absence that caused de Finetti to not solve his problem with correlated risks. A different explanation is proposed.

The two main results of the 1940s and early 1950s used in Markowitz (1956) were (a) The Kuhn-Tucker (1951) theorem, and (b) George Dantzig's simplex algorithm. (See Dantzig 1963) The former certifies that each of the portfolios produced by CLA is efficient. One obvious service (b) supplies is to provide a portfolio with maximum expected return; since the linear programming problem, which the simplex algorithm solves, is to maximize a linear function (like portfolio expected return) subject to the same kinds of constraints adopted by the "general" portfolio selection model. The simplex algorithm also renders various other essential services to CLA. It determines whether the model is feasible. If the model is feasible but contains one or more redundant equation, simplex supplies an equivalent model without this defect. The Markowitz (1956) handling of degenerate cases is patterned on the Dantzig, Orden and Wolfe (1955) solution to the same problem within linear programming. The simplex algorithm also alerts CLA if portfolio expected return is unbounded, or if it is bounded but the maximum expected return solution is not unique. In these cases CLA must proceed differently than simplex.

Since, as we will see shortly, CLA is an example of linear complementary programming, it might seem that work on the latter subject also contributed to CLA. In this case, however, the influence went in the other direction. Phil Wolfe relates that he was the anonymous referee of Markowitz (1956). In the process he noted that the steps of CLA were like the steps of the simplex algorithm, with one crucial exception. In CLA there are complementary pairs of variables -- $(X_i, \eta_i)$ for $i = 1, \ldots,$ n--such that if one is positive the other must be zero. This lead to Wolfe (1959) and linear complementary programming.

None of the ways in which the Kuhn-Tucker theorem and Dantzig's simplex algorithm contributed to CLA seem to be the missing ingredient needed by de Finetti to solve the reinsurance problem. Specifically, with (2) and (3) as the only constraints, there is no question that the model is feasible or has redundant equations. Also, assuming with de Finetti that all expected returns are positive, no computation is required to see that (1, 1, …, 1) is the unique, E-maximizing portfolio. As to the use of the Kuhn-Tucker theorem, since the constraint set has no equality constraints, only upper and lower bounds on variables, only a simple version of KT is required. De Finetti's treatment of the uncorrelated case shows that de Finetti had seen or had worked out this special case of Kuhn-Tucker.

The difference, then, between de Finetti circa 1940 and Markowitz in the 1950s was not some specific theorem or method. Rather, I believe, it was because by the 1950s quadratic programming was "in the air". For example, I mentioned that Wolfe refereed Markowitz (1956). About the same time I refereed what became Frank and Wolfe (1956). The existence of the Kuhn-Tucker theorem and the success of linear programming encouraged a presumption that a neat quadratic programming algorithm existed if we persisted in seeking it. De Finetti did not have the benefit of this environment.

**References**

Dantzig, G. B. (1963), "Linear Programming and Extensions", Princeton University Press, Princeton, New Jersey.

Dantzig, G. B., A. Orden and P. Wolfe (1955), *The Generalized Simplex Method for Minimizing a Linear Form Under Linear Inequality Restraints*. Pacific Journal of Mathematics, Vol. 5, No. 2, June.

De Finetti, Bruno, (1940) *"Il Problema dei ≪ pieni ≫"*, Giornale dell'Istituto Italiano degli Attuari, Vol. 11, No. 1, pp. 1-88. An English version of the first chapter, "The Problem in a Single Accounting Period," translated by Luca Barone in 2005, appears in the Journal of Investment Management, this issue.

Frank, M. and P. Wolfe (1956) *An Algorithm for Quadratic Programming*, Naval Research Logistics Quarterly, 3, pp.95-110.

Kuhn, H.W., and A.W. Tucker (1951) *Nonlinear Programming*, Proceedings of the Second Berkeley Symposium on Mathematical Statistics and Probability, edited by J. Neyman, University of California Press, Berkeley, pp. 481-492.

Markowitz, Harry M. (1952), *Portfolio Selection*, The Journal of Finance, 7, (1), March, pp. 77-91.

Markowitz, Harry M. (1956), *The Optimization of a Quadratic Function Subject to Linear Constraints*, Naval Research Logistics Quarterly, 3, pp. 111-33.

Markowitz, Harry M. (1959), "Portfolio Selection: Efficient Diversification of Investments", Cambridge, MA. Basil Blackwell, 1991, 2nd edition.

Markowitz, Harry M. (1987), "Mean-Variance Analysis in Portfolio Choice and Capital Markets", Basil Blackwell Ltd., Oxford.

Markowitz, Harry M. and Peter Todd (2000), "Mean-Variance Analysis in Portfolio Choice and Capital Markets", (revised reissue of Markowitz (1987) with chapter by Peter Todd) Frank J. Fabozzi Associates, New Hope, PA.

Perold, A. F. (1984), *Large-Scale Portfolio Optimization*, Management Science, 30 (10), October, pp. 1143-60.

Roy, A. D. (1952), *Safety First and the Holding of Assets*, Econometrica, 20, pp. 431-49.

Rubinstein, Mark ( ) *Bruno de Finetti and Mean-Variance Portfolio Selection*, Journal of Investment Management, this issue.

Sharpe, W. F. (1964), *Capital Asset Prices: A Theory of Market Equilibrium Under Conditions of Risk*, The Journal of Finance, 19 (3), September, pp. 425-442.

Tobin, James (1958), *Liquidity Preference as Behavior Toward Risk,* Review of Economic Studies, Vol. 25, pp. 65-87 February.

Williams, John Burr (1938), "The Theory of Investment Value", Harvard University Press, Cambridge, MA.

Wolfe (1959), *The Simplex Method for Quadratic Programming*, Econometrica, Vol. 27, No. 3, July, pp. 382-398

Endnote

1. The version of the Tobin Separation Theorem in Tobin (1958), as distinguished from that in Sharpe (1964), is a last segment theorem for the Markowitz (1952) model when $\sigma_{kk} = 0$ for some *k* ("cash"). It says that, in tracing out the efficient frontier from high E to low V, the first segment whose IN-set has *k* IN is the last segment. The Tobin Separation Theorem and the correct last segment theorem for the de Finetti model are special cases of a last segment theorem that applies to the "general" portfolio selection model As in M or MT we write the constraint set of this model as

$$AX = b \tag{N.1}$$

$$X \geq 0 \tag{N.2}$$

where A is $m \times n$. Inequalities, such as the upper bounds in the de Finetti model, and variables not required to be nonnegative, as in the Roy model, can be cast into the above format by introducing slack variables or by separating variables into their positive and negative parts. In general, a model of the above form may admit of no feasible portfolio, or may have feasible portfolios but no efficient portfolios. Except in these two cases--even when C is singular, A is rank deficient, or the model is degenerate--CLA generates a piecewise linear path that provides one and only one efficient portfolio for each efficient EV combination. Each segment of this set has an IN-set and an associated affine space $s_{IN}$ in which $X_i = 0$ for $i$ OUT, equations (N.1) are satisfied, and inequalities (N.2) are ignored. The efficient segment for this IN-set is a segment of the critical line $l_{IN}$ which is the locus of points in $s_{IN}$ that minimize $\frac{1}{2}V - \lambda_E E$ for various $\lambda_E$. The generalized final segment theorem as follows. Suppose that the model has at least one efficient portfolio. Also suppose that a risk-free *portfolio* is feasible. (This risk-free portfolio may or may not involve a risk-free security.) Then (obviously) a risk-free portfolio is efficient, and (less obviously) the first IN set whose $s_{IN}$ contains a feasible zero-variance portfolio is the last IN-set. A proof may be obtained from the author.

# CAPM Investors Do Not Get Paid for Bearing Risk: *A Linear Relation Does Not Imply Payment for Risk*

HARRY M. MARKOWITZ

HARRY M. MARKOWITZ is the president of Harry Markowitz Company, and is an adjunct professor at the Rady School of Management, University of California in San Diego, CA.
harryhmm@aol.com

The relation between the excess return of a security and its beta, where beta is defined as its regression against the return on the market portfolio, is linear in the Sharpe–Lintner (S–L) capital asset pricing model (CAPM). This linear relation is often interpreted to mean that CAPM investors are paid for bearing systematic risk.

I will show that this is not a correct interpretation, because two securities may have identical risk structures in terms of their covariances with other securities in the market, and yet have different excess returns. In fact, if the parameters of the CAPM are generated in a natural way, then securities with the same risk structure almost surely will have different expected returns.

## THE MOSSIN VERSION OF THE SHARPE–LINTNER MODEL

Although the premises and conclusions of the S–L CAPM were first presented in Sharpe [1964], I will use Mossin's [1966] systematic formulation of Sharpe's version of the S–L model in this analysis. Mossin states the premise of the Sharpe model explicitly, and draws valid conclusions from it, while Sharpe's version is vague on the statement of his premises and deduction of his conclusions, and one of his conclusions is incorrect.*

Inputs to Mossin's version of the S–L CAPM model (Mossin S–L model) include: the utility functions of many investors; the number of shares each investor first owns of each stock; and the expected returns per share (not per dollar) and covariances per share (not per dollar) upon which all investors agree. There is also an interest rate input that is used only in the excess return calculation.

Outputs of the Mossin S–L model include market clearing prices, expected returns, and covariances per dollar; the composition of the market portfolio; and the regressions against the market portfolio. Along the way, Mossin uses Tobin and Sharpe to prove various things about the S–L model. For example, the Mossin S–L model involves equations that include cash, expected returns, and a Lagrangian multiplier. With a little algebra, the equations can be expressed in excess returns rather than expected returns.

Mossin also notes that Sharpe's version of the Tobin separation theorem holds in equilibrium (Sharpe's version of the Tobin separation theorem involves borrowing as well as lending, while Tobin's version involves lending only). In Sharpe's version of the Tobin separation theorem, each investor's efficient set consists of one particular portfolio of risky securities as well as the ability to borrow or lend. This much of the theorem holds for any mean-variance investor who can borrow without limit (even if no one else can do so), or has the same beliefs, or even seeks mean-variance efficiency. When we assume that all investors have the same beliefs and all seek mean-variance efficiency, it follows that all investors mix the same portfolio of risky assets, and this must be the market portfolio.

The Mossin S– L model is like a giant television set: You set the dials (e.g., share availabilities), and the screen shows outputs (e.g., prices and expected returns per dollar invested). Like Mossin, I won't worry about the existence of the uniqueness of the solutions to the equations involved. I assume that when I flip on the switch, the TV works. If I set the dials the same way as yesterday, I get the same output as yesterday.

One noteworthy feature of the Mossin S–L television is that you can tell how the inputs are set by looking at the outputs. Outputs include prices, and with prices you can figure the original shares available as well as means and covariances per share.

Yet there are some conceivable outputs one will never see—namely, outputs that could happen only if some assumed given shares were negative (or zero for all investors because such securities would be dropped from the analysis). But these would be the securities with zero or negative percent demanded in the market portfolio.

We will see only output consistent with any positive vectors $X > 0$ in Equation (1), where $\nu_i$ and $\sigma_{ij}$ are the excess return and covariance per dollar invested:

$$\begin{pmatrix} \sigma_{11} \\ \sigma_{21} \\ \cdot \\ \cdot \\ \cdot \\ \sigma_{n1} \end{pmatrix} X_1 + \begin{pmatrix} \sigma_{12+} \\ \sigma_{22} \\ \cdot \\ \cdot \\ \cdot \\ \sigma_{n2} \end{pmatrix} X_2 + \cdots \begin{pmatrix} \sigma_{1n} \\ \sigma_{2n} \\ \cdot \\ \cdot \\ \cdot \\ \sigma_{nn} \end{pmatrix} X_n = h \begin{pmatrix} \nu_1 \\ \nu_2 \\ \cdot \\ \cdot \\ \cdot \\ \nu_n \end{pmatrix} \quad (1)$$

So, how do we find all possible outputs that can occur on the Mossin S–L television? The question practically answers itself if we write the basic optimization equations as in Equation (1).

We know that the covariance matrix $C$ per dollar invested is nonsingular because the covariance matrix per share is nonsingular—the only thing that Mossin tells us about the latter matrix. Given any nonsingular covariance matrix $C$, consider the set of excess return vectors, $\nu$, that result if $X$ in Equation (1) is nonnegative, $X \geq 0$ (later we consider only $X > 0$). These $\nu$ form the cone, $K_C$, generated by $C$, which is drawn as follows.

As in Equation (1), look at each column $\sigma^{(i)}$ of $C$ in turn as if it were a point in $\nu$-space. Draw a ray from the origin through $\sigma^{(i)}$ for $i = 1, \ldots, n$. Fill in the middle of the diagram by taking convex combinations of anything you already have. This gives you $K_C$, the cone generated by $C$.

But we want $X > 0$ only. We get this by throwing away the exterior of $K_C$ and keeping only the interior $K_C^0$ Even though we toss the exterior of $K_C$, there will be plenty of good stuff left as long as $|C| \neq 0$ as assumed.

I call $K_C^0$ the *compatible cone* (i.e., the cone full of $\nu$-vectors that are compatible with the given $C$ in the sense that only these will ever appear together on the Mossin S–L TV). In other words, *given any nonsingular $C$, pick any $\nu$-vector from $K_C^0$. This combination of $C$ and $\nu$ may appear on the Mossin S–L TV. These are the only $C$, $\nu$ combinations that can appear.*

Because in this model $C$ is all things risk and $\nu$ all things return, in an abstract way we now know the model's possible combinations of risk and return. But we do not have a concrete picture of what these $K_C^0$ look like. For starters, let's assume that returns are uncorrelated (this is not plausible, but easy to analyze). $C$ is then diagonal:

$$C = \begin{pmatrix} V_1 & 0 & 0 & 0 \\ 0 & V_2 & 0 & 0 \\ 0 & 0 & V_3 & 0 \\ 0 & 0 & 0 & V_4 \end{pmatrix}$$

The first column, considered as a $\nu$-vector, lies on the $X_1$-axis, the second on the $X_2$-axis, and so on. Because the location of a ray from the origin through a point does not depend on which point on the ray we choose, we have:

$$K_C = K_I$$

That is, $K_C$ is the same as the cone generated by the identity matrix $I$, the entire positive orthantant, specifically the entire first quadrant when $n = 2$. $K_C^0$ is the interior of this region.

Thus, if returns are uncorrelated, then any positive $V_1, V_2, \ldots, V_n$ and $\nu_1, \ldots, \nu_n$ are permitted. In particular, two securities can have identical $V$ but different $\nu$. So they can have identical risk structures, given our assumption of zero correlation, and yet have differing expected returns, therefore differing excess returns.

From Equation (1) with diagonal $C$ we see that when returns are uncorrelated, stocks with high means and low variances are a large part of the market:

$$X_i = k\nu_i / V_i \qquad (2)$$

Equation (2) may also be written as

$$X_i\, V_i = k\nu_i \qquad (3)$$

Because $\sigma_{ij} = 0$ for all $j \neq i$, the standard formula for the covariance between a security and a portfolio yields $X_i V_i$ as the covariance between the return on security i and the return on the investor's portfolio. Thus, in this case, Equation (3) says that the covariance of each security with its portfolio is proportional to the securities excess return. If you divide both sides of Equation (3) by the variance of the investor's portfolio you obtain:

$$\beta_i = \bar{k}\nu_i \qquad (4)$$

where $\beta_i$ is the regression against the portfolio. Here $\bar{k} = k/ Var\,(R_P)$ and should be given no other interpretation.

Equation (4) is a relation between a security and one individual's portfolio. But, if all investors are essentially the same, then Equation (4) carries over to the market as well.

It is useful to see where Equation (1)—and therefore Equation (4)—comes from. For this purpose, let us further simplify the Mossin S–L by assuming the form of utility function:

$$E - cV \qquad (5)$$

where $c$ may vary from one investor to the next. To maximize Equation (5), take partial derivatives and set the result to zero to get:

$$\frac{\partial V}{\partial X_1} = (1/c)\frac{\partial E}{\partial X_i} \quad \text{for } i = 1, \ldots, n \qquad (6)$$

In other words, each investor is advised to push each security into the investor's portfolio to the point where each security has the same ratio of marginal effect on portfolio variance to marginal effect on portfolio mean. This aggregates up to the portfolio level. Divide by market variance, and the well-known CAPM relation is created.

If returns are correlated, compatible cones are derived by plotting the columns of $C$ as if they were $\nu$ vectors, drawing straight lines from the origin through them, and then taking convex combinations of these lines using the interior of the resulting cone. This is easiest to do for $n = 2$.

Note that the cone tends to close up as correlation increases. If the two securities have a correlation of 1, then the two lines are identical. This is the law of one price. Identically distributed securities must have proportionate expected returns. As long as the $C$ matrix is nonsingular, therefore, $K_C^0$ is not empty, and two different points can be found in $K_C^0$ not on the same ray through the origin. In this case two securities can have identical risk structures $C$ but differing excess returns.

If the initial endowments of investors are drawn randomly from some continuous distribution, the probability distribution of $(C, \mu)$ cases will be continuous. Thus, there is a zero probability that two securities with the same covariances will have the same $\mu$.

## AFTERTHOUGHTS

Rather than use

$$(\sigma_{1j}, \sigma_{2j}, \ldots, \sigma_{nj})'$$

as a point in $\nu$-space to help draw a ray from the origin, we could use any other point on the ray, like:

$$(\sigma_{1j}/\sigma_{jj}, \sigma_{2j}/\sigma_{jj}, \ldots, \sigma_{nj}/\sigma_{jj})'$$

Recall that $\sigma_{ij}/\sigma_{jj}$, is the regression coefficient—the beta if you will—of the return of security $i$ against that of $j$.

Do not expect compatible cones to go away in multiperiod discrete- or continuous-time analysis. In particular, there is not much difference between a one-period optimization and a multiperiod optimization. Bellman [1957] tells us that to optimize a many-period problem you just optimize a sequence of one-period problems. It's a matter of getting the objectives right for the problem-within-a-problem. But often problems-within-a-problem are not all that different from the genuine one-period problem. Therefore, if the genuine one-period model has compatible cones, there should also be some lurking around the model's many-period version.

The continuous-time model is a special case of the discrete-time model. If you do not believe that, it is because you are stuck with an old-fashioned number system. The Greeks thought you could not squeeze any new numbers between the fractions (also called the rationals, because the Greeks thought that the new numbers were irrational). In fact, in some sense there are more irrationals than there are rationals.

Now we have infinitesimals between each real (rational or irrational) number. Gottfried Leibniz [1646–1716] thought he had a handle on the infinitesimals, but it had to wait for Abraham Robinson [1966] to figure it out rigorously in the mid-20th century. In a dynamic analysis, rather than have time travel along the real continuum, we can have it walk step-by-step along the "hyperfinite time line." The latter is as simple as a garden path, but with infinitesimal distances between successive stepping stones.

With the hyperfinite timeline, Brownian motion is what comes out at a macro level if at each infinitesimal time increment you flip a coin and record an infinitesimal gain or loss.

Thus, it is a small step to imagine a single-period Mossin S–L CAPM where the period is an infinitesimal tick long. After all, where does Mossin say how long the period has to be in a single-period CAPM? A day? A year? A microsecond? An infinitesimal increment? Because the analysis is independent of the time increment, it applies to infinitesimal steps and, therefore, to continuous models. Our conclusion is the same as in the discrete-time model; that possible combinations of excess returns that can go with a given covariance matrix will probably form some kind of cone, and that two securities with the same risk structure will probably have different expected returns. Therefore, one clearly cannot say that the CAPM investor is paid for bearing risk in either in the single-period, multiple-period, or continuous-time case.

## WHERE DOES THIS LEAVE THE CAPM?

The points I make do not change the major conclusions of the capital asset pricing model. Given the assumptions of the CAPM, the market is an efficient portfolio, and there is a linear relation between the expected return of each security and its regression against the market. But we must not interpret this as the bearing of risk.

Insofar as the world works like the CAPM at an aggregate level, this linear relation is useful. Someone who wishes to issue a new security could estimate its beta, which would indicate the expected return that the market would assign to this new security.

## ENDNOTE

[a]Sharpe [1964] asserts (with regard to his Exhibit 6) that as prices adjust, some combinations of risky assets can become perfectly correlated with each other. This is associated with a linear portion of the resulting expected return, standard deviation-efficient set considering risky securities only. Mossin [1966] assumes that the covariance matrix per share is nonsingular. It is not clear whether Sharpe assumes that this covariance matrix of returns is nonsingular, but he certainly permits this, so no matter what positive prices are present in the market, the covariance matrix of returns per dollar invested as well as the covariance matrix per share is nonsingular. This implies that no two distinct linear combinations of risky assets can be perfectly correlated, and there can be no linear segment in the set of mean-standard deviation-efficient combinations.

## REFERENCES

Bellman, Richard E. *Dynamic Programming.* Princeton, NJ: Princeton University Press, 1957.

Mossin, Jan. "Equilibrium in a Capital Asset Market." *Econometrica*, Vol. 34, No. 4 (October 1966), pp. 768–783.

Robinson, Abraham. *Nonstandard Analysis.* Amsterdam: North-Holland,1966.

Sharpe, William F. "Capital Asset Prices: A Theory of Market Equilibrium under Conditions of Risk." *Journal of Finance*, Vol. 19, No. 3 (September 1964), pp. 425–442.